www.brookscole.com

www.brookscole.com is the World Wide Web site for Thomson Brooks/Cole and is your direct source to dozens of online resources.

At *www.brookscole.com* you can find out about supplements, demonstration software, and student resources. You can also send email to many of our authors and preview new publications and exciting new technologies.

www.brookscole.com
Changing the way the world learns®

This book is dedicated

To my wife, Liz, for your love and encouragement,
and to my three sons, Kevin, Glenn, and Brandon,
who always bring me joy and make me proud

AST

And to five grandboys,
Daniel, Tyler, Spencer, Skyler, and Garrett

RDG

Books in the Tussy and Gustafson Series

In paperback:

Basic Mathematics for College Students Third Edition
Student edition: ISBN 0-534-42223-3
Instructor's edition: ISBN 0-534-42224-1

Prealgebra Third Edition
Student edition: ISBN 0-534-40280-1
Instructor's edition: ISBN 0-534-40281-X

Developmental Mathematics Second Edition
Student edition: ISBN 0-534-99776-7
Instructor's edition: ISBN 0-534-99775-5

Introductory Algebra Third Edition
Student edition: ISBN 0-534-40735-8
Instructor's edition: ISBN 0-534-40736-6

Intermediate Algebra Third Edition
Student edition: ISBN 0-534-49394-7
Instructor's edition: ISBN 0-534-49395-5

In hardcover:

Elementary Algebra Third Edition
Student edition: ISBN 0-534-41914-3
Instructor's edition: ISBN 0-534-41915-1

Intermediate Algebra Third Edition
Student edition: ISBN 0-534-41923-2
Instructor's edition: ISBN 0-534-41924-0

Elementary and Intermediate Algebra Third Edition
Student edition: ISBN 0-534-41932-1
Instructor's edition: ISBN 0-534-41933-X

THIRD EDITION

Prealgebra

Alan S. Tussy
Citrus College

R. David Gustafson
Rock Valley College

Australia • Canada • Mexico • Singapore • Spain • United Kingdom • United States

Prealgebra, Third Edition
Alan S. Tussy and R. David Gustafson

Executive Editor: Jennifer Laugier
Development Editor: Kirsten Markson
Assistant Editor: Rebecca Subity
Editorial Assistant: Christina Ho
Technology Project Manager: Sarah Woicicki
Marketing Manager: Greta Kleinert
Marketing Assistant: Jessica Bothwell
Advertising Project Manager: Bryan Vann
Project Manager, Editorial Production: Hal Humphrey
Art Director: Vernon Boes
Print/Media Buyer: Barbara Britton
Permissions Editor: Kiely Sisk

Printed in the United States of America
1 2 3 4 5 6 7 08 07 06 05

For more information about our products, contact us at:
Thomson Learning Academic Resource Center
1-800-423-0563

For permission to use material from this text or product, submit a request online at **http://www.thomsonrights.com.**

Any additional questions about permissions can be submitted by email to **thomsonrights@thomson.com.**

Library of Congress Control Number: 2004118199
Student Edition: ISBN 0-534-40280-1
Annotated Instructor's Edition: ISBN 0-534-40281-X

Production Service: Helen Walden
Text Designer: Diane Beasley
Art Editor: Helen Walden
Photo Researcher: Helen Walden
Copy Editor: Carol Reitz
Illustrator: Lori Heckelman
Cover Designer: Cheryl Carrington
Cover Image: Eric Bean/Getty Images
Cover Printer: Quebecor World/Dubuque
Compositor: G & S Typesetters, Inc.
Printer: Quebecor World/Dubuque
Study Skills Workshop photos from Getty Images

Thomson Higher Education
10 Davis Drive
Belmont, CA 94002-3098
USA

Asia (including India)
Thomson Learning
5 Shenton Way #01-01
UIC Building
Singapore 068808

Australia/New Zealand
Thomson Learning Australia
102 Dodds Street
Southbank, Victoria 3006
Australia

Canada
Thomson Nelson
1120 Birchmount Road
Toronto, Ontario M1K 5G4
Canada

UK/Europe/Middle East/Africa
Thomson Learning
High Holborn House
50/51 Bedford Row
London WC1R 4LR
United Kingdom

Latin America
Thomson Learning
Seneca, 53
Colonia Polanco
11560 Mexico D.F.
Mexico

Spain (including Portugal)
Thomson Paraninfo
Calle Magallanes, 25
28015 Madrid, Spain

CONTENTS

3 The Language of Algebra 155

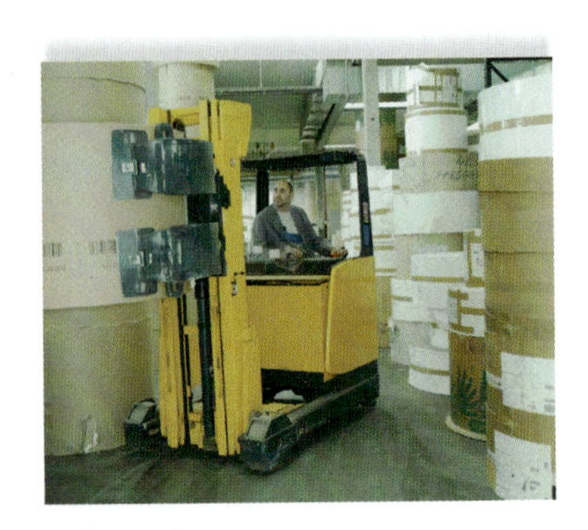

4 Fractions and Mixed Numbers 219

5 Decimals 307

6 Graphing, Exponents, and Polynomials 381

7 Percent 440

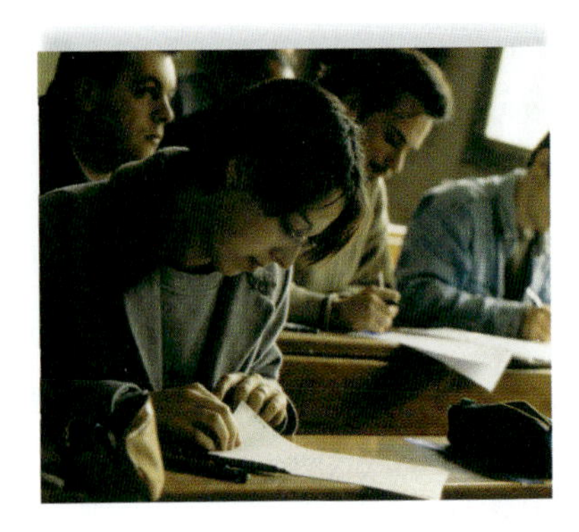

8 Ratio, Proportion, and Measurement 493

9 Introduction to Geometry 555

PREFACE

For the Instructor

The purpose of this textbook is to teach students how to read, write, and think mathematically using the language of mathematics. It is written for students studying prealgebra for the first time and for those who need a review of the basics. *Prealgebra,* Third Edition, employs a variety of instructional methods that reflect the recommendations of NCTM and AMATYC. You will find extensive opportunities for skill practice and the well-defined pedagogy that are hallmarks of a traditional approach to teaching mathematics. You will also find that we emphasize the reasoning, modeling, and communication skills that are a part of today's mathematics reform movement.

The third edition retains the basic philosophy of the second edition. However, we have made several improvements as a direct result of the comments and suggestions we received from instructors and students. Our goal has been to make the book more enjoyable to read, easier to understand, and more relevant.

New to This Edition

New features make this the most student relevant and engaging edition of this popular text.

New Chapter Openers with TLE Labs: TLE (The Learning Equation) is interactive courseware that uses a guided inquiry approach to teaching developmental math concepts. Each chapter opener integrates a TLE Lab with a real-world application, which offers an opportunity to seamlessly integrate TLE lessons into your course. For more information on TLE, instructors can see the Preview at the beginning of their book.

Check Your Knowledge: New pretests at the beginning of each chapter are helpful in gauging a student's knowledge base for the upcoming chapter. Instructors can assign the pretest to see how well students are prepared for the material in the chapter, and customize subsequent lessons based on the needs of their students. Students can also take the pretest themselves as a warm-up for the chapter and to structure review. Answers to the pretests appear at the back of the book.

Study Skills Workshop: This complete mini-course in mathematics study skills helps students and instructors tackle the problem of lack of preparedness and inadequate study habits in an organized series of lessons. Each chapter opens with a one-page Study Skills Workshop sequenced to address relevant study skills issues as the students move through the course. For example, students learn how to use a calendar to schedule study times in the first lesson, how to use study groups to prepare for a midterm exam, and how to effectively study for a final in one of the last lessons. This helpful reference can be used in the classroom or assigned as homework.

Think It Through: Each chapter contains one or two of these features, which make the connection between mathematics and student life. These problems are student-relevant and apply mathematics skills from the chapter to real-life situations. Topics include tuition costs, statistics about college life, job opportunities, and many more topics directly connected to the student experience.

A colorful new design visually organizes information on the page.

Proven Features of *Prealgebra*

- The authors' proven five-step problem-solving strategy teaches students to analyze the problem, form an equation, solve the equation, state the conclusion, and check the result. In a step-by-step manner, this approach clarifies the thought process and mathematical skills necessary to solve a wide variety of problems. As a result, students' confidence is increased and their problem-solving abilities are strengthened.
- STUDY SETS are found at the end of every section and feature a unique organization, tailored to improve students' ability to read, write, and communicate mathematical ideas, thereby approaching topics from a variety of perspectives. Each comprehensive STUDY SET is divided into seven parts: VOCABULARY, CONCEPTS, NOTATION, PRACTICE, APPLICATIONS, WRITING, and REVIEW.
  - VOCABULARY, NOTATION, and WRITING problems help students improve their ability to read, write, and communicate mathematical ideas.
  - The CONCEPT problems section reinforces major ideas through exploration and fosters independent thinking and the ability to interpret graphs and data.
  - PRACTICE problems provide the necessary drill for mastery while the APPLICATIONS provide opportunities for students to deal with real-life situations. Each STUDY SET concludes with a REVIEW section that consists of problems randomly selected from previous sections.
- SELF CHECK problems, adjacent to most worked examples, reinforce concepts and build confidence. The answer to each Self Check follows the problem to give students instant feedback.
- The KEY CONCEPT section is a one-page review, found at the end of each chapter, which revisits the importance of the role the concept plays in the overall picture.
- REAL-LIFE APPLICATIONS are presented from a number of disciplines, including science, business, economics, manufacturing, entertainment, history, art, music, and mathematics.
- CALCULATOR SNAPSHOT sections introduce keystrokes and show how scientific calculators can be used to solve application problems, for instructors who wish to integrate calculators into their course.
- CUMULATIVE REVIEW EXERCISES at the end of every chapter except Chapter 1 help students retain what they have learned in prior chapters.

For detailed information on the ancillary resources available for this text, instructors can see the Preview section at the beginning of their book.

Acknowledgments

We are grateful to the following people who reviewed this manuscript and the other manuscripts in the paperback series at various stages of development. They all had valuable suggestions that have been incorporated into the text.

The following people reviewed the first and second editions:

Linda Beattie
Western New Mexico University

Julia Brown
Atlantic Community College

Linda Clay
Albuquerque TVI

John Coburn
Saint Louis Community College–Florissant Valley

Sally Copeland
Johnson County Community College

Ben Cornelius
Oregon Institute of Technology

James Edmondson
Santa Barbara Community College

David L. Fama
Germanna Community College

Barbara Gentry
Parkland College

Laurie Hoecherl
Kishwaukee College

Judith Jones
Valencia Community College

Therese Jones
Amarillo College

Joanne Juedes
University of Wisconsin–Marathon County

Dennis Kimzey
Rogue Community College

Sally Lesik
Holyoke Community College

Elizabeth Morrison
Valencia Community College

Jan Alicia Nettler
Holyoke Community College

Scott Perkins
Lake–Sumter Community College

Angela Peterson
Portland Community College

J. Doug Richey
Northeast Texas Community College

Angelo Segalla
Orange Coast College

June Strohm
Pennsylvania State Community College–DuBois

Rita Sturgeon
San Bernardino Valley College

Jo Anne Temple
Texas Technical University

Sharon Testone
Onondaga Community College

Marilyn Treder
Rochester Community College

Thomas Vanden Eynden
Thomas More College

The following people reviewed the series in preparation for the third edition:

Cedric E. Atkins
Mott Community College

William D. Barcus
SUNY, Stony Brook

Kathy Bernunzio
Portland Community College

Girish Budhwar
United Tribes Technical College

Sharon Camner
Pierce College–Fort Steilacoom

Robin Carter
Citrus College

Ann Corbeil
Massasoit Community College

Carolyn Detmer
Seminole Community College

Maggie Flint
Northeast State Technical Community College

Charles Ford
Shasta College

Michael Heeren
Hamilton College

Barbara Hughes
San Jacinto College

Monica C. Kurth
Scott Community College

Sandra Lofstock
St. Petersberg College–Tarpon Springs Center

Marge Palaniuk
United Tribes Technical College

Jane Pinnow
University of Wisconsin–Parkside

Eric Sims
Art Institute of Dallas

Annette Squires
Palomar College

Lee Ann Spahr
Durham Technical Community College

John Strasser
Scottsdale Community College

Stuart Swain
University of Maine at Machias

Celeste M. Teluk
D'Youville College

Sven Trenholm
Herkeimer County Community College

Stephen Whittle
Augusta State University

Mary Lou Wogan
Klamath Community College

Without the talents and dedication of the editorial, marketing, and production staff of Brooks/Cole, this revision of *Prealgebra* could not have been so well accomplished. We express our sincere appreciation for the hard work of Bob Pirtle, Jennifer Laugier, Helen Walden, Lori Heckleman, Vernon Boes, Diane Beasley, Sarah Woicicki, Greta Kleinert, Jessica Bothwell, Bryan Vann, Kirsten Markson, Rebecca Subity, Hal Humphrey, Jolene Rhodes, Christine Davis, Diane Koenig, and G&S Typesetters for their help in creating the book. Special thanks to David Casey of Citrus College for his hard work on the pretests and to Sheila Pisa for writing the excellent Study Skills Workshops.

Alan S. Tussy
R. David Gustafson

For the Student

Success in Mathematics

To be successful in mathematics, you need to know how to study it. The following checklist will help you develop your own personal strategy to study and learn the material. The suggestions below require some time and self-discipline on your part, but it will be worth the effort. This will help you get the most out of the course.

As you read each of the following statements, place a check mark in the box if you can truthfully answer Yes. If you can't answer Yes, think of what you might do to make the suggestion part of your personal study plan. You should go over this checklist several times during the semester to be sure you are following it.

Preparing for the Class

- ❑ I have made a commitment to myself to give this course my best effort.
- ❑ I have the proper materials: a pencil with an eraser, paper, a notebook, a ruler, a calculator, and a calendar or day planner.
- ❑ I am willing to spend a minimum of two hours doing homework for every hour of class.
- ❑ I will try to work on this subject every day.
- ❑ I have a copy of the class syllabus. I understand the requirements of the course and how I will be graded.
- ❑ I have scheduled a free hour after the class to give me time to review my notes and begin the homework assignment.

Class Participation

- ❑ I know my instructor's name.
- ❑ I will regularly attend the class sessions and be on time.
- ❑ When I am absent, I will find out what the class studied, get a copy of any notes or handouts, and make up the work that was assigned when I was gone.

- ❑ I will sit where I can hear the instructor and see the board.
- ❑ I will pay attention in class and take careful notes.
- ❑ I will ask the instructor questions when I don't understand the material.
- ❑ When tests, quizzes, or homework papers are passed back and discussed in class, I will write down the correct solutions for the problems I missed so that I can learn from my mistakes.

Study Sessions

- ❑ I will find a comfortable and quiet place to study.
- ❑ I realize that reading a math book is different from reading a newspaper or a novel. Quite often, it will take more than one reading to understand the material.
- ❑ After studying an example in the textbook, I will work the accompanying Self Check.
- ❑ I will begin the homework assignment only after reading the assigned section.
- ❑ I will try to use the mathematical vocabulary mentioned in the book and used by my instructor when I am writing or talking about the topics studied in this course.
- ❑ I will look for opportunities to explain the material to others.
- ❑ I will check all my answers to the problems with those provided in the back of the book (or with the *Student Solutions Manual*) and resolve any differences.
- ❑ My homework will be organized and neat. My solutions will show all the necessary steps.
- ❑ I will work some review problems every day.
- ❑ After completing the homework assignment, I will read the next section to prepare for the coming class session.
- ❑ I will keep a notebook containing my class notes, homework papers, quizzes, tests, and any handouts—all in order by date.

Special Help

- ❑ I know my instructor's office hours and am willing to go in to ask for help.
- ❑ I have formed a study group with classmates that meets regularly to discuss the material and work on problems.
- ❑ When I need additional explanation of a topic, I use the tutorial videos and the interactive CD, as well as the Web site.
- ❑ I make use of extra tutorial assistance that my school offers for mathematics courses.
- ❑ I have purchased the *Student Solutions Manual* that accompanies this text, and I use it.

To follow each of these suggestions will take time. It takes a lot of practice to learn mathematics, just as with any other skill.

No doubt, you will sometimes become frustrated along the way. This is natural. When it occurs, take a break and come back to the material after you have had time to clear your thoughts. Keep in mind that the skills and discipline you learn in this course will help make for a brighter future. Good luck!

iLrn Tutorial Quick Start Guide

iLrn Can Help You Succeed in Math

iLrn™ is an online program that facilitates math learning by providing resources and practice to help you succeed in your math course. Your instructor chose to use iLrn because it provides online opportunities for learning (Explanations found by clicking

Read Book), practice (Exercises), and evaluating (Quizzes). It also gives you a way to keep track of your own progress and manage your assignments.

The mathematical notation in iLrn is the same as that you see in your textbooks, in class, and when using other math tools like a graphing calculator. iLrn can also help you run calculations, plot graphs, enter expressions, and grasp difficult concepts. You will encounter various problem types as you work through iLrn, all of which are designed to strengthen your skills and engage you in learning in different ways.

Logging in to 1Pass

Registering with the PIN Code on the 1Pass Card *Situation:* Your instructor has not given you a PIN code for an online course, but you have a textbook with a 1Pass PIN code. With 1Pass, you have one simple PIN code access to all media resources associated with your textbook. Please refer to your 1Pass card for a complete list of those resources.

Initial Log-in

To access your Web gateway through 1Pass:

1. Check the outside of your textbook to see if there is an additional 1Pass card.
2. Take this card (and the additional 1Pass card if appropriate) and go to http://1pass.thomson.com.
3. Type in your 1Pass access code (or codes).
4. Follow the directions on the screen to set up your personal username and password.
5. Click through to launch your personal portal.
6. Access the media resources associated with your text . . . all the resources are just one click away.
7. Record your username and password for future visits and be sure to use the same username for all Thomson Learning resources.

For tech support, contact us at 1 (800) 423-0563.

You will be asked to enter a valid e-mail address and password. Save your password in a safe place. You will need them to log in the next time you use 1Pass. Only your e-mail address and password will allow you to reenter 1Pass.

Subsequent Log-in

1. Go to **http://1pass.thomson.com.**
2. Type your e-mail address and password (see boxed information above) in the "Existing Users" box; then click on **Login.**

Navigating through Your iLrn Tutorial

To navigate between chapters and sections, use the drop-down menu below the top navigation bar. This will give you access to the study activities available for each section.

The view of a tutorial in iLrn looks like this.

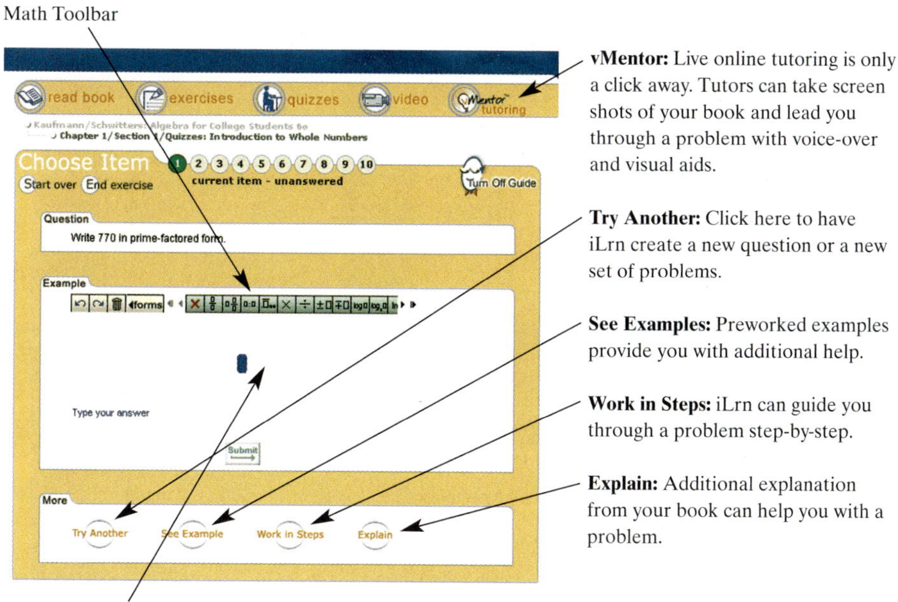

Type your answer here.

Online Tutoring with vMentor

Access to iLrn also means access to online tutors and support through vMentor™, which provides live homework help and tutorials. To access vMentor while you are working in the Exercises or Tutorial areas in iLrn, click on the **vMentor Tutoring** button at the top right of the navigation bar above the problem or exercise.

Next, click on the **vMentor** button; you will be taken to a Web page that lists the steps for entering a vMentor classroom. If you are a first-time user of vMentor, you might need to download Java software before entering the class for the first class. You can either take an Orientation Session or log in to a vClass from the links at the bottom of the opening screen.

All vMentor Tutoring is done through a vClass, an Internet-based virtual classroom that features two-way audio, a shared whiteboard, chat, messaging, and experienced tutors.

You can access vMentor Sunday through Thursday, as follows:

5 p.m. to 9 p.m. Pacific Time

6 p.m. to 10 p.m. Mountain Time

7 p.m. to 11 p.m. Central Time

8 p.m. to midnight Eastern Time

If you need additional help using vMentor, you can access the Participant Quick Reference Guide at this Web site: **http://www.elluminate.com/support/docs/Elive_Participant_Quick_Reference_Guide_6.0.pdf.**

Interact with TLE Online Labs

Use TLE Online Labs to explore and reinforce key concepts introduced in this text. These electronic labs give you access to additional instruction and practice problems, so you can explore each concept interactively, at your own pace. Not only will you be better prepared, but you will also perform better in the class overall.

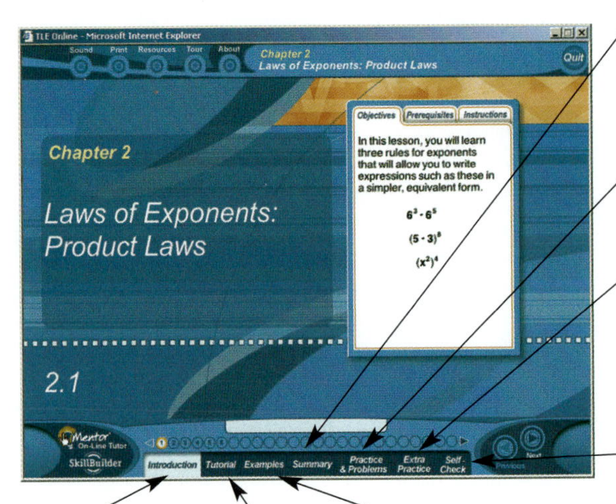

Introduction: Each lesson opens with objectives and prerequisites and provides brief instructions on using TLE.

Tutorial: The Tutorial provides the main instruction for the lesson. Hint and Success Tips teach strategies that can be used to solve the problem.

Examples: The examples expand on what you learned in the Tutorial. A hidden picture is progressively revealed as you complete each example.

Summary: The Summary revisits the problem presented in the Introduction and encourages you to apply the mathematics you learned in the Tutorial and Examples.

Practice & Problems: Practice & Problems presents up to 25 questions organized in four or five categories.

Extra Practice: Extra practice presents questions like those in the Examples. After each question you have the option to try again, see the answer, see a sample solution, or try another question of the same type.

Self-Check: Self-Check presents up to 10 dynamically generated questions. To complete a lesson you must obtain the minimum standard (about 70%).

APPLICATIONS INDEX

Examples that are applications are shown with boldface page numbers.
Exercises that are applications are shown with lightface page numbers.

CHAPTER 1

Whole Numbers

Getty Images

TLE Office managers play an important role in many businesses. They oversee the day-to-day activities of a company, making sure that the business runs smoothly and efficiently. To be an effective office manager, one needs excellent organizational, planning, and communication skills. Strong mathematical skills are also necessary to perform such job responsibilities as scheduling meetings, managing payroll and budgets, and designing office workspace layouts.

To learn more about the use of mathematics in the business world, visit The Learning Equation on the Internet at http://tle.brookscole.com. (The log-in instructions are in the Preface.) For Chapter 1, the online lessons are:

- *TLE* Lesson 1: Whole Numbers
- *TLE* Lesson 2: Order of Operations
- *TLE* Lesson 3: Solving Equations

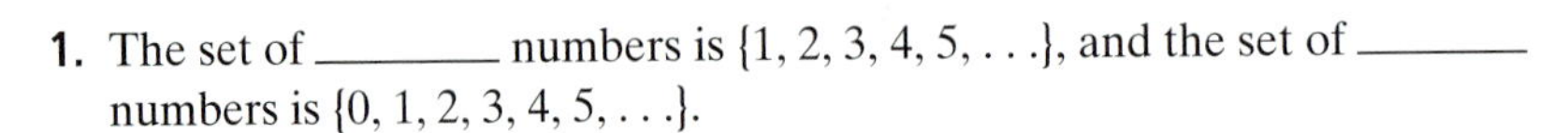

Check Your Knowledge

1. The set of ________ numbers is {1, 2, 3, 4, 5, . . .}, and the set of ________ numbers is {0, 1, 2, 3, 4, 5, . . .}.

2. Numbers that are to be multiplied are called ________. The result of a multiplication problem is called a ________. The answer to a division problem is called the ________.

3. The property that guarantees that we can add two numbers in either order is called the ____________ property of addition. The property that allows us to group numbers in an addition in any way we wish is called the ____________ property of addition.

4. A ________ number is a whole number, greater than 1, that has only 1 and itself as factors.

5. Write 7,343 in expanded notation.

6. Round 27,450 to the nearest hundred.

Number of persons / Blood types (A, B, O, AB)

PROBLEM 7

Refer to the data in the table.

Blood type	A	B	O	AB
Number of persons	5	7	9	4

7. Use the data to make a bar graph.

8. Use the data to make a line graph.

9. Place one of the symbols $<$ or $>$ in the blank to make a true statement: 91 ▢ 19

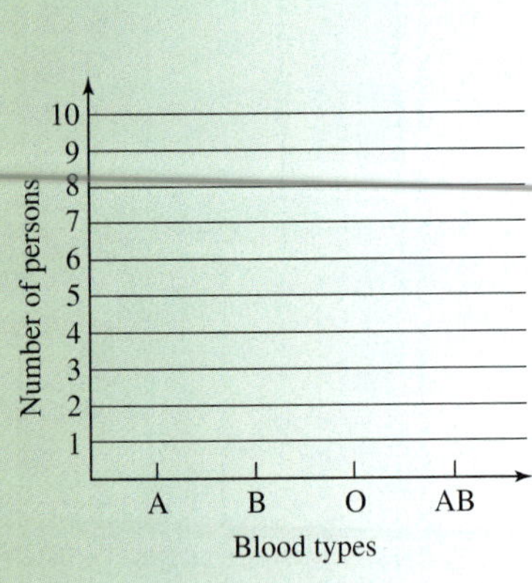

PROBLEM 8

Perform each operation.

10. Add: $\begin{array}{r} 3{,}742 \\ +\ 1{,}379 \\ \hline \end{array}$

11. Subtract 289 from 347.

12. On Wednesday, the high temperature was 72°. The high temperature rose 9° on Thursday and fell 12° on Friday. What was the high temperature on Friday?

13. Multiply: $\begin{array}{r} 432 \\ \times\ 57 \\ \hline \end{array}$

14. Divide: $79\overline{)4{,}537}$

15. Find the perimeter and the area of a rectangle that is 13 ft wide and 19 ft long.

16. Find the prime factorization of 950.

Evaluate the expressions.

17 $3 + 4 - 2$

18. $3 + 4 \cdot 25$

19. $\dfrac{(4^3 - 2) + 7}{5(2 + 4) - 7}$

20. $3 \cdot 7 - 2\,[10 - 3\,(5 - 2)]$

Solve each equation and check the result.

21. $x - 2 = 1$

22. $y + 3 = 5$

23. $4n = 64$

24. $\dfrac{z}{4} = 12$

25. Joan scored 95, 85, 73, 62, and 0 on five math quizzes. Find her mean (average) quiz score.

26. Let a variable represent the unknown quantity. Then write and solve an equation to answer the following question:

After moving his shop, a barber lost 24 regular customers. If he has 65 regular customers left, how many did he have before the move?

Study Skills Workshop

GET ORGANIZED!

Sometimes students who have had difficulty learning math in the past think that their problem is not being born with the ability to "do math." This isn't true! Learning math is a skill and, much like learning to play a musical instrument, takes daily, organized practice. Below are some strategies that will get you off to a good start.

Attend Class. Attending class every meeting is one of the most important things you can do to succeed. Your instructor explains material, gives examples to support the text, and information about topics that are not in your book, or may make announcements regarding homework assignments and test dates. Getting to know at least a few of your classmates is also important to your success. Find a classmate or two on whom you can depend for information, who can help with homework, or with whom you can form a study group.

Make a Calendar. Because daily practice is so important in learning math, it is a good idea to set up a calendar that lists all of your time commitments and time for studying and doing your homework. A general rule for how much study time to budget is to allot two hours outside of class for every lecture hour. If your class meets for three hours per week, plan on six hours per week for homework and study.

Gather Needed Materials. All math classes require textbooks, notebooks, pencils (with big erasers!), and usually as much scrap paper as you can gather. If you are not sure that you have everything you need, check with your instructor. Ideally, you should have your materials by your second class meeting and bring them to every class meeting after that. Additional materials that may be of use outside of class are the online tutorial program iLrn (www.iLrn.com) and the Video Skillbuilder CD-ROM that is packaged with your textbook.

What Does Your Instructor Expect From You? Your instructor's syllabus is documentation of his or her expectations. Many times your instructor will detail in the syllabus how your grade is determined, when office hours are held, and where you can get help outside of class. Read the syllabus thoroughly and make sure you understand all that is required.

ASSIGNMENT

1. Download a calendar online at series.brookscole.com/tussypaperback or make up your own calendar that includes class and study times for each course you are taking as well as times for work and other essential activities (e.g., church activities, social obligations, time with children). You may want to schedule additional time to study a week before a test. Also include time for physical exercise and rest — important for reducing the effects of stress that school brings.
2. Download and print out the Course Information Sheet online at series.brookscole.com/tussypaperback or make a list of the following items:
   a. Instructor name, office location and hours, phone number, e-mail address
   b. Test dates, if scheduled
   c. The work that determines your course grade and how grades are calculated
3. Write down names, phone numbers, and e-mail addresses of at least two classmates.
4. Does your school have tutorial services or a math lab/learning center? If so, where are they located? What are their hours of operation?

In this chapter, we begin our mathematics study by examining the procedures used to solve problems that involve whole numbers.

1.1 An Introduction to the Whole Numbers

- Sets of numbers • Place value • Expanded notation • Graphing on the number line
- Ordering of the whole numbers • Rounding whole numbers • Tables and graphs

In this section, we will discuss the natural numbers and the whole numbers. These numbers are used to answer questions such as How many?, How fast?, How heavy?, and How far?

- The movie *Saving Private Ryan* won 5 Academy Awards.
- The speed limit on interstate highways in Wisconsin is 65 mph.
- The Statue of Liberty weighs 225 tons.
- The driving distance between Chicago and Houston is 1,067 miles.

Sets of numbers

A **set** is a collection of objects. Two sets in mathematics are the natural numbers (the numbers that we count with) and the whole numbers. When writing a set, we use **braces** { } to enclose its **members** (or **elements**).

The set of natural numbers

$\{1, 2, 3, 4, 5, 6, 7, 8, 9, 10, 11, 12, \ldots\}$

The set of whole numbers

$\{0, 1, 2, 3, 4, 5, 6, 7, 8, 9, 10, 11, 12, \ldots\}$

The three dots at the ends of the lists above indicate that the sets continue on forever. There is no largest natural number or whole number.

Since every natural number is also a whole number, we say that the set of natural numbers is a **subset** of the set of whole numbers. However, not all whole numbers are natural numbers, because 0 is a whole number but not a natural number.

Place value

When we express a whole number with a *numeral* containing the *digits* 0, 1, 2, 3, 4, 5, 6, 7, 8, 9, we say that we have written the number in **standard notation.** The position of a digit in a numeral determines its value. In the numeral 325, the 5 is in the *ones column,* the 2 is in the *tens column,* and the 3 is in the *hundreds column.*

3 2 5

Hundreds column ↑ ↑ ↑ Ones column

Tens column

To make a numeral easy to read, we use commas to separate its digits into groups of three, called **periods.** Each period has a name, such as *ones, thousands, millions,* and so on. The following table shows the place value of each digit in the numeral 345,576,402,897,415, which is read as

three hundred forty-five trillion, five hundred seventy-six billion, four hundred two million, eight hundred ninety-seven thousand, four hundred fifteen

345 trillion			576 billion			402 million			897 thousand			4 hundred fifteen		
3	4	5	5	7	6	4	0	2	8	9	7	4	1	5
Trillions			Billions			Millions			Thousands			Ones		
Hundreds	Tens	Ones	Hundreds	Tens	Ones	Hundreds	Tens	Ones	Hundreds	Tens	Ones	Hundreds	Tens	Ones

As we move to the left in this table, the place value of each column is 10 times greater than the column to its right. This is why we call our number system a *base-10 number system.*

EXAMPLE 1 **TV news.** In 2003, there were 73,365,880 basic cable subscribers in the United States. Which digit in 73,365,880 tells the number of hundreds?

Solution In 73,365,**8**80, the hundreds column is the third column from the right. The digit 8 tells the number of hundreds.

Self Check 1
In 2003, there were 158,722,000 cellular telephone subscribers in the United States. Which digit in 158,722,000 tells the number of ten thousands?

Answer 2

Expanded notation

In the numeral 6,352, the digit 6 is in the thousands column, 3 is in the hundreds column, 5 is in the tens column, and 2 is in the ones (or units) column. The meaning of 6,352 becomes clear when we write it in **expanded notation.**

6 thousands + 3 hundreds + 5 tens + 2 ones

We read the numeral 6,352 as "six thousand, three hundred fifty-two."

EXAMPLE 2 Write each number in expanded notation: **a.** 63,427 and **b.** 1,251,609.

Solution

a. 6 ten thousands + 3 thousands + 4 hundreds + 2 tens + 7 ones

We read this number as "sixty-three thousand, four hundred twenty-seven."

b. 1 million + 2 hundred thousands + 5 ten thousands + 1 thousand + 6 hundreds + 0 tens + 9 ones

Since 0 tens is zero, the expanded notation can also be written as

1 million + 2 hundred thousands + 5 ten thousands + 1 thousand + 6 hundreds + 9 ones

We read this number as "one million, two hundred fifty-one thousand, six hundred nine."

Self Check 2
Write 808,413 in expanded notation.

Answer 8 hundred thousands + 8 thousands + 4 hundreds + 1 ten + 3 ones. Read as "eight hundred eight thousand, four hundred thirteen."

Self Check 3

Write seventy-six thousand three in standard notation.

Answer 76,003

EXAMPLE 3 Write twenty-three thousand forty in standard notation.

Solution In expanded notation, the number is written as

2 ten thousands + 3 thousands + 4 tens   There are 0 hundreds and 0 ones.

In standard notation, this is written as 23,040.

Graphing on the number line

Whole numbers can be represented by points on the **number line.** The number line is a horizontal or vertical line that is used to represent numbers graphically. Like a ruler, the number line is straight and has uniform markings. (See Figure 1-1.) To construct the number line, we begin on the left with a point on the line representing the number 0. This point is called the **origin.** We then proceed to the right, drawing equally spaced marks and labeling them with whole numbers that increase progressively in size. The arrowhead at the right indicates that the number line continues forever.

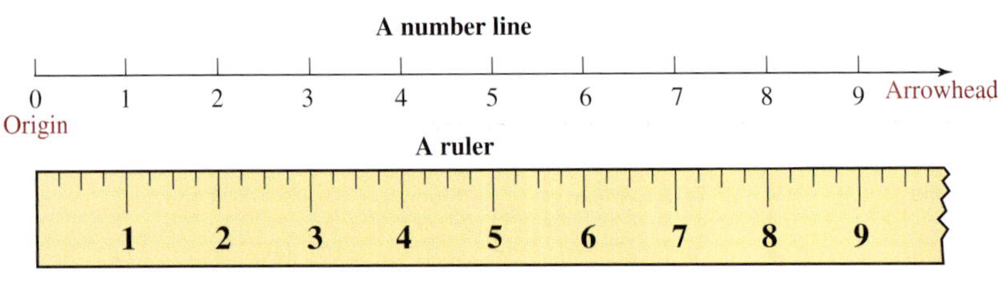

FIGURE 1-1

Using a process known as **graphing,** we can represent a single number or a set of numbers on the number line. *The graph of a number* is the point on the number line that corresponds to that number. *To graph a number* means to locate its position on the number line and then to highlight it using a dot. Figure 1-2 shows the graphs of the whole numbers 5 and 8.

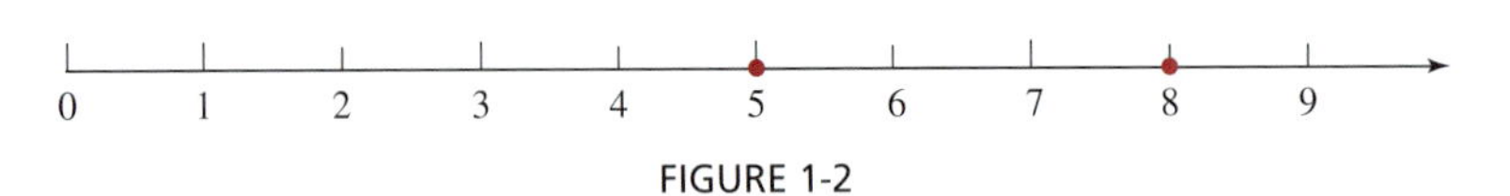

FIGURE 1-2

Ordering of the whole numbers

As we move to the right on the number line, the numbers get larger. Because 8 lies to the right of 5, we say that 8 is greater than 5. The **inequality symbol** $>$ ("is greater than") can be used to write this fact.

$8 > 5$ Read as "8 is greater than 5."

Since $8 > 5$, it is also true that $5 < 8$. (Read as "5 is less than 8.")

COMMENT To distinguish between these two inequality symbols, remember that the inequality symbol always points to the smaller of the two numbers involved.

$8 > 5$ $5 < 8$

Points to the smaller number

EXAMPLE 4 Place an < or > symbol in the box to make a true statement: **a.** 3 ☐ 7 and **b.** 18 ☐ 16.

Solution

a. Since 3 is to the left of 7 on the number line, $3 < 7$.

b. Since 18 is to the right of 16 on the number line, $18 > 16$.

Self Check 4

Place an < or > symbol in the box to make a true statement.

a. 12 ☐ 4 **b.** 7 ☐ 10

Answers **a.** >, **b.** <

Rounding whole numbers

When we don't need exact results, we often round numbers. For example, when a teacher with 36 students in his class orders 40 textbooks, he has rounded the actual number to the *nearest ten,* because 36 is closer to 40 than it is to 30.

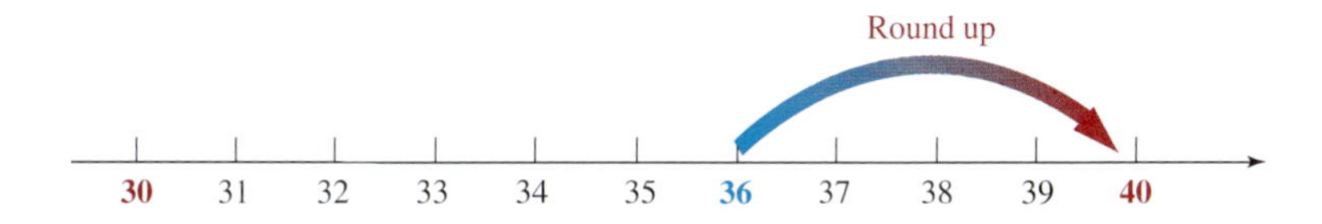

When a geologist says that the height of Alaska's Mount McKinley is "about 20,300 feet," she has rounded to the *nearest hundred,* because its actual height of 20,320 feet is closer to 20,300 than it is to 20,400.

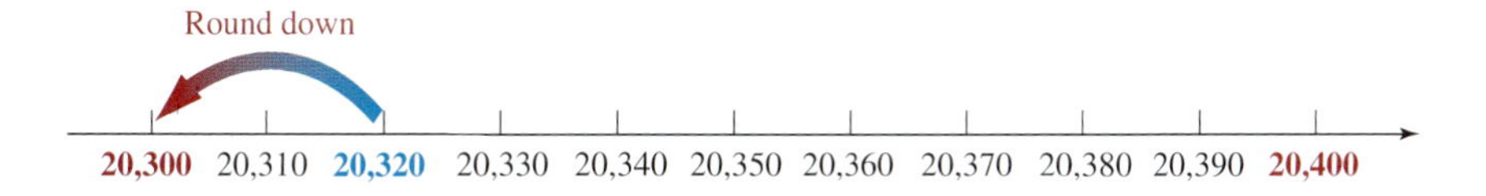

To round a whole number, we follow an established set of rules. To round a number to the nearest ten, for example, we begin by locating the **rounding digit** in the tens column. If the **test digit** to the right of that column (the digit in the ones column) is 5 or greater, we *round up* by increasing the tens digit by 1 and placing a 0 in the ones column. If the test digit is less than 5, we *round down* by leaving the tens digit unchanged and placing a 0 in the ones column.

EXAMPLE 5 Round each number to the nearest ten: **a.** 3,764 and **b.** 12,087.

Solution

a. We find the rounding digit in the tens column, which is 6.

Rounding digit

3,764

Test digit

We then look at the test digit to the right of 6, the 4 in the ones column. Since $4 < 5$, we round down by leaving the 6 unchanged and replacing the test digit with 0. The rounded answer is 3,760.

b. We find the rounding digit in the tens column, which is 8.

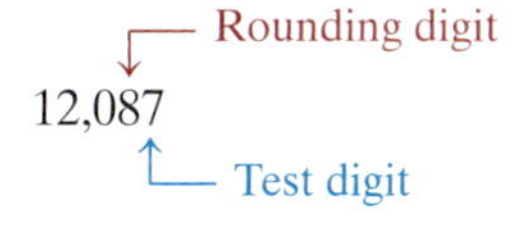

Self Check 5

Round each number to the nearest ten:

a. 35,642

b. 3,756

We then look at the test digit to the right of 8, the 7 in the ones column. Because $7 > 5$, we round up by adding 1 to 8 and replacing the test digit with 0. The rounded answer is 12,090.

Answers **a.** 35,640, **b.** 3,760

A similar procedure is used to round numbers to the nearest hundred, the nearest thousand, the nearest ten thousand, and so on.

Rounding a whole number

1. To round a number to a certain place, locate the rounding digit in that place.
2. Look at the test digit to the right of the rounding digit.
3. If the test digit is 5 or greater, round up by adding 1 to the rounding digit and changing all of the digits to the right of the rounding digit to 0.

   If the test digit is less than 5, round down by keeping the rounding digit and changing all of the digits to the right of the rounding digit to 0.

Self Check 6
Round 365,283 to the nearest hundred.

EXAMPLE 6 Round 7,960 to the nearest hundred.

Solution First, we find the rounding digit in the hundreds column. It is 9.

Rounding digit (9)
7,960
Test digit (6)

We then look at the 6 to the right of 9. Because $6 > 5$, we round up and increase 9 in the hundreds column by 1. Since the 9 in the hundreds column represents 900, increasing 9 by 1 represents increasing 900 to 1,000. Thus, we replace the 9 with a 0 and add 1 to the 7 in the thousands column. Finally, we replace the two rightmost digits with 0's. The rounded answer is 8,000.

Answer 365,300

Self Check 7
Round the elevation of Denver **a.** to the nearest hundred feet and **b.** to the nearest thousand feet.

EXAMPLE 7 **U.S. cities.** In 2003, Denver was the nation's 26th largest city. Round the 2003 population of Denver given in Figure 1-3 **a.** to the nearest thousand and **b.** to the nearest ten thousand.

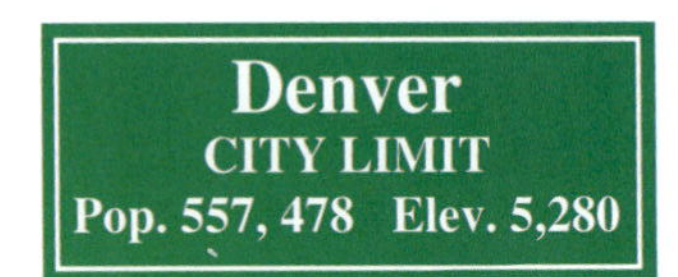

FIGURE 1-3

Solution

a. The rounding digit in the thousands column is 7. The test digit, 4, is less than 5, so we round down. To the nearest thousand, Denver's population in 2003 was 557,000.

b. The rounding digit in the ten thousands column is 5. The test digit, 7, is greater than 5, so we round up. To the nearest ten thousand, Denver's population in 2003 was 560,000.

Answers **a.** 5,300 ft, **b.** 5,000 ft

Tables and graphs

The table in Figure 1-4(a) on the next page is an example of the use of whole numbers. It shows the number of women elected to the United States House of Representatives in the congressional elections held every two years from 1996 to 2004.

Table

Year	1996	1998	2000	2002	2004
Number of women elected	54	56	59	60	65

(a)

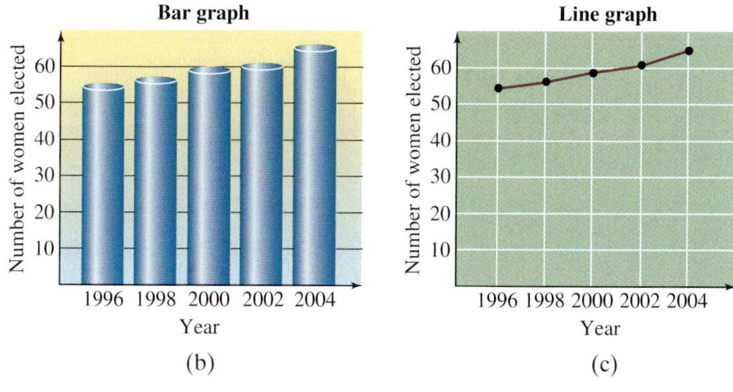

(b) (c)

FIGURE 1-4

In Figure 1-4(b), the election results are presented in a **bar graph.** The horizontal axis is labeled "Year" and scaled in units of 2 years. The vertical axis is labeled "Number of women elected" and scaled in units of 10. The bar directly over each year extends to a height indicating the number of women elected to Congress that year.

Another way to present the information in the table is with a **line graph.** Instead of using a bar to denote the number of women elected, we use a dot drawn at the correct height. After drawing data points for 1996, 1998, 2000, 2002, and 2004, we connect the points with line segments to create the line graph in Figure 1-4(c).

Re-entry Students

THINK IT THROUGH

"A re-entry student is considered one who is the age of 25 or older, or those students that have had a break in their academic work for 5 years or more. Nationally, this group of students is growing at an astounding rate."

Student Life and Leadership Department, University Union, Cal Poly University, San Luis Obispo

Some common concerns expressed by adult students considering returning to school are listed below in Column I. Match each concern to an encouraging reply in Column II.

Column I

1. I'm too old to learn.
2. I don't have the time.
3. I didn't do well in school the first time around. I don't think a college would accept me.
4. I'm afraid I won't fit in.
5. I don't have the money to pay for college.

Column II

a. Many students qualify for some type of financial aid.

b. Taking even a single class puts you one step closer to your educational goal.

c. There's no evidence that older students can't learn as well as younger ones.

d. More than 41% of the students in college are older than 25.

e. Typically, community colleges and career schools have an open admissions policy.

Adapted from *Common Concerns for Adult Students,* Minnesota Higher Education Services Office

Section 1.1 STUDY SET

VOCABULARY *Fill in the blanks.*

1. A ____ is a collection of objects.
2. The set of ______ numbers is {1, 2, 3, 4, 5, . . .}, and the set of ______ numbers is {0, 1, 2, 3, 4, 5, . . .}.
3. When 297 is written as 2 hundreds + 9 tens + 7 ones, it is written in ________ notation.
4. If we ______ 627 to the nearest ten, we get 630.
5. Using a process known as graphing, we can represent whole numbers as points on the ______ line.
6. The symbols > and < are ________ symbols.

CONCEPTS *Consider the numeral 57,634.*

7. What digit is in the tens column?
8. What digit is in the thousands column?
9. What digit is in the hundreds column?
10. What digit is in the ten thousands column?
11. What set of numbers is obtained when 0 is combined with the natural numbers?
12. Place the numbers 25, 17, 37, 15, 45 in order, from smallest to largest.
13. Graph: 1, 3, 5, and 7.

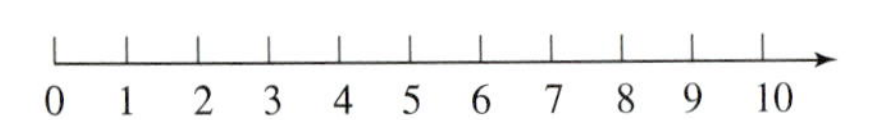

14. Graph: 0, 2, 4, 6, and 8.

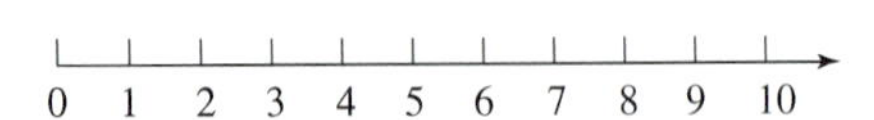

15. Graph: the whole numbers less than 6.

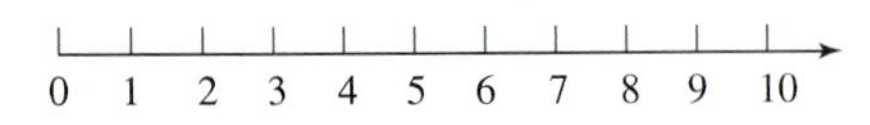

16. Graph: the whole numbers between 2 and 8.

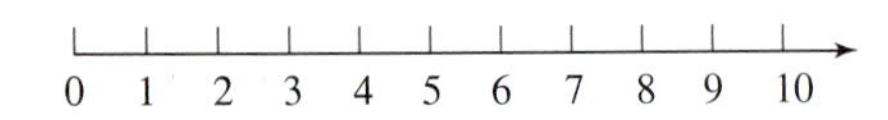

Place an > or < symbol in the box to make a true statement.

17. 47 ☐ 41
18. 53 ☐ 67
19. 309 ☐ 300
20. 841 ☐ 814
21. 2,052 ☐ 2,502
22. 999 ☐ 998
23. Since 4 < 7, it is also true that 7 ☐ 4.
24. Since 9 > 0, it is also true that 0 ☐ 9.

NOTATION *Fill in the blanks.*

25. The symbols { }, called ______, are used when writing a set.
26. The symbol > means ___ _____ _____, and the symbol < means ___ ____ ____.

PRACTICE *Write each number in expanded notation and then write it in words.*

27. 245
28. 508
29. 3,609
30. 3,960
31. 32,500
32. 73,009
33. 104,401
34. 570,003

Write each number in standard notation.

35. 4 hundreds + 2 tens + 5 ones
36. 7 hundreds + 7 tens + 7 ones
37. 2 thousands + 7 hundreds + 3 tens + 6 ones
38. 7 billions + 3 hundreds + 5 tens
39. Four hundred fifty-six
40. Three thousand seven hundred thirty-seven
41. Twenty-seven thousand five hundred ninety-eight
42. Seven million, four hundred fifty-two thousand, eight hundred sixty
43. Nine thousand one hundred thirteen
44. Nine hundred thirty
45. Ten million, seven hundred thousand, five hundred six
46. Eighty-six thousand four hundred twelve

Round 79,593 to the nearest . . .

47. ten

48. hundred

49. thousand

50. ten thousand

Round 5,925,830 to the nearest . . .

51. thousand

52. ten thousand

53. hundred thousand

54. million

Round $419,161 to the nearest . . .

55. $10

56. $100

57. $1,000

58. $10,000

APPLICATIONS

59. EATING HABITS The following list shows the ten countries with the largest per-person annual consumption of meat. Construct a two-column table that presents the data in order, beginning with the largest per-person consumption. (The abbreviation "lb" means "pounds.")

Australia: 239 lb *New Zealand: 259 lb*
Austria: 229 lb *Saint Lucia: 222 lb*
Canada: 211 lb *Spain: 211 lb*
Cyprus: 236 lb *Uruguay: 230 lb*
Denmark: 219 lb *United States: 261 lb*

60. PRESIDENTS The following list shows the ten youngest U.S. presidents and their ages (in years/days) when they took office. Construct a two-column table that presents the data in order, beginning with the youngest president.

C. Arthur 50 yr/350 days *U. Grant 46 yr/236 days*
G. Cleveland 47 yr/351 days *J. Kennedy 43 yr/236 days*
W. Clinton 46 yr/154 days *F. Pierce 48 yr/101 days*
M. Filmore 50 yr/184 days *J. Polk 49 yr/122 days*
J. Garfield 49 yr/105 days *T. Roosevelt 42 yr/322 days*

61. MISSIONS TO MARS The United States, Russia, Europe, and Japan have launched Mars space probes. The graph in the next column shows the success rate of the missions, by decade.

a. What decade had the greatest number of successful or partially successful missions? How many?

b. What decade had the greatest number of unsuccessful missions? How many?

c. Which decade had the greatest number of missions? How many?

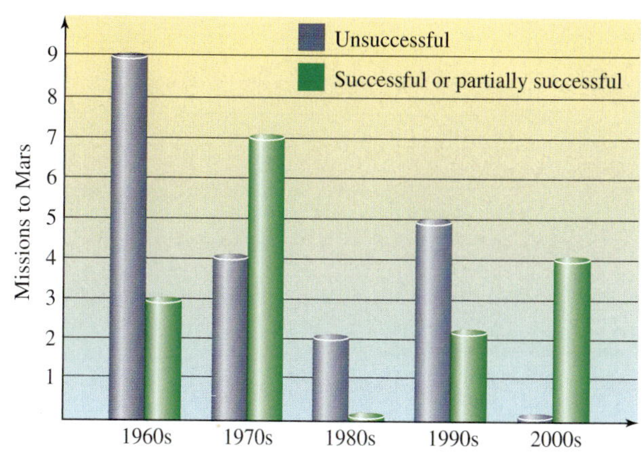

Source: The Planetary Society

62. BANKING The illustration shows the number of banks that were closed or taken over by federal agencies during the years 1935–1995.

a. During what two time spans was there an upsurge in bank failures?

b. In what year were there the most bank failures? Estimate the number of banks that failed that year.

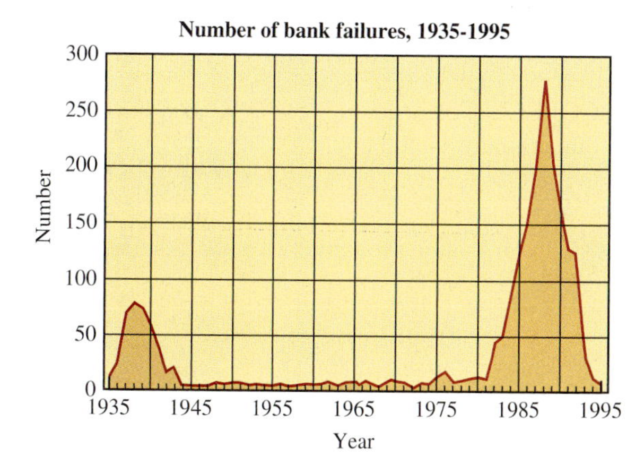

Source: FDIC Division of Research and Statistics

63. ENERGY RESERVES Construct a bar graph (see the next page) using the data shown in the table.

NATURAL GAS RESERVES, 2003 (IN TRILLION CUBIC FEET)	
United States	187
Venezuela	148
Canada	60
Argentina	27
Mexico	9

Source: *Oil and Gas Journal*

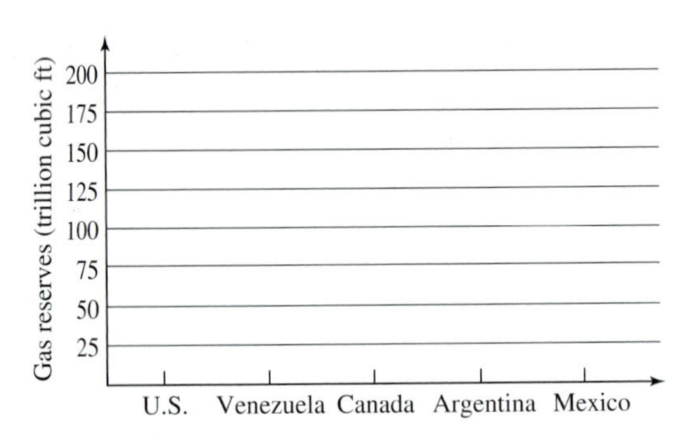

64. ENERGY RESERVES Construct a line graph using the data in the table on the previous page.

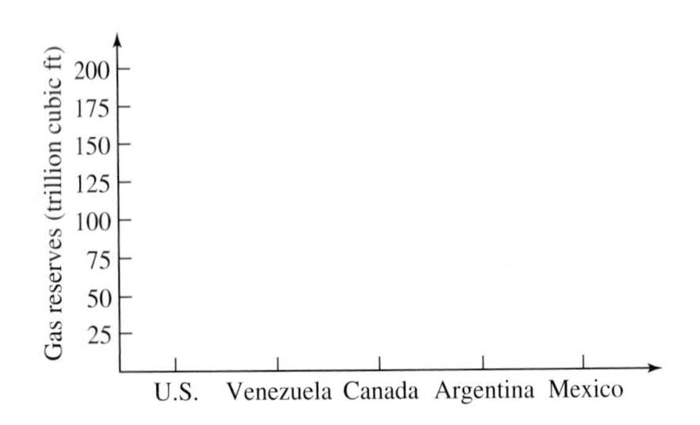

65. COFFEE Construct a line graph using the data in the table.

STARBUCKS LOCATIONS

Year	Number
1997	1,412
1998	1,886
1999	2,135
2000	3,501
2001	4,709
2002	5,886
2003	7,225
2004	8,337

Source: Starbucks Company

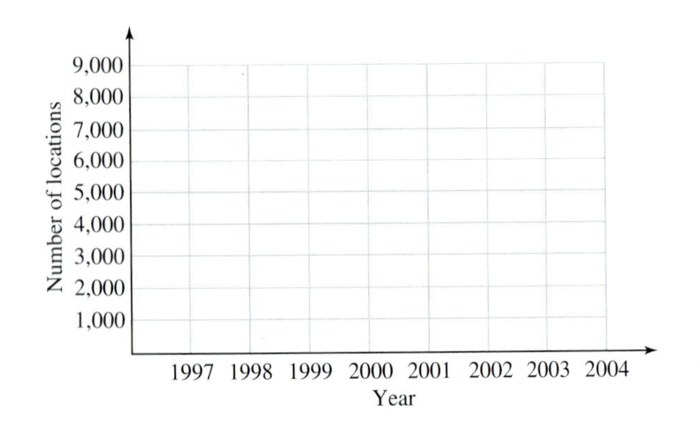

66. COFFEE Construct a bar graph using the data in the table in Exercise 65.

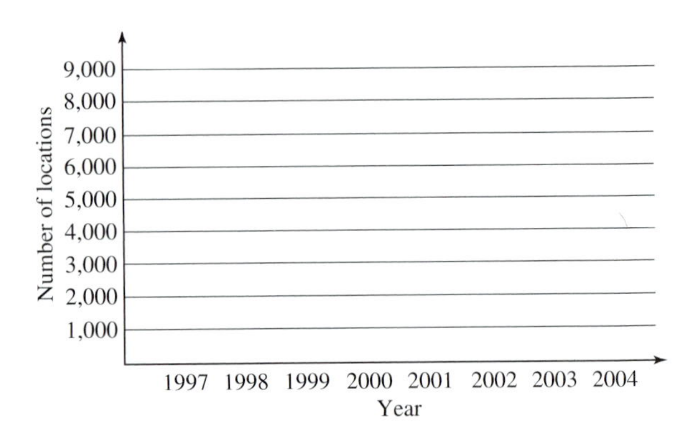

67. Complete each check by writing the amount in words on the proper line.

a.

No. 201 March 9, 20 05

Payable to Davis Chevrolet $ 15,601.00

____________________ DOLLARS

45-365-02 Don Smith

b.

No. 7890 Aug. 12, 20 05

Payable to Dr. Anderson $ 3,433.00

____________________ DOLLARS

45-828-02 Juan Decito

68. ANNOUNCEMENTS One style used when printing formal invitations and announcements is to write all numbers in words. Use this style to write each of the following phrases.

a. This diploma awarded this 27th day of June 2004.

b. The suggested contribution for the fundraiser is $850 a plate, or an entire table may be purchased for $5,250.

69. EDITING Edit this excerpt from a history text by circling all numbers written in words and rewriting them using digits.

Abraham Lincoln was elected with a total of one million, eight hundred sixty-five thousand, five hundred ninety-three votes — four hundred eighty-two thousand, eight hundred eighty more than the runner-up, Stephen Douglas. He was assassinated after having served a total of one thousand five hundred three days in office. Lincoln's Gettysburg Address, a mere two hundred sixty-nine words long, was delivered at the battle site where forty-three thousand four hundred forty-nine casualties occurred.

70. READING METERS The amount of electricity used in a household is measured by a meter in kilowatt-hours (kwh). Determine the reading on the meter shown in the illustration. (When the pointer is between two numbers, read the lower number.)

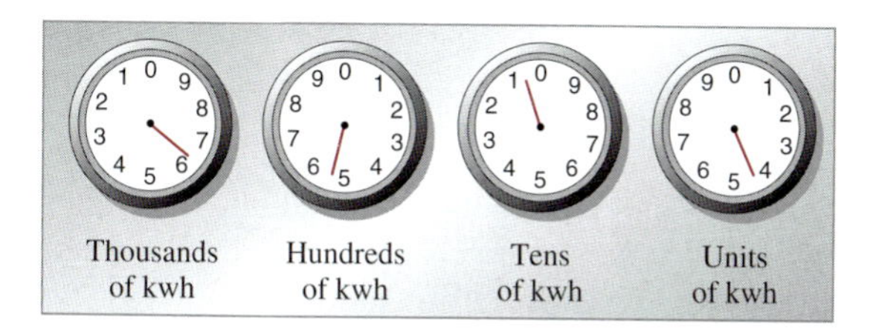

71. SPEED OF LIGHT The speed of light in a vacuum is 299,792,458 meters per second. Round this number

a. to the nearest hundred thousand meters per second.

b. to the nearest million meters per second.

72. CLOUDS Draw a vertical number line scaled from 0 to 40,000 feet, in units of 5,000 feet. Graph each cloud type given in the table at the proper altitude.

Cloud type	Altitude (ft)
Altocumulus	21,000
Cirrocumulus	37,000
Cirrus	38,000
Cumulonimbus	15,000
Cumulus	8,000
Stratocumulus	9,000
Stratus	4,000

WRITING

73. Explain why the natural numbers are called the counting numbers.

74. Explain how you would round 687 to the nearest ten.

75. The houses in a new subdivision are priced "in the low 130s." What does this mean?

76. A million is a thousand thousands. Explain why this is so.

1.2 Adding and Subtracting Whole Numbers

- Properties of addition
- Adding whole numbers
- Perimeter of a rectangle and a square
- Subtracting whole numbers
- Combinations of operations

Mastering addition and subtraction of whole numbers enables us to solve many problems from geometry, business, and science. For example, to find the distance around a rectangle, we need to add the lengths of its four sides. To prepare an annual budget, we need to add separate line items. To find the difference between two temperatures, we need to subtract one from the other.

Properties of addition

Addition is the process of finding the total of two (or more) numbers. It can be illustrated using the number line. (See Figure 1-5 on the next page.) For example, to compute $4 + 5$, we begin at 0 and draw an arrow 4 units long, extending to the right. This represents 4. From the tip of that arrow, we draw another arrow 5 units long, also extending to the right. The second arrow points to 9. This result corresponds to the addition fact $4 + 5 = 9$, where 4 and 5 are called **addends** or **terms,** and 9 is called the **sum.**

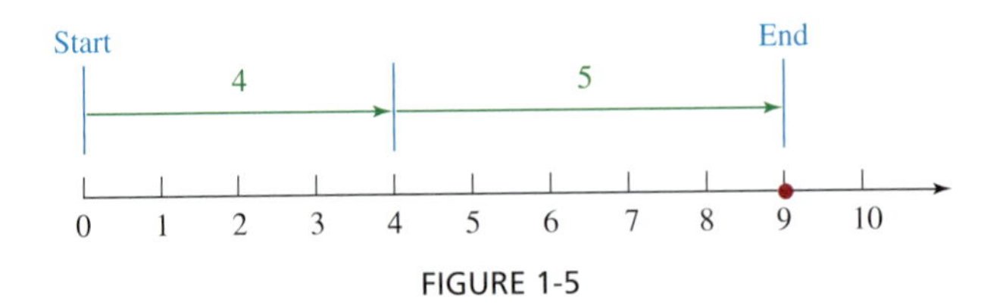

FIGURE 1-5

We have used a number line to find that $4 + 5 = 9$. If we add 4 and 5 in the opposite order, Figure 1-6 shows that we get the same result: $5 + 4 = 9$.

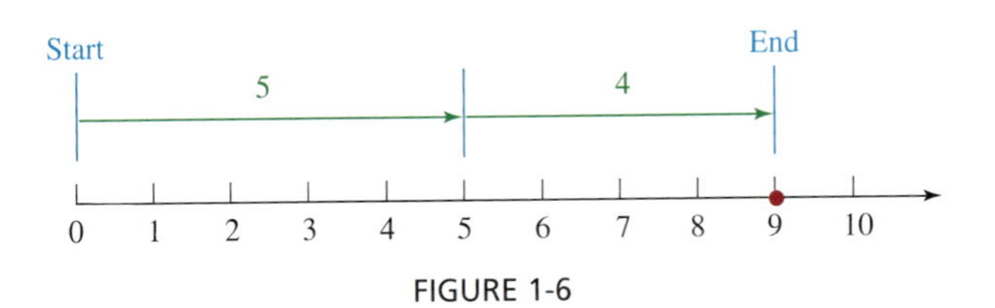

FIGURE 1-6

These examples illustrate that two whole numbers can be added in either order to get the same sum. The order in which we add two numbers does not affect the result. This property is called the **commutative property of addition.** To state the commutative property of addition concisely, we can use variables.

Variables

A **variable** is a letter that is used to stand for a number.

We now use the variables a and b to state the commutative property of addition.

Commutative property of addition

If a and b represent numbers, then

$$a + b = b + a$$

To find the sum of three whole numbers, we add two of them and then add the third to that result. In the following examples, we add $3 + 4 + 2$ in two ways. We will use the **grouping symbols** ( ), called **parentheses,** to show this. We must perform the operation within parentheses first.

Method 1: Group 3 + 4

$(\mathbf{3 + 4}) + 2 = \mathbf{7} + 2$ Because of the parentheses, add 3 and 4 first to get 7.

$= 9$ Then add 7 and 2 to get 9.

Method 2: Group 4 + 2

$3 + (\mathbf{4 + 2}) = 3 + \mathbf{6}$ Because of the parentheses, add 4 and 2 to get 6.

$= 9$ Then add 3 and 6 to get 9.

Either way, the sum is 9. It does not matter how we group (or associate) numbers in addition. This property is called the **associative property of addition.**

Associative property of addition

If a, b, and c represent numbers, then

$$(a + b) + c = a + (b + c)$$

Whenever we add 0 to a number, the number remains the same. For example,

$$3 + 0 = 3, \quad 5 + 0 = 5, \quad \text{and} \quad 9 + 0 = 9$$

These examples suggest the **addition property of 0.**

Addition property of 0

If a represents any number, then

$$a + 0 = a \quad \text{and} \quad 0 + a = a$$

EXAMPLE 1 Find each sum: **a.** $8 + 9$ and $9 + 8$, **b.** $5 + (1 + 8)$ and $(5 + 1) + 8$, and **c.** $(3 + 0) + 4$.

Solution

a. $8 + 9 = 17$ and $9 + 8 = 17$ The results are the same.

b. In each case, we perform the addition within parentheses first.

$$5 + (1 + 8) = 5 + 9 = 14 \qquad (5 + 1) + 8 = 6 + 8 = 14$$

The results are the same.

c. $(3 + 0) + 4 = 3 + 4 = 7$ Perform the addition within parentheses first: $3 + 0 = 3$.

Self Check 1

Find each sum:

a. $6 + 7$ and $7 + 6$

b. $2 + (6 + 3)$ and $(2 + 6) + 3$

c. $3 + (0 + 4)$

Answers **a.** 13, 13, **b.** 11, 11, **c.** 7

Adding whole numbers

We can add whole numbers greater than 10 by using a vertical format that adds digits with the same place value. Because the additions in each column often exceed 9, it is sometimes necessary to *carry* the excess to the next column to the left. For example, to add 27 and 15, we write the numerals with the digits of the same place value aligned vertically.

$$\begin{array}{r} 27 \\ +15 \\ \hline \end{array}$$

This is called vertical format.

We begin by adding the digits in the ones column: $7 + 5 = 12$. Because $12 =$ 1 ten and 2 ones, we place a 2 in the ones column of the answer and carry 1 to the tens column.

$$\begin{array}{r} \scriptstyle 1 \\ 27 \\ +15 \\ \hline 2 \end{array}$$

Add the digits in the ones column: $7 + 5 = 12$. Carry 1 (shown in blue) to the tens column.

Then we add the digits in the tens column.

$$\begin{array}{r} \scriptstyle 1 \\ 27 \\ +15 \\ \hline 42 \end{array}$$

Add 1, 2, and 1. Place the result, 4, in the tens column of the answer.

Thus, $27 + 15 = 42$.

Self Check 2
Add: 675 + 1,497.

EXAMPLE 2 Add: 9,834 + 692.

Solution We write the numerals with their corresponding digits aligned vertically. Then we add the numbers, one column at a time, working from right to left.

$$\begin{array}{r} 9{,}834 \\ +\ \ \ 692 \\ \hline 6 \end{array}$$

Add the digits in the ones column and place the result in the ones column of the answer.

$$\begin{array}{r} \scriptstyle 1\ \ \\ 9{,}834 \\ +\ \ \ 692 \\ \hline 26 \end{array}$$

Add the digits in the tens column. The result, 12, exceeds 9. Place the 2 in the tens column of the answer and carry 1 (shown in blue) to the hundreds column.

$$\begin{array}{r} \scriptstyle 1\ 1\ \ \ \ \\ 9{,}834 \\ +\ \ \ 692 \\ \hline 526 \end{array}$$

Add the digits in the hundreds column. Since the result, 15, exceeds 9, place the 5 in the hundreds column of the answer and carry 1 (shown in green) to the thousands column.

$$\begin{array}{r} \scriptstyle 1\ 1\ \ \ \ \\ 9{,}834 \\ +\ \ \ 692 \\ \hline 10{,}526 \end{array}$$

Since the sum of the digits in the thousands column is 10, write 0 in the thousands column and 1 in the ten thousands column of the answer.

Thus, 9,834 + 692 = 10,526.

Answer 2,172

To see if the result in Example 2 is reasonable, we can **estimate** the answer. We know that 9,834 is a little less than 10,000, and 692 is a little less than 700. We estimate that the answer will be a little less than 10,000 + 700, or 10,700. An answer of 10,526 is reasonable. Estimation is discussed in more detail later in this chapter.

Words such as *increase, gain, credit, up, forward, rise, in the future, and to the right* are used to indicate addition.

EXAMPLE 3 **Calculating temperatures.** At noon, the temperature in Helena, Montana, was 31°. By 1:00 P.M., the temperature had increased 5°, and by 2:00 P.M., it had risen another 7°. Find the temperature at 2:00 P.M.

Solution To the temperature at noon, we add the two increases.

$$31 + 5 + 7$$

The two additions are done working from left to right.

$$\begin{aligned} 31 + 5 + 7 &= 36 + 7 \\ &= 43 \end{aligned}$$

The temperature at 2:00 P.M. was 43°.

Self Check 4
By 1700, the populations of the four colonies were New Hampshire 5,000, New York 19,100, Massachusetts 55,900, and Virginia 58,600. Find the total population.

EXAMPLE 4 **History.** The populations of four American colonies in 1630 are shown in Figure 1-7 on the next page. Find the total population.

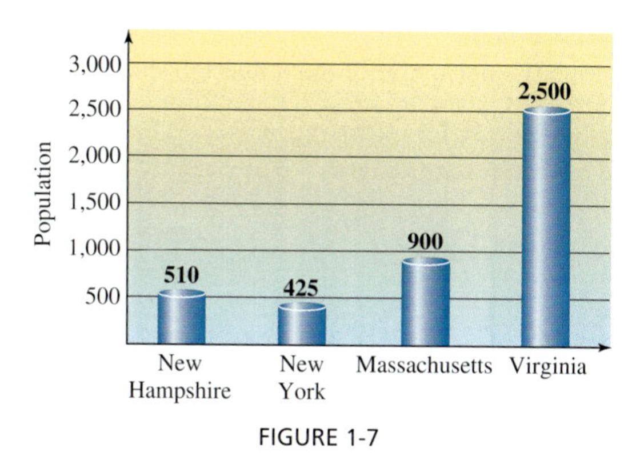

FIGURE 1-7

Solution The word *total* indicates that we must add the populations of the colonies.

$$\begin{array}{r} \overset{2}{5}10 \\ 425 \\ 900 \\ +2{,}500 \\ \hline 4{,}335 \end{array}$$

Align the numerals vertically. Add the digits, one column at a time, working from right to left.

The total population was 4,335.

Answer 138,600

Perimeter of a rectangle and a square

Figure 1-8(a) is an example of a four-sided figure called a **rectangle.** Either of the longer sides of a rectangle is called its **length** and either of the shorter sides is called its **width.** Together, the length and width are called the **dimensions** of the rectangle. For any rectangle, opposite sides have the same measure.

When all four of the sides of a rectangle have the same measure, we call the rectangle a **square.** An example of a square is shown in Figure 1-8(b).

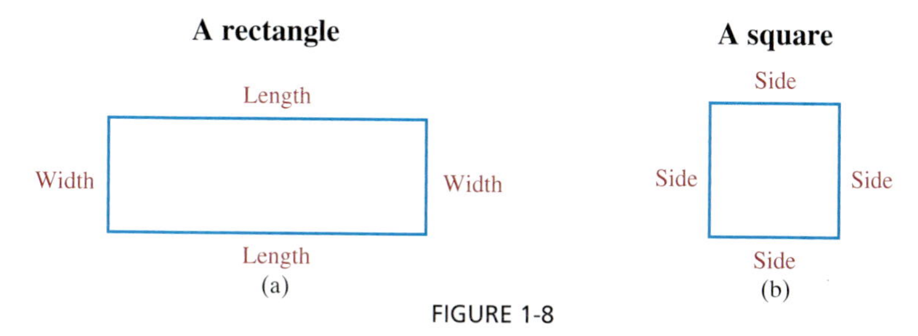

FIGURE 1-8

The distance around a rectangle or a square is called its **perimeter.** To find the perimeter of a rectangle, we add the lengths of its four sides.

The perimeter of a rectangle = length + length + width + width

To find the perimeter of a square, we add the lengths of its four sides.

The perimeter of a square = side + side + side + side

Self Check 5
A Monopoly game board is square, with sides 19 inches long. Find the perimeter of the board.

Answer 76 in.

EXAMPLE 5 Find the perimeter of the dollar bill shown in Figure 1-9.

Solution To find the perimeter of the rectangular-shaped bill, we add the lengths of its four sides.

$$\begin{array}{r} \overset{2\,2}{156} \\ 156 \\ 65 \\ +\ 65 \\ \hline 442 \end{array}$$

FIGURE 1-9

The perimeter is 442 mm.

To see whether this result is reasonable, we estimate the answer. Because the rectangle is about 150 mm by 70 mm, its perimeter is approximately 150 + 150 + 70 + 70, or 440 mm. An answer of 442 mm is reasonable.

Subtracting whole numbers

Subtraction is the process of finding the difference between two numbers. It can be illustrated using the number line. (See Figure 1-10.) For example, to compute $9 - 4$, we begin at 0 and draw an arrow 9 units long, extending to the right. From the tip of that arrow, we draw another arrow 4 units long, but extending to the left. (This represents taking away 4.) The second arrow points to 5, indicating that $9 - 4 = 5$. In this subtraction fact, 9 is called the **minuend,** 4 is called the **subtrahend,** and 5 is called the **difference.**

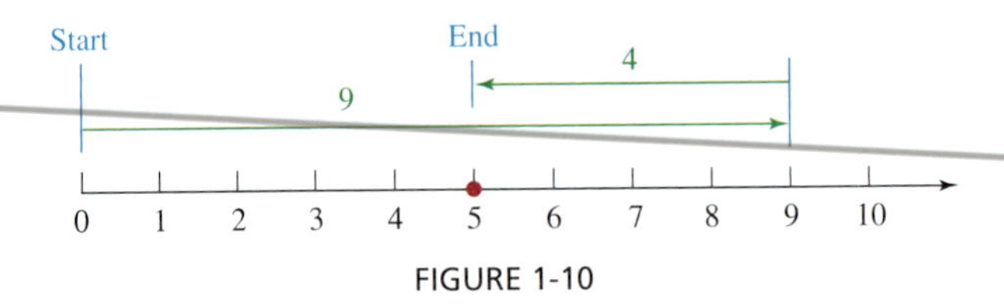

FIGURE 1-10

With whole numbers, we cannot subtract in the opposite order and find the difference $4 - 9$, because we cannot take away 9 objects from 4 objects. Since subtraction of whole numbers cannot be done in either order, **subtraction is not commutative.**

Subtraction is not associative either, because if we group in different ways, we get different answers. For example,

$$(9 - 5) - 1 = 4 - 1 \quad \text{but} \quad 9 - (5 - 1) = 9 - 4$$
$$= 3 \qquad\qquad\qquad = 5$$

Self Check 6
Perform the subtractions: $8 - (5 - 2)$ and $(8 - 5) - 2$.

Answers 5, 1

EXAMPLE 6 Perform the subtractions: $9 - (6 - 3)$ and $(9 - 6) - 3$.

Solution In each case, we perform the subtraction within parentheses first.

$$9 - (6 - 3) = 9 - 3 \qquad (9 - 6) - 3 = 3 - 3$$
$$= 6 \qquad\qquad\qquad = 0$$

Note that the results are different.

Whole numbers can be subtracted using a vertical format. Because subtractions often require subtracting a larger digit from a smaller digit, we may need to *borrow*. For example, to subtract 15 from 32, we write the minuend, 32, and the subtrahend, 15, in a vertical format, aligning the digits with the same place value.

$$\begin{array}{r} 32 \\ -15 \\ \hline \end{array}$$

Write the numerals in a column, with corresponding digits aligned vertically.

Since 5 can't be subtracted from 2, we borrow from the tens column of 32.

$$\begin{array}{r} \overset{2}{\not3}\,\overset{12}{\not2} \\ -1\;5 \\ \hline 7 \end{array}$$

To subtract in the ones column, borrow 1 ten from the tens column. We show this by drawing a slash through the 3 and writing a 2 above it. Add 10 to the 2 in the ones column, which gives 12. Then subtract: 12 − 5 = 7.

$$\begin{array}{r} \overset{2}{\not3}\,\overset{12}{\not2} \\ -1\;5 \\ \hline 1\;7 \end{array}$$

Subtract in the tens column: 2 − 1 = 1.

Thus, 32 − 15 = 17. To check the result, we add the difference, 17, and the subtrahend, 15. We should obtain the minuend, 32.

Check:

$$\begin{array}{r} \overset{1}{1}7 \\ +15 \\ \hline 32 \end{array}$$

EXAMPLE 7 Subtract 576 from 2,021.

Solution The number to be subtracted is 576. This sentence translates to 2,021 − 576. In vertical format, we have

$$\begin{array}{r} 2{,}021 \\ -\;576 \\ \hline \end{array}$$

Write the numerals in a column, with the digits of the same place value aligned vertically.

$$\begin{array}{r} 2{,}0\,\overset{1}{\not2}\,\overset{11}{\not1} \\ -\;5\;7\;6 \\ \hline 5 \end{array}$$

To subtract in the ones column, borrow 1 ten from the tens column and add it to the ones column. Then subtract: 11 − 6 = 5.

Since we can't subtract 7 from 1 in the tens column, we borrow. Because there is a 0 in the hundreds column of 2,021, we must borrow from the thousands column. We can take 1 thousand from the thousands column (leaving 1 thousand behind) and write it as 10 hundreds, placing a 10 in the hundreds column. From these 10 hundreds, we take 1 hundred (leaving 9 hundreds behind) and think of it as 10 tens. We add these 10 tens to the 1 ten that is already in the tens column to get 11 tens. From these 11 tens, we subtract 7 tens: 11 − 7 = 4.

$$\begin{array}{r} \overset{1}{\not2},\overset{\overset{9}{\not{10}}}{\not0}\,\overset{11}{\not2}\,\overset{11}{\not1} \\ -\;5\;7\;6 \\ \hline 4\;5 \end{array}$$

To subtract in the tens column, borrow 10 hundreds from the thousands digit and add it to the hundreds digit. Borrow 10 tens from the hundreds digit and add it to the tens digit. Then subtract: 11 − 7 = 4.

$$\begin{array}{r} \overset{1}{\not2},\overset{\overset{9}{\not{10}}}{\not0}\,\overset{11}{\not2}\,\overset{11}{\not1} \\ -\;5\;7\;6 \\ \hline 4\;4\;5 \end{array}$$

Subtract in the hundreds column: 9 − 5 = 4.

Self Check 7

Subtract 1,445 from 2,021. Then check the result using addition.

$$\begin{array}{r} \overset{1}{2},\overset{\overset{9}{10}}{\cancel{0}}\,\overset{11}{\cancel{2}}\,\overset{11}{\cancel{1}} \\ -\quad 5\ 7\ 6 \\ \hline 1,4\ 4\ 5 \end{array}$$

Subtract in the thousands column: $1 - 0 = 1$.

Answer 576; $576 + 1,445 = 2,021$

Thus, $2,021 - 576 = 1,445$. Check the result using addition.

Words such as *minus, decrease, loss, debit, down, backward, fall, reduce, in the past,* and *to the left* indicate subtraction.

EXAMPLE 8 Vehicle crashes. In 2000, the number of motor vehicle traffic crashes in the United States was 6,394,000. That number declined in 2001, dropping by 71,000. In 2002, it fell by an additional 7,000. How many motor vehicle traffic crashes were there in 2002?

Solution The words *dropping* and *fell* indicate subtraction. We can show the calculations necessary to solve this example in a single expression, as shown below. The two subtractions are done working from left to right.

$$\begin{aligned} 6,394,000 - 71,000 - 7,000 &= 6,323,000 - 7,000 \\ &= 6,316,000 \end{aligned}$$

In 2002, there were 6,316,000 motor vehicle traffic crashes in the United States.

To answer questions about *how much more* or *how many more,* subtraction can be used.

Combinations of operations

Additions and subtractions often appear in the same problem. It is important to read the problem carefully, locate the useful information, and organize it correctly.

Self Check 9

One share of ABC Corporation stock cost $75. The price fell $7 per share. However, it recovered and rose $13 per share. What is its current price?

Answer $81

EXAMPLE 9 Bus passengers. Twenty-seven people were riding a bus on Route 47. At the Seventh Street stop, 16 riders got off the bus and 5 got on. How many riders were left on the bus?

Solution The route and street number are not important. The phrase *got off the bus* indicates subtraction, and the phrase *got on* indicates addition. The number of riders on the bus can be found by calculating $27 - 16 + 5$. Working from left to right, we have

$$\begin{aligned} 27 - 16 + 5 &= 11 + 5 \\ &= 16 \end{aligned}$$

There were 16 riders left on the bus.

! COMMENT When making the calculation in Example 9, we must perform the subtraction first. If the addition is done first, we obtain an incorrect answer of 6. For expressions containing addition and subtraction, perform them as they occur from *left to right.*

$$\begin{aligned} 27 - 16 + 5 &= 27 - 21 \\ &= 6 \end{aligned}$$

Calculators

CALCULATOR SNAPSHOT

A calculator can be helpful when you are checking an answer or performing a tedious computation. Before making regular use of one, make sure that you have mastered the fundamentals of arithmetic.

Several brands of calculators are available. For specific details about the operation of your calculator, please consult the owner's manual.

To check the addition done in Example 4 using a scientific calculator, we enter these numbers and press these keys.

510 [+] 425 [+] 900 [+] 2500 [=] 4335

The display shows that in 1630, the total population of the four colonies was 4,335.

We can use a scientific calculator to check the subtraction performed in Example 8 by entering these numbers and pressing these keys.

17126 [−] 937 [−] 253 [=] 15936

The display shows that there were 15,936 alcohol-related traffic deaths in 1998.

Section 1.2 STUDY SET

VOCABULARY *Fill in the blanks.*

1. When two numbers are added, the result is called a ______. The numbers that are to be added are called ______.
2. A ______ is a letter that stands for a number.
3. When two numbers are added, the result is called a ______.
4. The figure on the left is an example of a ______. The figure on the right is an example of a ______.

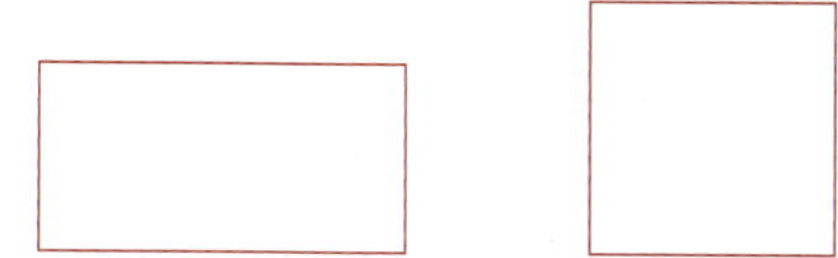

5. When two numbers are subtracted, the result is called a ______. In a subtraction problem, the ______ is subtracted from the ______.
6. The property that guarantees that we can add two numbers in either order and get the same sum is called the ______ property of addition.
7. The property that allows us to group numbers in an addition in any way we want is called the ______ property of addition.
8. The distance around a rectangle (or a square) is called its ______.

CONCEPTS *What property of addition is shown?*

9. $3 + 4 = 4 + 3$
10. $(3 + 4) + 5 = 3 + (4 + 5)$
11. $7 + (8 + 2) = (7 + 8) + 2$
12. $(8 + 5) + 1 = 1 + (8 + 5)$
13. a. Use the variables x and y to write the commutative property of addition.

    b. Use the variables x, y, and z to write the associative property of addition.
14. Show how to check the result:

$$\begin{array}{r} 74 \\ -29 \\ \hline 45 \end{array}$$

15. Fill in the blank: Any number added to ☐ stays the same.
16. a. In calculating $12 + (8 + 5)$, which numbers should be added first?

    b. In calculating $60 - 15 + 4$, which operation should be performed first?
17. What addition fact is illustrated below?

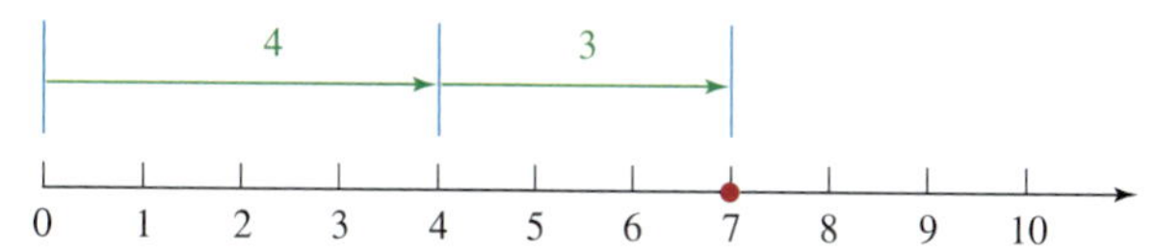

18. What subtraction fact is illustrated below?

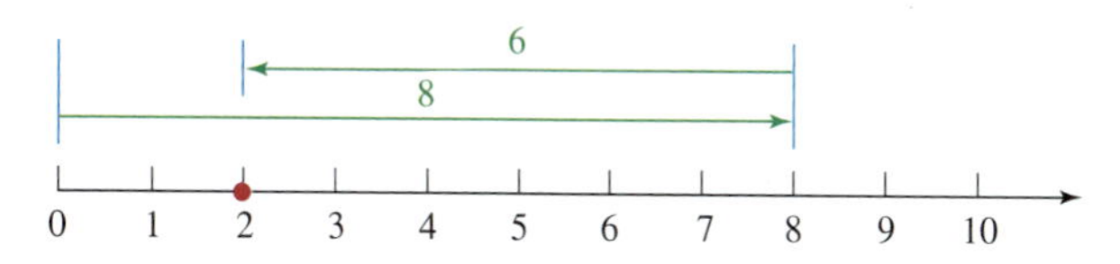

NOTATION *Fill in the blanks.*

19. The grouping symbols ( ) are called ___________.

20. The minus sign − means ___________.

Complete each solution.

21. $(36 + 11) + 5 = \square + 5$
$= \square$

22. $12 + (15 + 2) = 12 + \square$
$= \square$

PRACTICE *Perform each addition.*

23. 25 + 13

24. 47 + 12

25. 156 + 305

26. 647 + 38

27. 19 + 39 + 53

28. 27 + 16 + 48

29. (95 + 16) + 39

30. 832 + (97 + 27)

31. 25 + (321 + 17)

32. (4,231 + 213) + 5,234

33. $\begin{array}{r} 632 \\ +347 \\ \hline \end{array}$

34. $\begin{array}{r} 423 \\ +570 \\ \hline \end{array}$

35. $\begin{array}{r} 1,372 \\ +\ \ 613 \\ \hline \end{array}$

36. $\begin{array}{r} 2,477 \\ +\ \ 693 \\ \hline \end{array}$

37. $\begin{array}{r} 6,427 \\ +3,573 \\ \hline \end{array}$

38. $\begin{array}{r} 3,567 \\ +8,778 \\ \hline \end{array}$

39. $\begin{array}{r} 8,539 \\ +7,368 \\ \hline \end{array}$

40. $\begin{array}{r} 5,799 \\ +6,879 \\ \hline \end{array}$

41. $\begin{array}{r} 1,246 \\ 578 \\ +\ \ \ 37 \\ \hline \end{array}$

42. $\begin{array}{r} 4,689 \\ 3,422 \\ +\ \ \ 26 \\ \hline \end{array}$

43. $\begin{array}{r} 3,156 \\ 1,578 \\ +\ \ 578 \\ \hline \end{array}$

44. $\begin{array}{r} 2,379 \\ 4,779 \\ +2,339 \\ \hline \end{array}$

Find the perimeter of each rectangle or square.

45.

46.

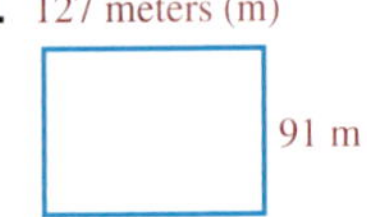

47. 17 inches (in.)

17 in. 17 in.

17 in.

48.

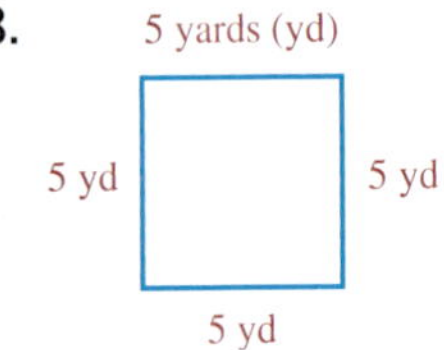

Perform each subtraction.

49. 17 − 14

50. 42 − 31

51. 39 − 14

52. 45 − 32

53. 174 − 71

54. 257 − 155

55. 633 − (598 − 30)

56. 600 − (497 − 60)

57. 160 − 15 − 4

58. 498 − 17 − 162

59. 29 − 17 − 12

60. 53 − 26 − 27

57. 160 − 15 − 4

58. 498 − 17 − 162

59. 29 − 17 − 12

60. 53 − 26 − 27

61. Subtract 343 from 367.

62. Subtract 122 from 224.

63. Subtract 305 from 423.

64. Subtract 270 from 330.

65. $\begin{array}{r} 1,537 \\ -\ \ 579 \\ \hline \end{array}$

66. $\begin{array}{r} 2,470 \\ -\ \ 863 \\ \hline \end{array}$

67. $\begin{array}{r} 4,267 \\ -2,578 \\ \hline \end{array}$

68. $\begin{array}{r} 7,356 \\ -3,578 \\ \hline \end{array}$

69. $\begin{array}{r} 17,246 \\ -\ \ 6,789 \\ \hline \end{array}$

70. $\begin{array}{r} 34,510 \\ -27,593 \\ \hline \end{array}$

71. $\begin{array}{r} 15,700 \\ -15,397 \\ \hline \end{array}$

72. $\begin{array}{r} 35,021 \\ -23,999 \\ \hline \end{array}$

Perform the computations.

73. 43 − 12 + 9

74. 59 − 16 + 2

75. 120 + 30 − 40

76. 600 + 99 − 54

APPLICATIONS

77. TAXIS For a 17-mile trip, Wanda paid the taxi driver $23. If $5 was a tip, how much was the fare?

78. SPACE FLIGHTS Astronaut Walter Schirra's first space flight orbited the Earth 6 times and lasted 9 hours. His second flight orbited the Earth 16 times and lasted 26 hours. How long was Schirra in space?

79. DOW JONES AVERAGE How much did the Dow rise on the day described by the graph?

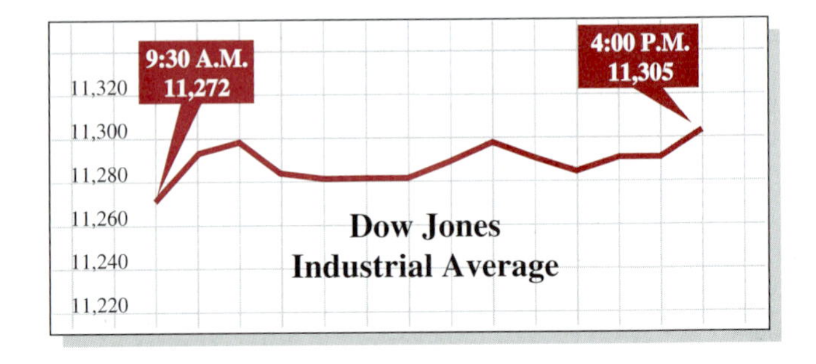

80. JEWELRY Gold melts at about 1,947° F. The melting point of silver is 183° F lower. What is the melting point of silver?

81. BANKING A savings account contained $370. After a deposit of $40 and a withdrawal of $197, how much is in the account?

82. TRAVEL A student wants to make the 2,221-mile trip from Detroit to Seattle in three days. If she drives 751 miles on the first day and 875 miles on the second day, how far must she travel on the third day?

83. TAX DEDUCTIONS For tax purposes, a woman kept the mileage records shown on the right. Find the total number of miles that she drove in the first 6 months of the year.

Month	Miles driven
January	2,345
February	1,712
March	1,778
April	445
May	1,003
June	2,774

84. COMPANY BUDGETS A department head prepared an annual budget with the line items shown on the right. Find the projected number of dollars to be spent.

Line item	Amount
Equipment	$17,242
Utilities	$5,443
Travel	$2,775
Supplies	$10,553
Development	$3,225
Maintenance	$1,075

Refer to the following table. To use this salary schedule, note that the annual salary of a third-year teacher with 15 units of course work beyond a Bachelor's degree is $30,887 (Step 3/Column 2).

Teachers' Salary Schedule			
Years teaching	Column 1: B.D.	Column 2: B.D. + 15	Column 3: B.D. + 30
Step 1	$26,785	$28,243	$29,701
Step 2	$28,107	$29,565	$31,023
Step 3	$29,429	$30,887	$32,345
Step 4	$30,751	$32,209	$33,667
Step 5	$32,073	$33,531	$34,989

85. INCOME How much money will a new teacher make in his first five years of teaching if he begins at

a. Step 1/Column 1?

b. Step 1/Column 3?

86. PAY INCREASES If a teacher is now on Step 2/Column 2, how much more money will she make next year when she

a. gains one year of teaching experience?

b. completes 15 units of course work?

87. BLUEPRINTS Find the length of the house shown in the blueprint.

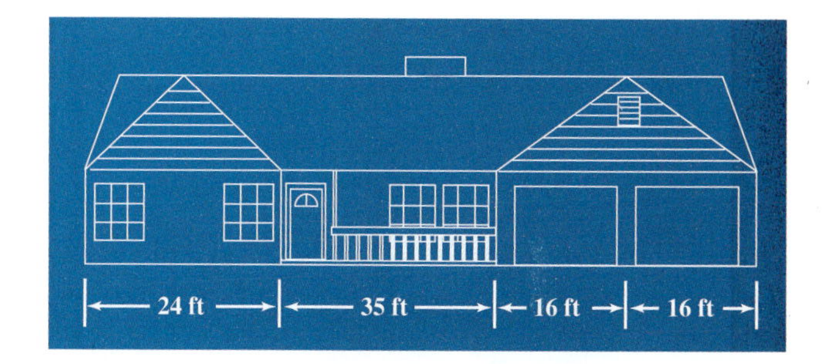

88. MACHINERY Find the length of the motor on the machine shown below (cm means centimeters).

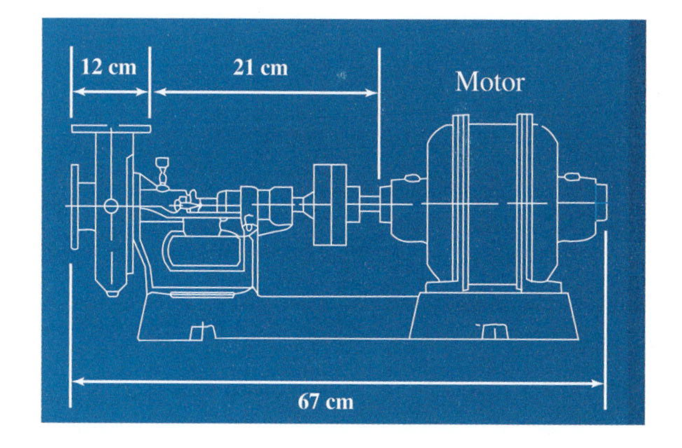

89. CANDY The graph below shows U.S. candy sales in 2003 during four holiday periods. Find the sum of these seasonal candy sales.

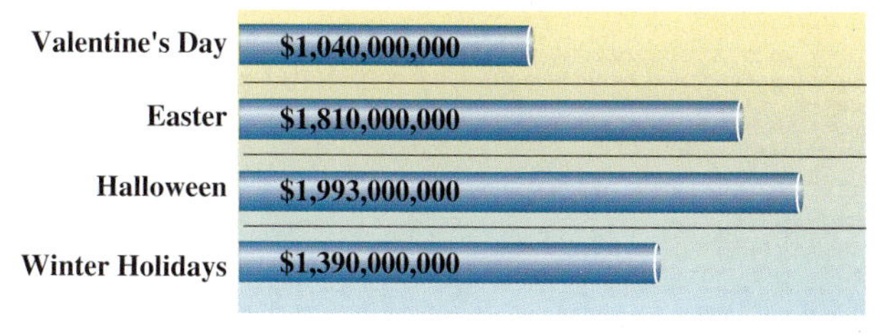

Source: National Confectioners Association

90. DALMATIANS See the graph below. Between which two years was the drop in registrations the greatest? What was that drop?

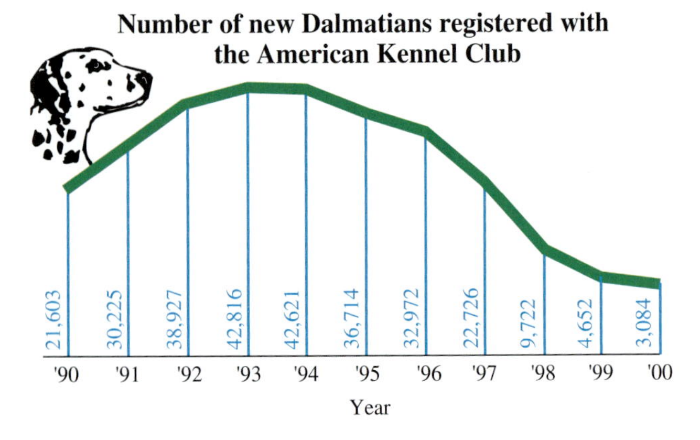

91. CITY FLAGS To decorate a city flag, yellow fringe is to be sewn around its outside edges, as shown. The fringe comes on long spools and is sold by the inch. How many inches of fringe must be purchased to complete the project?

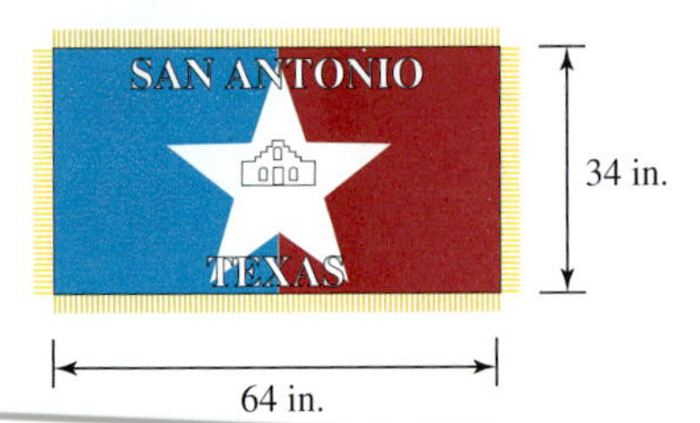

92. BOXING How much padded rope is needed to create the square boxing ring below if each side is 24 feet long?

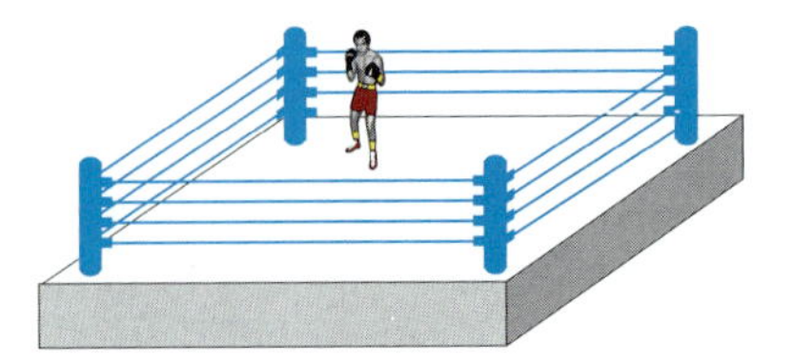

WRITING

93. Explain why the operation of addition is commutative.

94. Explain how addition can be used to check subtraction.

REVIEW *Write each numeral in expanded notation.*

95. 3,125

96. 60,037

Round 6,354,784 to the specified place.

97. Nearest ten

98. Nearest hundred

99. Nearest ten thousand

100. Nearest hundred thousand

1.3 Multiplying and Dividing Whole Numbers

- Properties of multiplication
- Multiplying whole numbers
- Finding the area of a rectangle
- Division
- Properties of division
- Dividing whole numbers

Mastering multiplication and division of whole numbers enables us to find areas of geometric figures and to solve many business and transportation problems. For example, to find the area of a rectangle, we need to multiply its length by its width. To figure a paycheck, we need to multiply the number of hours worked by the hourly rate of pay. To calculate the fuel economy of a bus, we need to divide the miles it travels by the number of gallons of gas that are used.

Properties of multiplication

Three symbols are used to indicate multiplication.

Symbols that are used for multiplication

Symbol		**Example**
$\times$	times symbol	4×5 or $\begin{array}{r} 4 \\ \times 5 \\ \hline \end{array}$
$\cdot$	raised dot	$4 \cdot 5$
( )	parentheses	$(4)(5)$ or $4(5)$ or $(4)5$

Recall that a variable is a letter that stands for a number. We often multiply a variable by another number or multiply a variable by another variable. When we do this, we don't need to use a symbol for multiplication.

$5a$ means $5 \cdot a$, ab means $a \cdot b$, and xyz means $x \cdot y \cdot z$

! COMMENT In this book, we seldom use the $\times$ symbol, because it can be confused with the letter x.

Multiplication is repeated addition. For example, $4 \cdot 5$ means the sum of four 5's.

$$4 \cdot 5 = \overbrace{5 + 5 + 5 + 5}^{\text{The sum of four 5's}}$$
$$= 20$$

In the above multiplication, the result of 20 is called a **product.** The numbers that were multiplied (4 and 5) are called **factors.**

$$\begin{array}{ccccc} \text{Factor} & & \text{Factor} & & \text{Product} \\ \downarrow & & \downarrow & & \downarrow \\ 4 & \cdot & 5 & = & 20 \end{array}$$

The multiplication $5 \cdot 4$ means the sum of five 4's.

$$5 \cdot 4 = \overbrace{4 + 4 + 4 + 4 + 4}^{\text{The sum of five 4's}}$$
$$= 20$$

We see that $4 \cdot 5 = 20$ and $5 \cdot 4 = 20$. The results are the same. These examples illustrate that the order in which we multiply two numbers does not affect the result. This property is called the **commutative property of multiplication.**

Commutative property of multiplication

If a and b represent numbers, then

$a \cdot b = b \cdot a$ or, more simply, $ab = ba$

Table 1-1 on the next page summarizes the basic multiplication facts.

- To find the product of 6 and 8 using the table, we find the intersection of the 6th row and the 8th column. The product is 48.
- To find the product of 8 and 6, we find the intersection of the 8th row and the 6th column. Once again, the product is 48.

In the table, we see that the set of answers above and the set of answers below the diagonal line in bold print are identical. This further illustrates that multiplication is commutative.

·	0	1	2	3	4	5	6	7	8	9
0	**0**	0	0	0	0	0	0	0	0	0
1	0	**1**	2	3	4	5	6	7	8	9
2	0	2	**4**	6	8	10	12	14	16	18
3	0	3	6	**9**	12	15	18	21	24	27
4	0	4	8	12	**16**	20	24	28	32	36
5	0	5	10	15	20	**25**	30	35	40	45
6	0	6	12	18	24	30	**36**	42	48	54
7	0	7	14	21	28	35	42	**49**	56	63
8	0	8	16	24	32	40	48	56	**64**	72
9	0	9	18	27	36	45	54	63	72	**81**

TABLE 1-1

From the table, we see that whenever we multiply a number by 0, the product is 0. For example,

$0 \cdot 5 = 0$, $0 \cdot 8 = 0$, and $9 \cdot 0 = 0$

We also see that whenever we multiply a number by 1, the number remains the same. For example,

$3 \cdot 1 = 3$, $7 \cdot 1 = 7$, and $1 \cdot 9 = 9$

These examples suggest the multiplication properties of 0 and 1.

Multiplication properties of 0 and 1

If a represents any number, then

$a \cdot 0 = 0$ and $0 \cdot a = 0$

$a \cdot 1 = a$ and $1 \cdot a = a$

Application problems that involve repeated addition are often more easily solved using multiplication.

Self Check 1

At a rate of \$8 per hour, how much will a school bus driver earn if she works from 8:00 A.M. until noon?

Answer \$32

EXAMPLE 1 Computing daily wages.

Raul worked an 8-hour day at an hourly rate of \$9. How much money did he earn?

Solution For each of the 8 hours, Raul earned \$9. His total pay for the day is the sum of eight 9's: $9 + 9 + 9 + 9 + 9 + 9 + 9 + 9$. This repeated addition can be calculated by multiplication.

$$\begin{aligned} \text{Total wages} &= 8 \cdot 9 \\ &= 72 \quad \text{See the multiplication table.} \end{aligned}$$

Raul earned \$72.

To multiply three numbers, we first multiply two of them and then multiply that result by the third number. In the following examples, we multiply $3 \cdot 2 \cdot 4$ in two ways. The parentheses show us which multiplication to perform first.

Method 1: Group 3 · 2

$$\begin{aligned} (3 \cdot 2) \cdot 4 &= 6 \cdot 4 \quad \text{Multiply 3 and 2 to get 6.} \\ &= 24 \quad \text{Then multiply 6 and 4 to get 24.} \end{aligned}$$

Method 2: Group 2 · 4

$$\begin{aligned} 3 \cdot (2 \cdot 4) &= 3 \cdot 8 && \text{Multiply 2 and 4 to get 8.} \\ &= 24 && \text{Then multiply 3 and 8 to get 24.} \end{aligned}$$

The answers are the same. This illustrates that changing the grouping when multiplying numbers does not affect the result. This property is called the **associative property of multiplication.**

> **Associative property of multiplication**
>
> If a, b, and c represent numbers, then
>
> $(a \cdot b) \cdot c = a \cdot (b \cdot c)$ or, more simply, $(ab)c = a(bc)$

Multiplying whole numbers

To find the product $8 \cdot 47$, it is inconvenient to add up eight 47's. Instead, we find the product by a multiplication process.

$$\begin{array}{r} 47 \\ \times \quad 8 \\ \hline \end{array}$$

Write the factors in a column, with the corresponding digits aligned vertically.

$$\begin{array}{r} {\scriptstyle 5} \\ 47 \\ \times \quad 8 \\ \hline 6 \end{array}$$

Multiply 7 by 8. The product is 56. Place 6 in the ones column of the answer and carry 5 (in blue) to the tens column.

$$\begin{array}{r} {\scriptstyle 5} \\ 47 \\ \times \quad 8 \\ \hline 376 \end{array}$$

Multiply 4 by 8. The product is 32. To the 32, add the carried 5 to get 37. Place the 7 in the tens column and the 3 in the hundreds column of the answer.

The product is 376.

To find the product $23 \cdot 435$, we use the multiplication process. Because $23 = 20 + 3$, we multiply 435 by 20 and by 3 and then add the products. To do this, we write the factors in a column, with the corresponding digits aligned vertically. We then begin the process by multiplying 435 by 3.

$$\begin{array}{r} {\scriptstyle 1} \\ 435 \\ \times \quad 23 \\ \hline 5 \end{array}$$

Multiply 5 by 3. The product is 15. Place 5 in the ones column and carry 1 (in blue) to the tens column.

$$\begin{array}{r} {\scriptstyle 1\,1} \\ 435 \\ \times \quad 23 \\ \hline 05 \end{array}$$

Multiply 3 by 3. The product is 9. To the 9, add the carried 1 to get 10. Place the 0 in the tens column and carry the 1 (in green) to the hundreds column.

$$\begin{array}{r} {\scriptstyle 1\,1} \\ 435 \\ \times \quad 23 \\ \hline 1305 \end{array}$$

Multiply 4 by 3. The product is 12. Add the 12 to the carried 1 to get 13. Write 13.

We continue by multiplying 435 by 2 tens, or 20.

$$\begin{array}{r} {\scriptstyle 1} \\ 435 \\ \times \quad 23 \\ \hline 1305 \\ 0 \end{array}$$

Multiply 5 by 2. The product is 10. Write 0 in the tens column and carry 1 (in purple).

$$\begin{array}{r} {\scriptstyle 1} \\ 435 \\ \times \quad 23 \\ \hline 1305 \\ 70 \end{array}$$

Multiply 3 by 2. The product is 6. Add 6 to the carried 1 to get 7. Write the 7. There is no carry.

$$\begin{array}{r} \overset{1}{4}35 \\ \times\ 23 \\ \hline 1305 \\ 870 \end{array}$$

Multiply 4 by 2. The product is 8. There is no carry to add. Write the 8.

$$\begin{array}{r} 435 \\ \times\ 23 \\ \hline 1305 \\ 870 \\ \hline 10005 \end{array}$$

Draw another line beneath the two completed rows. Add the two rows. This sum gives the product of 435 and 23.

Thus, $23 \cdot 435 = 10{,}005$.

Self Check 2

For highway driving, how far can the Explorer travel on a tank of gas?

Answer 399 mi

EXAMPLE 2 **Mileage.** Specifications for a Ford Explorer 4 × 4 are shown in the table below. For city driving, how far can it travel on a tank of gas? (The abbreviation "mpg" means "miles per gallon.")

Engine	4.0 L V6
Fuel capacity	21 gal
Fuel economy (mpg)	15 city/19 hwy

Solution For city driving, each of the 21 gallons of gas that the tank holds enables the Explorer to go 15 miles. The total distance it can travel is the sum of twenty-one 15's. This can be calculated by multiplication: $21 \cdot 15$.

$$\begin{array}{r} \overset{1}{1}5 \\ \times 21 \\ \hline 15 \\ 30 \\ \hline 315 \end{array}$$

For city driving, the Explorer can go 315 miles on a tank of gas.

EXAMPLE 3 **Calculating production.** The labor force of an electronics firm works two 8-hour shifts each day and manufactures 53 television sets each hour. Find how many sets will be manufactured in 5 days.

Solution The number of TV sets manufactured in 5 days is given by the product.

$$\underset{\text{per day}}{\overset{\text{2 shifts}}{\downarrow}}\ \ \underset{\text{per shift}}{\overset{\text{8 hr}}{\downarrow}}\ \ \underset{\text{per hr}}{\overset{\text{53 sets}}{\downarrow}}\ \ \underset{\text{days}}{\overset{5}{\downarrow}}$$

$$2 \cdot 8 \cdot 53 \cdot 5$$

This could also be written 2(8)(53)(5).

We perform the multiplications working from left to right.

$$\begin{aligned} 2 \cdot 8 \cdot 53 \cdot 5 &= 16 \cdot 53 \cdot 5 && \text{Multiply 2 and 8.} \\ &= 848 \cdot 5 && \text{Multiply 16 and 53.} \\ &= 4{,}240 \end{aligned}$$

So 4,240 television sets will be manufactured in 5 days.

Checking an answer

CALCULATOR SNAPSHOT

We can use a scientific calculator to check the multiplication performed in Example 3. To find the product $2 \cdot 8 \cdot 53 \cdot 5$, we enter these numbers and press these keys.

2 [×] 8 [×] 53 [×] 5 [=]   4240

The display verifies that the multiplication was done correctly in Example 3.

We can use multiplication to count objects arranged in rectangular patterns. For example, the following display on the left below shows a rectangular array consisting of 5 rows of 7 stars. The product $5 \cdot 7$, or 35, indicates the total number of stars.

Because multiplication is commutative, the array on the right below, consisting of 7 rows of 5 stars, contains the same number of stars.

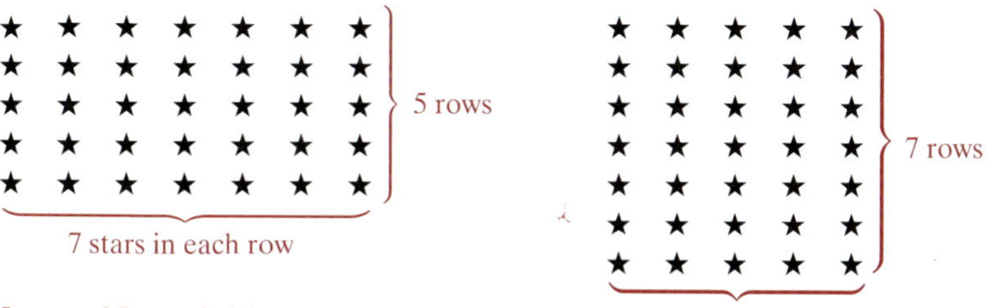

EXAMPLE 4 **Computer science.** To draw graphics on a computer screen, a computer controls each *pixel* (one dot on the screen). See Figure 1-11. If a computer graphics image is 800 pixels wide and 600 pixels high, how many pixels does the computer control?

FIGURE 1-11

Solution The graphics image is a rectangular array of pixels. Each of its 600 rows consists of 800 pixels. The total number of pixels is the product of 600 and 800.

$600 \cdot 800 = 480{,}000$ This could be written as 600(800).

The computer controls 480,000 pixels.

Self Check 4
On a color monitor, each of the pixels can be red, green, or blue. How many colored pixels does the computer control?

Answer 1,440,000

Finding the area of a rectangle

One important application of multiplication is finding the area of a rectangle. The **area of a rectangle** is the measure of the amount of surface it encloses. Area is measured in square units, such as square inches (denoted as in.^{2}) or square centimeters (denoted as cm^2). (See Figure 1-12 on the next page.)

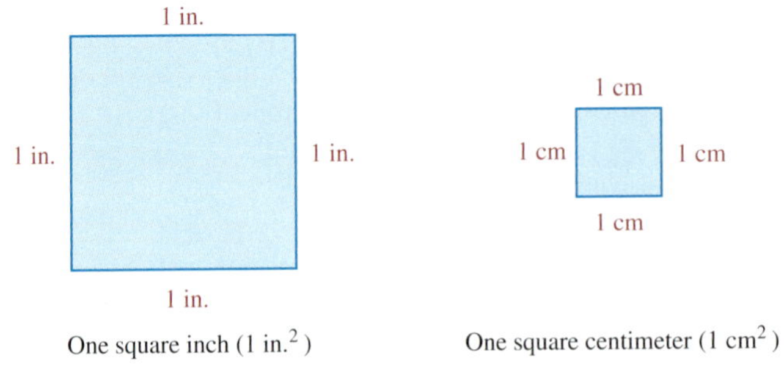

FIGURE 1-12

The rectangle in Figure 1-13 has a length of 5 centimeters and a width of 3 centimeters. Each small square covers an area of 1 square centimeter (1 cm^2). The small squares form a rectangular pattern, with 3 rows of 5 squares.

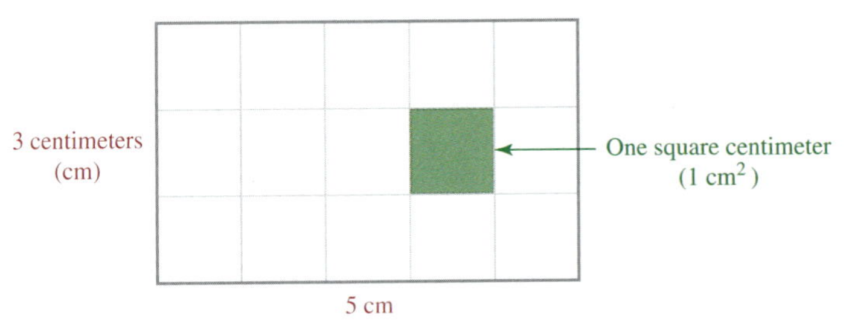

FIGURE 1-13

Because there are $5 \cdot 3$, or 15, small squares, the area of the rectangle is 15 cm^2. This suggests that the area of any rectangle is the product of its length and its width.

Area of a rectangle = length · width

Using the variables l and w to represent the length and width, we can write this formula in simpler form.

Area of a rectangle

The area A of a rectangle is the product of the rectangle's length l and its width w.

$A = l \cdot w$ or, more simply, $A = lw$

Self Check 5

Find the area of a 9-inch-by-12-inch sheet of paper.

Answer 108 in.^{2}

EXAMPLE 5 Gift wrap.

When completely unrolled, a long sheet of gift wrapping paper has the dimensions shown in Figure 1-14. How many square feet of gift wrap are on the roll?

FIGURE 1-14

Solution To find the number of square feet of paper, we need to find the area of the rectangle shown in the figure.

$A = lw$ This is the formula for the area of a rectangle.

$= 12 \cdot 3$ Replace l with 12 and w with 3.

$= 36$ Perform the multiplication.

There are 36 square feet (ft^2) of wrapping paper on the roll.

! COMMENT Remember that the perimeter of a rectangle is the distance around it. The area of a rectangle is a measure of the surface it encloses.

Division

If \$12 is distributed equally among 4 people, we must divide to see that each person receives \$3.

$$4\overline{)12}\ \text{with quotient } 3$$

Three symbols can be used to indicate division.

Symbols that are used for division

Symbol		Example
$\div$	division symbol	$12 \div 4$
$\overline{)\quad}$	long division	$4\overline{)12}$
—	fraction bar	$\frac{12}{4}$

In a division, the number that is being divided is called the **dividend,** the number that we are dividing by is called the **divisor,** and the answer is called the **quotient.**

$$\text{Dividend} \div \text{divisor} = \text{quotient} \qquad \text{Divisor}\overline{)\text{dividend}} \text{ (quotient on top)} \qquad \frac{\text{Dividend}}{\text{Divisor}} = \text{quotient}$$

Division can be thought of as repeated subtraction. To divide 12 by 4 is to ask, How many 4's can be subtracted from 12? Exactly three 4's can be subtracted from 12 to get 0.

$$12 \underbrace{- 4 - 4 - 4}_{\text{Three 4's}} = 0$$

Thus, $12 \div 4 = 3$.

Division is also related to multiplication.

$$\frac{12}{4} = 3 \quad \text{because} \quad 4 \cdot 3 = 12 \qquad \text{and} \qquad \frac{20}{5} = 4 \quad \text{because} \quad 5 \cdot 4 = 20$$

Properties of division

We will now consider three types of division that involve zero. In the first case, we will examine a division of zero; in the second, a division by zero; in the third, a division of zero by zero.

Division statement	Related multiplication statement	Result
$\frac{0}{2} = ?$	$2(?) = 0$ ↑ This must be 0 if the product is to be 0.	$\frac{0}{2} = 0$

$\frac{2}{0} = ?$ $0(?) = 2$ — There is no quotient.

↑ There is no number that gives 2 when multiplied by 0.

$\frac{0}{0} = ?$ $0(?) = 0$ — Any number can be the quotient.

↑ Any number times 0 is 0.

We see that $\frac{0}{2} = 0$. This suggests that the quotient of 0 divided by any nonzero number is 0. Since $\frac{2}{0}$ does not have a quotient, we say that division of 2 by 0 is *undefined.* In general, division of any nonzero number by 0 is undefined. Since $\frac{0}{0}$ can be any number, we say that $\frac{0}{0}$ is undetermined.

Division with 0

1. If a represents any nonzero number, $\frac{0}{a} = 0$.
2. If a represents any nonzero number, $\frac{a}{0}$ is undefined.
3. $\frac{0}{0}$ is undetermined.

The example $\frac{12}{1} = 12$ illustrates that *any number divided by 1 is the number itself.* The example $\frac{12}{12} = 1$ illustrates that *any number (except 0) divided by itself is 1.*

Division properties

If a represents any number,

$$\frac{a}{1} = a \quad \text{and} \quad \frac{a}{a} = 1 \text{ (provided that } a \neq 0)$$

Read ≠ as "is not equal to."

Dividing whole numbers

We can use a process called **long division** to divide whole numbers. To divide 832 by 23, for example, we proceed as follows.

Quotient →
Divisor → $23\overline{)832}$ — Place the divisor and the dividend as indicated. The quotient will appear above the long division symbol.
Dividend (832)

We will find the quotient using the following division process.

$$\begin{array}{r} 4 \\ 23\overline{)832} \end{array}$$

Ask: How many times will 23 divide 83? Because an estimate is 4, place 4 in the tens column of the quotient.

$$\begin{array}{r} 4 \\ 23\overline{)832} \\ 92 \end{array}$$

Multiply $23 \cdot 4$ and place the answer, 92, under the 83. Because 92 is larger than 83, our estimate of 4 for the tens column of the quotient was too large.

$$\begin{array}{r} 3 \\ 23\overline{)832} \\ \underline{69\downarrow} \\ 142 \end{array}$$

Revise the estimate of the quotient to be 3. Multiply $23 \cdot 3$ to get 69, place 69 under the 83, draw a line, and subtract.

Bring down the 2 in the ones column.

$$\begin{array}{r} 37 \\ 23\overline{)832} \\ \underline{69} \\ 142 \\ \underline{161} \end{array}$$

Ask: How many times will 23 divide 142? The answer is approximately 7. Place 7 in the ones column of the quotient. Multiply $23 \cdot 7$ to get 161. Place 161 under 142. Because 161 is larger than 142, the estimate of 7 is too large.

$$\begin{array}{r} 36 \\ 23\overline{)832} \\ \underline{69} \\ 142 \\ \underline{138} \\ 4 \end{array}$$

Revise the estimate of the quotient to be 6. Multiply: $23 \cdot 6 = 138$.

Place 138 under 142 and subtract.

The quotient is 36, and the leftover 4 is the **remainder.** We can write this result as 36 R 4.

To check the result of a division, we multiply the divisor by the quotient and then add the remainder. The result should be the dividend.

Check: Quotient · divisor + remainder = dividend

$$36 \cdot 23 + 4 = 832$$
$$828 + 4 = 832$$
$$832 = 832$$

Applications that involve forming groups can often be solved using division.

EXAMPLE 6 **Soup kitchens.** A soup kitchen plans to feed 1,990 people. Because of space limitations, only 165 people can be served at one time. How many seatings will be necessary to feed everyone? How many will be served at the last seating?

Solution The 1,990 people can be fed 165 at a time. To find the number of seatings, we divide.

$$\begin{array}{r} 12 \\ 165\overline{)1{,}990} \\ \underline{1\,65\downarrow} \\ 340 \\ \underline{330} \\ 10 \end{array}$$

The quotient is 12, and the remainder is 10. Thirteen seatings will be needed: 12 full-capacity seatings and one partial seating to serve the remaining 10 people.

Self Check 6

Each gram of fat in a meal provides 9 calories. A fast-food meal contains 243 calories from fat. How many grams of fat does the meal contain?

Answer 27

Retailing

CALCULATOR SNAPSHOT

A salesperson sold a number of calculators for \$17 each, and her total sales were \$1,819. To find the number of calculators she sold, we must divide the total sales by the cost of each calculator. We can use a calculator to evaluate $1{,}819 \div 17$ by entering these numbers and pressing these keys.

1819 [÷] 17 [=] 107

The salesperson sold 107 calculators.

Section 1.3 STUDY SET

VOCABULARY *Fill in the blanks.*

1. ____________ is repeated addition.
2. Numbers that are to be multiplied are called ________. The result of a multiplication is called a ________.
3. The statement $ab = ba$ expresses the ____________ property of multiplication. The statement $(ab)c = a(bc)$ expresses the __________ property of multiplication.
4. If a square measures 1 inch on each side, its area is 1 ________ ________.
5. In a division, the dividend is divided by the ________. The result of a division is called a ________.
6. The ______ of a rectangle is the amount of surface it encloses.

CONCEPTS

7. Write $8 + 8 + 8 + 8$ as a multiplication.
8. **a.** Use the variables x and y to write the commutative property of multiplication.

   b. Use the variables x, y, and z to write the associative property of multiplication.
9. How do we find the amount of surface enclosed by a rectangle?
10. Determine whether *perimeter* or *area* is the concept that should be applied to find each of the following.

    a. The amount of floor space to be carpeted

    b. The amount of clear glass to be tinted

    c. The amount of lace needed to trim the sides of a handkerchief
11. Perform each multiplication.

    a. $1 \cdot 25$ **b.** $62(1)$

    c. $10 \cdot 0$ **d.** $0(4)$
12. Perform each division.

    a. $25 \div 1$ **b.** $\frac{7}{1}$

    c. $\frac{0}{1}$ **d.** $\frac{5}{0}$

    e. $\frac{0}{0}$ **f.** $\frac{0}{2{,}757}$

13. Write a multiplication statement that finds the number of red squares.

14. Consider

$$\begin{array}{r} 12 \\ 15\overline{)182} \\ \underline{15} \\ 32 \\ \underline{30} \\ 2 \end{array}$$

Fill in the blanks: $12 \cdot \_\_ + \_\_ = \_\_\_$.

NOTATION

15. **a.** Write three symbols that are used for multiplication.

    b. Write three symbols that are used for division.
16. Write each multiplication in simpler form.

    a. $8 \cdot x$ **b.** $l \cdot w$
17. What does ft^2 mean?
18. Draw a figure having an area of 1 square inch.

PRACTICE *Perform each multiplication.*

19. $12 \cdot 7$
20. $15 \cdot 8$
21. $27(12)$
22. $35(17)$
23. $9 \cdot (4 \cdot 5)$
24. $(3 \cdot 5) \cdot 12$
25. $5 \cdot 7 \cdot 3$
26. $7 \cdot 6 \cdot 8$
27. 99×77
28. 73×59
29. 20×53
30. 78×20
31. 112×23
32. 232×53
33. 207×97
34. 768×70
35. $13{,}456 \cdot 217$
36. $17{,}456 \cdot 257$
37. $3{,}302 \cdot 358$
38. $123{,}112 \cdot 46$

Find the area of each rectangle or square.

39.

40.

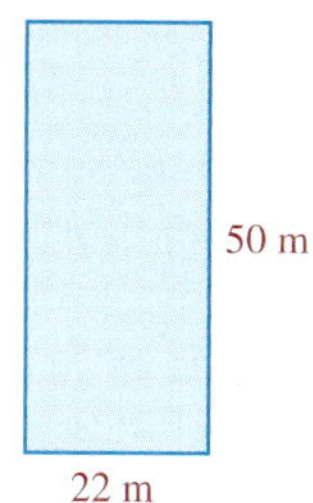

41.

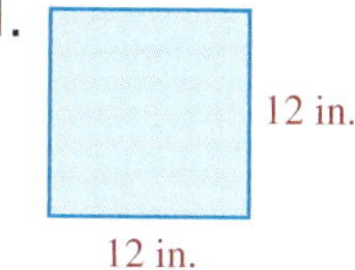

42.

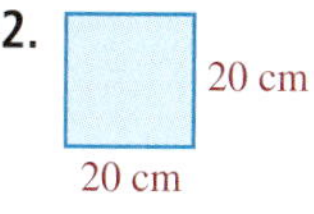

Perform each division.

43. $40 \div 5$

44. $40 \div 8$

45. $42 \div 14$

46. $65 \div 13$

47. $132 \div 11$

48. $132 \div 12$

49. $\frac{221}{17}$

50. $\frac{221}{13}$

51. $13\overline{)949}$

52. $73\overline{)949}$

53. $33\overline{)1{,}353}$

54. $41\overline{)1{,}353}$

55. $39\overline{)7{,}995}$

56. $71\overline{)7{,}313}$

57. $29\overline{)6{,}090}$

58. $13\overline{)7{,}410}$

Perform each division and give the quotient and the remainder.

59. $31\overline{)273}$

60. $25\overline{)290}$

61. $37\overline{)743}$

62. $79\overline{)931}$

63. $42\overline{)1{,}273}$

64. $83\overline{)3{,}280}$

65. $57\overline{)1{,}795}$

66. $99\overline{)9{,}876}$

APPLICATIONS

67. WAGES A cook worked 12 hours at $11 per hour. How much did she earn?

68. CHESSBOARDS A chessboard consists of 8 rows, with 8 squares in each row. How many squares are on a chessboard?

69. FINDING DISTANCE A car with a tank that holds 14 gallons of gasoline goes 29 miles on 1 gallon. How far can the car go on a full tank?

70. RENTING APARTMENTS Mia owns an apartment building with 18 units. Each unit generates a monthly income of $450. Find her total monthly income.

71. CONCERTS A jazz quartet gave two concerts in each of 37 cities. Approximately 1,700 fans attended each concert. How many people heard the group?

72. CEREAL A cereal maker advertises "Two cups of raisins in every box." Find the number of cups of raisins in a case of 36 boxes of cereal.

73. ORANGE JUICE It takes 13 oranges to make one can of orange juice. Find the number of oranges used to make a case of 24 cans.

74. ROOM CAPACITY A college lecture hall has 17 rows of 33 seats. A sign on the wall reads, "Occupancy by more than 570 persons is prohibited." If the seats are filled and there is one instructor, is the college breaking the rule?

75. ELEVATORS There are 14 people in an elevator with a capacity of 2,000 pounds. If the average weight of a person on the elevator is 150 pounds, is the elevator overloaded?

76. CHANGING UNITS There are 12 inches in 1 foot. How many inches are in 80 feet?

77. WORD PROCESSING A student used the option shown in the illustration when typing a report. How many entries will the table hold?

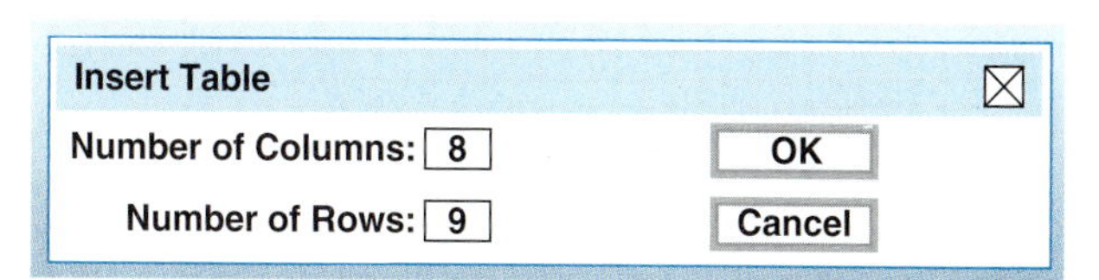

78. PRESCRIPTIONS How many tablets should a pharmacist put in the container shown on the right?

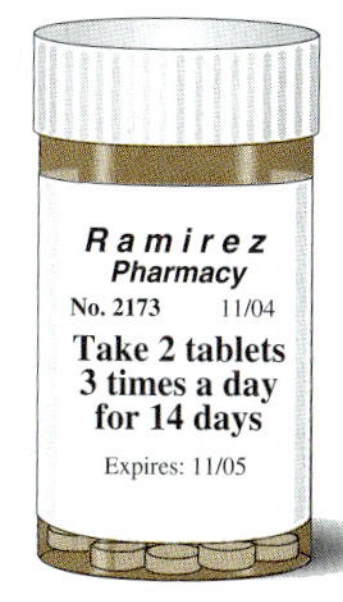

79. DISTRIBUTING MILK A first-grade class received 73 half-pint cartons of milk to distribute evenly to the 23 students. How many cartons were left over?

80. LIFT SYSTEMS If the bus shown below weighs 58,000 pounds, how much weight is on each jack?

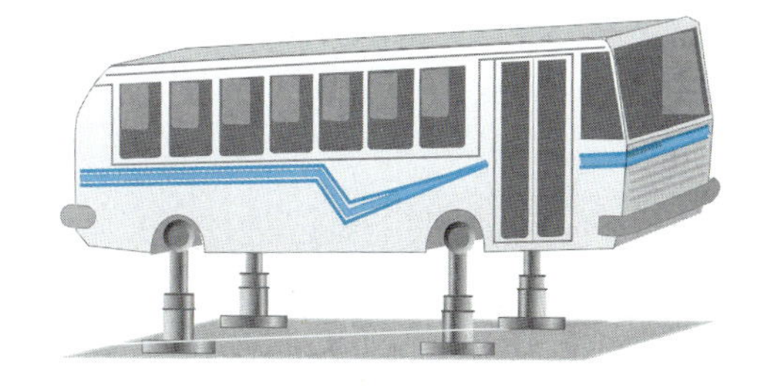

81. MILEAGE A touring rock group travels in a bus that has a range of 700 miles on one tank (140 gallons) of gasoline. How far can the bus travel on 1 gallon of gas?

82. RUNNING Brian runs 7 miles each day. In how many days will Brian run 371 miles?

83. How many feet more than 2 miles is 11,000 feet? (*Hint:* 5,280 feet = 1 mile.)

84. DOUGHNUTS How many dozen doughnuts must be ordered for a meeting if 156 people are expected to attend and each person will be served one doughnut?

85. PRICE OF TEXTBOOKS An author knows that her publisher received $954,193 on the sale of 23,273 textbooks. What is the price of each book?

86. WATER DISCHARGES The Susquehanna River discharges 38,200 cubic feet of water per second into the Chesapeake Bay. How long will it take for the river to discharge 1,719,000 cubic feet?

87. VOLLEYBALL LEAGUES A total of 216 girls tried out for a city volleyball program. How many girls should be put on each team roster if the following requirements must be met?

- All the teams are to have the same number of players.
- A reasonable number of players on a team is 7 to 10.
- For scheduling purposes, there must be an even number of teams.

88. AREA OF WYOMING The state of Wyoming is approximately rectangular-shaped with dimensions 360 miles long and 270 miles wide. Find its perimeter and its area.

89. COMPARING ROOMS Which has the greater area: a rectangular room that is 14 feet by 17 feet or a square room that is 16 feet on each side? Which has the greater perimeter?

90. MATTRESSES A queen-size mattress measures 60 inches by 80 inches, and a full-size mattress measures 54 inches by 75 inches. How much more sleeping surface is there on a queen-size mattress?

91. GARDENING A rectangular garden is 27 feet long and 19 feet wide. A path in the garden uses 125 square feet of space. How many square feet are left for planting?

92. TENNIS See the illustration.

a. Find the number of square feet of court area a singles tennis player must defend.

b. Do the same for a doubles player.

c. What is the difference between the two results?

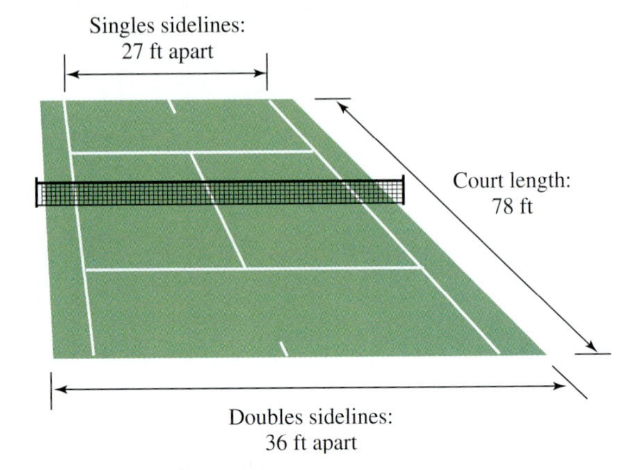

WRITING

93. Explain why the division of two numbers is not commutative.

94. Explain the difference between what perimeter measures and what area measures.

95. Explain the difference between 1 foot and 1 square foot.

96. When two numbers are multiplied, the result is 0. What conclusion can be drawn about the numbers?

REVIEW

97. Consider 372,856. What digit is in the hundreds column?

98. Round 45,995 to the nearest thousand.

99. Add: 357, 39, and 476.

100. DISCOUNTS A car, originally priced at $17,550, is being sold for $13,970. By how many dollars has the price been decreased?

ESTIMATION

In the previous two sections, we have used **estimation** as a means of checking the reasonableness of an answer. We now take a more in-depth look at the process of estimating.

Estimation is used to find an *approximate* answer to a problem. Estimates can be helpful in two ways. First, they serve as an accuracy check that can detect major computational errors. If an answer does not seem reasonable when compared to the estimate, the original problem should be reworked. Second, some situations call for only an approximate answer rather than the exact answer.

There are several ways to estimate, but there is one overriding theme of all the methods: The numbers in the problem are simplified so that the computation can be made easily and quickly. The first method we will study uses what is called **front-end rounding.** Each number is rounded to its largest place value, so that all but the first digit of each number is 0.

EXAMPLE 1 Estimating sums, differences, and products.

a. Estimate the sum: 3,714 + 2,489 + 781 + 5,500 + 303.

Solution Use front-end rounding.

$$\begin{array}{r} 3{,}714 \\ 2{,}489 \\ 781 \\ 5{,}500 \\ +\ 303 \\ \hline \end{array} \quad \begin{array}{c} \longrightarrow \\ \longrightarrow \\ \longrightarrow \\ \longrightarrow \\ \longrightarrow \end{array} \quad \begin{array}{r} 4{,}000 \\ 2{,}000 \\ 800 \\ 6{,}000 \\ +\ 300 \\ \hline 13{,}100 \end{array}$$

Each number is rounded to its largest place value. All but the first digit is 0.

The estimate is 13,100.

If we compute 3,714 + 2,489 + 781 + 5,500 + 303, the sum is 12,787. We can see that our estimate is close; it's just 313 more than 12,787. This example illustrates the tradeoff when using estimation: The calculations are easier to perform and they take less time, but the answers are not exact.

b. Estimate the difference: 46,721 − 13,208.

Solution Use front-end rounding.

$$\begin{array}{r} 46{,}721 \\ -13{,}208 \\ \hline \end{array} \quad \begin{array}{c} \longrightarrow \\ \longrightarrow \end{array} \quad \begin{array}{r} 50{,}000 \\ -10{,}000 \\ \hline 40{,}000 \end{array}$$

Only the first digit is nonzero.

The estimate is 40,000.

c. Estimate the product: 334 · 59.

Solution Use front-end rounding.

$$\begin{array}{r} 334 \\ \times\ 59 \\ \hline \end{array} \quad \begin{array}{c} \longrightarrow \\ \longrightarrow \end{array} \quad \begin{array}{r} 300 \\ \times\ 60 \\ \hline 18{,}000 \end{array}$$

334 rounds to 300, 59 rounds to 60.

The estimate is 18,000.

Self Check 1

a. Estimate the sum:

$$\begin{array}{r} 6{,}780 \\ 3{,}278 \\ 566 \\ 4{,}230 \\ +1{,}923 \\ \hline \end{array}$$

b. Estimate the difference:

$$\begin{array}{r} 89{,}070 \\ -15{,}331 \\ \hline \end{array}$$

c. Estimate the product:

$$\begin{array}{r} 707 \\ \times 251 \\ \hline \end{array}$$

Answers **a.** 16,600, **b.** 70,000, **c.** 210,000

To estimate quotients, we will use a method that approximates both the dividend and the divisor so that they will divide easily. With this method, some insight and intuition are needed. There is one rule of thumb for this method: If possible, round both numbers up or both numbers down.

Self Check 2
Estimate: 33,642 ÷ 42.

Answer 800

EXAMPLE 2 Estimating quotients. Estimate the quotient: 170,715 ÷ 57.

Solution Both numbers are rounded up. The division can then be done in your head.

170,715 ÷ 57 → 180,000 ÷ 60 = 3,000

(The dividend is approximately 180,000; the divisor is approximately 60.)

The estimate is 3,000.

STUDY SET *Use front-end rounding to find an estimate to check the reasonableness of each answer. Write yes if it appears reasonable and no if it does not.*

1.
```
  25,405
  11,222
   8,909
   1,076
  14,595
 +33,999
  73,206
```

2.
```
  568,334
 − 31,225
  497,109
```

3.
```
   451
 ×  73
 39,923
```

4.
```
   616
 ×  98
 60,368
```

Use estimation to check the reasonableness of each answer.

5. 57,238 ÷ 28 = 200

6. $322\overline{)13{,}202}$ with quotient 41

Use an estimation procedure to answer each problem.

7. CAMPAIGNING The number of miles flown each day by a politician on a campaign swing are shown here. Estimate the number of miles she flew during this time.

Day 1	3,546 miles
Day 2	567
Day 3	1,203
Day 4	342
Day 5	2,699

8. SHOPPING MALLS The total sales income for a downtown mall in its first three years in operation are shown here.

2000	$5,234,301
2001	$2,898,655
2002	$6,343,433

Estimate the difference in income for 2001 and 2002 as compared to the first year, 2000.

9. GOLF COURSES Estimate the number of bags of grass seed needed to plant a fairway whose area is 86,625 square feet if the seed in each bag covers 2,850 square feet.

10. CENSUS Estimate the total population of the ten largest counties in the United States as of 2003.

LARGEST COUNTIES, BY POPULATION	
1. Los Angeles, CA	9,871,506
2. Cook, IL	5,351,552
3. Harris, TX	3,596,086
4. Maricopa, AZ	3,389,260
5. Orange, CA	2,957,766
6. San Diego, CA	2,930,886
7. Kings, NY	2,472,523
8. Miami-Dade, FL	2,341,167
9. Dallas, TX	2,284,096
10. Queens, NY	2,225,486

Source: Bureau of the Census

11. CURRENCY Estimate the number of $5 bills in circulation as of June 30, 2004, if the total value of the currency was $9,373,288,075.

12. CORPORATIONS In 2003, General Motors Corporation had sales of $185,524,000,000. Approximately how many times larger was this than the 2003 sales of IBM, which were $89,000,000,000?

1.4 Prime Factors and Exponents

- Factoring whole numbers • Even and odd whole numbers • Prime numbers
- Composite numbers • Finding prime factorizations with the tree method
- Exponents • Finding prime factorizations with the division method

In this section, we will describe how to represent whole numbers in alternative forms. The procedures used to find these forms involve multiplication and division. We will then discuss exponents, a shortcut way to represent repeated multiplication.

Factoring whole numbers

The statement $3 \cdot 2 = 6$ has two parts: the numbers that are being multiplied, and the answer. The numbers that are being multiplied are *factors,* and the answer is the *product.* We say that 3 and 2 are factors of 6.

Factors

Numbers that are multiplied together are called **factors.**

EXAMPLE 1 Find the factors of 12.

Solution We need to find the possible ways that we can multiply two whole numbers to get a product of 12.

$1 \cdot 12 = 12, \quad 2 \cdot 6 = 12, \quad \text{and} \quad 3 \cdot 4 = 12$

In order from least to greatest, the factors of 12 are 1, 2, 3, 4, 6, and 12.

Self Check 1
Find the factors of 20.

Answer 1, 2, 4, 5, 10, and 20

Example 1 shows that 1, 2, 3, 4, 6, and 12 are the factors of 12. This observation was established by using multiplication facts. Each of these factors is related to 12 by division as well. Each of them divides 12, leaving a remainder of 0. Because of this fact, we say that 12 is *divisible* by each of its factors. When a division ends with a remainder of 0, we say that the division comes out even or that one of the numbers divides the other *exactly*.

Divisibility

One number is divisible by another if, when we divide them, the remainder is 0.

When we say that 3 is a factor of 6, we are using the word *factor* as a noun. The word *factor* is also used as a verb.

Factoring a whole number

To **factor** a whole number means to express it as the product of other whole numbers.

Self Check 2
Factor 18 using **a.** two factors and **b.** three factors.

Answers **a.** $1 \cdot 18, 2 \cdot 9, 3 \cdot 6$, **b.** $2 \cdot 3 \cdot 3$

EXAMPLE 2 Factor 40 using **a.** two factors and **b.** three factors.

Solution

a. There are several possibilities.

$$40 = 1 \cdot 40, \quad 40 = 2 \cdot 20, \quad 40 = 4 \cdot 10, \quad \text{or} \quad 40 = 5 \cdot 8$$

b. Again, there are several possibilities. Two of them are

$$40 = 5 \cdot 4 \cdot 2 \quad \text{and} \quad 40 = 2 \cdot 2 \cdot 10$$

Even and odd whole numbers

Even and odd whole numbers

If a whole number is divisible by 2, it is called an **even** number.

If a whole number is not divisible by 2, it is called an **odd** number.

The even whole numbers are the numbers

0, 2, 4, 6, 8, 10, 12, 14, 16, 18, . . .

The odd whole numbers are the numbers

1, 3, 5, 7, 9, 11, 13, 15, 17, 19, . . .

There are infinitely many even and infinitely many odd whole numbers.

Prime numbers

Self Check 3
Find the factors of 23.

Answer 1 and 23

EXAMPLE 3 Find the factors of 17.

Solution

$$1 \cdot 17 = 17$$

The only factors of 17 are 1 and 17.

In Example 3 and its Self Check, we saw that the only factors of 17 are 1 and 17, and the only factors of 23 are 1 and 23. Numbers that have only two factors, 1 and the number itself, are called **prime numbers.**

Prime numbers

A **prime number** is a whole number, greater than 1, that has only 1 and itself as factors.

The prime numbers are the numbers

2, 3, 5, 7, 11, 13, 17, 19, 23, 29, 31, . . .

The dots at the end of the list indicate that there are infinitely many prime numbers.

Note that the only even prime number is 2. Any other even whole number is divisible by 2 and thus has 2 as a factor, in addition to 1 and itself. Also note that not all odd whole numbers are prime numbers. For example, since 15 has factors of 1, 3, 5, and 15, it is not a prime number.

Composite numbers

The set of whole numbers contains many prime numbers. It also contains many numbers that are not prime.

> **Composite numbers**
>
> The **composite numbers** are whole numbers, greater than 1, that are not prime.

The composite numbers are the numbers

4, 6, 8, 9, 10, 12, 14, 15, 16, 18, . . .

The three dots at the end of the list indicate that there are infinitely many composite numbers.

EXAMPLE 4 **a.** Is 37 a prime number? **b.** Is 45 a prime number?

Solution

a. Since 37 is a whole number greater than 1 and its only factors are 1 and 37, it is prime.

b. The factors of 45 are 1, 3, 5, 9, 15, and 45. Since there are factors other than 1 and 45, 45 is not prime. It is a composite number.

Self Check 4

a. Is 57 a prime number?

b. Is 39 a prime number?

Answers **a.** no, **b.** no

! COMMENT The numbers 0 and 1 are neither prime nor composite, because neither is a whole number greater than 1.

Finding prime factorizations with the tree method

Every composite number can be formed by multiplying a specific combination of prime numbers. The process of finding that combination is called **prime factorization.**

> **Prime factorization**
>
> To find the **prime factorization** of a whole number means to write it as the product of only prime numbers.

One method for finding the prime factorization of a number is called the **tree method.** We use the tree method in the diagrams below to find the prime factorization of 90 in two ways.

1. Factor 90 as $9 \cdot 10$.
2. Factor 9 and 10.
3. The process is complete when only prime numbers appear.

90

9 10

3 3 2 5

1. Factor 90 as $6 \cdot 15$.
2. Factor 6 and 15.
3. The process is complete when only prime numbers appear.

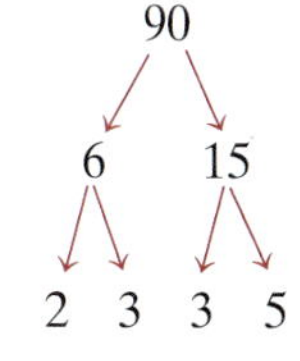

In either case, the prime factors are $2 \cdot 3 \cdot 3 \cdot 5$. Thus, the prime-factored form of 90 is $2 \cdot 3 \cdot 3 \cdot 5$. As we have seen, it does not matter how we factor 90. We will always get the same set of prime factors. No other combination of prime factors will multiply together and produce 90. This example illustrates an important fact about composite numbers.

Fundamental theorem of arithmetic

Any composite number has exactly one set of prime factors.

Self Check 5

Use a factor tree to find the prime factorization of 120.

Answer $2 \cdot 2 \cdot 2 \cdot 3 \cdot 5$

EXAMPLE 5 Use a factor tree to find the prime factorization of 210.

Solution

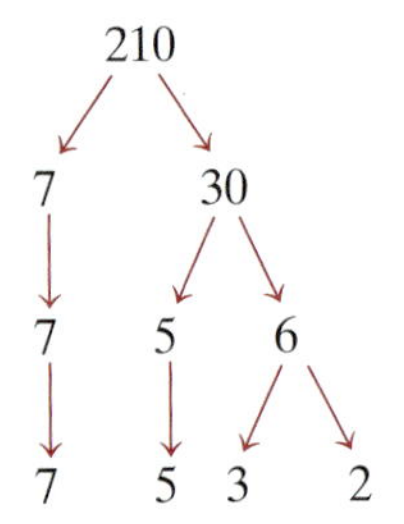

Factor 210 as $7 \cdot 30$.

Bring down the 7. Factor 30 as $5 \cdot 6$.

Bring down the 7 and the 5. Factor 6 as $3 \cdot 2$.

The prime factorization of 210 is $7 \cdot 5 \cdot 3 \cdot 2$. Writing the prime factors in order, from least to greatest, we have $210 = 2 \cdot 3 \cdot 5 \cdot 7$.

Exponents

In the Self Check of Example 5, we saw that the prime factorization of 120 is $2 \cdot 2 \cdot 2 \cdot 3 \cdot 5$. Because this factorization has three factors of 2, we call 2 a *repeated factor.* To express a repeated factor, we can use an **exponent.**

Exponent and base

An **exponent** is used to indicate repeated multiplication. It tells how many times the **base** is used as a factor.

The exponent is 3.

$$\underbrace{2 \cdot 2 \cdot 2}_{\text{Repeated factors}} = 2^3$$

The base is 2.

Read 2^3 as "2 to the third power" or "2 cubed."

The prime factorization of 120 can be written in a more compact form using exponents: $2 \cdot 2 \cdot 2 \cdot 3 \cdot 5 = 2^3 \cdot 3 \cdot 5$.

In the **exponential expression** a^n, a is the base and n is the exponent. The expression is called a **power of *a*.**

Self Check 6

Use exponents to write each prime factorization:

a. $3 \cdot 3 \cdot 7$ **b.** $5(5)(7)(7)$

c. $2 \cdot 2 \cdot 2 \cdot 3 \cdot 3 \cdot 5$

Answers **a.** $3^2 \cdot 7$, **b.** $5^2(7^2)$, **c.** $2^3 \cdot 3^2 \cdot 5$

EXAMPLE 6 Use exponents to write each prime factorization: **a.** $5 \cdot 5 \cdot 5$, **b.** $7 \cdot 7 \cdot 11$, and **c.** $2(2)(2)(2)(3)(3)(3)$.

Solution

a. $5 \cdot 5 \cdot 5 = 5^3$ — 5 is used as a factor 3 times.

b. $7 \cdot 7 \cdot 11 = 7^2 \cdot 11$ — 7 is used as a factor 2 times.

c. $2(2)(2)(2)(3)(3)(3) = 2^4(3^3)$ — 2 is used as a factor 4 times, and 3 is used as a factor 3 times.

EXAMPLE 7 Find each power: **a.** 7^2, **b.** 2^5, **c.** 10^4, and **d.** 6^1.

Solution

a. $7^2 = 7 \cdot 7 = 49$ — Read 7^2 as "7 to the second power" or "7 squared." Write the base 7 as a factor two times.

b. $2^5 = 2 \cdot 2 \cdot 2 \cdot 2 \cdot 2 = 32$ — Read 2^5 as "2 to the fifth power." Write the base 2 as a factor five times.

c. $10^4 = 10 \cdot 10 \cdot 10 \cdot 10 = 10{,}000$ — Read 10^4 as "10 to the fourth power." Write the base 10 as a factor four times.

d. $6^1 = 6$ — Read 6^1 as "6 to the first power." Write the base 6 once.

Self Check 7
Which of the numbers 3^5, 4^4, and 5^3 is the largest?

Answer $4^4 = 256$

! COMMENT Note that 2^5 means $2 \cdot 2 \cdot 2 \cdot 2 \cdot 2$. It does not mean $2 \cdot 5$. That is, $2^5 = 32$ and $2 \cdot 5 = 10$.

EXAMPLE 8 The prime factorization of a number is $2^3 \cdot 3^4 \cdot 5$. What is the number?

Solution To find the number, we find the value of each power and then perform the multiplication.

$$2^3 \cdot 3^4 \cdot 5 = 8 \cdot 81 \cdot 5 \quad 2^3 = 8 \text{ and } 3^4 = 81.$$
$$= 648 \cdot 5 \quad \text{Perform the multiplications, working from left to right.}$$
$$= 3{,}240$$

The number is 3,240.

Self Check 8
The prime factorization of a number is $3 \cdot 5^2 \cdot 7$. What is the number?

Answer 525

Bacterial growth

CALCULATOR SNAPSHOT

At the end of one hour, a culture contains two bacteria. Suppose the number of bacteria doubles every hour thereafter. Use exponents to determine how many bacteria the culture will contain after 24 hours.

We can use Table 1-2 to help model the situation. From the table, we see a pattern developing: The number of bacteria in the culture after 24 hours will be 2^{24}. We can evaluate this exponential expression using the exponential key $\boxed{y^x}$ on a scientific calculator ($\boxed{x^y}$ on some models).

To find the value of 2^{24}, we enter these numbers and press these keys.

Time	Number of bacteria
1 hr	$2 = 2^1$
2 hr	$4 = 2^2$
3 hr	$8 = 2^3$
4 hr	$16 = 2^4$
24 hr	$? = 2^{24}$

TABLE 1-2

2 $\boxed{y^x}$ 24 $\boxed{=}$ $\boxed{16777216}$

Since $2^{24} = 16{,}777{,}216$, there will be 16,777,216 bacteria after 24 hours.

Finding prime factorizations with the division method

We can also find the prime factorization of a whole number by division. For example, to find the prime factorization of 363, we begin the division method by choosing the *smallest* prime number that will divide the given number exactly. We continue this "inverted division" process until the result of the division is a prime number.

Step 1: The prime number 2 doesn't divide 363 exactly, but 3 does. The result is 121, which is not prime. We continue the division process.

$$\begin{array}{r|l} 3 & \underline{363} \\ & 121 \end{array}$$

Step 2: Next, we choose the smallest prime number that will divide 121. The primes 2, 3, 5, and 7 don't divide 121 exactly, but 11 does. The result is 11, which is prime. We are done.

$$\begin{array}{r|l} 3 & \underline{363} \\ 11 & \underline{121} \\ & 11 \end{array}$$

$$363 = 3 \cdot 11 \cdot 11$$

Using exponents, we can write the prime factorization of 363 as $3 \cdot 11^2$.

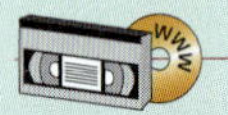

Self Check 9
Use the division method to find the prime factorization of 108. Use exponents to express the result.

Answer $2^2 \cdot 3^3$

EXAMPLE 9 Use the division method to find the prime factorization of 100. Use exponents to express the result.

Solution

2 divides 100 exactly. The result is 50, which is not prime. → $2\underline{|100}$
2 divides 50 exactly. The result is 25, which is not prime. → $2\underline{|50}$
5 divides 25 exactly. The result is 5, which is prime. We are done. → $5\underline{|25}$
5

The prime factorization of 100 is $2 \cdot 2 \cdot 5 \cdot 5$ or $2^2 \cdot 5^2$.

COMMENT In Example 9, it would be incorrect to begin the division process with

$10\underline{|100}$

because 10 is not a prime number.

Section 1.4 STUDY SET

VOCABULARY *Fill in the blanks.*

1. Numbers that are multiplied together are called ________.
2. One number is ________ by another if the remainder is 0 when they are divided. When a division ends with a remainder of 0, we say that one of the numbers divides the other ________.
3. To ________ a whole number means to express it as the product of other whole numbers.
4. A ________ number is a whole number, greater than 1, that has only 1 and itself as factors.
5. Whole numbers, greater than 1, that are not prime numbers are called ________ numbers.
6. An ________ whole number is exactly divisible by 2. An ________ whole number is not exactly divisible by 2.
7. To prime factor a number means to write it as a product of only ________ numbers.
8. An ________ is used to represent repeated multiplication.
9. In the exponential expression 6^4, 6 is called the ________ and 4 is called the ________.
10. Another way to say "5 to the second power" is 5 ________. Another way to say "7 to the third power" is 7 ________.

CONCEPTS

11. Write 27 as the product of two factors.
12. Write 30 as the product of three factors.

13. The complete list of the factors of a whole number is given. What is the number?

a. 2, 4, 22, 44, 11, 1

b. 20, 1, 25, 100, 2, 4, 5, 50, 10

14. **a.** Find the factors of 24.

b. Find the prime factorization of 24.

15. Find the factors of each number.

a. 11

b. 23

c. 37

d. From the results obtained in parts a–c, what can be said about 11, 23, and 37?

16. Suppose a number is divisible by 10. Is 10 a factor of the number?

17. If 4 is a factor of a whole number, will 4 divide the number exactly?

18. Give examples of whole numbers that have 11 as a factor.

The prime factorization of a whole number is given. Find the number.

19. $2 \cdot 3 \cdot 3 \cdot 5$

20. $3^3 \cdot 2$

21. $11^2 \cdot 5$

22. $2 \cdot 2 \cdot 2 \cdot 7$

23. Can we change the order of the base and the exponent in an exponential expression and obtain the same result? In other words, does $3^2 = 2^3$?

24. Find the prime factors of 20 and 35. What prime factor do they have in common?

25. Find the prime factors of 20 and 50. What prime factors do they have in common?

26. Find the prime factors of 30 and 165. What prime factors do they have in common?

27. Find the prime factors of 30 and 242. What prime factor do they have in common?

28. Find 1^2, 1^3, and 1^4. From the results, what can be said about any power of 1?

29. Finish the process of prime factoring 150. Compare the results.

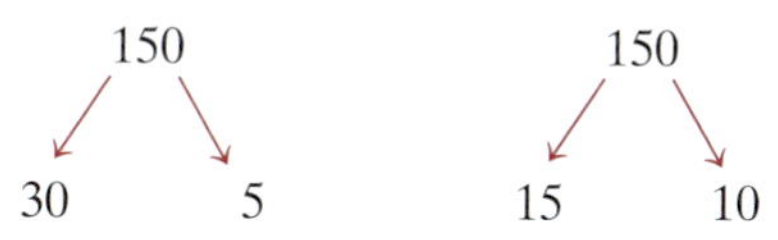

30. Find three whole numbers, less than 10, that would fit at the top of this tree diagram.

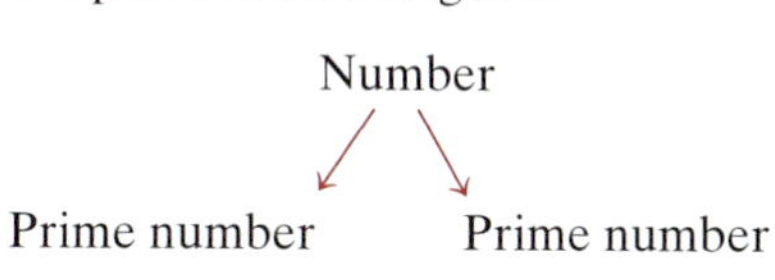

31. Complete the table.

Product of the factors of 12	Sum of the factors of 12
$1 \cdot 12$	
$2 \cdot 6$	
$3 \cdot 4$	

32. Consider 1, 4, 9, 16, 25, 36, 49, 64, 81, 100. Of the numbers listed, which is the *largest* factor of

a. 18 **b.** 24 **c.** 50

33. When using the division method to find the prime factorization of an even number, what is an obvious choice with which to start the division process?

34. When using the division method to find the prime factorization of a number ending in 5, what is an obvious choice with which to start the division process?

NOTATION *Write the repeated multiplication represented by each expression.*

35. 7^3

36. 8^4

37. 3^5

38. 4^6

39. $5^2(11)$

40. $2^3 \cdot 3^2$

Simplify each expression.

41. 10^1

42. 2^1

Use exponents to write each expression in simpler form.

43. $2 \cdot 2 \cdot 2 \cdot 2 \cdot 2$

44. $3 \cdot 3 \cdot 3 \cdot 3 \cdot 3 \cdot 3$

45. $5 \cdot 5 \cdot 5 \cdot 5$

46. $9 \cdot 9 \cdot 9$

47. $4(4)(5)(5)$

48. $12 \cdot 12 \cdot 12 \cdot 16$

PRACTICE *Find the factors of each whole number.*

49. 10

50. 6

51. 40

52. 75

53. 18

54. 32

55. 44

56. 65

57. 77

58. 81

59. 100

60. 441

Write each number in prime-factored form.

61. 39

62. 20

63. 99

64. 105

65. 162

66. 400

67. 220

68. 126

69. 64

70. 243

71. 147

72. 98

Evaluate each exponential expression.

73. 3^4

74. 5^3

75. 2^5

76. 10^5

77. 12^2

78. 7^3

79. 8^4

80. 9^5

81. $3^2(2^3)$

82. $3^3(4^2)$

83. $2^3 \cdot 3^3 \cdot 4^2$

84. $3^2 \cdot 4^3 \cdot 5^2$

85. 234^3

86. 51^4

87. $23^2 \cdot 13^3$

88. $12^3 \cdot 15^2$

APPLICATIONS

89. PERFECT NUMBERS A whole number is called a **perfect number** when the sum of its factors that are less than the number equals the number. For example, 6 is a perfect number, because $1 + 2 + 3 = 6$. Find the factors of 28. Then use addition to show that 28 is also a perfect number.

90. CRYPTOGRAPHY Information is often transmitted in code. Many codes involve writing products of large primes, because they are difficult to factor. To see how difficult, try finding two prime factors of 7,663. (*Hint:* Both primes are greater than 70.)

91. LIGHT The illustration shows that the light energy that passes through the first unit of area, 1 yard away from the bulb, spreads out as it travels away from the source. How much area does that energy cover 2 yards, 3 yards, and 4 yards from the bulb? Express each answer using exponents.

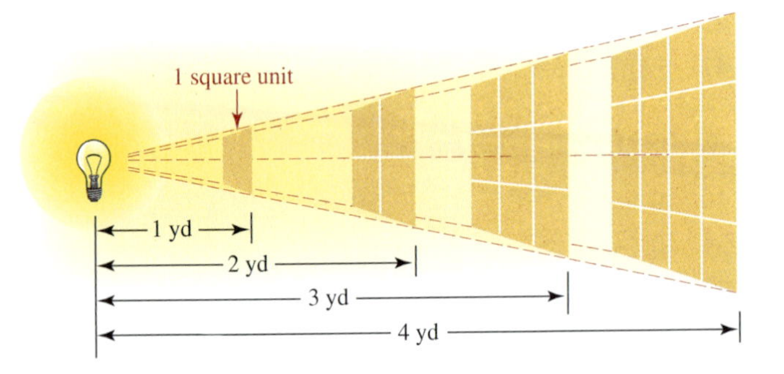

92. CELL DIVISION After one hour, a cell has divided to form another cell. In another hour, these two cells have divided so that four cells exist. In another hour, these four cells divide so that eight exist.

a. How many cells exist at the end of the fourth hour?

b. The number of cells that exist after each division can be found using an exponential expression. What is the base?

c. Use a calculator to find the number of cells after 12 hours.

WRITING

93. Explain how to test a number to see whether it is prime.

94. Explain how to test a number to see whether it is even.

95. Explain the difference between the *factors* of a number and the *prime factorization* of the number.

96. Explain why it would be incorrect to say that the area of the square shown on the right is 25^2 ft. How should we express its area?

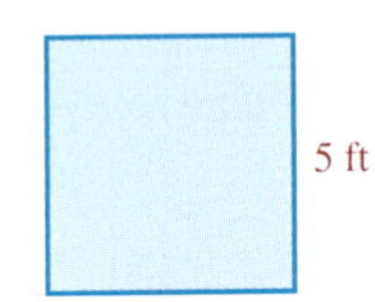

REVIEW

97. Round 230,999 to the nearest thousand.

98. Write the set of whole numbers.

99. What is $0 \div 15$?

100. Multiply: $15 \cdot (6 \cdot 9)$.

101. What is the formula for the area of a rectangle?

102. MARCHING BANDS When a university band lines up in eight rows of 15 musicians, there are five musicians left over. How many band members are there?

1.5 Order of Operations

- Order of operations
- Evaluating expressions with no grouping symbols
- Evaluating expressions with grouping symbols
- The arithmetic mean (average)

Punctuation marks, such as commas, quotations, and periods, serve an important purpose when writing compositions. They determine the way in which sentences are to be read and interpreted. To read and interpret mathematical expressions correctly, we must use an agreed-upon set of priority rules for the *order of operations*.

Order of operations

Suppose you are asked to contact a friend if you see a certain type of watch for sale while you are traveling in Europe. While in Switzerland, you spot the watch and send the following e-mail message.

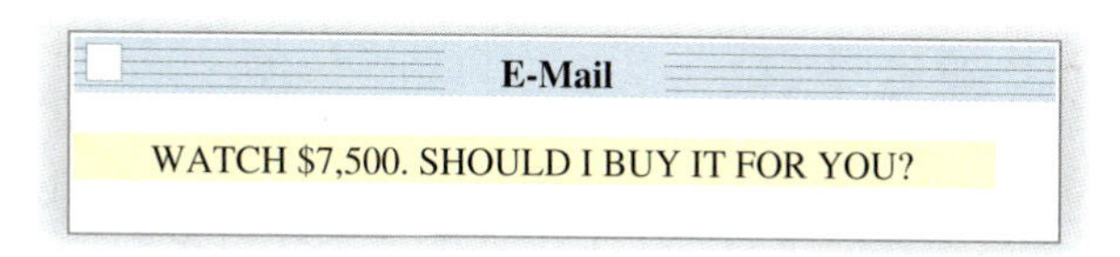

The next day, you get this response from your friend.

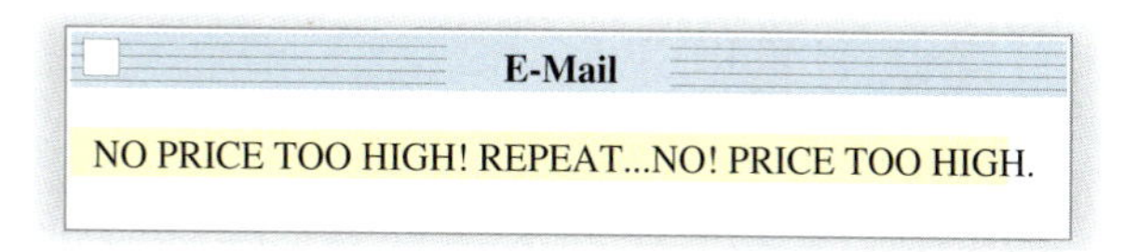

Something is wrong. One statement says to buy the watch at any price. The other says not to buy it, because it's too expensive. The placement of the exclamation point makes us read these statements differently, which results in different interpretations.

When we read a mathematical statement, the same kind of confusion is possible. For example, we consider

$$3 + 2 \cdot 5$$

The above expression contains two operations: addition and multiplication. We can *evaluate* it (find its value) in two ways. We can perform the addition first and then the multiplication. Or we can perform the multiplication first and then the addition. However, we get different results.

$\mathbf{3 + 2} \cdot 5 = \mathbf{5} \cdot 5$	Add first: $3 + 2 = 5$.	$3 + \mathbf{2 \cdot 5} = 3 + \mathbf{10}$	Multiply first: $2 \cdot 5 = 10$.
$= 25$	Multiply 5 and 5.	$= 13$	Add 3 and 10.

Different results

If we don't establish an order of operations, the expression $3 + 2 \cdot 5$ will have two different answers. To avoid this possibility, we evaluate expressions in the following order.

Order of operations

1. Perform all calculations within parentheses and other grouping symbols following the order listed in steps 2–4, working from the innermost pair to the outermost pair.
2. Evaluate all exponential expressions.
3. Perform all multiplications and divisions as they occur from left to right.
4. Perform all additions and subtractions as they occur from left to right.

When all grouping symbols have been removed, repeat steps 2–4 to complete the calculation.

If a fraction bar is present, evaluate the expression above the bar (called the **numerator**) and the expression below the bar (called the **denominator**) separately. Then perform the division indicated by the fraction bar, if possible.

To evaluate $3 + 2 \cdot 5$ correctly, we must apply the rules for the order of operations. Since there are no parentheses and there are no exponents, we perform the multiplication first and then the addition.

Ignore the addition for now → and perform the multiplication first: $2 \cdot 5 = 10$.

$$3 + \mathbf{2 \cdot 5} = 3 + \mathbf{10}$$
$$= 13 \quad \text{Now perform the addition.}$$

Using the rules for the order of operations, we see that the correct answer is 13.

Evaluating expressions with no grouping symbols

Self Check 1
Evaluate: $4 \cdot 3^3 - 6$.

Answer 102

EXAMPLE 1 Evaluate: $2 \cdot 4^2 - 8$.

Solution Since the expression does not contain any grouping symbols, we begin with step 2 of the rules for the order of operations.

$2 \cdot \mathbf{4^2} - 8 = 2 \cdot \mathbf{16} - 8$ Evaluate the exponential expression: $4^2 = 16$.
$= 32 - 8$ Perform the multiplication: $2 \cdot 16 = 32$.
$= 24$ Perform the subtraction.

Self Check 2
Evaluate: $10 - 2 \cdot 3 + 24$.

Answer 28

EXAMPLE 2 Evaluate: $8 - 3 \cdot 2 + 16$.

Solution Since the expression does not contain grouping symbols and since there are no powers to find, we look for multiplications or divisions to perform.

$8 - \mathbf{3 \cdot 2} + 16 = 8 - \mathbf{6} + 16$ Perform the multiplication: $3 \cdot 2 = 6$.
$= 2 + 16$ Working from left to right, perform the subtraction: $8 - 6 = 2$.
$= 18$ Perform the addition.

! COMMENT Some students incorrectly think that additions are always performed before subtractions. As Example 2 shows, this is not true. Working from left to right, we perform the additions or subtractions *in the order in which they occur.* The same is true for multiplications and divisions.

Self Check 3
Evaluate: $36 \div 9 + 4(2)3$.

Answer 28

EXAMPLE 3 Evaluate: $192 \div 6 - 5(3)2$.

Solution Although this expression contains parentheses, there are no calculations to perform within them. Since there are no powers, we perform multiplications and divisions as they are encountered from left to right.

$\mathbf{192 \div 6} - 5(3)2 = \mathbf{32} - 5(3)2$ Working from left to right, perform the division: $192 \div 6 = 32$.
$= 32 - 15(2)$ Working from left to right, perform the multiplication: $5(3) = 15$.
$= 32 - 30$ Perform the multiplication: $15(2) = 30$.
$= 2$ Perform the subtraction.

EXAMPLE 4 **Phone bills.** Figure 1–15 shows the rates for international telephone calls charged by a 10-10 long-distance company. A businesswoman calls Germany for 20 minutes, South Korea for 5 minutes, and Mexico City for 35 minutes. What is the total cost of the calls?

All rates are per minute.	
Canada	10¢
Germany	23¢
Jamaica	68¢
Mexico City	42¢
South Korea	29¢

FIGURE 1-15

Solution We can find the cost of a call (in cents) by multiplying the rate charged per minute by the length of the call (in minutes). To find the total cost, we add the costs of the three calls.

The cost of the call to Germany ↓ $23(20)$ + The cost of the call to South Korea ↓ $29(5)$ + The cost of the call to Mexico City ↓ $42(35)$

To evaluate this expression, we apply the rules for the order of operations.

$23(20) + 29(5) + 42(35) = 460 + 145 + 1{,}470$ Perform the multiplications.

$= 2{,}075$ Perform the additions.

The total cost of the calls is 2,075 cents, or $20.75.

Evaluating expressions with grouping symbols

Grouping symbols serve as mathematical punctuation marks. They help determine the order in which an expression is to be evaluated. Examples of grouping symbols are parentheses ( ), brackets [ ], and the fraction bar —.

In the next example, we have two similar-looking expressions. However, because of the parentheses, we evaluate them in a different order.

EXAMPLE 5 Evaluate each expression: **a.** $12 - 3 + 5$ and **b.** $12 - (3 + 5)$.

Solution

a. We perform the additions and subtractions as they occur, from left to right.

$\mathbf{12 - 3} + 5 = \mathbf{9} + 5$ Perform the subtraction: $12 - 3 = 9$.

$= 14$ Perform the addition.

b. This expression contains parentheses. We must perform the calculation within the parentheses first.

$12 - (\mathbf{3 + 5}) = 12 - \mathbf{8}$ Perform the addition: $3 + 5 = 8$.

$= 4$ Perform the subtraction.

Self Check 5

Evaluate each expression:

a. $20 - 7 + 6$

b. $20 - (7 + 6)$

Answer **a.** 19, **b.** 7

EXAMPLE 6 Evaluate: $(2 + 6)^3$.

Solution We begin by performing the calculation within the parentheses.

$(\mathbf{2 + 6})^3 = \mathbf{8}^3$ Perform the addition.

$= 512$ Evaluate the exponential expression: $8^3 = 8 \cdot 8 \cdot 8 = 512$.

Self Check 6

Evaluate: $(1 + 3)^4$.

Answer 256

Self Check 7
Evaluate: $50 - 4(12 - 5 \cdot 2)$.

Answer 42

EXAMPLE 7 Evaluate: $5 + 2(13 - 5 \cdot 2)$.

Solution This expression contains grouping symbols. We will apply the rules for the order of operations within the parentheses first, to evaluate $13 - 5 \cdot 2$.

$5 + 2(13 - \mathbf{5 \cdot 2}) = 5 + 2(13 - \mathbf{10})$	Perform the multiplication within the parentheses.
$= 5 + 2(3)$	Perform the subtraction within the parentheses.
$= 5 + 6$	Perform the multiplication: $2(3) = 6$.
$= 11$	Perform the addition.

Sometimes an expression contains two or more sets of grouping symbols. Since it can be confusing to read an expression such as $16 + 2(14 - 3(5 - 2))$, we often use brackets in place of the second pair of parentheses.

$$16 + 2[14 - 3(5 - 2)]$$

If an expression contains more than one pair of grouping symbols, we always begin by working within the innermost pair and then work to the outermost pair.

Innermost parentheses
↓ ↓
$$16 + 2[14 - 3(5 - 2)]$$
↑ ↑
Outermost brackets

Self Check 8
Evaluate: $140 - 7[4 + 3(6 - 2)]$.

Answer 28

EXAMPLE 8 Evaluate: $16 + 6[14 - 3(5 - 2)]$.

Solution

$16 + 6[14 - 3(\mathbf{5 - 2})] = 16 + 6[14 - 3(\mathbf{3})]$	Perform the subtraction within the parentheses.
$= 16 + 6(14 - 9)$	Perform the multiplication within the brackets. Since only one set of grouping symbols is needed, write $14 - 9$ within parentheses.
$= 16 + 6(5)$	Perform the subtraction within the parentheses.
$= 16 + 30$	Perform the multiplication: $6(5) = 30$.
$= 46$	Perform the addition.

Self Check 9
Evaluate: $\dfrac{3(14) - 6}{2(3^2)}$.

EXAMPLE 9 Evaluate: $\dfrac{2(13) - 2}{3(2^3)}$.

Solution A fraction bar is a grouping symbol. We evaluate the numerator and denominator separately and then perform the indicated division.

$$\frac{2(13) - 2}{3(2^3)} = \frac{26 - 2}{3(8)}$$ In the numerator, perform the multiplication. In the denominator, perform the calculation within the parentheses.

$$= \frac{24}{24}$$ In the numerator, perform the subtraction. In the denominator, perform the multiplication.

$$= 1$$ Perform the division.

Answer 2

The arithmetic mean (average)

The **arithmetic mean,** or **average,** of several numbers is a value around which the numbers are grouped. It gives you an indication of the center of the set of numbers. When finding the mean of a set of numbers, we usually need to apply the rules for the order of operations.

Finding an arithmetic mean

To find the mean of a set of scores, divide the sum of the scores by the number of scores.

EXAMPLE 10 **Basketball.** In 1998, the Lady Vols of the University of Tennessee won the women's basketball championship, capping a perfect 39-0 season. Find their average margin of victory in their last four tournament games shown below.

Regional	Regional final	Semifinal	Championship
Beat Rutgers by 32 points	Beat North Carolina by 6 points	Beat Arkansas by 28 points	Beat Louisiana Tech by 18 points

Solution To find the average margin of victory, add the margins of victory and divide by 4.

$$\text{Average} = \frac{32 + 6 + 28 + 18}{4}$$

$$= \frac{84}{4}$$

$$= 21$$

Their average margin of victory was 21 points.

Self Check 10
Syracuse University won the 2003 NCAA men's basketball championship. Find their average margin of victory in their six tournament games, which they won by 11, 12, 1, 16, 11, and 3 points.

Answer 9 points

Order of operations and parentheses CALCULATOR SNAPSHOT

Scientific calculators have the rules for order of operations built in. Even so, some evaluations require the use of a left parenthesis key (and a right parenthesis key). For example, to evaluate $\frac{240}{20 - 15}$, we enter these numbers and press these keys.

240 ÷ (20 − 15) =   48

THINK IT THROUGH Preparing for Class

"Only about 13% of full-time students spend more than 25 hours a week preparing for class, the approximate number that faculty members say is needed to do well in college." The National Survey of Student Engagement Annual Report 2003

The National Survey of Student Engagement 2003 Annual Report questioned thousands of full-time college students about their weekly activities. Use the given clues to determine the results of the survey shown below.

Full-time Student Time Usage per Week

Activity	Time per week
Preparing for class	14 hours
Working on-campus or off-campus	Four hours less than the time spent preparing for class
Participating in co-curricular activities	Half as many hours as the time spent working on-campus or off-campus
Relaxing and socializing	Two hours more than twice the time spent participating in co-curricular activities
Providing care for dependents	Three hours less than one-half of the time spent relaxing and socializing
Commuting to class	One hour more than the time spent providing care for dependents

Section 1.5 STUDY SET

VOCABULARY *Fill in the blanks.*

1. The grouping symbols ( ) are called ____________, and the symbols [ ] are called __________.
2. The expression above a fraction bar is called the __________. The expression below a fraction bar is called the ____________.
3. To ________ the expression $2 + 5 \cdot 4$ means to find its value.
4. To find the ______________ of several values, we add the values and divide by the number of values.

CONCEPTS

5. Consider $5(2)^2 - 1$. How many operations need to be performed to evaluate the expression? List them in the order in which they should be performed.
6. Consider $15 - 3 + (5 \cdot 2)^3$. How many operations need to be performed to evaluate this expression? List them in the order in which they should be performed.
7. Consider $\frac{5 + 5(7)}{2 + (8 - 4)}$. In the numerator, what operation should be done first? In the denominator, what operation should be done first?
8. In the expression $\frac{3 - 5(2)}{5(2) + 4}$, the bar is a grouping symbol. What does it separate?
9. Explain the difference between $2 \cdot 3^2$ and $(2 \cdot 3)^2$.
10. Use brackets to write $2(12 - (5 + 4))$ in better form.

NOTATION *Complete each solution to evaluate the expression.*

11. $28 - 5(2)^2 = 28 - 5(\square)$
    $= 28 - \square$
    $= \square$
12. $2 + (5 + 6 \cdot 2) = 2 + (5 + \square)$
    $= 2 + \square$
    $= \square$

13. $[4(2+7)] - 6 = [4(9)] - 6$
$= \square - 6$
$= \square$

14. $\dfrac{5(3)+12}{9-6} = \dfrac{\square + 12}{\square}$
$= \dfrac{\square}{\square}$
$= \square$

PRACTICE *Evaluate each expression.*

15. $7 + 4 \cdot 5$
16. $10 - 2 \cdot 2$
17. $2 + 3(0)$
18. $5(0) + 8$
19. $20 - 10 + 5$
20. $80 - 5 + 4$
21. $25 \div 5 \cdot 5$
22. $6 \div 2 \cdot 3$
23. $7(5) - 5(6)$
24. $4 \cdot 2 + 2 \cdot 4$
25. $4^2 + 3^2$
26. $12^2 - 5^2$
27. $2 \cdot 3^2$
28. $3^3 \cdot 5$
29. $3 + 2 \cdot 3^4 \cdot 5$
30. $3 \cdot 2^3 \cdot 4 - 12$
31. $5 \cdot 10^3 + 2 \cdot 10^2 + 3 \cdot 10^1 + 9$
32. $8 \cdot 10^3 + 0 \cdot 10^2 + 7 \cdot 10^1 + 4$
33. $3(2)^2 - 4(2) + 12$
34. $5(1)^3 + (1)^2 + 2(1) - 6$
35. $(8-6)^2 + (4-3)^2$
36. $(2+1)^2 + (3+2)^2$
37. $60 - \left(6 + \dfrac{40}{8}\right)$
38. $7 + \left(5^3 - \dfrac{200}{2}\right)$
39. $6 + 2(5+4)$
40. $3(5+1) + 7$
41. $3 + 5(6-4)$
42. $7(9-2) - 1$
43. $(7-4)^2 + 1$
44. $(9-5)^3 + 8$
45. $6^3 - (10+8)$
46. $5^2 - (9+3)$
47. $50 - 2(4)^2$
48. $30 + 2(3)^3$
49. $16^2 - 4(2)(5)$
50. $8^2 - 4(3)(1)$
51. $39 - 5(6) + 9 - 1$
52. $15 - 3(2) - 4 + 3$
53. $(18-12)^3 - 5^2$
54. $(9-2)^2 - 3^3$
55. $2(10 - 3^2) + 1$
56. $1 + 3(18 - 4^2)$
57. $6 + \dfrac{25}{5} + 6(3)$
58. $15 - \dfrac{24}{6} + 8 \cdot 2$
59. $3\left(\dfrac{18}{3}\right) - 2(2)$
60. $2\left(\dfrac{12}{3}\right) + 3(5)$
61. $(2 \cdot 6 - 4)^2$
62. $2(6-4)^2$
63. $4[50 - (3^3 - 5^2)]$
64. $6[15 + (5 \cdot 2^2)]$
65. $80 - 2[12 - (5+4)]$
66. $15 + 5[12 - (2^2 + 4)]$
67. $2[100 - (5+4)] - 45$
68. $8[6(6) - 6^2] + 4(5)$
69. $\dfrac{10+5}{6-1}$
70. $\dfrac{18+12}{2(3)}$
71. $\dfrac{5^2+17}{6-2^2}$
72. $\dfrac{3^2-2^2}{(3-2)^2}$
73. $\dfrac{(3+5)^2+2}{2(8-5)}$
74. $\dfrac{25-(2 \cdot 3-1)}{2 \cdot 9-8}$
75. $\dfrac{(5-3)^2+2}{4^2-(8+2)}$
76. $\dfrac{(4^3-2)+7}{5(2+4)-7}$
77. $12{,}985 - (1{,}800 + 689)$
78. $\dfrac{897-655}{88-77}$
79. $3{,}245 - 25(16-12)^2$
80. $\dfrac{24^2-4^2}{22+58}$

APPLICATIONS *In Problems 81–86, write an expression to solve each problem. Then evaluate the expression.*

81. BUYING GROCERIES At the supermarket, Carlos has 2 cases of soda, 4 bags of potato chips, and 2 cans of dip in his cart. Each case of soda costs \$6, each bag of chips costs \$2, and each can of dip costs \$1. Find the total cost of the groceries.

82. JUDGING The scores received by a junior diver are as follows.

5	2	4	6	3	4

The formula for computing the overall score for the dive is as follows:

1. Throw out the lowest score.
2. Throw out the highest score.
3. Divide the sum of the remaining scores by 4.

Find the diver's score.

83. BANKING When a customer deposits cash, a teller must complete a "currency count" on the back of the deposit slip. In the illustration, what is the total amount of cash being deposited?

Currency count, for financial use only			
24	x 1's		
—	x 2's		
6	x 5's		
10	x 10's		
12	x 20's		
2	x 50's		
1	x 100's		
	TOTAL \$		

84. WRAPPING GIFTS How much ribbon is needed to wrap the package shown if 15 inches of ribbon are needed to make the bow?

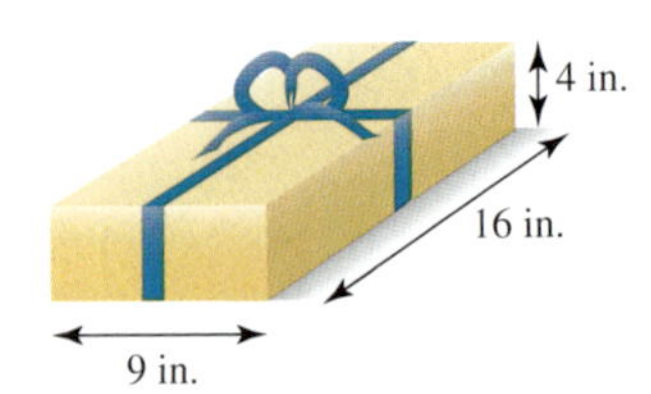

85. SCRABBLE Illustration (a) shows part of the game board before and Illustration (b) shows it after the words *brick* and *aphid* were played. Determine the scoring for each word. (The number on each tile gives the point value of the letter.)

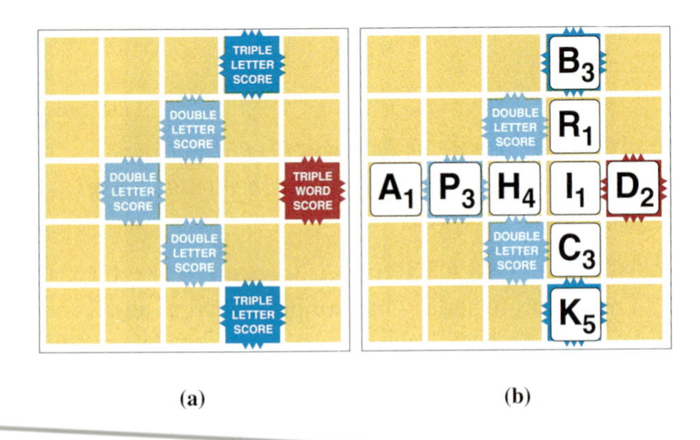

86. THE GETTYSBURG ADDRESS Here is an excerpt from Abraham Lincoln's Gettysburg Address.

Fourscore and seven years ago, our fathers brought forth on this continent a new nation, conceived in liberty, and dedicated to the proposition that all men are created equal.

Lincoln's comments refer to the year 1776, when the United States declared its independence. If a score is 20 years, in what year did Lincoln deliver the Gettysburg Address?

87. CLIMATE One December week, the temperatures in Honolulu, Hawaii, were 75°, 80°, 83°, 80°, 77°, 72°, and 86°. Find the week's average (mean) temperature.

88. GRADES In a psychology class, a student had test scores of 94, 85, 81, 77, and 89. He also overslept, missed the final exam, and received a 0 on it. What was his test average in the class?

89. NATURAL NUMBERS What is the average (mean) of the first nine natural numbers?

90. ENERGY USAGE Find the average number of therms of natural gas used per month. Then draw a dashed line across the graph in the next column showing the average.

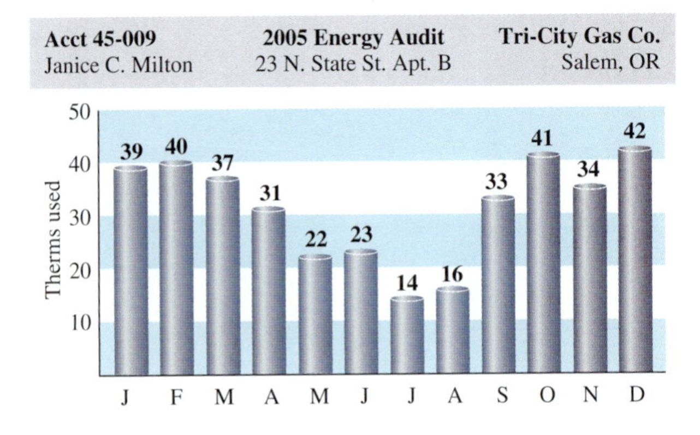

91. FAST FOOD The table shows the sandwiches Subway advertises as its low-fat menu. What is the average (mean) number of calories for the group of sandwiches?

6-inch subs	Calories	Fat (g)
Veggie Delite	237	3
Turkey Breast	289	4
Turkey Breast & Ham	295	5
Ham	302	5
Roast Beef	303	5
Subway Club	312	5
Roasted Chicken Breast	348	6

92. TV RATINGS The list below shows the number of people watching *Who Wants to Be a Millionaire*? on five weeknights in November 1999. How large was the average audience?

Monday	26,800,000
Tuesday	24,900,000
Wednesday	22,900,000
Thursday	25,900,000
Friday	21,900,000

WRITING

93. Explain why rules for the order of operations are necessary.

94. Explain the difference between the steps used to evaluate $5 \cdot 2^3$ and $(5 \cdot 2)^3$.

95. Explain the process of finding the mean of a large group of numbers. What does an average tell you?

96. What does it mean when we say to perform all additions and subtractions *as they occur from left to right*?

REVIEW *Perform the operations.*

97. $\begin{array}{r} 4{,}029 \\ +3{,}271 \\ \hline \end{array}$

98. $\begin{array}{r} 4{,}263 \\ -3{,}764 \\ \hline \end{array}$

99. $\begin{array}{r} 417 \\ \times\ 23 \\ \hline \end{array}$

100. $82\overline{)50{,}430}$

1.6 Solving Equations by Addition and Subtraction

- Equations
- Checking solutions
- Solving equations
- Problem solving with equations

The language of mathematics is *algebra.* The word *algebra* comes from the title of a book written by the Arabian mathematician Al-Khowarazmi around A.D. 800. Its title, *Ihm aljabr wa'l muqabalah,* means restoration and reduction, a process then used to solve equations. In this section, we will begin discussing equations, one of the most powerful ideas in algebra.

Equations

An **equation** is a statement indicating that two expressions are equal. Some examples of equations are

$$x + 5 = 21, \qquad 16 + 5 = 21, \qquad \text{and} \qquad 10 + 5 = 21$$

Equations

Equations are mathematical sentences that contain an = symbol.

In the equation $x + 5 = 21$, the expression $x + 5$ is called the **left-hand side,** and 21 is called the **right-hand side.** The letter x is the **variable** (or the **unknown**).

An equation can be true or false. For example, $16 + 5 = 21$ is a true equation, whereas $10 + 5 = 21$ is a false equation. An equation containing a variable can be true or false, depending upon the value of the variable. If x is 16, the equation $x + 5 = 21$ is true, because

$16 + 5 = 21$ Substitute 16 for x.

However, this equation is false for all other values of x.

Any number that makes an equation true when substituted for its variable is said to *satisfy* the equation. Such numbers are called **solutions.** Because 16 is the only number that satisfies $x + 5 = 21$, it is the only solution of the equation.

Checking solutions

EXAMPLE 1 Verify that 18 is a solution of the equation $x - 3 = 15$.

Solution We substitute 18 for x in the equation and verify that both sides of the equation are equal.

Self Check 1
Is 8 a solution of $x + 17 = 25$?

$x - 3 = 15$ This is the given equation.

$\mathbf{18} - 3 \stackrel{?}{=} 15$ Substitute 18 for x. Read $\stackrel{?}{=}$ as "is possibly equal to."

$15 = 15$ Perform the subtraction.

Since $15 = 15$ is a true equation, 18 is a solution of $x - 3 = 15$.

Answer yes

Self Check 2
Is 5 a solution of $20 = y - 17$?

Answer no

EXAMPLE 2 Is 23 a solution of $32 = y + 10$?

Solution We substitute 23 for y and simplify.

$32 = y + 10$ This is the given equation.

$32 \stackrel{?}{=} \mathbf{23} + 10$ Substitute 23 for y.

$32 \neq 33$ Perform the addition.

Since the left-hand and right-hand sides are not equal, 23 is not a solution of $32 = y + 10$.

Solving equations

Since the solution of an equation is usually not given, we must develop a process to find it. This process is called *solving the equation.* To develop an understanding of the properties and procedures used to solve an equation, we will examine $x + 2 = 5$ and make some observations as we solve it in a practical way.

We can think of the scales shown in Figure 1-16(a) as representing the equation $x + 2 = 5$. The weight (in grams) on the left-hand side of the scales is $x + 2$, and the weight (in grams) on the right-hand side is 5. Because these weights are equal, the scales are in balance. To find x, we need to isolate it. That can be accomplished by removing 2 grams from the left-hand side of the scales. Common sense tells us that we must also remove 2 grams from the right-hand side if the scales are to remain in balance. In Figure 1-16(b), we can see that x grams are balanced by 3 grams. We say that we have *solved* the equation and that the *solution* is 3.

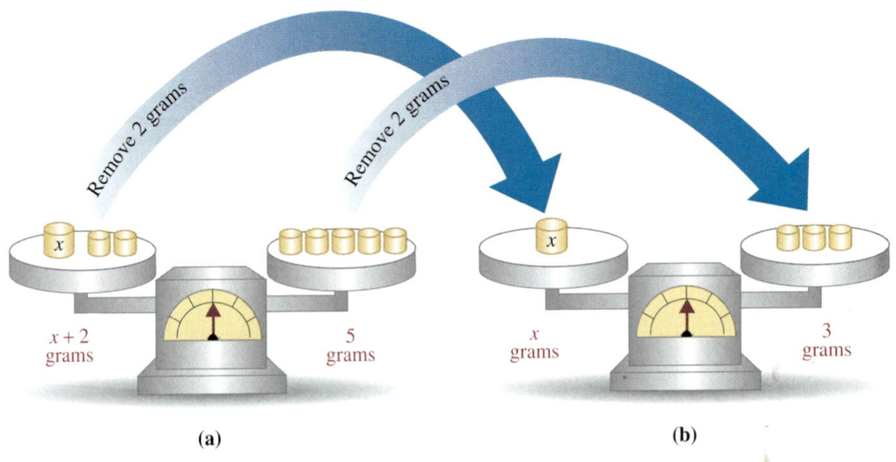

FIGURE 1-16

From this example, we can make observations about solving an equation.

- To find the value of x, we needed to isolate it on the left-hand side of the scales.
- To isolate x, we had to undo the addition of 2 grams. This was accomplished by subtracting 2 grams from the left-hand side.

- We wanted the scales to remain in balance. When we subtracted 2 grams from the left-hand side, we subtracted the same amount from the right-hand side.

The observations suggest a property of equality: *If the same quantity is subtracted from equal quantities, the results will be equal quantities.* We can express this property in symbols.

Subtraction property of equality

Let a, b, and c represent numbers.

If $a = b$, then $a - c = b - c$.

When we use this property, the resulting equation will be equivalent to the original equation.

Equivalent equations

Two equations are **equivalent equations** when they have the same solutions.

In the previous example, we found that $x + 2 = 5$ is equivalent to $x = 3$. This is true because these equations have the same solution, 3.

We now show how to solve $x + 2 = 5$ using an algebraic approach.

EXAMPLE 3 Solve: $x + 2 = 5$.

Solution To isolate x on the left-hand side of the equation, we undo the addition of 2 by subtracting 2 from both sides of the equation.

$x + 2 = 5$	This is the equation to solve.
$x + 2 - 2 = 5 - 2$	Subtract 2 from both sides.
$x = 3$	On the left-hand side, subtracting 2 undoes the addition of 2 and leaves x. On the right-hand side, $5 - 2 = 3$.

We check by substituting 3 for x in the original equation and simplifying. If 3 is the solution, we will obtain a true statement.

Check:	$x + 2 = 5$	This is the original equation.
	$3 + 2 \stackrel{?}{=} 5$	Substitute 3 for x.
	$5 = 5$	Perform the addition.

Since the resulting equation is true, 3 is the solution.

Self Check 3

Solve $x + 7 = 14$ and check the result.

Answer 7

A second property that we will use to solve equations involves addition. It is based on the following idea: *If the same quantity is added to equal quantities, the results will be equal quantities.* In symbols, we have the following property.

Addition property of equality

Let a, b, and c represent numbers.

If $a = b$, then $a + c = b + c$.

We can think of the scales shown in Figure 1-17(a) on the next page as representing the equation $x - 2 = 3$. To find x, we need to use the addition property of equality and add 2 grams of weight to each side. The scales will remain in balance. From the scales in Figure 1-17(b), we can see that x grams are balanced by 5 grams. The solution of $x - 2 = 3$ is therefore 5.

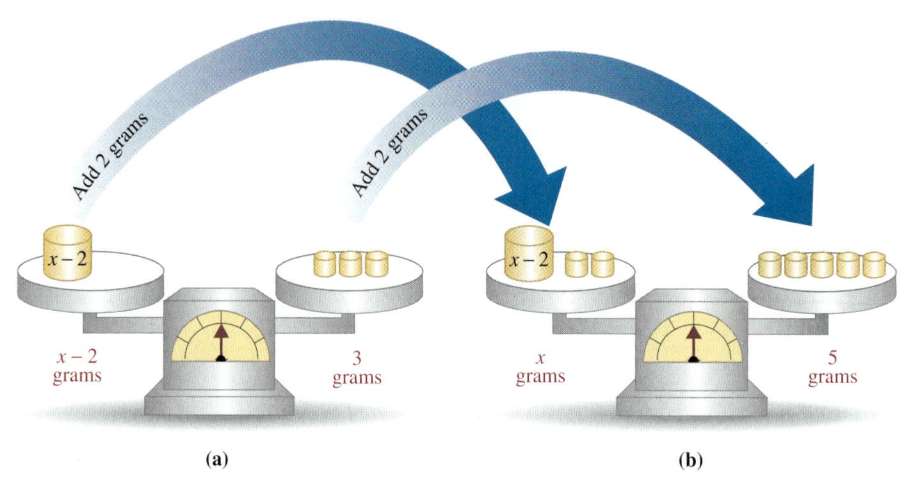

FIGURE 1-17

To solve $x - 2 = 3$ algebraically, we apply the addition property of equality. We can isolate x on the left-hand side of the equation by adding 2 to both sides.

$x - 2 = 3$	This is the equation to solve.
$x - 2 + \mathbf{2} = 3 + \mathbf{2}$	To undo the subtraction of 2, add 2 to both sides.
$x = 5$	On the left-hand side, adding 2 undoes the subtraction of 2 and leaves x. On the right-hand side, $3 + 2 = 5$.

To check this result, we substitute 5 for x in the original equation and simplify.

Check: $x - 2 = 3$	This is the original equation.
$\mathbf{5} - 2 \stackrel{?}{=} 3$	Substitute 5 for x.
$3 = 3$	Perform the subtraction.

Since this is a true statement, 5 is the solution.

Self Check 4
Solve $75 = b - 38$ and check the result.

EXAMPLE 4 Solve: $19 = y - 7$.

Solution To isolate the variable y on the right-hand side, we use the addition property of equality. We can undo the subtraction of 7 by adding 7 to both sides.

$19 = y - 7$	This is the equation to solve.
$19 + \mathbf{7} = y - 7 + \mathbf{7}$	Add 7 to both sides.
$26 = y$	On the left-hand side, $19 + 7 = 26$. On the right-hand side, adding 7 undoes the subtraction of 7 and leaves y.
$y = 26$	When we state a solution, it is common practice to write the variable first. If $26 = y$, then $y = 26$.

We check by substituting 26 for y in the original equation and simplifying.

Check: $19 = y - 7$	This is the original equation.
$19 \stackrel{?}{=} \mathbf{26} - 7$	Substitute 26 for y.
$19 = 19$	Perform the subtraction.

Since this is a true statement, 26 is the solution.

Answer 113

Problem solving with equations

The key to problem solving is to understand the problem and then to devise a plan for solving it. The following list of steps provides a good strategy to follow.

Strategy for problem solving

1. **Analyze the problem** by reading it carefully to understand the given facts. What information is given? What vocabulary is given? What are you asked to find? Often a diagram will help you visualize the facts of a problem.
2. **Form an equation** by picking a variable to represent the quantity to be found. Then express all other unknown quantities as expressions involving that variable. Key words or phrases can be helpful. Finally, translate the words of the problem into an equation.
3. **Solve the equation.**
4. **State the conclusion.**
5. **Check the result** in the words of the problem.

We will now use this five-step strategy to solve problems. The purpose of the following examples is to help you learn the strategy, even though you can probably solve these examples without it.

EXAMPLE 5 **Financial data.** Figure 1-18 shows the 1999 quarterly net income for Nike, the athletic shoe company. What was the company's total net income for 1999?

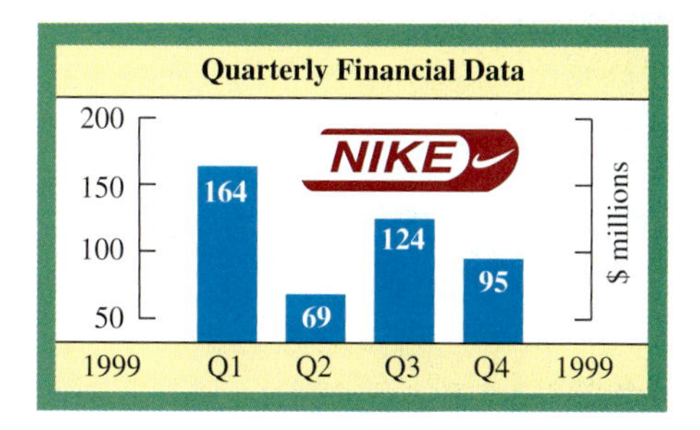

FIGURE 1-18

Analyze the problem

- We are given the net income for each quarter.
- We are asked to find the total net income.

Form an equation

Let $n =$ the total net income for 1999. To form an equation involving n, we look for a key word or phrase in the problem.

Key word: *total* **Translation:** *addition*

Now we translate the words of the problem into an equation.

Total net income	is	1st qtr. net income	plus	2nd qtr. net income	plus	3rd qtr. net income	plus	4th qtr. net income.
n	$=$	164	$+$	69	$+$	124	$+$	95

Solve the equation

$$n = 164 + 69 + 124 + 95$$ We are working in millions of dollars.

$$= 452$$ Perform the additions.

State the conclusion

Nike's total net income was $452 million.

Check the result

We can check the result by estimation. To estimate, we round the net income from each quarter and add.

$$160 + 70 + 120 + 100 = 450$$

The answer, 452, is reasonable.

EXAMPLE 6 Small business. Last year a hairdresser lost 17 customers who moved away. If he now has 73 customers, how many did he have originally?

Analyze the problem

- We know that he started with some unknown number of customers, and after 17 moved away, 73 were left.
- We are asked to find the number of customers he had before any moved away.

Form an equation

We can let c = the original number of customers. To form an equation involving c, we look for a key word or phrase in the problem.

Key phrase: *moved away* **Translation:** *subtraction*

Now we translate the words of the problem into an equation.

The original number of customers	minus	17	is	the remaining number of customers.
c	$-$	17	$=$	73

Solve the equation

$$c - 17 = 73$$

$$c - 17 + \mathbf{17} = 73 + \mathbf{17}$$ To undo the subtraction of 17, add 17 to both sides.

$$c = 90$$ Simplify each side of the equation.

State the conclusion

He originally had 90 customers.

Check the result

The hairdresser had 90 customers. After losing 17, he has $90 - 17$, or 73 left. The answer, 90, checks.

EXAMPLE 7 **Buying a house.** Sue wants to buy a house that costs \$87,000. Since she has only \$15,000 for a down payment, she will have to borrow some additional money by taking a mortgage. How much will she have to borrow?

Analyze the problem

- The house costs \$87,000.
- Sue has \$15,000 for a down payment.
- We must find how much money she needs to borrow.

Form an equation

We can let x = the money that she needs to borrow. To form an equation involving x, we look for a key word or phrase in the problem.

Key phrase: *borrow some additional money* **Translation:** *addition*

Now we translate the words of the problem into an equation.

The amount Sue has	plus	the amount she borrows	is	the total cost of the house.
15,000	$+$	x	$=$	87,000

Solve the equation

$$15{,}000 + x = 87{,}000$$

$$15{,}000 + x - \mathbf{15{,}000} = 87{,}000 - \mathbf{15{,}000}$$ To undo the addition of 15,000, subtract 15,000 from both sides.

$$x = 72{,}000$$ Perform the subtractions.

State the conclusion

She must borrow \$72,000.

Check the result

With a \$72,000 mortgage, she will have \$15,000 + \$72,000, which is the \$87,000 that is necessary to buy the house. The answer, 72,000, checks.

Section 1.6 STUDY SET

VOCABULARY *Fill in the blanks.*

1. An equation is a statement that two expressions are ________. An equation contains an ☐ symbol.
2. A ________ of an equation is a number that satisfies the equation.
3. To ________ the solution of an equation, we substitute the value for the variable in the original equation and see whether the result is a true statement.
4. A letter that is used to represent a number is called a ________.
5. ________ equations have exactly the same solutions.
6. To solve an equation, we ________ the variable on one side of the equals symbol.

CONCEPTS *Complete the rules.*

7. If $x = y$ and c is any number, then $x + c =$ ☐ + ☐.
8. If $x = y$ and c is any number, then $x - c =$ ☐ − ☐.

9. In $x + 6 = 10$, what operation is performed on the variable? How do we undo that operation to isolate the variable?

10. In $9 = y - 5$, what operation is performed on the variable? How do we undo that operation to isolate the variable?

NOTATION *Complete each solution to solve the given equation.*

11.
$$x + 8 = 24$$
$$x + 8 - \square = 24 - \square$$
$$x = 16$$
Check: $x + 8 = 24$
$$\square + 8 \;\square\; 24$$
$$\square = 24$$
So $\square$ is the solution.

12.
$$x - 8 = 24$$
$$x - 8 + \square = 24 + \square$$
$$x = 32$$
Check: $x - 8 = 24$
$$\square - 8 \stackrel{?}{=} 24$$
$$\square = 24$$
So $\square$ is the solution.

PRACTICE *Determine whether each statement is an equation.*

13. $x = 2$
14. $y - 3$
15. $7x < 8$
16. $7 + x = 2$
17. $x + y = 0$
18. $3 - 3y > 2$
19. $1 + 1 = 3$
20. $5 = a + 2$

For each equation, is the given number a solution?

21. $x + 2 = 3;\ 1$
22. $x - 2 = 4;\ 6$
23. $a - 7 = 0;\ 7$
24. $x + 4 = 4;\ 0$
25. $8 - y = y;\ 5$
26. $10 - c = c;\ 5$
27. $x + 32 = 0;\ 16$
28. $x - 1 = 0;\ 4$
29. $z + 7 = z;\ 7$
30. $n - 9 = n;\ 9$
31. $x = x;\ 0$
32. $x = 2;\ 0$

Use the addition or subtraction property of equality to solve each equation. **Check each answer.**

33. $x - 7 = 3$
34. $y - 11 = 7$
35. $a - 2 = 5$
36. $z - 3 = 9$
37. $1 = b - 2$
38. $0 = t - 1$
39. $x - 4 = 0$
40. $c - 3 = 0$
41. $y - 7 = 6$
42. $a - 2 = 4$
43. $70 = x - 5$
44. $66 = b - 6$
45. $312 = x - 428$
46. $x - 307 = 113$
47. $x - 117 = 222$
48. $y - 27 = 317$
49. $x + 9 = 12$
50. $x + 3 = 9$
51. $y + 7 = 12$
52. $c + 11 = 22$
53. $t + 19 = 28$
54. $s + 45 = 84$
55. $23 + x = 33$
56. $34 + y = 34$
57. $5 = 4 + c$
58. $41 = 23 + x$
59. $99 = r + 43$
60. $92 = r + 37$
61. $512 = x + 428$
62. $x + 307 = 513$
63. $x + 117 = 222$
64. $y + 38 = 321$
65. $3 + x = 7$
66. $b - 4 = 8$
67. $y - 5 = 7$
68. $z + 9 = 23$
69. $4 + a = 12$
70. $5 + x = 13$
71. $x - 13 = 34$
72. $x - 23 = 19$

APPLICATIONS *Complete each solution.*

73. ARCHAEOLOGY A 1,700-year-old manuscript is 425 years older than the clay jar in which it was found. How old is the jar?

Analyze the problem

- The manuscript is ________ old.
- The manuscript is ________ older than the jar.
- We are asked to find ______________.

Form an equation Since we want to find the age of the jar, we can let x = ______________. Now we look for a key word or phrase in the problem.

Key phrase: ________

Translation: ________

Now we translate the words of the problem into an equation.

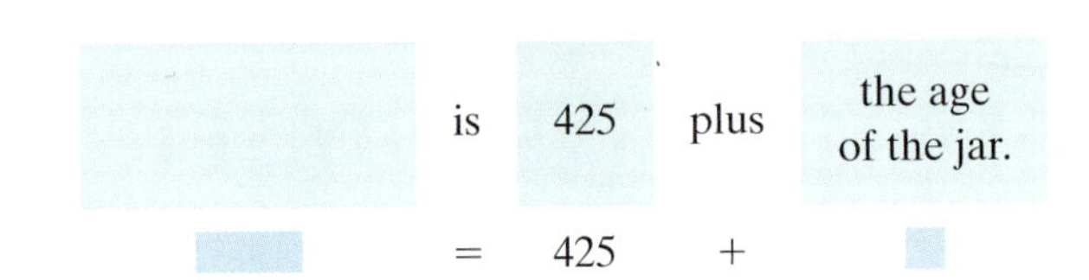

Solve the equation

$$\square = 425 + x$$
$$1{,}700 - \square = 425 + x - \square$$
$$\square = x$$

State the conclusion ______________.

Check the result If the jar is $\square$ years old, then the manuscript is $1{,}275 + 425 = \square$ years old. The answer checks.

74. BANKING After a student wrote a $1,500 check to pay for a car, he had a balance of $750 in his account. How much did he have in the account before he wrote the check?

Analyze the problem

- A ________ check was written.
- The balance became ______.
- We are asked to find ______________________________.

Form an equation Since we want to find his balance before he wrote the check, we let $x =$ ______________________________

Now we look for a key word or phrase in the problem.

Key phrase: ______________

Translation: ____________

Now we translate the words of the problem into an equation.

The original balance in the account	minus	$1,500	is	$750.
x	$-$		$=$	750

Solve the equation

$$\square - 1{,}500 = 750$$
$$x - 1{,}500 + \square = 750 + \square$$
$$x = \square$$

State the conclusion ______________________________

Check the result The original balance was ______. After writing a check, he had a balance of $2,250 − $1,500, or ______. The answer checks.

Let a variable represent the unknown quantity. Then write and solve an equation to answer the question.

75. ELECTIONS The illustration shows the votes received by the three major candidates running for President of the United States in 1996. Find the total number of votes cast for them.

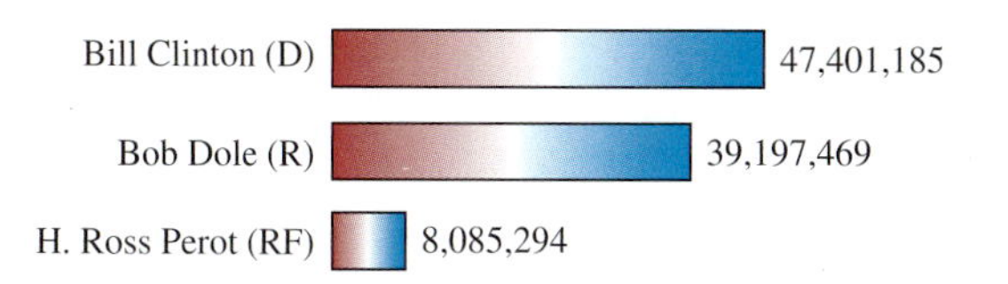

76. HIT RECORDS The oldest artist to have a number 1 single was Louis Armstrong at age 67, with *Hello Dolly.* The youngest artist to have the number 1 single was 12-year-old Jimmy Boyd, with *I Saw Mommy Kissing Santa Claus.* What is the difference in their ages?

77. PARTY INVITATIONS Three of Mia's party invitations were lost in the mail, but 59 were delivered. How many invitations did she send?

78. HEARING PROTECTION The sound intensity of a jet engine is 110 decibels. What noise level will an airplane mechanic experience if the ear plugs she is wearing reduce the sound intensity by 29 decibels?

79. FAST FOODS The franchise fee and startup costs for a Taco Bell restaurant are $287,000. If an entrepreneur has $68,500 to invest, how much money will she need to borrow to open her own Taco Bell restaurant?

80. BUYING GOLF CLUBS A man needs $345 for a new set of golf clubs. How much more money does he need if he now has $317?

81. CELEBRITY EARNINGS *Forbes* magazine estimates that in 2003, Celine Dion earned $28 million. If this was $152 million less than Oprah Winfrey's earnings, how much did Oprah earn in 2003?

82. HELP WANTED From the following ad from the classified section of a newspaper, determine the value of the benefit package. ($45 K means $45,000.)

★ACCOUNTS PAYABLE★
2-3 yrs exp as supervisor. Degree a +. High vol company. Good pay, $45K & xlnt benefits; total compensation worth $52K. Fax resume.

83. POWER OUTAGES The electrical system in a building automatically shuts down when the meter shown reads 85. By how much must the current reading increase to cause the system to shut down?

84. VIDEO GAMES After a week of playing Sega's *Sonic Adventure,* a boy scored 11,053 points in one game — an improvement of 9,485 points over the very first time he played. What was his score for his first game?

85. AUTO REPAIRS A woman paid \$29 less to have her car fixed at a muffler shop than she would have paid at a gas station. At the gas station, she would have paid \$219. How much did she pay to have her car fixed?

86. RIDING BUSES A man had to wait 20 minutes for a bus today. Three days ago, he had to wait 15 minutes longer than he did today, because four buses passed by without stopping. How long did he wait three days ago?

WRITING

87. Explain what it means for a number to satisfy an equation.

88. Explain how to tell whether a number is a solution of an equation.

89. Explain what Figure 1-16 (page 56) is trying to show.

90. Explain what Figure 1-17 (page 58) is trying to show.

91. When solving equations, we *isolate* the variable. Write a sentence in which the word *isolate* is used in a different context.

92. Think of a number. Add 8 to it. Now subtract 8 from that result. Explain why we will always obtain the original number.

REVIEW

93. Round 325,784 to the nearest ten

94. Find the power: 1^5.

95. Evaluate: $2 \cdot 3^2 \cdot 5$.

96. Represent $4 + 4 + 4$ as a multiplication.

97. Evaluate: $8 - 2(3) + 1^3$.

98. Write 1,055 in words.

1.7 Solving Equations by Division and Multiplication

- The division property of equality
- The multiplication property of equality
- Problem solving with equations

In the previous section, we solved equations of the forms

$$x - 4 = 10 \qquad \text{and} \qquad x + 5 = 16$$

by using the addition and subtraction properties of equality. In this section, we will learn how to solve equations of the forms

$$2x = 8 \qquad \text{and} \qquad \frac{x}{3} = 25$$

by using the division and multiplication properties of equality.

The division property of equality

To solve many equations, we must divide both sides of the equation by the same nonzero number. The resulting equation will be equivalent to the original one. This idea is summed up in the division property of equality: *If equal quantities are divided by the same nonzero quantity, the results will be equal quantities.*

Division property of equality

Let a, b, and c represent numbers and c is not 0.

$$\text{If } a = b, \text{ then } \frac{a}{c} = \frac{b}{c}.$$

We will now consider how to solve the equation $2x = 8$. Recall that $2x$ means $2 \cdot x$. Therefore, the given equation can be rewritten as $2 \cdot x = 8$. We can think of the scales in Figure 1-19(a) as representing the equation $2 \cdot x = 8$. The weight (in grams) on the left-hand side of the scales is $2 \cdot x$, and the weight (in grams) on the right-hand side is 8. Because these weights are equal, the scales are in balance. To find x, we need to isolate it. That can be accomplished by using the division property of equality to remove half of the weight from each side. The scales will remain in balance. From the scales shown in Figure 1-19(b), we see that x grams are balanced by 4 grams.

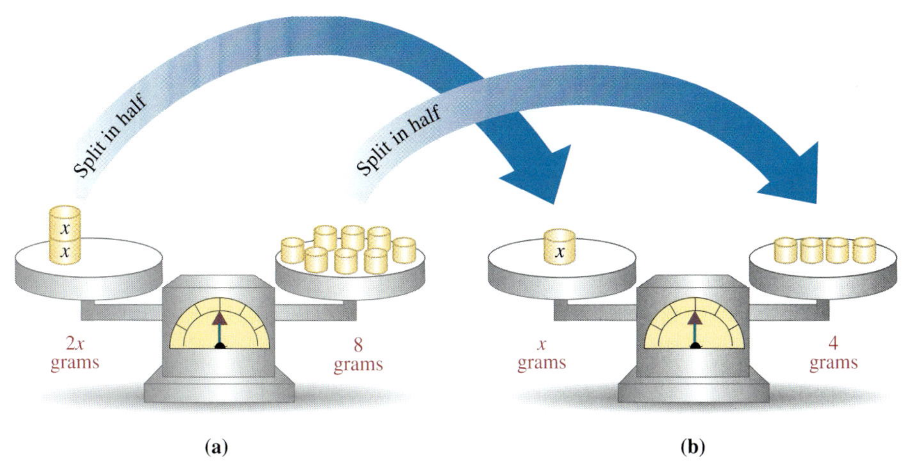

FIGURE 1-19

We now show how to solve $2x = 8$ using an algebraic approach.

EXAMPLE 1 Solve: $2x = 8$.

Solution Recall that $2x = 8$ means $2 \cdot x = 8$. To isolate x on the left-hand side of the equation, we undo the multiplication by 2 by dividing both sides of the equation by 2.

$2x = 8$ This is the equation to solve.

$\frac{2x}{2} = \frac{8}{2}$ To undo the multiplication by 2, divide both sides by 2.

$x = 4$ When x is multiplied by 2 and that product is then divided by 2, the result is x. Perform the division: $8 \div 2 = 4$.

To check this result, we substitute 4 for x in $2x = 8$.

Check: $2x = 8$ This is the original equation.

$2 \cdot 4 \stackrel{?}{=} 8$ Substitute 4 for x.

$8 = 8$ Perform the multiplication.

Since $8 = 8$ is a true statement, 4 is the solution.

Self Check 1
Solve $17x = 153$ and check the result.

Answer 9

The multiplication property of equality

We can also multiply both sides of an equation by the same nonzero number to get an equivalent equation. This idea is summed up in the multiplication property of equality: *If equal quantities are multiplied by the same nonzero quantity, the results will be equal quantities.*

Multiplication property of equality

Let a, b, and c represent numbers and c is not 0.

If $a = b$, then $c \cdot a = c \cdot b$ or, more simply, $ca = cb$.

We can think of the scales shown in Figure 1-20(a) as representing the equation $\frac{x}{3} = 25$. The weight on the left-hand side of the scales is $\frac{x}{3}$ grams, and the weight on the right-hand side is 25 grams. Because these weights are equal, the scales are in balance. To find x, we can use the multiplication property of equality to triple (or multiply by 3) the weight on each side. The scales will remain in balance. From the scales shown in Figure 1-20(b), we can see that x grams are balanced by 75 grams.

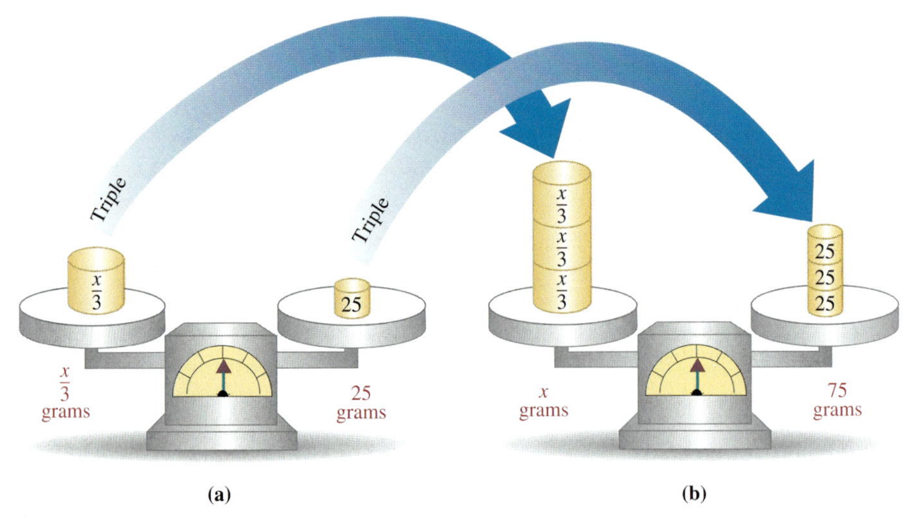

FIGURE 1-20

We now show how to solve $\frac{x}{3} = 25$ using an algebraic approach.

Self Check 2

Solve $\frac{x}{12} = 24$ and check the result.

Answer 288

EXAMPLE 2 Solve: $\frac{x}{3} = 25$.

Solution To isolate x on the left-hand side of the equation, we undo the division of the variable by 3 by multiplying both sides by 3.

$\frac{x}{3} = 25$ This is the equation to solve.

$3 \cdot \frac{x}{3} = 3 \cdot 25$ To undo the division by 3, multiply both sides by 3.

$x = 75$ When x is divided by 3 and that quotient is then multiplied by 3, the result is x. Perform the multiplication: $3 \cdot 25 = 75$.

Check: $\frac{x}{3} = 25$ This is the original equation.

$\frac{75}{3} \stackrel{?}{=} 25$ Substitute 75 for x.

$25 = 25$ Perform the division: $75 \div 3 = 25$. The answer checks.

Problem solving with equations

As before, we can use equations to solve problems. Remember that the purpose of these early examples is to help you learn the strategy, even though you can probably solve the examples without it.

EXAMPLE 3 **Buying electronics.** The owner of an apartment complex bought six television sets that were on sale for $499 each. What was the total cost?

Analyze the problem

- 6 television sets were bought.
- They cost $499 each.
- We are asked to find the total cost.

Form an equation

We let $c =$ the total cost of the TVs. To form an equation, we look for a key word or phrase in the problem. We can add 499 six times or multiply 499 by 6. Since it is easier, we will multiply.

Key phrase: *six TVs, each costing $499* **Translation:** *multiplication*

Now we translate the words of the problem into an equation.

The total number of TVs	multiplied by	the cost of each TV	is	the total cost.
6	$\cdot$	499	$=$	c

Solve the equation

$$6 \cdot 499 = c$$

$$2{,}994 = c \quad \text{Perform the multiplication.}$$

State the conclusion

The total cost will be $2,994.

Check the result

We can check by estimation. Since each TV costs a little less than $500, we would expect the total cost to be a little less than 6 · $500, or $3,000. An answer of $2,994 is reasonable.

EXAMPLE 4 **Splitting an inheritance.** If seven brothers inherit $343,000 and split the money evenly, how much will each brother get?

Analyze the problem

- There are 7 brothers.
- They split $343,000 evenly.
- We are asked to find how much each brother will get.

Form an equation

We can let $g =$ the number of dollars each brother will get. To form an equation, we look for a key word or phrase in the problem.

Key phrase: *split the money evenly* **Translation:** *division*

Now we translate the words of the problem into an equation.

The total amount of the inheritance	divided by	the number of brothers	is	the share each brother will get.
343,000	$\div$	7	$=$	g

Solve the equation

$$\frac{343{,}000}{7} = g \quad 343{,}000 \div 7 \text{ can be written as } \frac{343{,}000}{7}.$$

$$49{,}000 = g \quad \text{Perform the division.}$$

State the conclusion

Each brother will get \$49,000.

Check the result

If we multiply \$49,000 by 7, we get \$343,000.

EXAMPLE 5 **Traffic violations.** For a speeding ticket, a motorist had to pay a fine of \$592. The violation occurred on a stretch of highway posted with special signs like the one shown in Figure 1-21. What would the fine have been if such signs were not posted?

TRAFFIC FINES DOUBLED IN CONSTRUCTION ZONE

FIGURE 1-21

Analyze the problem

- The motorist was fined \$592.
- The fine was double what it would normally have been.
- We are asked to find what the fine would have been, had the area not been posted.

Form an equation

We can let $f =$ the amount that the fine would normally have been. To form an equation, we look for a key word or phrase in the problem or analysis.

Key word: *double* **Translation:** *multiply by 2*

Now we translate the words of the problem into an equation.

Two	times	the normal speeding fine	is	the new fine.
2	$\cdot$	f	$=$	592

Solve the equation

$$2f = 592 \quad \text{Write } 2 \cdot f \text{ as } 2f.$$

$$\frac{2f}{2} = \frac{592}{2} \quad \text{To undo the multiplication by 2, divide both sides by 2.}$$

$$f = 296 \quad \text{Perform the division: } 592 \div 2 = 296.$$

State the conclusion

The fine would normally have been \$296.

Check the result

If we double \$296, we get 2(\$296) = \$592. The answer checks.

EXAMPLE 6 **Entertainment costs.** A five-piece band worked on New Year's Eve. If each player earned \$120, what fee did the band charge?

Analyze the problem

- There were 5 players in the band.
- Each player made \$120.
- We are asked to find the band's fee. We know that the fee divided by the number of players will give each person's share.

Form an equation

We can let f = the band's fee. To form an equation, we look for a key word or phrase. In this case, we find it in the analysis of the problem.

Key phrase: *divided by* **Translation:** *division*

Now we translate the words of the problem into an equation.

The band's fee	divided by	the number in the band	is	each person's share.
f	$\div$	5	$=$	120

Solve the equation

$$\frac{f}{5} = 120 \quad \text{Write } f \div 5 \text{ as } \frac{f}{5}.$$

$$\mathbf{5} \cdot \frac{f}{5} = \mathbf{5} \cdot 120 \quad \text{To undo the division by 5, multiply both sides by 5.}$$

$$f = 600 \quad \text{Perform the multiplication: } 5 \cdot 120 = 600.$$

State the conclusion

The band's fee was \$600.

Check the result

If we divide \$600 by 5, we get each person's share: \$120.

Section 1.7 STUDY SET

VOCABULARY *Fill in the blanks.*

1. According to the ________ property of equality: If equal quantities are divided by the same nonzero quantity, the results will be equal quantities.

2. According to the ____________ property of equality: If equal quantities are multiplied by the same nonzero quantity, the results will be equal quantities.

CONCEPTS *Fill in the blanks.*

3. If we multiply x by 6 and then divide that product by 6, the result is ___.

4. If we divide x by 8 and then multiply that quotient by 8, the result is ___.

5. If $x = y$, then $\frac{x}{z} = \frac{\square}{\square}$ where $z \neq 0$.

6. If $x = y$, then $zx =$ ___ where $z \neq 0$.

7. In the equation $4t = 40$, what operation is being performed on the variable? How do we undo it?

8. In the equation $\frac{t}{15} = 1$, what operation is being performed on the variable? How do we undo it?

9. Name the first step in solving each of the following equations.

   a. $x + 5 = 10$ b. $x - 5 = 10$

   c. $5x = 10$ d. $\frac{x}{5} = 10$

10. For each of the following equations, check the given possible solution.

    a. $16 = t - 8;\ 33$

    b. $16 = t + 8;\ 8$

    c. $16 = 8t;\ 128$

    d. $16 = \frac{t}{8};\ 2$

NOTATION *Complete each solution.*

11. $3x = 12$

$\frac{3x}{\square} = \frac{12}{\square}$

$x = 4$

Check: $3x = 12$

$3 \cdot \square \stackrel{?}{=} 12$

$\square = 12$

So ___ is the solution.

12. $\frac{x}{5} = 9$

$\square \cdot \frac{x}{5} = \square \cdot 9$

$x = 45$

Check: $\frac{x}{5} = 9$

$\frac{\square}{5} \stackrel{?}{=} 9$

$\square = 9$

So ___ is the solution.

PRACTICE *Use the division or the multiplication property of equality to solve each equation.* **Check each answer.**

13. $3x = 3$
14. $5x = 5$
15. $2x = 192$
16. $4x = 120$
17. $17y = 51$
18. $19y = 76$
19. $34y = 204$
20. $18y = 90$
21. $100 = 100x$
22. $35 = 35y$
23. $16 = 8r$
24. $44 = 11m$
25. $\frac{x}{7} = 2$
26. $\frac{x}{12} = 4$
27. $\frac{y}{14} = 3$
28. $\frac{y}{13} = 5$
29. $\frac{a}{15} = 5$
30. $\frac{b}{25} = 5$
31. $\frac{c}{13} = 3$
32. $\frac{d}{100} = 11$
33. $1 = \frac{x}{50}$
34. $1 = \frac{x}{25}$
35. $7 = \frac{t}{7}$
36. $4 = \frac{m}{4}$
37. $9z = 90$
38. $3z = 6$
39. $7x = 21$
40. $13x = 52$
41. $86 = 43t$
42. $288 = 96t$
43. $21s = 21$
44. $31x = 155$
45. $\frac{d}{20} = 2$
46. $\frac{x}{16} = 4$
47. $400 = \frac{t}{3}$
48. $250 = \frac{y}{2}$

APPLICATIONS *Complete each solution.*

49. THE NOBEL PRIZE In 1998, three Americans, Louis Ignarro, Robert Furchgott, and Dr. Fred Murad, were awarded the Nobel Prize for Medicine. They shared the prize money. If each person received \$318,500, what was the Nobel Prize cash award?

Analyze the problem

- ___ people shared the cash award.
- Each person received ______.
- We are asked to find the ______________________.

Form an equation

Since we want to find what the Nobel Prize cash award was, we let c = ______________________. To form an equation, we look for a key word or phrase in the problem.

Key phrase: ______________

Translation: ________

Now we translate the words of the problem into an equation.

The Nobel Prize cash award	divided by	the number of recipients	was	$318,500.
___	$\div$	3	=	318,500

Solve the equation

$$\frac{x}{3} = 318{,}500$$

$$\square \cdot \frac{x}{3} = \square \cdot 318{,}500$$

$$x = \square$$

State the conclusion

____

Check the result

If we divide the Nobel Prize cash award by 3, we have

$\frac{\square}{3} = \square$. This was the amount each person received. The answer checks.

50. INVESTING An investor has watched the value of his portfolio double in the last 12 months. If the current value of his portfolio is $274,552, what was its value one year ago?

Analyze the problem

- The value of the portfolio ______ in 12 months.
- The current value is ______.
- We must find ______________.

Form an equation

We can let x = ______________________.

We now look for a key word or phrase in the problem.

Key phrase: ________

Translation: __________

Now we translate the words of the problem into an equation.

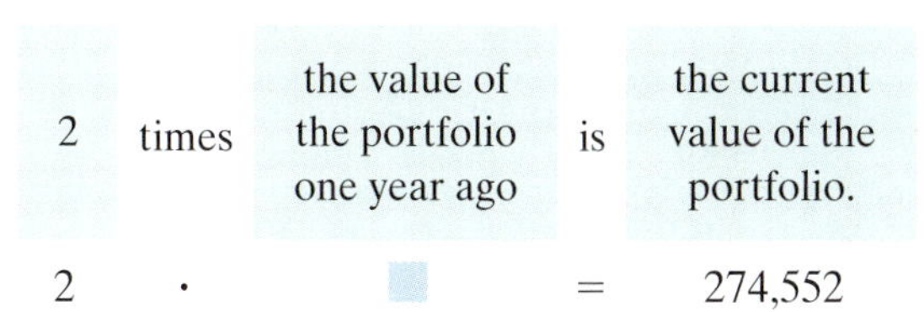

2	times	the value of the portfolio one year ago	is	the current value of the portfolio.
2	$\cdot$	___	=	274,552

Solve the equation

$$2x = \square$$

$$\frac{2x}{\square} = \frac{274{,}552}{\square}$$

$$x = \square$$

State the conclusion

____

Check the result

If the value of the portfolio one year ago was ______ and it doubled, its current value would be ______. The answer checks.

Let a variable represent the unknown quantity. Then write and solve an equation to answer the question.

51. SPEED READING An advertisement for a speed reading program claimed that successful completion of the course could triple a person's reading rate. If Alicia can currently read 130 words a minute, at what rate can she expect to read after taking the classes?

52. COST OVERRUNS Lengthy delays and skyrocketing costs caused a rapid-transit construction project to go over budget by a factor of 10. The final audit showed the project costing $540 million. What was the initial cost estimate?

53. STAMPS Large sheets of commemorative stamps honoring Marilyn Monroe are to be printed. On each sheet, there are 112 stamps, with 8 stamps per row. How many rows of stamps are on a sheet?

54. SPREADSHEETS The grid shown below is a Microsoft Excel spreadsheet. The rows are labeled with numbers, and the columns are labeled with letters. Each empty box of the grid is called a *cell.* Suppose a certain project calls for a spreadsheet with 294 cells, using columns A through F. How many rows will need to be used?

Microsoft Excel-Book 1

File Edit View Insert Format Tools

	A	B	C	D	E	F
1						
2						
3						
4						
5						
6						
7						
8						

Sheet 1 Sheet 2 Sheet 3 Sheet 4 Sheet 5

55. PHYSICAL EDUCATION A high school PE teacher had the students in her class form three-person teams for a basketball tournament. Thirty-two teams participated in the tournament. How many students were in the PE class?

56. LOTTO WINNERS The grocery store employees listed below pooled their money to buy $120 worth of lottery tickets each week, with the understanding they would split the prize equally if they happened to win. One week they did have the winning ticket and won $480,000. What was each employee's share of the winnings?

Sam M. Adler	Ronda Pellman	Manny Fernando
Lorrie Jenkins	Tom Sato	Sam Lin
Kiem Nguyen	H. R. Kinsella	Tejal Neeraj
Virginia Ortiz	Libby Sellez	Alicia Wen

57. ANIMAL SHELTERS The number of phone calls to an animal shelter quadrupled after the evening news aired a segment explaining the services the shelter offered. Before the publicity, the shelter received 8 calls a day. How many calls did the shelter receive each day after being featured on the news?

58. OPEN HOUSES The attendance at an elementary school open house was only half of what the principal had expected. If 120 people visited the school that evening, how many had she expected to attend?

59. GRAVITY The weight of an object on Earth is 6 times greater than what it is on the moon. The following weighing situation took place on the Earth. If it took place on the moon, what weight would the scale register?

60. INFOMERCIALS The number of orders received each week by a company selling skin care products increased fivefold after a Hollywood celebrity was added to the company's infomercial. After adding the celebrity, the company received abou 175 orders each week. How many orders were received each week before the celebrity took part?

WRITING

61. Explain what Figure 1-19 (page 65) is trying to show.

62. Explain what Figure 1-20 (page 66) is trying to show.

63. What does it mean to solve an equation?

64. Think of a number. Double it. Now divide it by 2. Explain why you always obtain the original number.

REVIEW

65. Find the perimeter of a rectangle with sides measuring 8 cm and 16 cm.

66. Find the area of a rectangle with sides measuring 23 inches and 37 inches.

67. Find the prime factorization of 120.

68. Find the prime factorization of 150.

69. Evaluate: $3^2 \cdot 2^3$.

70. Evaluate: $5 + 6 \cdot 3$.

71. FUEL ECONOMY Five basic models of automobiles made by Saturn have city mileage ratings of 24, 22, 28, 29, and 27 miles per gallon. What is the average (mean) city mileage for the five models?

72. Solve: $x - 4 = 20$.

KEY CONCEPT

Variables

One of the objectives of this course is for you to become comfortable working with **variables.** You will recall that a variable is a letter that stands for a number.

The application problems of Sections 1.6 and 1.7 were solved with the help of a variable. In these problems, we let the variable represent an unknown quantity such as the number of customers a hairdresser used to have, the age of a jar, and the cash award given a Nobel Prize winner. We then wrote an equation to describe the situation mathematically and solved the equation to find the value represented by the variable.

Suppose that you are going to solve the following problems. What quantity should be represented by a variable? State your response in the form "Let $x = \ldots$."

1. The monthly cost to lease a van is $120 less than the monthly cost to buy it. To buy it, the monthly payments are $290. How much does it cost to lease the van each month?
2. One piece of pipe is 10 feet longer than another. Together, their lengths total 24 feet. How long is the shorter piece of pipe?
3. The length of a rectangular field is 50 feet. What is its width if it has a perimeter of 200 feet?
4. If one hose can fill a vat in 2 hours and another can fill it in 3 hours, how long will it take to fill the vat if both hoses are used?
5. Find the distance traveled by a motorist in 3 hours if her average speed was 55 mph.
6. In what year was a couple married if their 50th anniversary was in 2004?

Variables can also be used to state properties of mathematics in a concise, "shorthand" notation. State each property using mathematical symbols and the given variable(s).

7. Use the variables a and b to state that two numbers can be added in either order to get the same sum.
8. Use the variable x to state that when 0 is subtracted from a number, the result is the same number.
9. Use the variable b to state that the result when dividing a number by 1 is the same number.
10. Use the variable x to show that the sum of a number and 1 is greater than the number.
11. Using the variable n, state the fact that when 1 is subtracted from any number, the difference is less than the number.
12. State the fact that the product of any number and 0 is 0, using the variable a.
13. Use the variables r, s, and t to state that the way we group three numbers when adding them does not affect the answer.
14. Using the variable n, state the fact that when a number is multiplied by 1, the result is the number.

ACCENT ON TEAMWORK

SECTION 1.1

PLACE VALUE Have each student in your group bring a calculator to class so that you can examine several different models. For each model, determine the largest number (if there is one) that can be entered on the display of the calculator. Then press the appropriate calculator keys to add 1 to that number. What does the display show?

LARGE NUMBERS Bill Gates, founder of Microsoft Corporation, is said to be a billionaire. How many millions make 1 billion?

SECTION 1.2

READING THE PROBLEM In reading Example 9 of Section 1.2, you will notice that it contains several facts that are not used in the solution of the problem. Have each person in your group write a similar problem that requires careful reading to extract the useful information. Then have each person share his or her problem with the other students in the group.

SECTION 1.3

DIVISIBILITY TESTS Certain tests can help us decide whether one whole number is divisible by another.

- A number is divisible by 2 if the last digit of the number is 0, 2, 4, 6, or 8.
- A number is divisible by 3 if the sum of the digits is divisible by 3.
- A number is divisible by 4 if the number formed by the last two digits is divisible by 4.
- A number is divisible by 5 if the last digit of the number is 0 or 5.
- A number is divisible by 6 if the last digit of the number is 0, 2, 4, 6, or 8 and the sum of the digits is divisible by 3.
- A number is divisible by 8 if the number formed by the last three digits is divisible by 8.
- A number is divisible by 9 if the sum of the digits is divisible by 9.
- A number is divisible by 10 if the last digit of the number is 0.
- Determine whether each number is divisible by 2, 3, 4, 5, 6, 8, 9, and/or 10.

a. 660 **b.** 2,526
c. 11,523 **d.** 79,503
e. 135,405 **f.** 4,444,440

SECTION 1.4

COMMON FACTORS The prime factorizations of 36 and 126 are shown below. The prime factors that are common to 36 and 126 (highlighted in color) are 2, 3, and 3.

$$36 = \mathbf{2} \cdot 2 \cdot \mathbf{3} \cdot \mathbf{3}$$
$$126 = \mathbf{2} \cdot \mathbf{3} \cdot \mathbf{3} \cdot 7$$

Find the common prime factors for each of the following pairs of numbers.

a. 25, 45 **b.** 24, 60
c. 18, 45 **d.** 40, 112
e. 180, 210 **f.** 242, 198

SECTION 1.5

ORDER OF OPERATIONS Consider the expression

$$5 + 8 \cdot 2^3 - 3 \cdot 2$$

Insert a set of parentheses somewhere in the expression so that, when it is evaluated, you obtain

a. 63 **b.** 132
c. 21 **d.** 127

SECTION 1.6

SOLVING EQUATIONS Borrow a scale and some weights from the chemistry department. Use them as part of a class presentation to explain how the subtraction property of equality is used to solve the equation $x + 2 = 5$. See the discussion and Figure 1-16 on page 56 for some suggestions on how to do this.

SECTION 1.7

FORMING EQUATIONS Reread Example 4 in Section 1.7. This problem could have been solved by forming an equation involving the operation of multiplication instead of the operation of division.

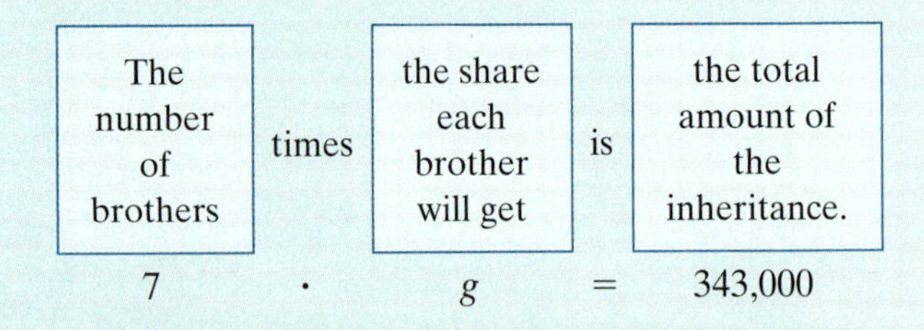

For Examples 5 and 6 in Section 1.7, write another equation that could be used to solve the problem. Then solve the equation and state the result.

CHAPTER REVIEW

SECTION 1.1 An Introduction to the Whole Numbers

CONCEPTS

A *set* is a collection of objects. The set of *natural numbers* is

{1, 2, 3, 4, 5, . . .}

The set of *whole numbers* is

{0, 1, 2, 3, 4, 5, . . .}

Whole numbers are often used in tables, bar graphs, and line graphs.

REVIEW EXERCISES

Graph each set.

1. The natural numbers less than 5

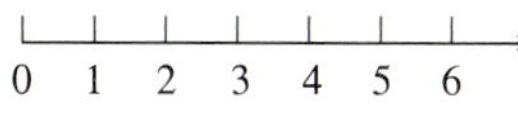

2. The whole numbers between 0 and 3

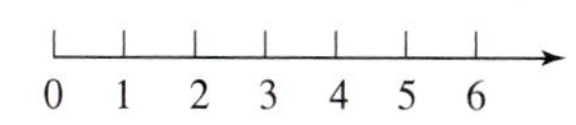

FARMING The table below shows the size of the average U.S. farm (in acres) for the period 1940–2000, in 20-year increments.

Year	1940	1960	1980	2000
Average size (acres)	174	297	426	432

3. Construct a bar graph of the data.

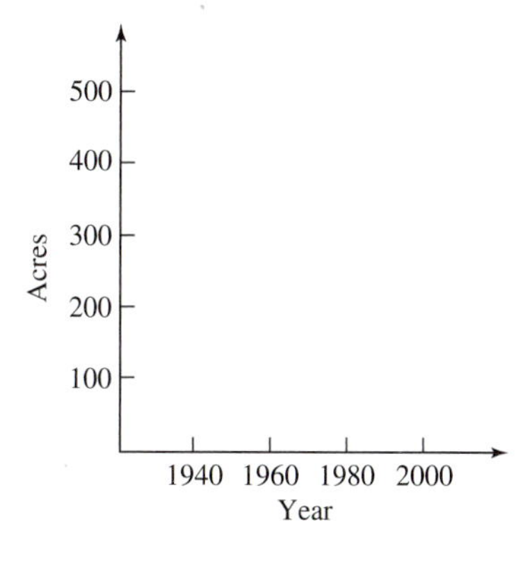

4. Construct a line graph of the data.

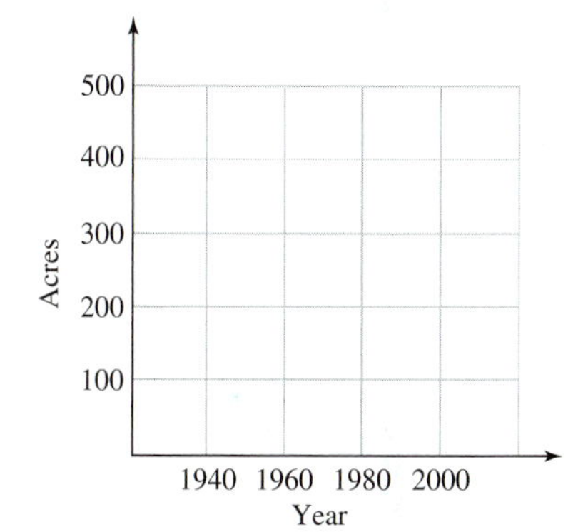

The digits in a whole number have *place value.*

Consider the number 2,365,720. Which digit is in the given column?

5. Ten thousands

6. Hundreds

A whole number is written in *expanded notation* when its digits are written with their place values.

Write each number in expanded notation.

7. 570,302

8. 37,309,054

We use the digits 0, 1, 2, 3, 4, 5, 6, 7, 8, and 9 to write a number in *standard notation.*

Write each number in standard notation.

9. 3 thousands + 2 hundreds + 7 ones

10. Twenty-three million, two hundred fifty-three thousand, four hundred twelve

11. Sixteen billion

The symbol $<$ means "is less than." The symbol $>$ means "is greater than."

Place an $<$ or $>$ symbol in the box to make a true statement.

12. $9 \ \square \ 7$

13. $3 \ \square \ 5$

To give approximate answers, we often use *rounded numbers.*

Round 2,507,348 to the specified place.

14. Nearest hundred

15. Nearest ten thousand

16. Nearest ten

17. Nearest hundred thousand

SECTION 1.2 Adding and Subtracting Whole Numbers

Addition is the process of finding the total of two (or more) numbers. Do additions within parentheses first.

Find each sum.

18. $56 + 22$

19. $137 + 0$

20. $15 + (27 + 13)$

21. $82 + 17 + 50$

22. $(111 + 222) + 444$

23. $0 + 2{,}332$

The *commutative* and *associative properties of addition:*

$$a + b = b + a$$

$$(a + b) + c = a + (b + c)$$

Perform each addition.

24. $\begin{array}{r} 236 \\ +\underline{282} \end{array}$

25. $\begin{array}{r} 5{,}345 \\ +\underline{\ \ \ 655} \end{array}$

26. $135 + 213 + 615 + 47$

27. $4{,}447 + 7{,}478 + 13{,}061$

What property of addition is shown?

28. $12 + 8 = 8 + 12$

29. $12 + (8 + 2) = (12 + 8) + 2$

The *perimeter* of a rectangle or a square is the distance around it.

30. Find the perimeter of square with sides 24 inches long.

Subtraction is the process of finding the difference between two numbers.

Perform each subtraction.

31. $18 - 5$

32. $9 - (7 - 2)$

33. $22 - 5 - 6$

34. Subtract 5,177 from 5,231.

35. $\begin{array}{r} 343 \\ -\underline{269} \end{array}$

36. $\begin{array}{r} 17{,}800 \\ -\underline{15{,}725} \end{array}$

Give the addition or subtraction fact that is illustrated by each figure.

37.

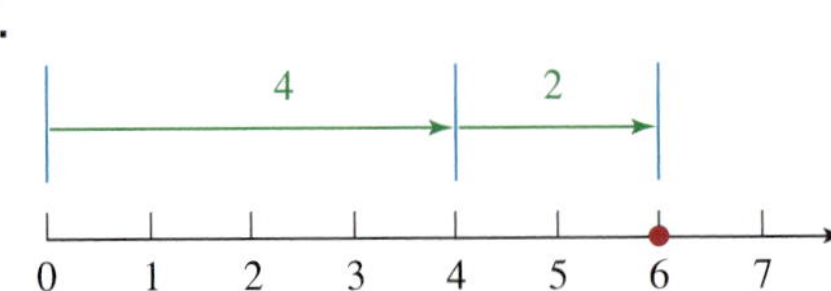

38.

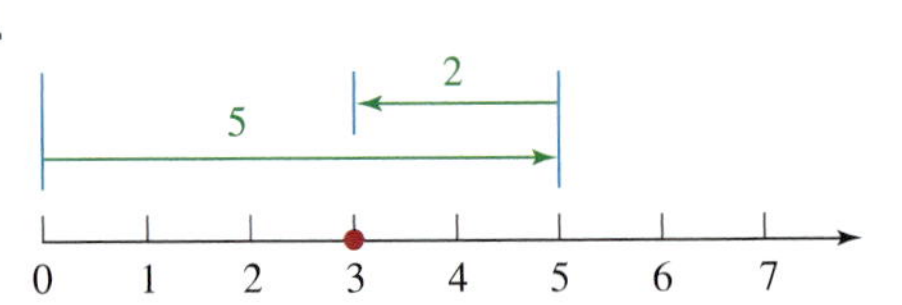

39. TRAVEL A direct flight from Omaha to San Francisco costs $237. Another flight with one stop in Reno costs $192. How much can be saved by taking the less expensive flight?

40. SAVINGS ACCOUNTS A savings account contains $931. If the owner deposits $271 and makes withdrawals of $37 and $380, find the final balance.

41. REBATES The price of a new Honda Civic was advertised in a newspaper as $21,991*. A note at the bottom of the ad read, "*Reflects $1,550 factory rebate." What was the car's original sticker price?

Multiplication is repeated addition. For example,

$$4 \cdot 6 = \overbrace{6 + 6 + 6 + 6}^{\text{The sum of four 6's}} = 24$$

The result, 24, is called the *product*, and the 4 and 6 are called *factors*.

The *commutative* and *associative properties of multiplication:*

$$a \cdot b = b \cdot a$$

$$(a \cdot b) \cdot c = a \cdot (b \cdot c)$$

The *area A of a rectangle* is the product of its length l and its width w.

$$A = l \cdot w$$

Perform each multiplication.

42. $8 \cdot 7$

43. $7(8)$

44. $8 \cdot 0$

45. $7 \cdot 1$

46. $10 \cdot 8 \cdot 7$

47. $5 \cdot (7 \cdot 6)$

48. $157 \cdot 21$

49. $3{,}723 \cdot 48$

50. $\begin{array}{r} 356 \\ \times\ \ 89 \\ \hline \end{array}$

51. $\begin{array}{r} 5{,}624 \\ \times\ \ \ 81 \\ \hline \end{array}$

What property of multiplication is shown?

52. $12 \cdot (8 \cdot 2) = (12 \cdot 8) \cdot 2$

53. $12 \cdot 8 = 8 \cdot 12$

54. WAGES If a math tutor worked for 38 hours and was paid $9 per hour, how much did she earn?

55. HORSESHOES Find the perimeter and the area of the rectangular horseshoe court shown below.

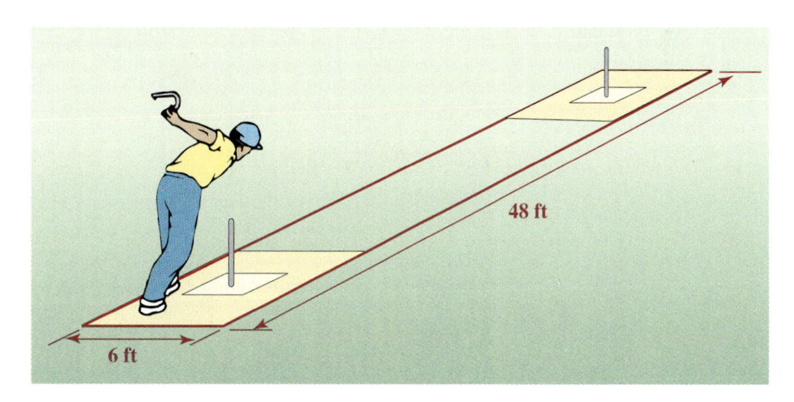

56. PACKAGING There are 12 eggs in one dozen, and 12 dozen in one gross. How many eggs are in a shipment of 5 gross?

Division is an operation that determines how many times a number (the *divisor*) is contained in another number (the *dividend*). *Remember that you can never divide by 0.*

Perform each division, if possible.

57. $\frac{6}{3}$

58. $\frac{15}{1}$

59. $73 \div 0$

60. $\frac{0}{8}$

61. $357 \div 17$

62. $1{,}443 \div 39$

63. $21\overline{)405}$

64. $54\overline{)1{,}269}$

65. TREATS If 745 candies are divided equally among 45 children, how many will each child receive? How many candies will be left over?

66. COPIES An elementary school teacher had copies of a 3-page social studies test made at Quick Copy Center. She was charged for 84 sheets of paper. How many copies of the test were made?

SECTION 1.4 *Prime Factors and Exponents*

Numbers that are multiplied together are called *factors.*

Find all of the factors of each number.

67. 18

68. 25

A *prime number* is a whole number, greater than 1, that has only 1 and itself as factors. Whole numbers greater than 1 that are not prime are called *composite numbers.*

Identify each number as a prime, composite, or neither.

69. 31

70. 100

71. 1

72. 0

73. 125

74. 47

Whole numbers divisible by 2 are *even* numbers. Whole numbers not divisible by 2 are *odd* numbers.

Identify each number as an even or odd number.

75. 171

76. 214

77. 0

78. 1

The *prime factorization* of a whole number is the product of its prime factors.

Find the prime factorization of each number.

79. 42

80. 375

An *exponent* is used to indicate repeated multiplication. In the *exponential expression* a^n, a is the base and n is the exponent.

Write each expression using exponents.

81. $6 \cdot 6 \cdot 6 \cdot 6$

82. $5 \cdot 5 \cdot 5 \cdot 13 \cdot 13$

Evaluate each expression.

83. 5^3

84. 11^2

85. $2^3 \cdot 5^2$

86. $2^2 \cdot 3^3 \cdot 5^2$

SECTION 1.5 *Order of Operations*

Perform mathematical operations in the following order:

1. Perform all calculations within parantheses and other grouping symbols.
2. Evaluate all exponential expressions.
3. Perform all multiplications and divisions in order from left to right.
4. Perform all additions and subtractions in order from left to right.

To evaluate an expression containing grouping symbols, perform all calculations within each pair of grouping symbols, working from the innermost pair to the outermost pair.

Evaluate each expression.

87. $13 + 12 \cdot 3$

88. $35 - 15 \div 5$

89. $(13 + 12)3$

90. $(8 - 2)^2$

91. $8 \cdot 5 - 4 \div 2$

92. $8 \cdot (5 - 4 \div 2)$

93. $2 + 3(10 - 4 \cdot 2)$

94. $4(20 - 5 \cdot 3 + 2) - 4$

95. $3^3\left(\frac{12}{6}\right) - 1^4$

96. $\frac{12 + 3 \cdot 7}{5^2 - 14}$

97. $7 + 3[10 - 3(4 - 2)]$

98. $5 + 2[(15 - 3 \cdot 4) - 2]$

99. DICE GAMES Write an expression that finds the total on all the dice shown below. Then evaluate the expression.

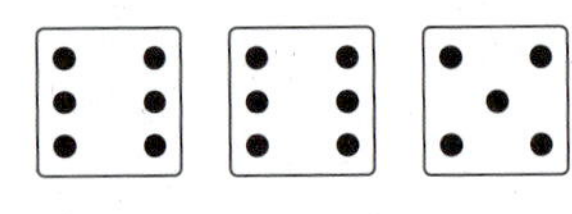

The *arithmetic mean* (average) is a value around which numbers are grouped.

100. YAHTZEE Find the player's average (mean) score for the 6 games.

Yahtzee SCORE CARD

Game #1	Game #2	Game #3	Game #4	Game #5	Game #6
159	244	184	240	166	213

SECTION 1.6 Solving Equations by Addition and Subtraction

An *equation* is a statement that two expressions are equal.

A *variable* is a letter that stands for a number.

Two equations with exactly the same solutions are called *equivalent equations.*

To solve an equation, isolate the variable on one side of the equation by undoing the operation performed on it.

If the same number is added to (or subtracted from) both sides of an equation, an equivalent equation results.

If $a = b$, then $a + c = b + c$.

If $a = b$, then $a - c = b - c$.

Problem-solving strategy:

1. Analyze the problem.
2. Form an equation.
3. Solve the equation.
4. State the conclusion.
5. Check the result.

Determine whether the given number is a solution of the equation. Explain.

101. $x + 2 = 13$; 5

102. $x - 3 = 1$; 4

Identify the variable in each equation.

103. $y - 12 = 50$

104. $114 = 4 - t$

Solve the equation and check the result.

105. $x - 7 = 2$

106. $x - 11 = 20$

107. $225 = y - 115$

108. $101 = p - 32$

109. $x + 9 = 18$

110. $b + 12 = 26$

111. $175 = p + 55$

112. $212 = m + 207$

113. $x - 7 = 0$

114. $x + 15 = 1{,}000$

Let a variable represent the unknown quantity. Then write and solve an equation to answer the question.

115. FINANCING A newly married couple made a $25,500 down payment on a $122,750 house. How much did they need to borrow?

116. DOCTOR'S PATIENTS After moving his office, a doctor lost 13 patients. If he had 172 patients left, how many did he have originally?

SECTION 1.7 *Solving Equations by Division and Multiplication*

If both sides of an equation are divided by (or multiplied by) the same nonzero number, an equivalent equation results.

If $a = b$, then $\frac{a}{c} = \frac{b}{c}$

If $a = b$, then $a \cdot c = b \cdot c$

Solve the equation and check the result.

117. $3x = 12$

118. $15y = 45$

119. $105 = 5r$

120. $224 = 16q$

121. $\frac{x}{7} = 3$

122. $\frac{a}{3} = 12$

123. $15 = \frac{s}{21}$

124. $25 = \frac{d}{17}$

125. $12x = 12$

126. $\frac{x}{12} = 12$

Let a variable represent the unknown quantity. Then write and solve an equation to answer the question.

127. CARPENTRY If you cut a 72-inch board into three equal pieces, how long will each piece be? Disregard any loss due to cutting.

128. JEWELRY Four sisters split the cost of a gold chain evenly. How much did the chain cost if each sister's share was $32?

CHAPTER 1 TEST

1. Graph the whole numbers less than 5.

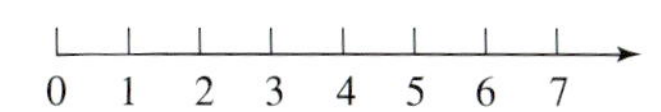

2. Write "five thousand two hundred sixty-six" in expanded notation.

3. Write "7 thousands + 5 hundreds + 7 ones" in standard notation.

4. Round 34,752,341 to the nearest million.

In Problems 5–6, refer to the data in the table.

Lot number	1	2	3	4
Defective bolts	7	10	5	15

5. Use the data to make a bar graph.

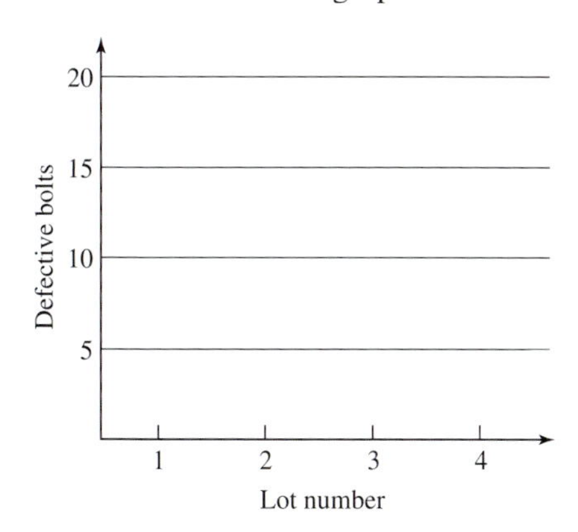

6. Use the data to make a line graph.

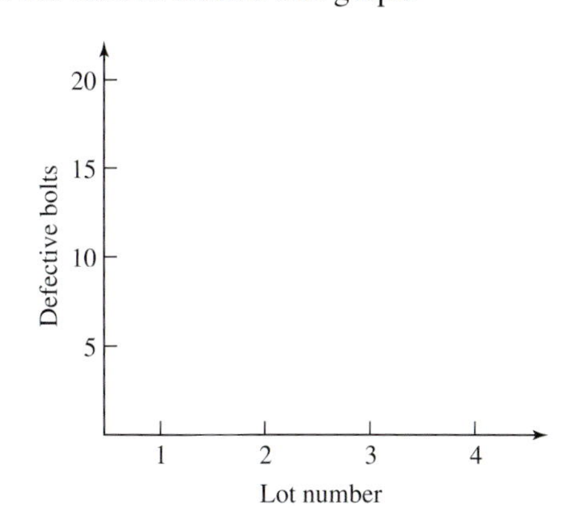

Place an $<$ or $>$ symbol in the box to make a true statement.

7. 15 ☐ 10

8. 1,247 ☐ 1,427

9. Add: 327 + 435 + 123 + 606.

10. Subtract 287 from 535.

11. Add: $\begin{array}{r} 44{,}526 \\ +13{,}579 \\ \hline \end{array}$

12. Subtract: $\begin{array}{r} 4{,}521 \\ -3{,}579 \\ \hline \end{array}$

13. STOCKS On Tuesday, a share of KBJ Company was selling at \$73. The price rose \$12 on Wednesday and fell \$9 on Thursday. Find its price on Thursday.

14. List the factors of 20 in order, from least to greatest.

Perform each operation.

15. Multiply: $\begin{array}{r} 53 \\ \times\ 8 \\ \hline \end{array}$

16. Multiply: 367(73).

17. Divide: $63\overline{)4{,}536}$.

18. Divide $73\overline{)8{,}379}$.

19. Perform each operation, if possible.

a. $15 \cdot 0$ **b.** $\frac{0}{15}$

20. What property is illustrated by each statement?

a. $18 \cdot (9 \cdot 40) = (18 \cdot 9) \cdot 40$

b. $23{,}999 + 1 = 1 + 23{,}999$

21. FURNITURE SALES See the following ad. Find the perimeter of the rectangular space under the tent. Then fill in the blank in the advertisement.

22. PRINTOUTS A computer printout can list 74 student names on one page of paper. If 3,451 student names are to be printed, how many names will appear on the last page of the printout?

23. COLLECTIBLES There are 12 baseball cards in every pack. There are 24 packs in every box. There are 12 boxes in every case. How many cards are in a case?

24. Find the prime factorization of 252.

25. Evaluate: $9 + 4 \cdot 5$.

26. Evaluate: $\dfrac{3 \cdot 4^2 - 2^2}{(2-1)^3}$.

27. Evaluate: $10 + 2[12 - 2(6 - 4)]$.

28. GRADES A student scored 73, 52, and 70 on three exams and received 0 on two missed exams. Find his average (mean) score.

29. Is 3 a solution of the equation $x + 13 = 16$? Explain why or why not.

Solve each equation. **Check the result.**

30. $100 = x + 1$

31. $y - 12 = 18$

32. $5t = 55$

33. $\dfrac{q}{3} = 27$

Let a variable represent the unknown quantity. Then write and solve an equation to answer the question.

34. PARKING After many student complaints, a college decided to commit funds to double the number of parking spaces on campus. This increase would bring the total number of spaces up to 6,200. How many parking spaces does the college have at this time?

35. LIBRARIES A library building is 6 years shy of its 200th birthday. How old is the building at this time?

36. Explain what it means to *solve* an equation.

CHAPTER 2

The Integers

Getty Images

TLE There are few things more breathtaking than a star-filled sky on a clear night. Stars come in different colors, sizes, shapes, and ages. You have probably noticed that some stars are bright and others are faint. Astronomers have developed a scale to classify the relative brightness of stars. The most brilliant stars (including the sun) are assigned negative number magnitudes while positive number magnitudes are assigned to the dimmer stars (and planets).

To learn more about positive and negative numbers, visit The Learning Equation on the Internet at http://tle.brookscole.com. (The log-in instructions are in the Preface.) For Chapter 2, the online lessons are:

- *TLE* Lesson 4: An Introduction to Integers
- *TLE* Lesson 5: Subtracting Integers

Check Your Knowledge

1. The ________ value of a number is the distance between the number and zero on a number line.
2. When 0 is added to a number, the number remains the same. We call 0 the additive ________.
3. Two numbers that are the same distance from 0 on the number line, but on opposite sides of it, are called ________________.
4. The product of two integers with ________ signs is negative.
5. Insert one of the symbols $>$ or $<$ in the blank to make the statement true: $-15 \; \square \; -16$.
6. Find the mean (average) of the temperatures shown in the table on the left.

Daily high temperatures	
Sunday	1
Monday	7
Tuesday	−3
Wednesday	1
Thursday	0
Friday	−1
Saturday	2

7. Add:
   a. $-27 + 13$
   b. $12 + (-12)$
   c. $(-2) + (-2) + (-2) + (-2)$
   d. $(-4 + 7) + [1 + (-6) + 4]$
8. Subtract:
   a. $7 - 13$ b. $3 - (-3)$ c. $0 - 5 - 7$
9. Find each product.
   a. $3(-3)$ b. $-5(-20)$ c. $(-3)(-2)(-4)$
10. Write the related multiplication statement for $\frac{-18}{3} = -6$.
11. Find each quotient, if possible.
   a. $\frac{36}{-9}$ b. $\frac{-900}{-30}$ c. $\frac{-3}{2 + 1 - 3}$
12. Evaluate each expression.
   a. $-(-7)$ b. $-|-7|$ c. $3|-6 + 2|$
13. Evaluate each power.
   a. -3^2 b. $(-3)^2$ c. $(3 - 4)^2$
14. Evaluate each expression.
   a. $24 \div 3 \cdot 2$ b. $6 + \frac{12}{-4} - 4^2$
   c. $-5 - 2[7 - (-3)(-2)^3]$ d. $\frac{-3^3 + (-4)(-6)}{-3(3 - 5)^2 + 9}$
15. The price of a certain computer dropped from \$620 to \$500 in six months. How much did the price drop per month?
16. On one lie detector test, a burglar scored −19, which indicates deception. However, on a second test, he scored −4, which is inconclusive. Find the difference in the scores.
17. Let a variable represent the unknown quantity. Then write and solve an equation to answer the following question.
   After Michelle deposited \$175 in her checking account, it was still \$55 overdrawn. What was the account balance before the deposit?
18. Explain what is meant when we say that subtraction is the same as adding the opposite.

Study Skills Workshop

ATTITUDES, PAST EXPERIENCE, AND LEARNING STYLES

Your personality will affect the way you learn math; some traits will work toward your benefit and some may not. The key is to maximize your strengths, minimize weaknesses, and find learning methods that will best suit your personal style.

Attitudes. How do you feel about math? How do you feel about taking control of your learning environment? What do you think of your chance for success in this class? If you respond to these questions with answers like: "Ugh! I'll never learn math!" or "My math teachers just never taught me well," you may be setting yourself up for failure. Try to alter your attitude slightly. Constructive thinking and positive self-talk may bring good results. Instead of telling yourself "I'm just stupid when it comes to math," try saying "Math is not my strongest subject, but I will work hard at it and learn new things this time." Know that in truly learning any new concept there are bound to be frustrations, but they can be overcome with support, strategy, and hard work.

Past Experience. Good or bad experiences in previous math classes may affect the way you view math now. Many students who have had negative math experiences in the past may feel anxiety and stress in the present. If you have high levels of stress or feel helpless when dealing with numbers, you may have math anxiety. These counterproductive feelings can be overcome with extra preparation, support services, relaxation techniques, and even hypnotherapy.

Learning Styles. What type of learner are you? The answer to this question will help you determine how you study, how you do your homework, and maybe even where you choose to sit in class. For example, visual-verbal learners learn best by reading and writing, so a good strategy for them in studying is to rewrite notes and examples. However, audio learners learn best by listening, so making audiotapes of important concepts may be their best study strategy. Kinesthetic learners like to move and do things with their hands, so incorporating the use of games or puzzles (often called manipulatives) in studying may be helpful.

ASSIGNMENT

1. Take a learning skills inventory test such as that found online at http://www.metamath.com/multiple/multiple_choice_questions.cgi to determine what type of learner you are. Once you have determined your learning style, write a plan on how will you use this information to help you succeed in your class.
2. Describe your past experiences in math courses. Have they generally been good or bad? Why? If you feel you have math anxiety, visit your school's counseling office and schedule an appointment to find ways of dealing with this problem.

In this chapter, we introduce the concept of negative number and explore an extension of the set of whole numbers, called the integers.

2.1 An Introduction to the Integers

- The integers • Extending the number line • More on inequality
- Absolute value • The opposite of a number • The – symbol

We have seen that whole numbers can be used to describe many situations that arise in everyday life. For example, we can use whole numbers to express temperatures above zero, the balance in a checking account, or an altitude above sea level. However, we cannot use whole numbers to express temperatures below zero, the balance in a checking account that is overdrawn, or how far an object is below sea level. (See Figure 2-1.)

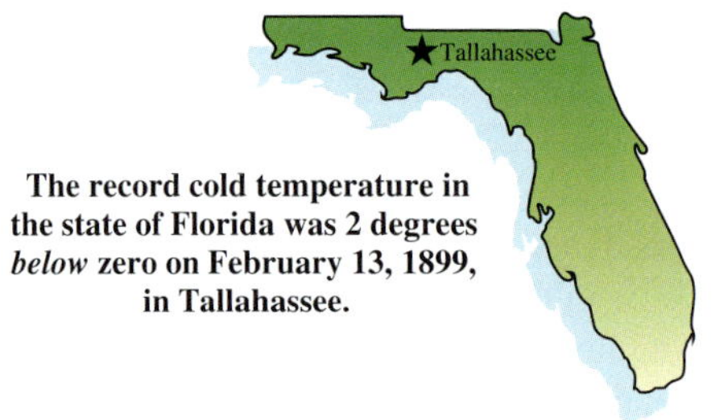

The record cold temperature in the state of Florida was 2 degrees *below* zero on February 13, 1899, in Tallahassee.

RECORD ALL CHARGES OR CREDITS THAT AFFECT YOUR ACCOUNT

NUMBER	DATE	DESCRIPTION OF TRANSACTION	PAYMENT/DEBIT (–)	√ T	FEE (IF ANY) (–)	DEPOSIT/CREDIT (+)	BALANCE
							$ 450 00
1207	5/2	Wood's Auto Repair Transmission	$ 500 00		$	$	

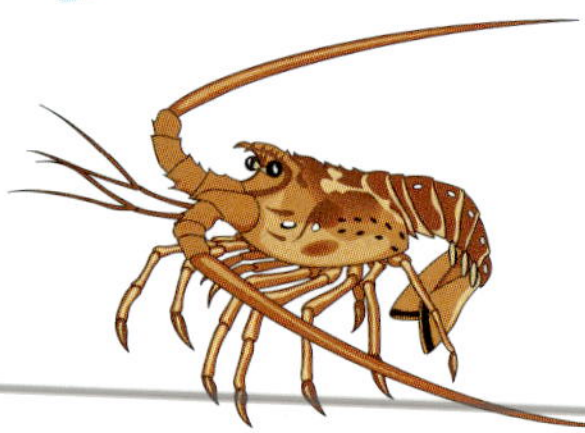

A check for $500 was written when there was only $450 in the account. The checking account is *overdrawn*.

The American lobster is found off the East Coast of North America at depths as much as 600 feet *below* sea level.

FIGURE 2-1

In this section, we see how negative numbers can be used to describe these three situations as well as many others.

The integers

To describe a temperature of 2 degrees (2°) above zero, a balance of $50, or 600 feet above sea level, we can use numbers called **positive numbers.** All positive numbers are greater than 0, and we can write them using a **positive sign** + or no sign at all.

In words	In symbols	Read as
2 degrees above zero	+2 or 2	positive two
A balance of $50	+50 or 50	positive fifty
600 feet above sea level	+600 or 600	positive six hundred

To describe a temperature of 2 degrees below zero, $50 overdrawn, or 600 feet below sea level, we need to use negative numbers. **Negative numbers** are numbers less than 0, and they are written using a **negative sign** −.

In words	In symbols	Read as
2 degrees below zero	-2	negative two
\$50 overdrawn	-50	negative fifty
600 feet below sea level	-600	negative six hundred

Positive and negative numbers

Positive numbers are greater than 0. **Negative numbers** are less than 0.

! COMMENT Zero is neither positive nor negative.

We often call positive and negative numbers **signed numbers.** The first three of the following signed numbers are positive, and the last three are negative.

$$+12, \quad +26, \quad 515, \quad -12, \quad -26, \quad \text{and} \quad -515$$

The collection of positive whole numbers, the negatives of the whole numbers, and 0 is called the set of **integers** (read as "in-ti-jers").

The set of integers

$$\{\ldots, -5, -4, -3, -2, -1, 0, 1, 2, 3, 4, 5, \ldots\}$$

Since every natural number is an integer, we say that the set of natural numbers is a subset of the integers. See Figure 2-2. Since every whole number is an integer, we say that the set of whole numbers is a **subset** of the integers.

The set of natural numbers

The set of integers → $\{\ldots, -7, -6, -5, -4, -3, -2, -1, 0, 1, 2, 3, 4, 5, 6, 7, \ldots\}$

The set of whole numbers

FIGURE 2-2

! COMMENT Note that the negative integers and 0 are not natural numbers. Also note that negative integers are not whole numbers.

Extending the number line

In Section 1.1, we introduced the number line. We can use an extension of the number line to learn about negative numbers.

Negative numbers can be represented on a number line by extending the line to the left. Beginning at the origin (the 0 point), we move to the left, marking equally spaced points as shown in Figure 2-3. As we move to the right on the number line, the values of the numbers increase. As we move to the left, the values of the numbers decrease.

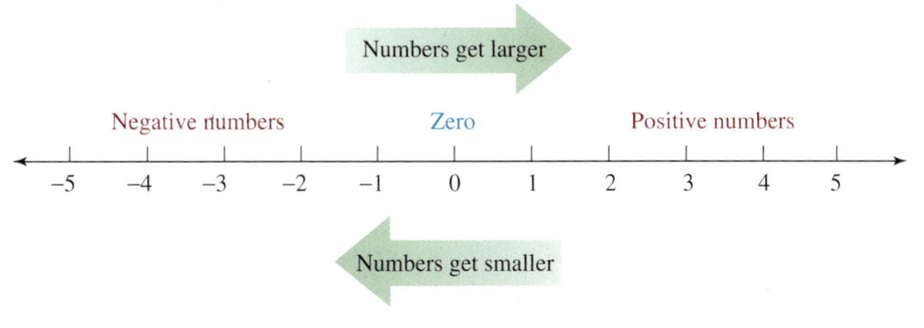

FIGURE 2-3

The thermometer shown in Figure 2-4(a) is an example of a vertical number line. It is scaled in degrees and shows a temperature of $-10°$. The time line shown in Figure 2-4(b) is an example of a horizontal number line. It is scaled in increments of 500 years.

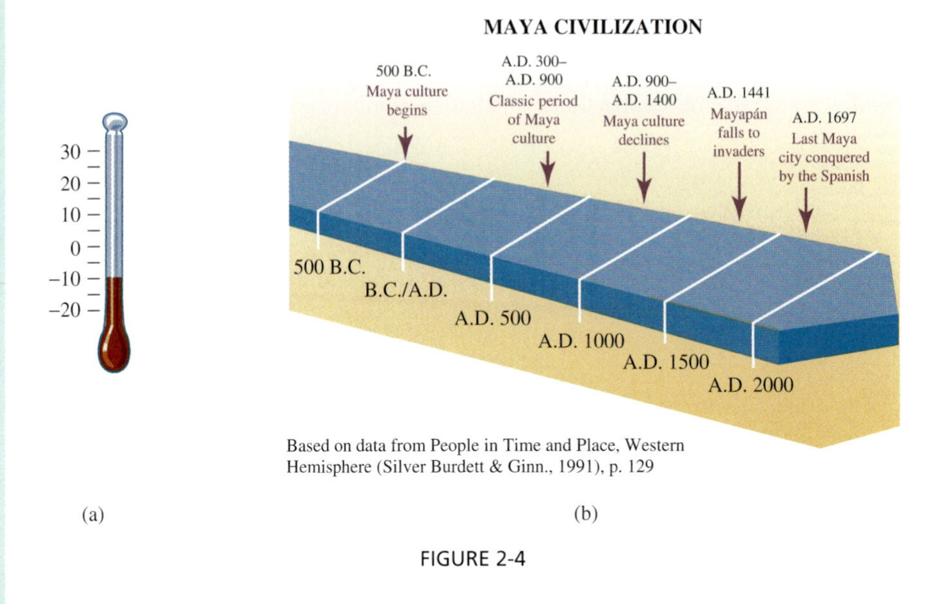

(a) (b)

FIGURE 2-4

Self Check 1

On the number line, graph -4, -2, 1, and 3.

Answer

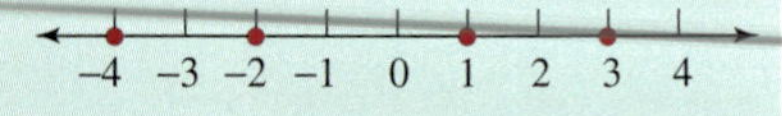

EXAMPLE 1 On the number line, graph -3, -1, 2, and 4.

Solution To graph each integer, we locate its position on the number line and draw a dot.

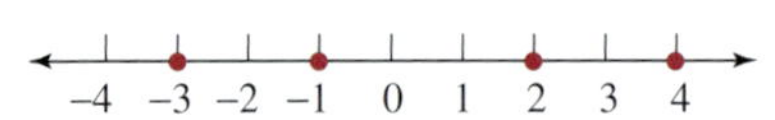

By extending the number line to include negative numbers, we can represent more situations using bar graphs and line graphs. For example, the bar graph shown in Figure 2-5 illustrates the annual profits *and losses* of Toys R Us over a nine-year period. Note that the profit in 2004 was \$88 million and that the loss in 1999 was \$132 million.

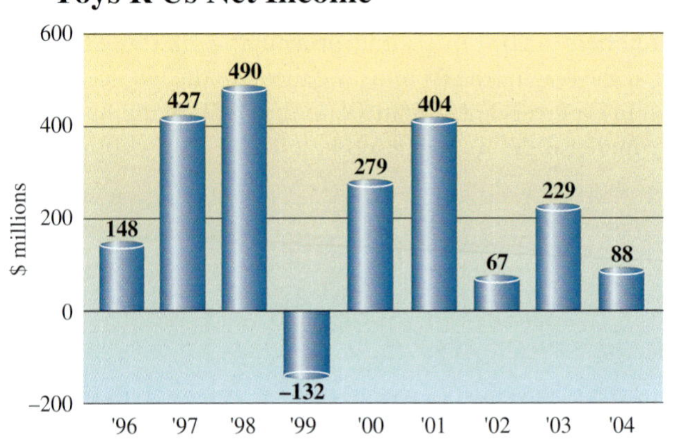

Source: Morningstar.com

FIGURE 2-5

Credit Card Debt

THINK IT THROUGH

"The most dangerous pitfall for many college students is the overuse of credit cards. Many banks do their best to entice new card holders with low or zero-interest cards." Gary Schatsky, certified financial planner

Which numbers on the credit card statement below are actually debts and, therefore, could be represented using negative numbers?

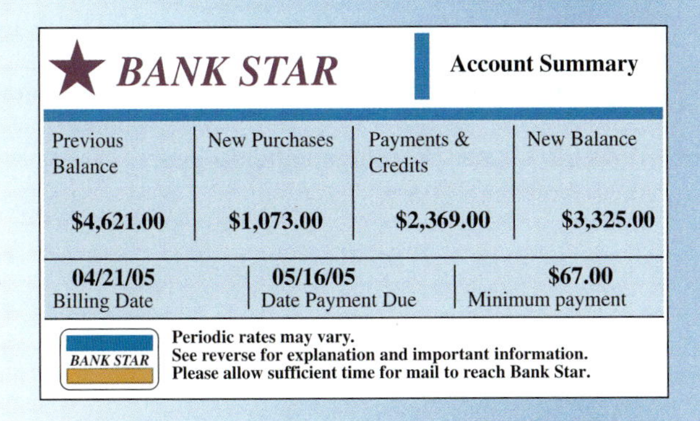

More on inequality

Recall that the symbol $<$ means "is less than" and that $>$ means "is greater than." Figure 2-6 shows the graph of the integers -2 and 1. Since -2 is to the left of 1 on the number line, $-2 < 1$. Since $-2 < 1$, it is also true that $1 > -2$.

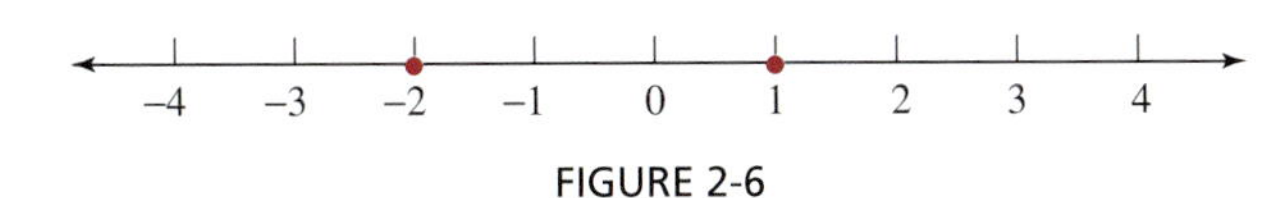

FIGURE 2-6

EXAMPLE 2 Use one of the symbols $>$ or $<$ to make each statement true:
a. $4 \;\square\; -5$ and **b.** $-4 \;\square\; -2$.

Solution

a. Since 4 is to the right of -5 on the number line, $4 > -5$.

b. Since -4 is to the left of -2 on the number line, $-4 < -2$.

Self Check 2

Use one of the symbols $>$ or $<$ to make each statement true:

a. $6 \;\square\; -6$

b. $-6 \;\square\; -5$

Answers **a.** $>$, **b.** $<$

Three other commonly used inequality symbols are the "is not equal to" symbol $\neq$, the "is less than or equal to" symbol $\leq$, and the "is greater than or equal to" symbol $\geq$.

$5 \neq 2$	Read as "5 is not equal to 2."
$6 \leq 10$	Read as "6 is less than or equal to 10." This statement is true, because $6 < 10$.
$12 \leq 12$	Read as "12 is less than or equal to 12." This statement is true, because $12 = 12$.
$17 \geq 15$	Read as "17 is greater than or equal to 15." This statement is true, because $17 > 15$.
$20 \geq 20$	Read as "20 is greater than or equal to 20." This statement is true, because $20 = 20$.

Self Check 3
Use an inequality symbol to write "30 is less than or equal to 35."

Answer $30 \leq 35$

EXAMPLE 3 Use an inequality symbol to write **a.** "8 is not equal to 5" and **b.** "50 is greater than or equal to 40."

Solution

a. $8 \neq 5$

b. $50 \geq 40$

Absolute value

Using the number line, we can see that the numbers 3 and −3 are both a distance of 3 units away from 0, as shown in Figure 2-7.

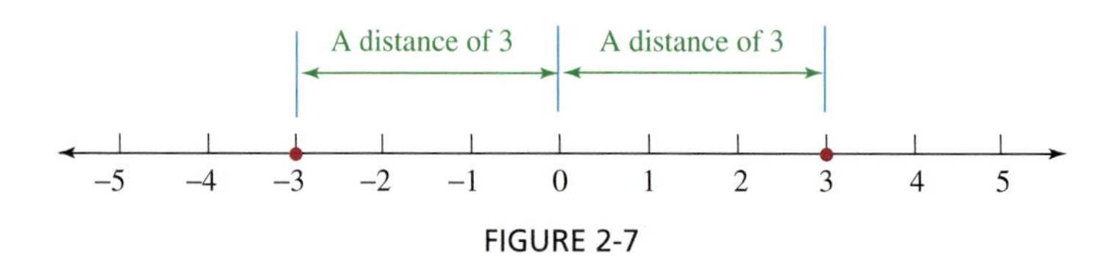

FIGURE 2-7

The **absolute value** of a number gives the distance between the number and 0 on the number line. To indicate absolute value, the number is inserted between two vertical bars, called the **absolute value symbol.** For example, we can write $|-3| = 3$. This is read as "The absolute value of negative 3 is 3," and it tells us that the distance between −3 and 0 on the number line is 3 units. From Figure 2-7, we also see that $|3| = 3$.

Absolute value

The **absolute value** of a number is the distance on the number line between the number and 0.

! COMMENT Absolute value expresses distance. The absolute value of a number is always positive or 0. It is never negative.

Self Check 4
Evaluate each expression:

a. $|-9|$

b. $|4|$

Answers **a.** 9, **b.** 4

EXAMPLE 4 Evaluate each expression: **a.** $|8|$, **b.** $|-5|$, and **c.** $|0|$.

Solution

a. On the number line, the distance between 8 and 0 is 8. Therefore,

$$|8| = 8$$

b. On the number line, the distance between −5 and 0 is 5. Therefore,

$$|-5| = 5$$

c. On the number line, the distance between 0 and 0 is 0. Therefore,

$$|0| = 0$$

The opposite of a number

Opposites or negatives

Two numbers that are the same distance from 0 on the number line, but on opposite sides of it, are called **opposites** or **negatives.**

Figure 2-8 shows that for each natural number on the number line, there is a corresponding natural number, called its *opposite,* to the left of 0. For example, we see that 3 and -3 are opposites, as are -5 and 5. Note that 0 is its own opposite.

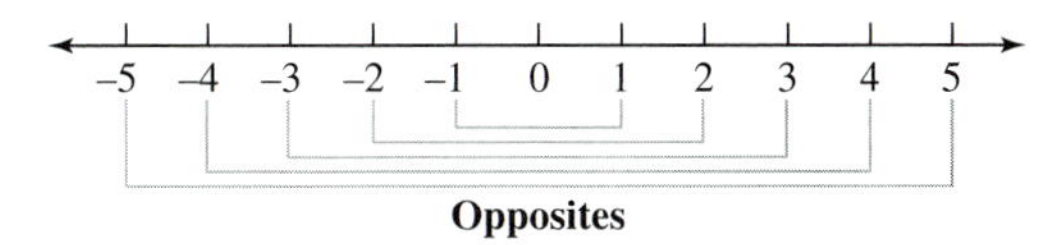

FIGURE 2-8

To write the opposite of a number, a $-$ symbol is used. For example, the opposite of 5 is -5 (read as "negative 5"). Parentheses are needed to express the opposite of a negative number. The opposite of -5 is written as $-(-5)$. Since 5 and -5 are the same distance from 0, the opposite of -5 is 5. Therefore, $-(-5) = 5$. This leads to the following conclusion.

The double negative rule

If a is any number, then

$$-(-a) = a$$

In words, this rule says that *the opposite of the negative of a number is that number.*

Number	Opposite	
57	-57	Read as "negative fifty-seven."
-8	$-(-8) = 8$	Read as "the opposite of negative eight." Apply the double negative rule.
0	$-0 = 0$	The opposite of 0 is 0.

The concept of opposite can also be applied to an absolute value. For example, the opposite of the absolute value of -8 can be written as $-|-8|$. Think of this as a two-step process. Find the absolute value first, and then attach a $-$ to that result.

First, find the absolute value.

$$-|-8| = -8$$

Then attach a $-$ sign.

EXAMPLE 5 Simplify each expression: **a.** $-(-44)$ and **b.** $-|-225|$.

Solution

a. $-(-44)$ means the opposite of -44. Since the opposite of -44 is 44, we write

$$-(-44) = 44$$

b. $-|-225|$ means the opposite of $|-225|$. Since $|-225| = 225$, and the opposite of 225 is -225, we write

$$-|-225| = -225$$

Self Check 5

Simplify each expression:
a. $-(-1)$ and **b.** $-|-99|$.

Answers **a.** 1, **b.** -99

The − symbol

The − symbol is used to indicate a negative number, the opposite of a number, and the operation of subtraction. The key to interpreting the − symbol correctly is to examine the context in which it is used.

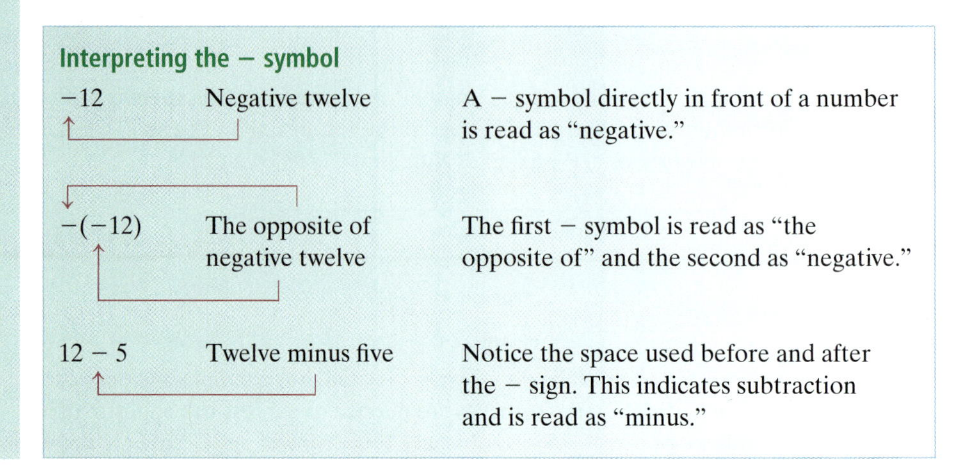

Section 2.1 STUDY SET

VOCABULARY *Fill in the blanks.*

1. Numbers can be represented by points equally spaced on the number ______.
2. The point on the number line representing 0 is called the ________.
3. To _______ a number means to locate it on the number line and highlight it with a dot.
4. The graph of a number is the point on the number ______ that represents that number.
5. The symbols > and < are called __________ symbols.
6. __________ numbers are less than 0.
7. The __________ _______ of a number is the distance between the number and 0 on the number line.
8. Two numbers that are the same distance from 0 on the number line, but on opposite sides of it, are called __________.
9. $\{\ldots, -3, -2, -1, 0, 1, 2, 3, \ldots\}$ is called the set of ________.
10. The double negative rule states that the negative of the ____________________ of a number is that number. If a is any number, then $-(-a) = \square$.

CONCEPTS

11. Refer to each graph and use an inequality symbol > or < to make a true statement.

a.

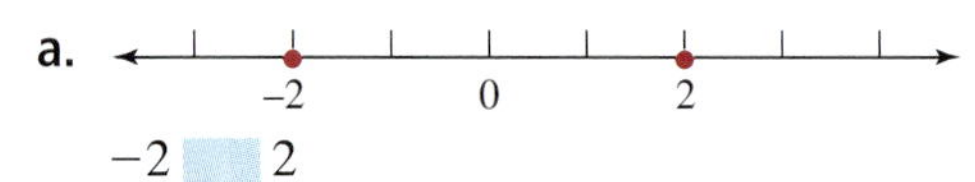

$-2 \;\square\; 2$

b.

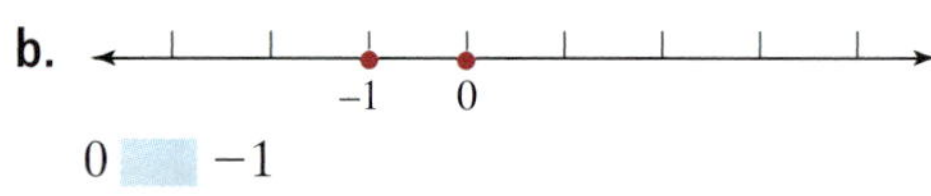

$0 \;\square\; -1$

c.

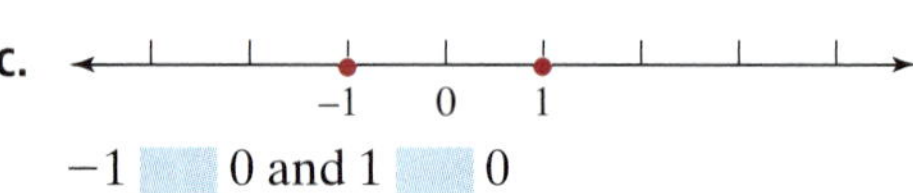

$-1 \;\square\; 0$ and $1 \;\square\; 0$

12. Determine what is wrong with each number line.

a.

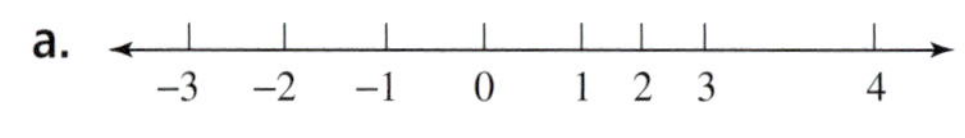

b. (−3, −2, −1, 0, 2, 4, 6, 8)

c. (−3, −2, −1, 1, 2, 3, 4, 5)

d.

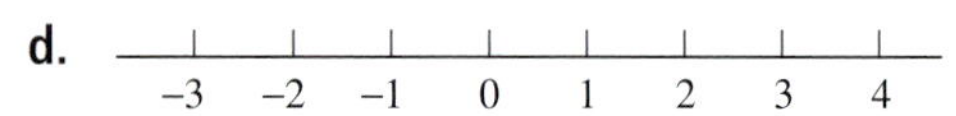

13. Does every number on the number line have an opposite?

14. Is the absolute value of a number always positive?

15. Which of the following contains a minus sign: $15 - 8$, $-(-15)$, or -15?

16. Is there a number that is both greater than 10 and less than 10 at the same time?

17. Express the fact $12 < 15$ using the $>$ symbol.

18. Express the fact $5 > 4$ using the $<$ symbol.

19. Represent each of these situations using a signed number.

 a. $225 overdrawn

 b. 10 seconds before liftoff

 c. 3 degrees below normal

 d. A deficit of $12,000

 e. A racehorse finished 2 lengths behind the leader.

20. Represent each of these situations using a signed number, and then describe its opposite in words.

 a. A trade surplus of $3 million

 b. A bacteria count 70 more than the standard

 c. A profit of $67

 d. A business $1 million in the "black"

 e. 20 units over their quota

21. If a number is less than 0, what type of number must it be?

22. If a number is greater than 0, what type of number must it be?

23. On the number line, what number is 3 units to the right of -7?

24. On the number line, what number is 4 units to the left of 2?

25. Name two numbers on the number line that are a distance of 5 away from -3.

26. Name two numbers on the number line that are a distance of 4 away from 3.

27. Which number is closer to -3 on the number line: 2 or -7?

28. Which number is farther from 1 on the number line: -5 or 8?

29. Give examples of the $-$ symbol used in three different ways.

30. What is the opposite of 0?

NOTATION

31. Translate each phrase to mathematical symbols.

 a. The opposite of negative eight

 b. The absolute value of negative eight

 c. Eight minus eight

 d. The opposite of the absolute value of negative eight

32. Write the set of integers.

PRACTICE *Simplify each expression.*

33. $|9|$

34. $|12|$

35. $|-8|$

36. $|-1|$

37. $|-14|$

38. $|-85|$

39. $-|20|$

40. $-|110|$

41. $-|-6|$

42. $|0|$

43. $|203|$

44. $-|-11|$

45. -0

46. $-|0|$

47. $-(-11)$

48. $-(-1)$

49. $-(-4)$

50. $-(-9)$

51. $-(-12)$

52. $-(-25)$

Graph each set of numbers on the number line.

53. $\{-3, 0, 3, 4, -1\}$

54. $\{-4, -1, 2, 5, 1\}$

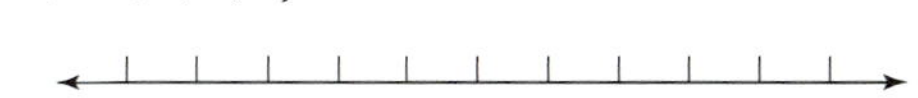

55. The opposite of -3, the opposite of 5, and the absolute value of -2

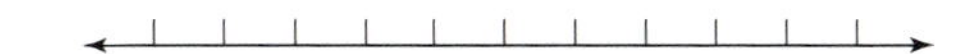

56. The absolute value of 3, the opposite of 3, and the number that is 1 less than -3

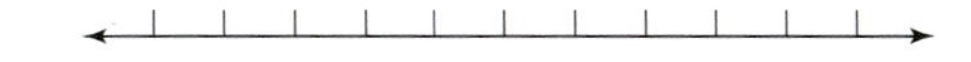

Insert one of the symbols $>$, $<$, or $=$ in the blank to make a true statement.

57. -5 ___ 5

58. 0 ___ -1

59. -12 ___ -6

60. -6 ___ -7

61. -10 ___ -11

62. -11 ___ -20

63. $|-2|$ ___ 0

64. $|-30|$ ___ -40

Insert one of the symbols $\geq$ or $\leq$ in the blank to make a true statement.

65. -14 ___ -15
66. -77 ___ -76
67. 210 ___ 210
68. 37 ___ 37
69. $-1{,}255$ ___ $-(-1{,}254)$
70. 0 ___ -3
71. $-|-3|$ ___ 4
72. $-|-163|$ ___ -150

APPLICATIONS

73. FLIGHT OF A BALL A boy throws a ball from the top of a building, as shown. At the instant he does this, his friend starts a stopwatch and keeps track of the time as the ball rises to a peak and then falls to the ground. Use the vertical number line to complete the table by finding the position of the ball at each specified time.

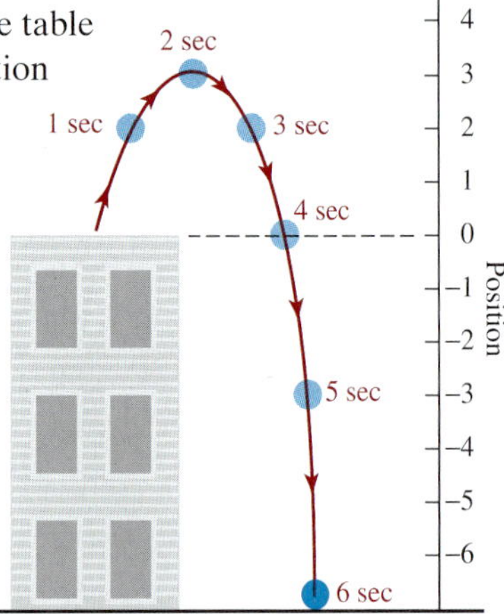

Time	Position of ball
1 sec	
2 sec	
3 sec	
4 sec	
5 sec	
6 sec	

74. SHOOTING GALLERIES At an amusement park, a shooting gallery contains moving ducks. The path of one duck is shown, along with the time it takes the duck to reach certain positions on the gallery wall. Complete the table using the horizontal number line in the illustration.

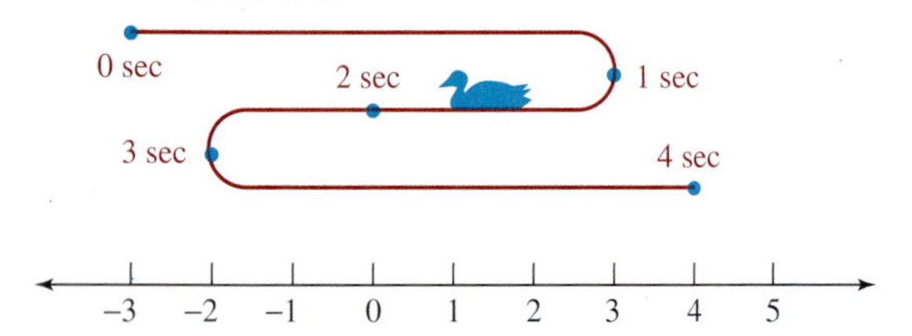

Time	Position of duck
0 sec	
1 sec	
2 sec	
3 sec	
4 sec	

75. TECHNOLOGY The readout from a testing device is shown. It is important to know the height of each of the three peaks and the depth of each of the three valleys. Use the vertical number line to find these numbers.

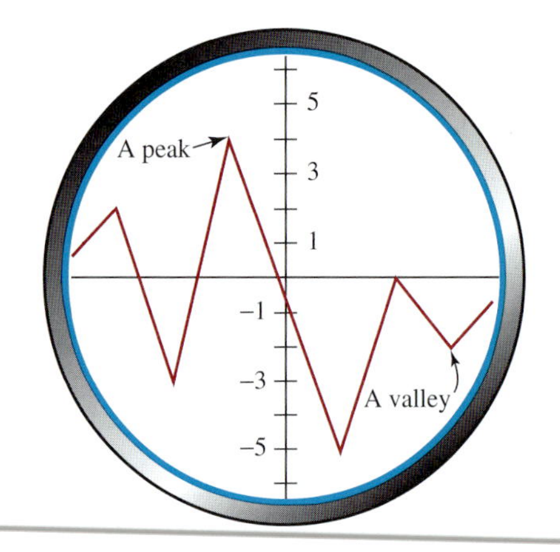

76. FLOODING A week of daily reports listing the height of a river in comparison to flood stage is given in the table. Complete the bar graph in the illustration.

Flood stage report	
Sun.	2 ft below
Mon.	3 ft over
Tue.	4 ft over
Wed.	2 ft over
Thu.	1 ft below
Fri.	3 ft below
Sat.	4 ft below

77. GOLF In golf, *par* is the standard number of strokes considered necessary on a given hole. A score of -2 indicates that a golfer used 2 strokes less than par. A score of $+2$ means 2 more strokes than par were used. In the illustration, each golf ball represents the score of a professional golfer on the 16th hole of a certain course.

a. What score was shot most often on this hole?

b. What was the best score on this hole?

c. Explain why this hole appears to be too easy for a professional golfer.

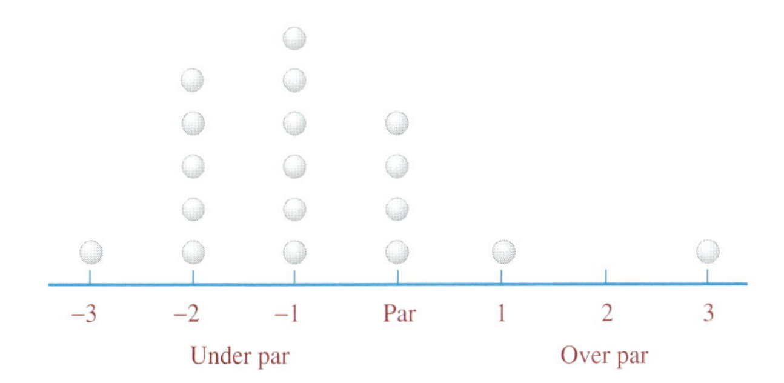

78. PAYCHECKS Examine the items listed on the following paycheck stub. Then write two columns on your paper — one headed "positive" and the other "negative." List each item under the appropriate heading.

Tom Dryden Dec. 04		Christmas bonus	$100
Gross pay	$2,000	**Reductions**	
Overtime	$300	Retirement	$200
Deductions		**Taxes**	
Union dues	$30	Federal withholding	$160
U.S. Bonds	$100	State withholding	$35

79. WEATHER MAPS The illustration shows the predicted Fahrenheit temperatures for a day in mid-January.

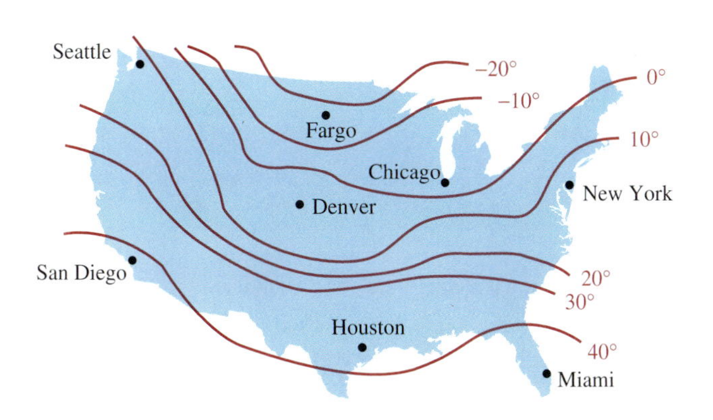

a. What is the temperature range for the region including Fargo, North Dakota?

b. According to the prediction, what is the warmest it should get in Houston?

c. According to this prediction, what is the coldest it should get in Seattle?

80. PROFITS/LOSSES The graph in the illustration shows the net income of Apple Computer Inc. for the years 2000–2004.

a. In what year did the company suffer a loss? Estimate each loss.

b. In what year did Apple have the greatest profit? Estimate it.

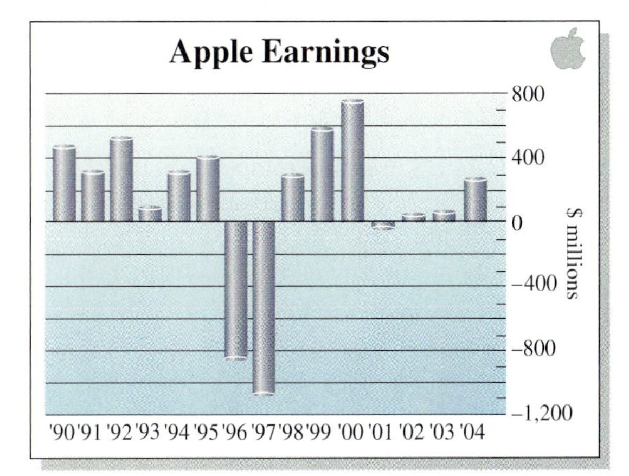

Source: Morningstar.com

81. HISTORY Number lines can be used to display historical data. Some important world events are shown on the time line in the illustration.

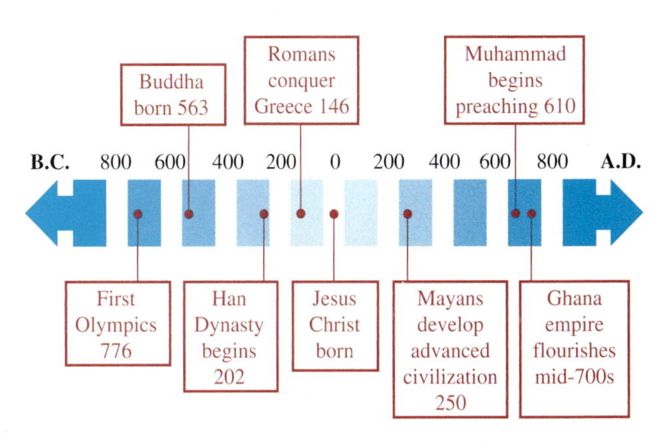

a. What basic unit is used to scale this time line?

b. What can be thought of as positive numbers?

c. What can be thought of as negative numbers?

d. What important event distinguishes the positive from the negative numbers?

82. ASTRONOMY Astronomers use a type of number line called the *apparent magnitude scale* to denote the brightness of objects in the sky. The brighter an object appears to an observer on Earth, the more negative is its apparent magnitude. Graph each of the following on the scale in the illustration.

- Full moon -12
- Pluto $+15$
- Sirius (a bright star) -2
- Sun -26
- Venus -4
- Visual limit of binoculars $+10$
- Visual limit of large telescope $+20$
- Visual limit of naked eye $+6$

83. LINE GRAPHS Each thermometer in the illustration gives the daily high temperature in degrees Fahrenheit. Plot each daily high temperature on the grid and then construct a line graph.

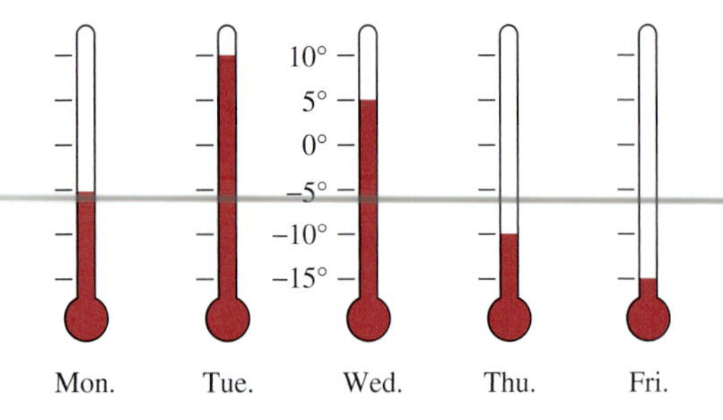

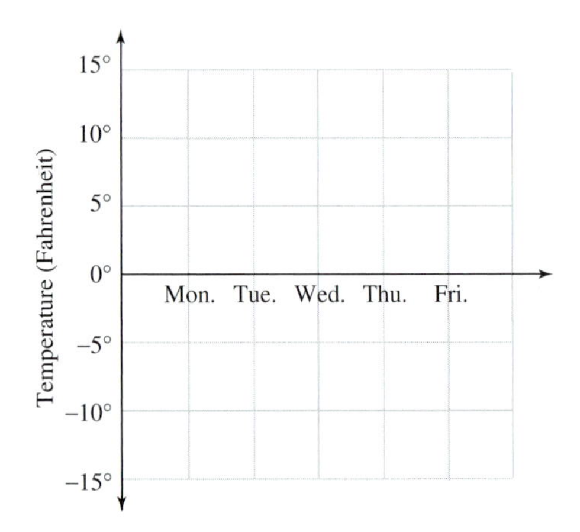

84. GARDENING The illustration in the next column shows the depths at which the bottoms of various types of flower bulbs should be planted. (The symbol ″ represents inches.)

a. At what depth should a tulip bulb be planted?

b. How much deeper are hyacinth bulbs planted than gladiolus bulbs?

c. Which bulb must be planted the deepest? How deep?

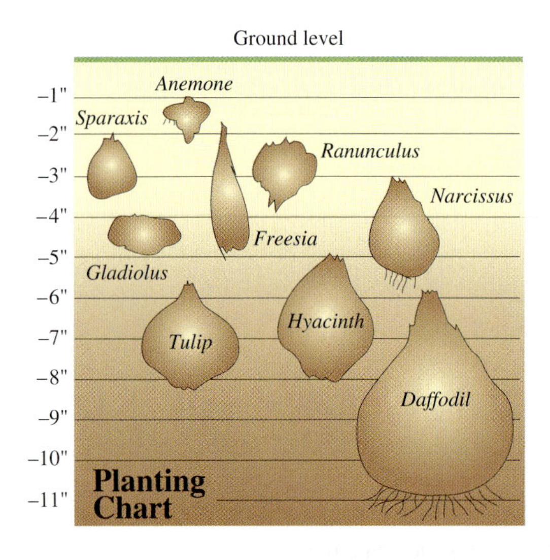

WRITING

85. Explain the concept of the opposite of a number.

86. What real-life situation do you think gave rise to the concept of a negative number?

87. Explain why the absolute value of a number is never negative.

88. Give an example of the use of the number line that you have seen in another course.

89. DIVING Divers use the terms *positive buoyancy, neutral buoyancy,* and *negative buoyancy* as shown. What do you think each of these terms means?

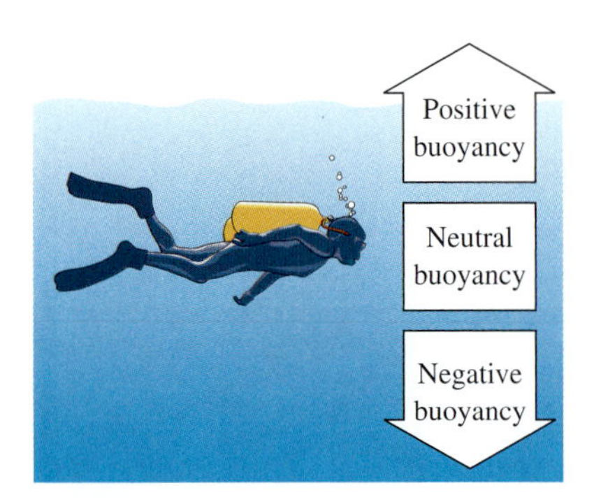

90. NEW ORLEANS The city of New Orleans, Louisiana, lies largely below sea level. Find out why the city is not under water.

REVIEW

91. Round 23,456 to the nearest hundred.

92. Evaluate: $19 - 2 \cdot 3$.

93. Subtract 2,081 from 2,842.

94. Divide 345 by 15.

95. Give the name of the property illustrated here:

$$(13 \cdot 2) \cdot 5 = 13 \cdot (2 \cdot 5)$$

96. Write four times five using three different notations.

2.2 Adding Integers

- Adding two integers with the same sign
- Adding two integers with different signs
- The addition property of zero
- The additive inverse of a number

A dramatic change in temperature occurred in 1943 in Spearfish, South Dakota. On January 22, at 7:30 A.M., the temperature was -4°F. In just two minutes, the temperature rose 49 degrees! To calculate the temperature at 7:32 A.M., we need to add 49 to -4.

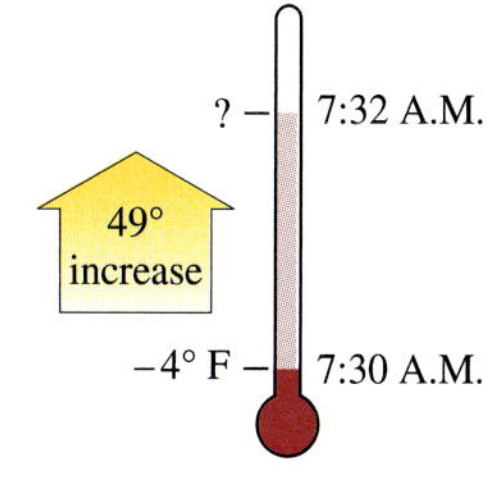

$$-4 + 49$$

To perform this addition, we must know how to add positive and negative integers. In this section, we develop rules to help us make such calculations.

Adding two integers with the same sign

4 + 3
both positive

To explain addition of signed numbers, we can use the number line. (See Figure 2-9.) To compute $4 + 3$, we begin at the **origin** (the point labeled 0) and draw an arrow 4 units long, pointing to the right. This represents positive 4. From that point, we draw an arrow 3 units long, pointing to the right, to represent positive 3. The second arrow points to the answer. Therefore, $4 + 3 = 7$.

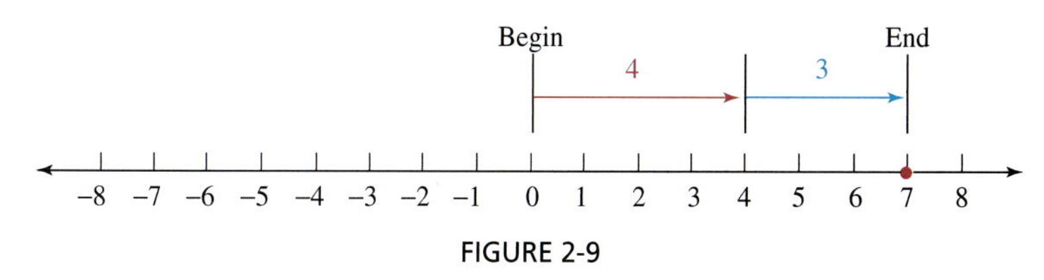

FIGURE 2-9

To check our work, let's think of the problem in terms of money. If you had \$4 and earned \$3 more, you would have a total of \$7.

−4 + (−3)
both negative

To compute $-4 + (-3)$, we begin at the origin and draw an arrow 4 units long, pointing to the left. (See Figure 2-10.) This represents -4. From there, we draw an arrow 3 units long, pointing to the left, to represent -3. The second arrow points to the answer: $-4 + (-3) = -7$.

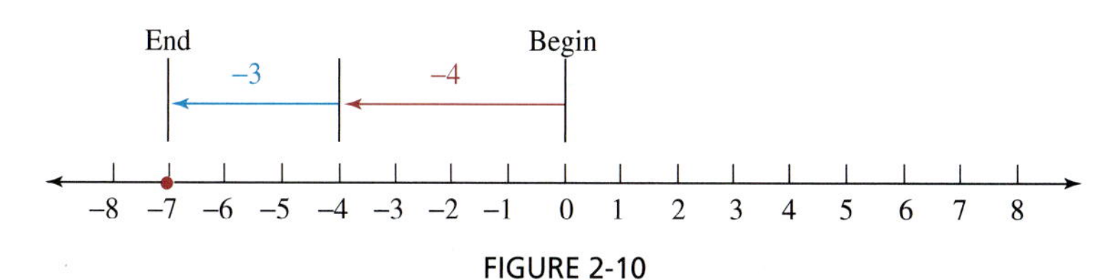

FIGURE 2-10

Let's think of this problem in terms of money. If you had a debt of \$4 (negative 4) and then took on \$3 more debt (negative 3), you would be in debt \$7 (negative 7).

Here are some observations about the process of adding two numbers that have the same sign, using the number line.

- Both arrows point in the same direction and "build" on each other.
- The answer has the same sign as the two numbers being added.

4	+	3	=	7		−4	+	(−3)	=	−7
Positive	+	positive	=	positive answer		Negative	+	negative	=	negative answer

These observations suggest the following rule.

Adding two integers with the same sign

To add two integers with the same sign, add their absolute values and attach their common sign to the sum. If both integers are positive, their sum is positive. If both integers are negative, their sum is negative.

! COMMENT When writing additions that involve signed numbers, write negative numbers within parentheses to separate the negative sign − from the plus sign +.

$9 + (-4)$ ~~$9 + -4$~~ and $-9 + (-4)$ ~~$-9 + -4$~~

EXAMPLE 1 Find the sum: $-9 + (-4)$.

Solution

Step 1: To add two integers with the same sign, we first add the absolute values of each of the integers. Since $|-9| = 9$ and $|-4| = 4$, we begin by adding 9 and 4.

$$9 + 4 = 13$$

Step 2: We then attach the common sign (which is negative) to this result. Therefore,

$$-9 + (-4) = -13$$

Make the answer negative.

After some practice, you will be able to do this kind of problem in your head. It will not be necessary to show all the steps as we have done here.

Self Check 2

Find the sum:

a. $7 + 5$

b. $-300 + (-100)$

Answers **a.** 12, **b.** −400

EXAMPLE 2 Find the sum: **a.** $6 + 4$ and **b.** $-80 + (-60)$.

Solution

a. Since both integers are positive, the answer is positive.

$$6 + 4 = 10$$

b. Since both integers are negative, the answer is negative.

$$-80 + (-60) = -140$$

Adding two integers with different signs

$4 + (-3)$
one positive, one negative

To compute $4 + (-3)$, we start at the origin and draw an arrow 4 units long, pointing to the right. (See Figure 2-11.) This represents positive 4. From there, we draw an arrow 3 units long, pointing to the left, to represent -3. The second arrow points to the answer: $4 + (-3) = 1$.

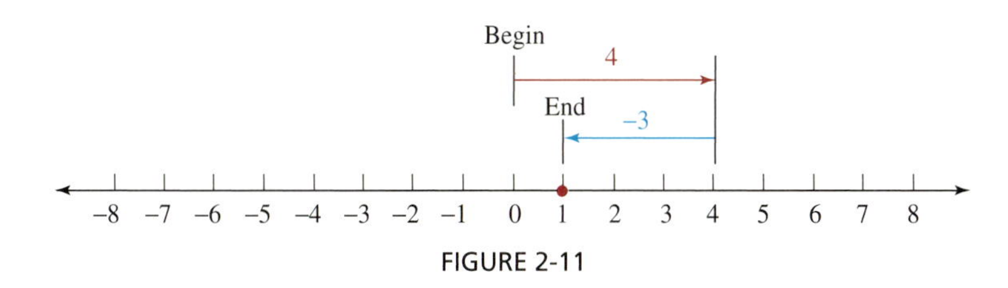

FIGURE 2-11

In terms of money, if you had \$4 (positive 4) and then took on a debt of \$3 (negative 3), you would have \$1 (positive 1).

$-4 + 3$
one positive, one negative

The problem $-4 + 3$ can be illustrated by drawing an arrow 4 units long from the origin, pointing to the left. (See Figure 2-12.) This represents -4. From there, we draw an arrow 3 units long, pointing to the right, to represent positive 3. The second arrow points to the answer: $-4 + 3 = -1$.

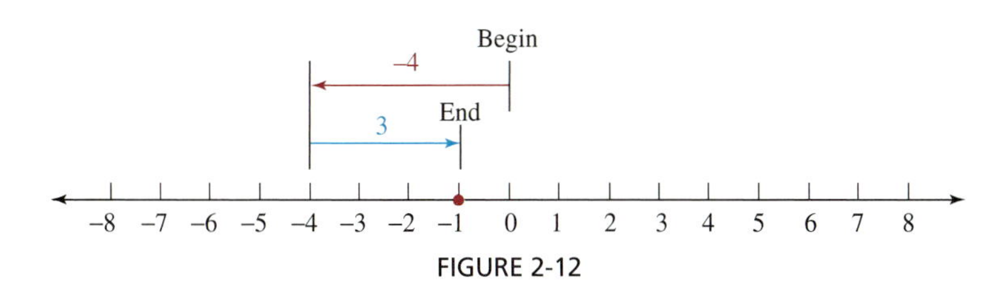

FIGURE 2-12

This problem can be thought of as owing \$4 (negative 4) and then paying back \$3 (positive 3). You will still owe \$1 (negative 1).

The last two examples lead us to some observations about adding two integers with different signs, using the number line.

- The arrows representing the integers point in opposite directions.
- The longer of the two arrows determines the sign of the answer.

These observations suggest the following rule.

Adding two integers with different signs

To add two integers with different signs, subtract their absolute values, the smaller from the larger. Then attach to that result the sign of the integer with the larger absolute value.

CALCULATOR SNAPSHOT — Entering negative numbers

Nigeria is the United States' second largest trading partner in Africa. To calculate the 2002 U.S. trade balance with Nigeria, we add the $1,057,700,000 worth of exports to Nigeria (considered positive) to the $5,945,300,000 worth of imports from Nigeria (considered negative). We can use a scientific calculator to perform the addition: $1,057,700,000 + (-5,945,300,000)$.

- We do not have to do anything special to enter a positive number. When we key in 1,057,700,000, a positive number is entered.
- To enter $-5,945,300,000$, we press the change-of-sign key [+/−] *after entering* 5,945,300,000. Note that the change-of-sign key is different from the subtraction key [−].

1057700000 [+] 5945300000 [+/−] [=]   [-4887600000]

In 2002, the United States had a trade balance of −$4,887,600,000 with Nigeria. Because the result is negative, it is called a *trade deficit*.

The addition property of zero

When 0 is added to a number, the number remains the same. For example, $5 + 0 = 5$, and $0 + (-4) = -4$. Because of this, we call 0 the **additive identity.**

Addition property of 0

For any number a,

$$a + 0 = a \quad \text{and} \quad 0 + a = a$$

In words, this property states that *the sum of any number and 0 is that number.*

The additive inverse of a number

A second fact concerning 0 and the operation of addition can be demonstrated by considering the sum of a number and its opposite. To illustrate this, we use the number line in Figure 2-13 to add 6 and its opposite, −6. We see that $6 + (-6) = 0$.

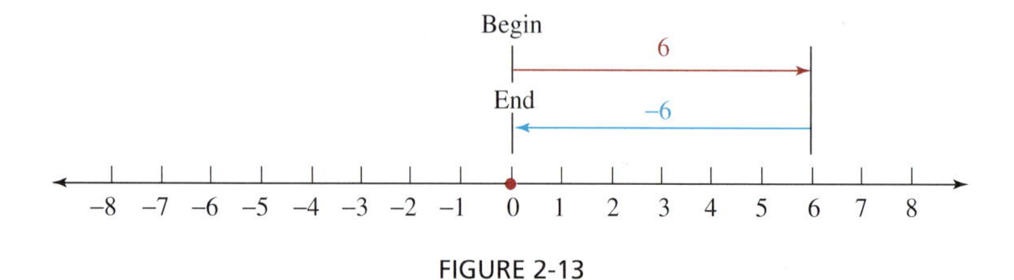

FIGURE 2-13

If the sum of two numbers is 0, the numbers are said to be **additive inverses** of each other. Since $6 + (-6) = 0$, we say that 6 and −6 are additive inverses.

We can now classify a pair of numbers such as 6 and −6 in three ways: as opposites, negatives, or additive inverses.

The additive inverse of a number

For any numbers a and b, if $a + b = 0$, then a and b are called **additive inverses.** That is, two numbers are said to be additive inverses if their sum is 0.

EXAMPLE 9 What is the additive inverse of -3? Justify your result.

Solution The additive inverse of -3 is its opposite, 3. To justify the result, we add and show that the sum is 0.

$$-3 + 3 = 0$$

Self Check 9
What is the additive inverse of 12? Justify your result.

Answer -12; $12 + (-12) = 0$

Section 2.2 STUDY SET

VOCABULARY *Fill in the blanks.*

1. When 0 is added to a number, the number remains the same. We call 0 the additive ________.
2. Since $-5 + 5 = 0$, we say that 5 is the additive ________ of -5. We can also say that 5 and -5 are ________.

CONCEPTS *Find each answer using the number line.*

3. $-3 + 6$
4. $-3 + (-2)$
5. $-5 + 3$
6. $-1 + (-3)$
7. **a.** Is the sum of two positive integers always positive?
   b. Is the sum of two negative integers always negative?
8. **a.** What is the sum of a number and its additive inverse?
   b. What is the sum of a number and its opposite?
9. Find each absolute value.
   a. $|-7|$ **b.** $|10|$
10. If the sum of two numbers is 0, what can be said about the numbers?

Fill in the blanks.

11. To add two integers with unlike signs, ________ their absolute values, the smaller from the larger. Then attach to that result the sign of the number with the ________ absolute value.
12. To add two integers with like signs, add their ________ values and attach their common ______ to the sum.

NOTATION *Complete each solution to evaluate the expression.*

13. $-16 + (-2) + (-1) = \square + (-1)$
    $= \square$
14. $-8 + (-2) + 6 = \square + 6$
    $= \square$
15. $(-3 + 8) + (-3) = \square + (-3)$
    $= \square$
16. $-5 + [2 + (-9)] = -5 + (\square)$
    $= \square$
17. Explain why the expression $-6 + -5$ is not written correctly. How should it be written?
18. What mathematical symbol is suggested when the word *sum* is used?

PRACTICE *Find the additive inverse of each number.*

19. -11
20. 9
21. -23
22. -43
23. 0
24. 1
25. 99
26. 250

Find each sum.

27. $-6 + (-3)$
28. $-2 + (-3)$
29. $-5 + (-5)$
30. $-8 + (-8)$
31. $-6 + 7$
32. $-2 + 4$
33. $-15 + 8$
34. $-18 + 10$
35. $20 + (-40)$
36. $25 + (-10)$
37. $30 + (-15)$
38. $8 + (-20)$
39. $-1 + 9$
40. $-2 + 7$
41. $-7 + 9$
42. $-3 + 6$
43. $5 + (-15)$
44. $16 + (-26)$
45. $24 + (-15)$
46. $-4 + 14$
47. $35 + (-27)$
48. $46 + (-73)$
49. $24 + (-45)$
50. $-65 + 31$

Evaluate each expression.

51. $-2 + 6 + (-1)$

52. $4 + (-3) + (-2)$

53. $-9 + 1 + (-2)$

54. $5 + 4 + (-6)$

55. $6 + (-4) + (-13) + 7$

56. $8 + (-5) + (-10) + 6$

57. $9 + (-3) + 5 + (-4)$

58. $-3 + 7 + 1 + (-4)$

59. Find the sum of -6, -7, and -8.

60. Find the sum of -11, -12, and -13.

Find each sum.

61. $-7 + 0$

62. $6 + 0$

63. $9 + 0$

64. $0 + (-15)$

65. $-4 + 4$

66. $18 + (-18)$

67. $2 + (-2)$

68. $-10 + 10$

69. What number must be added to -5 to obtain 0?

70. What number must be added to 8 to obtain 0?

Evaluate each expression.

71. $2 + (-10 + 8)$

72. $(-9 + 12) + (-4)$

73. $(-4 + 8) + (-11 + 4)$

74. $(-12 + 6) + (-6 + 8)$

75. $[-3 + (-4)] + (-5 + 2)$

76. $[9 + (-10)] + (-7 + 9)$

77. $[6 + (-4)] + [8 + (-11)]$

78. $[5 + (-8)] + [9 + (-15)]$

79. $-2 + [-8 + (-7)]$

80. $-8 + [-5 + (-2)]$

81. $789 + (-9{,}135)$

82. $2{,}701 + (-4{,}089)$

83. $-675 + (-456) + 99$

84. $-9{,}750 + (-780) + 2{,}345$

APPLICATIONS *Use signed numbers to help answer each question.*

85. G FORCES As a fighter pilot dives and loops, different forces are exerted on the body, just like the forces you experience when riding on a roller coaster. Some of the forces, called G's, are positive and some are negative. The force of gravity, 1G, is constant. Complete the diagram in the next column.

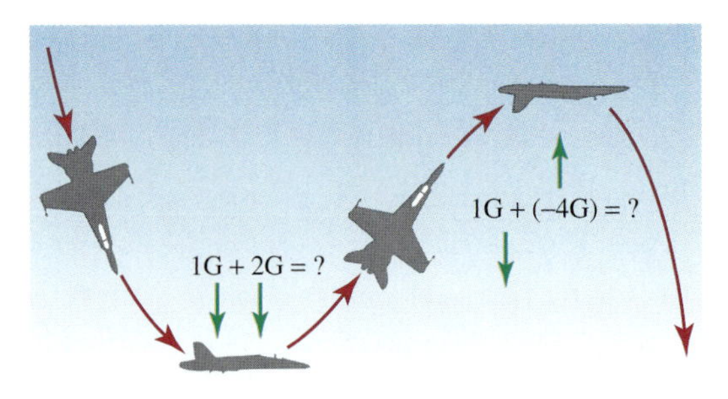

86. CHEMISTRY The first several steps of a chemistry lab experiment are listed here. The experiment begins with a compound that is stored at $-40°$ F.

Step 1: Raise the temperature of the compound 200°.

Step 2: Add sulfur and then raise the temperature 10°.

Step 3: Add 10 milliliters of water, stir, and raise the temperature 25°.

What is the resulting temperature of the mixture after step 3?

87. CASH FLOW The maintenance costs, utilities, and taxes on a duplex are \$900 per month. The owner of the apartments receives monthly rental payments of \$450 and \$380. Does this investment produce a positive cash flow each month?

88. JOGGING A businessman's lunchtime workout includes jogging up ten stories of stairs in his high-rise office building. If he starts on the fourth level below ground in the underground parking garage, on what story of the building will he finish his workout?

89. HEALTH Find the point total for the six risk factors (in blue) on the medical questionnaire below. Then use the table at the bottom of the form to determine the risk of contracting heart disease for the man whose responses are shown.

Age		**Total Cholesterol**	
Age	Points	Reading	Points
35	*–4*	*280*	*3*
Cholesterol		**Blood Pressure**	
HDL	Points	Systolic/Diastolic	Points
62	*–3*	*124/100*	*3*
Diabetic		**Smoker**	
	Points		Points
Yes	*4*	*Yes*	*2*

10-Year Heart Disease Risk

Total Points	Risk	Total Points	Risk
–2 or less	1%	5	4%
–1 to 1	2%	6	6%
2 to 3	3%	7	6%
4	4%	8	7%

Source: National Heart, Lung, and Blood Institute

90. SPREADSHEETS Monthly rain totals for four counties are listed in the spreadsheet shown below. The **−1** entered in cell B1 means that the rain total for Suffolk County for a certain month was 1 inch below average. We can analyze this data by asking the computer to perform various operations.

a. To ask the computer to add the numbers in cells C1, C2, C3, and C4, we type SUM(C1:C4). Find this sum.

b. Find SUM(B4:F4).

	A	B	C	D	E	F
1	Suffolk	**−1**	−1	0	+1	+1
2	Marin	0	−2	+1	+1	−1
3	Logan	−1	+1	+2	+1	+1
4	Tipton	−2	−2	+1	−1	−3

91. ATOMS An atom is composed of protons, neutrons, and electrons. A proton has a positive charge (represented by +1), a neutron has no charge, and an electron has a negative charge (−1). Two simple models of atoms are shown. What is the net charge of each atom?

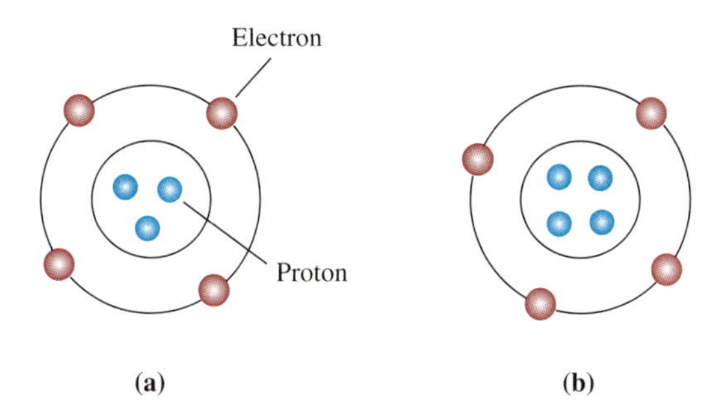

92. POLITICAL POLLS Six months before a general election, the incumbent senator found himself trailing the challenger by 18 points. To overtake his opponent, the campaign staff decided to use a four-part strategy. Each part of this plan is shown below, with the anticipated point gain.

1. Intense TV ad blitz +10

2. Ask for union endorsement +2

3. Voter mailing +3

4. Get-out-the-vote campaign +1

With these gains, will the incumbent overtake the challenger on election day?

93. FLOODING After a heavy rainstorm, a river that had been 4 feet under flood stage rose 11 feet in a 48-hour period. Find the height of the river after the storm in comparison to flood stage.

94. MILITARY SCIENCE During a battle, an army retreated 1,500 meters, regrouped, and advanced 3,500 meters. The next day, it had to retreat 1,250 meters. Find the army's net gain.

95. AIRLINES The graph below shows the annual net income for Delta Air Lines during the years 2000–2003. Estimate the company's total net income over this span of four years.

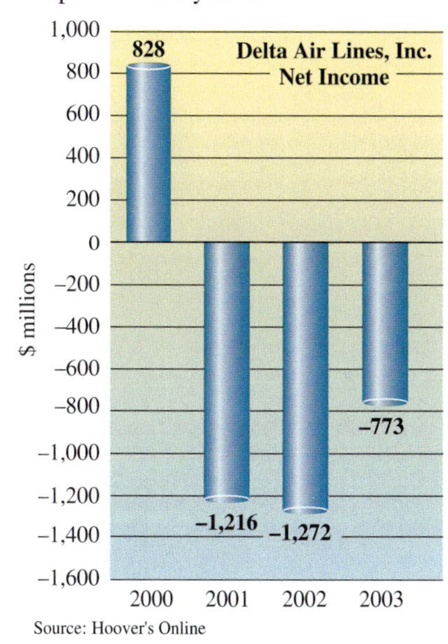

Source: Hoover's Online

96. ACCOUNTING On a financial balance sheet, debts (considered negative numbers) are denoted within parentheses. Assets (considered positive numbers) are written without parentheses. What is the 2004 fund balance for the preschool whose financial records are shown?

Community Care Preschool Balance Sheet, June 2004	
Fund balances	
Classroom supplies	$ 5,889
Emergency needs	927
Holiday program	(2,928)
Insurance	1,645
Janitorial	(894)
Licensing	715
Maintenance	(6,321)
BALANCE	?

WRITING

97. Is the sum of a positive and a negative number always positive? Explain why or why not.

98. How do you explain the fact that when asked to *add* −4 and 8, we must actually *subtract* to obtain the result?

99. Why is the sum of two negative numbers a negative number?

100. Write an application problem that will require adding -50 and -60.

REVIEW

101. Find the area of the rectangle shown below.

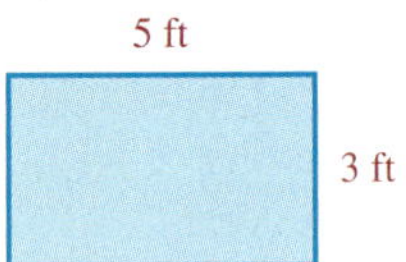

102. Find the perimeter of the rectangle in Exercise 101.

103. A car with a tank that holds 15 gallons of gasoline goes 25 miles on 1 gallon. How far can it go on a full tank?

104. What property is illustrated by the statement $5 \cdot 15 = 15 \cdot 5$?

105. Prime factor 125. Use exponents to express the result.

106. Perform the division: $\frac{144}{12}$.

2.3 Subtracting Integers

• Adding the opposite • Order of operations • Applications of subtraction

In this section, we study another way to think about subtraction. This new procedure is helpful when subtraction problems involve negative numbers.

Adding the opposite

The subtraction problem $6 - 4$ can be thought of as taking away 4 from 6. We can use the number line to illustrate this. (See Figure 2-14.) Beginning at the origin, we draw an arrow of length 6 units in the positive direction. From that point, we move back 4 units to the left. The answer, called the **difference,** is 2.

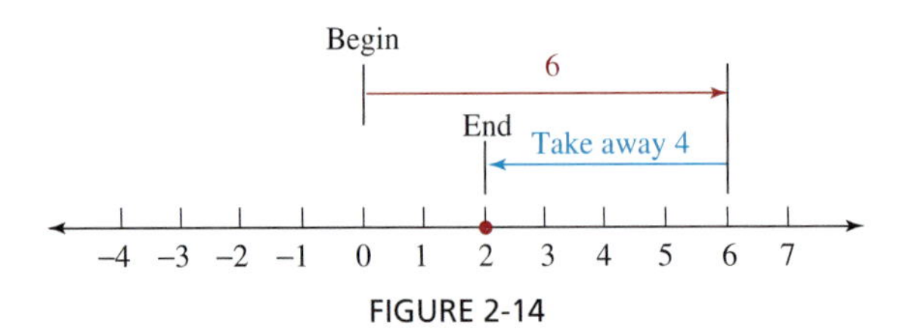

FIGURE 2-14

The work shown in Figure 2-14 looks like the illustration for the *addition* problem $6 + (-4) = 2$, shown in Figure 2-15.

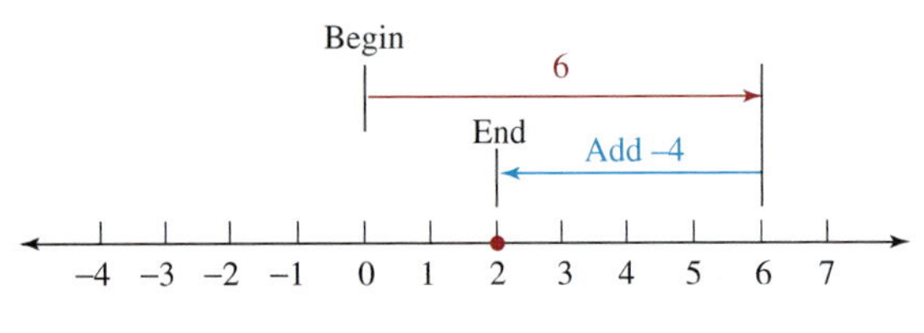

FIGURE 2-15

In the first problem, $6 - 4$, we subtracted 4 from 6. In the second, $6 + (-4)$, we added -4 (which is the opposite of 4) to 6. In each case, the result was 2.

Subtracting → $6 - 4 = 2$

Adding the opposite of 4 → $6 + (-4) = 2$

The same result

This observation helps to justify the following rule for subtraction.

Rule for subtraction

If a and b are any numbers, then

$$a - b = a + (-b)$$

In words, this rule states that *to subtract two integers, add the opposite of the second integer to the first integer.*

The rule for subtraction is also stated as follows: *subtraction is the same as adding the opposite of the number to be subtracted.*

You won't need to use this rule for every subtraction problem. For example, $6 - 4$ is obviously 2; it does not need to be rewritten as adding the opposite. But for more complicated problems such as $-6 - 4$ or $3 - (-5)$, where the result is not obvious, the subtraction rule will be quite helpful.

EXAMPLE 1 Subtract: $-6 - 4$.

Solution The number to be subtracted is 4. Applying the subtraction rule, we write

$-6 - 4 = -6 + (-4)$ Write the subtraction as an addition of the opposite of 4, which is -4. Write -4 within parentheses.

$= -10$ To add -6 and -4, apply the rule for adding two negative numbers.

To check the result, we add the difference, -10, and the subtrahend, 4. We should obtain the minuend, -6.

$$-10 + 4 = -6$$

The answer, -10, checks.

Self Check 1

Find $-2 - 3$ and check the result.

Answer -5

EXAMPLE 2 Subtract: $3 - (-5)$.

Solution The number being subtracted is -5.

$3 - (-5) = 3 + 5$ Write the subtraction as an addition of the opposite of -5, which is 5.

$= 8$ Perform the addition.

Self Check 2

Subtract: $3 - (-2)$.

Answer 5

Self Check 3
Subtract -8 from -3.

Answer 5

EXAMPLE 3 Subtract -3 from -8.

Solution The number being subtracted is -3, so we write it after -8.

$-8 - (-3) = -8 + 3$ Add the opposite of -3, which is 3.

$= -5$ Perform the addition.

Remember that any subtraction problem can be rewritten as an equivalent addition. We just add the opposite of the number that is to be subtracted.

Subtraction can be written as addition . . .

$$4 - 8 = 4 + (-8) = -4$$
$$4 - (-8) = 4 + 8 = 12$$
$$-4 - 8 = -4 + (-8) = -12$$
$$-4 - (-8) = -4 + 8 = 4$$

of the opposite of the number to be subtracted.

Order of operations

Expressions can contain repeated subtraction or subtraction in combination with grouping symbols. To work these problems, we apply the rules for the order of operations, listed on page 47.

Self Check 4
Evaluate: $-3 - 5 - (-1)$.

Answer -7

EXAMPLE 4 Evaluate: $-1 - (-2) - 10$.

Solution This problem involves two subtractions. We work from left to right, rewriting each subtraction as an addition of the opposite.

$-1 - (-2) - 10 = -1 + 2 + (-10)$ Add the opposite of -2, which is 2. Add the opposite of 10, which is -10. Write -10 in parentheses.

$= 1 + (-10)$ Work from left to right. Add $-1 + 2$.

$= -9$ Perform the addition.

Self Check 5
Evaluate: $-2 - (-6 - 5)$.

Answer 9

EXAMPLE 5 Evaluate: $-8 - (-2 - 2)$.

Solution We must perform the subtraction within the parentheses first.

$-8 - (-2 - 2) = -8 - [-2 + (-2)]$ Add the opposite of 2, which is -2. Since -2 must be written within parentheses, we write $-2 + (-2)$ within brackets.

$= -8 - (-4)$ Add -2 and -2. Since only one set of grouping symbols is now needed, we write -4 within parentheses.

$= -8 + 4$ Add the opposite of -4, which is 4.

$= -4$ Perform the addition.

Applications of subtraction

Things are constantly changing in our daily lives. The temperature, the amount of money we have in the bank, and our ages are examples. In mathematics, the operation of subtraction is used to measure change.

COMMENT In general, to find the change in a quantity, we subtract the earlier value from the later value.

EXAMPLE 6 Change of water level. On Monday, the water level in a city storage tank was 6 feet above normal. By Friday, the level had fallen to a mark 4 feet below normal. Find the change in the water level from Monday to Friday. (See Figure 2-16.)

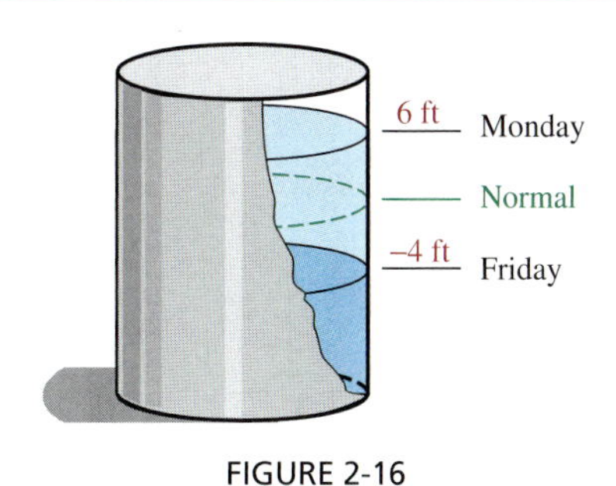

FIGURE 2-16

Solution We use subtraction to find the amount of change. The water levels of 4 feet below normal (the later value) and 6 feet above normal (the earlier value) can be represented by -4 and 6, respectively.

The water level Friday	minus	the water level Monday	is	the change in the water level.

$-4 - 6 = -4 + (-6)$ Add the opposite of 6, which is -6.

$= -10$ Perform the addition. The negative result indicates that the water level fell.

The water level fell 10 feet from Monday to Friday.

In the next example, the number line serves as a mathematical model of a real-life situation. You will see how the operation of subtraction can be used to find the distance between two points on the number line.

EXAMPLE 7 Artillery. In a practice session, an artillery group fired two rounds at a target. The first landed 65 yards short of the target, and the second landed 50 yards past it. (See Figure 2-17.) How far apart were the two impact points?

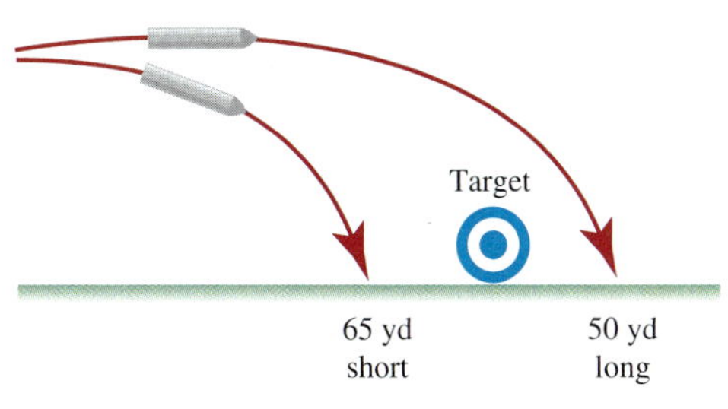

FIGURE 2-17

Solution We can use the number line to model this situation. The target is the origin. The words *short of the target* indicate a negative number, and the words *past it* indicate a positive number. Therefore, we graph the impact points at −65 and 50 in Figure 2-18.

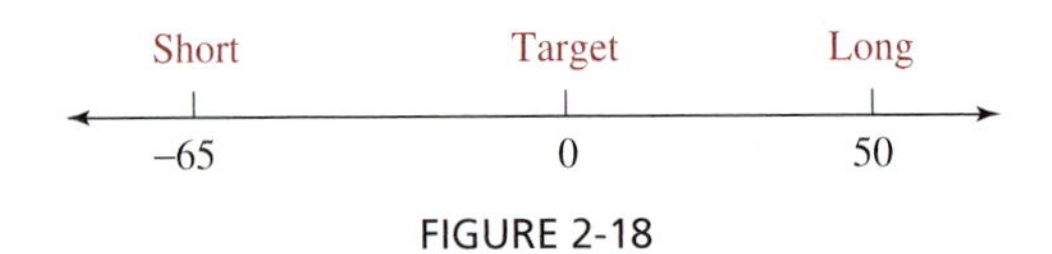

FIGURE 2-18

The phrase *how far apart* tells us to subtract.

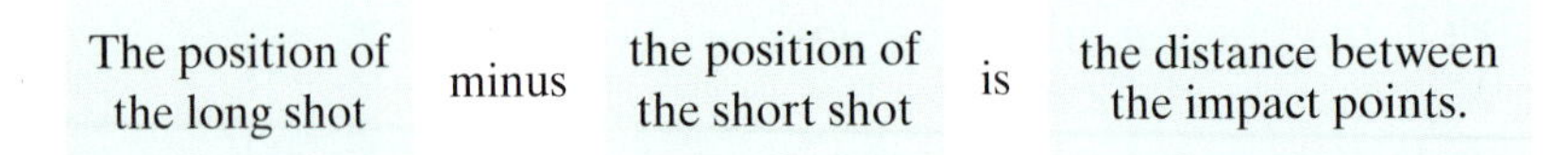

The position of the long shot	minus	the position of the short shot	is	the distance between the impact points.

$50 - (-65) = 50 + 65$ Add the opposite of −65.

$= 115$ Perform the addition.

The impact points are 115 yards apart.

CALCULATOR SNAPSHOT

Subtraction with negative numbers

The world's highest peak is Mount Everest in the Himalayas. The greatest ocean depth yet measured lies in the Mariana Trench near the island of Guam in the western Pacific. (See Figure 2-19) To find the range between the highest peak and the greatest depth, we must subtract.

$$29{,}035 - (-36{,}025)$$

To perform this subtraction using a scientific calculator, we enter these numbers and press these keys.

29035 [−] 36025 [+/−] [=] 65060

The range is 65,060 feet between the highest peak and the lowest depth. We could have applied the subtraction rule to write 29,035 − (−36,025) as 29,035 + 36,025 before using the calculator.

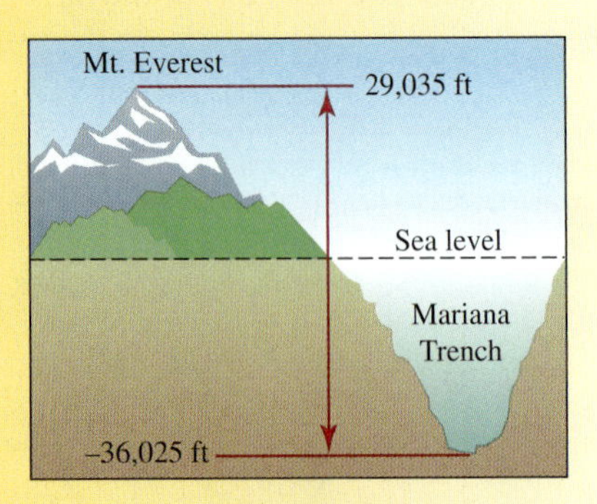

FIGURE 2-19

Section 2.3 STUDY SET

VOCABULARY *Fill in the blanks.*

1. The answer to a subtraction problem is called the ________.
2. Two numbers that are the same distance from 0 on the number line, but on opposite sides of it, are called ________.

CONCEPTS *Fill in the blanks.*

3. Subtraction is the same as ______ the ________ of the number to be subtracted.
4. Subtracting 3 is the same as adding ▢.
5. Subtracting -6 is the same as adding ▢.
6. The opposite of -8 is ▢.
7. For any numbers a and b, $a - b = a +$ ▢.
8. a. $2 - 7 = 2 +$ ▢
   b. $2 - (-7) = 2 +$ ▢
   c. $-2 - 7 = -2 +$ ▢
   d. $-2 - (-7) = -2 +$ ▢
9. After using parentheses as grouping symbols, if another set of grouping symbols is needed, we use ________.
10. We can find the ______ in a quantity by subtracting the earlier value from the later value.
11. Write this problem using mathematical symbols: negative eight minus negative four.
12. Write this problem using mathematical symbols: negative eight subtracted from negative four.
13. Find the distance between -4 and 3 on the number line.
14. Find the distance between -10 and 1 on the number line.
15. Is subtracting 3 from 8 the same as subtracting 8 from 3? Explain.
16. Evaluate each expression.
   a. $-2 - 0$ b. $0 - (-2)$

NOTATION *Complete each solution to evaluate each expression.*

17. $1 - 3 - (-2) = 1 + ($▢$) + 2$
   $= -2 +$ ▢
   $=$ ▢
18. $-6 + 5 - (-5) = -6 + 5 +$ ▢
   $=$ ▢ $+ 5$
   $=$ ▢
19. $(-8 - 2) - (-6) = [-8 + ($▢$)] - (-6)$
   $=$ ▢ $- (-6)$
   $= -10 +$ ▢
   $=$ ▢
20. $-5 - (-1 - 4) = -5 - [-1 + ($▢$)]$
   $= -5 - ($▢$)$
   $= -5 +$ ▢
   $=$ ▢

PRACTICE *Find each difference.*

21. $8 - (-1)$
22. $3 - (-8)$
23. $-4 - 9$
24. $-7 - 6$
25. $-5 - 5$
26. $-7 - 7$
27. $-5 - (-4)$
28. $-9 - (-1)$
29. $-1 - (-1)$
30. $-4 - (-3)$
31. $-2 - (-10)$
32. $-6 - (-12)$
33. $0 - (-5)$
34. $0 - 8$
35. $0 - 4$
36. $0 - (-6)$
37. $-2 - 2$
38. $-3 - 3$
39. $-10 - 10$
40. $4 - 4$
41. $9 - 9$
42. $4 - (-4)$
43. $-3 - (-3)$
44. $-5 - (-5)$

Evaluate each expression.

45. $-4 - (-4) - 15$
46. $-3 - (-3) - 10$
47. $-3 - 3 - 3$
48. $-1 - 1 - 1$
49. $5 - 9 - (-7)$
50. $6 - 8 - (-4)$
51. $10 - 9 - (-8)$
52. $16 - 14 - (-9)$
53. $-1 - (-3) - 4$
54. $-2 - 4 - (-1)$
55. $-5 - 8 - (-3)$
56. $-6 - 5 - (-1)$
57. $(-6 - 5) - 3$
58. $(-2 - 1) - 5$
59. $(6 - 4) - (1 - 2)$
60. $(5 - 3) - (4 - 6)$
61. $-9 - (6 - 7)$
62. $-3 - (6 - 12)$
63. $-8 - [4 - (-6)]$
64. $-1 - [5 - (-2)]$
65. $[-4 + (-8)] - (-6)$
66. $[-5 + (-4)] - (-2)$

67. Subtract -3 from 7.

68. Subtract 8 from -2.

69. Subtract -6 from -10.

70. Subtract -4 from -9.

Use a calculator to perform each subtraction.

71. $-1{,}557 - 890$

72. $-345 - (-789)$

73. $20{,}007 - (-496)$

74. $-979 - (-44{,}879)$

75. $-162 - (-789) - 2{,}303$

76. $-787 - 1{,}654 - (-232)$

APPLICATIONS *Use signed numbers to help answer each question.*

77. SCUBA DIVING A diver jumps from his boat into the water and descends 50 feet. He pauses to check his equipment and then descends an additional 70 feet. Use a signed number to represent the diver's final depth.

78. TEMPERATURE CHANGE Rashawn flew from his New York home to Hawaii for a week of vacation. He left blizzard conditions and a temperature of $-6°$, and stepped off the airplane into 85° weather. What temperature change did he experience?

79. READING PROGRAMS In a state reading test administered at the start of a school year, an elementary school's performance was 23 points below the county average. The principal immediately began a special tutorial program. At the end of the school year, retesting showed the students to be only 7 points below the average. How many points did the school's reading score improve over the year?

80. SUBMARINES A submarine was traveling 2,000 feet below the ocean's surface when the radar system warned of an impending collision with another sub. The captain ordered the navigator to dive an additional 200 feet and then level off. Find the depth of the submarine after the dive.

81. AMPERAGE During normal operation, the ammeter on a car reads +5. If the headlights, which draw a current of 7 amps, and the radio, which draws a current of 6 amps, are both turned on, what number will the ammeter register?

82. GIN RUMMY After a losing round, a card player must subtract the value of each of the cards left in his hand from his previous point total of 21. If face cards are counted as 10 points, what is his new score?

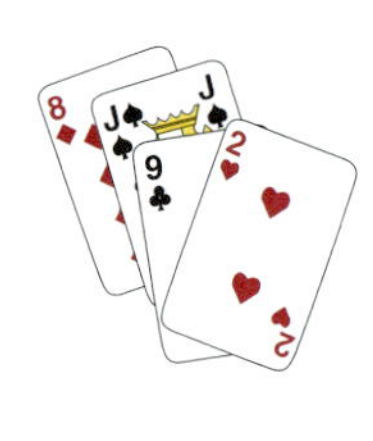

83. GEOGRAPHY Death Valley, California, is the lowest land point in the United States, at 282 feet below sea level. The lowest land point on the Earth is the Dead Sea, which is 1,348 feet below sea level. How much lower is the Dead Sea than Death Valley?

84. LIE DETECTOR TESTS On one lie detector test, a burglar scored -18, which indicates deception. However, on a second test, he scored -1, which is inconclusive. Find the difference in the scores.

85. FOOTBALL A college football team records the outcome of each of its plays during a game on a stat sheet. Find the net gain (or loss) after the 3rd play.

Down	Play	Result
1st	Run	Lost 1 yd
2nd	Pass — sack!	Lost 6 yd
Penalty	Delay of game	Lost 5 yd
3rd	Pass	Gained 8 yd
4th	Punt	—

86. ACCOUNTING Complete the balance sheet below. Then determine the overall financial condition of the company by subtracting the total liabilities from the total assets.

Walker Corporation
Balance Sheet 2001

Assets				
Cash	$11	1	0	9
Supplies	7	8	6	2
Land	67	5	4	3
Total assets	$			
Liabilities				
Accounts payable	$79	0	3	7
Income taxes	20	1	8	1
Total liabilities	$			

87. DIVING A diver jumps from a platform. After she hits the water, her momentum takes her to the bottom of the pool.

a. Use the number line and signed numbers to model this situation. Show the top of the platform, the water level as 0, and the bottom of the pool.

b. Find the total length of the dive from the top of the platform to the bottom of the pool.

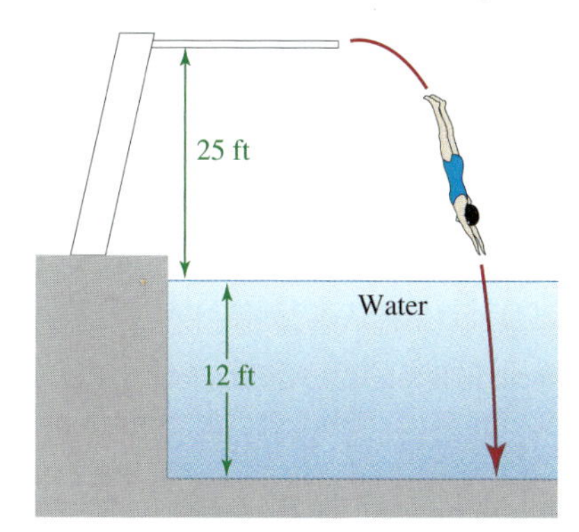

88. TEMPERATURE EXTREMES The highest and lowest temperatures ever recorded in several cities are shown below. List the cities in order, from the largest to smallest range in temperature extremes.

	Extreme temperatures	
City	**Highest**	**Lowest**
Atlantic City, NJ	106	−11
Barrow, AK	79	−56
Kansas City, MO	109	−23
Norfolk, VA	104	−3
Portland, ME	103	−39

89. CHECKING ACCOUNTS Michael has $1,303 in his checking account. Can he pay his car insurance premium of $676, his utility bills of $121, and his rent of $750 without having to make another deposit? Explain.

90. HISTORY Two of the greatest Greek mathematicians were Archimedes (287–212 B.C.) and Pythagoras (569–500 B.C.). How many years apart were they born?

WRITING

91. Explain what is meant when we say that subtraction is the same as addition of the opposite.

92. Give an example showing that it is possible to subtract something from nothing.

93. Explain how to check the result: $-7 - 4 = -11$.

94. Explain why students don't need to change every subtraction they encounter to an addition of the opposite. Give some examples.

REVIEW

95. Round 5,989 to the nearest ten.

96. Round 5,999 to the nearest hundred.

97. List the factors of 20.

98. It takes 13 oranges to make one can of orange juice. Find the number of oranges used to make 12 cans.

99. Evaluate: $12^2 - (5 - 4)^2$.

100. What property does the following illustrate?

$$15 + 12 = 12 + 15$$

101. Solve: $5x = 15$.

102. Solve: $x + 5 = 15$.

2.4 Multiplying Integers

- Multiplying two positive integers • Multiplying a positive and a negative integer
- Multiplying a negative and a positive integer • Multiplying by zero
- Multiplying two negative integers • Powers of integers

We now turn our attention to multiplication of integers. When we multiply two nonzero integers, the first factor can be positive or negative. The same is true for the second factor. This means that there are four possible combinations to consider.

Positive · positive	Positive · negative
Negative · positive	Negative · negative

In this section, we discuss these four combinations and use our observations to establish rules for multiplying two integers.

Multiplying two positive integers

4(3)
like signs
both positive

We begin by considering the product of two positive integers, 4(3). Since both factors are positive, we say that they have *like* signs. In Chapter 1, we learned that multiplication is repeated addition. Therefore, 4(3) represents the sum of four 3's.

$4(3) = 3 + 3 + 3 + 3$ Multiplication is repeated addition. Write 3 as an addend four times.

$4(3) = 12$ The result is 12, which is a positive number.

This result suggests that *the product of two positive integers is positive.*

Multiplying a positive and a negative integer

4(−3)
unlike signs
one positive, one negative

Next, we consider 4(−3). This is the product of a positive and a negative integer. The signs of these factors are *unlike.* According to the definition of multiplication, 4(−3) means that we are to add −3 four times.

$4(-3) = (-3) + (-3) + (-3) + (-3)$ Use the definition of multiplication. Write −3 as an addend four times.

$4(-3) = (-6) + (-3) + (-3)$ Work from left to right. Apply the rule for adding two negative numbers.

$4(-3) = (-9) + (-3)$ Work from left to right. Apply the rule for adding two negative numbers.

$4(-3) = -12$ Perform the addition.

This result is −12, which suggests that *the product of a positive integer and a negative integer is negative.*

Multiplying a negative and a positive integer

−3(4)
unlike signs
one negative, one positive

To develop a rule for multiplying a negative and a positive integer, we consider −3(4). Notice that the factors have *unlike* signs. Because of the commutative property of multiplication, the answer to −3(4) will be the same as the answer to 4(−3). We know that 4(−3) = −12 from the previous discussion, so −3(4) = −12. This suggests that *the product of a negative integer and a positive integer is negative.*

Putting the results of the last two cases together leads us to the rule for multiplying two integers with unlike signs.

Multiplying two integers with unlike signs

To multiply a positive integer and a negative integer, or a negative integer and a positive integer, multiply their absolute values. Then make the answer negative.

EXAMPLE 1 Find each product: **a.** $7(-5)$, **b.** $20(-8)$, and **c.** $-8 \cdot 5$.

Solution To multiply integers with unlike signs, we multiply their absolute values and make the product negative.

a. $7(-5) = -35$ Multiply the absolute values, 7 and 5, to get 35. Then make the answer negative.

b. $20(-8) = -160$ Multiply the absolute values, 20 and 8, to get 160. Then make the answer negative.

c. $-8 \cdot 5 = -40$ Multiply the absolute values, 8 and 5, to get 40. Then make the answer negative.

Self Check 1
Find each product:
a. $2(-6)$
b. $30(-2)$
c. $-15 \cdot 2$

Answers **a.** -12, **b.** -60, **c.** -30

! COMMENT When writing multiplication involving signed numbers, do not write a negative sign – next to a raised dot · (the multiplication symbol). Instead, use parentheses to show the multiplication.

$6(-2)$ ~~$6 \cdot -2$~~ and $-6(-2)$ ~~$-6 \cdot -2$~~

Multiplying by zero

Before we can develop a rule for multiplying two negative integers, we need to examine multiplication by 0. If 4(3) means that we are to find the sum of four 3's, then $0(-3)$ means that we are to find the sum of zero -3's. Obviously, the sum is 0. Thus, $0(-3) = 0$.

The commutative property of multiplication guarantees that we can change the order of the factors in the multiplication problem without affecting the result.

$$(-3)(0) = 0(-3) = 0$$

↑ ↑ Change the order of the factors. ↑ The result is still 0.

We see that the order in which we write the factors 0 and -3 doesn't matter — their product is 0. This example suggests that *the product of any number and 0 is 0.*

Multiplying by 0
If a is any number, then

$$a \cdot 0 = 0 \qquad 0 \cdot a = 0$$

EXAMPLE 2 Find $-12 \cdot 0$.

Solution Since the product of any number and 0 is 0, we have

$$-12 \cdot 0 = 0$$

Self Check 2
Find $0(-56)$.

Answer 0

Multiplying two negative integers

$-3(-4)$
like signs
both negative

To develop a rule for multiplying two negative integers, we consider the pattern displayed on the next page. There, we multiply -4 by a series of factors that decrease by 1. After determining each product, we graph each product on the number line (Figure 2-20). See if you can determine the answers to the last three multiplication problems by examining the pattern of answers leading up to them.

This factor decreases by 1 as you read down the column. Look for a pattern here.

$$\begin{aligned}
4(-4) &= -16\\
3(-4) &= -12\\
2(-4) &= -8\\
1(-4) &= -4\\
0(-4) &= 0\\
-1(-4) &= ?\\
-2(-4) &= ?\\
-3(-4) &= ?
\end{aligned}$$

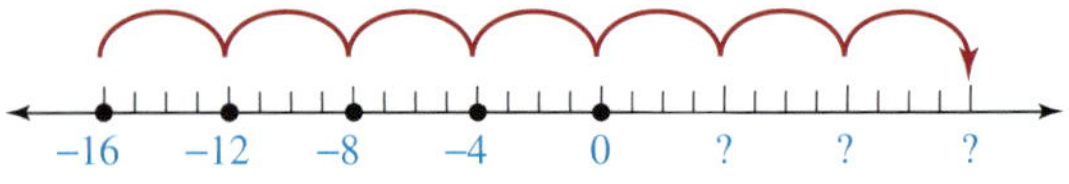

FIGURE 2-20

From the pattern, we see that

$$\begin{aligned}
-1(-4) &= 4\\
-2(-4) &= 8\\
-3(-4) &= 12
\end{aligned}$$

For two negative factors, the product is a positive.

These results suggest that *the product of two negative integers is positive.* Earlier in this section, we saw that the product of two positive integers is also positive. This leads to the following conclusion.

Multiplying two integers with like signs

To multiply two positive integers, or two negative integers, multiply their absolute values. The answer is positive.

Self Check 3

Find each product:

a. $-9(-7)$

b. $-12(-2)$

Answers **a.** 63, **b.** 24

EXAMPLE 3 Find each product: **a.** $-5(-9)$ and **b.** $-8(-10)$.

Solution To multiply two negative integers, we multiply their absolute values and make the result positive.

a. $-5(-9) = 45$ Multiply the absolute values, 5 and 9, to get 45. The answer is positive.

b. $-8(-10) = 80$ Multiply the absolute values, 8 and 10, to get 80. The answer is positive.

We now summarize the rules for multiplying two integers.

Multiplying two integers

To multiply two integers, multiply their absolute values.

1. The product of two integers with *like* signs is positive.
2. The product of two integers with *unlike* signs is negative.

We can use a calculator to multiply signed numbers.

Multiplication with negative numbers

CALCULATOR SNAPSHOT

At Thanksgiving time, a large supermarket chain offered customers a free turkey with every grocery purchase of $100 or more. Each turkey cost the store $8, and 10,976 people took advantage of the offer. Since each of the 10,976 turkeys given away represented a loss of $8 (which can be expressed as -8 dollars), the company lost $10{,}976(-8)$ dollars. To find this product, we enter these numbers and press these keys on a scientific calculator.

10976 [×] 8 [+/−] [=]   -87808

The negative result indicates that with this promotion, the supermarket chain gave away $87,808 in turkeys.

EXAMPLE 4 Multiply: **a.** $-3(-2)(-6)(-5)$ and **b.** $-2(-4)(-5)$.

Solution

a. $-3(-2)(-6)(-5) = 6(-6)(-5)$ Work from left to right: $-3(-2) = 6$.

$= -36(-5)$ Work from left to right: $6(-6) = -36$.

$= 180$

b. $-2(-4)(-5) = 8(-5)$ Work from left to right: $-2(-4) = 8$.

$= -40$

Self Check 4

Multiply: **a.** $-1(-2)(-5)$ and **b.** $-2(-7)(-1)(-2)$.

Answers **a.** -10, **b.** 28

Example 4, part a, illustrates that

A product is positive when there are an even number of negative factors.

Part b illustrates that

A product is negative when there are an odd number of negative factors.

Powers of integers

Recall that exponential expressions are used to represent repeated multiplication. For example, 2 to the third power, or 2^3, is a shorthand way of writing $2 \cdot 2 \cdot 2$. In this expression, 3 is the exponent and the base is positive 2. In the next example, we evaluate exponential expressions with bases that are negative numbers.

EXAMPLE 5 Find each power: **a.** $(-2)^4$ and **b.** $(-5)^3$.

Solution

a. $(-2)^4 = (-2)(-2)(-2)(-2)$ Write -2 as a factor 4 times.

$= 4(-2)(-2)$ Work from left to right. Multiply -2 and -2 to get 4.

$= -8(-2)$ Work from left to right. Multiply 4 and -2 to get -8.

$= 16$ Perform the multiplication.

b. $(-5)^3 = (-5)(-5)(-5)$ Write -5 as a factor 3 times.

$= 25(-5)$ Work from left to right. Multiply -5 and -5 to get 25.

$= -125$ Perform the multiplication.

Self Check 5

Find each power:

a. $(-3)^4$

b. $(-4)^3$

Answers **a.** 81, **b.** -64

In Example 5, part a, -2 was raised to an even power, and the answer was positive. In part b, another negative number, -5, was raised to an odd power, and the answer was negative. These results suggest a general rule.

Even and odd powers of a negative integer

When a negative integer is raised to an even power, the result is positive.

When a negative integer is raised to an odd power, the result is negative.

Self Check 6

Find the power: $(-1)^8$.

Answer 1

EXAMPLE 6 Find the power: $(-1)^5$.

Solution We have a negative integer raised to an odd power. The result will be negative.

$$(-1)^5 = (-1)(-1)(-1)(-1)(-1)$$
$$= -1$$

! COMMENT Although the expressions -3^2 and $(-3)^2$ look similar, they are not the same. In -3^2, the base is 3 and the exponent 2. The $-$ sign in front of 3^2 means the opposite of 3^2. In $(-3)^2$, the base is -3 and the exponent is 2. When we evaluate them, it becomes clear that they are not equivalent.

-3^2 represents *the opposite of* 3^2.

$-3^2 = -(3 \cdot 3)$	Write 3 as a factor 2 times.
$= -9$	Multiply within the parentheses first.

$(-3)^2$ represents $(-3)(-3)$.

$(-3)^2 = (-3)(-3)$	Write -3 as a factor 2 times.
$= 9$	The product of two negative numbers is positive.

Notice that the results are different.

Self Check 7

Evaluate: **a.** -4^2 and **b.** $(-4)^2$.

Answers **a.** -16, **b.** 16

EXAMPLE 7 Evaluate: a. -2^2 and b. $(-2)^2$.

Solution

a. $-2^2 = -(2 \cdot 2)$	Since 2 is the base, write 2 as a factor two times.
$= -4$	Perform the multiplication within the parentheses.
b. $(-2)^2 = (-2)(-2)$	The base is -2. Write it as a factor twice.
$= 4$	The signs are like, so the product is positive.

CALCULATOR SNAPSHOT **Raising a negative number to a power**

Negative numbers can be raised to a power using a scientific calculator. We use the change-of-sign key [+/−] and the power key [y^x] (on the some calculators, [x^y]). For example, to evaluate $(-5)^6$, we enter these numbers and press these keys.

5 [+/−] [y^x] 6 [=]   [15625]

The result is 15,625.

Some scientific calculators require parentheses when entering a negative base raised to a power.

Section 2.4 STUDY SET

VOCABULARY *Fill in the blanks.*

1. In the multiplication $-5(-4)$, the integers -5 and -4, which are being multiplied, are called ______. The answer, 20, is called the ______.
2. The definition of multiplication tells us that $3(-4)$ represents repeated ______ $-4 + (-4) + (-4)$.
3. In the expression -3^5, ___ is the base and 5 is the ______.
4. In the expression $(-3)^5$, ___ is the base and ___ is the exponent.

CONCEPTS *Fill in the blanks.*

5. The product of two integers with ______ signs is negative.
6. The product of two integers with like signs is ______.
7. The ______ property of multiplication implies that $-2(-3) = -3(-2)$.
8. The product of 0 and any number is ___.
9. Find $-1(9)$. In general, what is the result when we multiply a positive number by -1?
10. Find $-1(-9)$. In general, what is the result when we multiply a negative number by -1?
11. When we multiply two integers, there are four possible combinations of signs. List each of them.
12. When we multiply two integers, there are four possible combinations of signs. How can they be grouped into two categories?
13. If each of the following powers were evaluated, what would be the *sign* of the result?
    a. $(-5)^{13}$ b. $(-3)^{20}$
14. A student claimed, "A positive and a negative is negative." What is wrong with this statement?
15. Find each absolute value.
    a. $|-3|$ b. $|12|$
    c. $|-5|$ d. $|9|$
    e. $|10|$ f. $|-25|$
16. Find each product and then graph it on a number line. What is the distance between every two products?

    $2(-2)$, $1(-2)$, $0(-2)$, $-1(-2)$, $-2(-2)$

17. a. Complete the table.

Problem	Number of negative factors	Answer
$-2(-2)$		
$-2(-2)(-2)(-2)$		
$-2(-2)(-2)(-2)(-2)(-2)$		

    b. The answers entered in the table help to justify the following rule: The product of an ______ number of negative integers is positive.

18. a. Complete the table.

Problem	Number of negative factors	Answer
$-2(-2)(-2)$		
$-2(-2)(-2)(-2)(-2)$		
$-2(-2)(-2)(-2)(-2)(-2)(-2)$		

    b. The answers entered in the table help to justify the following rule: The product of an ______ number of negative integers is negative.

NOTATION *Complete each solution to evaluate the expression.*

19. $-3(-2)(-4) = \square(-4)$
    $= \square$
20. $(-3)^4 = (-3)(-3)(-3)\square$
    $= \square(-3)(-3)$
    $= \square(-3)$
    $= \square$
21. Explain why the expression below is not written correctly. How should it be written?

    $-6 \cdot -5$

22. Translate to mathematical symbols.
    a. the product of negative three and negative two
    b. negative five, squared
    c. the opposite of five squared

PRACTICE *Find each product.*

23. $-9(-6)$
24. $-5(-5)$
25. $-3 \cdot 5$
26. $-6 \cdot 4$

27. $12(-3)$
28. $11(-4)$
29. $(-8)(-7)$
30. $(-9)(-3)$
31. $(-2)10$
32. $(-3)8$
33. $-40 \cdot 3$
34. $-50 \cdot 2$
35. $-8(0)$
36. $0(-27)$
37. $-1(-6)$
38. $-1(-8)$
39. $-7(-1)$
40. $-5(-1)$
41. $1(-23)$
42. $-35(1)$

Evaluate each expression.

43. $-6(-4)(-2)$
44. $-3(-2)(-3)$
45. $5(-2)(-4)$
46. $3(-3)(3)$
47. $2(3)(-5)$
48. $6(2)(-2)$
49. $6(-5)(2)$
50. $4(-2)(2)$
51. $(-1)(-1)(-1)$
52. $(-1)(-1)(-1)(-1)$
53. $-2(-3)(3)(-1)$
54. $5(-2)(3)(-1)$
55. $3(-4)(0)$
56. $-7(-9)(0)$
57. $-2(0)(-10)$
58. $-6(0)(-12)$
59. Find the product of -6 and the opposite of 10.
60. Find the product of the opposite of 9 and the opposite of 8.

Find each power.

61. $(-4)^2$
62. $(-6)^2$
63. $(-5)^3$
64. $(-6)^3$
65. $(-2)^3$
66. $(-4)^3$
67. $(-9)^2$
68. $(-10)^2$
69. $(-1)^5$
70. $(-1)^6$
71. $(-1)^8$
72. $(-1)^9$

Evaluate each expression.

73. $(-7)^2$ and -7^2
74. $(-5)^2$ and -5^2
75. -12^2 and $(-12)^2$
76. -11^2 and $(-11)^2$

Use a calculator to evaluate each expression.

77. $-76(787)$
78. $407(-32)$
79. $(-81)^4$
80. $(-6)^5$
81. $(-32)(-12)(-67)$
82. $(-56)(-9)(-23)$
83. $(-25)^4$
84. $(-41)^5$

APPLICATIONS *Use signed numbers to help solve each problem.*

85. DIETING After giving a patient a physical exam, a physician felt that the patient should begin a diet. Two options were discussed.

	Plan #1	Plan #2
Length	10 weeks	14 weeks
Daily exercise	1 hr	30 min
Weight loss per week	3 lb	2 lb

a. Find the expected weight loss from each diet plan. Express each answer as a signed number.

b. With which plan should the patient expect to lose more weight? Explain why the patient might not choose it.

86. INVENTORIES A spreadsheet is used to record inventory losses at a warehouse. The items, their cost, and the number missing are listed in the table.

	A	B	C	D
1	Item	Cost	Number of units	$ losses
2	CD	$5	−11	
3	TV	$200	−2	
4	Radio	$20	−4	

a. What instruction should be given to find the total losses for each type of item? Find each of those losses and fill in column D.

b. What instruction should be given to find the *total* inventory losses for the warehouse? Find this number.

87. MAGNIFICATION Using an electronic testing device, a mechanic can check the emissions of a car. The results of the test are displayed on a screen.

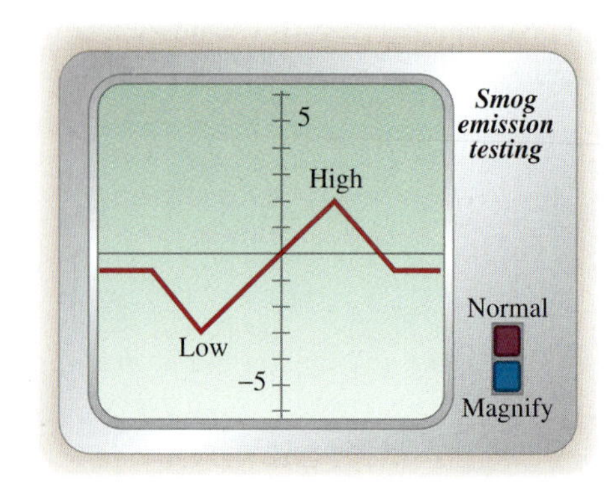

a. Find the high and low values for this test as shown on the screen.

b. By switching a setting on the monitor, the picture on the screen can be magnified. What would be the new high and new low if every value were doubled?

88. LIGHT Sunlight is a mixture of all colors. When sunlight passes through water, the water absorbs different colors at different rates, as shown.

a. Use a signed number to represent the depth to which red light penetrates water.

b. Green light penetrates 4 times deeper than red light. How deep is this?

c. Blue light penetrates 3 times deeper than orange light. How deep is this?

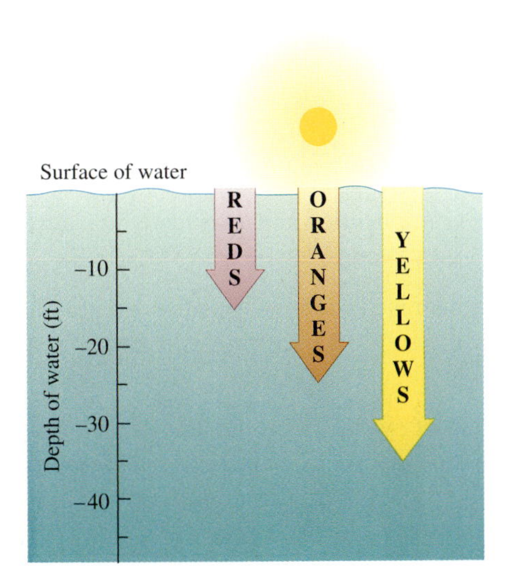

89. TEMPERATURE CHANGE A farmer, worried about his fruit trees suffering frost damage, calls the weather service for temperature information. He is told that temperatures will be decreasing approximately 4° every hour for the next five hours. What signed number represents the total change in temperature expected over the next five hours?

90. DEPRECIATION For each of the last four years, a businesswoman has filed a \$200 depreciation allowance on her income tax return, for an office computer system. What signed number represents the total amount of depreciation written off over the four-year period?

91. EROSION A levee protects a town in a low-lying area from flooding. According to geologists, the banks of the levee are eroding at a rate of 2 feet per year. If something isn't done to correct the problem, what signed number indicates how much of the levee will erode during the next decade?

92. DECK SUPPORTS After a winter storm, a homeowner has an engineering firm inspect his damaged deck. Their report concludes that the original pilings were not anchored deep enough, by a factor of 3. What signed number represents the depth to which the pilings should have been sunk?

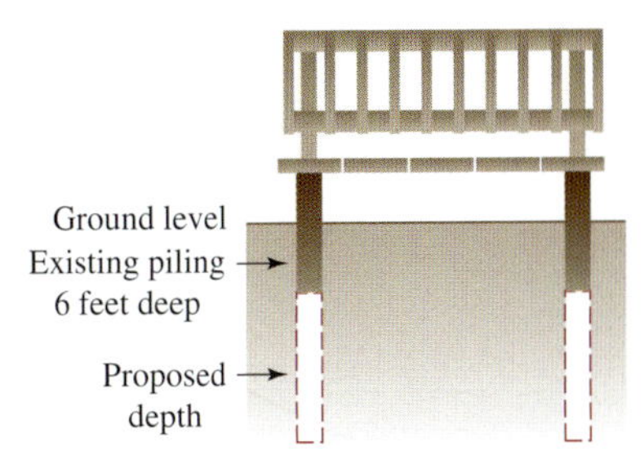

93. WOMEN'S NATIONAL BASKETBALL ASSOCIATION The average attendance for the WNBA Houston Comets is 8,110 a game. Suppose the team gives a sports bag, costing \$3, to everyone attending a game. What signed number expresses the financial loss from this promotional giveaway?

94. HEALTH CARE A health care provider for a company estimates that 75 hours per week are lost by employees suffering from stress-related or preventable illness. In a 52-week year, how many hours are lost? Use a signed number to answer.

WRITING

95. If a product contains an even number of negative factors, how do we know that the result will be positive?

96. Explain why the product of a positive number and a negative number is negative, using $5(-3)$ as an example.

97. Explain why the result is the opposite of the original number when a number is multiplied by -1.

98. Can you think of any number that yields a nonzero result when it is multiplied by 0? Explain your response.

REVIEW

99. The prime factorization of a number is $3^2 \cdot 5$. What is the number?

100. Solve: $\frac{y}{8} = 10$.

101. The enrollment at a college went from 10,200 to 12,300 in one year. What was the increase in enrollment?

102. Find the perimeter of a square with sides 6 yards long.

103. What does the symbol $<$ mean?

104. List the first ten prime numbers.

2.5 Dividing Integers

- The relationship between multiplication and division
- Rules for dividing integers
- Division and zero

In this section, we develop rules for division of integers, just as we did for multiplication of integers. We also consider two types of division involving 0.

The relationship between multiplication and division

Every division fact containing three numbers has a related multiplication fact involving the same three numbers. For example,

$$\frac{6}{3} = 2 \quad \text{because} \quad 3(2) = 6$$

Remember that in the division statement, 6 is the *dividend,* 3 is the *divisor,* and 2 is the *quotient.*

$$\frac{20}{5} = 4 \quad \text{because} \quad 5(4) = 20$$

Rules for dividing integers

We now use the relationship between multiplication and division to help develop rules for dividing integers. There are four cases to consider.

Case 1: In the first case, a positive integer is divided by a positive integer. From years of experience, we already know that the result is positive. Therefore, *the quotient of two positive integers is positive.*

Case 2: Next, we consider the quotient of two negative integers. As an example, consider the division $\frac{-12}{-2} = ?$. We can do this division by examining its related multiplication statement, $-2(?) = -12$. Our objective is to find the number that should replace the question mark. To do this, we use the rules for multiplying integers, introduced in the previous section.

Multiplication statement	**Division statement**
$-2(?) = -12$	$\frac{-12}{-2} = ?$
This must be *positive* 6 if the product is to be *negative* 12.	So the quotient is *positive* 6.

Therefore, $\frac{-12}{-2} = 6$. From this example, we can see that *the quotient of two negative integers is positive.*

Case 3: The third case we examine is the quotient of a positive integer and a negative integer. Let's consider $\frac{12}{-2} = ?$. Its equivalent multiplication statement is $-2(?) = 12$.

Multiplication statement	**Division statement**
$-2(?) = 12$	$\frac{12}{-2} = ?$
This must be -6 if the product is to be *positive* 12.	So the quotient is -6.

Therefore, $\frac{12}{-2} = -6$. This result shows that *the quotient of a positive integer and a negative integer is negative.*

Case 4: Finally, to find the quotient of a negative integer and a positive integer, let's consider $\frac{-12}{2} = ?$. Its equivalent multiplication statement is $2(?) = -12$.

Multiplication statement	**Division statement**
$2(?) = -12$	$\frac{-12}{2} = ?$
This must be -6 if the product is to be -12.	So the quotient is -6.

Therefore, $\frac{-12}{2} = -6$. From this example, we can see that *the quotient of a negative integer and a positive integer is negative.*

We now summarize the results from the previous discussion.

Dividing two integers

To divide two integers, divide their absolute values.

1. The quotient of two integers with *like* signs is positive.
2. The quotient of two integers with *unlike* signs is negative.

The rules for dividing integers are similar to those for multiplying integers.

EXAMPLE 1 Find each quotient: **a.** $\frac{-35}{7}$ and **b.** $\frac{20}{-5}$.

Solution To divide integers with unlike signs, we find the quotient of their absolute values and make the quotient negative.

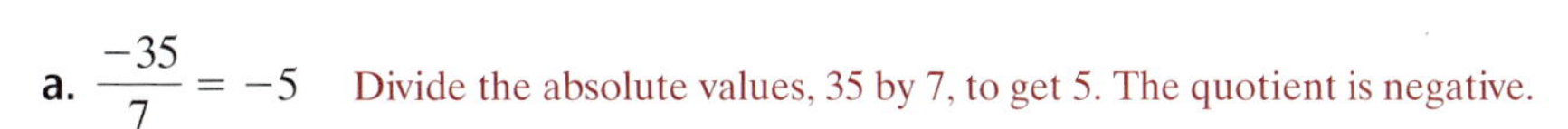

a. $\frac{-35}{7} = -5$ Divide the absolute values, 35 by 7, to get 5. The quotient is negative.

To check the result, we multiply the divisor, 7, and the quotient, -5. We should obtain the dividend, -35.

$7(-5) = -35$

The answer, -5, checks.

b. $\frac{20}{-5} = -4$ Divide the absolute values, 20 by 5, to get 4. The quotient is negative.

Self Check 1

Find each quotient and check the result:

a. $\frac{-45}{5}$

b. $\frac{60}{-20}$

Answers **a.** -9, **b.** -3

EXAMPLE 2 Divide: $\frac{-12}{-3}$.

Solution The integers have like signs. The quotient will be positive.

$\frac{-12}{-3} = 4$ Divide the absolute values, 12 by 3, to get 4. The quotient is positive.

Self Check 2

Divide: $\frac{-21}{-3}$.

Answer 7

EXAMPLE 3 **Price reductions.** Over the course of a year, a retailer reduced the price of a television set by an equal amount each month, because it was not selling. By the end of the year, the cost was \$132 less than at the beginning of the year. How much did the price fall each month?

Solution We label the drop in price of \$132 for the year as -132. It occurred in 12 equal reductions. This indicates division.

$$\frac{-132}{12} = -11$$ The quotient of a negative number and a positive number is negative.

The drop in price each month was \$11.

Division and zero

To review the concept of division of 0, we look at $\frac{0}{-2} = ?$. The related multiplication statement is $-2(?) = 0$.

Multiplication statement	Division statement
$-2(?) = 0$	$\frac{0}{-2} = ?$

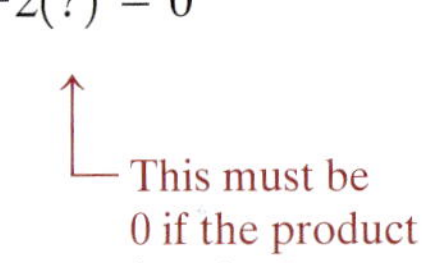

So the quotient is 0.

Therefore, $\frac{0}{-2} = 0$. This example suggests that *the quotient of 0 divided by any nonzero number is 0.*

To review division by 0, let's look at $\frac{-2}{0} = ?$. The related multiplication statement is $0(?) = -2$.

Multiplication statement	Division statement
$0(?) = -2$	$\frac{-2}{0} = ?$

There is no number that gives -2 when multiplied by 0.

There is no quotient.

Therefore, $\frac{-2}{0}$ does not have an answer. We say that division by 0 is undefined. This example suggests that *the quotient of any number divided by 0 is undefined.*

Division with 0

1. If a represents any nonzero number, $\frac{0}{a} = 0$.
2. If a represents any nonzero number, $\frac{a}{0}$ is undefined.
3. $\frac{0}{0}$ is undetermined.

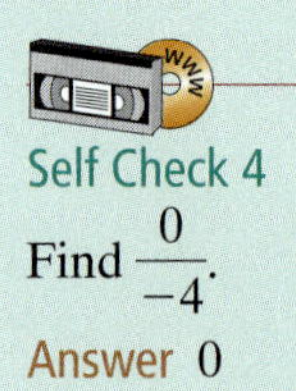

Self Check 4
Find $\frac{0}{-4}$.
Answer 0

EXAMPLE 4 Find $\frac{-4}{0}$, if possible.

Solution Since $\frac{-4}{0}$ is division by 0, the division is undefined.

Division with negative numbers — CALCULATOR SNAPSHOT

The Bureau of Labor Statistics estimated that the United States lost 146,400 jobs in the manufacturing sector of the economy in 2003. Because the jobs were lost, we write this as $-146,400$. To find the average number of manufacturing jobs lost each month, we divide: $\frac{-146,400}{12}$. To perform this division using a scientific calculator, we enter these numbers and press these keys.

146400 [+/−] [÷] 12 [=]   -12200

The average number of manufacturing jobs lost each month in 2003 was 12,200.

Section 2.5 STUDY SET

VOCABULARY *Fill in the blanks.*

1. In $\frac{-27}{3} = -9$, the number -9 is called the ________, and the number 3 is the ________.
2. Division by 0 is ________. Division ____ 0 by a nonzero number is 0.
3. The ________ ________ of a number is the distance between it and 0 on the number line.
4. $\{\ldots, -4, -3, -2, -1, 0, 1, 2, 3, 4, \ldots\}$ is the set of ________.
5. The quotient of two negative integers is ________.
6. The quotient of a negative integer and a positive integer is ________.

CONCEPTS

7. Write the related multiplication statement for $\frac{-25}{5} = -5$.
8. Write the related multiplication statement for $\frac{0}{-15} = 0$.
9. Show that there is no answer for $\frac{-6}{0}$ by writing the related multiplication statement.
10. Find the value of $\frac{0}{5}$.
11. Write a related division statement for $5(-4) = -20$.
12. How do the rules for multiplying integers compare with the rules for dividing integers?
13. Determine whether each statement is always true, sometimes true, or never true.
    a. The product of a positive integer and a negative integer is negative.
    b. The sum of a positive integer and a negative integer is negative.
    c. The quotient of a positive integer and a negative integer is negative.
14. Determine whether each statement is always true, sometimes true, or never true.
    a. The product of two negative integers is positive.
    b. The sum of two negative integers is negative.
    c. The quotient of two negative integers is negative.

PRACTICE *Find each quotient, if possible.*

15. $\frac{-14}{2}$
16. $\frac{-10}{5}$
17. $\frac{-8}{-4}$
18. $\frac{-12}{-3}$
19. $\frac{-25}{-5}$
20. $\frac{-36}{-12}$
21. $\frac{-45}{-15}$
22. $\frac{-81}{-9}$
23. $\frac{40}{-2}$
24. $\frac{35}{-7}$
25. $\frac{50}{-25}$
26. $\frac{80}{-40}$

27. $\frac{0}{-16}$

28. $\frac{0}{-6}$

29. $\frac{-6}{0}$

30. $\frac{-8}{0}$

31. $\frac{-5}{1}$

32. $\frac{-9}{1}$

33. $-5 \div (-5)$

34. $-11 \div (-11)$

35. $\frac{-9}{9}$

36. $\frac{-15}{15}$

37. $\frac{-10}{-1}$

38. $\frac{-12}{-1}$

39. $\frac{-100}{25}$

40. $\frac{-100}{50}$

41. $\frac{75}{-25}$

42. $\frac{300}{-100}$

43. $\frac{-500}{-100}$

44. $\frac{-60}{-30}$

45. $\frac{-200}{50}$

46. $\frac{-500}{100}$

47. Find the quotient of -45 and 9.

48. Find the quotient of -36 and -4.

49. Divide 8 by -2.

50. Divide -16 by -8.

 *Use a calculator to perform each division.*

51. $\frac{-13{,}550}{25}$

52. $\frac{-3{,}876}{-19}$

53. $\frac{272}{-17}$

54. $\frac{-6{,}776}{-77}$

APPLICATIONS *Use signed numbers to help solve each problem.*

55. TEMPERATURE DROP During a five-hour period, the temperature steadily dropped as shown. What was the average change in the temperature per hour over this five-hour time span?

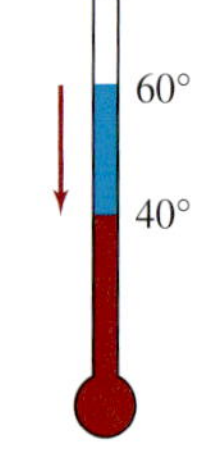

56. PRICE DROPS Over a three-month period, the price of a DVR steadily fell as shown. What was the average monthly change in the price of the DVR over this period?

57. SUBMARINE DIVES In a series of three equal dives, a submarine is programmed to reach a depth of 3,030 feet below the ocean surface. What signed number describes how deep each of the three dives will be?

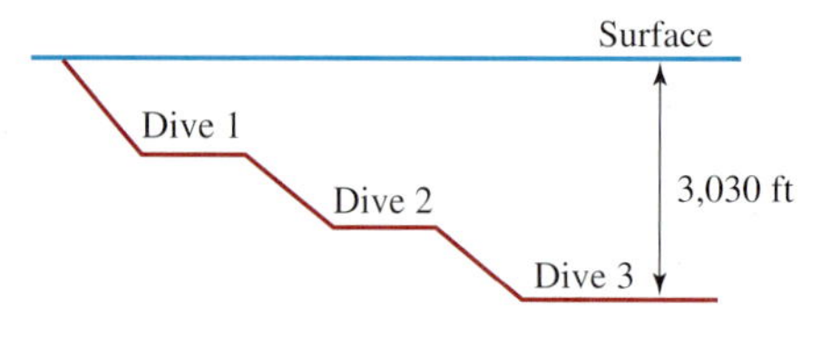

58. GRAND CANYON A mule train is to travel from a stable on the rim of the Grand Canyon to a camp on the canyon floor, approximately 5,000 feet below the rim. If the guide wants the mules to be rested after every 1,000 feet of descent, how many stops will be made on the trip?

59. BASEBALL TRADES At the midway point of the season, a baseball team finds itself 12 games behind the league leader. Team management decides to trade for a talented hitter, in hopes of making up at least half of the deficit in the standings by the end of the year. Where in the league standings does management expect to finish at season's end?

60. BUDGET DEFICITS A politician proposed a two-year plan for cutting a county's $20-million budget deficit, as shown. If this plan is put into effect, what will be the change in the county's financial status in two years?

	Plan	Prediction
1st year	Raise taxes, drop subsidy programs	Will cut deficit in half
2nd year	Search out waste and fraud	Will cut remaining deficit in half

61. MARKDOWNS The owner of a clothing store decides to reduce the price on a line of jeans that are not selling. She feels she can afford to lose $300 of projected income on these pants. By how much can she mark down each of the 20 pairs of jeans?

62. WATER RESERVOIRS Over a week's time, engineers at a city water reservoir released enough water to lower the water level 35 feet. On average, how much did the water level change each day during this period?

63. PAY CUTS In a cost-cutting effort, a business decides to lower expenditures on salaries by $9,135,000. To do this, all of the 5,250 employees will have their salaries reduced by an equal dollar amount. How big a pay cut will each employee experience?

64. STOCK MARKET On Monday, the value of Maria's 255 shares of stock was at an all-time high. By Friday, the value had fallen \$4,335. What was her per-share loss that week?

WRITING

65. Explain why the quotient of two negative numbers is positive.

66. Think of a real-life situation that could be represented by $\frac{0}{4}$. Explain why the answer would be 0.

67. Using a specific example, explain how multiplication can be used as a check for division.

68. Explain what it means when we say that division by 0 is undefined.

REVIEW

69. Evaluate: $3\left(\frac{18}{3}\right)^2 - 2(2)$.

70. List the set of whole numbers.

71. Find the prime factorization of 210.

72. The statement $(4 + 8) + 10 = 4 + (8 + 10)$ illustrates what property?

73. Is the following true? $17 \geq 17$

74. Does $8 - 2 = 2 - 8$?

75. Solve: $99 = r - 43$.

76. Sharif has scores of 55, 70, 80, and 75 on four mathematics tests. What is his mean (average) score?

2.6 Order of Operations and Estimation

- Order of operations • Absolute value • Estimation

In this section, we evaluate expressions involving more than one operation. To do this, we apply the rules for the order of operations and the rules for working with integers. We also continue the discussion of estimating an answer. Estimation can be used when you need a quick indication of the size of the actual answer to a calculation.

Order of operations

In Section 1.5, we introduced the following rules for the order of operations: an agreed-upon sequence of steps for completing the operations of arithmetic.

Order of operations

1. Perform all calculations within parentheses and other grouping symbols in the following order listed in steps 2–4, working from innermost pair to outermost pair.
2. Evaluate all exponential expressions.
3. Perform all multiplications and divisions as they occur from left to right.
4. Perform all additions and subtractions as they occur from left to right.

When all grouping symbols have been removed, repeat steps 2–4 to complete the calculation.

If a fraction bar is present, evaluate the expression above the bar (the *numerator*) and the expression below the bar (the *denominator*) separately. Then perform the division indicated by the fraction bar, if possible.

EXAMPLE 1 Evaluate: $-4(-3)^2 - (-2)$.

Solution This expression contains the operations of multiplication, raising to a power, and subtraction. Since there are no calculations to be made *within* parentheses, we begin by evaluating the exponential expression.

Self Check 1
Evaluate: $-5(-2)^2 - (-6)$.

$-4(-3)^2 - (-2) = -4(9) - (-2)$	Evaluate the exponential expression: $(-3)^2 = 9$.
$= -36 - (-2)$	Perform the multiplication: $-4(9) = -36$.
$= -36 + 2$	To subtract, add the opposite of -2.
$= -34$	Perform the addition.

Answer -14

Self Check 2
Evaluate: $4(2) + (-4)(-3)(-2)$.

Answer -16

EXAMPLE 2 Evaluate: $2(3) + (-5)(-3)(-2)$.

Solution This expression contains the operations of multiplication and addition. By the rules for the order of operations, we perform the multiplications first.

$2(3) + (-5)(-3)(-2) = 6 + (-30)$	Perform the multiplications from left to right.
$= -24$	Perform the addition.

Self Check 3
Evaluate: $45 \div (-5)3$.

Answer -27

EXAMPLE 3 Evaluate: $40 \div (-4)5$.

Solution This expression contains the operations of division and multiplication. We perform the divisions and multiplications as they occur from left to right.

$40 \div (-4)5 = -10 \cdot 5$	Perform the division first: $40 \div (-4) = -10$.
$= -50$	Perform the multiplication.

Self Check 4
Evaluate: $-3^2 - (-3)^2$.

Answer -18

EXAMPLE 4 Evaluate: $-2^2 - (-2)^2$.

Solution This expression contains the operations of raising to a power and subtraction. We are to find the powers first. (Recall that -2^2 means the *opposite* of 2^2.)

$-2^2 - (-2)^2 = -4 - 4$	Find the powers: $-2^2 = -4$ and $(-2)^2 = 4$.
$= -8$	Perform the subtraction.

Self Check 5
Evaluate: $-18 + 4(-7 + 9)$.

Answer -10

EXAMPLE 5 Evaluate: $-15 + 3(-4 + 7)$.

Solution

$-15 + 3(-4 + 7) = -15 + 3(3)$	Perform the addition within the parentheses.
$= -15 + 9$	Perform the multiplication: $3(3) = 9$.
$= -6$	Perform the addition.

Self Check 6
Evaluate: $-2 + (9 - 3)^2$.

Answer 34

EXAMPLE 6 Evaluate: $-10 + (8 - 4)^2$.

Solution By the rules for the order of operations, we must perform the operation within the parentheses first.

$-10 + (8 - 4)^2 = -10 + (4)^2$	Perform the subtraction within the parentheses: $8 - 4 = 4$.
$= -10 + 16$	Evaluate the exponential expression: $(4)^2 = 16$.
$= 6$	Perform the addition.

EXAMPLE 7 Evaluate: $\frac{-20 + 3(-5)}{(-4)^2 - 21}$.

Solution We first evaluate the expressions in the numerator and the denominator, separately.

$\frac{-20 + 3(-5)}{(-4)^2 - 21} = \frac{-20 + (-15)}{16 - 21}$ In the numerator, perform the multiplication: $3(-5) = -15$. In the denominator, evaluate the power: $(-4)^2 = 16$.

$= \frac{-35}{-5}$ In the numerator, add: $-20 + (-15) = -35$. In the denominator, subtract: $16 - 21 = -5$.

$= 7$ Perform the division.

Self Check 7

Evaluate: $\frac{-9 + 6(-4)}{(-5)^2 - 28}$.

Answer 11

EXAMPLE 8 Evaluate: $-5[-1 + (2 - 8)^2]$.

Solution We begin by working within the innermost pair of grouping symbols and work to the outermost pair.

$-5[-1 + (2 - 8)^2] = -5[-1 + (-6)^2]$ Perform the subtraction within the parentheses.

$= -5(-1 + 36)$ Evaluate the power within the brackets.

$= -5(35)$ Perform the addition within the parentheses.

$= -175$ Perform the multiplication.

Self Check 8

Evaluate: $-4[-2 + (5 - 9)^2]$.

Answer -56

Absolute value

You will recall that the absolute value of a number is the distance between the number and 0 on the number line. Earlier in this chapter, we evaluated simple absolute value expressions such as $|-3|$ and $|10|$. Absolute value symbols are also used in combination with more complicated expressions, such as $|-4(3)|$ and $|-6 + 1|$. When we apply the rules for the order of operations to evaluate these expressions, *the absolute value symbols are considered to be grouping symbols,* and any operations within them are to be completed first.

EXAMPLE 9 Find each absolute value: **a.** $|-4(3)|$ and **b.** $|-6 + 1|$.

Solution We perform the operations within the absolute value symbols first.

a. $|-4(3)| = |-12|$ Perform the multiplication within the absolute value symbol: $-4(3) = -12$.

$= 12$ Find the absolute value of -12.

b. $|-6 + 1| = |-5|$ Perform the addition within the absolute value symbol: $-6 + 1 = -5$.

$= 5$ Find the absolute value of -5.

Self Check 9

Find each absolute value:

a. $|(-6)(5)|$

b. $|-3 + (-26)|$

Answers **a.** 30, **b.** 29

! COMMENT Just as $-5(8)$ means $-5 \cdot 8$, the expression $-5|8|$ (read as "negative 5 times the absolute value of 8") means $-5 \cdot |8|$. To evaluate such an expression, we find the absolute value first and then multiply.

$$-5|8| = -5 \cdot 8 \quad \text{Find the absolute value: } |8| = 8.$$
$$= -40 \quad \text{Perform the multiplication.}$$

Self Check 10
Evaluate: $7 - 5|-1 - 6|$.

Answer -28

EXAMPLE 10 Evaluate: $8 - 4|-6 - 2|$.

Solution We perform the operation within the absolute value symbol first.

$$8 - 4|\mathbf{-6 - 2}| = 8 - 4|\mathbf{-8}| \quad \text{Perform the subtraction within the absolute value symbol: } -6 - 2 = -8.$$
$$= 8 - 4(8) \quad \text{Find the absolute value: } |-8| = 8.$$
$$= 8 - 32 \quad \text{Perform the multiplication: } 4(8) = 32.$$
$$= -24 \quad \text{Perform the subtraction.}$$

Estimation

Recall that the idea behind estimation is to simplify calculations by using rounded numbers that are close to the actual values in the problem. When an exact answer is not necessary and a quick approximation will do, we can use estimation.

EXAMPLE 11 **The stock market.** The Dow Jones Industrial Average is announced at the end of each trading day to give investors an indication of how the New York Stock Exchange performed. A positive number indicates good performance, while a negative number indicates poor performance. Estimate the net gain or loss of points in the Dow for the week shown in Figure 2-21.

FIGURE 2-21

Solution We will approximate each of these numbers. For example, -139 is close to -140, and $+124$ is close to $+120$. To estimate the net gain or loss, we add the approximations.

$$-140 + (-360) + 120 + 130 + 270 = -500 + 520 \quad \text{Add positive and negative numbers separately to get subtotals.}$$
$$= 20 \quad \text{Perform the addition.}$$

This estimate tells us that there was a gain of approximately 20 points in the Dow.

Section 2.6 STUDY SET

VOCABULARY *Fill in the blanks.*

1. When asked to evaluate expressions that contain more than one operation, we should apply the rules for the ______ of operations.
2. In situations where an exact answer is not needed, an approximation or ________ is a quick way of obtaining a rough idea of the size of the actual answer.
3. Absolute value symbols, parentheses, and brackets are types of ________ symbols.
4. If an expression involves two sets of grouping symbols, always begin working within the ________ symbols and then work to the ________.

CONCEPTS

5. Consider $5(-2)^2 - 1$. How many operations need to be performed to evaluate this expression? List them in the order in which they should be performed.
6. Consider $15 - 3 + (-5 \cdot 2)^3$. How many operations need to be performed to evaluate this expression? List them in the order in which they should be performed.
7. Consider $\dfrac{5 + 5(7)}{2 + (4 - 8)}$. In the numerator, what operation should be performed first? In the denominator, what operation should be performed first?
8. In the expression $4 + 2(-7 - 1)$, how many operations need to be performed? List them in the order in which they should be performed.
9. Explain the difference between -3^2 and $(-3)^2$.
10. In the expression $-2 \cdot 3^2$, what operation should be performed first?

NOTATION *Complete each solution to evaluate the expression.*

11. $-8 - 5(-2)^2 = -8 - 5(\square)$
$= -8 - \square$
$= -8 + (\square)$
$= \square$

12. $2 + (5 - 6 \cdot 2) = 2 + (5 - \square)$
$= 2 + [5 + (\square)]$
$= 2 + (\square)$
$= \square$

13. $[-4(2 + 7)] - 6 = [-4(\square)] - 6$
$= \square - 6$
$= \square$

14. $\dfrac{|-9 + (-3)|}{9 - 6} = \dfrac{|\square|}{3}$
$= \dfrac{\square}{3}$
$= \square$

PRACTICE *Evaluate each expression.*

15. $(-3)^2 - 4^2$
16. $-7 + 4 \cdot 5$
17. $3^2 - 4(-2)(-1)$
18. $2^3 - 3^3$
19. $(2 - 5)(5 + 2)$
20. $-3(2)^2 4$
21. $-10 - 2^2$
22. $-50 - 3^3$
23. $\dfrac{-6 - 8}{2}$
24. $\dfrac{-6 - 6}{-2 - 2}$
25. $\dfrac{-5 - 5}{2}$
26. $\dfrac{-7 - (-3)}{2 - 4}$
27. $-12 \div (-2)2$
28. $-60(-2) \div 3$
29. $-16 - 4 \div (-2)$
30. $-24 + 4 \div (-2)$
31. $|-5(-6)|$
32. $|-7 - 9|$
33. $|-4 - (-6)|$
34. $|-2 + 6 - 5|$
35. $5|3|$
36. $5|4|$
37. $-6|-7|$
38. $-6|-4|$
39. $(7 - 5)^2 - (1 - 4)^2$
40. $5^2 - (-9 - 3)$
41. $-1(2^2 - 2 + 1^2)$
42. $(-7 - 4)^2 - (-1)$
43. $-50 - 2(-3)^3$
44. $(-2)^3 - (-3)(-2)$
45. $-6^2 + 6^2$
46. $-9^2 + 9^2$
47. $3\left(\dfrac{-18}{3}\right) - 2(-2)$
48. $2\left(\dfrac{-12}{3}\right) + 3(-5)$
49. $6 + \dfrac{25}{-5} + 6 \cdot 3$
50. $-5 - \dfrac{24}{6} + 8(-2)$

51. $\frac{1 - 3^2}{-2}$

52. $\frac{-3 - (-7)}{2^2 - 3}$

53. $\frac{-4(-5) - 2}{-6}$

54. $\frac{(-6)^2 - 1}{-4 - 3}$

55. $-3\left(\frac{32}{-4}\right) - (-1)^5$

56. $-5\left(\frac{16}{-4}\right) - (-1)^4$

57. $6(2^3)(-1)$

58. $2(3^3)(-2)$

59. $2 + 3[5 - (1 - 10)]$

60. $12 - 2[1 - (-8 + 2)]$

61. $-7(2 - 3 \cdot 5)$

62. $-4(1 + 3 \cdot 5)$

63. $-[6 - (1 - 4)^2]$

64. $-[9 - (9 - 12)^2]$

65. $15 + (-3 \cdot 4 - 8)$

66. $11 + (-2 \cdot 2 + 3)$

67. $|-3 \cdot 4 + (-5)|$

68. $|-8 \cdot 5 - 2 \cdot 5|$

69. $|(-5)^2 - 2 \cdot 7|$

70. $|8 \div (-2) - 5|$

71. $-2 + |6 - 4^2|$

72. $-3 - 4|6 - 7|$

73. $2|1 - 8| \cdot |-8|$

74. $2(5) - 6(|-3|)^2$

75. $-2(-34)^2 - (-605)$

75. $11 - (-15)(24)^2$

77. $-60 - \frac{1{,}620}{-36}$

78. $\frac{2^5 - 4^6}{-42 + 58}$

79. $-30 + (7 - 2)^2$

80. $-11 + (4 - 1)^2$

81. $(3 - 4)^2 - (3 - 9)^2$

82. $(5 - 9)^2 - (2 - 10)^2$

Make an estimate.

83. $-379 + (-103) + 287$

84. $\frac{-67 - 9}{-18}$

85. $-39 \cdot 8$

86. $-568 - (-227)$

87. $-3{,}887 + (-5{,}106)$

88. $-333(-4)$

89. $\frac{6{,}267}{-5}$

90. $-36 + (-78) + 59 + (-4)$

APPLICATIONS

91. TESTING In an effort to discourage her students from guessing on multiple-choice tests, a professor uses the grading scale shown in the table in the next column. If unsure of an answer, a student does best to skip the question, because incorrect responses are penalized very heavily. Find the test score of a student who gets 12 correct and 3 wrong and leaves 5 questions blank.

Response	Value
Correct	+3
Incorrect	−4
Left blank	−1

92. THE FEDERAL BUDGET See below. Suppose you were hired to write a speech for a politician who wanted to highlight the improvement in the federal government's finances during 1990s. Would it be better for the politician to refer to the average budget deficit/surplus for the last half, or for the last four years of that decade? Explain your reasoning.

U.S. Budget Deficit/Surplus
($ billions)

Deficit	Year	Surplus
−164	1995	
−107	1996	
−22	1997	
	1998	+70
	1999	+123

93. SCOUTING REPORTS The illustration shows a football coach how successful his opponent was running a "28 pitch" the last time the two teams met. What was the opponent's average gain with this play?

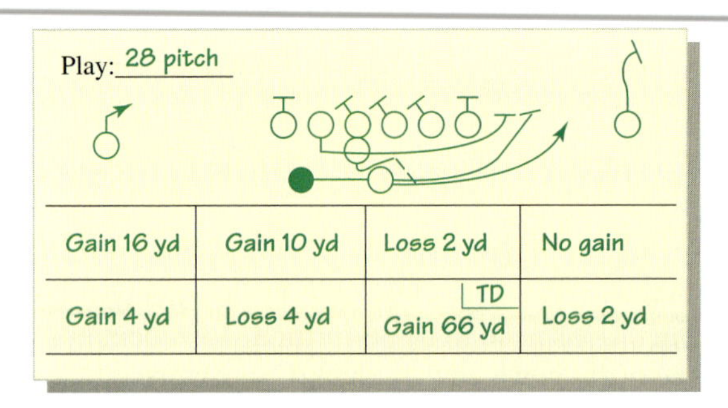

94. SPREADSHEETS The table shows the data from a chemistry experiment in spreadsheet form. To obtain a result, the chemist needs to add the values in row 1, double that sum, and then divide that number by the smallest value in column C. What is the final result of these calculations?

	A	B	C	D
1	12	−5	6	−2
2	15	4	5	−4
3	6	4	−2	8

Estimate the answer to each question.

95. OIL PRICES The price per barrel of crude oil fluctuates with supply and demand. It can rise and fall quickly. The line graph shows how many cents the price per barrel rose or fell each day for a week. For example, on Monday the price rose 68 cents, and on Tuesday it rose an additional 91 cents. Estimate the net gain or loss in the value of a barrel of crude oil for the week.

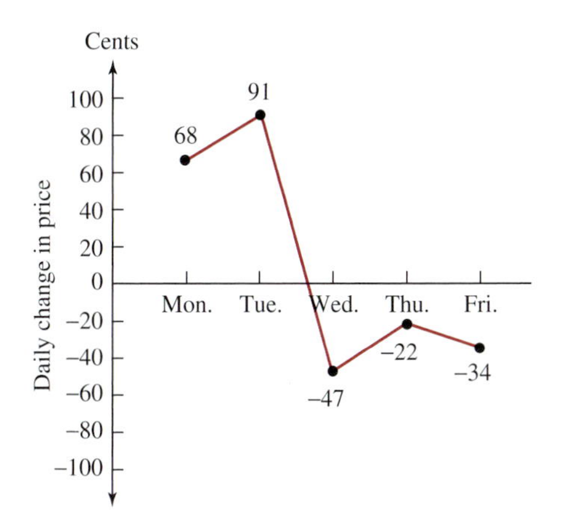

96. ESTIMATION Quickly determine a reasonable estimate of the exact answer in each of the following situations.

a. A diver, swimming at a depth of 34 feet below sea level, spots a sunken ship beneath him. He dives down another 57 feet to reach it. What is the depth of the sunken ship?

b. A dental hygiene company offers a money-back guarantee on its tooth whitener kit. When the kit is returned by a dissatisfied customer, the company loses the $11 it cost to produce it, because it cannot be resold. How much money has the company lost because of this return policy if 56 kits have been mailed back by customers?

c. A tram line makes a 7,561-foot descent from a mountaintop in 18 equal stages. How much does it descend in each stage?

WRITING

97. When evaluating expressions, why are rules for the order of operations necessary?

98. In the rules for the order of operations, what does the phrase *as they occur from left to right* mean?

99. Name a situation in daily life where you use estimation.

100. List some advantages and some disadvantages of the process of estimation.

REVIEW

101. Round 5,456 to the nearest thousand.

102. Evaluate: $6^2 - (10 - 8)$.

103. How do we find the perimeter of a rectangle?

104. Is this statement true or false? "When measuring perimeter, we use square units."

105. An elevator has a weight capacity of 1,000 pounds. Seven people, with an average weight of 140 pounds, are in it. Is it overloaded?

106. List the factors of 36.

2.7 Solving Equations Involving Integers

- The properties of equality
- Solving equations involving $-x$
- Combinations of operations
- Problem solving with equations

In this section, we revisit the topic of solving equations. The equations that we will be solving involve negative numbers, and some of the solutions are negative numbers as well. The section concludes with an application problem.

The properties of equality

Recall that the addition property of equality states: *If the same number is added to equal quantities, the results will be equal quantities.* When working with negative numbers, we can use this property in a new way.

To solve $x + (-8) = -10$, we need to isolate x on the left-hand side of the equation. We can do this by adding 8 to both sides.

$$x + (-8) = -10$$

$$x + (-8) \mathbf{+ 8} = -10 \mathbf{+ 8}$$ Use the addition property of equality. To undo the addition of -8, add 8 to both sides.

$$x = -2$$ Perform the additions: $(-8) + 8 = 0$ and $-10 + 8 = -2$.

From this example, we see that to undo addition, we can *add the opposite* of the number that is added to the variable.

Self Check 1
Solve $-6 + y = -8$ and check the result.

Answer -2

EXAMPLE 1 Solve: $-4 + x = -12$.

Solution We want to isolate x on one side of the equation. We can do this by adding the opposite of -4 to both sides.

$$-4 + x = -12$$

$$-4 + x \mathbf{+ 4} = -12 \mathbf{+ 4}$$ To undo the addition of -4, add 4 to both sides.

$$x = -8$$ Perform the additions: $-4 + 4 = 0$ and $-12 + 4 = -8$.

To check, we substitute -8 for x in the original equation and then simplify.

$$-4 + \mathbf{x} = -12$$ This is the original equation.

$$-4 + (\mathbf{-8}) \stackrel{?}{=} -12$$ Substitute -8 for x.

$$-12 = -12$$ Perform the addition: $-4 + (-8) = -12$.

Since the statement $-12 = -12$ is true, the solution is -8.

Recall that the subtraction property of equality states: *If the same quantity is subtracted from equal quantities, the results will be equal quantities.*

Self Check 2
Solve: $c + 4 = -3$.

Answer -7

EXAMPLE 2 Solve: $x + 16 = -8$.

Solution

$$x + 16 = -8$$

$$x + 16 \mathbf{- 16} = -8 \mathbf{- 16}$$ To undo the addition of 16, subtract 16 from both sides.

$$x = -8 + (-16)$$ Simplify the left side of the equation. On the right side, write the subtraction as addition of the opposite.

$$x = -24$$ Perform the addition.

Check the result.

Self Check 3
Solve $-2 + 8 = y + 3(-4)$ and check the result.

EXAMPLE 3 Solve: $-3 + 7 = h + 11(-2)$.

Solution The expressions on either side of the equation should be simplified before we use the properties of equality. There is an addition to perform on the left-hand side and a multiplication to perform on the right-hand side of the equation.

$$-3 + 7 = h + 11(-2)$$

$$4 = h + (-22)$$ Perform the addition: $-3 + 7 = 4$. Perform the multiplication: $11(-2) = -22$.

$$4 \mathbf{+ 22} = h + (-22) \mathbf{+ 22}$$ To undo the addition of -22, add 22 to both sides.

$$26 = h$$ Simplify: $4 + 22 = 26$ and $(-22) + 22 = 0$.

$$h = 26$$ Since $26 = h$, $h = 26$.

Check:

$-3 + 7 = h + 11(-2)$	This is the original equation.
$-3 + 7 \stackrel{?}{=} 26 + 11(-2)$	Substitute 26 for h.
$4 \stackrel{?}{=} 26 + (-22)$	Simplify both sides.
$4 = 4$	Perform the addition.

Answer 18

Recall that the division property of equality states: *If equal quantities are divided by the same nonzero quantity, the results will be equal quantities.*

EXAMPLE 4 Solve: **a.** $-3x = 15$ and **b.** $-16 = -4y$.

Solution

a. Recall that $-3x$ indicates multiplication: $-3 \cdot x$. We must undo the multiplication of x by -3. To do this, we divide both sides of the equation by -3.

$-3x = 15$	This is the original equation.
$\frac{-3x}{-3} = \frac{15}{-3}$	Divide both sides by -3.
$x = -5$	-3 times x, divided by -3, is x. On the right-hand side, perform the division: $15 \div (-3) = -5$.

Check:

$-3x = 15$	This is the original equation.
$-3(-5) \stackrel{?}{=} 15$	Substitute -5 for x.
$15 = 15$	Perform the multiplication: $-3(-5) = 15$.

b. $-16 = -4y$

$\frac{-16}{-4} = \frac{-4y}{-4}$	To undo the multiplication by -4, divide both sides by -4.
$4 = y$	Perform the divisions.
$y = 4$	If $4 = y$, then $y = 4$.

Check the result.

Self Check 4

Solve each equation and check the result:

a. $-7k = 28$

b. $-40 = -8f$

Answers **a.** -4, **b.** 5

Recall that the multiplication property of equality states: *If equal quantities are multiplied by the same nonzero quantity, the results will be equal quantities.*

EXAMPLE 5 Solve: $\frac{x}{-5} = -10$.

Solution In this equation, x is being divided by -5. To undo this division, we multiply both sides of the equation by -5.

$\frac{x}{-5} = -10$	
$-5\left(\frac{x}{-5}\right) = -5(-10)$	Use the multiplication property of equality. Multiply both sides by -5.
$x = 50$	When x is divided by -5 and then multiplied by -5, the result is x. Perform the multiplication: $-5(-10) = 50$.

Self Check 5

Solve $\frac{t}{-3} = 4$ and check the result.

Check:

$$\frac{x}{-5} = -10 \quad \text{This is the original equation.}$$

$$\frac{\mathbf{50}}{-5} \stackrel{?}{=} -10 \quad \text{Substitute 50 for } x.$$

$$-10 = -10 \quad \text{Perform the division.}$$

The solution is 50.

Answer -12

Solving equations involving $-x$

Consider the equation $-x = 3$. The variable x is not isolated, because there is a $-$ sign in front of it. The notation $-x$ means -1 times x. Therefore, the equation $-x = 3$ can be rewritten as $-1x = 3$. To isolate the variable, we can either multiply both sides by -1 or divide both sides by -1.

$-x = 3$		$-x = 3$	
$-1x = 3$	$-x = -1x$.	$-1x = 3$	$-x = -1x$.
$(\mathbf{-1})(-1x) = (\mathbf{-1})3$	Multiply both sides by -1.	$\frac{-1x}{\mathbf{-1}} = \frac{3}{\mathbf{-1}}$	Divide both sides by -1.
$1x = -3$	$-1(-1) = 1$.	$x = -3$	$-1x \div (-1) = x$; $3 \div (-1) = -3$.
$x = -3$	$1x = x$.		

Self Check 6
Solve $-h = -10$ and check the result.

EXAMPLE 6 Solve: $-x = -9$.

Solution

$$-x = -9$$

$$-1x = -9 \quad -x = -1x.$$

$$\mathbf{-1}(-1x) = \mathbf{-1}(-9) \quad \text{Multiply both sides by } -1.$$

$$x = 9 \quad \text{Perform the multiplications: } -1(-1x) = x \text{ and } -1(-9) = 9.$$

Check:

$$-\boldsymbol{x} = -9 \quad \text{This is the original equation.}$$

$$-(\mathbf{9}) \stackrel{?}{=} -9 \quad \text{Substitute 9 for } x.$$

$$-9 = -9$$

The solution is 9.

Answer 10

Combinations of operations

In the previous examples, each equation was solved by using a single property of equality. If the equation is more complicated, it is usually necessary to use several properties of equality to solve it. For example, consider the equation $2x + 5 = 9$ and the operations performed on x.

$$2x + 5 = 9$$

The variable is multiplied by 2. (points to $2x$)
Then 5 is added. (points to $+5$)

To solve this equation, we use the rules for the order of operations in reverse.

- First, use the subtraction property of equality to undo the addition of 5.
- Second, apply the division property of equality to undo the multiplication by 2.

$2x + 5 = 9$

$2x + 5 - 5 = 9 - 5$ — To undo the addition of 5, subtract 5 from both sides.

$2x = 4$ — Perform the subtraction on each side: $5 - 5 = 0$ and $9 - 5 = 4$.

$\frac{2x}{2} = \frac{4}{2}$ — To undo the multiplication by 2, divide both sides by 2.

$x = 2$ — Perform the divisions.

EXAMPLE 7 Solve: $-4x - 5 = 15$.

Solution The operations performed on x are multiplication by -4 and subtraction of 5. We undo these operations in the reverse order.

$-4x - 5 = 15$

$-4x - 5 + 5 = 15 + 5$ — Add 5 to both sides.

$-4x = 20$ — Perform the addition on each side: $-5 + 5 = 0$ and $15 + 5 = 20$.

$\frac{-4x}{-4} = \frac{20}{-4}$ — Divide both sides by -4.

$x = -5$ — Perform the divisions.

Check:

$-4x - 5 = 15$ — This is the original equation.

$-4(-5) - 5 \stackrel{?}{=} 15$ — Substitute -5 for x.

$20 - 5 \stackrel{?}{=} 15$ — Perform the multiplication: $-4(-5) = 20$.

$15 = 15$ — Perform the subtraction.

The solution is -5.

Self Check 7

Solve $-6b - 1 = 11$ and check the result.

Answer -2

EXAMPLE 8 Solve: $2 - 3p = -1$.

Solution We begin by writing the subtraction on the left-hand side as addition of the opposite.

$2 - 3p = -1$

$-3p + 2 = -1$ — Rewrite the left-hand side of the equation.

$-3p + 2 - 2 = -1 - 2$ — To undo the addition of 2 on the left-hand side of the equation, subtract 2 from both sides.

$-3p = -3$ — Simplify.

$\frac{-3p}{-3} = \frac{-3}{-3}$ — To undo the multiplication by -3, divide both sides by -3.

$p = 1$ — Perform the divisions.

Check this result in the *original* equation.

Self Check 8

Solve: $6 - 8k = -34$.

Answer 5

EXAMPLE 9 Solve: $\frac{y}{-2} - 6 = -18$.

Solution The operations performed on y are division by -2 and subtraction of 6. We undo these operations in the opposite order.

Self Check 9

Solve: $\frac{m}{-8} - 10 = -14$.

$$\frac{y}{-2} - 6 = -18$$

$$\frac{y}{-2} - 6 + 6 = -18 + 6$$ To undo the subtraction of 6, add 6 to both sides.

$$\frac{y}{-2} = -12$$ Simplify both sides of the equation.

$$-2\left(\frac{y}{-2}\right) = -2(-12)$$ To undo the division by −2, multiply both sides by −2.

$$y = 24$$ Perform the multiplications.

Check this result in the original equation.

Answer 32

Problem solving with equations

EXAMPLE 10 **Water management.** One day, enough water was released from a reservoir to lower the water level 17 feet to a reading of 33 feet below capacity. What was the water level reading before the release?

Analyze the problem

Figure 2-22 illustrates the given information and what we are asked to find.

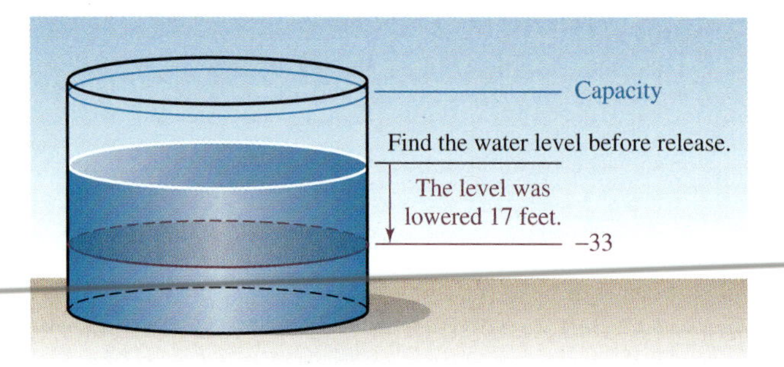

FIGURE 2-22

Form an equation

Let x = the water level reading before the release.

The water level reading before the release	minus	the number of feet the water level was lowered	is	the new water level reading.
x	$-$	17	$=$	-33

Solve the equation

$$x - 17 = -33$$

$$x - 17 + 17 = -33 + 17$$ To undo the subtraction of 17, add 17 to both sides.

$$x = -16$$ Perform the additions: $-17 + 17 = 0$ and $-33 + 17 = -16$.

State the conclusion

The water level reading before the release was −16 feet.

Check the result

If the water level reading was initially −16 feet and was then lowered 17 feet, the new reading would be $-16 - 17 = -16 + (-17) = -33$ feet. The answer checks.

Section 2.7 STUDY SET

VOCABULARY *Fill in the blanks.*

1. To ______ an equation, we isolate the variable on one side of the = symbol.
2. When solving an ________, the objective is to find all values of the variable that will make the equation true.

CONCEPTS

3. If we multiply x by -3 and then divide that product by -3, what is the result?
4. If we divide x by -4 and then multiply that quotient by -4, what is the result?
5. In the equation $x + 3 = 10$, we can isolate x in two ways. Find the missing numbers.
   a. $x + 3 - 3 = 10 - \square$
   b. $x + 3 + (-3) = 10 + \square$
6. In the equation $12 + c = 10$, we can isolate c in two ways. Find the missing numbers.
   a. $12 + c - \square = 10 - \square$
   b. $12 + c + (\square) = 10 + (\square)$
7. Determine what operations are performed on the variable x and the order in which they occur.
   a. $-2x = -100$
   b. $-6 + x = -9$
   c. $-4x - 8 = 12$
   d. $-1 = -6 + (-5x)$
8. Determine what operations are performed on the variable x and the order in which they occur.
   a. $\frac{x}{-4} - 8 = 50$
   b. $-16 = -5 + \frac{x}{-3}$

Fill in the blanks.

9. When solving the equation $t - 4 = -8 - 2$, it is best to ________ the right-hand side of the equation first before undoing any operations performed on the variable.
10. To solve the equation $-2x - 4 = 6$, we first undo the __________ of 4. Then we undo the ____________ by -2.
11. When solving an equation, we isolate the variable by undoing the operations performed on it in the ________ order.
12. To solve $-x = 6$, we can multiply or divide both sides of the equation by $\square$.
13. When solving each of these equations, which operation should you undo first?
    a. $-2x - 3 = -19$
    b. $-6 + \frac{h}{-3} = -14$
14. When solving each of these equations, which operation should you undo first?
    a. $5 + (-9x) = -1$
    b. $-16 = -9 + \frac{t}{7}$

NOTATION *Complete each solution to solve the equation.*

15.
$$\begin{aligned} y + (-7) &= -16 + 3 \\ y + (-7) &= \square \\ y + (-7) + 7 &= -13 + \square \\ y &= -6 \end{aligned}$$

16.
$$\begin{aligned} x - (-4) &= -1 - 5 \\ x + 4 &= -1 + \square \\ x + 4 &= -6 \\ x + 4 - 4 &= -6 - \square \\ x &= -6 + (-4) \\ x &= -10 \end{aligned}$$

17.
$$\begin{aligned} -13 &= -4y - 1 \\ -13 + \square &= -4y - 1 + \square \\ \square &= -4y \\ \frac{-12}{\square} &= \frac{-4y}{\square} \\ \square &= y \\ y &= 3 \end{aligned}$$

18.
$$\begin{aligned} 1 &= \frac{m}{-5} + 6 \\ 1 - \square &= \frac{m}{-5} + 6 - \square \\ -5 &= \frac{m}{-5} \\ \square(-5) &= \square\left(\frac{m}{-5}\right) \\ \square &= m \\ m &= 25 \end{aligned}$$

19. What does $-10x$ mean?
20. What does $\frac{x}{-8}$ mean?

PRACTICE *Determine whether the given number is a solution of the equation.*

21. $-3x - 4 = 2;\ -2$

22. $\frac{x}{-2} + 5 = -10;\ 20$

23. $-x + 8 = -4;\ 4$

24. $-3 + 2x = -3;\ 0$

Solve each equation. **Check each result.**

25. $x + 6 = -12$

26. $y + 1 = -4$

27. $-6 + m = -20$

28. $-12 + r = -19$

29. $-5 + 3 = -7 + f$

30. $-10 + 4 = -9 + t$

31. $h - 8 = -9$

32. $x - 1 = -7$

33. $0 = y + 9$

34. $0 = t + 5$

35. $r - (-7) = -1 - 6$

36. $x - (-1) = -4 - 3$

37. $t - 4 = -8 - (-2)$

38. $r - 1 = -3 - (-4)$

39. $x - 5 = -5$

40. $r - 4 = -4$

41. $-2s = 16$

42. $-3t = 9$

43. $-5t = -25$

44. $-6m = -60$

45. $-2 + (-4) = -3n$

46. $-10 + (-2) = -4x$

47. $-9h = -3(-3)$

48. $-6k = -2(-3)$

49. $\frac{t}{-3} = -2$

50. $\frac{w}{-4} = -5$

51. $0 = \frac{y}{8}$

52. $0 = \frac{h}{7}$

53. $\frac{x}{-2} = -6 + 3$

54. $\frac{a}{-5} = -7 + 6$

55. $\frac{x}{4} = -5 - 8$

56. $\frac{r}{2} = -5 - 1$

57. $2y + 8 = -6$

58. $5y + 1 = -9$

59. $-21 = 4h - 5$

60. $-22 = 7l - 8$

61. $-3v + 1 = 16$

62. $-4e + 4 = 24$

63. $8 = -3x + 2$

64. $15 = -2x + (-11)$

65. $-35 = 5 - 4x$

66. $12 = -9 - 3x$

67. $4 - 5x = 34$

68. $15 - 6x = 21$

69. $-5 - 6 - 5x = 4$

70. $-7 - 5 - 7x = 16$

71. $4 - 6x = -5 - 9$

72. $8 - 2d = -5 - 5$

73. $\frac{h}{-6} + 4 = 5$

74. $\frac{p}{-3} + 3 = 8$

75. $-2(4) = \frac{t}{-6} + 1$

76. $-2(5) = \frac{y}{-3} + 3$

77. $0 = 6 + \frac{c}{-5}$

78. $0 = -6 + \frac{s}{-3}$

79. $-1 = -8 + \frac{h}{-2}$

80. $-5 = 4 + \frac{g}{-4}$

81. $2x + 3(0) = -6$

82. $3x - 4(0) = -12$

83. $2(0) - 2y = 4$

84. $5(0) - 2y = 10$

85. $-x = 8$

86. $-y = 12$

87. $-15 = -k$

88. $-4 = -p$

APPLICATIONS *Complete each solution.*

89. SHARKS During a research project, a diver inside a shark cage made observations at a depth of 120 feet. For a second set of observations, the cage was raised to a depth of 75 feet. How many feet was the cage raised between observations?

Analyze the problem

- The first observations were at -120 ft.
- The next observations were at ____ ft.
- We must find ____________________.

Form an equation

Let x = ____________________

Key word: *raised* **Translation:** ________

Translate the words of the problem into an equation.

The first position of the cage	plus	the amount the cage was raised	is	the second position of the cage.
-120	$+$	____	$=$	____

Solve the equation

$$\_\_\_\_ + x = \_\_\_\_$$
$$-120 + x + \_\_\_\_ = -75 + \_\_\_\_$$
$$x = 45$$

State the conclusion

____

Check the result

If we add the number of feet the cage was raised to the first position, we get $-120 + \_\_\_\_ = -75$. The answer checks.

90. TRACK TIMES In one race, an athlete's time for the mile was 7 seconds under the school record. In a second race, her time continued to drop, to 16 seconds under the old school record. How much time did she drop between the first and second races?

Analyze the problem

- The 1st race was 7 sec under the record.
- The 2nd race was ____ sec under the record.
- We must find ____________________

Form an equation

Let x = ______________________________

Key phrase: *dropped* **Translation:** ___________

Translate the words of the problem into an equation.

First race perfor-mance	minus	amount of time dropped	is	second race perfor-mance.
-7	$-$	___	$=$	___

Solve the equation

$$\_\_\_ - x = \_\_\_$$
$$-7 - x + \_\_\_ = -16 + \_\_\_$$
$$-x = \_\_\_$$
$$x = 9$$

State the conclusion

____

Check the result

If we subtract the time she dropped in the second race from the time dropped in the first race, we get $-7 - \_\_\_ = -16$. The answer checks.

Let a variable represent the unknown quantity. Then write and solve an equation to answer the question.

91. MARKET SHARE After its first year of business, a manufacturer of smoke detectors found its market share 43 points behind the industry leader. Five years later, it trailed the leader by only 9 points. How many points of market share did the company pick up over this five-year span?

92. WEATHER FORECASTS The weather forecast for Fairbanks, Alaska warned listeners that the daytime high temperature of 2° below zero would drop to a nighttime low of 28° below. What was the overnight change in temperature?

93. CHECKING ACCOUNTS After he made a deposit of \$220, a student's account was still \$215 overdrawn. What was his checking account balance before the deposit?

94. FOOTBALL During the first half of a football game, a team ran for a total of 43 yards. After a dismal second half, they ended the game with a total of -8 yards rushing. What was their rushing total in the second half?

95. POLLS Six months before an election, a political candidate was 31 points behind in the polls. Two days before the election, polls showed that his support had skyrocketed; he was now only 2 points behind. How much support had he gained over the six-month period?

96. HORSE RACING At the midway point of a 6-furlong horse race, the long shot was 3 lengths ahead of the pre-race favorite. In the last half of the race, the long shot lost ground and eventually finished 6 lengths behind the favorite. By how many lengths did the long shot lose to the favorite during the last half of the race?

97. PRICE REDUCTIONS Over the past year, the price of a video game player has dropped each month. If the price fell \$60 this year, how much did the price drop each month on average?

98. REBATES A store decided that it could afford to lose some money to promote a new line of sunglasses. A \$9 rebate was offered to each customer purchasing these sunglasses. If the rebate program resulted in a loss of \$225 for the store, how many customers took advantage of the offer?

99. DREDGING A HARBOR In order to handle larger vessels, port officials are having a harbor deepened by dredging. The harbor bottom is already 47 feet below sea level. After the dredging, the bottom will be 65 feet below sea level. How many feet must be dredged out?

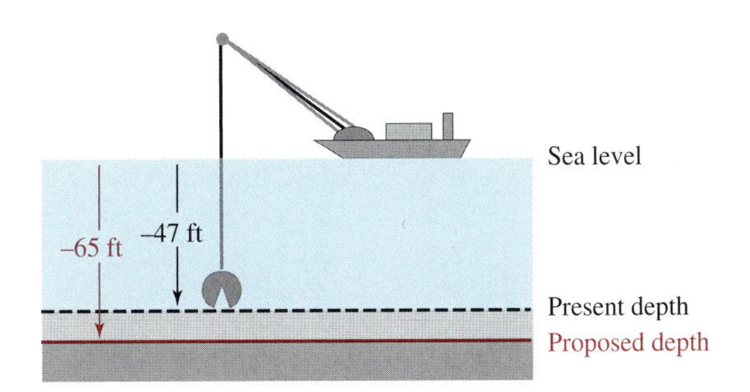

100. ROLLER COASTERS The end of a roller coaster ride consists of a steep plunge from a peak 145 feet above ground level. The car then comes to a screeching halt in a cave that is 25 feet below ground level. How many feet does the roller coaster drop at the end of the ride shown in the illustration below?

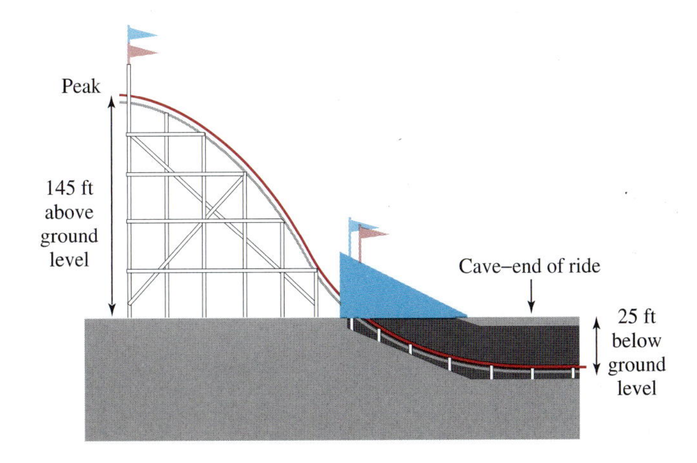

101. INTERNATIONAL TIME ZONES The world is divided into 24 times zones. Each zone is one hour ahead of or behind its neighboring zones. In the portion of the world time zone map shown, we see that Tokyo is in zone +9. What time zone is Seattle in if it is 17 hours behind Tokyo?

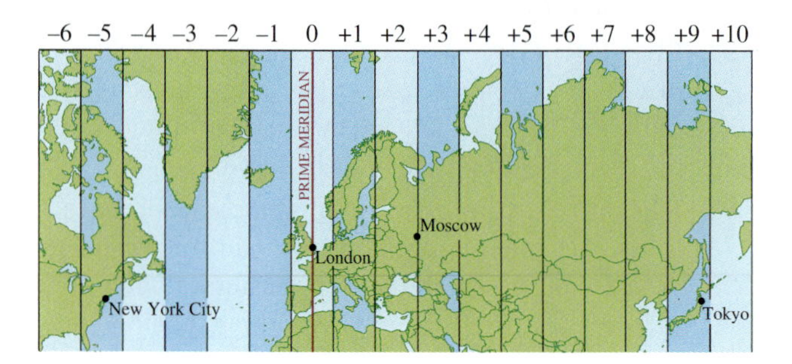

102. PROFITS AND LOSSES In its first year of business, a nursery suffered a loss due to frost damage, ending the year $11,560 in the red. In the second year, it made a sizable profit. If the total profit for the first two years in business was $32,090, how much profit was made the second year?

WRITING

103. Explain why the variable is not isolated in the equation $-x = 10$.

104. Explain how to check the result after solving an equation.

REVIEW

105. Write 5^6 without using exponents.

106. Give the definition of an even whole number.

107. Solve: $7 + 3y = 43$.

108. How can the addition $2 + 2 + 2 + 2 + 2$ be represented using multiplication?

109. Write $16 \div 8$ using a fraction bar.

110. To evaluate the expression $5(6) - 3 + 2$, list the operations in the order in which they must be performed.

KEY CONCEPT

Signed Numbers

In algebra, we work with both positive and negative numbers. We study negative numbers because they are necessary to describe many situations in daily life.

Represent each of these situations using a signed number.

1. Stocks fell 5 points.
2. The river was 12 feet over flood stage.
3. 30 seconds before going on the air
4. A business \$6 million in the red
5. 10 degrees above normal
6. The year 2000 B.C.
7. \$205 overdrawn
8. 14 units under their quota

The number line can be used to illustrate positive and negative numbers.

9. On this number line, label the location of the positive integers and the negative integers.

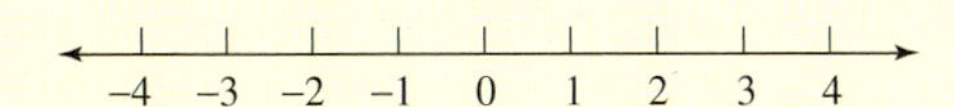

10. On this number line, graph 2 and its opposite.

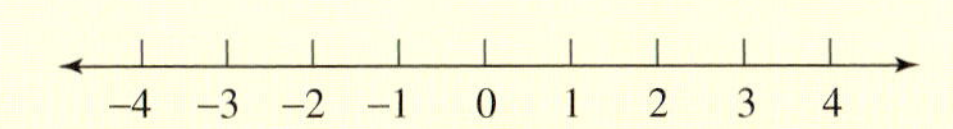

11. Two numbers, x and y, are graphed on this number line. What can you say about their relative sizes?

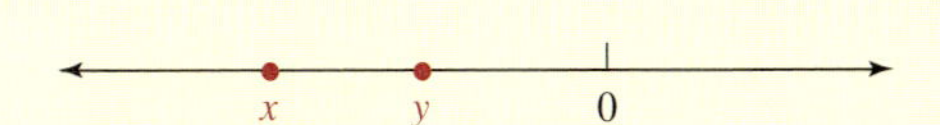

12. The absolute value of -3, written $|-3|$, is the distance between -3 and 0 on a number line. Show this distance on the number line.

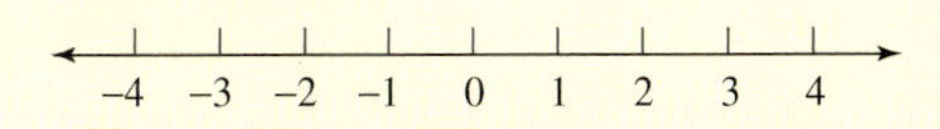

In the space provided, summarize how addition, multiplication, and division are performed with two integers having like signs and with two integers having unlike signs. Then explain the method that is used to subtract integers.

13. **Addition**

    Like signs:

    Unlike signs:

14. **Multiplication**

    Like signs:

    Unlike signs:

15. **Division**

    Like signs:

    Unlike signs:

16. **Subtraction with integers**

The opposite of the opposite of a number is that number.

Find each of the following.

18. $-(-12)$

19. The opposite of 8

20. The opposite of -8

21. -0

SECTION 2.2 *Adding Integers*

To add two integers with *like signs,* add their absolute values and attach their common sign to that sum.

To add two integers with *unlike signs,* subtract their absolute values, the smaller from the larger. Attach the sign of the number with the larger absolute value to that result.

Use a number line to find each sum.

22. $4 + (-2)$

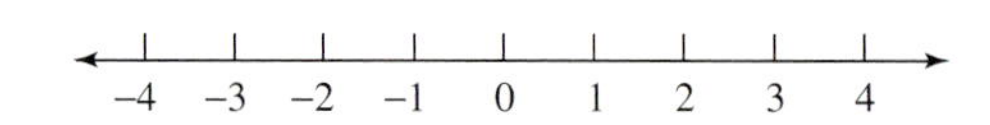

23. $-1 + (-3)$

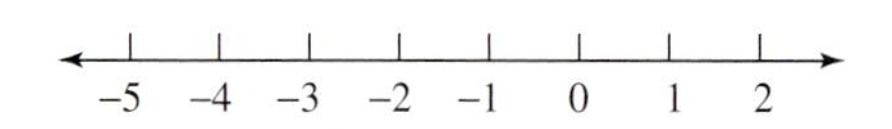

Add.

24. $-6 + (-4)$

25. $-23 + (-60)$

26. $-1 + (-4) + (-3)$

27. $-4 + 3$

28. $-28 + 140$

29. $9 + (-20)$

30. $3 + (-2) + (-4)$

31. $(-2 + 1) + [(-5) + 4]$

Addition property of 0: If a is any number, then

$$a + 0 = a \quad \text{and} \quad 0 + a = a$$

If $a + b = 0$, then a and b are called *additive inverses.*

Add.

32. $-4 + 0$

33. $0 + (-20)$

34. $-8 + 8$

35. $73 + (-73)$

Give the additive inverse of each number.

36. -11

37. 4

38. DROUGHT During a drought, the water level in a reservoir fell to a point 100 feet below normal. After two rainy months, it rose a total of 35 feet. How far below normal was the water level after the rain?

SECTION 2.3 *Subtracting Integers*

Rule for subtraction: If a and b are any numbers, then

$$a - b = a + (-b)$$

Subtract.

39. $5 - 8$

40. $-9 - 12$

41. $-4 - (-8)$

42. $-6 - 106$

43. $-8 - (-2)$

44. $7 - 1$

45. $0 - 37$

46. $0 - (-30)$

47. Fill in the blanks: Subtracting a number is the same as ________ the ________ of that number.

Evaluate each expression.

48. $-9 - 7 + 12$

49. $7 - [(-6) - 2]$

50. $1 - (2 - 7)$

51. $-12 - (6 - 10)$

52. Subtract 27 from -50.

53. Evaluate: $2 - [-(-3)]$.

54. GOLD MINING Some miners discovered a small vein of gold at a depth of 150 feet. This prompted them to continue their exploration. After descending another 75 feet, they came upon a much larger find. Use a signed number to represent the depth of the second discovery.

To find the *change* in a quantity, subtract the earlier value from the later value.

55. RECORD TEMPERATURES The lowest and highest recorded temperatures for Alaska and Virginia are shown here. For each state, find the difference in temperature between the record high and low.

Alaska: Low $-80°$ Jan. 23, 1971; High 100° June 27, 1915

Virginia: Low $-30°$ Jan. 22, 1985; High 110° July 15, 1954

SECTION 2.4 Multiplying Integers

The product of two integers with *like signs* is positive. The product of two integers with *unlike signs* is negative.

Multiply.

56. $-9 \cdot 5$

57. $-3(-6)$

58. $7(-2)$

59. $(-8)(-47)$

60. $-20 \cdot 5$

61. $-1(-1)$

62. $-1(25)$

63. $(5)(-30)$

64. $(-6)(-2)(-3)$

65. $4(-3)3$

66. $0(-7)$

67. $(-1)(-1)(-1)(-1)$

68. TAX DEFICITS A state agency's prediction of a tax shortfall proved to be two times worse than the actual deficit of \$3 million. The federal prediction of the same shortfall was even more inaccurate—three times the amount of the actual deficit. Complete the illustration, which summarizes these incorrect forecasts.

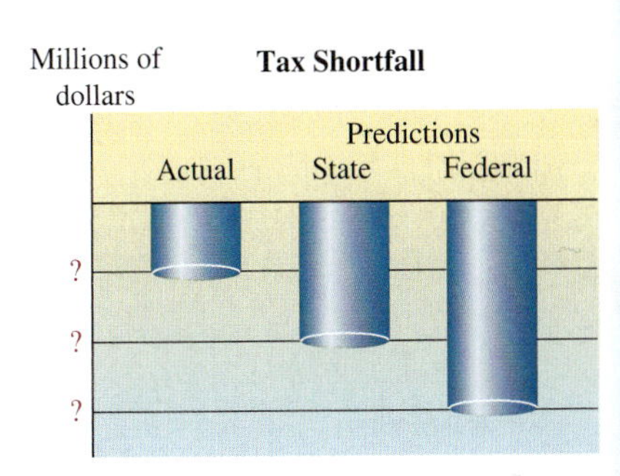

An *exponent* is used to represent repeated multiplication.

Find each power.

69. $(-5)^2$

70. $(-2)^5$

71. $(-8)^2$

72. $(-4)^3$

When a negative integer is raised to an *even* power, the result is positive. When it is raised to an *odd* power, the result is negative.

73. When $(-5)^9$ is evaluated, will the result be positive or negative?

74. Explain the difference between -2^2 and $(-2)^2$ and then evaluate each.

SECTION 2.5 Dividing Integers

75. Fill in the blanks: We know that $\frac{-15}{5} = -3$ because ___(___) = ___.

The quotient of two numbers with *like* signs is positive.

The quotient of two integers with *unlike* signs is negative.

Divide.

76. $\frac{-14}{7}$

77. $\frac{25}{-5}$

78. $-64 \div 8$

79. $\frac{-202}{-2}$

If a is any nonzero number,

$$\frac{0}{a} = 0$$

Division by 0 is undefined.

Find each quotient, if possible.

80. $\frac{0}{-5}$

81. $\frac{-4}{0}$

82. $\frac{-673}{-673}$

83. $\frac{-10}{-1}$

84. PRODUCTION TIME Because of improved production procedures, the time needed to produce an electronic component dropped by 12 minutes over the past six months. If the drop in production time was uniform, how much did it change each month over this period?

SECTION 2.6 Order of Operations and Estimation

The rules for the order of operations:

1. Perform all calculations within parentheses and other grouping symbols.
2. Evaluate all exponential expressions.
3. Perform all the multiplications and divisions, working from left to right.
4. Perform all the additions and subtractions, working from left to right.

Always work from the *innermost* set of grouping symbols to the *outermost* set.

A fraction bar is a grouping symbol.

Evaluate each expression.

85. $2 + 4(-6)$

86. $7 - (-2)^2 + 1$

87. $2 - 5(4) + (-25)$

88. $-3(-2)^3 - 16$

89. $-2(5)(-4) + \frac{|-9|}{3^2}$

90. $-4^2 + (-4)^2$

91. $-12 - (8 - 9)^2$

92. $7|-8| - 2(3)(4)$

93. $-4\left(\frac{15}{-3}\right) - 2^3$

94. $-20 + 2(12 - 5 \cdot 2)$

95. $-20 + 2[12 - (-7 + 5)^2]$

96. $8 - |-3 \cdot 4 + 5|$

97. $\frac{10 + (-6)}{-3 - 1}$

98. $\frac{3(-6) - 11 + 1}{4^2 - 3^2}$

An *estimation* is an approximation that gives a quick idea of what the actual answer would be.

Estimate each answer.

99. $-89 + 57 + (-42)$

100. $\frac{-507}{-26}$

101. $(-681)(9)$

102. $317 - (-775)$

SECTION 2.7 *Solving Equations Involving Integers*

To *solve an equation* means to find all values of the variable that, when substituted into the original equation, make the equation a true statement.

Is −4 a solution of the equation? Explain why or why not.

103. $2x + 6 = -2$

104. $6 + \frac{x}{2} = -4$

To solve an equation, we undo the operations in the reverse order from that in which they were performed on the variable. The objective is to isolate the variable.

Solve each equation. **Check the result.**

105. $t + (-8) = -18$

106. $\frac{x}{-3} = -4$

107. $y + 8 = 0$

108. $-7m = -28$

109. $-x = -15$

110. $4 = -y$

111. $-5t + 1 = -14$

112. $3(2) = 2 - 2x$

113. $\frac{x}{-4} - 5 = -1 - 1$

114. $c - (-5) = 5$

The five-step *problem-solving strategy* can be used when solving application problems.

1. Analyze the problem.
2. Form an equation.
3. Solve the equation.
4. State the conclusion.
5. Check the result.

Let a variable represent the unknown quantity. Then write and solve an equation to answer the question.

115. CHECKING ACCOUNTS After Lee made a deposit of \$85 into his checking account, it was still \$47 overdrawn. What was his account balance before making the deposit?

116. WIND-CHILL FACTOR If the wind is blowing at 25 miles per hour, an air temperature of 5° below zero will feel like 51° below zero. Find the perceived change in temperature that is caused by the wind.

117. CREDIT CARD PROMOTIONS During the holidays, a store offered an \$8 gift certificate to any customer applying for its credit card. If this promotion cost the company \$968, how many customers applied for credit?

118. BANK FAILURES When a group of 7 investors decided to acquire a failing bank, each had to assume an equal share of the bank's total indebtedness, which was \$57,400. How much debt did each investor assume?

CHAPTER 2 TEST

1. Insert one of the symbols > or < in the blank to make the statement true.
   a. -8 ☐ -9
   b. -8 ☐ $|-8|$
   c. The opposite of 5 ☐ 0

2. List the integers.

3. SCHOOL ENROLLMENTS According to the projections in the table, which high school will face the greatest shortage in the year 2010?

High schools with shortage of classroom seats by 2010	
Sylmar	−669
San Fernando	−1,630
Monroe	−2,488
Cleveland	−350
Canoga Park	−586
Polytechnic	−2,379
Van Nuys	−1,690
Reseda	−462
North Hollywood	−1,004
Hollywood	−774

4. Use a number line to find the sum: $-3 + (-2)$.

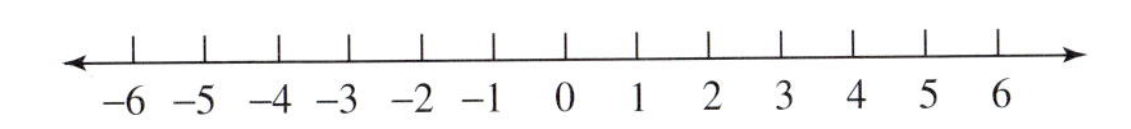

5. Add.
   a. $-65 + 31$
   b. $-17 + (-17)$
   c. $[6 + (-4)] + [-6 + (-4)]$

6. Subtract.
   a. $-7 - 6$
   b. $-7 - (-6)$
   c. $0 - 15$
   d. $-60 - 50 - 40$

7. Find each product.
   a. $-10 \cdot 7$
   b. $-4(-2)(-6)$
   c. $(-2)(-2)(-2)(-2)$
   d. $-55(0)$

8. Write the related multiplication statement for $\frac{-20}{-4} = 5$.

9. Find each quotient, if possible.
   a. $\frac{-32}{4}$
   b. $\frac{8}{6 - 6}$
   c. $\frac{-5}{1}$
   d. $\frac{0}{-6}$

10. BUSINESS TAKEOVERS Six businessmen are contemplating taking over a company that has potential, but they must pay off the debt taken on by the company over the past three quarters. If they plan equal ownership, how much will each have to contribute to pay off the debt?

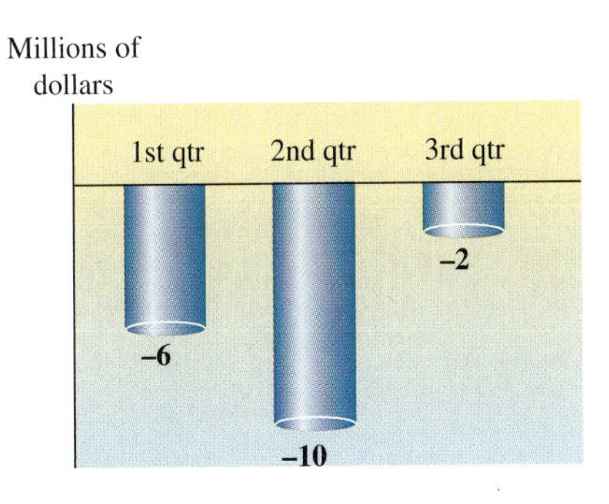

11. GEOGRAPHY The lowest point on the African continent is the Qattarah Depression in the Sahara Desert, 436 feet below sea level. The lowest point on the North American continent is Death Valley, California, 282 feet below sea level. Find the difference in these elevations.

12. Evaluate each expression.
    a. $-(-6)$
    b. $|-7|$
    c. $-|-9 + 3|$
    d. $2|-66|$

13. Find each power.
  a. $(-4)^2$ b. -4^2
  c. $(-4 + 3)^5$

Evaluate each expression.

14. $-18 \div 2 \cdot 3$

15. $4 - (-3)^2 + 6$

16. $-3 + \left(\frac{-16}{4}\right) - 3^3$

17. $-10 + 2[6 - (-2)^2(-5)]$

18. $\frac{4(-6) - 2^2}{-3 - 4}$

19. TEMPERATURE CHANGE In a lab, the temperature of a fluid was reduced 6° per hour for 12 hours. What signed number represents the change in temperature?

20. GAUGES Many automobiles have an ammeter like that shown. If the headlights, which draw a current of 8 amps, and the radio, which draws a current of 4 amps, are both turned on, which way will the arrow move? What will be the new reading?

21. CARD GAMES After the first round of a card game, Tommy had a score of 8. When he lost the second round, he had to deduct the value of the cards left in his hand from his first-round score. (See the illustration below.) What was his score after two rounds of the game? For scoring, face cards are counted as 10 points and aces as 1 point.

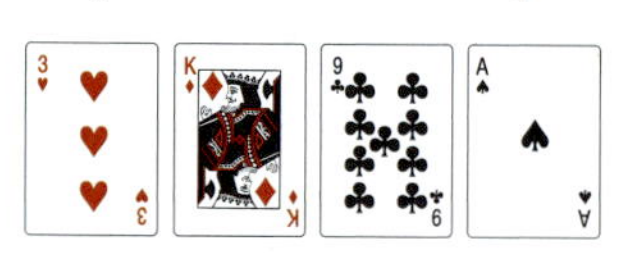

22. Is -10 a solution of $\frac{x}{5} - 16 = -14$?

Solve each equation.

23. $c - (-7) = -8$

24. $6 - x = -10$

25. $\frac{x}{-4} = 10$

26. $-6x = 0$

27. $3x + (-7) = -11 + (-11)$

28. $-5 = -6a + 7$

29. $\frac{x}{-2} + 3 = (-2)(-6)$

Let a variable represent the unknown quantity. Then write and solve an equation to answer each question.

30. CHECKING ACCOUNTS After making a deposit of \$225, a student's account was still \$19 overdrawn. What was her balance before the deposit?

31. HOSPITAL CAPACITY One morning, the number of beds occupied by patients in a hospital was 3 under capacity. By afternoon, the number of unoccupied beds was 21. If no new patients were admitted, how many patients were released to go home?

32. Multiplication means repeated addition. Use this fact to show that the product of a positive and a negative number, such as $5(-4)$, is negative.

33. Explain why the absolute value of a number can never be negative.

34. Is the inequality $12 \geq 12$ true or false? Explain why.

CHAPTERS 1–2 CUMULATIVE REVIEW EXERCISES

Consider the numbers in the set $\{-2, -1, 0, 1, 2\frac{3}{4}, 5, 9\}$.

1. List each natural number.

2. List each whole number.

3. List each negative number.

4. List each integer.

Consider the number 7,326,549.

5. Which digit is in the thousands column?

6. Which digit is in the hundred thousands column?

7. Round to the nearest hundred.

8. Round to the nearest ten thousand.

9. BIDS A school district received the bids shown in the table. Which company should be awarded the contract?

Citrus Unified School District Bid 02-9899 CABLING AND CONDUIT INSTALLATION	
Datatel	\$2,189,413
Walton Electric	\$2,201,999
Advanced Telecorp	\$2,175,081
CRF Cable	\$2,174,999
Clark & Sons	\$2,175,801

10. NUCLEAR POWER The table gives the number of nuclear power plants in the United States for the years 1978–2003, in five-year increments. Construct a bar graph using the data.

Year	1978	1983	1988	1993	1998	2003
Plants	70	80	108	109	104	104

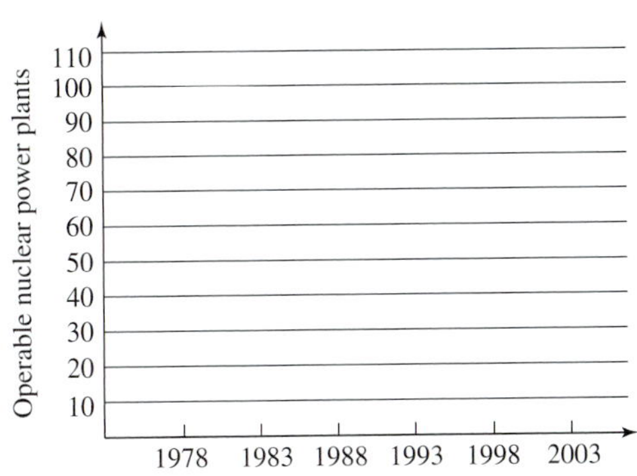

Source: *The World Almanac, 2005*

Perform each operation.

11. $237 + 549$

12. $6,375 - 2,569$

13. $\begin{array}{r} 5,369 \\ -\ \ 685 \\ \hline \end{array}$

14. $\begin{array}{r} 7,899 \\ +5,237 \\ \hline \end{array}$

15. Find the perimeter and the area of the rectangular garden.

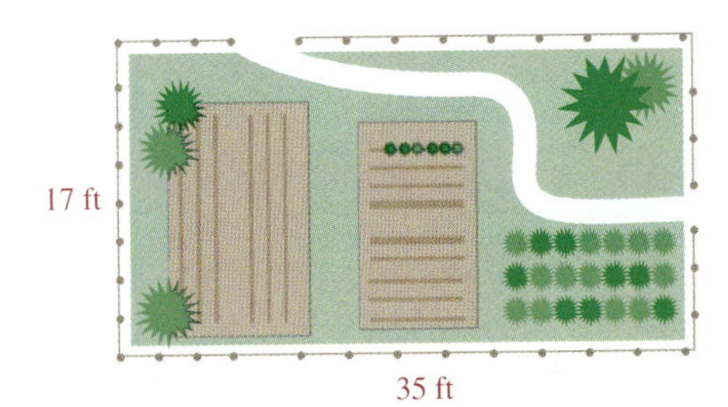

16. In a shipment of 147 pieces of furniture, 27 pieces were sofas, 55 were leather chairs, and the rest were wooden chairs. Find the number of wooden chairs.

Perform each operation.

17. $435 \cdot 27$

18. $1,261 \div 97$

19. $\begin{array}{r} 4,587 \\ \times\ \ \ 67 \\ \hline \end{array}$

20. $38\overline{)17,746}$

21. SHIPPING There are 12 tennis balls in one dozen, and 12 dozen in one gross. How many tennis balls are there in a shipment of 12 gross?

22. Find all of the factors of 18.

Identify each number as a prime, a composite, an even, or an odd.

23. 17

24. 18

25. 0

26. 1

27. Find the prime factorization of 504.

28. Write the expression $11 \cdot 11 \cdot 11 \cdot 11$ using an exponent.

Evaluate each expression.

29. $5^2 \cdot 7$

30. $16 + 2[14 - 3(5 - 4)^2]$

31. $25 + 5 \cdot 5$

32. $\frac{16 - 2 \cdot 3}{2 + (9 - 6)}$

33. SPEED CHECKS A traffic officer monitored several cars on a city street. She found that the speeds of the cars were as follows:

38, 42, 36, 38, 48, 44

On average, were the drivers obeying the 40-mph speed limit?

34. Determine whether 6 is a solution of the equation $3x - 2 = 16$. Explain why or why not.

Solve each equation and check the result.

35. $50 = x + 37$

36. $a - 12 = 41$

37. $5p = 135$

38. $\frac{y}{8} = 3$

Graph each set on the number line.

39. $\{-2, -1, 0, 2\}$

−3 −2 −1 0 1 2 3

40. The integers greater than -4 but less than 2

−4 −3 −2 −1 0 1 2

41. True or false: $-17 < -16$.

42. Evaluate: 3^2 and -3^2.

Evaluate each expression.

43. $-2 + (-3)$

44. $-15 + 10 + (-9)$

45. $-3 - 5$

46. $-15^2 - 2|-3|$

47. $(-8)(-3)$

48. $5(-7)^3$

49. $\frac{-14}{-7}$

50. $\frac{450}{-9}$

51. $5 + (-3)(-7)$

52. $-20 + 2[12 - 5(-2)(-1)]$

53. $\frac{10 - (-5)}{1 - 2 \cdot 3}$

54. $\frac{3(-6) - 10}{3^2 - 4^2}$

Solve each equation. Check the result.

55. $-5t + 1 = -14$

56. $\frac{x}{-3} - 2 = -2(-2)$

57. BUYING A BUSINESS When 12 investors decided to buy a bankrupt company, they agreed to assume equal shares of the company's debt of $1,512,444. How much was each person's share?

58. THE MOON The difference in the maximum and the minimum temperatures on the moon's surface is 540° F. The maximum temperature, which occurs at lunar noon, is 261° F. Find the minimum temperature, which occurs just before lunar dawn.

CHAPTER 3

The Language of Algebra

Getty Images

TLE Most manufacturing companies have a production process that uses different machines, materials, and workers. For example, printing presses, binders, and packaging machines are needed to produce textbooks. Paper, ink, glue, and covers must be ordered in the correct quantities. And teams of workers must be scheduled to arrive at the proper times.

Planners often use algebra to organize the entire production process. They use mathematics to make sure production runs are efficient and, most important, profitable.

To learn more about the use of algebra by production planners, visit The Learning Equation on the Internet at http://tle.brookscole.com. (The log-in instructions are in the Preface.) For Chapter 3, the online lessons are:

- *TLE* Lesson 6: Translating Written Phrases
- *TLE* Lesson 7: Writing and Evaluating Expressions

Check Your Knowledge

1. An algebraic __________ is a combination of variables, numbers, and the operation symbols for addition, subtraction, multiplication, and division.
2. To evaluate an algebraic expression, we __________ specific numbers for the variables in the expression and apply the rules for the order of operations.
3. We __________ expressions, and we ________ equations.
4. Terms with exactly the same variables and exponents are called ______ terms.
5. In $3(x - 5)$, to remove parentheses means to apply the __________ property.
6. Translate each phrase into mathematical symbols.
   a. 7 more than x
   b. 3 less than y
   c. $\frac{3}{4}$ of z
7. George had $500 in his bank account when he wrote a check for x dollars. Write an expression that represents the remaining balance in George's account.
8. Write an algebraic expression for
   a. the value of q quarters.
   b. the value of f five-dollar bills.
9. Evaluate each expression.
   a. $b^2 - 4ac$ for $a = 1$, $b = 1$, and $c = -6$
   b. $\frac{-(x - 1)^2}{1 - x}$ for $x = -5$
10. Joan left for a bicycle ride at 6:00 A.M. and returned at 9:00 A.M. Find the distance that Joan rode if she averaged 12 miles per hour.
11. Remove parentheses.
    a. $3(2x + 7)$
    b. $-(3x - 7)$
    c. $-3(2x - 3)$
12. Determine whether x is term or a factor.
    a. $5x$
    b. $x + 2y$
13. Combine like terms.
    a. $2x - 2y + x + 2y$
    b. $x + x - x - 2x$
14. Simplify.
    a. $-5x(-3)$
    b. $0 - 5x$
    c. $3(x - 3) - 2(4x - 5)$

Solve each equation. Check each result.

15. $3x - 3 = 2x + 3$
16. $4y + 7 - y = 10$
17. $3(2x - 5) = 5(x - 1)$
18. $12 - (x + 4) = 6$
19. To receive a safety certificate, a crossing guard must volunteer 78 hours of service by working 6-hour shifts at a local elementary school. If a crossing guard has already completed 42 hours of service, how many more 6-hour shifts must she work to receive the safety certificate?
20. The sum of two numbers is 17. Find the numbers if one number is one more than three times the other.

Study Skills Workshop

LISTENING IN CLASS AND TAKING NOTES

Attending all of your class meetings is crucial to your success in a math course. Your instructor will be giving explanations and examples that may not be found in your textbook, as well as other information about your course (test dates, homework assignments, etc.). Because this information is not found anywhere else, and because it is normal to forget material shortly after it is presented, it is important that you keep a written record of what occurred in your class. (*Note:* Auditory learners may also want to keep a recorded version of class notes. Ask your instructor for permission to record lectures.)

Listening in Class. Listening in class is different from listening to your favorite CDs or MP3 files because it requires that you be an *active* listener. It is usually impossible to write down everything that your instructor says in class, but you want to get what's important. In order to catch the significant material, you should be waiting for cues from your instructor: pauses in lectures and statements such as "This is really important" or "This is a question that shows up frequently on tests" are indications that you should be paying special attention. You should be listening with a pencil in hand, ready to record these examples, definitions, or concepts.

Taking Notes. Usually when giving a lecture, your instructor will tell you the important chapter(s) and section(s) in your textbook. Bring your textbook to every class and open it to the section that your instructor is covering. If possible, find out what section you will be covering before your next lecture and read this section *before* class. As you read, try to identify which terms and definitions your instructor will think are important. When your instructor refers to definitions that are found in the text, it is not necessary for you to write them down in entirety, but just jot the term being defined and the page number on which it appears. If topics are being discussed that are not in the text, record them using abbreviations and phrases rather than complete sentences.

Don't worry about making your class notes really neat; you should be "reworking" your notes after class (this will be discussed in more detail in the next Study Skills Workshop). However, you should organize your notes as much as possible as you write them. Write down the examples your instructor gives in step-by-step detail. Label examples and definitions that your instructor covers in detail as "key concepts." If you miss a step, or if you don't understand a step in the example, ask your instructor to explain. The examples should look a good deal like the problems that you will see in your homework assignment, so it is really important that you write them down as completely as possible. As soon as possible after class, sit down with a classmate and compare notes, to see whether either one of you missed anything.

ASSIGNMENT

1. Find out from your instructor which section(s) will be covered in your next class. Pre-read the section(s) and make a list of terms and definitions that you predict your instructor will think are most important.
2. In your next class, bring your textbook and keep it open to the sections being covered. If your instructor mentions a term or definition that is found in your text, write the term and the page number on which it appears in your notes. Write every example that your instructor gives, making sure that you have included all the steps.
3. Find at least one classmate with whom you can review notes. Make an appointment to compare your class notes as soon as possible after the class. Did you find differences in your notes?

Algebra is the language of mathematics. In this chapter, you will learn more about thinking and writing in this language, using its most important component — a variable.

3.1 Variables and Algebraic Expressions

- Algebraic expressions • Translating from English to mathematical symbols
- Writing algebraic expressions to represent unknown quantities
- Looking for hidden operations • Expressions involving more than one operation

In Chapter 1, we introduced the following strategy for problem solving.

1. Analyze the problem.
2. Form an equation.
3. Solve the equation.
4. State the conclusion.
5. Check the result.

In order to form equations, you must represent unknown quantities with variables. Success at doing this depends on your ability to translate English words and phrases into mathematical symbols.

Algebraic expressions

In the equation $x + 5 = 10$, the notation $x + 5$ is called an **algebraic expression** or, more simply, an **expression.**

Algebraic expressions

Variables and/or numbers can be combined with the operations of addition, subtraction, multiplication, and division to create **algebraic expressions.**

Here are some examples of algebraic expressions.

$5(2a)$ — This algebraic expression is a combination of the numbers 5 and 2, the variable a, and the operation of multiplication.

$x + 2x + 3x$ — This algebraic expression involves the variable x, the numbers 2 and 3, and the operations of addition and multiplication.

$\frac{10 - y}{-3}$ — This algebraic expression is a combination of the numbers 10 and -3, the variable y, and the operations of subtraction and division.

$5(r - 6)$ — This algebraic expression is a combination of the numbers 5 and 6, the variable r, and the operations of subtraction and multiplication.

Algebraic expressions can contain more than one variable. For example, the expressions $3x + 4y$ and $-6m^2n(mn)$ each contain two variables.

Translating from English to mathematical symbols

In order to solve application problems, which are almost always given in words, we must translate those words into mathematical symbols. The tables on the next page list some *key words* and *key phrases* that are used to represent the operations of addition, subtraction, multiplication, and division.

Addition

The phrase	Translates to the algebraic expression
the *sum* of p and 15	$p + 15$
10 *plus* c	$10 + c$
5 *added to* a	$a + 5$
4 *more than* r	$r + 4$
8 *greater than* A	$A + 8$
S *increased by* 100	$S + 100$
exceeds L *by* 20	$L + 20$

Subtraction

The phrase	Translates to the algebraic expression
the difference of 30 and k	$30 - k$
1,000 *minus* R	$1{,}000 - R$
15 *less than* w	$w - 15$
r *decreased by* 5	$r - 5$
T *reduced by* 80	$T - 80$
7 *subtracted from* s	$s - 7$
2,000 *less* c	$2{,}000 - c$

Multiplication

The phrase	Translates to the algebraic expression
the *product* of 60 and h	$60h$
10 *times* A	$10A$
twice w	$2w$
$\frac{1}{2}$ *of* t	$\frac{1}{2}t$

Division

The phrase	Translates to the algebraic expression
the *quotient* of B and 5	$\frac{B}{5}$
T *divided by* 50	$\frac{T}{50}$
the *ratio* of h to 3	$\frac{h}{3}$
n *split into* 8 equal parts	$\frac{n}{8}$

! COMMENT The phrase *greater than* is used to indicate addition. The phrase *is greater than* refers to the symbol >. A similar comment applies to the phrases *less than* and *is less than*.

EXAMPLE 1 Write each phrase as an algebraic expression: **a.** the sum of n and 12, **b.** the product of z and 60, and **c.** 6 less than z.

Solution

a. **Key word:** *sum* **Translation:** add

The phrase translates to

$n + 12$

b. **Key word:** *product* **Translation:** multiply

The variable z is to be multiplied by 60. So we have: $z \cdot 60$. This can be written as

$60z$

c. **Key phrase:** *less than* **Translation:** subtract

Since z is to be made less, we will subtract 6 from it. When we translate to mathematical symbols, we must reverse the order in which 6 and x appear in the given phrase.

6 less than z

$z - 6$

Self Check 1

Write each phrase as an algebraic expression:

a. $\frac{3}{4}$ of k

b. A split into 5 equal parts

c. Eighty subtracted from n

Answers **a.** $\frac{3}{4}k$, **b.** $\frac{A}{5}$, **c.** $n - 80$

Self Check 7
Complete the table. Then use that information to determine how many yards are in f feet.

Number of feet	Number of yards
3	
6	
9	
f	

Answers 1, 2, 3; $\frac{f}{3}$

EXAMPLE 7 How many eggs are there in d dozen?

Solution Since there are no key words, we must carefully analyze the problem to write an expression that gives the number of eggs in d dozen. It is often helpful to consider some specific cases. For example, let's calculate the number of eggs in 1 dozen, 2 dozen, and 3 dozen. When we write the results in a table, a pattern becomes apparent.

Number of dozen	Number of eggs
1	$12 \cdot 1 = 12$
2	$12 \cdot 2 = 24$
3	$12 \cdot 3 = 36$
d	$12 \cdot d = 12d$

We multiply the number of dozen by 12 to find the number of eggs.

Therefore, if d = the number of dozen eggs, the number of eggs is $12d$.

Expressions involving more than one operation

In the previous examples, the algebraic expressions contained one operation. We now examine expressions involving two operations.

Self Check 8
On the second day of her trip, Tamiko drove 100 miles less than twice as far as the first day. Choose a variable to represent the number of miles driven on the first day. Then write an expression that represents the number of miles driven on the second day.

Answers m = miles driven on first day, miles driven on second day = $2m - 100$

EXAMPLE 8 As Figure 3-3 shows, Alaska is much larger than Vermont. To be exact, the area of Alaska is 380 square miles more than 50 times that of Vermont. Choose a variable to represent one area. Then write an expression that represents the other area.

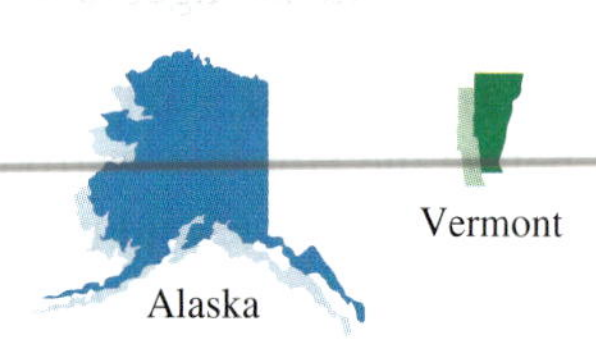

FIGURE 3-3

Solution Since the area of Alaska is expressed in terms of the area of Vermont, we let v = the area of Vermont.

Key phrase: *more than* **Translation:** add
Key word: *times* **Translation:** multiply

The area of Alaska = $50v + 380$ square miles.

Section 3.1 STUDY SET

VOCABULARY *Fill in the blanks.*

1. An algebraic ________ is a combination of variables, numbers, and the operation symbols for addition, subtraction, multiplication, and division.
2. The answer to an addition problem is called the ______. The answer to a subtraction problem is called the ________.
3. A ________ is a letter that is used to stand for a number.

4. The answer to a multiplication problem is called the ________. The answer to a division problem is called the ________.

CONCEPTS

5. Write two different algebraic expressions that contain the numbers 10 and 3 and the variable x.

6. Write an equation with one side containing an algebraic expression and the other side the number 20.

7. The illustration shows the commute to work (in miles) for two men who work in the same office. Who lives farther from the office? How much farther?

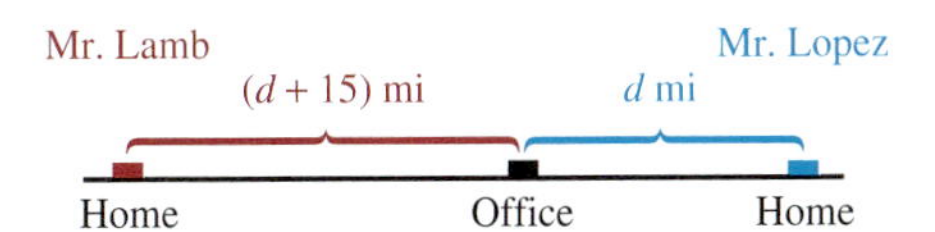

8. See the illustration below.
   a. If we let b represent the height of the birch tree, write an algebraic expression for the height of the elm tree.
   b. If we let e stand for the height of the elm tree, write an algebraic expression for the height of the birch tree.

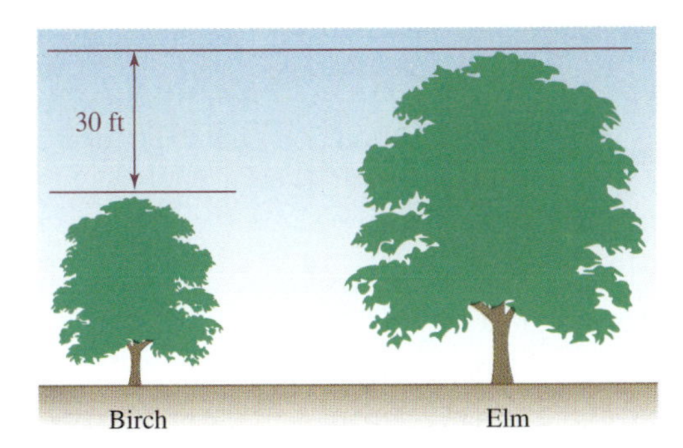

9. In 2002, the business profits of a video rental store were double those of the previous year. In 2003, the profits were triple those of 2001. If the profits in 2001 were p dollars, write algebraic expressions to complete the bar graph.

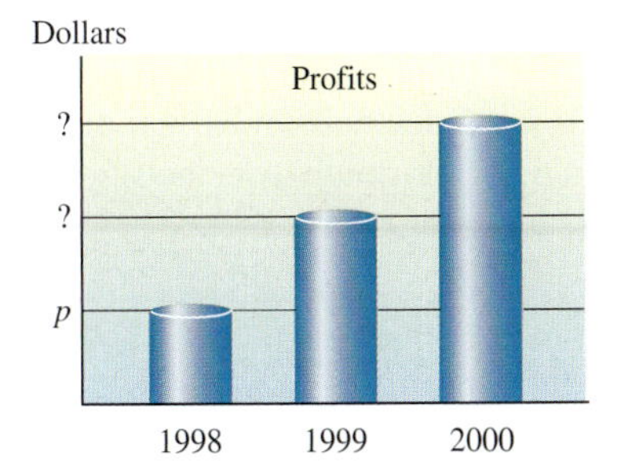

10. The following table shows the ages of three family members.

	Age (years)
Matthew	x
Sarah	$x - 8$
Joshua	$x + 2$

   a. Who is the youngest person shown in the table?
   b. Who is the oldest person listed in the table?
   c. On whose age are the ages in the table based?

11. On a flight from Dallas to Miami, a jet airliner, which flies at 500 mph in still air, experiences a tail wind of x mph. The tail wind increases the speed of the jet. On the immediate return flight to Dallas, the airliner flies into a head wind of the same strength. The head wind decreases the speed of the jet. Use this information to complete the table.

Wind conditions	Speed of jet (mph)
In still air	
With the tail wind	
Against the head wind	

12. The weights of two mixtures, measured in ounces, are compared on a balance, as shown.

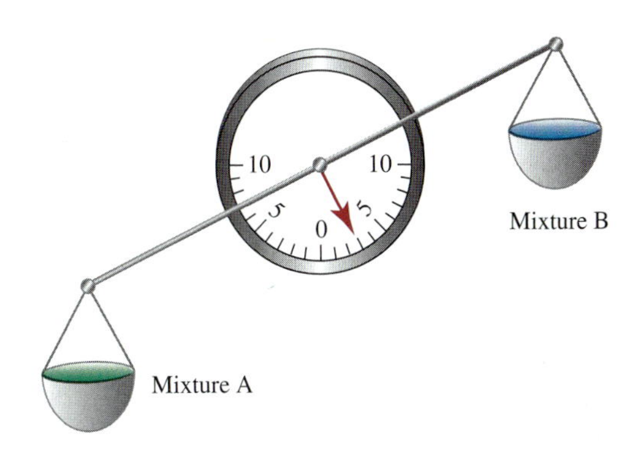

   a. Which mixture is heavier? By how much?
   b. Let x = the weight of mixture B. Then write an expression that represents the weight of mixture A.

13. A student figures that she has h hours to study for a government final. She wants to spread the studying evenly over a four-day period. Write an expression for how many hours she should study each day.

14. After all c children complete a Little League tryout, the league officials decide that they have enough players for 8 teams of equal size. Express the number of players on each team.

15. If x inches of tape have been used off the roll shown, how many inches of tape are left on the roll?

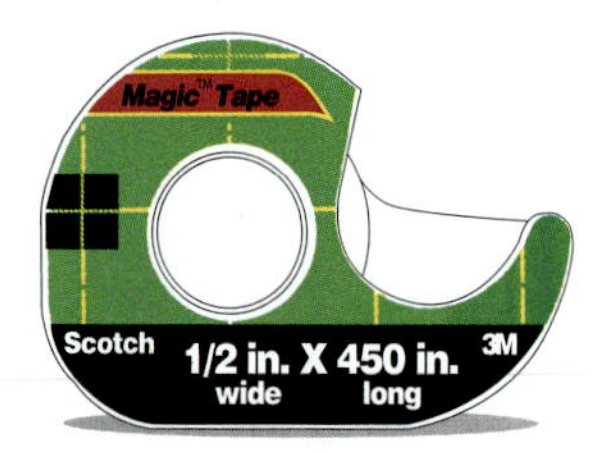

16. Complete each table.

a.

Number of decades	Number of years
1	
2	
3	
d	

b.

Number of inches	Number of feet
12	
24	
36	
i	

NOTATION *Write each expression in another algebraic form.*

17. $x \cdot 8$

18. $5(t)$

19. $10 \div g$

20. $h \div 16$

PRACTICE *Translate each phrase into an algebraic expression.*

21. The difference of x and 9

22. The sum of 12 and p

23. Two-thirds of the population p

24. The product of x and 34

25. r added to six

26. The ratio of i to 100

27. 15 less than d

28. Forty increased by w

29. s subtracted from 1

30. Sixteen minus a

31. Twice the price p

32. T reduced by 50

33. Exceeds the standard s by 14

34. The cost c split five equal ways

35. 35 divided by b

36. The total of 5 and 12 and q

37. x decreased by 2

38. 7 more than the average a

Write a word description of each algebraic expression. (Answers may vary.)

39. $c + 7$

40. $7 - c$

41. $c - 7$

42. $7c$

43. **a.** How many seconds are there in m minutes?
b. In h hours?

44. A man sleeps x hours per day.
a. How many hours does he sleep in a week?
b. In a year?

45. A secretary earns an annual salary of s dollars.
a. Express her salary per month.
b. Express her salary per week.

46. A store manager earns d dollars an hour.
a. How much money will he earn in an 8-hour day?
b. In a 40-hour week?

47. A rope is f feet long.
a. Express its length in inches.
b. Express its length in yards.

48. A chain is y yards long.
a. Express its length in feet.
b. Express its length in inches.

Write an expression that represents the unknown quantity.

49. The highest decibel reading during a rock concert registered just 5 decibels shy of that of a jet engine. If a jet engine is normally j decibels, what was the decibel reading for the concert?

50. A couple needed to purchase 21 presents for friends and relatives on their holiday gift list. If the husband purchased g presents, how many presents did the wife need to buy?

51. A restaurant owner purchased s six-packs of cola. How many individual cans would this be?

52. The height of a hedge was f feet before a gardener cut 2 feet off the top. What was the height of the trimmed hedge?

53. A pad of yellow legal paper contains p pages. If a lawyer uses 15 pages every day, how many days will one pad last?

54. The projected cost c (in dollars) of a freeway was too low by a factor of 10! What was the actual cost of the freeway?

55. In a recycling drive, a campus ecology club collected t tons of newspaper. A Boy Scout troop then contributed an additional 2 tons. How many tons of newspaper were collected?

56. A graduating class of x people took buses that held 40 students each to an all-night graduation party. How many buses were needed to transport the class?

Choose a variable to represent one unknown. Then write an expression that represents the other unknown.

57. The rectangle is 6 units longer than it is wide. Express the length and width of this rectangle.

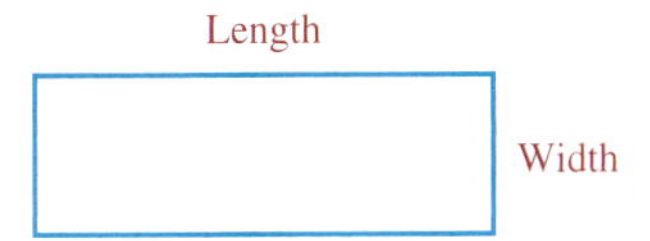

58. The smaller pipe in the illustration takes three times longer to fill the tank than does the larger pipe. Express how long it takes each pipe to fill the tank.

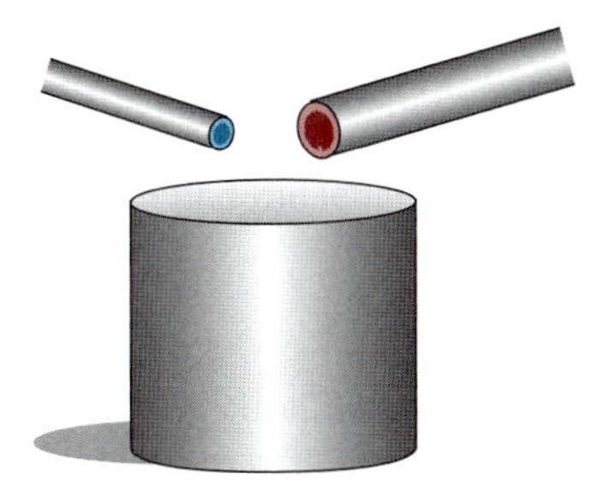

59. A car radiator originally contained 6 gallons of coolant before some was drained out. Express the amount that was drained out and the amount that remains in the radiator.

60. During a sale, the regular price of a CD was reduced by \$2. Express the regular price and the new sale price.

Translate each phrase to an algebraic expression.

61. Five more than triple x

62. The quotient of 5 and t is reduced by four.

63. The product of a and 10 is increased by 12.

64. Thirty more than the difference of 78 and d

APPLICATIONS *Choose a variable to represent one unknown. Then write an expression that represents the other unknown. (Answers may vary.)*

65. ELECTIONS In 1960, John F. Kennedy was elected President of the United States with a popular vote of only 118,550 votes more than that of Richard M. Nixon. Express how many votes each candidate received.

66. EARTHQUAKES An earthquake with a reading of 7.0 on the Richter scale releases ten times as much energy as an earthquake that registers 6.0 on the scale. Express the amount of energy released by an earthquake of each magnitude.

67. THE BEATLES According to music historians, sales of the Beatles' second most popular single, *Hey Jude,* trail the sales of their most popular single, *I Want to Hold Your Hand,* by 2,000,000 copies. Express the sales of each single.

68. YOUTH SPORTS The following illustration shows how participation in organized soccer and volleyball changes as girls and boys enter their teen years.

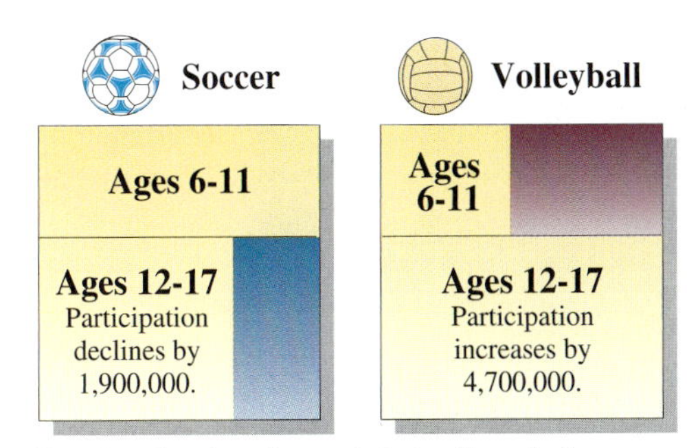

Based on data from Soccer Industry Council of America and American Sports Data

a. Express the number of soccer participants for both age groups.

b. Express the number of volleyball participants for both age groups.

b. $\frac{-x - 15}{6} = \frac{-(3) - 15}{6}$ Substitute 3 for x. Use parentheses.

$= \frac{-3 - 15}{6}$ $-(3) = -3$.

$= \frac{-3 + (-15)}{6}$ Add the opposite of 15.

$= \frac{-18}{6}$ Perform the addition: $-3 + (-15) = -18$.

$= -3$ Perform the division.

Answers **a.** 21, **b.** −4

Self Check 2
Evaluate each expression for $t = -3$:
a. $-2t + 4t^2$
b. $-t + 2(t + 1)$
c. $-t^2 + 16$

EXAMPLE 2 Evaluate each expression for $a = -2$: **a.** $-3a + 4a^2$, **b.** $-a + 3(1 + a)$, and **c.** $-a^2 - 5$.

Solution

a. $-3a + 4a^2 = -3(-2) + 4(-2)^2$ Substitute −2 for a. Use parentheses.

$= -3(-2) + 4(4)$ Evaluate the exponential exprsesion: $(-2)^2 = 4$.

$= 6 + 16$ Perform the multiplications.

$= 22$ Perform the addition.

b. $-a + 3(1 + a) = -(-2) + 3[1 + (-2)]$ Substitute −2 for a. Use parentheses.

$= -(-2) + 3(-1)$ Perform the addition within the brackets.

$= 2 + (-3)$ Simplify: $-(-2) = 2$. Perform the multiplication: $3(-1) = -3$.

$= -1$ Perform the addition.

c. $-a^2 - 5 = -(-2)^2 - 5$ Substitute −2 for a. Use parentheses.

$= -(4) - 5$ Evaluate the exponential expression: $(-2)^2 = 4$.

$= -4 - 5$ Simplify: $-(4) = -4$.

$= -9$ Subtract: $-4 - 5 = -4 + (-5) = -9$.

Answers **a.** 42, **b.** −1, **c.** 7

To evaluate algebraic expressions containing two or more variables, we need to know the value of each variable.

Self Check 3
Evaluate $(5rs + 4s)^2$ for $r = -1$ and $s = 5$.

EXAMPLE 3 Evaluate $(8hg + 6g)^2$ for $h = -1$ and $g = 5$.

Solution

$(8hg + 6g)^2 = [8(-1)(5) + 6(5)]^2$ Substitute −1 for h and 5 for g.

$= (-40 + 30)^2$ Perform the multiplications within the brackets.

$= (-10)^2$ Perform the addition within the parentheses.

$= 100$ Evaluate the power.

Answer 25

Formulas

A **formula** is an equation that is used to state a relationship between two or more variables. Formulas are used in many fields: economics, physical education, anthropology, biology, automotive repair, and nursing, just to name a few. In this section, we will consider several formulas from business, science, and mathematics.

Formulas from business

A Formula to Find the Sale Price. If a car that usually sells for $12,000 is discounted $1,500, you can find the sale price using the formula

Sale price = original price − discount

Using the variables s to represent the sale price, p the original price, and d the discount, we can write this formula as

$$s = p - d$$

To find the sale price of the car, we substitute 12,000 for p, substitute 1,500 for d, and evaluate the right-hand side of the equation.

$$\begin{aligned} s &= p - d \\ &= 12{,}000 - 1{,}500 && \text{Substitute 12,000 for } p \text{ and 1,500 for } d. \\ &= 10{,}500 && \text{Perform the subtraction.} \end{aligned}$$

The sale price of the car is $10,500.

A Formula to Find the Retail Price. To make a profit, a merchant must sell a product for more than he paid for it. The price at which he sells the product, called the *retail price,* is the *sum* of what the item cost him and the markup.

Retail price = cost + markup

Using the variables r to represent the retail price, c the cost, and m the markup, we can write this formula as

$$r = c + m$$

As an example, suppose that a store owner buys a lamp for $35 and then marks up the cost $20 before selling it. We can find the retail price of the lamp using this formula.

$$\begin{aligned} r &= c + m \\ &= 35 + 20 && \text{Substitute 35 for } c \text{ and 20 for } m. \\ &= 55 && \text{Perform the addition.} \end{aligned}$$

The retail price of the lamp is $55.

A Formula to Find Profit. The profit a business makes is the *difference* of the revenue (the money it takes in) and the costs.

Profit = revenue − costs

Using the variables p to represent the profit, r the revenue, and c the costs, we have the formula

$$p = r - c$$

As an example, suppose that a charity telethon took in $14 million in donations but had expenses totaling $2 million. We can find the profit made by the charity by subtracting the expenses (costs) from the donations (revenue).

$$\begin{aligned} p &= r - c \\ &= 14 - 2 \quad \text{Substitute 14 for } r \text{ and 2 for } c. \\ &= 12 \quad \text{Perform the subtraction.} \end{aligned}$$

The charity made a profit of $12 million from the telethon.

Formulas from science

A Formula to Find the Distance Traveled. If we know the rate (speed) at which we are traveling and the time we will be moving at that rate, we can find the distance traveled using the formula

Distance = rate · time

Using the variables d to represent the distance, r the rate, and t the time, we have the formula

$$d = rt$$

Self Check 4

Nevada's highway speed limit for trucks is 75 mph. How far would a truck travel in 3 hours?

Answer 225 mi

EXAMPLE 4 Highway speed limits. Three state speed limits for trucks are shown below. At each of these speeds, how far would a truck travel in 3 hours?

Solution To find the distance traveled by a truck in Ohio, we write

$$\begin{aligned} d &= rt \\ &= 55(3) \quad \text{55 mph is the rate } r \text{, and 3 hours is the time } t. \\ &= 165 \quad \text{Perform the multiplication. The units of the answer are miles.} \end{aligned}$$

At 55 mph, a truck would travel 165 miles in 3 hours. We can use a table to display the calculations for each state.

	r ·	t =	d
Ohio	55	3	165
Indiana	60	3	180
Kentucky	65	3	195

This column gives the distance traveled, in miles.

! COMMENT When using $d = rt$ to find distance, make sure that the units are similar. For example, if the rate is given in miles per hour, the time must be expressed in hours.

A Formula for Converting Degrees Fahrenheit to Degrees Celsius. Electronic message boards in front of some banks flash two temperature readings. This is because temperature can be measured using the Fahrenheit or the Celsius scale. The Fahrenheit scale is used in the American system of measurement and the Celsius scale in the metric system. The two scales are shown on the thermometers in Figure 3-6. This should help you to see how the two scales are related. There is a formula to convert a Fahrenheit reading F to a Celsius reading C.

$$C = \frac{5}{9}(F - 32)$$

or

$$C = \frac{5(F - 32)}{9}$$

Later we will see that there is a formula to convert a Celsius reading to a Fahrenheit reading.

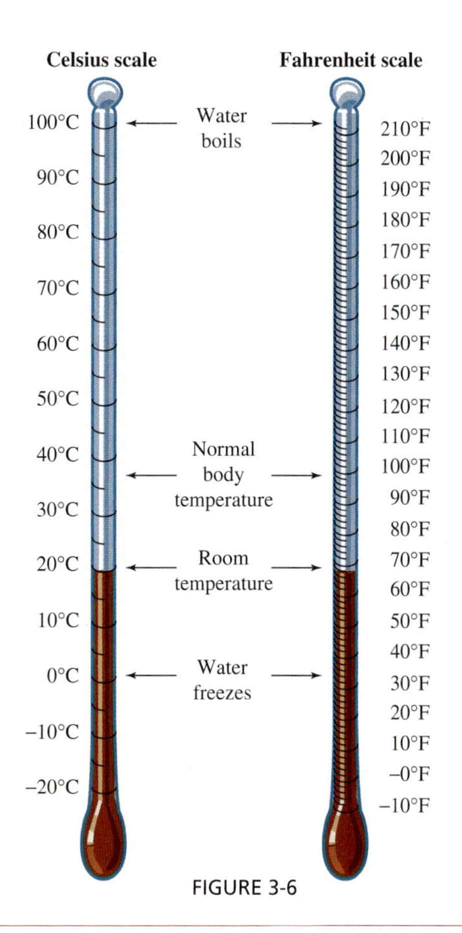

FIGURE 3-6

EXAMPLE 5 The thermostat in an office building was set at 77° F. Convert this setting to degrees Celsius.

Solution

$$C = \frac{5}{9}(F - 32)$$

$= \frac{5}{9}(77 - 32)$ Substitute the Fahrenheit temperature, 77, for F.

$= \frac{5}{9}(45)$ Perform the operation within parentheses first: $77 - 32 = 45$.

$= \frac{225}{9}$ Perform the multiplication: $\frac{5}{9}\left(\frac{45}{1}\right) = \frac{225}{9}$.

$= 25$ Perform the division.

The thermostat is set at 25° C. We would obtain the same result if we had used the other form of the formula.

Self Check 5

The record high temperature for New Mexico is 122° F, on June 27, 1994. Convert this temperature to degrees Celsius.

Answer 50° C

A Formula to Find the Distance an Object Falls. The distance an object falls (in feet) when it is dropped from a height is related to the time (in seconds) that it has been falling by the formula

Distance fallen = 16 · $(\text{time})^2$

Using the variables d to represent the distance and t the time, we have

$$d = 16t^2$$

Self Check 6
Find the distance a rock fell in 3 seconds if it was dropped over the edge of the Grand Canyon.

Answer 144 ft

EXAMPLE 6 **Balloon rides.**

Find the distance a camera fell in 6 seconds if it was dropped overboard by a vacationer taking a hot-air balloon ride.

Solution We can use the formula $d = 16t^2$ to find the distance the camera fell.

$d = 16t^2$

$= 16(6)^2$ The camera fell for 6 seconds. Substitute 6 for t.

$= 16(36)$ Evaluate the exponential expression: $6^2 = 36$.

$= 576$ Perform the multiplication.

The camera fell 576 feet.

Formulas from mathematics

A Formula to Find the Arithmetic Mean (Average). The arithmetic mean, or average, of a set of numbers is a value around which the numbers are grouped. To find the arithmetic mean (average), we divide the *sum* of all the values by the *number* of values. Writing this as a formula, we get

$$\text{Mean} = \frac{\text{sum of the values}}{\text{number of values}}$$

Using the variables A to represent the mean (average), S the sum, and n the number of values, we have

$$A = \frac{S}{n}$$

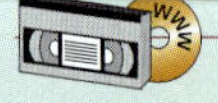

EXAMPLE 7 **Response time to 911 calls.**

To measure its effectiveness, a police department recorded the length of time between incoming 911 calls and the arrival of a police unit at the scene. The response times for an entire week are listed in Table 3-2. Find the average response time.

Response times	Occurrences
2 min	3
3 min	16
4 min	52
5 min	22

TABLE 3-2

Solution The average response time can be found using the formula $A = \frac{S}{n}$. To find S, we need to find the sum of all the response times. Since the response time of 2 minutes occurred 3 times, we need to add 2 + 2 + 2. Since the response time of 3 minutes occurred 16 times, we need to add sixteen 3's, and so on. More simply, we can find S by multiplying each response time by the number of occurrences and then adding the results.

$S = 2(3) + 3(16) + 4(52) + 5(22)$ 2 min occurred 3 times, 3 min occurred 16 times, 4 min occurred 52 times, and 5 min occurred 22 times.

$= 6 + 48 + 208 + 110$ Perform the multiplications.

$= 372$ Perform the additions.

To find n, the number of response times, we add the number of occurrences.

$n = 3 + 16 + 52 + 22 = 93$

Replacing S with 372 and n with 93, we have

$$A = \frac{S}{n}$$

$$= \frac{372}{93} \quad \text{Substitute 372 for } S \text{ and 93 for } n.$$

$$= 4 \quad \text{Perform the division.}$$

The average response time was 4 minutes.

Study Time — THINK IT THROUGH

"Your success in school is dependent on your ability to study effectively and efficiently. The results of poor study skills are wasted time, frustration, and low or failing grades." Effective Study Skills, Dr. Bob Kizlik, 2004

For a course that meets for h hours each week, the formula $H = 2h$ gives the suggested number of hours H that a student should study the course outside of class each week. If a student expects difficulty in a course, the formula can be adjusted upward to $H = 3h$. Use the formulas to complete the table at right.

If a course meets for:	Suggested study time (hours per week)	Expanded study time (hours per week)
2 hours per week		
3 hours per week		
4 hours per week		
5 hours per week		

Section 3.2 STUDY SET

VOCABULARY *Fill in the blanks.*

1. A ________ is an equation that states a relationship between two or more variables.
2. An algebraic __________ is a combination of variables, numbers, and the operation symbols for addition, subtraction, multiplication, and division.
3. To evaluate an algebraic expression, we __________ specific numbers for the variables in the expression and apply the rules for the order of operations.
4. The arithmetic mean or ________ of a group of numbers is a value around which the numbers are grouped.

CONCEPTS

5. Show the misunderstanding that occurs if we don't write parentheses around -8 when evaluating the expression $2x + 10$ for $x = -8$.
6. **a.** Which of the formulas studied in this section involve *a difference* of two quantities?

   b. Which of the formulas studied in this section involve a *product?*
7. The plans for building a children's swing set are shown.

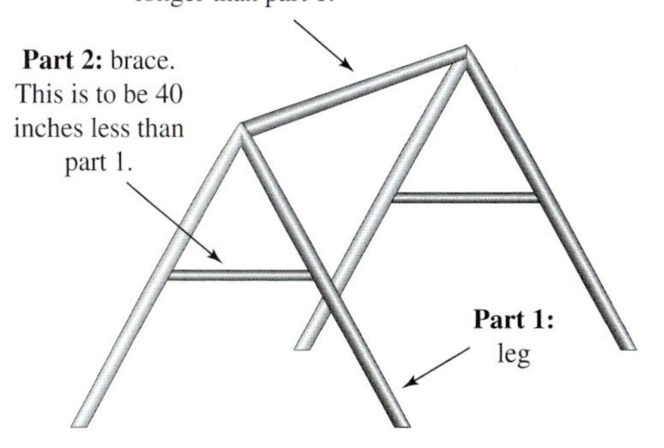

   a. Choose a variable to represent the length of one part of the swing set. Then write expressions that represent the lengths of the other two parts.

   b. If the builder chooses to have part 1 be 60 inches long, how long should parts 2 and 3 be?

8. A television studio art department plans to construct a series of set decorations out of plywood, using the plan shown.

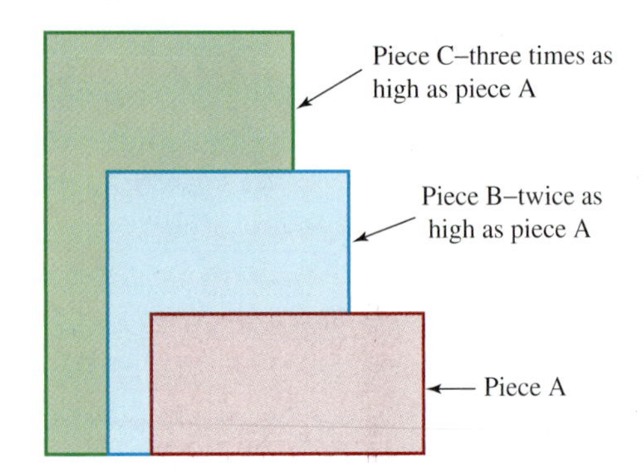

a. Choose a variable to represent the height of one piece of plywood. Then write algebraic expressions that represent the heights of the other two pieces.

b. For the foreground, designers will make piece A 15 inches high. How high should pieces B and C be?

c. For the background, piece A will be 30 inches high. How high should pieces B and C be?

9. A ticket outlet adds a service charge of \$2 to the price of every ticket it sells.

a. Complete the pricing chart.

Ticket price	Service charge	Total cost
20	2	
25	2	
p	2	

b. Write a formula for the total cost T if the price of a ticket is p dollars.

10. Explain why the following instruction is incomplete.

Evaluate the algebraic expression $3a^2 - 4$.

11. Complete the table below by finding the distance traveled in each instance.

	Rate (mph) ·	time (hr) =	distance (mi)
Bike	12	4	
Walking	3	t	
Car	x	3	

12. DASHBOARDS The illustration shows part of a dashboard. Explain what each instrument measures. How are these measurements mathematically related?

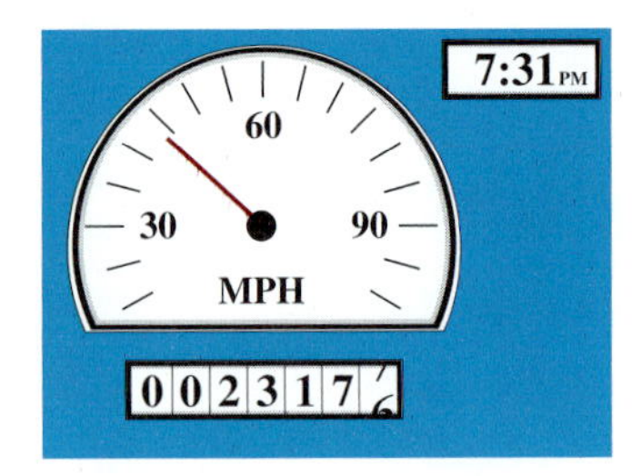

13. What occupation might use a formula that finds

a. target heart rate after a workout.

b. gas mileage of a car.

c. age of a fossil.

d. equity in a home.

e. dose to administer.

f. cost-of-living index.

14. A car travels at a rate of 65 mph for 15 minutes. What is wrong with the following thinking?

$$\begin{aligned} d &= rt \\ &= 65(15) \\ &= 975 \end{aligned}$$

The car travels 975 miles in 15 minutes.

NOTATION

15. Use variables to write the formula that relates each of the quantities listed below.

a. Rate, distance, time

b. Centigrade temperature, Fahrenheit temperature

c. Time, the distance an object falls when dropped

16. Use variables to write the formula that relates each of the quantities listed below.

a. Original price, sale price, discount

b. Number of values, average, sum of values

c. Cost, profit, revenue

d. Markup, retail price, cost

PRACTICE *Evaluate each expression for the given value of the variable.*

17. $3x + 5$ for $x = 4$
18. $1 + 7a$ for $a = 2$
19. $-p$ for $p = -4$
20. $-j$ for $j = -9$
21. $-4t$ for $t = -10$
22. $-12m$ for $m = -6$
23. $\frac{x-8}{2}$ for $x = -4$
24. $\frac{-10+y}{-4}$ for $y = -6$
25. $2(p + 9)$ for $p = -12$
26. $3(r - 20)$ for $r = 15$
27. $x^2 - x - 7$ for $x = -5$
28. $a^2 + 3a - 9$ for $a = -3$
29. $-s^3 + 8s$ for $s = -2$
30. $r^3 + 5r$ for $r = 1$
31. $4x^2$ for $x = 5$
32. $3f^2$ for $f = 3$
33. $-b^2 + 3b$ for $b = -4$
34. $-a^2 + 5a$ for $a = -3$
35. $\frac{24+k}{3k}$ for $k = 3$
36. $\frac{4-h}{h-4}$ for $h = -1$
37. $|6 - x|$ for $x = 50$
38. $|3c - 1|$ for $c = -1$
39. $-2|x| - 7$ for $x = -7$
40. $|x^2 - 7^2|$ for $x = 7$

Evaluate each algebraic expression for the given values of the variables.

41. $\frac{x}{y}$ for $x = 30$ and $y = -10$
42. $\frac{e}{3f}$ for $e = 24$ and $f = -8$
43. $-x - y$ for $x = -1$ and $y = 8$
44. $-a - 5b$ for $a = -9$ and $b = 6$
45. $x(5h - 1)$ for $x = -2$ and $h = 2$
46. $c(2k - 7)$ for $c = -3$ and $k = 4$
47. $b^2 - 4ac$ for $b = -3$, $a = 4$, and $c = -1$
48. $3r^2h$ for $r = 4$ and $h = 2$
49. $x^2 - y^2$ for $x = 5$ and $y = -2$
50. $x^3 - y^3$ for $x = -1$ and $y = 2$
51. $\frac{50-6s}{-t}$ for $s = 5$ and $t = 4$
52. $\frac{7v-5r}{-r}$ for $v = 8$ and $r = 4$
53. $-5abc + 1$ for $a = -2$, $b = -1$, and $c = 3$
54. $-rst + 2t$ for $r = -3$, $s = -1$, and $t = -2$
55. $5s^2t$ for $s = -3$ and $t = -1$
56. $-3k^2t$ for $k = -2$ and $t = -3$
57. $|a^2 - b^2|$ for $a = -2$ and $b = -5$
58. $-|2x - 3y + 10|$ for $x = 0$ and $y = -4$

Use the appropriate formula to answer each question.

59. It costs a snack bar owner 20 cents to make a snow cone. If the markup is 50 cents, what is the price of a snow cone?
60. Find the distance covered by a jet if it travels for 3 hours at 550 mph.
61. A school carnival brought in revenues of $13,500 and had costs of $5,300. What was the profit?
62. For the month of June, a florist's cost of doing business was $3,795. If June revenues totaled $5,115, what was her profit for the month?
63. A jewelry store buys bracelets for $18 and marks them up $5. What is the retail price of a bracelet?
64. A shopkeeper marks up the cost of every item she carries by the amount she paid for the item. If a fan costs her $27, what does she charge for the fan?
65. Find the distance covered by a car traveling 60 miles per hour for 5 hours.
66. Find the sale price of a pair of skis that usually sells for $200 but is discounted $35.
67. Find the Celsius temperature reading if the Fahrenheit reading is 14°.
68. Find the Celsius temperature reading if the Fahrenheit reading is 113°.
69. Find the average for a bowler who rolled scores of 254, 225, and 238.
70. On its first night of business, a pizza parlor brought in $445. The owner estimated his costs that night to be $295. What was the profit?
71. Find the distance a ball has fallen 2 seconds after being dropped from a tall building.
72. A store owner buys a pair of pants for $25 and marks them up $15 for sale. What is the retail price of the pants?

APPLICATIONS

73. FINANCIAL STATEMENTS Use the following data to complete the financial statement for Avon Products, Inc. on the next page.

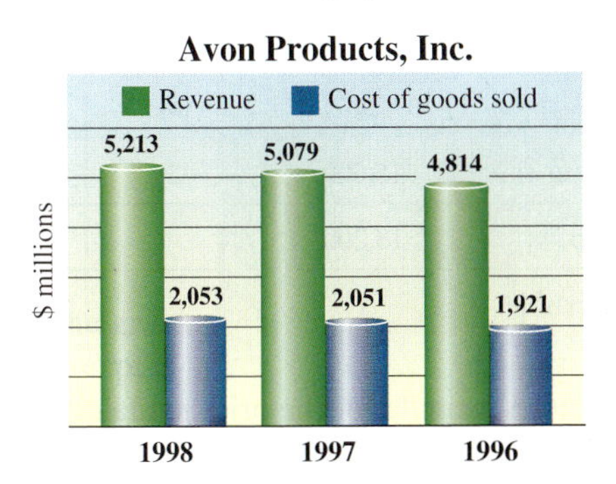

Section 3.3 STUDY SET

VOCABULARY *Fill in the blanks.*

1. The __________ property tells us how to multiply $5(x + 7)$. After doing the multiplication to obtain $5x + 35$, we say that the parentheses have been __________.
2. To __________ an algebraic expression means to use algebraic properties to write it in simpler form.
3. When an algebraic expression is simplified, the result is an __________ expression.
4. We __________ expressions, and we __________ equations.

CONCEPTS

5. State the distributive property using the variables x, y, and z.
6. Use the variables r, s, and t to state the distributive property of subtraction.
7. The following expressions are examples of the *right* and *left* distributive properties.

   $5(w + 7)$ and $(w + 7)5$

   Which of the two do you think would be termed the right distributive property?
8. For each of the following expressions, determine whether the distributive property applies.

   a. $2(5t)$ b. $2(t + 5)$
   c. $5(2 \cdot t)$ d. $(2t)5$
   e. $(2)(-t)5$ f. $(5 - t)(-2)$
9. The distributive property can be demonstrated using the following illustration. Fill in the blanks: Two groups of 6 plus three groups of 6 is ___ groups of 6. Therefore,

   $\_\_ \cdot 2 + \_\_ \cdot 3 = 6(\_\_ + \_\_)$.

10. a. Simplify: $2(5x)$.
    b. Remove parentheses: $2(5 + x)$.
11. Write an equivalent expression for $-(y + 9)$ without using parentheses.
12. Explain what the arrows are illustrating.

    $-9(y - 7)$

NOTATION *Complete each solution.*

13. $-5(7n) = (\_\_ \cdot 7)n$
    $= -35n$
14. $6y(-9) = 6(\_\_)y$
    $= [6(\_\_)]y$
    $= -54y$
15. $-9(-4 - 5y) = (\_\_)(-4) - (\_\_)(5y)$
    $= 36 - (\_\_)$
    $= 36 + 45y$
16. $4(2a + b - 1) = 4(\_\_) + 4(\_\_) - \_\_(1)$
    $= 8a + 4b - 4$
17. Write each expression in simpler form, using fewer mathematical symbols.

    a. $-(-x)$
    b. $x - (-5)$
    c. $5x - 10y + (-15)$
    d. $5 \cdot x$
18. Determine what number is to be distributed.

    a. $-6(x - 2)$ b. $(t + 1)(-5)$
    c. $(a + 24)8$ d. $-(z - 16)$

PRACTICE *Simplify each algebraic expression.*

19. $2(6x)$
20. $4(7b)$
21. $-5(6y)$
22. $-12(6t)$
23. $-10(-10t)$
24. $-8(-6k)$
25. $(4s)3$
26. $(9j)7$
27. $2c \cdot 7$
28. $11f \cdot 9$
29. $-5 \cdot 8h$
30. $-8 \cdot 4d$
31. $-7x(6y)$
32. $13a(-2b)$
33. $4r \cdot 4s$
34. $7x \cdot 7y$
35. $2x(5y)(3)$
36. $4(3z)(4)$
37. $5r(2)(-3b)$
38. $4d(5)(-3e)$
39. $5 \cdot 8c \cdot 2$
40. $3 \cdot 6j \cdot 2$
41. $(-1)(-2e)(-4)$
42. $(-1)(-5t)(-1)$

Use the distributive property to remove parentheses.

43. $4(x + 1)$
44. $5(y + 3)$
45. $4(4 - x)$
46. $5(7 + k)$
47. $-2(3e + 3)$
48. $-5(7t + 2)$
49. $-8(2q - 6)$
50. $-5(3p - 8)$

51. $-4(-3 - 5s)$
52. $-6(-1 - 3d)$
53. $(7 + 4d)6$
54. $(8r + 2)7$
55. $(5r - 6)(-5)$
56. $(3z - 7)(-8)$
57. $(-4 - 3d)6$
58. $(-4 - 2j)5$
59. $3(3x - 7y + 2)$
60. $5(4 - 5r + 8s)$
61. $-3(-3z - 3x - 5y)$
62. $-10(5e + 4a + 6t)$

Write each expression without using parentheses.

63. $-(x + 3)$
64. $-(5 + y)$
65. $-(4t + 5)$
66. $-(8x + 4)$
67. $-(-3w - 4)$
68. $-(-6 - 4y)$
69. $-(5x - 4y + 1)$
70. $-(6r - 5f + 1)$

Each expression is the result of an application of the distributive property. What was the original algebraic expression?

71. $2(4x) + 2(5)$
72. $3(3y) + 3(7)$
73. $-4(5) - 3x(5)$
74. $-8(7) - (4s)(7)$
75. $-3(4y) - (-3)(2)$
76. $-5(11s) - (-5)(11t)$
77. $3(4) - 3(7t) - 3(5s)$
78. $2(7y) + 2(8x) - 2(4)$

WRITING

79. Explain what it means to simplify an algebraic expression. Give an example.
80. Explain the commutative and associative properties of multiplication.
81. Explain how to apply the distributive property.
82. Explain why the distributive property applies to $2(3 + x)$ but does not apply to $2(3x)$.

REVIEW

83. Evaluate: $|-6 + 1|$.
84. Subtract: $-1 - (-4)$.
85. Identify the operation associated with each word: *product, quotient difference, sum.*
86. What steps are used to find the mean (average) of a set of scores?
87. Insert the proper inequality symbol: $-6 \quad -7$.
88. Fill in the blank: To factor a number means to express it as the ________ of other whole numbers.
89. Which of the following involve area: carpeting a room, fencing a yard, walking around a lake, painting a wall?
90. Write seven squared and seven cubed.

3.4 Combining Like Terms

- Terms of an algebraic expression
- Coefficients of a term
- Terms and factors
- Like terms
- Combining like terms
- Perimeter formulas

In this section, we will show how the distributive property can be used to simplify algebraic expressions that involve addition and subtraction. We will also review the concept of perimeter and write the formulas for the perimeter of a rectangle and a square using variables.

Terms of an algebraic expression

Addition signs break algebraic expressions into smaller parts called **terms.** The expression $3x + 8$ contains two terms, $3x$ and 8.

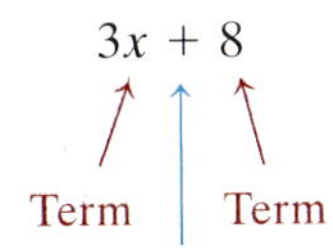

The addition sign breaks the expression into two terms.

5. $2(x + 3) = 2x + 2(3)$ is an example of the use of the __________ property.
6. Terms with exactly the same variables and exponents are called _______ terms.
7. Simplifying the _______ (or difference) of like terms is called combining like terms.
8. The numbers multiplied together to form a product are called __________.

CONCEPTS

9. Determine whether x is used as a factor or as a term.
   a. $12 + x$
   b. $7x$
   c. $12y + 12x - 6$
   d. $-36xy$
10. Determine whether $6y$ is used as a factor or as a term.
    a. $6yz$ b. $10 + 6y$
    c. $9xy + 6y$ d. $6y - 18$
11. What is the coefficient of each term?
    a. $11x$ b. $8t$
    c. $-4x^2$ d. a
    e. $-x$ f. $102xy$
12. What is the coefficient of the second term of each expression?
    a. $5x^2 + 6x + 7$
    b. $xy - x + y + 10$
    c. $9y^2 + y + 8$
    d. $5x^3 - 4x^2 + 3x + 1$
13. Complete the table.

Term	Coefficient	Variable part
$6m$		
$-75t$		
w		
$4bh$		

14. Simplify each pair of expressions, if possible.
    a. $5(2x)$ and $5 + 2x$
    b. $6(-7x)$ and $6 - 7x$
    c. $2(3x)(3)$ and $2 + 3x + 3$
    d. $x \cdot x$ and $x + x$
15. When simplifying an algebraic expression, some students use underlining.

$$\underline{\underline{3y}} + \underline{4} + \underline{\underline{5y}} + \underline{8}$$

What purpose does the underlining serve?

16. Determine whether each statement is true or false.
    a. $x = 1x$ b. $2x + 0 = 2x$
    c. $-y = -1y$ d. $0 - 4c = 4c$
17. The illustration shows the distance (in miles) that two men live from the office. Find the total distance the men travel from home to office.

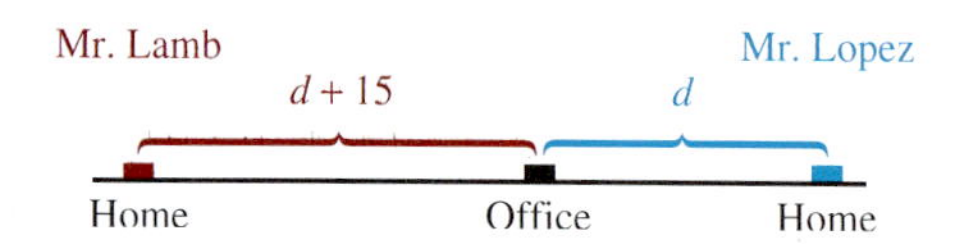

18. The heights of two trees are shown below. Find the sum of their heights.

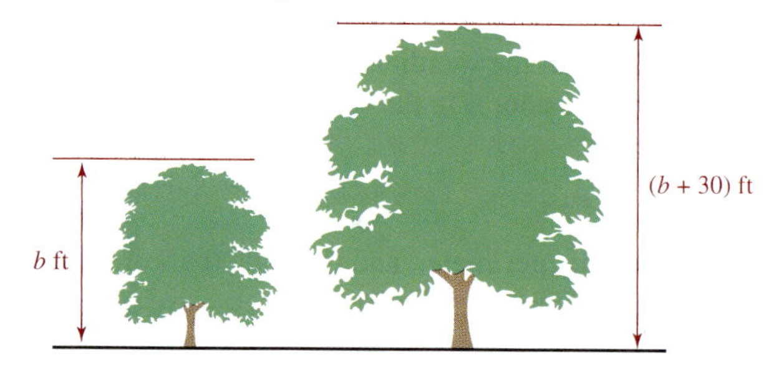

19. a. What does this diagram illustrate?

$$9x + 5x = 14x$$

    b. What does this diagram illustrate?

$$12k - 4k = 8k$$

20. For each expression, identify any like terms.
    a. $3a + 8 + 2a$ b. $10 - 13h + 12$
    c. $3x^2 + 3x + 3$ d. $9y^2 - 9m - 8y^2$

NOTATION *Complete each solution.*

21. $5x + 7x = (5 + \square)x$
    $= 12x$
22. $12w - 16w = (\square - 16)w$
    $= -4w$
23. $2(x - 1) + 3x = 2x - \square + 3x$
    $= \square - 2$
24. $-3(1 - b) - b = \square + \square - b$
    $= \square - 3$
25. In the formula $P = 2l + 2w$,
    a. what does P represent?
    b. what does $2l$ mean?
    c. what does $2w$ mean?
26. In the formula $P = 4s$,
    a. what does P represent?
    b. what does $4s$ mean?

PRACTICE *Identify the terms of each expression.*

27. $3x^2 + 5x + 4$
28. $y^2 + 12y + 6$
29. $5 + 5t - 8t + 4$
30. $3x - y - 5x + y$

What exponent must appear in each box to make the terms like terms?

31. $3x^{\square}, -6x^2$
32. $7a^3, 21a^{\square}$
33. $-8h^5, -5h^{\square}$
34. $25n^4, -15n^{\square}$

Simplify by combining like terms, if possible.

35. $6t + 9t$
36. $7r + 5r$
37. $5s - s$
38. $8y - y$
39. $-5x + 6x$
40. $-8m + 6m$
41. $-5d + 9d$
42. $-4a + 12a$
43. $3e - 7e$
44. $2s - 4s$
45. $h - 7$
46. $j - 8$
47. $4z - 10z$
48. $3w - 18w$
49. $-3x - 4x$
50. $-7y - 9y$
51. $2t - 2t$
52. $7r - 7r$
53. $-6s + 6s$
54. $19c + (-19c)$
55. $x + x + x + x$
56. $s - s - s$
57. $2x + 2y$
58. $5a - 5b$
59. $0 - 2y$
60. $0 - 7x$
61. $3a - 0$
62. $10t + 0$
63. $6t + 9 + 5t + 3$
64. $5x + 3 + 5x + 4$
65. $3w - 4 - w - 1$
66. $6y + 6 - y - 1$
67. $-4r + 8R + 2R - 3r + R$
68. $12a - A - a - 8A - a$
69. $-45d - 12a - 5d + 12a$
70. $-m - n - 8m + n$
71. $4x - 3y - 7 + 4x - 2 - y$
72. $2a + 8 - b - 5 + 5a - 9b$

Simplify each expression.

73. $4(x + 1) + 5(6 + x)$
74. $7(1 + y) + 8(2y + 3)$
75. $5(3 - 2s) + 4(2 - 3s)$
76. $6(t - 3) + 9(2 - t)$
77. $-4(6 - 4e) + 3(e + 1)$
78. $-5(7 - 4t) + 3(2 + 5t)$
79. $3t - (t - 8)$
80. $6n - (4n + 1)$
81. $-2(2 - 3x) - 3(x - 4)$
82. $-3(1 - y) - 5(2y - 6)$
83. $-4(-4y + 5) - 6(y + 2)$
84. $-3(-6y - 8) - 4(5 - y)$

APPLICATIONS

85. MOBILE HOMES The design of a mobile home calls for a six-inch-wide strip of stained pine around the outside of each exterior wall, as shown. If the strip costs 80¢ a linear foot, how much will be spent on the pine used for the trim?

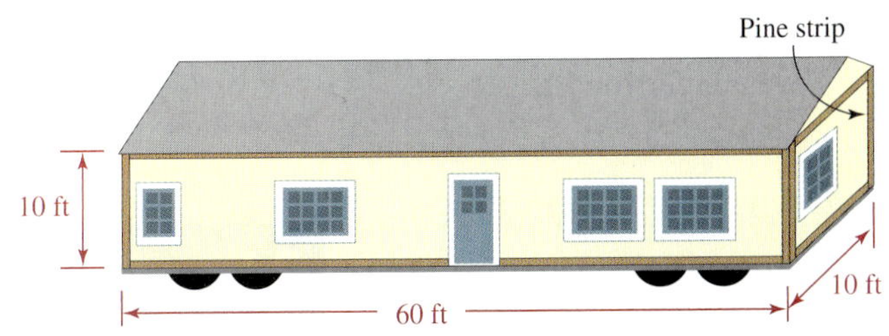

86. LANDSCAPING A landscape architect has designed a planter surrounding two birch trees, as shown. The planter is to be outlined with redwood edging in the shape of a rectangle and two squares. If the material costs 17¢ a running foot, how much will the redwood cost for this project?

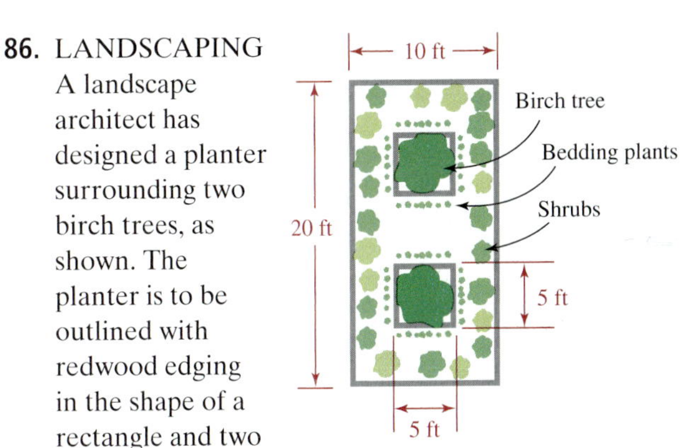

87. PARTY PREPARATIONS The appropriate size of a dance floor for a given number of dancers can be determined from the table shown. Find the perimeter of each of the dance floors listed.

Slow dancers	Fast dancers	Size of floor (in feet)
8	5	9 × 9
14	9	12 × 12
22	15	15 × 15
32	20	18 × 18
50	30	21 × 21

Self Check 6

A fitness club has 150 members. Monthly membership fees are \$25 for nonseniors and \$15 for senior citizens. Find the income the club receives from nonseniors and from seniors each month. Use the table below to present your results. Let m represent the number of non-seniors.

Member's age	Number ·	Fee =	Total income

Answers $25m$ dollars from non-seniors, $15(150 - m)$ dollars from seniors

EXAMPLE 6 Ninety-five people attended a movie matinee. Ticket prices were \$6 for adults and \$4 for children. Write expressions that represent the income received from the sale of children's tickets and from the sale of adults' tickets.

Solution If 95 people attended and if we let c = the number of children's tickets sold, then $95 - c$ = the number of adult tickets sold.

The value of a child's ticket is 4 dollars. Therefore, the income from the sale of c children's tickets is the number times the value: $c \cdot 4$ dollars = $4c$ dollars.

The value of an adult ticket is 6 dollars. The income from the sale of $(95 - c)$ adult tickets is $(95 - c) \cdot 6$ dollars or $6(95 - c)$ dollars.

Type of ticket	Number ·	Value =	Total value
Child	c	4	$4c$
Adult	$95 - c$	6	$6(95 - c)$

This column is number · value.

EXAMPLE 7 **Basketball.** On a night when they scored 110 points, a basketball team made only 5 free throws (worth 1 point each). The remainder of their points came from two- and three-point baskets. If the number of baskets from the field totaled 45, how many two-point and how many three-point baskets did they make?

Analyze the problem

- The team scored 110 points.
- They made 5 free throws (1 point each).
- They made 45 baskets from the field.
- Find the number of two-point and three-point baskets made.

Form an equation

The number of two- and three-point baskets totaled 45. If we let x = the number of three-point baskets made, then $45 - x$ = the number of two-point baskets made. We can now organize the data in a table. We must remember to multiply the *number* of each type of basket by its point *value.*

Type of basket	Number ·	Value =	Total value
Three-point	x	3	$3x$
Two-point	$45 - x$	2	$2(45 - x)$
Free throw	5	1	5

We can express the total number of points scored in two ways.

3 ·	the number of three-point baskets	plus	2 ·	the number of two-point baskets	plus	the number of free throws	is	110.
3 ·	x	+	2 ·	$(45 - x)$	+	5	=	110

Solve the equation

$$\begin{aligned} 3x + 2(45 - x) + 5 &= 110 \\ 3x + 90 - 2x + 5 &= 110 && \text{Distribute the multiplication by 2.} \\ x + 95 &= 110 && \text{Combine like terms.} \\ x + 95 \mathbf{- 95} &= 110 \mathbf{- 95} && \text{Subtract 95 from both sides.} \\ x &= 15 && \text{Combine like terms.} \end{aligned}$$

We can substitute 15 for x in $45 - x$ to find the number of two-point baskets made.

$$45 - x = 45 - \mathbf{15} = 30$$

State the conclusion

The basketball team made 15 three-point baskets and 30 two-point baskets.

Check the result

If we multiply the number of three-point baskets by their value, we get $15 \cdot 3 = 45$ points. If we multiply the number of two-point baskets by their value, we get $30 \cdot 2 = 60$ points. If we add the number of made free throws to these two subtotals, we get $45 + 60 + 5 = 110$ points. The answers check.

Section 3.6 STUDY SET

VOCABULARY *Fill in the blanks.*

1. The words *increased by, longer, taller, higher, total,* and *more than* indicate that the operation of ________ should be used.
2. The words *shorter, less than, fewer than, difference,* and *decreased by* indicate that the operation of __________ should be used.

CONCEPTS

3. BUSINESS ACCOUNTS Every month, a salesman adds five new accounts. How many new accounts will he add in x months?
4. ANTIQUE COLLECTING Every year, a woman purchases four antique spoons to add to her collection. How many spoons will she purchase in x years?
5. SERVICE STATIONS See the illustration. How many gallons does the smaller tank hold?

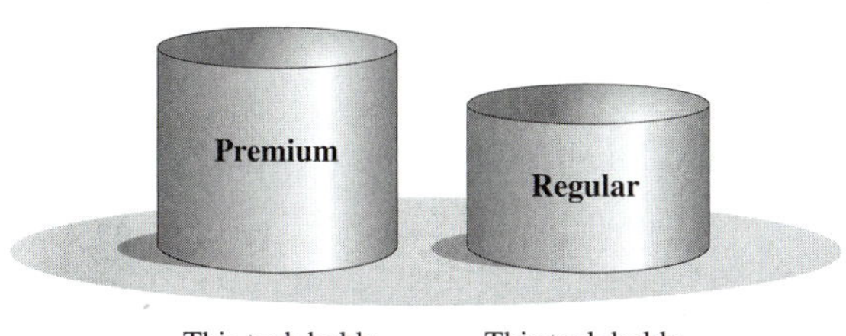

This tank holds g gallons. This tank holds 100 gallons less than the premium tank.

6. SCHOLARSHIPS See the illustration. How many scholarships were awarded this year?

Last year, s scholarships were awarded. Six more scholarships were awarded this year than last year.

7. OCEAN TRAVEL See the illustration. How many miles did the passenger ship travel?

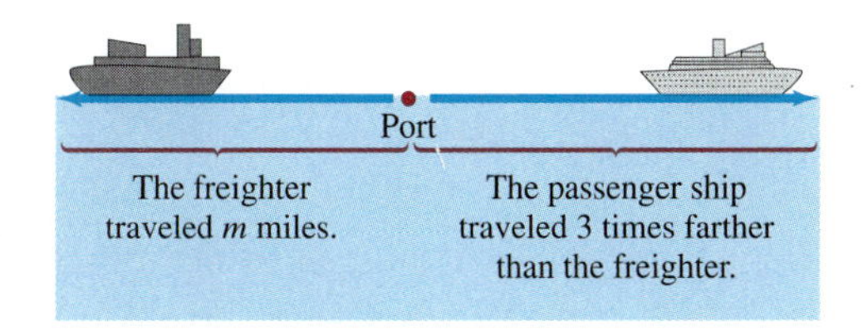

8. TAX REFUNDS See the illustration. How much of the tax refund did the husband get?

A husband and wife received a tax refund of $\$d$. The couple split the refund equally.

9. The length of a rectangle is twice its width. Write an expression that represents the length of the rectangle.

10. Complete this statement about the perimeter of the rectangle shown.

$2 \cdot \square + 2 \cdot \square = 240$

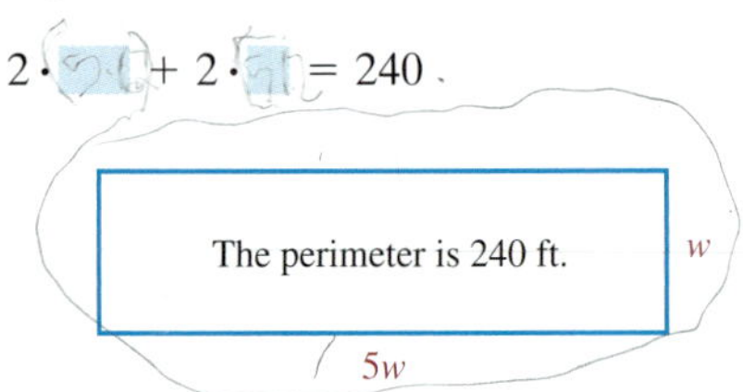

11. COMMISSIONS A shoe salesman receives a commission for every pair of shoes he sells. Complete the table.

Type of shoe	Number sold	Commission per shoe ($)	Total commission ($)
Dress	10	3	
Athletic	12	2	
Child's	x	5	
Sandal	$9 - x$	4	

12. Complete the table.

Type of coin	Number	Value (¢)	Total value (¢)
Nickel	12		
Dime	d		
Quarter	$q + 2$		

13. The illustration shows a rack that contains dress shoes and athletic shoes.

a. How many pairs of shoes are stored in the rack?

b. If there are d pairs of dress shoes in the rack, how many pairs of athletic shoes are there in the rack?

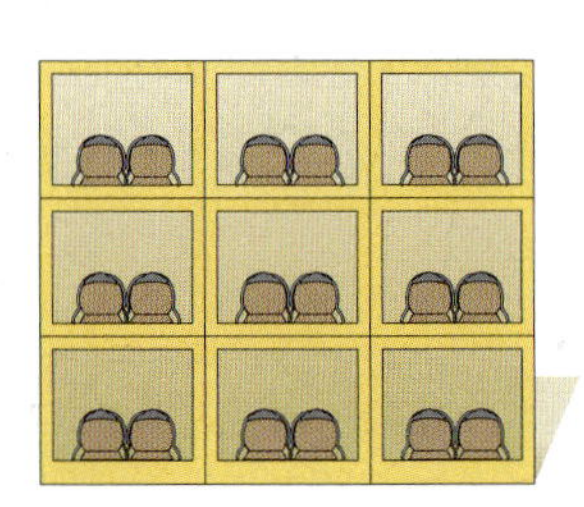

14. The answers to a prealgebra quiz are shown below.

a. How many questions were on the quiz?

b. If the student answered c questions correctly, how many did she answer incorrectly?

PREALGEBRA QUIZ
CHAPTER 3

1. 44	6. 250 ft
2. 376	7. 165 mi
3. equal	8. no
4. $9 - x$	9. yes
5. $4x$	10. simplify

15. AIRLINE SEATING An 88-seat passenger plane has ten times as many economy seats as first-class seats. Find the number of first-class seats.

Analyze the problem

- There are ☐ seats on the plane.
- There are ☐ times as many economy as first-class seats.
- Find ________________.

Form an equation Since the number of economy seats is related to the number of first-class seats, we let $x =$ ________________.

To represent the number of economy seats, look for a key phrase in the problem.

Key phrase: ten times as many

Translation: ________

So ☐ = the number of economy seats.

The number of first-class seats	plus	the number of economy seats	is	88.
x	$+$	☐	$=$	88

Solve the equation

$x + 10x = \square$

$\square = 88$

$x = 8$

State the conclusion There are ☐ first-class seats.

Check the result If there are 8 first-class seats, there are ☐ $\cdot 8 = 80$ economy seats. Adding ☐ and ☐, we get 88. The answer checks.

16. COUPONS A shopper used some 20-cents-off and some 40-cents-off coupons at the supermarket to get a reduction of \$2.60 from her grocery bill. If she used a total of eight coupons, how many of each type did she redeem at the checkout stand?

Analyze the problem

- ____ ¢ and ____ ¢ coupons were redeemed.
- The coupons saved her \$2.60, which is ____ ¢.
- ____ coupons were used.
- Find ____________________.

Form an equation The total number of coupons used was 8. If we let x = the number of 20¢ coupons used, then $8 - x$ = ____________________.

20	·	the number of 20¢ coupons used	plus	40	·	the number of 40¢ coupons used	is	260.
		$20x$	+			____	=	260

Solve the equation

$$20x + 40(8 - x) = \_\_\_\_$$
$$20x + \_\_\_\_ - 40x = 260$$
$$-20x + \_\_\_\_ = 260$$
$$-20x = \_\_\_\_$$
$$x = 3$$

If ____ of the 20¢ coupons were used, then $8 - 3 =$ ____ of the 40¢ coupons were used.

State the conclusion ____ of the 20¢ coupons and ____ of the 40¢ coupons were redeemed.

Check the result The value of ____ of the 20¢ coupons is $3 \cdot 20 =$ ____ ¢. The value of ____ of the 40¢ coupons is $5 \cdot 40 =$ ____ ¢. Adding these two subtotals, we get 260¢, which is \$2.60. The answers check.

Form an equation and solve it to answer each question.

17. BUSINESS After beginning a new position with 15 established accounts, a salesman made it his objective to add 5 new accounts every month. His goal was to reach 100 accounts. At this rate, how many months would it take to reach his goal?

18. LOANS A student plans to pay back a \$600 loan with monthly payments of \$30. How many payments has she made if the debt has been reduced to \$420?

19. ANTIQUES A woman purchases 4 antique spoons each year. She now owns 56 spoons. In how many years will she have 100 spoons in her collection?

20. CONSTRUCTION To get a heavy-equipment operator's certificate, 48 hours of on-the-job training are required. If a woman has completed 24 hours, and the training sessions last for 6 hours, how many more sessions must she take to get the certificate?

21. RENTALS In renting an apartment with two other friends, Enrique agreed to pay the security deposit of \$100 himself. The three of them agreed to contribute equally toward the monthly rent. Enrique's first check to the apartment owner was for \$425. What was the monthly rent for the apartment?

22. TAX REFUNDS After receiving their tax refund, a husband and wife split the refunded money equally. The husband then gave \$50 of his money to charity, leaving him with \$70. What was the amount of the tax refund check?

23. BOTTLED WATER DELIVERY A truck driver left the plant carrying 300 bottles of drinking water. His delivery route consisted of office buildings, each of which was to receive 3 bottles of water. The driver returned to the plant at the end of the day with 117 bottles of water on the truck. To how many office buildings did he deliver?

24. CORPORATE DOWNSIZING In an effort to cut costs, a corporation has decided to lay off 5 employees every month until the number of employees totals 465. If 510 people are now employed, how many months will it take to reach the employment goal?

25. NUMBER PROBLEMS Ten less than five times a number is the same as the number increased by six. What is the number?

26. NUMBER PROBLEMS Four less than seven times a number is the same as the number increased by eight. What is the number?

27. SERVICE STATIONS At a service station, the underground tank storing regular gas holds 100 gallons less than the tank storing premium gas. If the total storage capacity of the tanks is 700 gallons, how much does the premium gas tank hold?

28. SCHOLARSHIPS Because of increased giving, a college scholarship program awarded six more scholarships this year than last year. If a total of 20 scholarships were awarded over the last two years, how many were awarded last year?

29. OCEAN TRAVEL At noon, a passenger ship and a freighter left a port traveling in opposite directions. By midnight, the passenger ship was 3 times farther from port than the freighter was. How far was the freighter from port if the distance between the ships was 84 miles?

30. RADIO STATIONS The daily listening audience of an AM radio station is four times as large as that of its FM sister station. If 100,000 people listen to these two radio stations, how many listeners does the FM station have?

31. INTERIOR DECORATING As part of redecorating, crown molding was installed around the ceiling of a room. Sixty feet of molding was needed for the project. Find the width of the room if its length is twice the width.

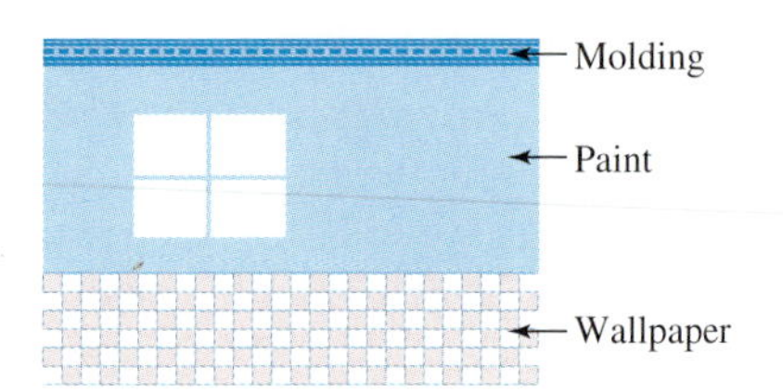

32. SPRINKLER SYSTEMS A landscaper buried a water line around a rectangular lawn to serve as a supply line for a sprinkler system. The length of the lawn is 5 times its width. If 240 feet of pipe was used to do the job, what is the width of the lawn?

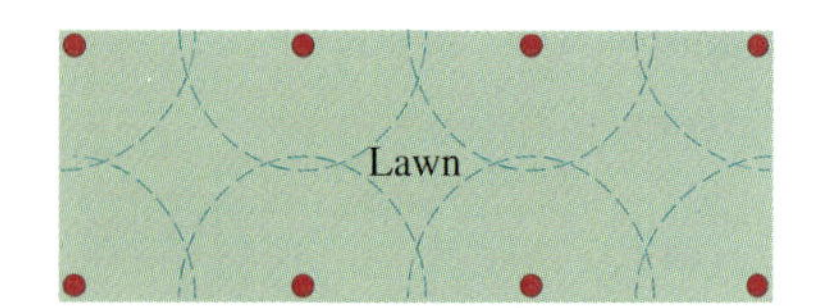

33. COMMERCIALS During a 30-minute television show, a viewer found that the actual program aired a total of 18 minutes more than the time devoted to commercials. How many minutes of commercials were there?

34. CLASS TIME In a biology course, students spend a total of 250 minutes in lab and lecture each week. The lab time is 50 minutes shorter than the lecture time. How many minutes do the students spend in lecture per week?

Form an equation and then solve it to answer each question. Make a table to organize the data.

35. COMMISSIONS A salesman receives a commission of \$3 for every pair of dress shoes he sells. He is paid \$2 for every pair of athletic shoes he sells. After selling 9 pairs of shoes in a day, his commission was \$24. How many pairs of each kind of shoe did he sell that day?

36. GRADING SCALES For every problem answered correctly on an exam, 3 points are awarded. For every incorrect answer, 4 points are deducted. In a 10-question test, a student scored 16 points. How many correct and incorrect answers did he have on the exam?

37. MOVER'S PAY SCALE A part-time mover's regular pay rate is \$60 an hour. If the work involves going up and down stairs, his rate increases to \$90 an hour. In one week, he earned \$1,380 and worked 20 hours. How many hours did he work at each rate?

38. PRESCHOOL ENROLLMENTS A preschool charges \$8 for a child to attend its morning session or \$10 to attend the afternoon session. No child can attend both. Thirty children are enrolled in the preschool. If the daily receipts are \$264, how many children attend each session?

WRITING

39. Explain what should be accomplished in each of the five steps of the problem-solving strategy studied in this section.

40. Use an example to explain the difference between the number of quarters a person has and the value of those quarters.

41. Write a problem that could be represented by the following equation.

Age of father	plus	age of son	is	50.
x	$+$	$x - 20$	$=$	50

42. Write a problem that could be represented by the following equation.

2 ·	length of a field	plus	2 ·	width of a field	is	600 ft.
2 ·	$4x$	$+$	2 ·	x	$=$	600

REVIEW

43. What property is illustrated?

$$(2 + 9) + 1 = 2 + (9 + 1)$$

44. Solve: $4 - x = -8$.

45. Evaluate: -10^2.

46. List the factors of 18.

47. Fill in the blank: Subtraction of a number is the same as ________ of the opposite of that number.

48. Round 123,808 to the nearest ten thousand.

49. Write this prime factorization using exponents: $2 \cdot 2 \cdot 2 \cdot 5 \cdot 5$.

50. The value of a stock dropped \$3 a day for 6 consecutive days. What was the change in the value of the stock over this period?

KEY CONCEPT

Order of Operations

One of the major objectives of this course is that you thoroughly understand the rules for the order of operations.

The following priority list gives the order in which operations must be performed. Fill in the blanks.

1. Perform all calculations within ____________ and other grouping symbols, following the order listed in steps 2–4 below and working from the __________ pair to the __________ pair.
2. Evaluate all ____________ expressions.
3. Perform all ______________ and _________ as they occur from left to right.
4. Perform all __________ and ____________ as they occur from left to right.

When all the grouping symbols have been removed, repeat steps 2–4 to complete the calculation.

If a fraction bar is present, evaluate the numerator and the denominator __________. Then do the division indicated by the fraction bar, if possible.

In algebra, we have to apply the rules for the order of operations in many different settings.

Evaluating numerical expressions

5. Evaluate: $-10 + 4 - 3^2$.
6. Evaluate: $\frac{-30}{6} - (-4)3$.
7. Evaluate: $-2(-3) - 12 \div 6 \cdot 3$.
8. Evaluate: $2(-3)^3(4) + (-6)$.
9. Evaluate: $2(4 + 3 \cdot 2)^2 - (-6)$.
10. Evaluate: $-1^2 + 3[6 - (1 - 5)]$.

Evaluating algebraic expressions

11. Evaluate $1 - 2|8w - w^3|$ for $w = -2$.
12. Evaluate $\frac{50 - 6s}{-t}$ for $s = 5$ and $t = -4$.

Working with formulas

13. Use the formula $P = 2l + 2w$ to find the perimeter of a rectangle that has a length of 30 feet and a width of 16 feet.
14. WINDOW WASHING Use the formula $d = 16t^2$ to determine the distance a squeegee fell if it took 6 seconds for it to hit the ground after a window washer dropped it.

Simplifying algebraic expressions

15. Simplify: $(3x)4 - 2(5x) + x$.
16. Simplify: $2(y + 3) - 3(y - 4)$.

Checking a solution of an equation

17. Is -3 a solution of the equation $15 - 3x = 23$?
18. Is -2 a solution of the equation $-4(2c + 5) - 4(2c - 5) = 32$?

Solving equations

When solving equations, we use the rules for the order of operations in reverse.

19. Consider $2x - 3 = 13$. In what order should we undo the operations so that we can isolate x on the left side?
20. Consider $-10 = 5 + \frac{m}{-6}$. In what order should we undo the operations so that we can isolate m on the right side?

ACCENT ON TEAMWORK

SECTION 3.1

TRANSLATING In column 1, write in all the words or phrases you can think of that indicate addition. In column 2, write in all the words or phrases you can think of that indicate subtraction. Continue in this way for each of the remaining columns.

Addition	Subtraction	Multiplication	Division	Equals

SECTION 3.2

AREA OF A SQUARE What patterns do you see as the following squares increase in size? Draw the next four squares of this sequence, labeling them in a similar way.

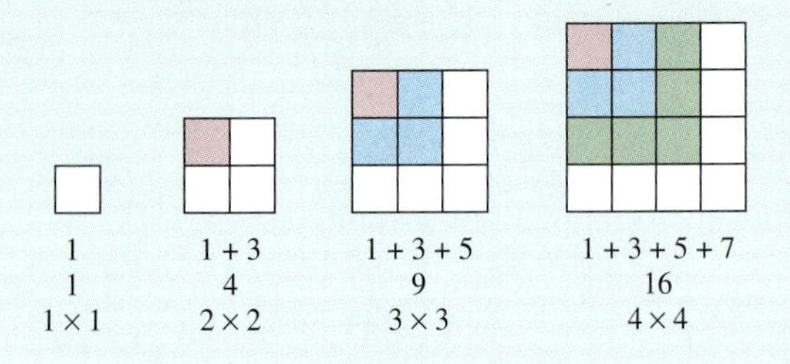

SECTION 3.3

THE DISTRIBUTIVE PROPERTY The illustration shows three rectangles that are divided into squares. Since the area of the rectangle on the left-hand side of the equals sign can be found by multiplying its width by its length, its area is $4(5 + 3)$ square units. Evaluating this expression, we see that the area shaded in blue is $4(8) = 32$ square units.

The area on the right-hand side of the equals sign is the sum of the areas of the two rectangles: $4(5) + 4(3)$. Evaluating this expression, we see that the area shaded in red is also 32 square units: $4(5) + 4(3) = 20 + 12 = 32$. Therefore,

$$4(5 + 3) = 4(5) + 5(3)$$

Create a similar demonstration of the distributive property using rectangles with dimensions different from those in the illustration.

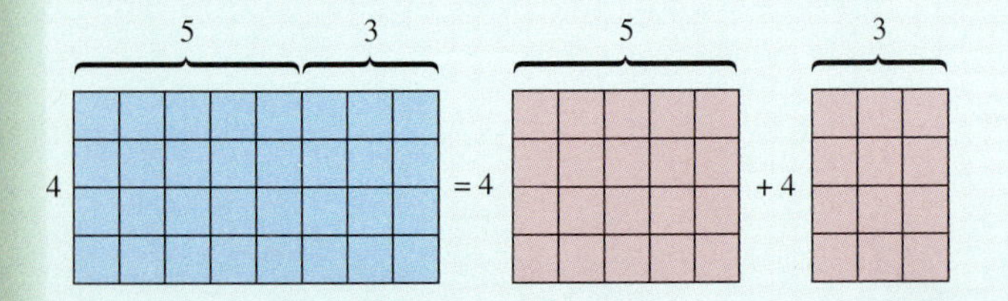

SECTION 3.4

LIKE TERMS If we are to add or subtract objects, they must be similar. Simplify each of the following expressions by combining like terms. You will have to change some of the units so that you are working with like terms. To do so, use the following conversion facts.

- There are 12 inches in one foot.
- There are 36 inches in one yard.
- There are 3 feet in one yard.
- There are 5,280 feet in one mile.

a. 1 foot + 6 inches
b. 1 yard + 11 inches
c. 1 mile − 1 foot
c. 12 feet − 1 yard
e. 1 yard + 1 foot + 5 inches
f. 2 yards + 2 feet − 2 inches
g. 6 inches + 3 feet − 4 inches + 2 feet

SECTION 3.5

SOLVING EQUATIONS In Sections 1.5 and 1.6, we saw how scales can be used to illustrate the steps used to solve an equation.

a. What equation is being solved in the illustration? What is the solution?

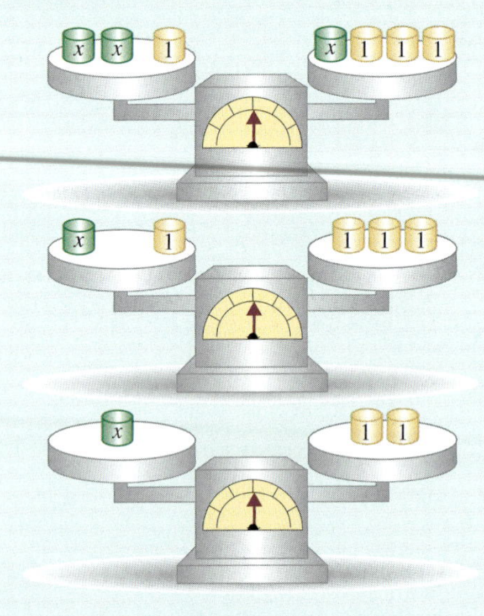

b. Draw a similar series of pictures showing the solution of each of the following equations.

1. $3x = 2x + 4$
2. $3x + 1 = 2x + 4$
3. $2x + 6 = 4x$
4. $2x + 6 = 4x + 2$

SECTION 3.6

PROBLEM SOLVING Write a number–value problem that involves two quantities that have different values. Then solve it. See Exercises 33–36 in Study Set 3.6 for some examples.

CHAPTER REVIEW

SECTION 3.1 Variables and Algebraic Expressions

CONCEPTS

A *variable* is used to represent an unknown quantity.

Variables and numbers can be combined with the operations of addition, subtraction, multiplication, and division to create *algebraic expressions.*

Key words and *key phrases* are used to represent the operations of addition, subtraction, multiplication, and division.

Sometimes you must rely on common sense and insight to find *hidden operations.*

REVIEW EXERCISES

1. The illustration shows the distances from two towns to an airport. Which town is closer to the airport? How much closer is it?

2. See the illustration. Let h represent the height of the ladder, and write an algebraic expression for the height of the ceiling in feet.

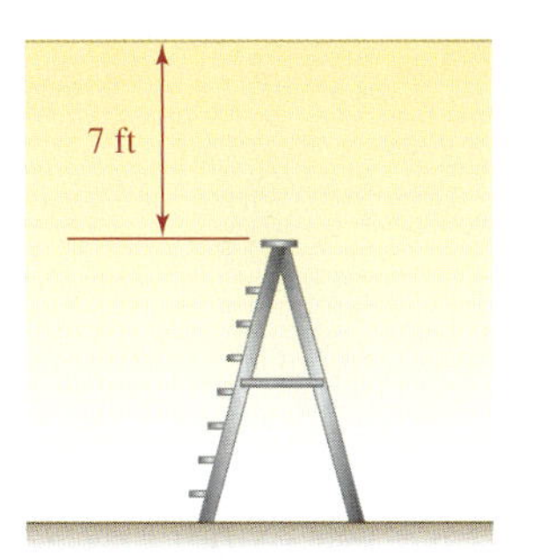

Translate each of the following phrases to mathematical symbols.

3. Five less than n
4. The product of 7 and x
5. The quotient of six and p
6. The sum of s and -15
7. Twice the length l
8. D reduced by 100
9. Two more than r
10. 45 divided by x

11. A child care center has six rooms, and the same number of children are in each room. If c children attend the center, how many will be in each room?
12. A used car, originally advertised for \$1,000, did not sell. The owner decided to drop the price \$$x$. What is the new price of the car?
13. On a cross-country vacation, a husband drove twice as many hours as his wife. Let x = the hours the wife drove. Write an expression that represents the number of hours driven by the husband.
14. The length of a rectangle is 3 units more than its width. Let w = the width of the rectangle. Write an expression that represents the length of the rectangle.

15. How many eggs are in x dozen?
16. d days is how many weeks?

SECTION 3.2 Evaluating Algebraic Expressions and Formulas

When we replace the variable, or variables, in an algebraic expression with a specific number and then apply the rules for the order of operations, we are *evaluating the algebraic expression.*

RETAINING WALLS The illustration shows the design for a retaining wall. The relationships between the lengths of its important parts are given in words.

17. Choose a variable to represent the height of the wall. Write algebraic expressions to represent the lengths of the upper and lower bases.

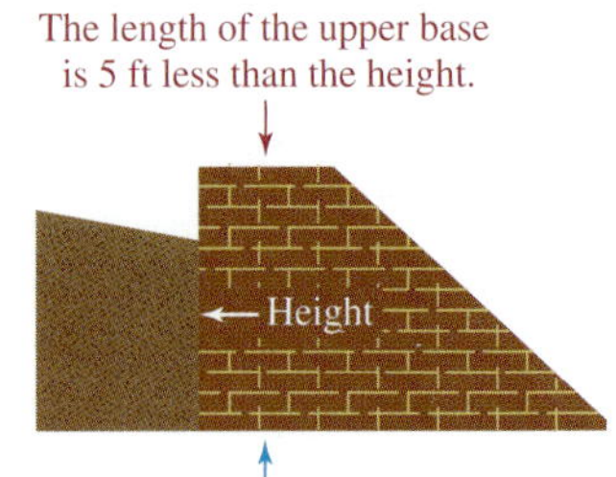

18. Suppose engineers determine that a 10-foot-high wall is needed. Find the lengths of the upper and lower bases.

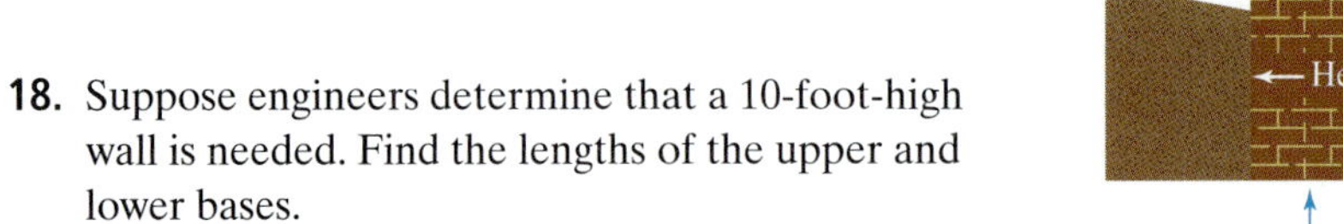

Evaluate each algebraic expression.

19. $-2x + 6$ for $x = -3$

20. $\dfrac{6 - a}{1 + a}$ for $a = -2$

21. $b^2 - 4ac$ for $a = 4$, $b = 6$, and $c = -4$

22. $\dfrac{-2k^3}{1 - 2 - 3}$ for $k = -2$

A *formula* is a general rule that describes a known relationship between two or more variables.

Distance = rate · time

23. DISTANCE TRAVELED Complete the table by finding the distance traveled for a given time at a given rate.

	Rate (mph)	Time (hr)	Distance traveled (mi)
Monorail	65	2	
Subway	38	3	
Train	x	6	
Bus	55	t	

Formulas from business:

Sale price = original price − discount

Retail price = cost + markup

Profit = revenue − cost

24. SALE PRICE Find the sale price of a trampoline that usually sells for \$315 if a \$37 discount is being offered.

25. RETAIL PRICE Find the retail price of a car if the dealer pays \$14,505 and the markup is \$725.

ANNUAL PROFIT The bar graph shows the revenue and costs for a company for the years 2002 to 2004, in millions of dollars.

26. In which year was there the most revenue?

27. Which year had the largest profit?

28. What can you say about costs over this three-year span?

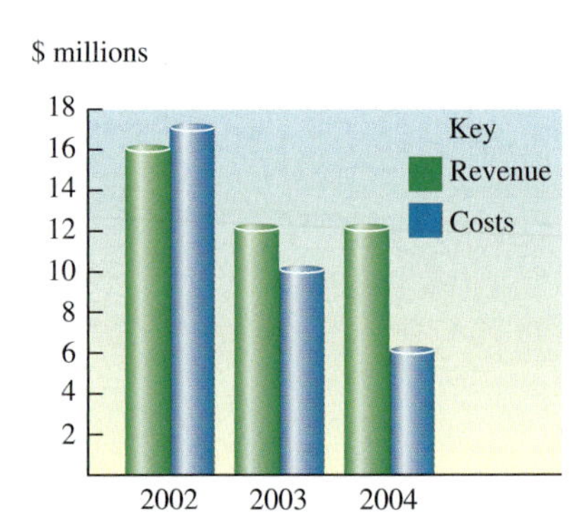

Formulas from science:

$$C = \frac{5}{9}(F - 32)$$

Distance fallen = $16 \cdot (\text{time})^2$

29. TEMPERATURE CONVERSION At a summer resort, visitors can relax by taking a dip in a swimming pool or a lake. The pool water is kept at a constant temperature of 77° F. The water in the lake is 23° C. Which water is warmer, and by how many degrees Celsius?

30. DISTANCE FALLEN A steelworker accidentally dropped his hammer while working atop a new high-rise building. How far will the hammer fall in 3 seconds?

Formulas from mathematics:

$$\text{Mean} = \frac{\text{sum of values}}{\text{number of values}}$$

31. AVERAGE YEARS OF EXPERIENCE Three generations of Smiths now operate a family-owned real estate office. The grandparents, who started the business, have been realtors for 40 years. Their son and daughter-in-law joined the company as realtors 18 years ago. Their grandson has worked as a realtor for 4 years. What is the average number of years a member of the Smith family has worked at Smith Realty?

SECTION 3.3 *Simplifying Algebraic Expressions and the Distributive Property*

To *simplify an algebraic expression,* we use properties of algebra to write the expression in simpler form.

Simplify each expression.

32. $-2(5x)$ **33.** $-7x(-6y)$ **34.** $4d \cdot 3e \cdot 5$ **35.** $(4s)8$

36. $-1(-e)(2)$ **37.** $7x \cdot 7y$ **38.** $4 \cdot 3k \cdot 7$ **39.** $(-10t)(-10)$

The *distributive property:* If a, b, and c are numbers, then

$$a(b + c) = ab + ac$$
$$(b + c)a = ba + ca$$
$$a(b - c) = ab - ac$$
$$(b - c)a = ba - ca$$
$$a(b + c + d) = ab + ac + ad$$

Remove parentheses.

40. $4(y + 5)$ **41.** $-5(6t + 9)$

42. $(-3 - 3x)7$ **43.** $-3(4e - 8x - 1)$

Write an equivalent expression without parentheses.

44. $-(6t - 4)$ **45.** $-(5 + x)$ **46.** $-(6t - 3s + 1)$ **47.** $-(-5a - 3)$

SECTION 3.4 *Combining Like Terms*

A *term* is a number or a product of a number and one or more variables.

Identify the second term and the coefficient of the third term.

48. $5x^2 - 4x + 8$ **49.** $7y - 3y + x - y$

In a term that is a product of a number and one or more variables, the factor that is a number is called the *coefficient* of the term.

Determine whether x is used as a factor or a term.

50. $5x - 6y^2$ **51.** $x + 6$ **52.** $6xy$ **53.** $-36 - x + b$

Like terms, or *similar terms,* are terms with exactly the same variables and exponents.

Determine whether the following are like terms.

54. $4x, -5x$ **55.** $4x, 4x^2$

56. $3xy, xy$ **57.** $-5b^2c, -5bc^2$

CHAPTER 3 TEST

1. Translate each phrase to mathematical symbols.
   a. 2 less than r
   b. The product of 3, x, and y
   c. x increased by 100

2. Together, a couple earns $51,000 a year. If the wife earns e dollars a year, write an expression that represents the husband's yearly earnings.

3. Evaluate each expression.
   a. $\frac{x - 16}{x}$ for $x = 4$
   b. $2t^2 - 3(t - s)$ for $t = -2$ and $s = 4$
   c. $-a^2 + 10$ for $a = -3$

4. DISTANCE TRAVELED Find the distance traveled by a motorist who departed from home at 9:00 A.M. and arrived at his destination at noon, traveling at a rate of 55 miles per hour.

5. PROFITS A craft show promoter had revenues and costs as shown. Find the profit.

Revenues	Costs
Ticket sales: $40,000	Supplies: $13,000
Booth rental: $15,000	Facility rental fee: $5,000

6. FALLING OBJECTS If a tennis ball was dropped from the top of a 200-foot-tall building, would it hit the ground after falling for 3 seconds? If not, how far short of the ground would it be?

7. METER READINGS Every hour between 8 A.M. and 5 P.M., a technician noted the value registered by a meter in a power plant and recorded that number on a line graph. Find the average meter value reading for this period.

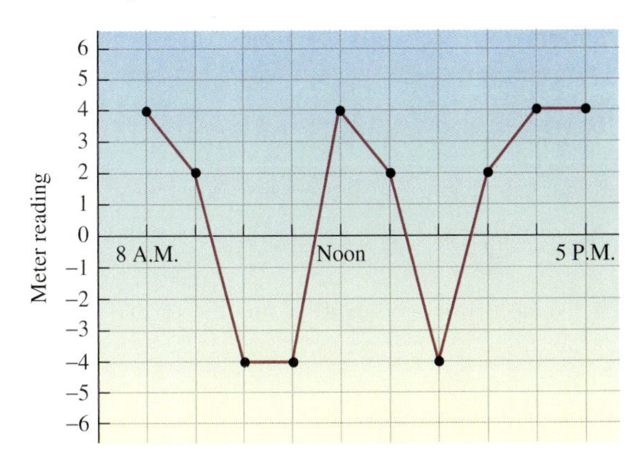

8. LANDMARKS Overlund College is going to construct a gigantic block letter O on a foothill slope near campus. The outline of the letter is to be done using redwood edging. How many feet of edging will be needed?

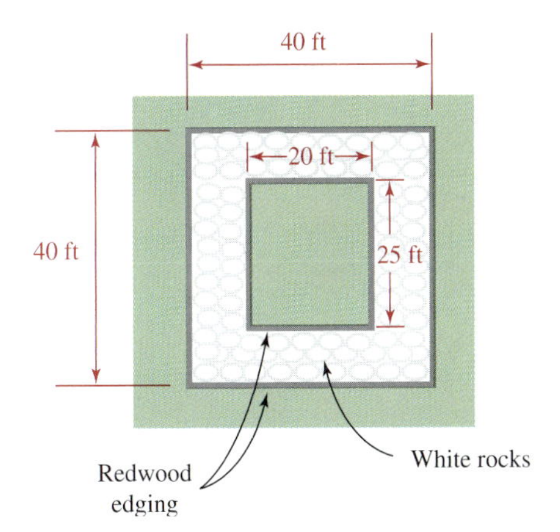

9. AIR CONDITIONING After the air conditioner in a classroom was accidentally left on all night, the room's temperature in the morning was a cool 59°F. What was the temperature in degrees Celsius?

10. Simplify by removing parentheses.

a. $5(5x + 1)$ **b.** $-6(7 - x)$

c. $-(6y + 4)$ **d.** $3(2a + 3b - 7)$

11. Determine whether x is used as a factor or as a term.

a. $5xy$ **b.** $8y + x + 6$

12. Simplify by combining like terms.

a. $-20y + 6 - 8y + 4$

b. $-t - t - t$

13. **a.** Identify each term in this algebraic expression: $8x^2 - 4x - 6$.

b. What is the coefficient of the first term?

14. Simplify each expression.

a. $7x + 4x$ **b.** $3c \cdot 4e \cdot 2$

c. $6x - x$ **d.** $-5y(-6)$

e. $0 - 7x$ **f.** $0 + 9y$

15. Simplify: $4(y + 3) - 5(2y + 3)$.

16. Determine whether -5 is a solution of $6x - 8 = 12(x - 3)$.

Solve each equation.

17. $5x - 3x = -18$

18. $6r = 2r - 12$

19. $-45 = 3(1 - 4t)$

20. $6 - (y - 3) = 19$

21. $8 + 2(3x - 4) = -60$

22. **a.** What is the value (in cents) of k dimes?

b. What is the value of $p + 2$ twenty-dollar bills?

Form an equation and then solve it to answer each question.

23. DRIVING SCHOOLS A driver's training program requires students to attend regularly scheduled classroom sessions each afternoon on Monday through Thursday. On Friday, the students take a 2-hour final exam. If the entire program requires 14 hours of a student's time, how long is each classroom session?

24. CABLE TELEVISION In order to receive its broadcasting license, a cable television station was required to broadcast locally produced shows in addition to its nationally syndicated programming. During a typical 24-hour period, the national shows aired for 8 hours more than the local shows. How many hours of local shows were broadcast each day?

25. What are like terms?

26. Let a variable represent the length of one of the fish shown. Then write an expression that represents the length of the other fish. Give two possible sets of answers.

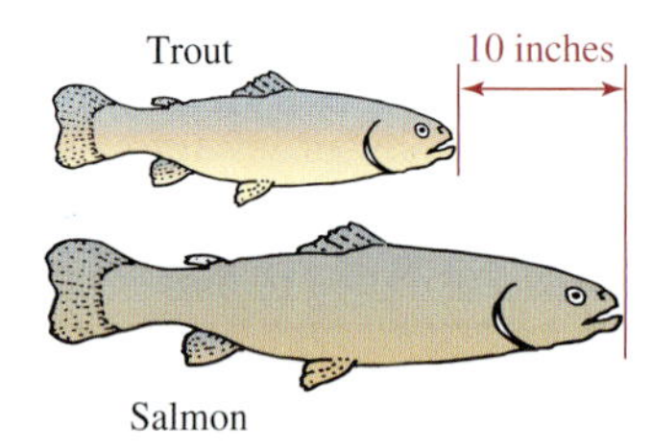

27. Do the instructions *simplify* and *solve* mean the same thing? Explain.

28. Give an example of an application of the distributive property.

CHAPTERS 1–3 CUMULATIVE REVIEW EXERCISES

1. GASOLINE In 1999, gasoline consumption in the United States was three hundred fifty-eight million, six hundred thousand gallons a day. Write this number in standard notation.

2. Round 49,999 to the nearest thousand.

Perform each operation.

3. $\begin{array}{r} 38{,}908 \\ +15{,}696 \\ \hline \end{array}$

4. $\begin{array}{r} 9{,}700 \\ -5{,}491 \\ \hline \end{array}$

5. $\begin{array}{r} 345 \\ \times\ 67 \\ \hline \end{array}$

6. $23\overline{)2{,}001}$

7. Explain how to check the following result using addition.

$$\begin{array}{r} 1{,}142 \\ -\quad 459 \\ \hline 683 \end{array}$$

8. VIETNAMESE CALENDAR An animal represents each Vietnamese lunar year. Recent Years of the Cat are listed below. If the cycle continues, what year will be the next Year of the Cat?

1915 1927 1939 1951 1963 1975 1987 1999

9. Consider the multiplication statement $4 \cdot 5 = 20$. Show that multiplication is repeated addition.

10. ROOM DIVIDERS Four pieces of plywood, each 22 inches wide and 62 inches high, are to be covered with fabric, front and back, to make the room divider shown. How many square inches of fabric will be used?

11. **a.** Find the factors of 18.

    b. Find the prime factorization of 18.

12. List the first ten prime numbers.

13. Why isn't 27 a prime number?

14. Evaluate: $(9 - 2)^2 - 3^3$.

15. Solve: $250 = \frac{y}{2}$.

16. Simplify: $-(-6)$.

17. Graph the integers greater than -3 but less than 4.

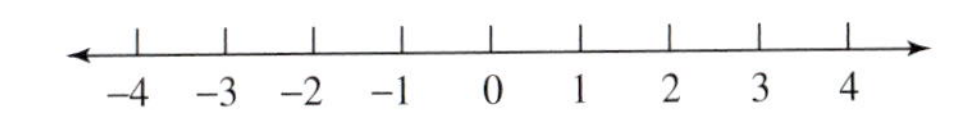

18. Find the absolute value: $|-5|$.

19. Is the statement $-12 > -10$ true or false?

20. NET INCOME Use the following data for the Polaroid Corporation to construct a line graph.

Year	1995	1996	1997	1998	1999
Total net income ($ millions)	−139	15	−127	−51	9

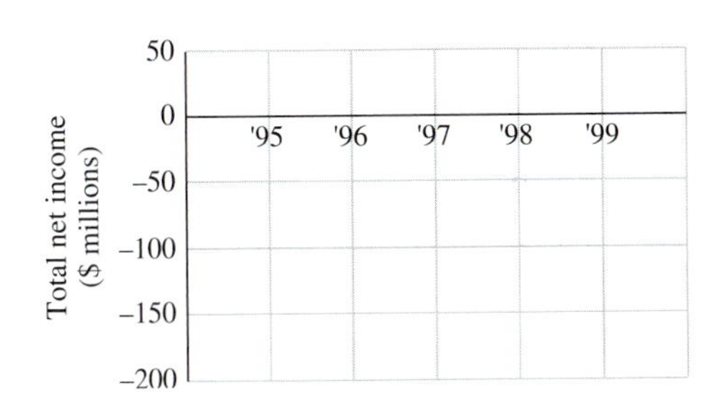

Perform each operation.

21. $-25 + 5$

22. $25 - (-5)$

23. $-25(5)(-1)$

24. $\dfrac{-25}{-5}$

25. Evaluate: $\dfrac{(-6)^2 - 1^5}{-4 - 3}$.

26. Evaluate: $-3 + 3[-4 - 4 \cdot 2]^2$.

27. Evaluate: -3^2 and $(-3)^2$.

28. PLANETS Mercury orbits closer to the sun than any other planet. Temperatures on Mercury can get as high as 810° F and as low as −290° F. What is the temperature range?

Solve each equation.

29. $-4x + 4 = -24$

30. $-y = 10$.

31. Write an expression illustrating division by 0 and an expression illustrating division of 0. Which is undefined?

32. What property allows us to rewrite $x \cdot 5$ as $5x$?

33. Translate to mathematical symbols: h increased by 12.

34. TENNIS Write an algebraic expression that represents the length of the handle of the tennis racket.

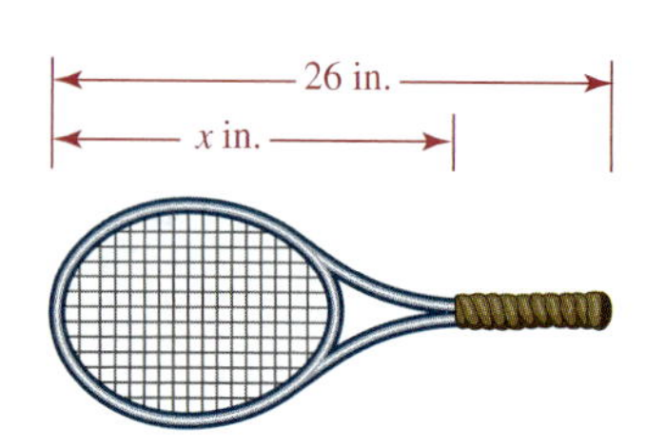

35. Explain the difference between x^2 and $2x$.

36. Evaluate $x^2 - 2x + 1$ for $x = -5$.

37. Complete the table.

	Rate (mph)	Time (hr)	Distance traveled (mi)
Truck	55	4	

38. Remove the parentheses: $5(2x - 7)$.

39. Simplify: $-6(-4t)$.

40. Complete the table.

Term	Coefficient
$4a$	
$-2y^2$	
x	
$-m$	

41. Write an expression in which x is used as a term. Then write an expression in which x is used as a factor.

42. What is the value (in cents) of q quarters?

43. Simplify: $5b + 8 - 6b - 7$.

Solve each equation and check the result.

44. $8p + 2p - 1 = -11$

45. $7 + 2x = 2 - (4x + 7)$.

Form an equation and then solve it to answer the following question.

46. CLASS TIME In a chemistry course, students spend a total of 300 minutes in lab and lecture each week. The time spent in lab is 50 minutes less than the time spent in lecture. How many minutes do the students spend in lecture each week?

CHAPTER 4

Fractions and Mixed Numbers

Getty Images

It has been said that "A penny saved is a penny earned." Smart shoppers certainly know this is true. A lot of money can be saved by waiting for big-ticket items, such as appliances, furniture, and jewelry, to go on sale before purchasing them. Many stores advertise from 1/3 to 1/2 off on these items at various times during the year. In such cases, an informed shopper needs a working knowledge of fractions to calculate discounts and sale prices.

To learn more about fractions and discount shopping, visit The Learning Equation on the Internet at http://tle.brookscole.com. (The log-in instructions are in the Preface.) For Chapter 4, the online lessons are:

- *TLE* Lesson 8: Adding and Subtracting Fractions
- *TLE* Lesson 9: Multiplying Fractions and Mixed Numbers

Check Your Knowledge

1. For the fraction $\frac{3}{4}$, 3 is the __________ and 4 is the __________.
2. A fraction is said to be in ________ terms if the only factor common to the numerator and denominator is 1.
3. Fractions that represent the same amount, such as $\frac{2}{3}$ and $\frac{4}{6}$, are called __________ fractions.
4. Two numbers are called __________ if their product is 1.
5. A ________ number such as $5\frac{3}{4}$ contains a whole-number part and a fractional part.
6. What fractional part of a day is 10 hours? What is the remaining fractional part of the day?
7. Simplify each fraction.
   a. $\frac{24}{30}$ b. $\frac{144y}{192y^3}$
8. Multiply.
   a. $\frac{1}{2}\left(\frac{3}{4}\right)$ b. $-\frac{x}{12}\left(\frac{20}{3x}\right)$
9. Divide.
   a. $-3\frac{3}{8} \div 2\frac{1}{4}$ b. $\frac{x^2}{2} \div 3x$
10. Add.
    a. $\frac{2}{5} + \frac{7}{8}$ b. $\frac{3}{5} + \frac{1}{x}$
11. Subtract.
    a. $\frac{2}{3} - \frac{1}{4}$ b. $1 - \frac{3}{8x}$
12. Express $\frac{5}{6}$ as an equivalent fraction with denominator $36x$.
13. Graph: $-\frac{9}{8}, \frac{1}{9}$, and $1\frac{1}{12}$.

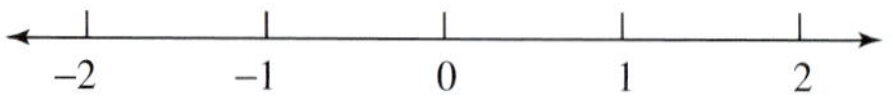

14. Find the area of a triangle with base of length $\frac{7}{4}$ cm and height $\frac{5}{7}$ cm.
15. Add: $123\frac{4}{5} + 189\frac{2}{3}$.
16. Subtract: $13\frac{1}{3} - 8\frac{7}{8}$.
17. Evaluate: $\left(\frac{5}{4}\right)^2 + \left(\frac{2}{3} - 2\frac{1}{6}\right)$.
18. Simplify each complex fraction.
    a. $\dfrac{\frac{3}{5}}{\frac{1}{3}}$ b. $\dfrac{\frac{1}{5} - \frac{1}{6}}{\frac{1}{4} - \frac{1}{3}}$
19. Solve each equation.
    a. $-\frac{x}{6} = -12$ b. $12 - (3x - 4) = 6$ c. $\frac{x}{2} - \frac{x}{3} = 1$
20. Four-fifths of the people questioned at a beach were using sun screen. Twelve of those surveyed were not using sun screen. How many people took part in the survey?
21. Josie and five friends purchased ten lottery tickets together. If one of their tickets is chosen, the prize is $\$10\frac{1}{2}$ million. What would Josie's share of the prize be if they won?
22. A corporate executive negotiated a seven-year employment contract providing for equal annual payments. The executive received 3 million for the first two years of the contract term. How much is she due for the rest of the contract term?

Study Skills Workshop

REWORKING NOTES AS A STUDY AID

In the last Study Skills Workshop, you learned the importance of taking notes. In this workshop, you will learn to rework your notes to use in studying. *Reworking notes* simply means that you look over class notes, fill in any missing information, and then put the information in a new, easy-to-study format.

Identify Concepts. As soon as possible after class, sit down with your notes and see whether you can identify the key concepts that were discussed in your lecture. Key concepts are usually identified by definitions, rules, or formulas.

Find Examples. Once you have identified a key concept, find all of the examples that were used to illustrate that concept, and make sure you understand them.

Choose a Format for Reworking. You can rework your notes using index cards, a study sheet divided into three columns, or audio form.

- *Index card format:* Make a tab for your cards that states a key concept. On the front of the first index card, write the concept. On the back of the card, write the definition or formula; you will want to memorize this. Using an index card for each example, write the example on the front of the card. On the back, work each step of the example, stating the reason for each step. Repeat this process for every concept learned. Store your index cards in a box, like those used for recipes, and group cards by concept.
- *Three-column study sheet:* Divide an $8\frac{1}{2}$-by-11-inch sheet of paper into three columns. Label the leftmost column "Concept," the middle column "Example," and the rightmost column "Steps and Reasons." Rewrite your notes, filling in the columns on your study sheet as labeled.
- *Audiotape or CD:* Using a recorder, state a key concept. Pause for a few seconds and then state the definition or formula. State an example and verbally explain each step. Continue to state examples and steps. Repeat for the remaining concepts.

Use the Reworked Notes for Studying.

- *Index cards:* Look at the concept on the first card in a group. Attempt to state the definition or the rule. If you are unable to do so, look at the back and read it. Keep doing this until you can recite from memory. Then see whether you can work the problem without any help. Continue to rework the problem until you can do it without looking at the solution.
- *Three-column study sheet:* Go through the concepts and examples, covering the work in the rightmost column while you attempt to work the problem in the middle. If you get stuck, uncover the work to look at the steps. Continue to rework the problem until you can do it without looking at the solution.
- *Audiotape:* In the pause between each phase of the problem, see whether you can recite the steps before they are revealed on the tape. Repeat until you can recite all of the steps on your own.

ASSIGNMENT

1. Determine which format is the best one for you to use to rework your notes. If you think of a different format, describe it.
2. Rework your notes from the last lecture in your chosen format.
3. Compare your reworked notes with a classmate who has chosen the same format. Then compare with someone who has chosen a different format. Are you satisfied with your reworked notes? If not, analyze what is not working and try to fix it.
4. Ask your instructor or a tutor for an evaluation of your reworked notes.

Whole numbers are used to count objects. When we need to represent parts of a whole, fractions can be used.

4.1 The Fundamental Property of Fractions

- Basic facts about fractions
- Equivalent fractions
- Simplifying a fraction
- Expressing a fraction in higher terms

There is no better place to start a study of fractions than with *the fundamental property of fractions*. This property is the foundation for two fundamental procedures that are used when working with fractions. But first, we review some basic facts about fractions.

Basic facts about fractions

1. A Fraction Can Be Used to Indicate Equal Parts of a Whole. In our everyday lives, we often deal with parts of a whole. For example, we talk about parts of an hour, parts of an inch, and parts of a pound.

half of an hour   three-eighths of an inch of rain   a quarter-pound hamburger

2. A Fraction Is Composed of a Numerator, a Denominator, and a Fraction Bar.

Fraction bar ⟶ $\frac{3}{4}$ ⟵ Numerator (3), ⟵ Denominator (4)

The denominator (in this case, 4) tells us that a whole was divided into four equal parts. The numerator tells us that we are considering three of those equal parts.

3. Fractions Can Be Proper or Improper. If the numerator of a fraction is less than its denominator, the fraction is called a **proper fraction.** A proper fraction is less than 1. Fractions whose numerators are greater than or equal to their denominators are called **improper fractions.**

Proper fractions	**Improper fractions**
$\frac{1}{4}, \frac{2}{3},$ and $\frac{98}{99}$	$\frac{7}{2}, \frac{98}{97}, \frac{16}{16},$ and $\frac{5}{1}$

4. The Denominator of a Fraction Cannot Be 0. $\frac{7}{0}, \frac{23}{0}$, and $\frac{0}{0}$ are meaningless expressions. (Recall that $\frac{7}{0}, \frac{23}{0}$, and $\frac{0}{0}$ represent *division* by 0, and a number cannot be divided by 0.) However, $\frac{0}{7} = 0$ and $\frac{0}{23} = 0$.

5. The Numerator and the Denominator of a Fraction Can Contain Variables. Since a variable is a letter that is used to stand for a number, a variable or a combination of variables can appear in the numerator or the denominator of a fraction. Here are several examples of such **algebraic fractions.**

$$\frac{x}{4}, \quad \frac{12}{b}, \quad \frac{x}{y}, \quad \frac{m^2}{2mn}, \quad \frac{2c + d}{3c^3d}$$

6. Fractions Can Be Negative. There are times when a negative fraction is needed to describe a quantity. For example, if an earthquake causes a road to sink one-half inch, the amount of movement can be represented by $-\frac{1}{2}$ inch.

Negative fractions can be written in three ways. The negative sign can appear in the numerator, in the denominator, or in front of the fraction.

$$\frac{-1}{2} = \frac{1}{-2} = -\frac{1}{2} \qquad \frac{-15}{8} = \frac{15}{-8} = -\frac{15}{8}$$

Negative fractions

If a and b represent positive numbers,

$$\frac{-a}{b} = \frac{a}{-b} = -\frac{a}{b}$$

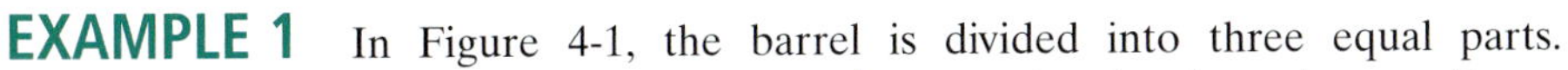

EXAMPLE 1 In Figure 4-1, the barrel is divided into three equal parts. **a.** What fractional part of the barrel is full? **b.** What fractional part is empty?

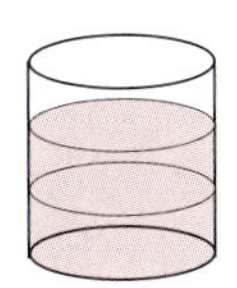

FIGURE 4-1

Solution

a. Two of the three parts are full. Therefore, the barrel is $\frac{2}{3}$ full.

b. One of the three equal parts is not filled. The barrel is $\frac{1}{3}$ empty.

The fractions $\frac{2}{3}$ and $\frac{1}{3}$ are both proper fractions.

Self Check 1

a. According to the calendar below, what fractional part of the month has passed? **b.** What fractional part remains?

DECEMBER

X	X	X	X	X	X	X
X	X	X	X	12	13	14
15	16	17	18	19	20	21
22	23	24	25	26	27	28
29	30	31				

Answers **a.** $\frac{11}{31}$, **b.** $\frac{20}{31}$

Fractions are often referred to as **rational numbers.** All integers are rational numbers, because every integer can be written as a fraction with a denominator of 1. For example,

$$2 = \frac{2}{1}, \quad -5 = \frac{-5}{1}, \quad \text{and} \quad 0 = \frac{0}{1}$$

Since every integer is also a rational number, the integers are a subset of the rational numbers.

! COMMENT Not all rational numbers are integers. For example, the rational number $\frac{7}{8}$ is not an integer.

Equivalent fractions

Fractions can look different but still represent the same number. For example, let's divide the rectangle in Figure 4-2(a) in two ways. In Figure 4-2(b), we divide it into halves (2 equal-sized parts). In Figure 4-2(c), we divide it into fourths (4 equal-sized parts). Notice that one-half of the figure is the same size as two-fourths of the figure.

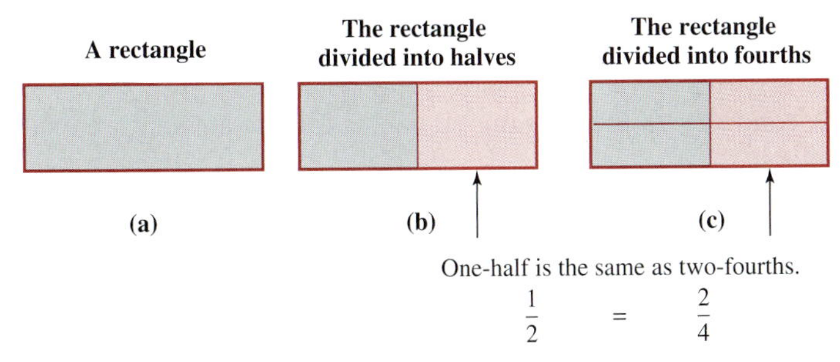

FIGURE 4-2

The fractions $\frac{1}{2}$ and $\frac{2}{4}$ look different, but Figure 4-2 shows that they represent the same amount. We say that they are **equivalent fractions.**

Equivalent fractions

Two fractions are **equivalent** if they represent the same number.

Simplifying a fraction

If we replace a fraction with an equivalent fraction that contains smaller numbers, we are **simplifying** or **reducing the fraction.** To simplify a fraction, we use the following fact.

The fundamental property of fractions

Multiplying or dividing the numerator and the denominator of a fraction by the same nonzero number does not change the value of the fraction. In symbols, for all numbers a, b, and x (provided b and x are not 0),

$$\frac{a}{b} = \frac{a \cdot x}{b \cdot x} \quad \text{and} \quad \frac{a}{b} = \frac{a \div x}{b \div x}$$

For example, we consider $\frac{24}{28}$. It is apparent that 24 and 28 have a common factor 4. By the fundamental property of fractions, we can divide the numerator and denominator of this fraction by 4.

$$\frac{24}{28} = \frac{24 \div \mathbf{4}}{28 \div \mathbf{4}}$$

Divide the numerator by 4.
Divide the denominator by 4.

$$= \frac{6}{7}$$

Perform each division: $24 \div 4 = 6$ and $28 \div 4 = 7$.

Thus, $\frac{24}{28} = \frac{6}{7}$. We say that $\frac{24}{28}$ and $\frac{6}{7}$ are equivalent fractions, because they represent the same number.

In practice, we show the previous simplification in a slightly different way.

$$\frac{24}{28} = \frac{\mathbf{4} \cdot 6}{\mathbf{4} \cdot 7}$$

Once you see a common factor of the numerator and the denominator, factor each of them so that it shows. In this case, 24 and 28 share a common factor 4.

$$= \frac{\overset{1}{\cancel{4}} \cdot 6}{\underset{1}{\cancel{4}} \cdot 7}$$

Divide the numerator and the denominator by 4 by drawing slashes through the common factors. Use small 1's to represent the result of each division of 4 by 4. Note that $\frac{4}{4} = 1$.

$$= \frac{6}{7}$$

Multiply in the numerator and in the denominator: $1 \cdot 6 = 6$ and $1 \cdot 7 = 7$.

In the second step of the previous simplification, we say that we *divided out the common factor 4.*

Simplifying a fraction

We can **simplify** a fraction by factoring its numerator and denominator and then dividing out all common factors in the numerator and denominator.

When a fraction can be simplified no further, we say that it is written in **lowest terms.**

Lowest terms

A fraction is in **lowest terms** if the only factor common to the numerator and denominator is 1.

EXAMPLE 2 Simplify to lowest terms: $\frac{25}{75}$.

Solution The numerator and the denominator have a common factor 25.

$$\frac{25}{75} = \frac{25 \cdot 1}{3 \cdot 25}$$ Factor 25 as 25 · 1 and 75 as 3 · 25.

$$= \frac{\overset{1}{\cancel{25}} \cdot 1}{3 \cdot \underset{1}{\cancel{25}}}$$ Divide out the common factor 25. Note that $\frac{25}{25} = 1$.

$$= \frac{1}{3}$$ Multiply in the numerator and in the denominator: $1 \cdot 1 = 1$ and $1 \cdot 3 = 3$.

Self Check 2

Simplify to lowest terms: $\frac{60}{80}$.

Answer $\frac{3}{4}$

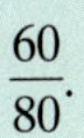

EXAMPLE 3 Simplify: $\frac{90}{126}$.

Solution To find the common factors that will divide out, we prime factor 90 and 126.

$$\frac{90}{126} = \frac{2 \cdot 3 \cdot 3 \cdot 5}{2 \cdot 3 \cdot 3 \cdot 7}$$ Use the tree method or the division method to prime factor 90 and 126: $90 = 2 \cdot 3 \cdot 3 \cdot 5$ and $126 = 2 \cdot 3 \cdot 3 \cdot 7$.

$$= \frac{\overset{1}{\cancel{2}} \cdot \overset{1}{\cancel{3}} \cdot \overset{1}{\cancel{3}} \cdot 5}{\underset{1}{\cancel{2}} \cdot \underset{1}{\cancel{3}} \cdot \underset{1}{\cancel{3}} \cdot 7}$$ Divide out the common factors 3, 3, and 2.

$$= \frac{5}{7}$$ Multiply in the numerator and in the denominator.

We can also simplify this fraction by noting that the numerator and denominator have a common factor 18.

$$\frac{90}{126} = \frac{5 \cdot \overset{1}{\cancel{18}}}{7 \cdot \underset{1}{\cancel{18}}}$$ Factor 90 as 5 · 18 and 126 as 7 · 18. Divide out the common factor 18.

$$= \frac{5}{7}$$

Self Check 3

Simplify: $\frac{42}{150}$.

Answer $\frac{7}{25}$

! COMMENT Negative fractions are simplified in the same way as positive fractions. Just remember to write a negative sign − in each step of the solution.

$$-\frac{45}{72} = -\frac{5 \cdot \overset{1}{\cancel{9}}}{8 \cdot \underset{1}{\cancel{9}}} = -\frac{5}{8}$$

EXAMPLE 4 Simplify: $\frac{24xy^2}{16x^2y^2}$.

Solution To simplify fractions that contain variables, we divide out common numerical factors and common variable factors.

Self Check 4

Simplify: $\frac{45ab^2}{36ab^3}$.

$$\frac{24xy^2}{16x^2y^2} = \frac{8 \cdot 3 \cdot xy^2}{8 \cdot 2 \cdot x^2y^2}$$ Since 24 and 16 have a common factor 8, factor 24 as $8 \cdot 3$ and 16 as $8 \cdot 2$.

$$= \frac{8 \cdot 3 \cdot x \cdot y \cdot y}{8 \cdot 2 \cdot x \cdot x \cdot y \cdot y}$$ Write the variable parts of the numerator and denominator in factored form: $y^2 = y \cdot y$ and $x^2 = x \cdot x$.

$$= \frac{\overset{1}{\cancel{8}} \cdot 3 \cdot \overset{1}{\cancel{x}} \cdot \overset{1}{\cancel{y}} \cdot \overset{1}{\cancel{y}}}{\underset{1}{\cancel{8}} \cdot 2 \cdot \underset{1}{\cancel{x}} \cdot x \cdot \underset{1}{\cancel{y}} \cdot \underset{1}{\cancel{y}}}$$ Divide out the common factors 8, x, and y^2.

$$= \frac{3}{2x}$$ Multiply in the numerator and in the denominator.

Answer $\frac{5}{4b}$

Expressing a fraction in higher terms

It is sometimes necessary to replace a fraction with an equivalent fraction that involves larger numbers or more complex terms. This is called **expressing the fraction in higher terms** or **building** the fraction.

For example, to write $\frac{3}{8}$ as an equivalent fraction with a denominator 40, we can use the fundamental property of fractions and multiply the numerator and denominator by 5.

$$\frac{3}{8} = \frac{3 \cdot 5}{8 \cdot 5}$$ Multiply the numerator by 5. Multiply the denominator by 5.

$$= \frac{15}{40}$$ Perform the multiplications in the numerator and in the denominator.

Therefore, $\frac{3}{8} = \frac{15}{40}$.

Self Check 5
Write $\frac{2}{3}$ as an equivalent fraction with a denominator 24.

Answer $\frac{16}{24}$

EXAMPLE 5 Write $\frac{5}{7}$ as an equivalent fraction with a denominator 28.

Solution We need to multiply the denominator by 4 to obtain 28. By the fundamental property of fractions, we must multiply the numerator by 4 as well.

$$\frac{5}{7} = \frac{5 \cdot 4}{7 \cdot 4}$$ Multiply the numerator and denominator by 4.

$$= \frac{20}{28}$$ Perform the multiplication in the numerator and in the denominator.

Self Check 6
Write 5 as a fraction with a denominator 3.

Answer $\frac{15}{3}$

EXAMPLE 6 Write 4 as a fraction with a denominator 6.

Solution First, express 4 as a fraction: $4 = \frac{4}{1}$. To obtain a denominator 6, we need to multiply the numerator and denominator by 6.

$$\frac{4}{1} = \frac{4 \cdot 6}{1 \cdot 6}$$

$$= \frac{24}{6}$$ Perform each multiplication: $4 \cdot 6 = 24$ and $1 \cdot 6 = 6$.

EXAMPLE 7 Write $\frac{3}{4}$ as an equivalent fraction with a denominator $28a$.

Solution We need to multiply the denominator of $\frac{3}{4}$ by $7a$ to obtain $28a$, so we must also multiply the numerator by $7a$.

$$\frac{3}{4} = \frac{3 \cdot 7a}{4 \cdot 7a} \quad \text{Multiply numerator and denominator by } 7a.$$

$$= \frac{21a}{28a} \quad \text{Perform each multiplication: } 3 \cdot 7a = 21a \text{ and } 4 \cdot 7a = 28a.$$

Self Check 7
Write $\frac{2}{5}$ as an equivalent fraction with a denominator $30x$.

Answer $\frac{12x}{30x}$

Section 4.1 STUDY SET

VOCABULARY *Fill in the blanks.*

1. For the fraction $\frac{7}{8}$, 7 is the ________ and 8 is the ________.
2. When we express 15 as $5 \cdot 3$, we say that we have ________ 15.
3. A ________ fraction is less than 1.
4. A fraction is said to be in ________ terms if the only factor common to the numerator and denominator is 1.
5. Two fractions are ________ if they have the same value.
6. A ________ can be used to indicate the number of equal parts of a whole.
7. Multiplying the numerator and denominator of a fraction by a number to obtain an equivalent fraction that involves larger numbers is called expressing the fraction in ________ terms or ________ the fraction.
8. We can ________ a fraction that is not in lowest terms by applying the fundamental property of fractions. We ________ out common factors of the numerator and denominator.

CONCEPTS

9. What common factor (other than 1) do the numerator and the denominator have?

   a. $\frac{2}{16}$ b. $\frac{6}{9}$ c. $\frac{10}{15}$ d. $\frac{14}{35}$

10. Given: $\frac{15}{35} = \frac{3 \cdot \overset{1}{\cancel{5}}}{\underset{1}{\cancel{5}} \cdot 7}$. In this work, what do the slashes and small 1's mean?

11. What concept studied in this section is shown?

12. Why can't we say that $\frac{2}{5}$ of the figure is shaded?

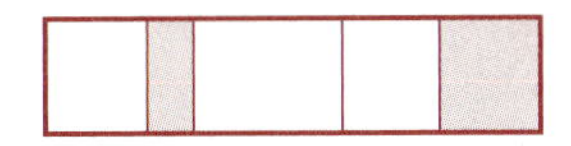

13. a. Explain the difference in the two approaches used to simplify $\frac{20}{28}$.

    $$\frac{\overset{1}{\cancel{4}} \cdot 5}{\underset{1}{\cancel{4}} \cdot 7} \quad \text{and} \quad \frac{\overset{1}{\cancel{2}} \cdot \overset{1}{\cancel{2}} \cdot 5}{\underset{1}{\cancel{2}} \cdot \underset{1}{\cancel{2}} \cdot 7}$$

    b. Are the results the same?

14. What concept studied in this section does this statement illustrate?

    $$\frac{5}{10} = \frac{4}{8} = \frac{3}{6} = \frac{2}{4} = \frac{1}{2}$$

15. Why isn't this a valid application of the fundamental property of fractions?

    $$\frac{10}{11} = \frac{2+8}{2+9} = \frac{\overset{1}{\cancel{2}} + 8}{\underset{1}{\cancel{2}} + 9} = \frac{9}{10}$$

16. Write the fraction $\frac{7}{-8}$ in two other ways.

17. Write as a fraction.

    a. 8 b. -25

    c. x d. $7a$

18. Fill in the blanks in the following solution to write $\frac{5}{9x}$ as an equivalent fraction with denominator $27x$.

    $$\frac{5 \cdot \square}{9x \cdot \square} = \frac{15}{27x}$$

NOTATION *Complete each solution to simplify each fraction.*

19. $\dfrac{18}{24} = \dfrac{2 \cdot \square \cdot 3}{2 \cdot 2 \cdot \square \cdot 3}$

$= \dfrac{\overset{1}{\not{\square}} \cdot 3 \cdot \overset{1}{\not{\square}}}{\underset{1}{\not{\square}} \cdot 2 \cdot 2 \cdot \underset{1}{\not{\square}}}$

$= \dfrac{3}{4}$

20. $\dfrac{60ab^2}{90a^2b} = \dfrac{\square \cdot 2 \cdot a \cdot b \cdot b}{\square \cdot 3 \cdot a \cdot a \cdot b}$

$= \dfrac{\overset{1}{\not{30}} \cdot \square \cdot \overset{1}{\not{a}} \cdot \square \cdot \overset{1}{\not{b}}}{\underset{1}{\not{30}} \cdot \square \cdot \underset{1}{\not{a}} \cdot \square \cdot \underset{1}{\not{b}}}$

$= \dfrac{2b}{3a}$

PRACTICE *Simplify each fraction to lowest terms, if possible.*

21. $\dfrac{3}{9}$ **22.** $\dfrac{5}{20}$

23. $\dfrac{7}{21}$ **24.** $\dfrac{6}{30}$

25. $\dfrac{20}{30}$ **26.** $\dfrac{12}{30}$

27. $\dfrac{15}{6}$ **28.** $\dfrac{24}{16}$

29. $-\dfrac{28}{56}$ **30.** $-\dfrac{45}{54}$

31. $-\dfrac{90}{105}$ **32.** $-\dfrac{26}{78}$

33. $\dfrac{60}{108}$ **34.** $\dfrac{75}{125}$

35. $\dfrac{180}{210}$ **36.** $\dfrac{76}{28}$

37. $\dfrac{55}{67}$ **38.** $\dfrac{41}{51}$

39. $\dfrac{36}{96}$ **40.** $\dfrac{48}{120}$

41. $\dfrac{25x^2}{35x}$ **42.** $\dfrac{16r^2}{20r}$

43. $\dfrac{12t}{15t}$ **44.** $\dfrac{10y}{15y}$

45. $\dfrac{6a}{7a}$ **46.** $\dfrac{4c}{5c}$

47. $\dfrac{7xy}{8xy}$ **48.** $\dfrac{10ab}{21ab}$

49. $-\dfrac{10rs}{30}$ **50.** $-\dfrac{14ab}{28}$

51. $\dfrac{15st^3}{25xt^3}$ **52.** $\dfrac{16wx^3}{24x^3y}$

53. $\dfrac{35r^2t}{28rt^2}$ **54.** $\dfrac{35m^3n^4}{25m^4n^3}$

55. $\dfrac{56p^4}{28p^6}$ **56.** $\dfrac{32q^2}{8q^5}$

Write each fraction as an equivalent fraction with the indicated denominator.

57. $\dfrac{7}{8}$, denominator 40 **58.** $\dfrac{3}{4}$, denominator 24

59. $\dfrac{4}{5}$, denominator 35 **60.** $\dfrac{5}{7}$, denominator 49

61. $\dfrac{5}{6}$, denominator 54 **62.** $\dfrac{11}{16}$, denominator 32

63. $\dfrac{1}{2}$, denominator 30 **64.** $\dfrac{1}{3}$, denominator 60

65. $\dfrac{2}{7}$, denominator $14x$ **66.** $\dfrac{3}{10}$, denominator $50a$

67. $\dfrac{9}{10}$, denominator $60t$ **68.** $\dfrac{2}{3}$, denominator $27t$

69. $\dfrac{5}{4s}$, denominator $20s$ **70.** $\dfrac{9}{4x}$, denominator $44x$

71. $\dfrac{2}{15}$, denominator $45y$ **72.** $\dfrac{5}{12}$, denominator $36n$

Write each number or algebraic expression as a fraction with the indicated denominator.

73. 3 as fifths **74.** 4 as thirds

75. 6 as eighths **76.** 3 as sixths

77. $4a$ as ninths **78.** $7x$ as fourths

79. $-2t$ as halves **80.** $-10c$ as ninths

APPLICATIONS *Use fractions to answer each question.*

81. COMMUTING How much of the commute from home to work has the motorist in the illustration made?

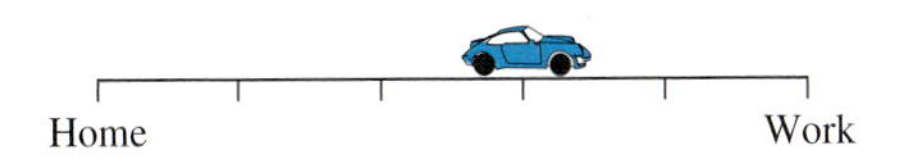

82. TIME CLOCKS For each clock, how much of the hour has passed?

a. 

b. 

c. 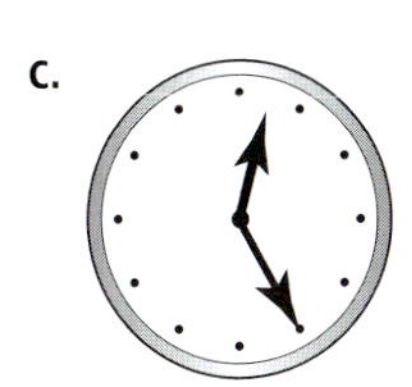

d. 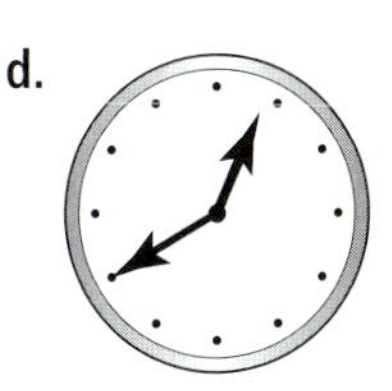

83. SINKHOLES The illustration shows a side view of a depression in the sidewalk near a sinkhole. Describe the movement of the sidewalk using a signed number. (On the tape measure, 1 inch is divided into 16 equal parts.)

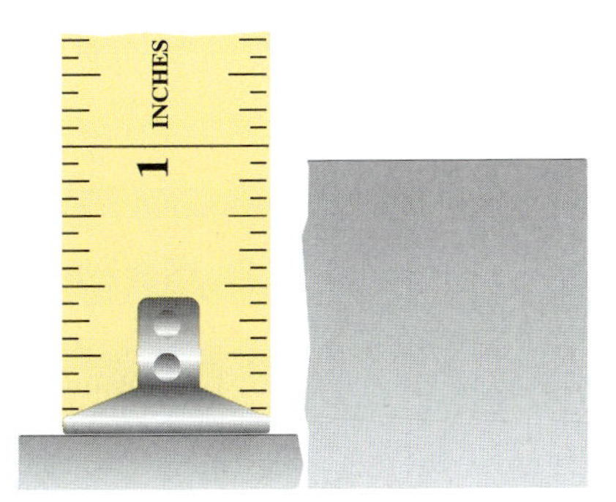

84. POLITICAL PARTIES The illustration in the next column shows the political party affiliation of the governors of the 50 states, as of January 1, 2000.

a. What fraction were Democrats?

b. What fraction were Republicans?

c. What fraction were neither Democrat nor Republican?

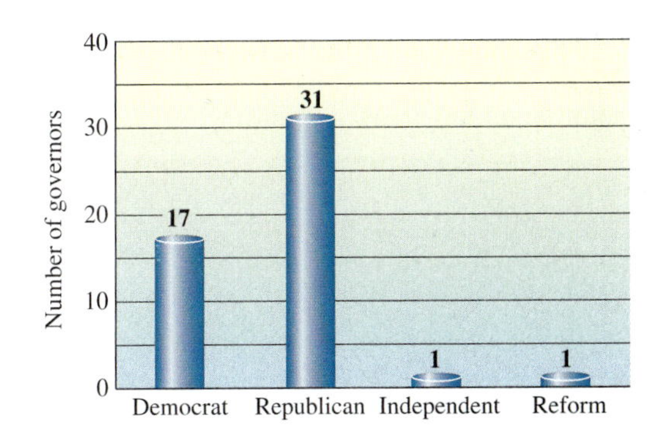

85. PERSONNEL RECORDS Complete the table by finding the amount of the job that will be completed by each person working alone for the given number of hours.

Name	Total time to complete the job alone	Time worked alone	Amount of job completed
Bob	10 hours	7 hours	
Ali	8 hours	1 hour	

86. GAS TANKS How full does the gauge indicate the gas tank is? How much of the tank has been used?

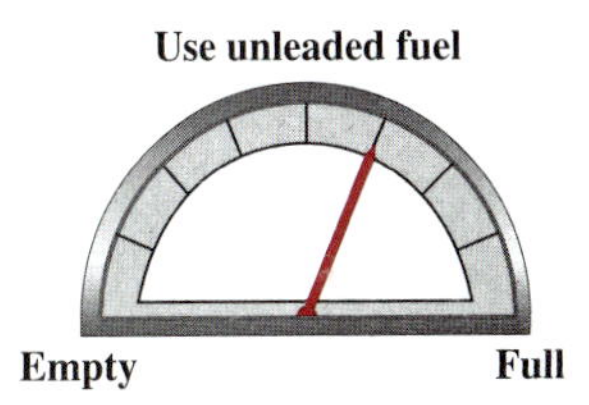

87. MUSIC The illustration shows the finger position needed to produce a length of string (from the bridge to the fingertip) that gives low C on a violin. To play other notes, fractions of that length are used. Locate these finger positions.

a. $\frac{1}{2}$ of the length gives middle C.

b. $\frac{3}{4}$ of the length gives F above low C.

c. $\frac{2}{3}$ of the length gives G.

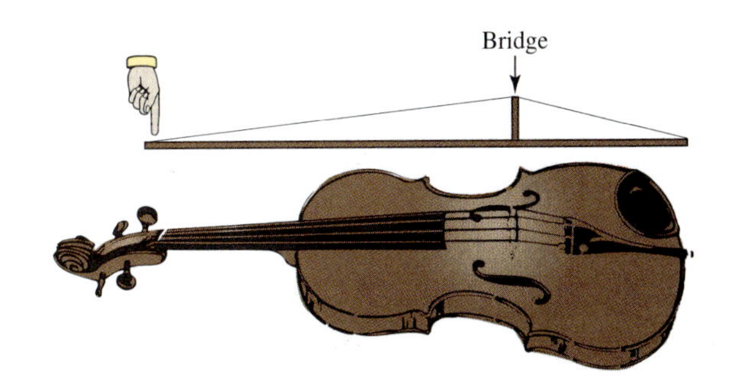

88. RULERS On the ruler, determine how many spaces are between the numbers 0 and 1. Then determine to what number the arrow is pointing.

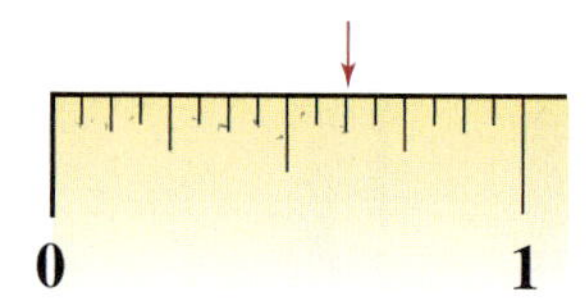

89. MACHINERY The operator of a machine is to turn the dial shown below from setting A to setting B. Express this instruction in two different ways, using fractions of one complete revolution.

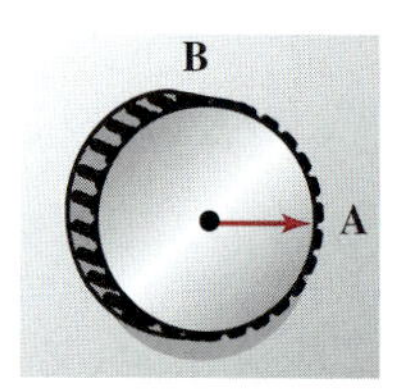

90. EARTH'S ROTATION The Earth rotates about its vertical axis once every 24 hours.

a. What is the significance of $\frac{1}{24}$ of a rotation to us on Earth?

b. What significance does $\frac{24}{24}$ of a revolution have?

91. SUPERMARKET DISPLAYS The amount of space to be given each type of snack food in a supermarket display case is expressed as a fraction. Complete the model of the display, showing where the adjustable shelves should be located, and label where each snack food should be stocked.

$\frac{3}{8}$: potato chips

$\frac{2}{8}$: peanuts

$\frac{1}{8}$: pretzels

$\frac{2}{8}$: tortilla chips

SNACKS

92. MEDICAL CENTERS Hospital designers have located a nurse's station at the center of a circular building. Use the circle graph in the next column to show how to divide the surrounding office space so that each medical department has the proper fractional amount allocated to it. Label each department.

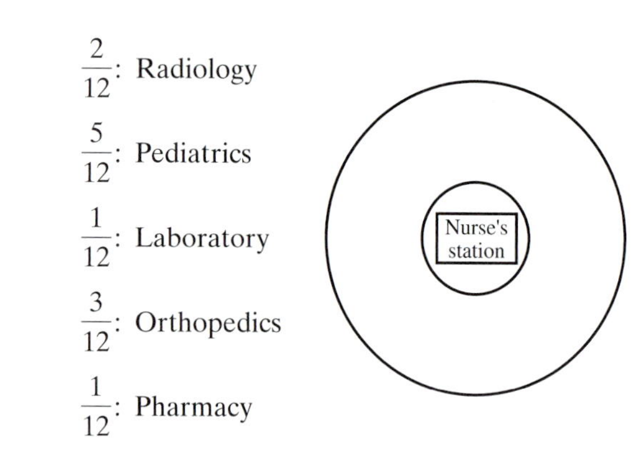

$\frac{2}{12}$: Radiology

$\frac{5}{12}$: Pediatrics

$\frac{1}{12}$: Laboratory

$\frac{3}{12}$: Orthopedics

$\frac{1}{12}$: Pharmacy

93. CAMERAS When the shutter of a camera stays open longer than $\frac{1}{125}$ second, any movement of the camera will probably blur the picture. With this in mind, if a photographer is taking a picture of a fast-moving object, should she select a shutter speed of $\frac{1}{60}$ or $\frac{1}{250}$?

94. GROSS DOMESTIC PRODUCT The GDP is the official measure of the size of the U.S. economy. It represents the market value of all goods and services that have been bought during a given period of time. The GDP for the second quarter of 2004 is listed below. What is meant by the phrase *second quarter of 2004?*

Second quarter of 2004	$11,657,000,000

WRITING

95. Explain the concept of equivalent fractions.

96. What does it mean for a fraction to be in lowest terms?

97. Explain the difference between three-fourths and three-fifths of a pizza.

98. Explain both parts of the fundamental property of fractions.

REVIEW

99. Solve: $-5x + 1 = 16$.

100. Simplify: $4t - 8 + t - 9$.

101. Round 564,112 to the nearest thousand.

102. Give the definition of a prime number.

103. Find the value (in cents) of d dimes.

104. A father is 24 years older than his son. If $s =$ the age of the son, write an algebraic expression that represents the age of the father.

4.2 Multiplying Fractions

- Multiplying fractions • Simplifying when multiplying fractions
- Multiplying algebraic fractions • Powers of a fraction • Applications

In the next three sections, we will discuss how to add, subtract, multiply, and divide fractions. We begin with the operation of multiplication.

Multiplying fractions

Suppose that a television network is going to take out a full-page ad to publicize its fall lineup of shows. The prime-time shows are to get $\frac{3}{5}$ of the ad space and daytime programming the remainder. Of the space devoted to prime time, $\frac{1}{2}$ is to be used to promote weekend programs. How much of the newspaper page will be used to advertise weekend prime-time programs?

The ad for the weekend prime-time shows will occupy $\frac{1}{2}$ of $\frac{3}{5}$ of the page. This can be expressed as $\frac{1}{2} \cdot \frac{3}{5}$. We can calculate $\frac{1}{2} \cdot \frac{3}{5}$ using a three-step process, illustrated below.

Step 1: We divide the page into fifths and shade three of them. This represents the fraction $\frac{3}{5}$, the amount of the page used to advertise prime-time shows.

Prime-time ad space

Step 2: Next, we find $\frac{1}{2}$ of the shaded part of the page by dividing the page into halves, using a vertical line.

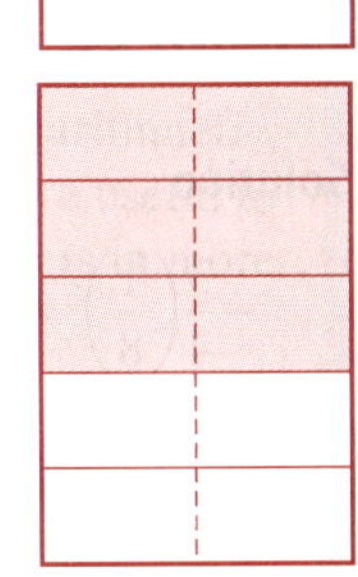

Step 3: Finally, we highlight (in purple) $\frac{1}{2}$ of the shaded parts determined in step 2. The highlighted parts are 3 out of 10 or $\frac{3}{10}$ of the page. They represent the amount of the page used to advertise the weekend prime-time shows. This leads us to the conclusion that $\frac{1}{2} \cdot \frac{3}{5} = \frac{3}{10}$.

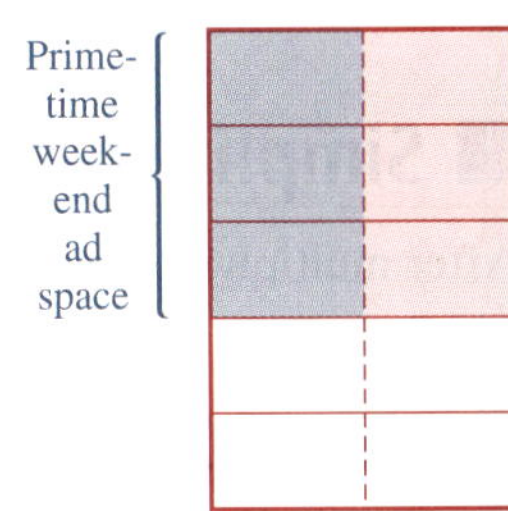

Two observations can be made from this result.

- The numerator of the answer is the product of the numerators of the original fractions.

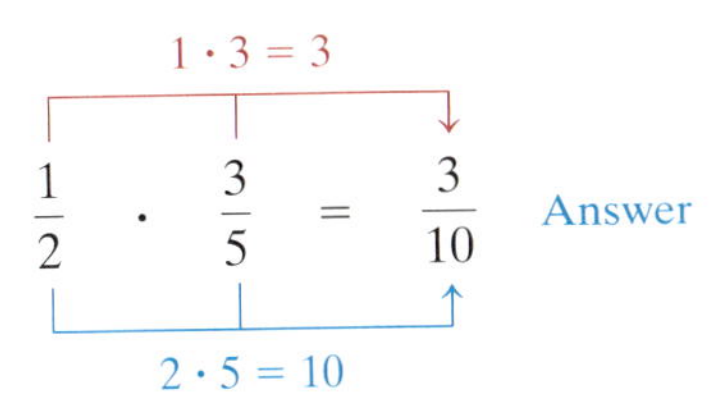

- The denominator of the answer is the product of the denominators of the original fractions.

These observations suggest the following rule for multiplying two fractions.

NOTATION *Complete each solution.*

17. $\frac{2}{5} + \frac{1}{3} = \frac{2 \cdot \square}{5 \cdot \square} + \frac{1 \cdot 5}{3 \cdot 5}$

$= \frac{\square}{15} + \frac{\square}{15}$

$= \frac{\square + \square}{15}$

$= \frac{11}{15}$

18. $\frac{7}{8} - \frac{2}{3} = \frac{7 \cdot 3}{\square \cdot 3} - \frac{2 \cdot 8}{\square \cdot 8}$

$= \frac{21}{\square} - \frac{16}{\square}$

$= \frac{21 - 16}{\square}$

$= \frac{5}{24}$

PRACTICE *The denominators of two fractions are given. Find the lowest common denominator.*

19. 18, 6
20. 15, 3
21. 8, 6
22. 10, 4
23. 8, 20
24. 14, 21
25. 15, 12
26. 25, 30

Perform each operation. Simplify when necessary.

27. $\frac{3}{7} + \frac{1}{7}$
28. $\frac{16}{25} - \frac{9}{25}$
29. $\frac{37}{103} - \frac{17}{103}$
30. $\frac{54}{53} - \frac{52}{53}$
31. $\frac{11}{25} - \frac{1}{25}$
32. $\frac{7}{8} - \frac{1}{8}$
33. $\frac{5}{d} + \frac{3}{d}$
34. $\frac{17}{x} - \frac{12}{x}$
35. $\frac{1}{4} + \frac{3}{8}$
36. $\frac{2}{3} + \frac{1}{6}$
37. $\frac{13}{20} - \frac{1}{5}$
38. $\frac{71}{100} - \frac{1}{10}$
39. $\frac{4}{5} + \frac{2}{3}$
40. $\frac{1}{4} + \frac{2}{3}$
41. $\frac{1}{8} + \frac{2}{7}$
42. $\frac{1}{6} + \frac{5}{9}$
43. $\frac{3}{4} - \frac{2}{3}$
44. $\frac{4}{5} - \frac{1}{6}$
45. $\frac{5}{6} - \frac{3}{4}$
46. $\frac{7}{8} - \frac{5}{6}$
47. $\frac{16}{25} - \left(-\frac{3}{10}\right)$
48. $\frac{3}{8} - \left(-\frac{1}{6}\right)$
49. $-\frac{7}{16} + \frac{1}{4}$
50. $-\frac{17}{20} + \frac{4}{5}$
51. $\frac{1}{12} - \frac{3}{4}$
52. $\frac{11}{60} - \frac{13}{20}$
53. $-\frac{5}{8} - \frac{1}{3}$
54. $-\frac{7}{20} - \frac{1}{5}$
55. $-3 + \frac{2}{5}$
56. $-6 + \frac{5}{8}$
57. $-\frac{3}{4} - 5$
58. $-2 - \frac{7}{8}$
59. $\frac{7}{8} - \frac{t}{7}$
60. $\frac{5}{6} + \frac{c}{7}$
61. $\frac{4}{5} - \frac{2b}{9}$
62. $\frac{3}{16} + \frac{4h}{8}$
63. $\frac{4}{7} - \frac{1}{r}$
64. $\frac{4}{m} + \frac{2}{7}$
65. $-\frac{5}{9} + \frac{1}{y}$
66. $-\frac{3}{5} + \frac{5}{x}$
67. $\frac{1}{3} + \frac{1}{4} + \frac{1}{5}$
68. $\frac{1}{10} + \frac{1}{8} + \frac{1}{5}$
69. $-\frac{2}{3} + \frac{5}{4} + \frac{1}{6}$
70. $-\frac{3}{4} + \frac{3}{8} + \frac{7}{6}$
71. $\frac{5}{24} + \frac{3}{16}$
72. $\frac{17}{20} - \frac{4}{15}$
73. $-\frac{11}{15} - \frac{2}{9}$
74. $-\frac{19}{18} - \frac{5}{12}$
75. $\frac{7}{25} + \frac{1}{15}$
76. $\frac{11}{20} - \frac{1}{8}$
77. $\frac{4}{27} + \frac{1}{6}$
78. $\frac{8}{9} - \frac{7}{12}$

79. Find the difference of $\frac{11}{60}$ and $\frac{2}{45}$.

80. Find the sum of $\frac{9}{48}$ and $\frac{7}{40}$.

81. Subtract $\frac{5}{12}$ from $\frac{2}{15}$.

82. Find the sum of $\frac{11}{24}$ and $\frac{7}{36}$ increased by $\frac{5}{48}$.

APPLICATIONS

83. BOTANY To assess the effects of smog on tree development, botanists cut down a pine tree and measured the width of the growth rings for the last two years (see the illustration on the next page).

a. What was the growth over this two-year period?

b. What is the difference in the widths of the two rings?

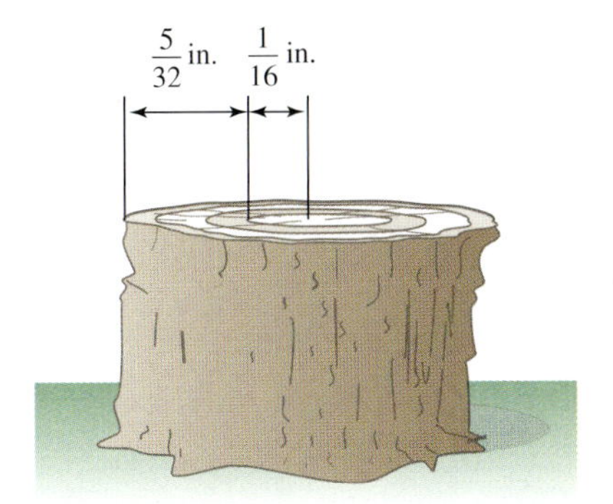

84. MAGAZINE LAYOUTS The page design for a magazine cover includes a blank strip at the top, called a header, and a blank strip at the bottom of the page, called a footer. In the illustration, how much page length is lost because of the header and footer?

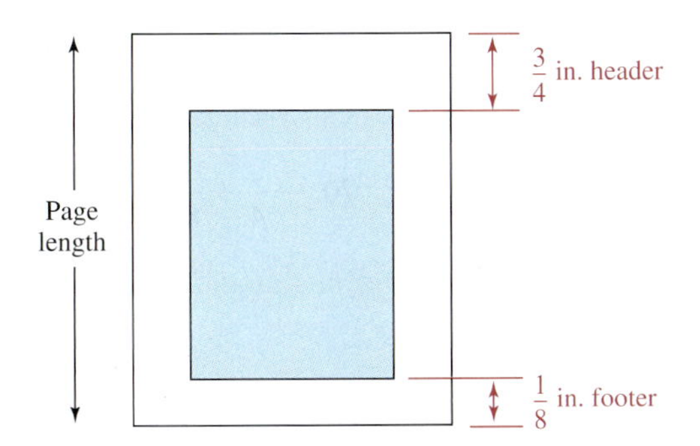

85. FAMILY DINNER A family bought two large pizzas for dinner. Several pieces of each pizza were not eaten, as shown. How much pizza was left? Could the family have been fed with just one pizza?

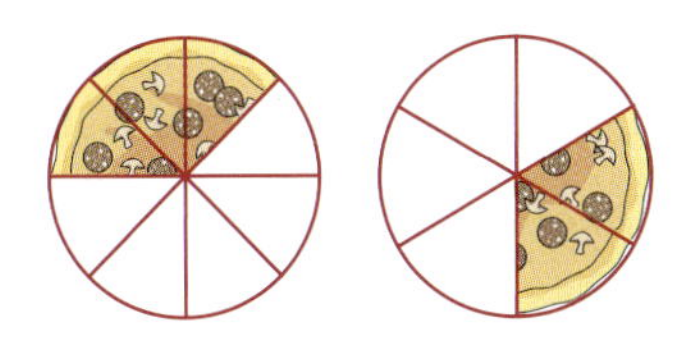

86. GASOLINE BARRELS The contents of two identical-sized barrels are shown. If they are poured into an empty third barrel that is the same size, how much of the third barrel will they fill?

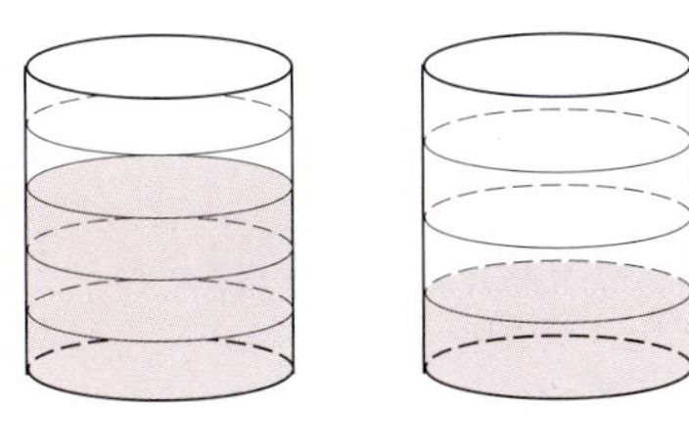

87. WEIGHTS AND MEASURES A consumer protection agency verifies the accuracy of butcher shop scales by placing a known three-quarter-pound weight on the scale and then comparing that to the scale's readout. According to the illustration, by how much is this scale off? Does it result in undercharging or overcharging customers on their meat purchases?

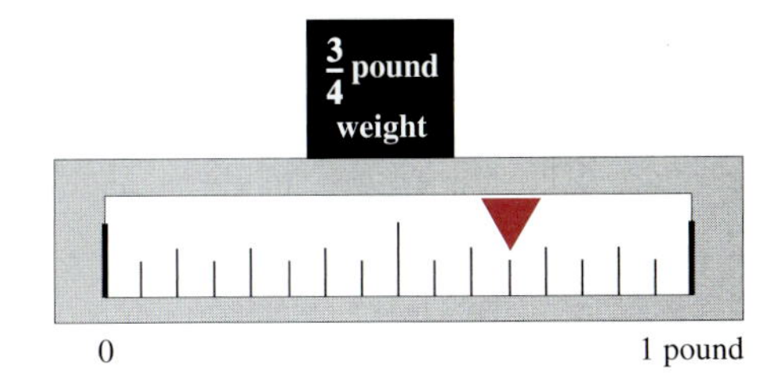

88. WRENCHES A mechanic hangs his wrenches above a tool bench in order of narrowest to widest. What is the proper order of the wrenches in the illustration?

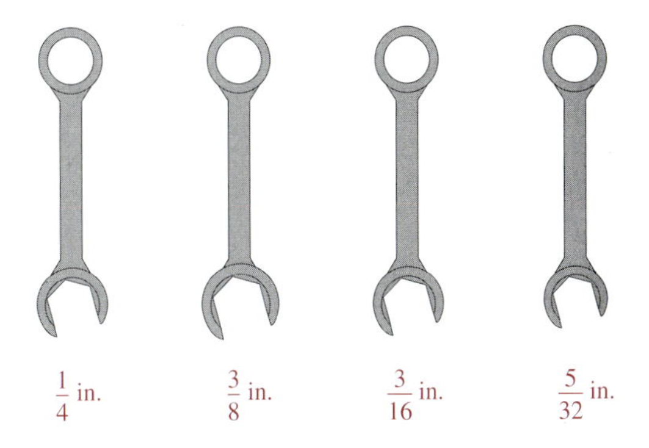

89. HIKING The illustration shows the length of each part of a three-part hike. Rank the lengths from longest to shortest.

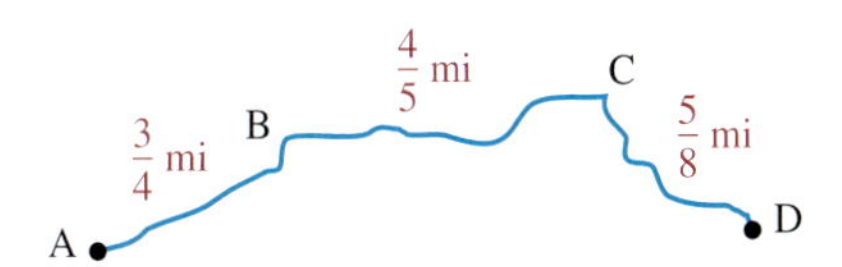

90. FIGURE DRAWING As an aid in drawing the human body, artists divide the body into three parts. Each part is then expressed as a fraction of the total body height. For example, the torso is $\frac{4}{15}$ of the body height. What fraction of body height is the head?

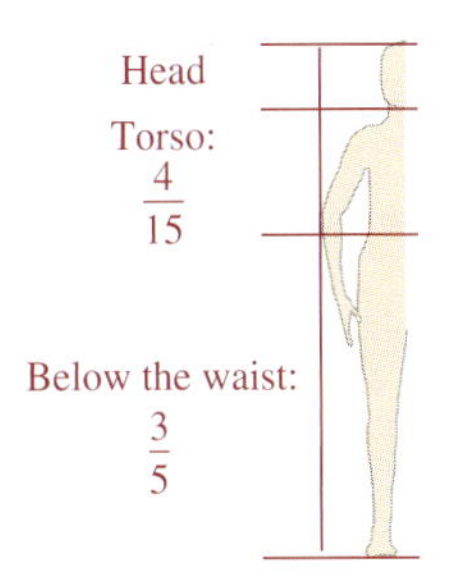

91. STUDY HABITS College students taking a full load were asked to give the average number of hours they studied each day. The results are shown in the pie chart. What fraction of the students study 2 hours or more daily?

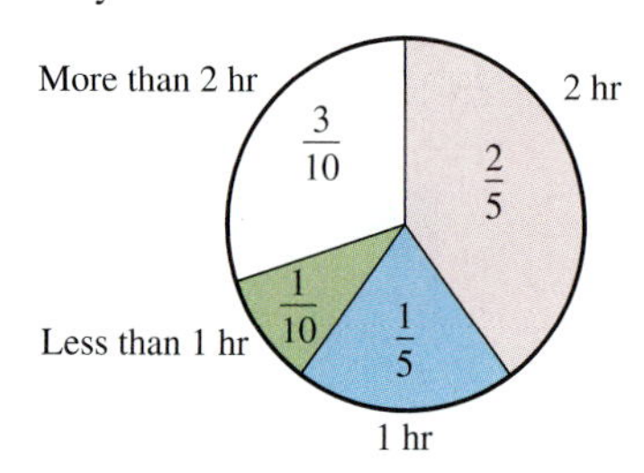

92. MUSICAL NOTES The notes used in music have fractional values. Their names and the symbols used to represent them are shown in Illustration (a). In common time, the values of the notes in each measure must add up to 1. Is the measure in Illustration (b) complete?

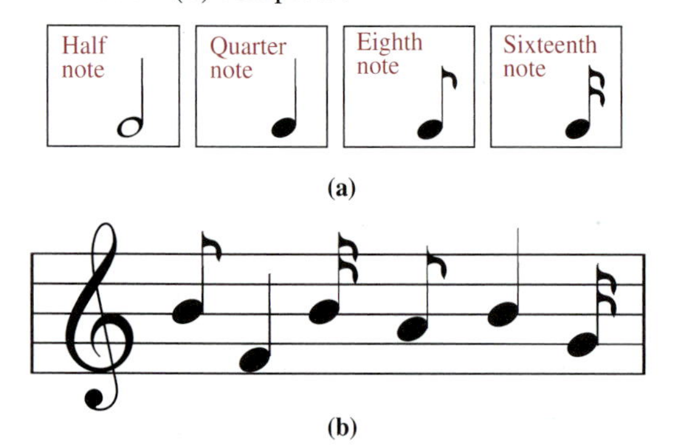

(a)

(b)

93. GARAGE DOOR OPENERS What is the difference in strength between a $\frac{1}{3}$-hp and a $\frac{1}{2}$-hp garage door opener?

94. DELIVERY TRUCKS A truck can safely carry a one-ton load. Should it be used to deliver one-half ton of sand, one-third ton of gravel, and one-fifth ton of cement in one trip to a job site?

WRITING

95. How are the procedures for expressing a fraction in higher terms and simplifying a fraction to lowest terms similar, and how are they different?

96. Given two fractions, how do we find their lowest common denominator?

97. How do we compare the sizes of two fractions with different denominators?

98. What is the difference between a common denominator and the lowest common denominator?

REVIEW

99. Simplify: $2(2 + x) - 3(x - 1)$.

100. A ball is dropped off the top of a building and falls for 3 seconds. How far does it travel in that time? (*Hint:* $d = 16t^2$.)

101. Translate to mathematical symbols: 5 less than x.

102. What is the formula for finding the area of a rectangle?

103. What is the formula for finding the perimeter of a rectangle?

104. Let $x = -4$. Find $2x^2 - x$.

The LCM and the GCF

As we have seen, the **multiples** of a number can be found by multiplying it successively by 1, 2, 3, 4, 5, and so on. The multiples of 4 and the multiples of 6 are shown below.

$1 \cdot 4 = 4$	$1 \cdot 6 = 6$	
$2 \cdot 4 = 8$	$2 \cdot 6 = \mathbf{12}$	
$3 \cdot 4 = \mathbf{12}$	$3 \cdot 6 = 18$	
$4 \cdot 4 = 16$	$4 \cdot 6 = \mathbf{24}$	Common multiples of 4 and 6 are highlighted in red.
$5 \cdot 4 = 20$	$5 \cdot 6 = 30$	
$6 \cdot 4 = \mathbf{24}$	$6 \cdot 6 = \mathbf{36}$	
$7 \cdot 4 = 28$	$7 \cdot 6 = 42$	
$8 \cdot 4 = 32$	$8 \cdot 6 = 48$	
$9 \cdot 4 = \mathbf{36}$	$9 \cdot 6 = 54$	

Because 12 is the smallest number that is a multiple of both 4 and 6, it is called the **least common multiple (LCM)** of 4 and 6.

Making lists like those shown above can be tedious. A more efficient method to find the least common multiple of several numbers is as follows.

Finding the least common multiple

1. Write each of the numbers in prime-factored form.
2. The least common multiple is a product of prime factors, where each prime factor is used the greatest number of times it appears in any one factorization found in step 1.

EXAMPLE 1 Least common multiple.

Find the LCM of 24 and 36.

Solution

Step 1: First, we find the prime factorizations of 24 and 36.

$$24 = 3 \cdot 2 \cdot 2 \cdot 2$$
$$36 = 3 \cdot 3 \cdot 2 \cdot 2$$

Step 2: The prime factorizations of 24 and 36 contain the prime factors 3 and 2. We use each of these factors the greatest number of times it appears in any one factorization.

The greatest number of times 3 appears in any one factorization is two times.

The greatest number of times 2 appears in any one factorization is three times.

$$\text{LCD} = 3 \cdot 3 \cdot 2 \cdot 2 \cdot 2 = 72$$

The least common multiple of 24 and 36 is 72.

Self Check 1

Find the LCM of 18 and 84.

Answer 252

Because 2 divides 36 exactly and because 2 divides 120 exactly, 2 is called a **common factor** of 36 and 120.

$$\frac{36}{2} = 18 \qquad \frac{120}{2} = 60$$

The numbers 36 and 120 have other common factors, such as 3 and 6. The **greatest common factor (GCF)** of 36 and 120 is the largest number that is a factor of both. We follow these steps to find the greatest common factor of several numbers.

Finding the greatest common factor

1. Write each of the numbers in prime-factored form.
2. The greatest common factor is the product of the prime factors that are common to the factorizations found in step 1. If the numbers have no factors in common, the GCF is 1.

EXAMPLE 2 Greatest common factor.

Find the GCF of 36 and 120.

Solution

Step 1: We find the prime factorizations of 36 and 120.

$$36 = 3 \cdot 3 \cdot 2 \cdot 2$$
$$120 = 5 \cdot 3 \cdot 2 \cdot 2 \cdot 2$$

Self Check 2

Find the GCF of 60 and 150.

NOTATION *Complete each solution.*

15. $-5\frac{1}{4}\cdot 1\frac{1}{7} = -\frac{21}{4}\cdot\frac{\square}{7}$

$= -\frac{21\cdot\square}{4\cdot 7}$

$= -\frac{\overset{1}{\cancel{7}}\cdot 3\cdot\overset{1}{\cancel{\square}}\cdot 2}{\underset{1}{\cancel{\square}}\cdot\underset{1}{\cancel{7}}}$

$= -\frac{\square}{1}$

$= -6$

16. $-5\frac{5}{6}\div 2\frac{1}{12} = -\frac{\square}{6}\div\frac{25}{12}$

$= -\frac{35}{6}\cdot\frac{12}{\square}$

$= -\frac{35\cdot 12}{6\cdot\square}$

$= -\frac{\overset{1}{\cancel{5}}\cdot\square\cdot\overset{1}{\cancel{6}}\cdot 2}{\underset{1}{\cancel{6}}\cdot\underset{1}{\cancel{5}}\cdot\square}$

$= -\frac{\square}{5}$

$= -2\frac{4}{5}$

PRACTICE *Write each improper fraction as a mixed number. Simplify the result, if possible.*

17. $\frac{15}{4}$

18. $\frac{41}{6}$

19. $\frac{29}{5}$

20. $\frac{29}{3}$

21. $-\frac{20}{6}$

22. $-\frac{28}{8}$

23. $\frac{127}{12}$

24. $\frac{197}{16}$

Write each mixed number as an improper fraction.

25. $6\frac{1}{2}$

26. $8\frac{2}{3}$

27. $20\frac{4}{5}$

28. $15\frac{3}{8}$

29. $-6\frac{2}{9}$

30. $-7\frac{1}{12}$

31. $200\frac{2}{3}$

32. $90\frac{5}{6}$

Graph each set of numbers on the number line.

33. $\left\{-2\frac{8}{9}, 1\frac{2}{3}, \frac{16}{5}\right\}$

−5 −4 −3 −2 −1 0 1 2 3 4 5

34. $\left\{-\frac{3}{4}, -3\frac{1}{4}, \frac{5}{2}\right\}$

−5 −4 −3 −2 −1 0 1 2 3 4 5

35. $\left\{3\frac{1}{7}, -\frac{98}{99}, -\frac{10}{3}\right\}$

−5 −4 −3 −2 −1 0 1 2 3 4 5

36. $\left\{-2\frac{1}{5}, \frac{4}{5}, -\frac{11}{3}\right\}$

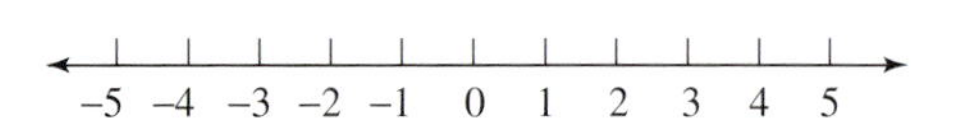

Multiply.

37. $1\frac{2}{3}\cdot 2\frac{1}{7}$

38. $2\frac{3}{5}\cdot 1\frac{2}{3}$

39. $-7\frac{1}{2}\left(-1\frac{2}{5}\right)$

40. $-4\frac{1}{8}\left(-1\frac{7}{9}\right)$

41. $3\frac{1}{16}\cdot 4\frac{4}{7}$

42. $5\frac{3}{5}\cdot 1\frac{11}{14}$

43. $-6\cdot 2\frac{7}{24}$

44. $-7\cdot 1\frac{3}{28}$

45. $2\frac{1}{2}\left(-3\frac{1}{3}\right)$

46. $\left(-3\frac{1}{4}\right)\left(1\frac{1}{5}\right)$

47. $2\frac{5}{8}\cdot\frac{5}{27}$

48. $3\frac{1}{9}\cdot\frac{3}{32}$

49. Find the product of $1\frac{2}{3}$, 6, and $-\frac{1}{8}$.

50. Find the product of $-\frac{5}{6}$, -8, and $-2\frac{1}{10}$.

Evaluate each power.

51. $\left(1\frac{2}{3}\right)^2$

52. $\left(3\frac{1}{2}\right)^2$

53. $\left(-1\frac{1}{3}\right)^3$

54. $\left(-1\frac{1}{5}\right)^3$

Divide.

55. $3\frac{1}{3} \div 1\frac{5}{6}$

56. $3\frac{3}{4} \div 5\frac{1}{3}$

57. $-6\frac{3}{5} \div 7\frac{1}{3}$

58. $-4\frac{1}{4} \div 4\frac{1}{2}$

59. $-20\frac{1}{4} \div \left(-1\frac{11}{16}\right)$

60. $-2\frac{7}{10} \div \left(-1\frac{1}{14}\right)$

61. $6\frac{1}{4} \div 20$

62. $4\frac{2}{5} \div 11$

63. $1\frac{2}{3} \div \left(-2\frac{1}{2}\right)$

64. $2\frac{1}{2} \div \left(-1\frac{5}{8}\right)$

65. $8 \div 3\frac{1}{5}$

66. $15 \div 3\frac{1}{3}$

67. Find the quotient of $-4\frac{1}{2}$ and $2\frac{1}{4}$.

68. Find the quotient of 25 and $-10\frac{5}{7}$.

APPLICATIONS

69. CALORIES A company advertises that its mints contain only $3\frac{1}{5}$ calories apiece. What is the calorie intake if you eat an entire package of 20 mints?

70. CEMENT MIXERS A cement mixer can carry $9\frac{1}{2}$ cubic yards of concrete. If it makes 8 trips to a job site, how much concrete will be delivered to the site?

71. SHOPPING In the illustration, what is the cost of buying the fruit in the scale?

72. FRAMES How much molding is needed to make the square picture frame below?.

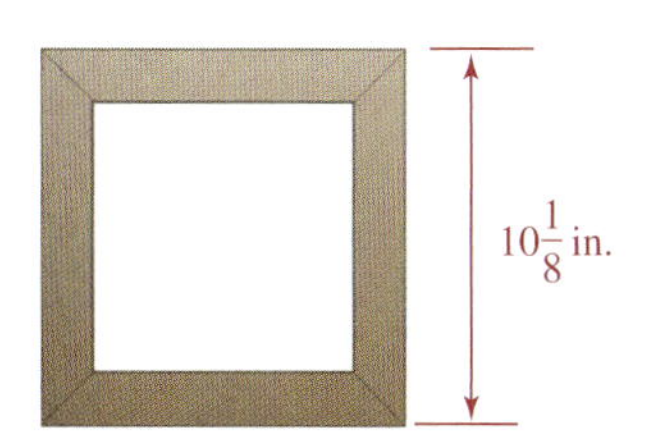

73. SUBDIVISIONS A developer donated to the county 100 of the 1,000 acres of land she owned. She divided the remaining acreage into $1\frac{1}{3}$-acre lots. How many lots were created?

74. CATERING How many people can be served $\frac{1}{3}$-pound hamburgers if a caterer purchases 200 pounds of ground beef?

75. GRAPH PAPER Mathematicians use specially marked paper, called *graph paper,* when drawing figures. It is made up of $\frac{1}{4}$-inch squares. Find the length and width of the following piece of graph paper.

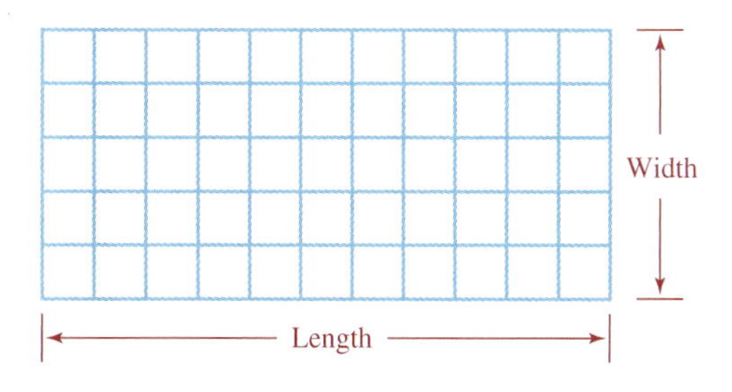

76. LUMBER As shown in the following illustration, 2-by-4's from the lumber yard do not really have dimensions of 2 inches by 4 inches. How wide and how high is the stack of 2-by-4's?

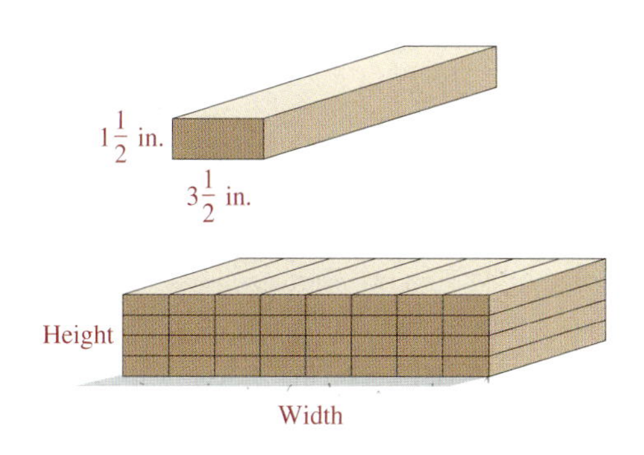

77. EMERGENCY EXIT The following sign marks the emergency exit on a school bus. Find the area of the sign.

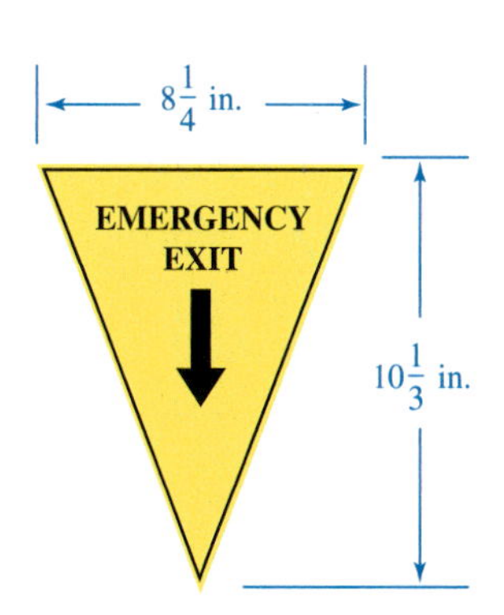

78. HORSE RACING The race tracks on which thoroughbred horses run are marked off in $\frac{1}{8}$-mile-long segments called furlongs. How many furlongs are there in a $1\frac{1}{16}$-mile race?

79. FIRE ESCAPES The fire escape stairway in an office building is shown. Each riser is $7\frac{1}{2}$ inches high. If each floor is 105 inches high and the building is 43 stories tall, how many steps are in the stairway?

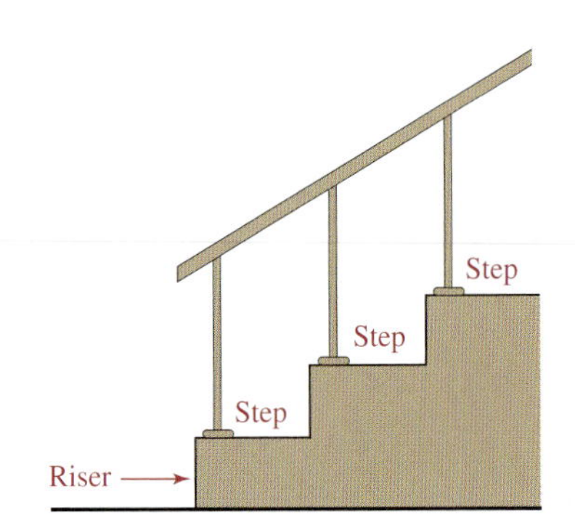

80. LICENSE PLATES Find the area of the license plate shown below.

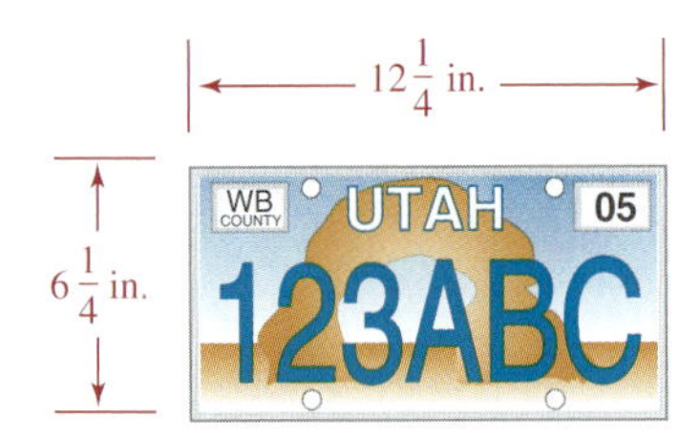

81. SHOPPING ON THE INTERNET A mother is ordering a pair of jeans for her daughter from the screen shown. If the daughter's height is $60\frac{3}{4}$ in. and her waist is $24\frac{1}{2}$ in., on what size and what cut should the mother point and click?

Girl's jeans- regular cut

Size	7	8	10	12	14	16
Height	50-52	52-54	54-56	$56\frac{1}{4}$-$58\frac{1}{2}$	59-61	61-62
Waist	$22\frac{1}{4}$-$22\frac{3}{4}$	$22\frac{3}{4}$-$23\frac{1}{4}$	$23\frac{3}{4}$-$24\frac{1}{4}$	$24\frac{3}{4}$-$25\frac{1}{4}$	$25\frac{3}{4}$-$26\frac{1}{4}$	$26\frac{1}{4}$-28

Girl's jeans- slim cut

Size	7	8	10	12	14	16
Height	50-52	52-54	54-56	$56\frac{1}{2}$-$58\frac{1}{2}$	59-61	61-62
Waist	$20\frac{3}{4}$-$21\frac{1}{4}$	$21\frac{1}{4}$-$21\frac{3}{4}$	$22\frac{1}{4}$-$22\frac{3}{4}$	$23\frac{1}{4}$-$23\frac{3}{4}$	$24\frac{1}{4}$-$24\frac{3}{4}$	25-$26\frac{1}{2}$

<u>To order:</u>
Point arrow to proper size/cut and click

82. SEWING Use the following table to determine the number of yards of fabric needed:

a. to make a size 16 top if the fabric to be used is 60 inches wide.

b. to make size 18 pants if the fabric to be used is 45 inches wide.

8767 Pattern
stitch'n save
by McCall's

Front

SIZES	8	10	12	14	16	18	20	
Top								
45"	$2\frac{1}{4}$	$2\frac{3}{8}$	$2\frac{3}{8}$	$2\frac{3}{8}$	$2\frac{1}{2}$	$2\frac{5}{8}$	$2\frac{3}{4}$	**Yds**
60"	2	2	$2\frac{1}{8}$	$2\frac{1}{8}$	$2\frac{1}{8}$	$2\frac{1}{8}$	$2\frac{1}{8}$	
Pants								
45"	$2\frac{5}{8}$	$2\frac{5}{8}$	$2\frac{5}{8}$	$2\frac{5}{8}$	$2\frac{5}{8}$	$2\frac{5}{8}$	$2\frac{5}{8}$	**Yds**
60"	$1\frac{3}{4}$	2	$2\frac{1}{4}$	$2\frac{1}{4}$	$2\frac{1}{4}$	$2\frac{1}{4}$	$2\frac{1}{2}$	

WRITING

83. Explain the difference between $2\frac{3}{4}$ and $2(\frac{3}{4})$.

84. Give three examples of how you use mixed numbers in daily life.

85. Explain the procedure used to write an improper fraction as a mixed number.

86. Explain the procedure used to multiply two mixed numbers.

REVIEW

87. Evaluate: $3^2 \cdot 2^3$.

88. If a represents a number, then $a \cdot 0 = \square$.

89. Write $8 + 8 + 8 + 8$ as a multiplication.

90. If a square measures 1 inch on each side, find its area?

91. Solve: $\frac{x}{2} = -12$.

92. In the formula $C = \frac{5}{9}(F - 32)$, what do C and F represent?

93. Explain why $3t$ and $3u$ are not like terms.

94. Add: $3x + 3x$.

4.6 Adding and Subtracting Mixed Numbers

- Adding mixed numbers
- Adding mixed numbers in vertical form
- Subtracting mixed numbers

In this section, we will discuss methods for adding and subtracting mixed numbers. The first method works well when the whole-number parts of the mixed numbers are small. The second method works well when the whole-number parts of the mixed numbers are large. The third method uses columns as a way to organize the work.

Adding mixed numbers

We can add mixed numbers by writing them as improper fractions. To do so, we follow these steps.

Adding mixed numbers: method 1

1. Write each mixed number as an improper fraction.
2. Write each improper fraction as an equivalent fraction with a denominator that is the LCD.
3. Add the fractions.
4. Change the result to a mixed number if desired.

EXAMPLE 1 Add: $4\frac{1}{6} + 2\frac{3}{4}$.

Solution

$$4\frac{1}{6} + 2\frac{3}{4} = \frac{25}{6} + \frac{11}{4}$$ Write each mixed number as an improper fraction: $4\frac{1}{6} = \frac{25}{6}$ and $2\frac{3}{4} = \frac{11}{4}$.

By inspection, we see that the common denominator is 12.

$$= \frac{25 \cdot 2}{6 \cdot 2} + \frac{11 \cdot 3}{4 \cdot 3}$$ Write each fraction as a fraction with a denominator of 12.

$$= \frac{50}{12} + \frac{33}{12}$$ Perform the multiplications in the numerators and denominators.

$$= \frac{83}{12}$$ Add the numerators: $50 + 33 = 83$. Write the sum over the common denominator 12.

$$= 6\frac{11}{12}$$ Write the improper fraction as a mixed number: $\frac{83}{12} = 6\frac{11}{12}$.

Self Check 1

Add: $3\frac{2}{3} + 1\frac{1}{5}$.

Answer $4\frac{13}{15}$

We can also add mixed numbers by adding their whole-number parts and their fractional parts. To do so, we follow these steps.

Adding mixed numbers: method 2

1. Write each mixed number as the sum of a whole number and a fraction.
2. Use the commutative property of addition to write the whole numbers together and the fractions together.
3. Add the whole numbers and the fractions separately.
4. Write the result as a mixed number if necessary.

Self Check 2

Find the sum: $275\frac{1}{6} + 81\frac{3}{5}$.

Answer $356\frac{23}{30}$

EXAMPLE 2 Find the sum: $168\frac{3}{4} + 85\frac{1}{5}$.

Solution

$$168\frac{3}{4} + 85\frac{1}{5} = 168 + \frac{3}{4} + 85 + \frac{1}{5}$$ Write each mixed number as the sum of a whole number and a fraction.

$$= 168 + 85 + \frac{3}{4} + \frac{1}{5}$$ Use the commutative property of addition to change the order of the addition.

$$= 253 + \frac{3}{4} + \frac{1}{5}$$ Add the whole numbers: $168 + 85 = 253$.

$$= 253 + \frac{3 \cdot 5}{4 \cdot 5} + \frac{1 \cdot 4}{5 \cdot 4}$$ Write each fraction as a fraction with denominator 20.

$$= 253 + \frac{15}{20} + \frac{4}{20}$$ Multiply in the numerators and denominators.

$$= 253 + \frac{19}{20}$$ Add the numerators and write the sum over the common denominator 20.

$$= 253\frac{19}{20}$$ Write the sum as a mixed number.

! COMMENT If we use method 1 to add the mixed numbers in Example 2, the numbers we encounter are cumbersome. As expected, the result is the same: $253\frac{19}{20}$.

$$168\frac{3}{4} + 85\frac{1}{5} = \frac{675}{4} + \frac{426}{5}$$ Write $168\frac{3}{4}$ and $85\frac{1}{5}$ as improper fractions.

$$= \frac{675 \cdot 5}{4 \cdot 5} + \frac{426 \cdot 4}{5 \cdot 4}$$ The LCD is 20.

$$= \frac{3{,}375}{20} + \frac{1{,}704}{20}$$

$$= \frac{5{,}079}{20}$$

$$= 253\frac{19}{20}$$

Generally speaking, the larger the whole-number parts of the mixed numbers get, the more difficult it becomes to add those mixed numbers using method 1.

Adding mixed numbers in vertical form

By working in columns, we can use a third method to add mixed numbers. The strategy is the same as in Example 2: Add whole numbers to whole numbers and fractions to fractions.

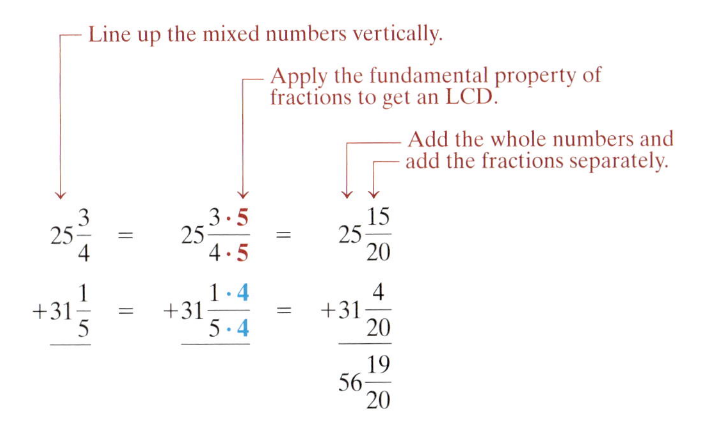

EXAMPLE 3 **Suspension bridges.** Find the total length of cable that must be ordered if cables a, d, and e of the suspension bridge in Figure 4-9 are to be replaced. (See the table on the right.)

Bridge Specifications

Cable	a	b	c
Length (feet)	$75\frac{1}{12}$	$54\frac{1}{6}$	$43\frac{1}{4}$

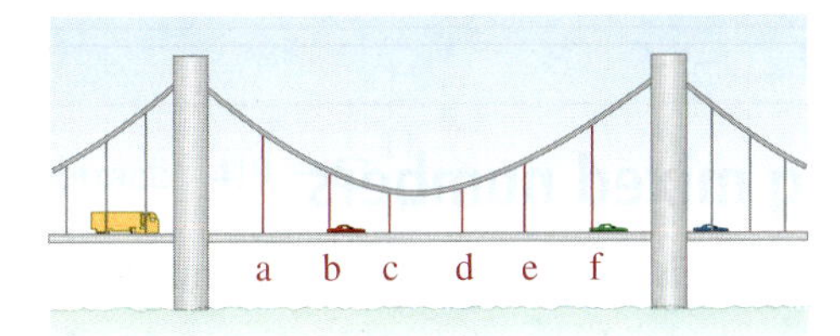

FIGURE 4-9

Solution To find the total length of cable to be ordered, we add the lengths of cables a, d, and e. Because of the symmetric design, cables e and b and cables d and c are the same length.

Length of cable a	plus	length of cable d (or cable c)	plus	length of cable e (or cable b)	equals	the total length needed.
$75\frac{1}{12}$	$+$	$43\frac{1}{4}$	$+$	$54\frac{1}{6}$	$=$	total length

We add the mixed numbers using a vertical format.

$$
\begin{aligned}
75\frac{1}{12} &= 75\frac{1}{12} = 75\frac{1}{12} \\
43\frac{1}{4} &= 43\frac{1\cdot 3}{4\cdot 3} = 43\frac{3}{12} \\
+54\frac{1}{6} &= +54\frac{1\cdot 2}{6\cdot 2} = +\ 54\frac{2}{12} \\
&\qquad 172\frac{6}{12} = 172\frac{1}{2} \quad \text{Simplify: } \frac{6}{12} = \frac{1}{2}.
\end{aligned}
$$

The total length of cable needed for the replacement is $172\frac{1}{2}$ feet.

Simplifying complex fractions

To *simplify* complex fractions means to express them as fractions in simplified form.

Simplifying a complex fraction: method 1

Write the numerator and the denominator of the complex fraction as single fractions. Then perform the indicated division of the two fractions and simplify.

Method 1 is based on the fact that the main fraction bar of the complex fraction indicates division.

$$\frac{\frac{1}{4}}{\frac{2}{5}} \longleftarrow \text{The main fraction bar means "divide the fraction in the numerator by the fraction in the denominator."} \longrightarrow \frac{1}{4} \div \frac{2}{5}$$

Self Check 5

Simplify: $\frac{\frac{1}{6}}{\frac{3}{8}}$.

Answer $\frac{4}{9}$

EXAMPLE 5 Simplify: $\frac{\frac{1}{4}}{\frac{2}{5}}$.

Solution Since the numerator and the denominator of this complex fraction are single fractions, we can do the indicated division.

$$\frac{\frac{1}{4}}{\frac{2}{5}} = \frac{1}{4} \div \frac{2}{5}$$ Express the complex fraction as an equivalent division problem.

$$= \frac{1}{4} \cdot \frac{5}{2}$$ Multiply by the reciprocal of $\frac{2}{5}$.

$$= \frac{1 \cdot 5}{4 \cdot 2}$$ Multiply the numerators and multiply the denominators.

$$= \frac{5}{8}$$

A second method is based on the fundamental property of fractions.

Simplifying a complex fraction: method 2

Multiply the numerator and the denominator of the complex fraction by the LCD of all the fractions that appear in its numerator and denominator. Then simplify.

EXAMPLE 6 Simplify: $\dfrac{-\frac{1}{4}+\frac{2}{5}}{\frac{1}{2}-\frac{4}{5}}$.

Solution Examine the numerator and the denominator of the complex fraction. The fractions involved have denominators of 4, 5, and 2. The LCD of these fractions is 20.

$$\frac{-\frac{1}{4}+\frac{2}{5}}{\frac{1}{2}-\frac{4}{5}}=\frac{20\left(-\frac{1}{4}+\frac{2}{5}\right)}{20\left(\frac{1}{2}-\frac{4}{5}\right)}$$

Use the fundamental property of fractions. Multiply the numerator and the denominator of the complex fraction by 20. Note how parentheses are used to show this.

$$=\frac{20\left(-\frac{1}{4}\right)+20\left(\frac{2}{5}\right)}{20\left(\frac{1}{2}\right)-20\left(\frac{4}{5}\right)}$$

Use the distributive property in the numerator and in the denominator.

$$=\frac{-5+8}{10-16}$$

Perform the multiplications by 20.

$$=\frac{3}{-6}$$

Perform the addition in the numerator and the subtraction in the denominator.

$$=-\frac{1}{2}$$

Simplify.

EXAMPLE 7 Simplify: $\dfrac{7-\frac{2}{3}}{4\frac{5}{6}}$.

Solution Examine the numerator and the denominator of the complex fraction. The fractions have denominators of 3 and 6. The LCD of these fractions is 6.

$$\frac{7-\frac{2}{3}}{4\frac{5}{6}}=\frac{7-\frac{2}{3}}{\frac{29}{6}}$$

Express $4\frac{5}{6}$ as an improper fraction.

$$=\frac{6\left(7-\frac{2}{3}\right)}{6\left(\frac{29}{6}\right)}$$

Use the fundamental property of fractions. Multiply the numerator and the denominator of the complex fraction by the LCD, 6.

$$=\frac{6(7)-6\left(\frac{2}{3}\right)}{6\left(\frac{29}{6}\right)}$$

Use the distributive property in the numerator. Distribute the multiplication by 6.

$$=\frac{42-4}{29}$$

Perform the multiplications by 6.

$$=\frac{38}{29}$$

Perform the subtraction in the numerator.

$$=1\frac{9}{29}$$

Write $\frac{38}{29}$ as a mixed number.

Self Check 7

Simplify: $\dfrac{5-\frac{3}{4}}{1\frac{7}{8}}$.

Answer $2\frac{4}{15}$

Section 4.7 STUDY SET

VOCABULARY *Fill in the blanks.*

1. $\dfrac{\frac{1}{2}}{\frac{3}{4}}$ is a ________ fraction.

2. To evaluate an algebraic expression, we ________ specific numbers for the variables in the expression and simplify.

CONCEPTS

3. What division is represented by the complex fraction?

$$\frac{\frac{2}{3}}{\frac{1}{5}}$$

4. Write the division as a complex fraction.

$$-\frac{7}{8} \div \frac{3}{4}$$

5. What is the common denominator of all the fractions in the complex fraction?

$$\frac{\frac{2}{3} - \frac{1}{5}}{\frac{1}{3} + \frac{4}{5}}$$

6. What is the common denominator of all the fractions in the complex fraction?

$$\frac{\frac{1}{8} - \frac{3}{16}}{-5\frac{3}{4}}.$$

7. When the complex fraction is simplified, will the result be positive or negative?

$$\frac{-\frac{2}{3}}{\frac{3}{4}}$$

8. What property is being applied?

$$\frac{1 + \frac{1}{11}}{\frac{1}{2}} = \frac{22\left(1 + \frac{1}{11}\right)}{22\left(\frac{1}{2}\right)}$$

9. What is the LCD of fractions with the denominators 6, 4, and 5?

10. What operations are involved in the numerical expression?

$$5\left(6\frac{1}{3}\right) + \left(-\frac{1}{4}\right)^2$$

NOTATION *Complete each solution to simplify the complex fraction.*

11. $\dfrac{\frac{1}{8}}{\frac{3}{4}} = \dfrac{1}{8} \div \square$

$= \dfrac{1}{8} \cdot \square$

$= \dfrac{1\,\square}{8 \cdot 3}$

$= \dfrac{1 \cdot \overset{1}{\cancel{4}}}{2 \cdot \underset{1}{\cancel{}} \cdot 3}$

$= \dfrac{1}{6}$

12. $\dfrac{\frac{1}{6} + \frac{1}{5}}{-\frac{1}{15}} = \dfrac{30\left(\frac{1}{6} + \frac{1}{5}\right)}{\square\left(-\frac{1}{15}\right)}$

$= \dfrac{\square\left(\frac{1}{6}\right) + \square\left(\frac{1}{5}\right)}{30\left(-\frac{1}{15}\right)}$

$= \dfrac{5 + 6}{\square}$

$= \dfrac{\square}{-2}$

$= -5\frac{1}{2}$

PRACTICE *Evaluate each expression.*

13. $\frac{2}{3}\left(-\frac{1}{4}\right) + \frac{1}{2}$

14. $-\frac{7}{8} - \left(\frac{1}{8}\right)\left(\frac{2}{3}\right)$

15. $\frac{4}{5} - \left(-\frac{1}{3}\right)^2$

16. $-\frac{3}{16} - \left(-\frac{1}{2}\right)^3$

17. $-4\left(-\frac{1}{5}\right) - \left(\frac{1}{4}\right)\left(-\frac{1}{2}\right)$

18. $(-3)\left(-\frac{2}{3}\right) - (-4)\left(-\frac{3}{4}\right)$

19. $1\frac{3}{5}\left(\frac{1}{2}\right)^2\left(\frac{3}{4}\right)$

20. $2\frac{3}{5}\left(-\frac{1}{3}\right)^2\left(\frac{1}{2}\right)$

21. $\frac{7}{8} - \left(\frac{4}{5} + 1\frac{3}{4}\right)$

22. $\left(\frac{5}{4}\right)^2 + \left(\frac{2}{3} - 2\frac{1}{6}\right)$

23. $\left(\frac{9}{20} \div 2\frac{2}{5}\right) + \left(\frac{3}{4}\right)^2$

24. $\left(1\frac{2}{3}\cdot 15\right) + \left(\frac{7}{9} \div \frac{7}{81}\right)$

25. $\left(-\frac{3}{4}\cdot\frac{9}{16}\right) + \left(\frac{1}{2} - \frac{1}{8}\right)$

26. $\left(\frac{8}{5} - 1\frac{1}{3}\right) - \left(-\frac{4}{5}\cdot 10\right)$

27. $\left|\frac{2}{3} - \frac{9}{10}\right| \div \left(-\frac{1}{5}\right)$

28. $\left|-\frac{3}{16} \div 2\frac{1}{4}\right| + \left(-2\frac{1}{8}\right)$

29. $\left(2 - \frac{1}{2}\right)^2 + \left(2 + \frac{1}{2}\right)^2$

30. $\left(1 - \frac{3}{4}\right)\left(1 + \frac{3}{4}\right)$

Find one-half of the given number and square that result. Express the answer as an improper fraction.

31. -7

32. -5

33. $\frac{11}{2}$

34. $\frac{7}{3}$

Evaluate each algebraic expression for $a = 1\frac{3}{4}$, $b = -\frac{1}{5}$, $r = -1\frac{2}{3}$, and $c = -\frac{2}{3}$.

35. $\frac{1}{3}b^2 + c$

36. $\left(-\frac{1}{2}c\right)^3$

37. $-1 - ar$

38. $ab - br$

Find the perimeter of each figure.

39.

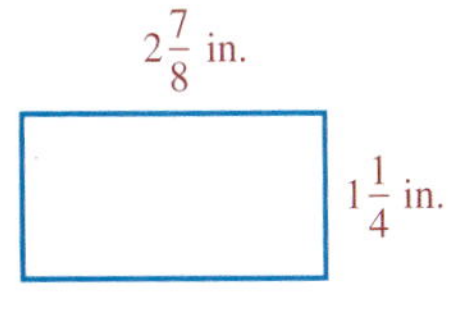

40.

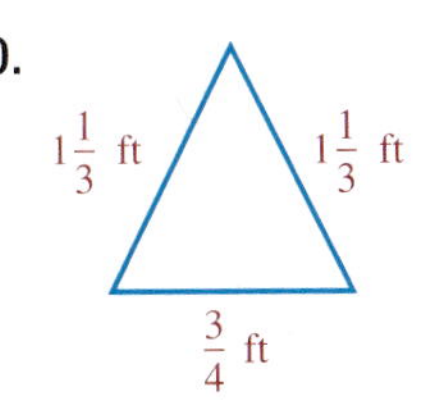

Simplify each complex fraction.

41. $\dfrac{\frac{2}{3}}{\frac{4}{5}}$

42. $\dfrac{\frac{3}{5}}{\frac{9}{25}}$

43. $\dfrac{-\frac{14}{15}}{\frac{7}{10}}$

44. $\dfrac{\frac{5}{27}}{-\frac{5}{9}}$

45. $\dfrac{5}{\frac{10}{21}}$

46. $\dfrac{6}{\frac{3}{8}}$

47. $\dfrac{-\frac{5}{6}}{-1\frac{7}{8}}$

48. $\dfrac{-\frac{4}{3}}{-2\frac{5}{6}}$

49. $\dfrac{\frac{1}{2} + \frac{1}{4}}{\frac{1}{2} - \frac{1}{4}}$

50. $\dfrac{\frac{1}{3} + \frac{1}{4}}{\frac{1}{3} - \frac{1}{4}}$

51. $\dfrac{\frac{3}{8} + \frac{1}{4}}{\frac{3}{8} - \frac{1}{4}}$

52. $\dfrac{\frac{2}{5} + \frac{1}{4}}{\frac{2}{5} - \frac{1}{4}}$

53. $\dfrac{\frac{1}{5} + 3}{-\frac{4}{25}}$

54. $\dfrac{-5 - \frac{1}{3}}{\frac{1}{6} + \frac{2}{3}}$

55. $\dfrac{5\frac{1}{2}}{-\frac{1}{4} + \frac{3}{4}}$

56. $\dfrac{4\frac{1}{4}}{\frac{2}{3} + \left(-\frac{1}{6}\right)}$

57. $\dfrac{\frac{1}{5} - \left(-\frac{1}{4}\right)}{\frac{1}{4} + \frac{4}{5}}$

58. $\dfrac{\frac{1}{8} - \left(-\frac{1}{2}\right)}{\frac{1}{4} + \frac{3}{8}}$

59. $\dfrac{\frac{1}{3} + \left(-\frac{5}{6}\right)}{1\frac{1}{3}}$

60. $\dfrac{\frac{3}{7} + \left(-\frac{1}{2}\right)}{1\frac{3}{4}}$

Evaluate each algebraic expression for $x = -\frac{3}{4}$ and $y = \frac{7}{8}$.

61. $\dfrac{x + y}{2}$

62. $\dfrac{x - y}{x + y}$

63. $\left|\dfrac{2x}{y - x}\right|$

64. $\left|\dfrac{y^2}{y - 2}\right|$

APPLICATIONS

65. SANDWICH SHOPS A sandwich shop sells a $\frac{1}{2}$-pound club sandwich, made up of turkey meat and ham. The owner buys the turkey in $1\frac{3}{4}$-pound packages and the ham in $2\frac{1}{2}$-pound packages. If he mixes a package of each of the meats together, how many sandwiches can he make from the mixture?

66. SKIN CREAMS Using a formula of $\frac{1}{2}$ ounce of sun block, $\frac{2}{3}$ ounce of moisturizing cream, and $\frac{3}{4}$ ounce of lanolin, a beautician mixes her own brand of skin cream. She packages it in $\frac{1}{4}$-ounce tubes. How many tubes can be produced using this formula?

67. PHYSICAL FITNESS Two people begin their workouts from the same point on a bike path and travel in opposite directions, as shown. How far apart are they in $1\frac{1}{2}$ hours? Use the table to help organize your work.

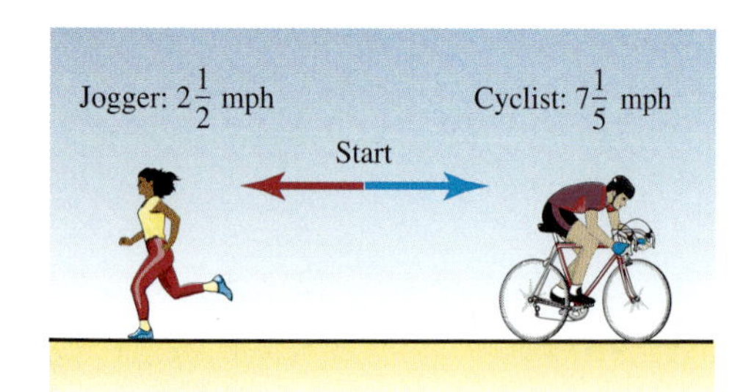

	Rate (mph)	Time (hr)	Distance (mi)
Jogger			
Cyclist			

68. SLEEP The illustration compares the amount of sleep a 1-month-old baby got to the $15\frac{1}{2}$-hour daily requirement recommended by Children's Hospital of Orange County, California. For the week, how far below the baseline was the baby's daily average?

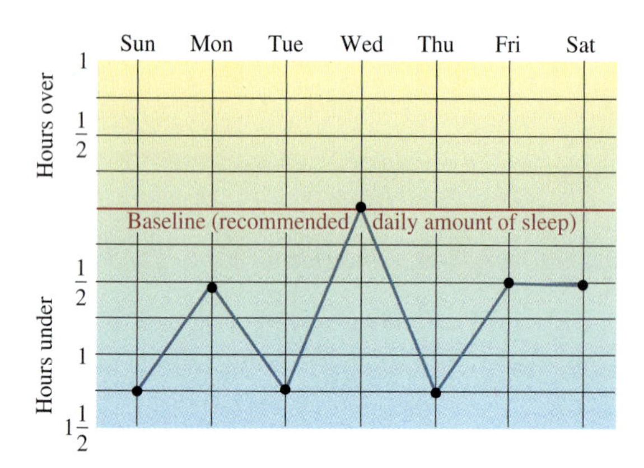

69. POSTAGE RATES Can the following ad package be mailed for the 1-ounce rate?

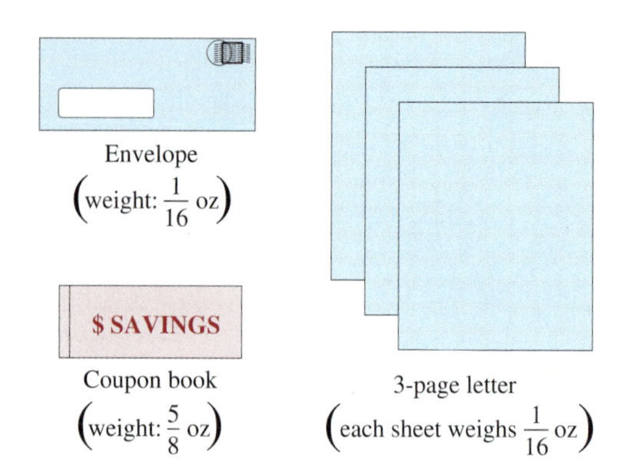

70. PLYWOOD To manufacture a sheet of plywood, several layers of thin laminate are glued together, as shown. Then an exterior finish is affixed to the top and bottom. How thick is the finished product?

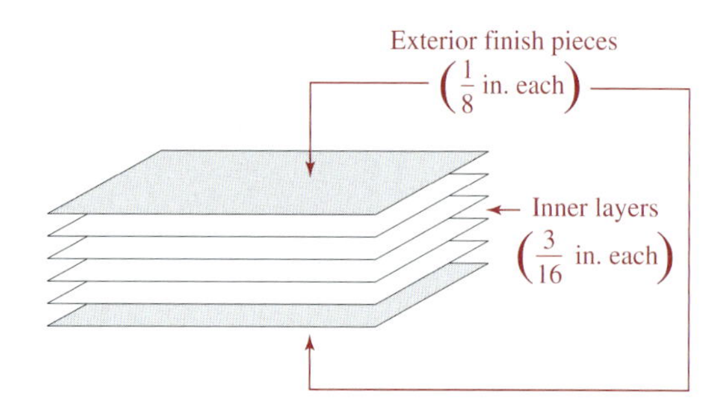

71. PHYSICAL THERAPY After back surgery, a patient undertook a walking program to rehabilitate her back muscles, as specified in the table. What was the total distance she walked over this three-week period?

Week	Distance per day
#1	$\frac{1}{4}$ mile
#2	$\frac{1}{2}$ mile
#3	$\frac{3}{4}$ mile

72. READING PROGRAMS To improve reading skills, elementary-school children read silently at the end of the school day for $\frac{1}{4}$ hour on Mondays and for $\frac{1}{2}$ hour on Fridays. For the month of January, how many total hours did the children read silently in class?

JANUARY						
S	M	T	W	T	F	S
	1	2	3	4	5	6
7	8	9	10	11	12	13
14	15	16	17	18	19	20
21	22	23	24	25	26	27
28	29	30	31			

73. AMUSEMENT PARKS At the end of a ride at an amusement park, a boat splashes into a pool of water. The time (in seconds) that it takes two pipes to refill the pool is given by

$$\frac{1}{\frac{1}{10}+\frac{1}{15}}$$

Find this time.

74. HIKING A scout troop plans to hike from the campground to Glenn Peak. Since the terrain is steep, they plan to stop and rest after every $\frac{2}{3}$ mile. With this plan, how many parts will there be to this hike?

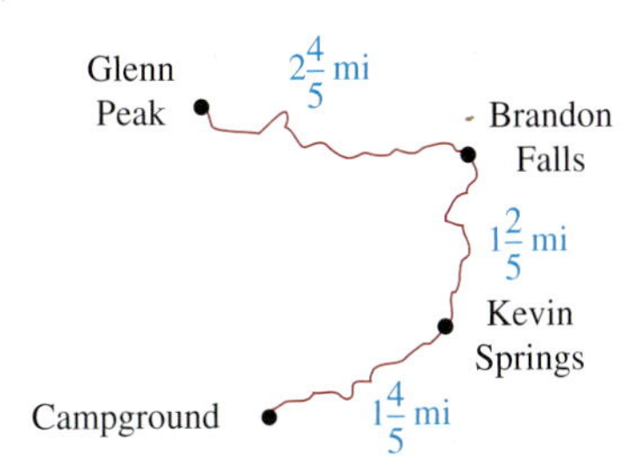

WRITING

75. What is a complex fraction?

76. Explain method 1 for simplifying complex fractions.

77. Write an application problem using a complex fraction, and then solve it.

78. Explain method 2 for simplifying complex fractions.

REVIEW

79. Simplify: $-4d-(-7d)$.

80. Solve: $8+2a=3-(4a+1)$.

81. Translate into mathematical symbols: the product of twice a number and its opposite.

82. List the factors of 24.

83. Evaluate: $2+3[-3-(-4-1)]$.

84. What is the sign of the quotient of two numbers with unlike signs?

85. Simplify: $3\cdot3\cdot3\cdot x\cdot x\cdot x\cdot x\cdot x$.

86. In the expression $-4x^2+3x-7$, what is the coefficient of the second term?

4.8 Solving Equations Containing Fractions

- Using reciprocals to solve equations
- An alternate method
- The addition and subtraction properties of equality
- Clearing an equation of fractions
- The steps to solve equations
- Problem solving with equations

In this section, we will discuss how to solve both equations containing fractions and equations whose solutions are fractions. We will make use of the properties of equality and several concepts from this chapter, including the reciprocal and the LCD.

Using reciprocals to solve equations

In the equation $\frac{3}{4}x=5$, the variable is multiplied by $\frac{3}{4}$. To undo this multiplication and isolate the variable, we can use the multiplication property of equality and multiply both sides of the equation by the reciprocal of $\frac{3}{4}$.

$\frac{4}{3}\left(\frac{3}{4}x\right)=\frac{4}{3}(5)$ Multiply both sides by the reciprocal of $\frac{3}{4}$, which is $\frac{4}{3}$.

$\left(\frac{4}{3}\cdot\frac{3}{4}\right)x=\frac{4}{3}\cdot\frac{5}{1}$ Use the associative property of multiplication to regroup the factors. Write 5 as $\frac{5}{1}$.

$\frac{\overset{1}{\cancel{4}}\cdot\overset{1}{\cancel{3}}}{\underset{1}{\cancel{3}}\cdot\underset{1}{\cancel{4}}}x=\frac{4\cdot5}{3\cdot1}$ Multiply the numerators and multiply the denominators. On the left, divide out the common factors 4 and 3.

$1x=\frac{20}{3}$ Multiply in the numerators and in the denominators.

$x=\frac{20}{3}$ Simplify: $1x=x$.

In algebra, we usually leave a solution to an equation as an improper fraction rather than converting it to a mixed number.

! COMMENT We can write expressions such as $\frac{4a}{5}$ and $\frac{-9h}{16}$ in an equivalent form so that the fractional coefficients are more evident.

$$\frac{4a}{5} = \frac{4}{5}a \qquad \frac{-9h}{16} = -\frac{9}{16}h$$

Self Check 1

Solve $-\frac{3}{2}t = 15$ and check the result.

Answer -10

EXAMPLE 1 Solve: $-\frac{7}{8}k = 21.$

Solution The coefficient of the variable is $-\frac{7}{8}$. To isolate k, we multiply both sides of the equation by the reciprocal of $-\frac{7}{8}$.

$$-\frac{7}{8}k = 21$$

$$-\frac{8}{7}\left(-\frac{7}{8}k\right) = -\frac{8}{7}(21)$$ Multiply both sides by the reciprocal of $-\frac{7}{8}$, which is $-\frac{8}{7}$.

$$1k = -\frac{8}{7}(21)$$ The product of a number and its reciprocal is 1: $-\frac{8}{7}(-\frac{7}{8}) = 1$.

$$k = -\frac{8}{7} \cdot \frac{21}{1}$$ Simplify: $1k = k$. Write 21 as $\frac{21}{1}$.

$$k = -\frac{8 \cdot 3 \cdot \overset{1}{\cancel{7}}}{\underset{1}{\cancel{7}} \cdot 1}$$ The product of two numbers with unlike signs is negative. Multiply the numerators and the denominators. Prime factor 21 and then divide out the common factor 7.

$$k = -24$$ Multiply in the numerator and the denominator.

Check the result by substituting -24 for k in the original equation.

An alternate method

Another method of solving equations such as $\frac{3}{4}x = 5$ uses two steps to isolate the variable. In this method, we consider the variable to be multiplied by 3 and divided by 4. Then, in reverse order, we undo these operations.

$$4\left(\frac{3}{4}x\right) = 4(5)$$ To undo the division by 4, multiply both sides by 4.

$$\left(4 \cdot \frac{3}{4}\right)x = 4(5)$$ Use the associative property to regroup the factors.

$$\left(\frac{\overset{1}{\cancel{4}} \cdot 3}{1 \cdot \underset{1}{\cancel{4}}}\right)x = 4(5)$$ Write 4 as $\frac{4}{1}$, multiply the numerators and the denominators, and divide out the common factor 4.

$$3x = 20$$ Multiply in the numerator and the denominator.

$$\frac{3x}{3} = \frac{20}{3}$$ To undo the multiplication by 3, divide both sides by 3.

$$x = \frac{20}{3}$$

EXAMPLE 2 Solve: $\frac{3}{5}h = -9$.

Solution

$$\frac{3}{5}h = -9$$

$$5\left(\frac{3}{5}h\right) = 5(-9)$$ To undo the division by 5, multiply both sides by 5.

$$3h = -45$$ Perform the multiplications.

$$\frac{3h}{3} = -\frac{45}{3}$$ To undo the multiplication by 3, divide both sides by 3.

$$h = -15$$ Perform the divisions.

Self Check 2

Solve: $\frac{5}{9}t = -10$.

Answer -18

The addition and subtraction properties of equality

The addition property of equality enables us to add the same number to both sides of an equation and obtain an equivalent equation. In the next example, we will use this property to help solve an equation that contains fractions.

EXAMPLE 3 Solve: $y - \frac{15}{32} = \frac{1}{32}$.

Solution To isolate y on the left-hand side, we need to undo the subtraction of $\frac{15}{32}$.

$$y - \frac{15}{32} = \frac{1}{32}$$

$$y - \frac{15}{32} + \frac{15}{32} = \frac{1}{32} + \frac{15}{32}$$ Add $\frac{15}{32}$ to both sides.

$$y = \frac{16}{32}$$ Simplify: $-\frac{15}{32} + \frac{15}{32} = 0$ and $\frac{1}{32} + \frac{15}{32} = \frac{16}{32}$.

$$y = \frac{1}{2}$$ Simplify the fraction: $\frac{16}{32} = \frac{\overset{1}{\cancel{16}} \cdot 1}{2 \cdot \underset{1}{\cancel{16}}} = \frac{1}{2}$.

Self Check 3

Solve: $\frac{11}{16} = a - \frac{1}{16}$.

Answer $\frac{3}{4}$

EXAMPLE 4 Solve: $x + \frac{1}{6} = \frac{3}{4}$.

Solution In this equation, $\frac{1}{6}$ is added to x. We undo this operation by subtracting $\frac{1}{6}$ from both sides.

$$x + \frac{1}{6} = \frac{3}{4}$$

$$x + \frac{1}{6} - \frac{1}{6} = \frac{3}{4} - \frac{1}{6}$$ Subtract $\frac{1}{6}$ from both sides.

$$x = \frac{3}{4} - \frac{1}{6}$$ Perform the subtraction: $\frac{1}{6} - \frac{1}{6} = 0$.

Self Check 4

Solve: $y + \frac{1}{5} = \frac{2}{3}$.

$$x = \frac{3 \cdot 3}{4 \cdot 3} - \frac{1 \cdot 2}{6 \cdot 2}$$ Use the fundamental property of fractions to write each fraction in terms of the LCD, which is 12.

$$x = \frac{9}{12} - \frac{2}{12}$$ Perform the multiplications in the numerators and denominators.

$$x = \frac{7}{12}$$ Subtract the fractions.

Answers $\frac{7}{15}$

Clearing an equation of fractions

In Example 4, we found an LCD so that we could subtract $\frac{1}{6}$ and $\frac{3}{4}$. We will now discuss a method in which we clear such an equation of fractions.

To clear $x + \frac{1}{6} = \frac{3}{4}$ of fractions, we multiply both sides by the LCD of all fractions that appear in the equation. In this case, the LCD of $\frac{1}{6}$ and $\frac{3}{4}$ is 12.

$$\mathbf{12}\left(x + \frac{1}{6}\right) = \mathbf{12}\left(\frac{3}{4}\right)$$ Multiply both sides by the LCD, 12.

$$12x + 12\left(\frac{1}{6}\right) = 12\left(\frac{3}{4}\right)$$ On the left-hand side, distribute the multiplication by 12.

$$12x + 2 = 9$$ Perform the multiplications: $12\left(\frac{1}{6}\right) = 2$ and $12\left(\frac{3}{4}\right) = 9$.

We note that the resulting equation, $12x + 2 = 9$, does not contain fractions. We now complete the solution.

$$12x + 2 \mathbf{- 2} = 9 \mathbf{- 2}$$ To undo the addition of 2, subtract 2 from both sides.

$$12x = 7$$ Perform the subtractions.

$$\frac{12x}{\mathbf{12}} = \frac{7}{\mathbf{12}}$$ To undo the multiplication by 12, divide both sides by 12.

$$x = \frac{7}{12}$$ Perform the divisions.

Self Check 5

Solve: $\frac{4}{5}p - \frac{1}{2} = \frac{3}{4}p$.

EXAMPLE 5 Solve: $\frac{3}{4}h - \frac{1}{2} = \frac{5}{8}h$.

Solution

$$\frac{3}{4}h - \frac{1}{2} = \frac{5}{8}h$$

$$\mathbf{8}\left(\frac{3}{4}h - \frac{1}{2}\right) = \mathbf{8}\left(\frac{5}{8}h\right)$$ To clear the equation of fractions, multiply both sides by the LCD of $\frac{3}{4}$, $\frac{1}{2}$ and $\frac{5}{8}$, which is 8.

$$8\left(\frac{3}{4}h\right) - 8\left(\frac{1}{2}\right) = 8\left(\frac{5}{8}h\right)$$ Distribute the multiplication by 8 on the left-hand side.

$$6h - 4 = 5h$$ Perform the multiplications: $8\left(\frac{3}{4}\right) = 6$, $8\left(\frac{1}{2}\right) = 4$, and $8\left(\frac{5}{8}\right) = 5$.

$$6h - 4 \mathbf{- 5h} = 5h \mathbf{- 5h}$$ To eliminate $5h$ from the right-hand side, subtract $5h$ from both sides.

$$h - 4 = 0$$ Combine like terms.

$$h - 4 \mathbf{+ 4} = 0 \mathbf{+ 4}$$ To undo the subtraction of 4, add 4 to both sides.

$$h = 4$$ Simplify.

Answer 10

The steps to solve equations

We can now complete the strategy for solving equations discussed earlier. You won't always have to use all six steps to solve a given equation. If a step doesn't apply, skip it and move to the next step.

Strategy for solving equations

Simplify the equation:

1. Clear the equation of fractions.
2. Use the distributive property to remove any parentheses.
3. Combine like terms on either side of the equation.

Isolate the variable:

4. Use the addition and subtraction properties of equality to get the variables on one side and the constant terms on the other.
5. Combine like terms when necessary.
6. Undo the operations of multiplication and division to isolate the variable.

Problem solving with equations

EXAMPLE 6 **Native Americans.** The United States Constitution requires a population count, called a *census,* to be taken every ten years. In the 2000 census, the population of the Navajo tribe was 298,000. This was about two-fifths of the population of the largest Native American tribe, the Cherokee. What was the population of the Cherokee tribe in 2000?

Analyze the problem

- In 2000, the population of the Navajo tribe was 298,000.
- The population of the Navajo tribe was $\frac{2}{5}$ the population of the Cherokee tribe.
- Find the population of the Cherokee tribe in 2000.

Form an equation

Let $x =$ the population of the Cherokee tribe. Next, we look for a key word or phrase in the problem.

Key phrase: two-fifths of **Translation:** multiply by $\frac{2}{5}$

The population of the Navajo tribe	was	$\frac{2}{5}$	of	the population of the Cherokee tribe.
298,000	$=$	$\frac{2}{5}$	$\cdot$	x

Solve the equation

$$298{,}000 = \frac{2}{5}x$$

$$\frac{5}{2}(298{,}000) = \frac{5}{2}\left(\frac{2}{5}x\right)$$ To isolate x on the right-hand side, multiply both sides by the reciprocal of $\frac{2}{5}$.

$$745{,}000 = x$$ On the left-hand side, $\frac{5}{2}(298{,}000) = \frac{1{,}490{,}000}{2} = 745{,}000$. On the right-hand side, $\frac{5}{2}(\frac{2}{5}x) = 1x = x$.

Solve the conclusion

In 2000, the population of the Cherokee tribe was about 745,000.

Check the result

Using a fraction to compare the two populations, we have

$$\frac{298{,}000}{745{,}000} = \frac{298}{745} = \frac{2 \cdot 149}{5 \cdot 149} = \frac{2}{5}$$

The answer checks.

EXAMPLE 7 **Filmmaking.** A movie director has sketched out a "storyboard" for a film that is in the planning stages. On the storyboard, he estimates the amount of time in the film that will be devoted to scenes involving dialogue, action scenes, and scenes that make a transition between the two. (See Figure 4-11.) From the information on the storyboard, how long will this film be, in minutes?

Storyboard		Film: "Terminating Force"
Dialogue	*Action scenes*	*Transition scenes*
One-half of film	One-third of film	20 minutes

FIGURE 4-11

Analyze the problem

- $\frac{1}{2}$ of the film is dialogue.
- $\frac{1}{3}$ of the film is action scenes.
- There are 20 minutes of transition scenes.
- How long is the film?

Form an equation

Let x = the length of the film in minutes. To represent the number of minutes for dialogue and action scenes, look for a key word or phrase.

Key phrases: one-half of, one-third of **Translation:** multiply

Therefore, $\frac{1}{2}x$ = the number of minutes for dialogue scenes and $\frac{1}{3}x$ = the number of minutes for action scenes.

The time for dialogue scenes	plus	the time for action scenes	plus	the time for transition scenes	is	the total length of the film.
$\frac{1}{2}x$	$+$	$\frac{1}{3}x$	$+$	20	$=$	x

Solve the equation

$$\frac{1}{2}x + \frac{1}{3}x + 20 = x$$

$$6\left(\frac{1}{2}x + \frac{1}{3}x + 20\right) = 6(x)$$ To clear the equation of fractions, multiply both sides by the LCD, 6.

$$6\left(\frac{1}{2}x\right) + 6\left(\frac{1}{3}x\right) + 6(20) = 6(x)$$ On the left-hand side, distribute the 6.

$$3x + 2x + 120 = 6x$$ Perform the multiplications: $6(\frac{1}{2}) = 3$ and $6(\frac{1}{3}) = 2$.

$$5x + 120 = 6x$$ Combine like terms.

$$5x + 120 - 5x = 6x - 5x$$ To eliminate $5x$ from the left-hand side, subtract $5x$ from both sides.

$$120 = x$$ Combine like terms.

State the conclusion

The length of the film will be 120 minutes.

Check the result

If $x = 120$, the time for dialogue scenes is $\frac{1}{2}x = \frac{1}{2} \cdot 120 = 60$ minutes. The time for action scenes is $\frac{1}{3}x = \frac{1}{3} \cdot 120 = 40$ minutes. The time for transition scenes is 20 minutes. Adding the three times, we get $60 + 40 + 20 = 120$ minutes. The answer checks.

Section 4.8 STUDY SET

VOCABULARY *Fill in the blanks.*

1. To find the __________ of a fraction, invert the numerator and the denominator.
2. In the term $\frac{5}{12}x$, the number $\frac{5}{12}$ is called the __________ of the term.
3. The ______ __________ ____________ of a set of fractions is the smallest number each denominator will divide exactly.
4. A ________ of an equation, when substituted into that equation, makes a true statement.

CONCEPTS

5. Is 40 a solution of $\frac{5}{8}x = 25$? Explain.
6. Give the reciprocal of each number.
   a. $\frac{7}{9}$
   b. $-\frac{1}{2}$
7. What is the result when a number is multiplied by its reciprocal?
8. Perform each multiplication.
   a. $\frac{3}{2}\left(\frac{2}{3}x\right)$
   b. $-\frac{16}{15}\left(-\frac{15}{16}t\right)$
   c. $25\left(\frac{2}{5}\right)$
   d. $16\left(\frac{3}{8}\right)$
9. Translate to mathematical symbols.
   a. Four-fifths of the population p
   b. One-quarter of the time t
10. What property is illustrated by the arrows?

$$12\left(y - \frac{1}{4}\right) = 12y - 3$$

11. Explain two ways in which the variable x can be isolated: $\frac{2}{3}x = -4$.

12. What is wrong with this portion of a solution?

$$\frac{x}{6} - \frac{3}{5} = 8$$
$$30\left(\frac{x}{6} - \frac{3}{5}\right) = 8$$
$$30\left(\frac{x}{6}\right) - 30\left(\frac{3}{5}\right) = 8$$

NOTATION *Complete each solution to solve the equation.*

13. $\frac{7}{8}x = 21$

$\square\left(\frac{7}{8}x\right) = \square(21)$

$x = 24$

14. $h + \frac{1}{2} = \frac{2}{3}$

$\square\left(h + \frac{1}{2}\right) = \square\left(\frac{2}{3}\right)$

$6\square + 6\left(\frac{1}{2}\right) = 6\left(\frac{2}{3}\right)$

$6h + \square = 4$

$6h + 3 - \square = 4 - \square$

$6h = 1$

$\frac{6h}{\square} = \frac{1}{\square}$

$h = \frac{1}{6}$

15. Determine whether each statement is true or false.

a. $\frac{1}{2}x = \frac{x}{2}$

b. $\frac{1}{8}y = 8y$

c. $-\frac{1}{2}x = \frac{-x}{2} = \frac{x}{-2}$

d. $\frac{7p}{8} = \frac{7}{8}p$

16. Write the product of $\frac{4}{7}$ and x in two ways.

PRACTICE *Solve each equation.*

17. $\frac{4}{7}x = 16$

18. $\frac{2}{3}y = 30$

19. $\frac{7}{8}t = -28$

20. $\frac{5}{6}c = -25$

21. $-\frac{3}{5}h = 4$

22. $-\frac{5}{6}f = -2$

23. $\frac{2}{3}x = \frac{4}{5}$

24. $\frac{5}{8}y = \frac{10}{11}$

25. $\frac{2}{5}y = 0$

26. $\frac{4}{9}x = 0$

27. $-\frac{5c}{6} = -25$

28. $-\frac{7t}{4} = -35$

29. $\frac{5f}{7} = -2$

30. $\frac{3h}{5} = -35$

31. $\frac{5}{8}y = \frac{1}{10}$

32. $\frac{1}{16}x = \frac{5}{24}$

33. $2x + 1 = 0$

34. $3y - 1 = 0$

35. $5x - 1 = 1$

36. $4c + 1 = -2$

37. $6x = 2x - 11$

38. $5t = t - 7$

39. $2(y - 3) = 7$

40. $3(r + 2) = 10$

41. $x - \frac{1}{9} = \frac{7}{9}$

42. $x + \frac{1}{3} = \frac{2}{3}$

43. $x + \frac{1}{9} = \frac{4}{9}$

44. $x - \frac{1}{6} = \frac{1}{6}$

45. $x - \frac{1}{6} = \frac{2}{9}$

46. $y - \frac{1}{3} = \frac{4}{5}$

47. $y + \frac{7}{8} = \frac{1}{4}$

48. $t + \frac{5}{6} = \frac{1}{8}$

49. $\frac{5}{4} + t = \frac{1}{4}$

50. $\frac{2}{3} + y = \frac{4}{3}$

51. $x + \frac{3}{4} = -\frac{1}{2}$

52. $y - \frac{5}{6} = \frac{1}{3}$

53. $\frac{-x}{4} + 1 = 10$

54. $\frac{-y}{6} - 1 = 5$

55. $2x - \frac{1}{2} = \frac{1}{3}$

56. $3y - \frac{2}{5} = \frac{1}{8}$

57. $\frac{1}{2}x - \frac{1}{9} = \frac{1}{3}$

58. $\frac{1}{4}y - \frac{2}{3} = \frac{1}{2}$

59. $5 + \frac{x}{3} = \frac{1}{2}$

60. $4 + \frac{y}{2} = \frac{3}{5}$

61. $\frac{2}{5}x + 1 = \frac{1}{3} + x$

62. $\frac{2}{3}y + 2 = \frac{1}{5} + y$

63. $\frac{x}{3} + \frac{x}{4} = -2$

64. $\frac{y}{6} + \frac{y}{4} = -1$

65. $4 + \frac{s}{3} = 8$

66. $6 + \frac{y}{5} = 1$

67. $\frac{5h}{6} - 8 = 12$

68. $\frac{6a}{7} - 1 = 11$

69. $-4 + 9 + \frac{5t}{12} = 0$

70. $-4 + 10 + \frac{3y}{8} = 0$

71. $-3 - 2 + \frac{4x}{15} = 0$

72. $-1 - 9 + \frac{2y}{15} = 0$

APPLICATIONS *Complete each solution.*

73. TRANSMISSION REPAIRS A repair shop found that $\frac{1}{3}$ of its customers with transmission problems needed a new transmission. If the shop installed 32 new transmissions last year, how many customers did the shop have last year?

Analyze the problem

- Only ___ of the customers needed new transmissions.
- The shop installed ___ new transmissions last year.
- Find the number of __________ the shop had last year.

Form an equation

Let x = ______________________________.

Key phrase: *one-third of*
Translation: _________

$\frac{1}{3}$ of the number of customers last year	was	32.
	=	32

Solve the equation

$$\frac{1}{3}x = 32$$

$$\square\left(\frac{1}{3}x\right) = \square(32)$$

$$x = \square$$

State the conclusion ______________________________

Check the result If we find $\frac{1}{3}$ of 96, we get ___. The answer checks.

74. CATTLE RANCHING A rancher is preparing to fence in a rectangular grazing area next to a $\frac{3}{4}$-mile-long lake. He has determined that $1\frac{1}{2}$ square miles of land are needed to ensure that overgrazing does not occur. How wide should this grazing area be?

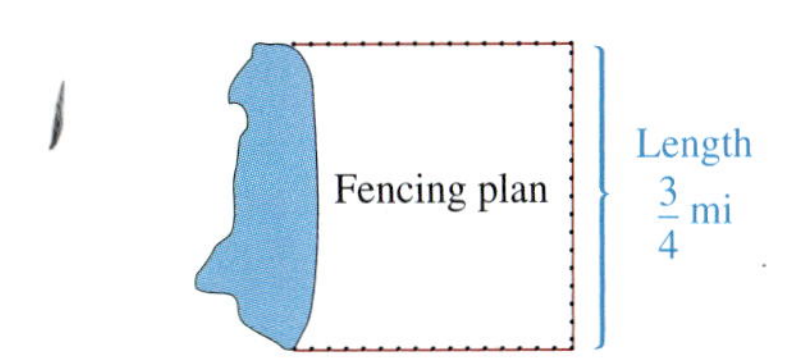

Analyze the problem

- The grazing area is ___ $= \frac{3}{2}$ square miles.
- The length of the rectangle is $\frac{3}{4}$ mile.
- Find the _______ of the grazing area.

Form an equation

Let w = __________________.

Key word: *area* **Translation:** $A =$ ___

The area of the rectangle	is	the length times the width.
	=	

Solve the equation

$$\frac{3}{2} = \frac{3}{4}w$$

$$\square\left(\frac{3}{2}\right) = \square\left(\frac{3}{4}w\right)$$

$$\square = w$$

State the conclusion

____

Check the result If we multiply the length and the width of the rectangular area, we get $\frac{3}{4} \cdot 2 = \frac{3}{2} = 1\frac{1}{2}$ square miles. The answer checks.

Choose a variable to represent the unknown. Then write and solve an equation to answer each question.

75. TOOTH DEVELOPMENT During a checkup, a pediatrician found that only four-fifths of a child's baby teeth had emerged. The mother counted 16 teeth in the child's mouth. How many baby teeth will the child eventually have?

76. GENETICS Bean plants with inflated pods were cross-bred with bean plants with constricted pods. Of the offspring plants, three-fourths had inflated pods and one-fourth had constricted pods. If 244 offspring plants had constricted pods, how many offspring plants resulted from the cross-breeding experiment?

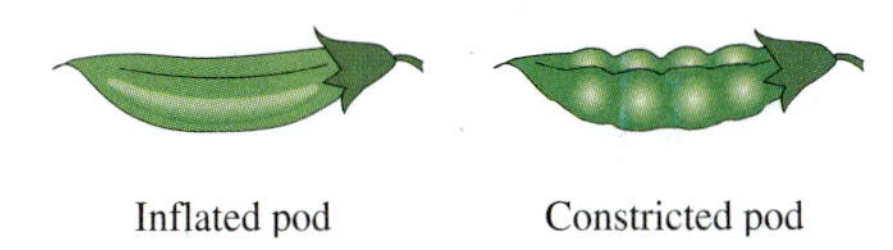

Inflated pod Constricted pod

77. TELEPHONE BOOKS A telephone book consists of the white pages and the yellow pages. Two-thirds of the book consists of the white pages; the white pages number 300. Find the total number of pages in the telephone book.

78. BROADWAY MUSICALS A theater usher at a Broadway musical finds that seven-eighths of the patrons attending a performance, which is 350 people, are in their seats by show time. If the show is always a complete sellout, how many seats does the theater have?

79. HOME SALES In less than a month, three-quarters of the homes in a new subdivision were purchased. This left only 9 homes to be sold. How many homes are there in the subdivision? (*Hint:* First determine what fractional part of the homes in the subdivision were not yet sold.)

80. WEDDING GUESTS Of those invited to a wedding, three-tenths were friends of the bride. The friends of the groom numbered 84. How many people were invited to the wedding? (*Hint:* First determine what fractional part of the people invited to the wedding were friends of the groom.)

81. SAFETY REQUIREMENTS In developing taillights for an automobile, designers must be aware of a safety standard that requires an area of 30 square inches to be visible from behind the vehicle. If the designers want the taillights to be $3\frac{3}{4}$ inches high, how wide must they be to meet safety standards?

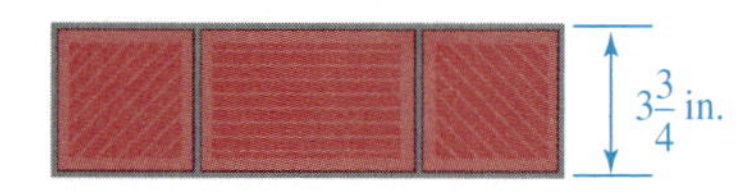

82. GRAPHIC ARTS A design for a yearbook is shown. The page is divided into 12 parts. The parts that are shaded will contain pictures, and the remainder of the squares will contain copy. If the pictures are to cover an area of 100 square inches, how many square inches are there on the page?

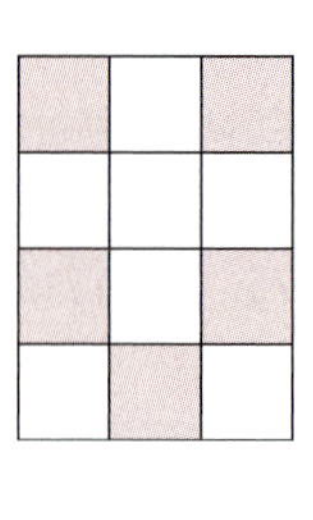

83. CPR CLASS The instructor for a course in CPR (cardiopulmonary resuscitation) has three segments in her lesson plan, as shown. How many minutes long is the CPR course?

Lecture on subject	Practicing CPR techniques	Legal responsibilities
One-fourth of class	Two-thirds of class	30 min

84. FIREFIGHTING A firefighting crew is composed of three elements, as shown. How many firefighters are in the crew?

A 50-man parachuting team is called the "Smoke Jumpers."

One-half of the crew comes from the National Forest Service.

One-third of the crew comes from county fire departments.

WRITING

85. What does it mean to isolate the variable when solving an equation?

86. What does it mean to clear an equation of fractions before solving the equation?

87. Which method, the reciprocal method or the two-step method, would you use to solve the equation $\frac{5}{16}t = 15$? Why?

88. Use an example to show why dividing by a number is the same as multiplying by its reciprocal.

REVIEW

89. Use the distributive property to fill in the blank.

$$a(b + c) = \underline{\qquad}$$

90. What is the value (in cents) of q quarters?

91. Convert 41° Fahrenheit to Celsius.

92. Evaluate: $\dfrac{-4 - 8}{3}$.

93. Solve: $5x - 3 = 2x + 12$.

94. Solve: $10 - (x - 5) = 40$.

95. Round 12,590,767 to the nearest million.

96. In the expression $(-4)^6$, what do we call -4 and what do we call 6?

KEY CONCEPT

The Fundamental Property of Fractions

The **fundamental property of fractions** states that multiplying or dividing the numerator and the denominator of a fraction by the same nonzero number does not change the value of the fraction. This property is used to simplify fractions and to express fractions in higher terms. The following problems review both procedures. Complete each solution.

1. Simplify: $\frac{15}{25}$.

Step 1: The numerator and the denominator share a common factor of $\square$.

Step 2: Apply the fundamental property of fractions. Divide the numerator and the denominator by the common factor $\square$.

$$\frac{15}{25} = \frac{15 \div \square}{25 \div \square}$$

Step 3: Perform the divisions to simplify the fraction.

$$= \frac{3}{\square}$$

2. In practice, we often show the simplifying process described in Problem 1 in a different form.

Step 1: Factor 15 as $3 \cdot \square$ and 25 as $\square \cdot 5$.

$$\frac{15}{25} = \frac{3 \cdot \square}{\square \cdot 5}$$

Step 2: The slashes and small 1's indicate that the numerator and the denominator have been divided by $\square$.

$$= \frac{3 \cdot \overset{1}{\cancel{5}}}{\underset{1}{\cancel{5}} \cdot 5}$$

Step 3: Multiply in the numerator and the denominator.

$$= \frac{\square}{5}$$

3. When adding or subtracting fractions and mixed numbers, we often need to express a fraction in higher terms. This is called building the fraction. Express $\frac{1}{5}$ as a fraction with denominator 35.

Step 1: We must multiply 5 by $\square$ to obtain 35.

Step 2: Use the fundamental property of fractions. Multiply the numerator and the denominator by $\square$.

$$\frac{1}{5} = \frac{1 \cdot \square}{5 \cdot \square}$$

Step 3: Multiply in the numerator and the denominator.

$$= \frac{\square}{35}$$

4. When adding or subtracting algebraic fractions, we often need to express a fraction in higher terms. Express $\frac{2}{3x}$ as a fraction with a denominator of $18x$.

Step 1: We must multiply $3x$ by $\square$ to obtain $18x$.

Step 2: Use the fundamental property of fractions. Multiply the numerator and the denominator by $\square$.

$$\frac{2}{3x} = \frac{2 \cdot \square}{3x \cdot \square}$$

Step 3: Perform the multiplication in the numerator and the denominator.

$$= \frac{12}{\square}$$

ACCENT ON TEAMWORK

SECTION 4.1

EQUIVALENT FRACTIONS Complete the labeling of each number line using fractions with the same denominator.

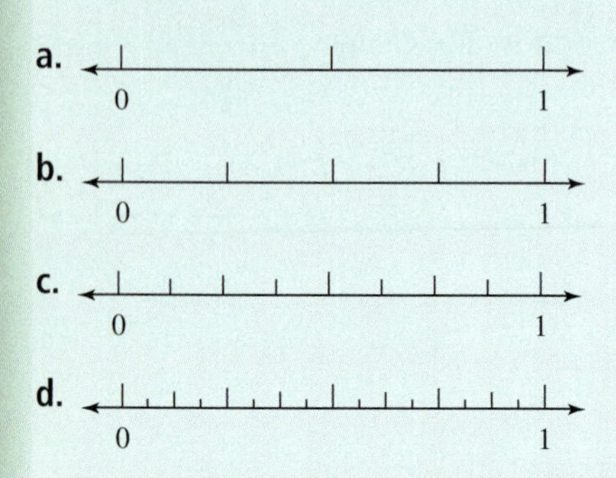

FRACTIONS Give everyone in your group a strip of paper that is the same length. Determine ways to fold the strip of paper into

a. fourths **b.** eighths
c. thirds **d.** sixths

SECTION 4.2

MULTIPLICATION When we multiply 2 and 4, the answer is greater than 2 and greater than 4. Is this always the case? Is the product of two numbers always greater than either of the two numbers? Explain your answer.

POWERS When we square the number 4, the answer is greater than 4. Is the square of a number always greater than the number? Explain your answer.

SECTION 4.3

DIVIDING SNACKS Devise a way to divide seven brownies equally among six people.

SECTION 4.4

ADDING FRACTIONS Without actually doing the addition, explain why $\frac{3}{7} + \frac{1}{4}$ must be less than 1 and why $\frac{4}{7} + \frac{3}{4}$ must be greater than 1.

COMPARING FRACTIONS

a. When 1 is added to the numerator of a fraction, is the result greater than or less than the original fraction? Explain your reasoning.

b. When 1 is added to the denominator of a fraction, is the result greater than or less than the original fraction? Explain your reasoning.

COMPARING FRACTIONS Think of a fraction. Add 1 to its numerator and add 1 to its denominator. Is the resulting fraction greater than, less than, or equal to the original fraction? Explain your reasoning.

SECTION 4.5

DIVISION WITH MIXED NUMBERS Division can be thought of as repeated subtraction. Use this concept to solve the following problem.

$5\frac{1}{4}$ yards of ribbon needs to be cut into pieces that are $\frac{3}{4}$ of a yard long to form bows. How many bows can be made?

SECTION 4.6

MIXED NUMBERS Two mixed numbers, A and B, are graphed below. Estimate where on the number line the graph of $A + B$ would lie.

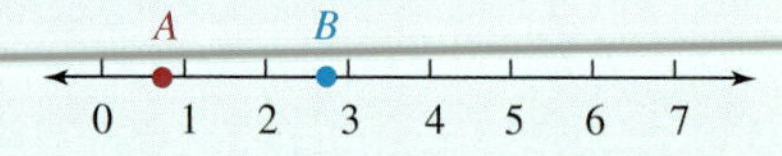

SECTION 4.7

COMPLEX FRACTIONS Write a problem that could be solved by simplifying the complex fraction.

$$\frac{\frac{7}{8}}{\frac{3}{4}}$$

SECTION 4.8

SOLVING EQUATIONS

a. Solve the equation $\frac{3}{4}x = 15$. Undo the multiplication by $\frac{3}{4}$ by dividing both sides by $\frac{3}{4}$.

b. Do the same for $-\frac{7}{8}x = 21$.

CHAPTER REVIEW

SECTION 4.1 The Fundamental Property of Fractions

CONCEPTS

Fractions can be used to indicate equal parts of a whole.

A fraction is composed of a *numerator, a denominator,* and a *fraction bar.*

If a and b are positive numbers,

$$\frac{-a}{b} = \frac{a}{-b} = -\frac{a}{b} \quad (b \neq 0)$$

Equivalent fractions represent the same number.

The *fundamental property of fractions:* Dividing the numerator and denominator of a fraction by the same nonzero number does not change the value of the fraction.

To *simplify* a fraction that is not in lowest terms, divide the numerator and denominator by the same number.

A fraction is in *lowest terms* if the only factor common to the numerator and denominator is 1.

The fundamental property of fractions: Multiplying the numerator and denominator of a fraction by a nonzero number does not change its value.

$$\frac{a}{b} = \frac{a \cdot x}{b \cdot x} \quad (b \neq 0, x \neq 0)$$

Expressing a fraction in higher terms results in an equivalent fraction that involves larger numbers or more complex terms.

REVIEW EXERCISES

1. If a woman gets seven hours of sleep each night, what part of a whole day does she spend sleeping?

2. In the illustration, why can't we say that $\frac{3}{4}$ of the figure is shaded?

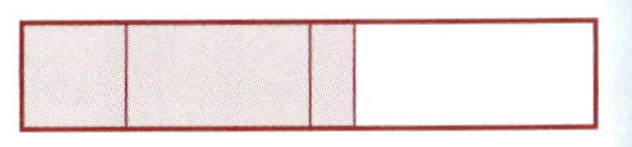

3. Write the fraction $\frac{2}{-3}$ in two other ways.

4. What concept about fractions does the illustration demonstrate?

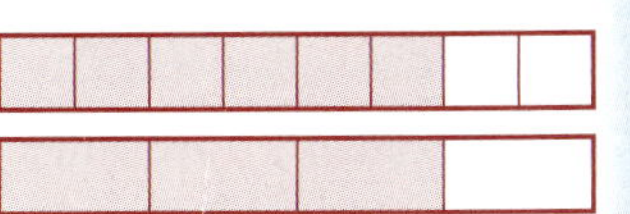

5. Explain the procedure shown here.

$$\frac{4}{6} = \frac{4 \div 2}{6 \div 2} = \frac{2}{3}$$

6. Explain what the slashes and the 1's mean.

$$\frac{4}{6} = \frac{\overset{1}{\cancel{2}} \cdot 2}{\underset{1}{\cancel{2}} \cdot 3} = \frac{2}{3}$$

Simplify each fraction to lowest terms.

7. $\frac{15}{45}$

8. $\frac{20}{48}$

9. $-\frac{63x^2}{84x}$

10. $\frac{66m^3n}{108m^4n}$

11. Explain what is being done and why it is valid.

$$\frac{5}{8} = \frac{5 \cdot 2}{8 \cdot 2} = \frac{10}{16}$$

Write each fraction or whole number with the indicated denominator (shown in red).

12. $\frac{2}{3}$, 18

13. $-\frac{3}{8}$, 16

14. $\frac{7}{15}$, $45a$

15. 4, 9

SECTION 4.2 *Multiplying Fractions*

To *multiply two fractions,* multiply their numerators and multiply their denominators.

$$\frac{a}{b} \cdot \frac{c}{d} = \frac{a \cdot c}{b \cdot d}$$

$(b \neq 0, d \neq 0)$

Multiply.

16. $\frac{1}{2} \cdot \frac{1}{3}$ **17.** $\frac{2}{5}\left(-\frac{7}{9}\right)$ **18.** $\frac{9}{16} \cdot \frac{20}{27}$ **19.** $\frac{5}{6} \cdot \frac{1}{3} \cdot \frac{18}{25}$

20. $\frac{3}{5} \cdot 7$ **21.** $-4\left(-\frac{9}{16}\right)$ **22.** $3\left(\frac{1}{3}\right)$ **23.** $-\frac{6}{7}\left(-\frac{7}{6}\right)$

Determine whether each statement is true or false.

24. $\frac{3}{4}x = \frac{3x}{4}$ **25.** $-\frac{5}{9}e = -\frac{5}{9e}$

Multiply.

26. $\frac{3t}{5} \cdot \frac{10}{27t}$ **27.** $-\frac{2}{3}\left(\frac{4}{7}s\right)$ **28.** $\frac{4d^2}{9} \cdot \frac{3}{28d}$ **29.** $9mn\left(-\frac{5}{81n^2}\right)$

An *exponent* indicates repeated multiplication.

Evaluate each power.

30. $\left(\frac{3}{4}\right)^2$ **31.** $\left(-\frac{5}{2}\right)^3$ **32.** $\left(\frac{x}{3}\right)^2$ **33.** $\left(-\frac{2c}{5}\right)^3$

In mathematics, the word *of* usually means multiply.

34. GRAVITY ON THE MOON Objects on the moon weigh only one-sixth as much as on Earth. How much will an astronaut weigh on the moon if he weighs 180 pounds on Earth?

The *area of a triangle:*

$$A = \frac{1}{2}bh$$

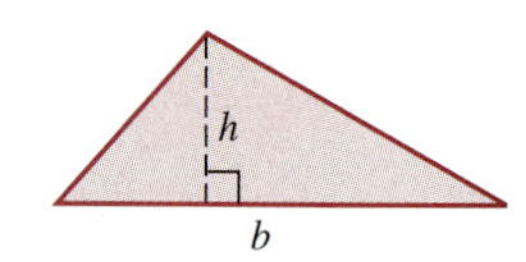

35. Find the area of the triangular sign.

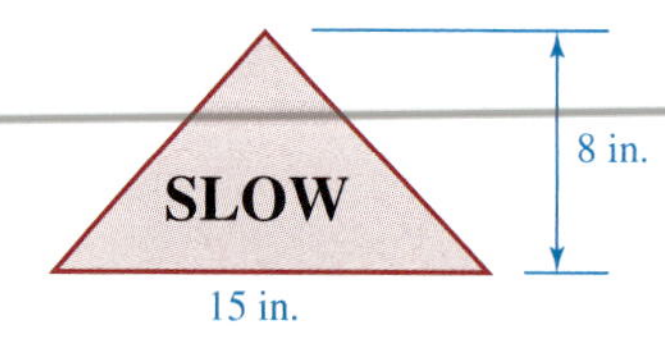

SECTION 4.3 *Dividing Fractions*

Two numbers are called *reciprocals* if their product is 1.

Find the reciprocal of each number.

36. $\frac{1}{8}$ **37.** $-\frac{11}{12}$ **38.** x **39.** $\frac{ab}{c}$

To *divide two fractions,* multiply the first by the reciprocal of the second.

$$\frac{a}{b} \div \frac{c}{d} = \frac{a}{b} \cdot \frac{d}{c}$$

$(b \neq 0, c \neq 0, d \neq 0)$

Divide.

40. $\frac{1}{6} \div \frac{11}{25}$ **41.** $-\frac{7}{8} \div \frac{1}{4}$ **42.** $-\frac{15}{16} \div (-10)$ **43.** $8 \div \frac{16}{5}$

44. $\frac{t}{8} \div \frac{1}{4}$ **45.** $\frac{4a}{5} \div \frac{a}{2}$ **46.** $-\frac{a}{b} \div \left(-\frac{b}{a}\right)$ **47.** $\frac{2}{3}x \div \left(-\frac{x^2}{9}\right)$

48. GOLD COINS How many $\frac{1}{16}$-ounce coins can be cast from a $\frac{3}{4}$-ounce bar of gold?

SECTION 4.4 Adding and Subtracting Fractions

To add (or subtract) fractions with like denominators, add (or subtract) their numerators and write the result over the common denominator.

$$\frac{a}{c} + \frac{b}{c} = \frac{a+b}{c} \quad (c \neq 0)$$

$$\frac{a}{c} - \frac{b}{c} = \frac{a-b}{c} \quad (c \neq 0)$$

Add or subtract.

49. $\frac{2}{7} + \frac{3}{7}$

50. $-\frac{3}{5} - \frac{3}{5}$

51. $\frac{3}{x} - \frac{1}{x}$

52. $\frac{7}{8} + \frac{t}{8}$

53. Explain why we cannot immediately add $\frac{1}{2} + \frac{2}{3}$ without doing some preliminary work.

The *LCD* must include the set of prime factors of each of the denominators.

54. Use prime factorization to find the least common denominator for fractions with denominators of 45 and 30.

To add or subtract fractions with unlike denominators, we must first express them as equivalent fractions with the same denominator, preferably the LCD.

Add or subtract.

55. $\frac{1}{6} + \frac{2}{3}$

56. $\frac{2}{5} + \left(-\frac{3}{8}\right)$

57. $-\frac{3}{8} - \frac{5}{6}$

58. $3 - \frac{1}{7}$

59. $\frac{x}{25} - \frac{3}{10}$

60. $\frac{1}{3} + \frac{7}{y}$

61. $\frac{13}{6} - 6$

62. $\frac{1}{3} + \frac{1}{4} + \frac{1}{5}$

63. MACHINE SHOPS How much must be milled off the $\frac{3}{4}$-inch-thick steel rod so that the collar will slip over the end of it?

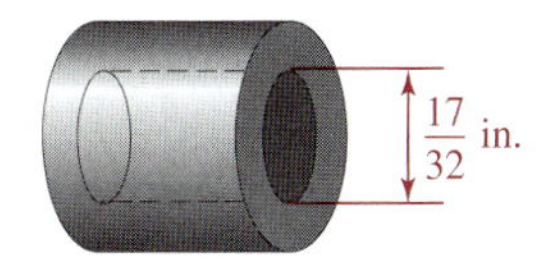

Steel rod

To *compare fractions,* write them as equivalent fractions with the same denominator. Then the fraction with the larger numerator will be the larger fraction.

64. TELEMARKETING In the first hour of work, a telemarketer made 2 sales out of 9 telephone calls. In the second hour, she made 3 sales out of 11 calls. During which hour was the rate of sales to calls better?

SECTION 4.5 Multiplying and Dividing Mixed Numbers

A *mixed number* is the sum of its whole-number part and its fractional part.

65. What mixed number is represented in the illustration?

66. What improper fraction is represented in the illustration?

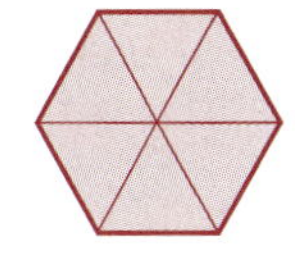
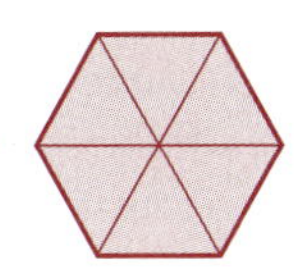
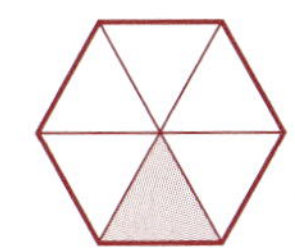

To change an *improper fraction* to a mixed number, divide the numerator by the denominator to obtain the whole-number part. Write the remainder over the denominator for the fractional part.

Express each improper fraction as a mixed number or a whole number.

67. $\frac{16}{5}$ **68.** $-\frac{47}{12}$ **69.** $\frac{6}{6}$ **70.** $\frac{14}{6}$

To change a mixed number to an improper fraction, multiply the whole number by the denominator and add the result to the numerator. Write this sum over the denominator.

Write each mixed number as an improper fraction.

71. $9\frac{3}{8}$ **72.** $-2\frac{1}{5}$ **73.** $100\frac{1}{2}$ **74.** $1\frac{99}{100}$

75. Graph: $-2\frac{2}{3}, \frac{8}{9}$, and $\frac{59}{24}$.

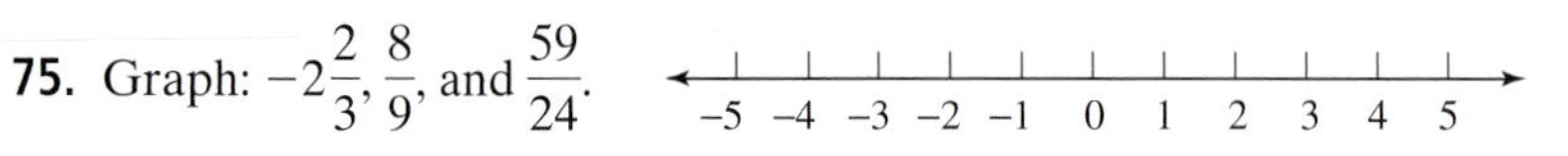

To *multiply* or *divide mixed numbers,* change the mixed numbers to improper fractions and then perform the operations as usual.

Multiply or divide. Write answers as mixed numbers when appropriate.

76. $-5\frac{1}{4} \cdot \frac{2}{35}$ **77.** $\left(-3\frac{1}{2}\right) \div \left(-3\frac{2}{3}\right)$

78. $\left(-6\frac{2}{3}\right)(-6)$ **79.** $-8 \div 3\frac{1}{5}$

80. CAMERA TRIPODS The three legs of a tripod can be extended to become $5\frac{1}{2}$ times their original length. If each leg is $8\frac{3}{4}$ inches long when collapsed, how long will a leg become when it is completely extended?

SECTION 4.6 Adding and Subtracting Mixed Numbers

To add (or subtract) mixed numbers, we can change each to an improper fraction and use the method of Section 4.4.

Add or subtract.

81. $1\frac{3}{8} + 2\frac{1}{5}$ **82.** $3\frac{1}{2} + 2\frac{2}{3}$

83. $2\frac{5}{6} - 1\frac{3}{4}$ **84.** $3\frac{7}{16} - 2\frac{1}{8}$

To add mixed numbers, we can add the whole numbers and the fractions separately.

85. PAINTING SUPPLIES In a project to restore a house, painters used $10\frac{3}{4}$ gallons of primer, $21\frac{1}{2}$ gallons of latex paint, and $7\frac{2}{3}$ gallons of enamel. Find the total number of gallons of paint used.

Vertical form can be used to add or subtract mixed numbers.

Add or subtract.

86. $\begin{array}{r} 133\frac{1}{9} \\ +\ 49\frac{1}{6} \\ \hline \end{array}$ **87.** $\begin{array}{r} 98\frac{11}{20} \\ +\ 14\frac{3}{5} \\ \hline \end{array}$

88. $\begin{array}{r} 50\frac{5}{8} \\ -19\frac{1}{6} \\ \hline \end{array}$ **89.** $\begin{array}{r} 375\frac{3}{4} \\ -\ 59 \\ \hline \end{array}$

If the fraction being subtracted is larger than the first fraction, we need to *borrow* from the whole number.

Subtract.

90. $23\frac{1}{3} - 2\frac{5}{6}$ **91.** $39 - 4\frac{5}{8}$

SECTION 4.7 Order of Operations and Complex Fractions

A *complex fraction* is a fraction whose numerator or denominator, or both, contain one or more fractions or mixed numbers.

To simplify a complex fraction, *Method 1:* The main fraction bar of a complex fraction indicates division.

Method 2: Multiply the numerator and denominator of the complex fraction by the LCD of all the fractions that appear in it.

Evaluate each numerical expression.

92. $\frac{3}{4} + \left(-\frac{1}{3}\right)^2\left(\frac{5}{4}\right)$

93. $\left(\frac{2}{3} \div \frac{16}{9}\right) - \left(1\frac{2}{3} \cdot \frac{1}{15}\right)$

Simplify each complex fraction.

94. $\dfrac{\frac{3}{5}}{-\frac{17}{20}}$

95. $\dfrac{\frac{2}{3} - \frac{1}{6}}{-\frac{3}{4} - \frac{1}{2}}$

Evaluate each expression for $c = -\frac{3}{4}$, $d = \frac{1}{8}$, and $e = -2\frac{1}{16}$.

96. $d^2 - 2c$

97. $-cd + e$

98. $e \div (cd)$

99. $\dfrac{c - d}{e}$

SECTION 4.8 Solving Equations Containing Fractions

To solve an equation,
Simplify the equation:

1. Clear the equation of fractions.
2. Remove parentheses using the distributive property.
3. Combine like terms.

Isolate the variable:

4. Get the variable on one side and constants on the other using the addition or subtraction properties of equality.
5. Combine like terms.
6. Undo multiplication and division.

Solve each equation. Check the result.

100. $\frac{2}{3}x = 16$

101. $-\frac{7s}{4} = -49$

102. $\frac{y}{5} = -\frac{1}{15}$

103. $2x - 3 = 8$

Solve each equation.

104. $\frac{c}{3} - \frac{c}{8} = 2$

105. $\frac{5h}{9} - 1 = -3$

106. $4 - \frac{d}{4} = 0$

107. $\frac{t}{10} - \frac{2}{3} = \frac{1}{5}$

108. TEXTBOOKS In writing a history text, the author decided to devote two-thirds of the book to events prior to World War II. The remainder of the book deals with history after the war. If pre-World War II history is covered in 220 pages, how many pages does the textbook have?

CHAPTER 4 TEST

1. See the illustration.

   a. What fractional part of the plant is above ground?

   b. What fractional part of the plant is below ground?

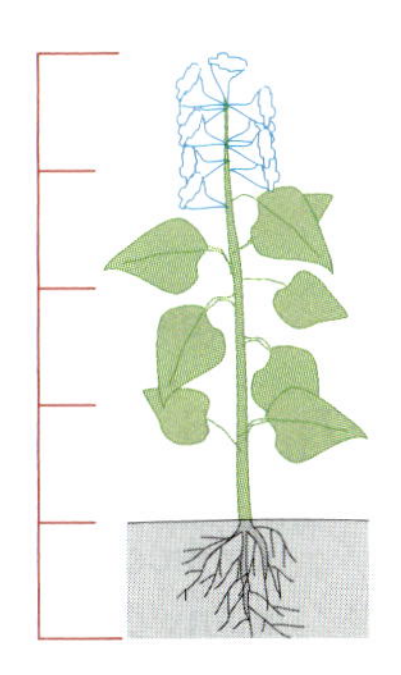

2. Simplify each fraction.

   a. $\frac{27}{36}$

   b. $\frac{72n^2}{180n}$

3. Multiply: $-\frac{3x}{4}\left(\frac{1}{5x^2}\right)$.

4. COFFEE DRINKERS Of 100 adults surveyed, $\frac{2}{5}$ said they started off their morning with a cup of coffee. Of the 100, how many would this be?

5. Divide: $\frac{4a}{3} \div \frac{a^2}{9}$.

6. Subtract: $\frac{x}{6} - \frac{4}{5}$.

7. Express $\frac{7}{8}$ as an equivalent fraction with denominator $24a$.

8. Graph: $2\frac{4}{5}$, $-1\frac{1}{7}$, and $\frac{7}{6}$.

   −2 −1 0 1 2 3

9. SPORTS CONTRACTS A basketball player signed a nine-year contract for $\$13\frac{1}{2}$ million. How much is this per year?

10. Evaluate $-2ct^2$ for $c = -2\frac{1}{12}$ and $t = \frac{2}{5}$.

11. Add: $157\frac{5}{9} + 103\frac{3}{4}$.

12. Subtract: $67\frac{1}{4} - 29\frac{5}{6}$.

13. BOXING When Oscar De La Hoya fought Pernell Whitaker, the "Tale of the Tape" shown below appeared in the sports section of many newspapers. What was the difference in the fighters'

    a. weights?

    b. chests (expanded)?

    c. waists?

Tale of the Tape

De La Hoya		Whitaker
24 yr	Age	33 yr
146 1/2 lb	Weight	146 1/2 lb
5-11	Height	5-6
72 in.	Reach	69 in.
39 in.	Chest (Normal)	37 in.
42 1/4 in.	Chest (Expanded)	39 1/2 in.
31 3/4 in.	Waist	28 in.

14. Add: $-\frac{3}{7} + 2$.

15. SEWING When cutting material for a $10\frac{1}{2}$-inch-wide placemat, a seamstress allows $\frac{5}{8}$ inch at each end for a hem. How wide should the material be cut?

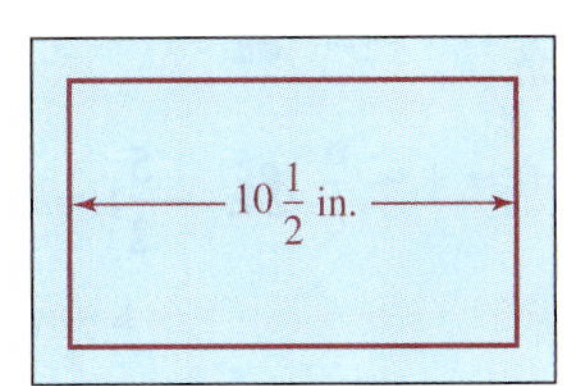

Reading a decimal

1. Look to the left of the decimal point and say the name of the whole number.
2. The decimal point is then read as "and."
3. Say the fractional part of the decimal as a whole number followed by the name of the place value column of the digit that is farthest to the right.

When we use this procedure, here is the other way to read 12.37.

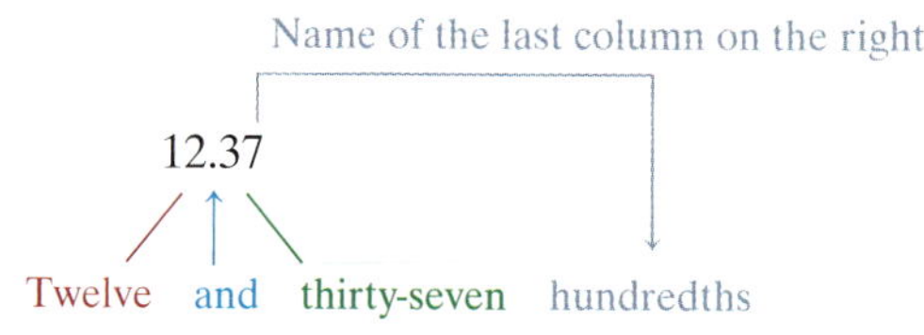

When we read a decimal in this way, it is easy to write it in words and as a mixed number.

Decimal	Words	Mixed number
12.37	Twelve and thirty-seven hundredths	$12\frac{37}{100}$

Self Check 1

Write each decimal in words and then as a mixed number.

a. Sputnik 1, the first artificial satellite, weighed 184.3 pounds.

b. The planet Mercury makes one revolution every 87.9687 days.

Answers a. One hundred eighty-four and three tenths, or $184\frac{3}{10}$ b. Eighty-seven and nine thousand six hundred eighty-seven ten-thousandths, or $87\frac{9,687}{10,000}$

EXAMPLE 1 World records. Write each decimal in words and then as a fraction or mixed number. **Do not simplify the fraction.**

a. According to the *Guinness Book of World Records*, the fastest qualifying speed at the Indianapolis 500 was 236.986 mph by Aire Luyendyk in 1996.

b. The smallest freshwater fish is the dwarf pygmy goby, found in the Philippines. Adult males weigh 0.00014 ounce.

Solution

a. The whole number part of 236.986 to the left of the decimal point is 236. The fractional part, stated as a whole number, is 986. The digit the farthest to the right is 6 and it is in the thousandths column. Thus,

236.986 is two hundred thirty-six and nine hundred eighty-six thousandths, or $236\frac{986}{1,000}$.

b. The whole number part of 0.00014 to the left of the decimal point is 0. The fractional part, stated as a whole number, is 14. The digit the farthest to the right is 4 and it is in the hundred-thousandths column. Thus,

0.00014 is fourteen hundred-thousandths, or $\frac{14}{100,000}$.

Decimals can be negative. For example, a record low temperature of −128.6°F was recorded in Vostok, Antarctica, on July 21, 1983. This is read as "negative one hundred twenty-eight and six tenths." Written as a mixed number, it is $-128\frac{6}{10}$.

Comparing decimals

The relative sizes of a set of decimals can be determined by scanning their place value columns from left to right, column by column, looking for a difference in the digits. For example,

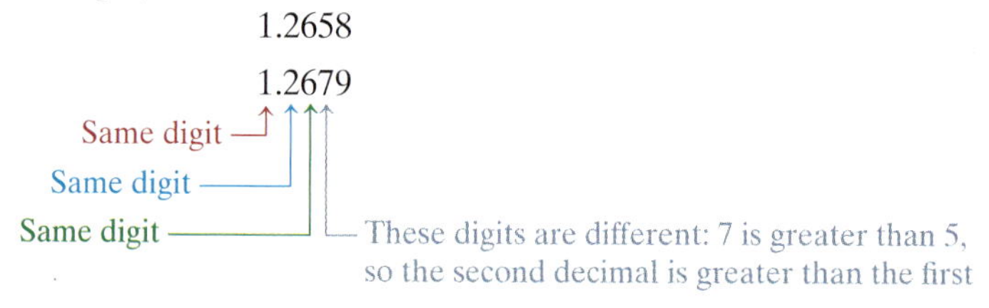

Thus, 1.2679 is greater than 1.2658. We write 1.2679 > 1.2658.

Comparing positive decimals

1. Make sure both numbers have the same number of decimal places to the right of the decimal point. Write any additional zeros necessary to achieve this.
2. Compare the digits of each decimal, column by column, working from left to right.
3. When two digits differ, the decimal with the greater digit is the greater number.

EXAMPLE 2 Which is greater: 54.9 or 54.929?

Solution

54.900 Write two zeros after 9 so that both decimals have the same number of digits to the right of the decimal point.

54.929
↑

Working from left to right, we find this is the first column in which the digits differ. Since 2 is greater than 0, we can conclude that $54.929 > 54.9$.

Self Check 2
Which is greater: 113.7 or 113.657?

Answer 113.7

Comparing negative decimals

1. Make sure both numbers have the same number of decimal places to the right of the decimal point. Write any additional zeros necessary to achieve this.
2. Compare the digits of each decimal, column by column, working from left to right.
3. When two digits differ, the decimal with the smaller digit is the greater number.

EXAMPLE 3 Which is greater: -10.45 or -10.419?

Solution

-10.450 Write a zero after 5 to help in the comparison.

-10.419
↑

As we work from left to right, this is the first column in which the digits differ. Since 1 is less than 5, we conclude that $-10.419 > -10.45$.

Self Check 3
Which is greater: -703.8 or -703.78?

Answer -703.78

Self Check 4

Graph: $-1.1, -0.6, 0.8$, and 1.9.

Answer

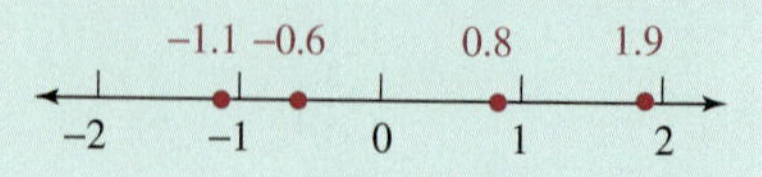

EXAMPLE 4 Graph: $-1.8, -1.23, -0.3$, and 1.89.

Solution To graph each decimal, we locate its position on the number line and draw a dot. Since -1.8 is to the left of -1.23, we can write $-1.8 < -1.23$.

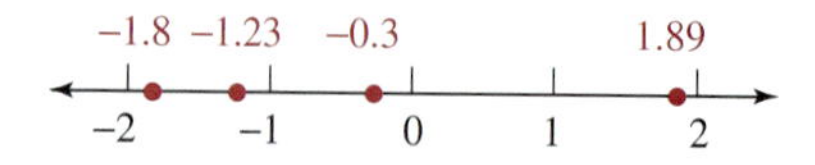

Rounding

When working with decimals, we often round answers to a specific number of decimal places.

Rounding a decimal

1. To round a decimal to a specified decimal place, locate the digit in that place. Call it the *rounding digit.*
2. Look at the *test digit* to the right of the rounding digit.
3. If the test digit is 5 or greater, round up by adding 1 to the rounding digit and dropping all the digits to its right. If the test digit is less than 5, round down by keeping the rounding digit and dropping all the digits to its right.

EXAMPLE 5 **Chemistry.** In a chemistry class, a student uses a balance to weigh a compound. The digital readout on the scale shows 1.2387 grams. Round this decimal to the nearest thousandth of a gram.

Solution We are asked to round to the nearest thousandth.

Add 1 to the 8.

1.2387

The rounding digit — The test digit is 5 or greater. Therefore, add 1 to the rounding digit and drop all other digits to its right.

The compound weighs approximately 1.239 grams.

Self Check 6

Round each decimal to the indicated place value:

a. -708.522 to the nearest tenth

b. 9.1198 to the nearest thousandth

EXAMPLE 6 Round each decimal to the indicated place value: **a.** -645.13 to the nearest tenth and **b.** 33.097 to the nearest hundredth.

Solution

a. -645.13

Rounding digit — Since the test digit is less than 5, drop it and all the digits to its right.

The result is -645.1.

b. 33.097

Rounding digit — Since the test digit is greater than 5, we add 1 to 9 and drop all the digits to the right.

33.09 (10) Adding a 1 to the 9 requires that we carry a 1 to the tenths column.

When we are asked to round to the nearest hundredth, we must have a digit in the hundredths column, even if it is a zero. Therefore, the result is 33.10.

Answers **a.** −708.5, **b.** 9.120

Section 5.1 STUDY SET

VOCABULARY *Fill in the blanks.*

1. Give the name of each place value column.

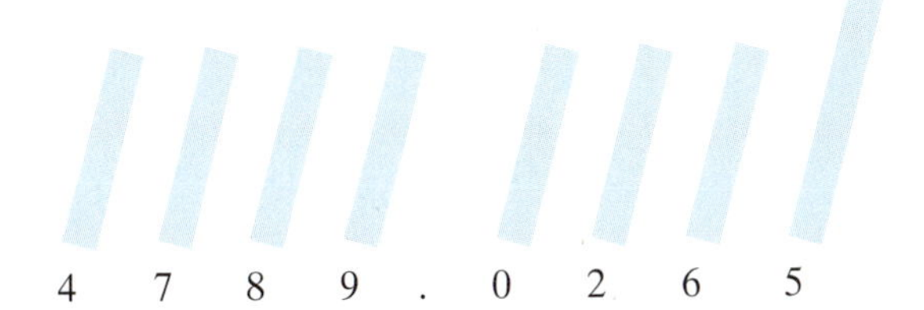

2. We can show the value represented by each digit of the decimal 98.6213 by using __________ notation.

$$98.6213 = 90 + 8 + \frac{6}{10} + \frac{2}{100} + \frac{1}{1{,}000} + \frac{3}{10{,}000}$$

3. We can approximate a decimal number using the process called __________.

4. When we read 2.37, the decimal point can be read as "_______" or "______."

CONCEPTS

5. Consider the decimal 32.415.

a. Write the decimal in words.

b. What is its whole-number part?

c. What is its fractional part?

d. Write the decimal in expanded notation.

6. Write $400 + 20 + 8 + \frac{9}{10} + \frac{6}{100}$ as a decimal.

7. Graph: $\frac{7}{10}$, -0.7, $-3\frac{1}{100}$, and 3.01.

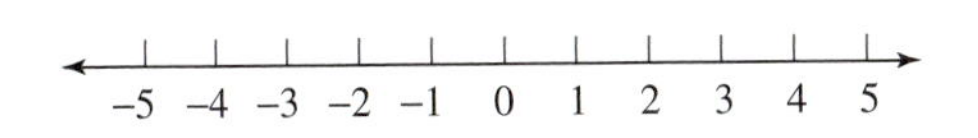

8. Graph: −1.21, −3.29, and −4.25.

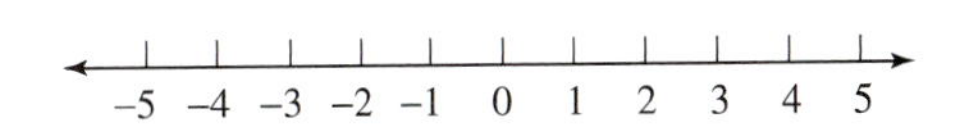

9. Determine whether the statement is true or false.

a. 0.9 = 0.90

b. 1.260 = 1.206

c. −1.2800 = −1.280

d. 0.001 = .0010

10. Write each fraction as a decimal.

a. $\frac{9}{10}$ **b.** $\frac{63}{100}$

c. $\frac{111}{1{,}000}$ **d.** $\frac{27}{10{,}000}$

11. Represent the shaded part of the square as a fraction and a decimal.

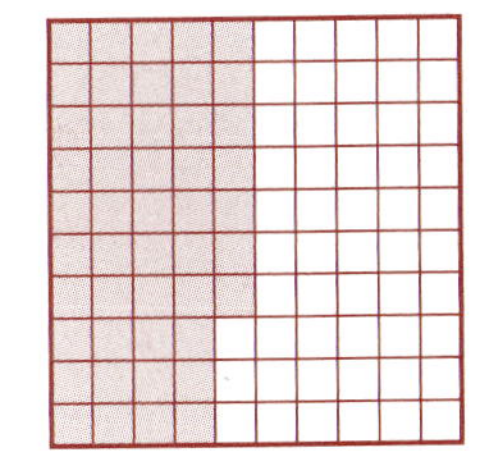

12. Represent the shaded part of the rectangle using a fraction and a decimal.

13. The line segment shown below is 1 inch long. Show a length of 0.3 inch on it.

14. Read the meter on the right. What decimal is indicated by the arrow?

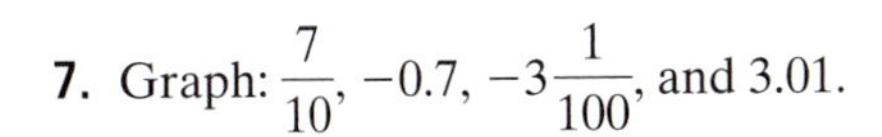

NOTATION

15. Construct a decimal number by writing
0 in the tenths column,
4 in the thousandths column,
1 in the tens column,
9 in the thousands column,
8 in the hundreds column,
2 in the hundredths column,
5 in the ten-thousandths column, and
6 in the ones column.

16. Represent each situation using a signed number.

a. A deficit of $15,600.55

b. A river 6.25 feet under flood stage

c. A state budget $6.4 million in the red

d. 3.9 degrees below zero

e. 17.5 seconds prior to liftoff

f. A checking account overdrawn by $33.45

PRACTICE *Write each decimal in words and as a fraction or mixed number.*

17. 50.1 **18.** 0.73

19. −0.0137 **20.** −76.09

21. 304.0003 **22.** 68.91

23. −72.493 **24.** −31.5013

Write each decimal using numbers.

25. Negative thirty-nine hundredths

26. Negative twenty-seven and forty-four hundredths

27. Six and one hundred eighty-seven thousandths

28. Ten and fifty-six ten-thousandths

Round each decimal to the nearest tenth.

29. 506.098 **30.** 0.441

31. 2.718218 **32.** 3,987.8911

Round each decimal to the nearest hundredth.

33. −0.137 **34.** −808.0897

35. 33.0032 **36.** 64.0059

Round each decimal to the nearest thousandth.

37. 3.14159 **38.** 16.0995

39. 1.414213 **40.** 2,300.9998

Round each decimal to the nearest whole number.

41. 38.901 **42.** 405.64

43. 2,988.399 **44.** 10,453.27

Round each amount to the value indicated.

45. $3,090.28

a. Nearest dollar

b. Nearest ten cents

46. $289.73

a. Nearest dollar

b. Nearest ten cents

Fill in the blanks with the proper symbol (<, >, or =).

47. −23.45 ☐ −23.1 **48.** −301.98 ☐ −302.45

49. −.065 ☐ −.066 **50.** −3.99 ☐ −3.9888

Arrange the decimals in order, from least to greatest.

51. 132.64, 132.6499, 132.6401

52. 0.007, 0.00697, 0.00689

APPLICATIONS

53. WRITING CHECKS Complete the check below by writing in the amount, using a decimal.

Ellen Russell
455 Santa Clara Ave.
Parker, CO 25413
April 14, 20 03

PAY TO THE ORDER OF Citicorp $ ________

One thousand twenty-five and $\frac{78}{100}$ DOLLARS

BA Downtown Branch
P.O. Box 2456
Colorado Springs, CO 23712
MEMO Mortgage
Ellen Russell
45-828-02-33-4660

54. MONEY We use a decimal point when working with dollars, but the decimal point is not necessary when working with cents. For each dollar amount in the table, give the equivalent amount expressed as cents.

Dollars	Cents
$0.50	
$0.05	
$0.55	
$5.00	
$0.01	

55. INJECTIONS A syringe is shown below. Use an arrow to show to what point the syringe should be filled if a 0.38-cc dose of medication is to be administered. ("cc" stands for "cubic centimeters.")

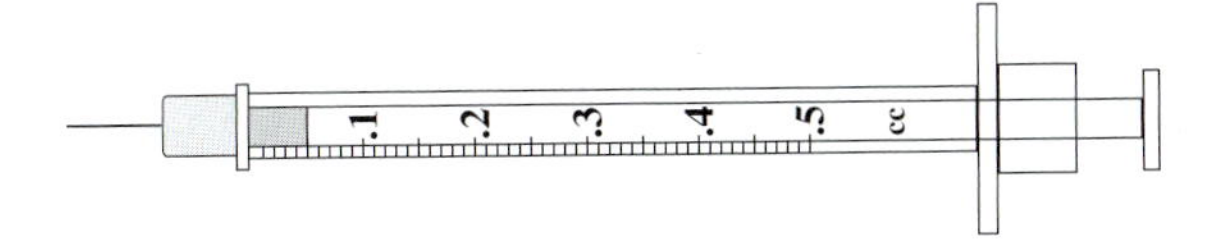

56. LASERS The laser used in laser vision correction is so precise that each pulse can remove 39 millionths of an inch of tissue in 12 billionths of a second. Write each of these numbers as decimals.

57. THE METRIC SYSTEM The metric system is widely used in science to measure length (meters), weight (grams), and capacity (liters). Round each decimal to the nearest hundredth.

a. 1 ft is 0.3048 meter.

b. 1 mi is 1,609.344 meters.

c. 1 lb is 453.59237 grams.

d. 1 gal is 3.785306 liters.

58. WORLD RECORDS As of October 2004, four American women held individual world records in swimming. Their times are given below in the form *minutes: seconds.* Round each to the nearest tenth of a second.

100-meter backstroke	Natalie Coughlin	0:59.58
200-meter breaststroke	Amanda Beard	2:22.44
400-meter freestyle	Janet Evans	4:03.85
800-meter freestyle	Janet Evans	8:16.22
1,500-meter freestyle	Janet Evans	15:52.10

59. GEOLOGY Geologists classify types of soil according to the grain size of the particles that make up the soil. The four major classifications are shown below. Complete the table by classifying each sample.

Clay	0.00008 in. and under
Silt	0.00008 in. to 0.002 in.
Sand	0.002 in. to 0.08 in.
Granule	0.08 in. to 0.15 in.

Sample	Location	Size (in.)	Classification
A	riverbank	0.009	
B	pond	0.0007	
C	NE corner	0.095	
D	dry lake	0.00003	

60. MICROSCOPES A microscope used in a lab is capable of viewing structures that range in size from 0.1 to 0.0001 centimeter. Which of the structures listed below would be visible through this microscope?

Structure	Size (in cm)
bacterium	0.00011
plant cell	0.015
virus	0.000017
animal cell	0.00093
asbestos fiber	0.0002

61. AIR QUALITY The following table shows the cities with the highest one-hour concentrations of ozone (in parts per million) during the summer of 1999. Rank the cities in order, beginning with the city with the highest reading.

Crestline, California	0.170
Galveston, Texas	0.176
Houston, Texas	0.202
Texas City, Texas	0.206
Westport, Connecticut	0.188
White Plains, New York	0.171

Source: *Los Angeles Times* (August 18, 1999)

62. DEWEY DECIMAL SYSTEM A system for classifying books in a library is the Dewey Decimal System. Books on the same subject are grouped together by number. For example, books about the arts are assigned numbers between 700 and 799. When stacked on the shelves, the books are to be in numerical order, from left to right. How should the titles in the illustration be rearranged to be in the proper order?

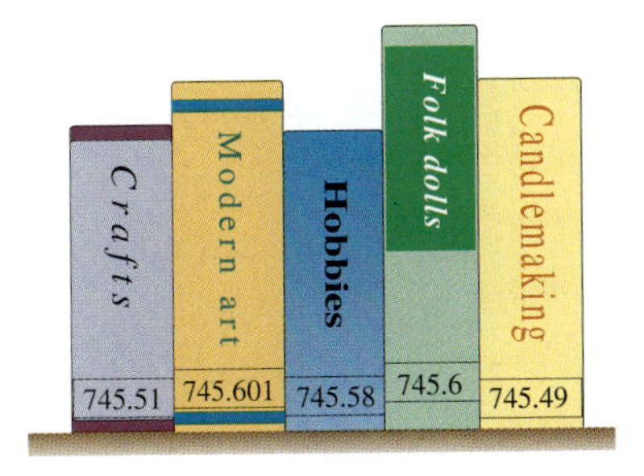

63. THE OLYMPICS The results of the women's all-around gymnastic competition in the 2004 Athens Olympic Games are shown in the following table. Which gymnasts won the gold, silver, and bronze medals?

Name	Country	Score
Nan Zhang	China	38.049
Ana Pavlova	Russia	38.024
Nicoleta Sofronie	Romania	37.948
Carly Patterson	U.S.A.	38.387
Svetlana Khorkina	Russia	38.211
Irina Yarotska	Ukraine	37.687

64. TUNEUPS The six spark plugs from the engine of a Nissan Quest were removed, and the spark plug gap was checked. If vehicle specifications call for the gap to be from 0.031 to 0.035 inch, which of the plugs should be replaced?

Cylinder 1: 0.035 in.
Cylinder 2: 0.029 in.
Cylinder 3: 0.033 in.
Cylinder 4: 0.039 in.
Cylinder 5: 0.031 in.
Cylinder 6: 0.032 in.

Spark plug gap

65. E-COMMERCE The gain (or loss) in value of one share of Amazon.com stock is shown in the graph below for eleven quarters. (For accounting purposes, a year is divided into four quarters.)

a. In what quarter, of what year, was there the greatest gain? Estimate the gain.

b. In what quarter, of what year, was there the greatest loss? Estimate the loss.

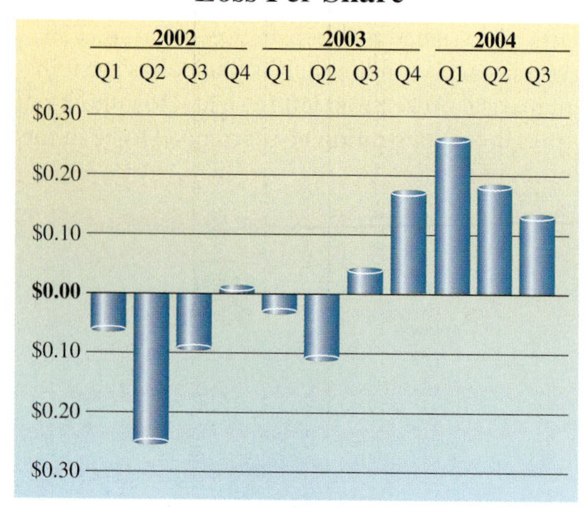

Source: Amazon.com

66. GASOLINE PRICES Refer to the data in the table. Then construct a line graph showing the annual national average retail price per gallon for unleaded regular gasoline for the years 1997 through 2003 (according to *The World Almanac 2005*).

Year	1997	1998	1999	2000	2001	2002	2003
Price (¢)	123.4	105.9	116.5	151.0	146.1	135.8	159.1

WRITING

67. Explain the difference between ten and a tenth.

68. "The more digits a number contains, the larger it is." Is this statement true? Explain your response.

69. How are fractions and decimals related?

70. Explain the benefits of a monetary system that is based on decimals instead of fractions.

71. The illustration shows an unusual notation many service stations use to express the price of a gallon of gasoline. Explain what this notation means.

REGULAR	UNLEADED	UNLEADED +
$1.79\frac{9}{10}$	$1.89\frac{9}{10}$	$1.99\frac{9}{10}$

72. Write a definition for each of these words.

decade *decathlon* *decimal*

REVIEW

73. Add: $75\frac{3}{4} + 88\frac{4}{5}$.

74. Multiply: $\frac{2x}{15}\left(-\frac{5}{4x^2}\right)$.

75. Simplify: $5R - 3(6 - R)$.

76. Solve: $6y + 8 = -y + 9 - y$.

77. Find the area of a triangle with base 16 in. and height 9 in.

78. Express the fraction $\frac{2}{3}$ as an equivalent fraction with a denominator of $3a$.

79. Add: $-2 + (-3) + 4$.

80. Subtract: $-15 - (-6)$.

5.2 Adding and Subtracting Decimals

- Adding decimals • Subtracting decimals • Adding and subtracting signed decimals

To add or subtract objects, they must be similar. The federal income tax form shown in Figure 5-4 has a vertical line to ensure that dollars are added to dollars and cents added to cents. In this section, we will show how decimals are added and subtracted using this type of vertical column format.

Form 1040EZ	Department of the Treasury—Internal Revenue Service **Income Tax Return for Single and Joint Filers With No Dependents 2003**			
Income **Attach Form(s) W-2 here.** Enclose but do not attach any payment.	1 Total wages, salaries, and tips. This should be shown in box 1 of your form(s) W-2.	1	21,056	89
	2 Taxable interest. If the total is over $1,500, you cannot use Form 1040EZ.	2	42	06
	3 Unemployment compensation and Alaska Permanent Fund dividends (see page 14).	3	200	00
	4 Add lines 1, 2, and 3. This is your **adjusted gross income.**	4	21,298	95

FIGURE 5-4

Adding decimals

When adding decimals, we line up the columns so that ones are added to ones, tenths are added to tenths, hundredths are added to hundredths, and so on. As an example, consider the following problem.

Line up the columns and the decimal points vertically. Then add the numbers.

$$\begin{array}{r} 12.140 \\ 3.026 \\ 4.000 \\ +\ 0.700 \\ \hline 19.866 \end{array}$$

Write the decimal point in the result directly under the decimal points in the problem.

Adding decimals

1. Line up the decimal points, using the vertical column format.
2. Add the numbers as you would add whole numbers.
3. Write the decimal point in the result directly below the decimal points in the problem.

EXAMPLE 1 Add: 1.903 + 0.6 + 8 + 0.78.

Solution

$$\begin{array}{r} \overset{2}{1}.903 \\ 0.600 \\ 8.000 \\ +\ 0.780 \\ \hline 11.283 \end{array}$$

To make the addition by columns easier, write two zeros after 6, a decimal point and three zeros after 8, and one zero after 0.78.

Carry a 2 (shown in blue) to the ones column.

The result is 11.283.

Self Check 1
Add: 0.07 + 35 + 0.888 + 4.1.

Answer 40.058

CALCULATOR SNAPSHOT

Preventing heart attacks

The bar graph in Figure 5-5 shows the number of grams of fiber in a standard serving of each of several foods. It is believed that men can significantly cut their risk of heart attack by eating at least 25 grams of fiber a day. Does this diet meet or exceed the 25-gram requirement?

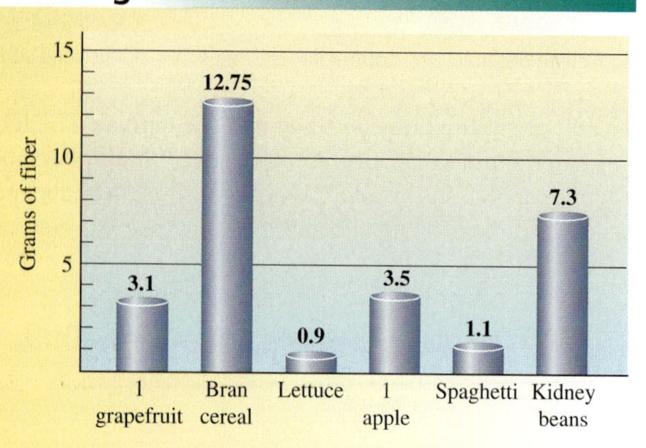

FIGURE 5-5

To find the total fiber intake, we will add the fiber content of each of the foods. We can use a scientific calculator to add the decimals.

3.1 [+] 12.75 [+] .9 [+] 3.5 [+] 1.1 [+] 7.3 [=]   28.65

Since $28.65 > 25$, this diet exceeds the daily fiber requirement of 25 grams.

Subtracting decimals

To subtract decimals, we line up the decimal points and corresponding columns so that we subtract like objects — tenths from tenths, hundredths from hundredths, and so on.

Subtracting decimals

1. Line up the decimal points using the vertical column format.
2. Subtract the numbers as you would subtract whole numbers.
3. Write the decimal point in the result directly below the decimal points in the problem.

Self Check 2

Subtract:

a. 382.5 − 227.1

b. 30.1 − 27.122

Answers **a.** 155.4, **b.** 2.978

EXAMPLE 2 Subtract: **a.** 279.6 − 138.7 and **b.** 15.4 − 13.059.

Solution

a.

$$\begin{array}{r} \overset{8}{\cancel{9}}\overset{16}{\cancel{6}} \\ 27\cancel{9}.\cancel{6} \\ -138.7 \\ \hline 140.9 \end{array}$$

To subtract in the tenths column, borrow 1 one in the form of 10 tenths from the ones column. Add 10 to the 6 in the tenths column, which gives 16 (shown in blue).

b.

$$\begin{array}{r} 15.\overset{3}{\cancel{4}}\,\overset{\overset{9}{\cancel{10}}}{\cancel{0}}\,\overset{10}{\cancel{0}} \\ -13.0\,5\,9 \\ \hline 2.3\,4\,1 \end{array}$$

Add two zeros to the right of 15.4 to make borrowing easier. First, borrow from the tenths column; then borrow from the hundredths column.

EXAMPLE 3 **Conditioning programs.** A 350-pound football player lost 15.7 pounds during the first week of practice. During the second week, he gained 4.9 pounds. Find his weight after the first two weeks of practice.

Solution The word *lost* indicates subtraction. The word *gained* indicates addition.

Beginning weight	minus	first week weight loss	plus	second week weight gain	equals	weight after two weeks of practice.

$350 - 15.7 + 4.9 = 334.3 + 4.9$ Working from left to right, perform the subtraction first: $350 - 15.7 = 334.3$.

$= 339.2$ Perform the addition.

The player's weight is 339.2 pounds after two weeks of practice.

Weather balloons

CALCULATOR SNAPSHOT

A giant weather balloon is made of neoprene, a flexible rubberized substance, that has an uninflated thickness of 0.011 inch. When the balloon is inflated with helium, the thickness becomes 0.0018 inch.

To find the change in thickness, we need to subtract. We can use a scientific calculator to subtract the decimals.

.011 [−] .0018 [=] 0.0092

After the balloon is inflated, the neoprene loses 0.0092 of an inch in thickness.

Adding and subtracting signed decimals

To add signed decimals, we use the same rules that we used for adding integers.

Adding two decimals

With like signs: Add their absolute values and attach their common sign to the sum.

With unlike signs: Subtract their absolute values (the smaller from the larger) and attach the sign of the number with the larger absolute value to the sum.

EXAMPLE 4 Add: $-6.1 + (-4.7)$.

Solution Since the decimals are both negative, we add their absolute values and attach a negative sign to the result.

$-6.1 + (-4.7) = -10.8$ Add the absolute values, 6.1 and 4.7, to get 10.8. Use their common sign.

Self Check 4
Add: $-5.04 + (-2.32)$.

Answer -7.36

EXAMPLE 5 Add: $5.35 + (-12.9)$.

Solution In this example, the signs are unlike. Since -12.9 has the larger absolute value, we subtract 5.35 from 12.9 to get 7.55, and attach a negative sign to the result.

$$5.35 + (-12.9) = -7.55$$

Self Check 5
Add: $-21.4 + 16.75$.

Answer -4.65

EXAMPLE 6 Subtract: $-4.3 - 5.2$.

Solution To subtract signed decimals, we can add the opposite of the decimal that is being subtracted.

$$-4.3 - 5.2 = -4.3 + (-5.2)$$ Add the opposite of 5.2, which is -5.2.

$$= -9.5$$ Add the absolute values, 4.3 and 5.2, to get 9.5. Attach a negative sign to the result.

Self Check 6
Subtract: $-1.18 - 2.88$

Answer -4.06

EXAMPLE 7 Subtract: $-8.37 - (-16.2)$.

Solution

$$-8.37 - (-16.2) = -8.37 + 16.2$$ Add the opposite of -16.2, which is 16.2.

$$= 7.83$$ Subtract the smaller absolute value from the larger, 8.37 from 16.2, to get 7.83. Since 16.2 has the larger absolute value, the result is positive.

Self Check 7
Subtract: $-2.56 - (-4.4)$.

Answer 1.84

EXAMPLE 8 Evaluate: $-12.2 - (-14.5 + 3.8)$.

Solution We perform the addition within the grouping symbols first.

$$-12.2 - (\mathbf{-14.5 + 3.8}) = -12.2 - (\mathbf{-10.7})$$ Perform the addition: $-14.5 + 3.8 = -10.7$.

$$= -12.2 + 10.7$$ Add the opposite of -10.7.

$$= -1.5$$ Perform the addition.

Self Check 8
Evaluate: $-4.9 - (-1.2 + 5.6)$.

Answer -9.3

Section 5.2 STUDY SET

VOCABULARY *Fill in the blanks.*

1. The answer to an addition problem is called the ______.
2. The answer to a subtraction problem is called the ______.

CONCEPTS

3. To subtract signed decimals, add the ________ of the decimal that is being subtracted.

4. a. Add: 0.3 + 0.17.

 b. Write 0.3 and 0.17 as fractions. Find a common denominator for the fractions and add them.

 c. Express your answer to part b as a decimal.

 d. Compare your answers from part a and part c.

NOTATION

5. Every whole number has an unwritten decimal _______ to its right.

6. In the subtraction problem below, we must borrow. How much is borrowed from the 3, and in what form is it borrowed?

$$\begin{array}{r} 29.\overset{2}{\not{3}}\,\overset{11}{\not{1}} \\ -25.1\;6 \\ \hline \end{array}$$

PRACTICE *Perform each operation.*

7. $\begin{array}{r} 32.5 \\ +\ 7.4 \\ \hline \end{array}$

8. $\begin{array}{r} 6.3 \\ +13.5 \\ \hline \end{array}$

9. $\begin{array}{r} 21.6 \\ +33.12 \\ \hline \end{array}$

10. $\begin{array}{r} 19.4 \\ +31.95 \\ \hline \end{array}$

11. 12 + 3.9

12. 0.01 + 3.6

13. 0.03034 + 0.2003

14. 19.9 + 19.9

15. 247.9 + 40 + 0.56

16. 0.0053 + 1.78 + 6

17. 45 + 9.9 + 0.12 + 3.02

18. 505.01 + 23 + 0.989 + 12.07

19. $\begin{array}{r} 12.98 \\ -\ 3.45 \\ \hline \end{array}$

20. $\begin{array}{r} 1.6 \\ -0.16 \\ \hline \end{array}$

21. $\begin{array}{r} 78.1 \\ -\ 7.81 \\ \hline \end{array}$

22. $\begin{array}{r} 202.234 \\ -\ 19.34 \\ \hline \end{array}$

23. 5 − 0.023

24. 30 − 11.98

25. 24 − 23.81

26. 7.001 − 5.9

27. −45.6 + 34.7

28. −19.04 + 2.4

29. 46.09 + (−7.8)

30. 34.7 + (−30.1)

31. −7.8 + (−6.5)

32. −5.78 + (−33.1)

33. −0.0045 + (−0.031)

34. −90.09 + (−0.087)

35. −9.5 − 7.1

36. −7.08 − 14.3

37. 30.03 − (−17.88)

38. 143.3 − (−64.01)

39. −2.002 − (−4.6)

40. −0.005 −(−8)

41. −7 − (−18.01)

42. −63.04 − (−8.911)

Evaluate each expression. Remember to perform the operations within grouping symbols first.

43. (3.4 − 6.6) + 7.3

44. 3.4 − (6.6 + 7.3)

45. (−9.1 − 6.05) − (−51)

46. −9.1 − (−6.05) + 51

47. 16 − (67.2 + 6.27)

48. −43 − (0.032 − 0.045)

49. (−7.2 + 6.3) − (−3.1 − 4)

50. 2.3 + [2.4 − (2.5 −2.6)]

51. |−14.1 + 6.9| + 8

52. 15 − |−2.3 + (−2.4)|

53. Find the sum of *two and forty-three hundredths* and *five and six tenths.*

54. Find the difference of *nineteen hundredths* and *six thousandths.*

APPLICATIONS

55. SPORTS PAGES In the sports pages of any newspaper, decimal numbers are used quite often.

 a. "German bobsledders set a world record today with a final run of 53.03, finishing ahead of the Italian team by only fourteen thousandths of a second." What was the time for the Italian bobsled team?

 b. "The women's figure skating title was decided by only thirty-three hundredths of a point." If the winner's point total was 102.71, what was the second-place finisher's total?

56. NURSING The following table shows a patient's health chart. A nurse failed to fill in certain portions. (98.6° Fahrenheit is considered normal.) Complete the table.

Day of week	Patient's temperature	How much above normal
Monday	99.7°	
Tuesday		2.5°
Wednesday	98.6°	
Thursday	100.0°	
Friday		0.9

57. VEHICLE SPECIFICATIONS Certain dimensions of a compact car are shown. Find the wheelbase of the car.

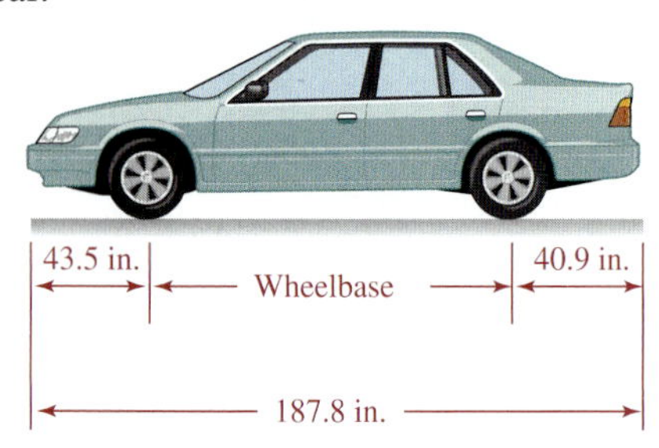

58. pH SCALE The pH scale shown below is used to measure the strength of acids and bases in chemistry. Find the difference in pH readings between

a. bleach and stomach acid.

b. ammonia and coffee.

c. blood and coffee.

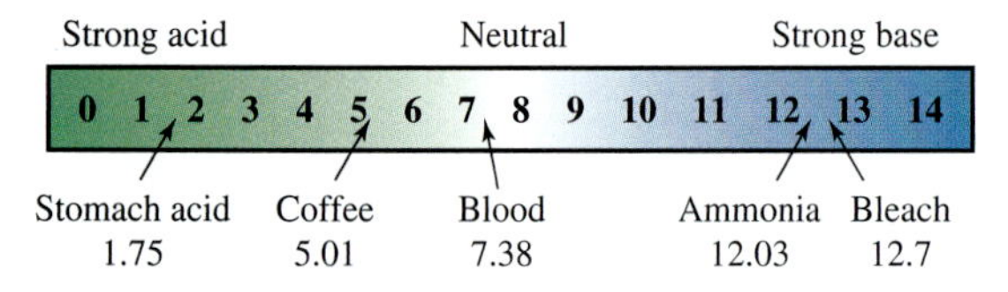

59. BAROMETRIC PRESSURES Barometric pressure readings are recorded on the following weather map. In a low-pressure area (L on the map), the weather is often stormy. The weather is usually fair in a high-pressure area (H). What is the difference in readings between the areas of highest and lowest pressure? In what part of the country would you expect the weather to be fair?

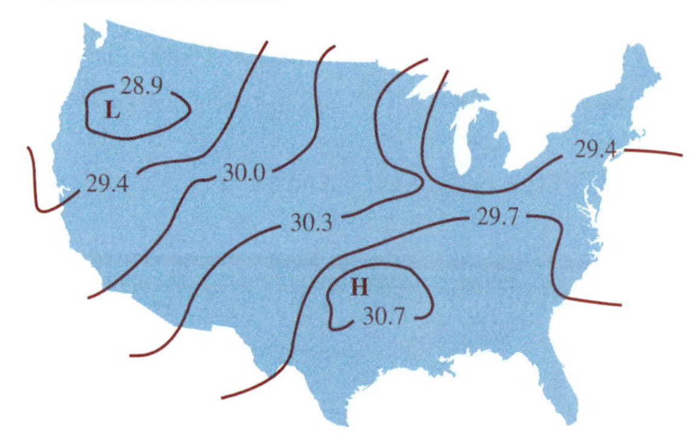

60. QUALITY CONTROL An electronics company has strict specifications for silicon chips used in a computer. The company will install only chips that are within 0.05 centimeter of the specified thickness. The table gives that specification for two types of chip. Fill in the blanks to complete the chart.

Chip type	Thickness specification	Acceptable range	
		Low	High
A	0.78 cm		
B	0.643 cm		

61. OFFSHORE DRILLING A company needs to construct a pipeline from an offshore oil well to a refinery located on the coast. Company engineers have come up with two plans for consideration, as shown. Use the information in the illustration to complete the table.

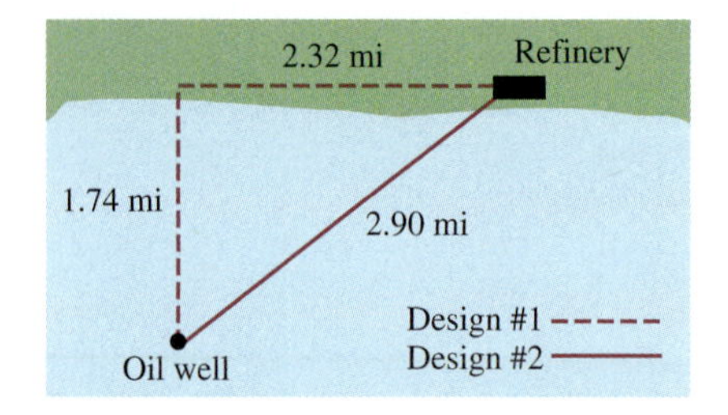

	Pipe underwater (mi)	Pipe underground (mi)	Total pipe (mi)
Design 1			
Design 2			

62. TELEVISION The following illustration shows the six most-watched television shows of all time (excluding Super Bowl games).

a. What was the combined total audience of all six shows?

b. How many more people watched the last episode of "MASH" than watched the last episode of "Seinfeld"?

c. How many more people would have had to watch the last "Seinfeld" to move it into a tie for fifth place?

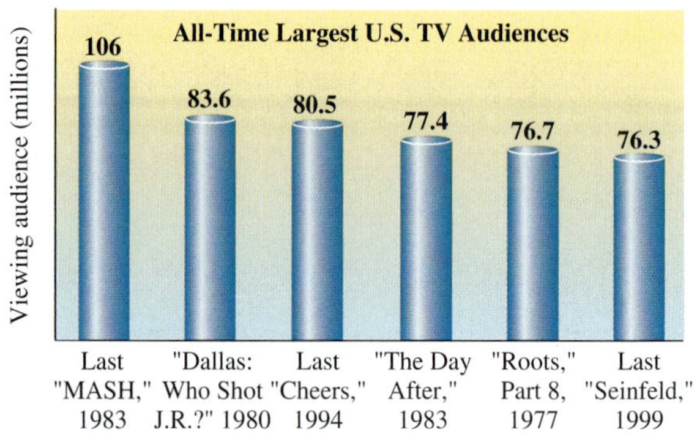

Source: Nielson Media Research

63. RECORDHOLDERS The late Florence Griffith-Joyner of the United States still holds the world record in the 100-meter sprint: 10.49 seconds. Jodie Henry of Australia holds the world record in the 100-meter freestyle swim. Jodie's time is 53.52. How much faster did Griffith-Joyner run the 100 meters than Henry swam it?

64. FLIGHT PATH See the illustration. Find the added distance a plane must travel to avoid flying through the storm.

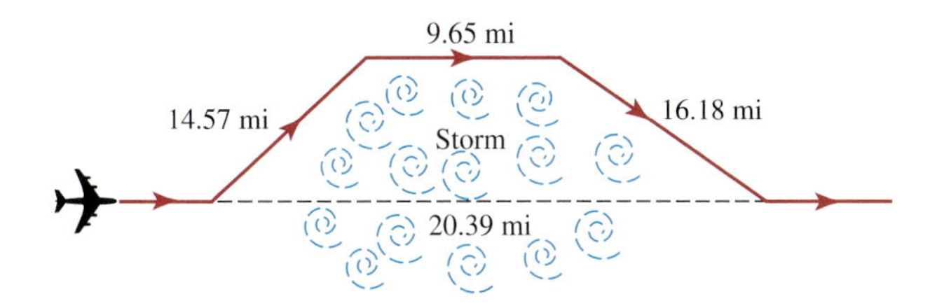

65. DEPOSIT SLIPS A deposit slip for a savings account is shown below. Find the subtotal and the total deposit.

Deposit

Cash	242	50
Checks (properly endorsed)	116	10
	47	93
Total from reverse side	359	16
Subtotal		
Less cash	25	00
Total deposit		

66. MOTION Forces such as water current or wind can increase or decrease the speed of an object in motion. Find the speed of each object.

a. An airplane's speed in still air is 450 mph, and it has a tail wind of 35.5 mph helping it along.

b. A man can paddle a canoe at 5 mph in still water, but he is going upstream. The speed of the current against him is 1.5 mph.

67. THE HOME SHOPPING NETWORK The illustration shows a description of a cookware set that was sold on television.

a. Find the difference between the manufacturer's suggested retail price (MSRP) and the sale price.

b. Including shipping and handling (S & H), how much will the cookware set cost?

Item 229-442

Continental 9-piece Cookware Set

Stainless steel

MSRP	\$149.79
HSN Price	\$59.85
On Sale	**\$47.85**
S & H	\$7.95

68. RETAILING Complete the table by filling in the retail price of each appliance, given its cost to the dealer and the store markup.

Item	Cost	Markup	Retail price
Refrigerator	\$510.80	\$105.00	
Washing machine	\$189.50	\$55.50	
Dryer	\$163.99	$\$x$	

Evaluate each expression.

69. 2,367.909 + 5,789.0253

70. 0.00786 + 0.3423

71. 9,000.09 − 7,067.445

72. 1 − 0.004999

73. 3,434.768 − (908 − 2.3 + .0098)

74. 12 − (0.723 + 3.05611)

WRITING

75. Explain why we line up the decimal points and corresponding columns when adding decimals.

76. Explain why we can write additional zeros to the right of a decimal such as 7.89 without affecting its value.

77. Explain what is wrong with the work shown below.

$$\begin{array}{r} 203.56 \\ 37 \\ +\ \ 0.43 \\ \hline 204.36 \end{array}$$

78. Consider the addition

$$\begin{array}{r} \overset{2}{2}3.7 \\ 41.9 \\ +12.8 \\ \hline 78.4 \end{array}$$

Explain the meaning of the small 2 written above the ones column.

REVIEW

79. Add: $44\frac{3}{8} + 66\frac{1}{5}$.

80. Simplify: $\dfrac{-\frac{3}{4}}{\frac{5}{16}}$.

81. Multiply: $\frac{-15}{26} \cdot 1\frac{4}{9}$.

82. Evaluate: $2 + 5[-2 - (6 + 1)]$.

5.3 Multiplying Decimals

- Multiplying decimals • Multiplying decimals by powers of 10
- Multiplying signed decimals • Order of operations

We now focus on the operation of multiplication. First, we develop a method used to multiply decimals. Then we use that method to evaluate expressions and to solve problems involving decimals.

Multiplying decimals

To show how to multiply decimals, we examine the multiplication 0.3 · 0.17, finding the product in a roundabout way. First, we will write 0.3 and 0.17 as fractions and multiply them. Then we will express the resulting fraction as a decimal.

$$0.3(0.17) = \frac{3}{10} \cdot \frac{17}{100}$$ Express 0.3 and 0.17 as fractions.

$$= \frac{3 \cdot 17}{10 \cdot 100}$$ Multiply the numerators and multiply the denominators.

$$= \frac{51}{1{,}000}$$ Multiply in the numerator and denominator.

$$= 0.051$$ Write $\frac{51}{1{,}000}$ as a decimal.

From this example, we can make observations about multiplying decimals.

- The digits in the answer are found by multiplying 3 and 17.

  $$0.3 \cdot 0.17 = 0.051$$

  3 · 17 = 51

- The answer has 3 decimal places. The *sum* of the number of decimal places in the factors 0.3 and 0.17 is also 3.

  $$0.3 \cdot 0.17 = 0.051$$

  1 decimal place, 2 decimal places, 3 decimal places

  These observations suggest the following rule for multiplying decimals.

Multiplying decimals

To multiply two decimals:

1. Multiply the decimals as if they were whole numbers.
2. Find the total number of decimal places in both factors.
3. Place the decimal point in the result so that the answer has the same number of decimal places as the total found in step 2.

Self Check 1

Mutiply: 2.74 · 4.3.

EXAMPLE 1 Multiply: 5.9 · 3.4.

Solution We temporarily ignore the decimal points and multiply the decimals as if they were whole numbers. Initially, we think of this problem as 59 times 34.

$$\begin{array}{r} 59 \\ \times\ 34 \\ \hline 236 \\ 177 \\ \hline 2006 \end{array}$$

To place the decimal point in the product, we find the total number of digits to the right of the decimal points in the factors.

$$\begin{array}{r} 5.9 \\ \times\ 3.4 \\ \hline 236 \\ 177 \\ \hline 20.06 \end{array}$$

5.9 ← 1 decimal place; 3.4 ← 1 decimal place — The answer will have 1 + 1 = 2 decimal places.

Locate the decimal point so that the answer has 2 decimal places.

Answer 11.782

When we multiply decimals, it is not necessary to line up the decimal points, as the next example illustrates.

EXAMPLE 2 Multiply: 1.3(0.005).

Solution We multiply 13 by 5.

$$\begin{array}{r} 1.3 \\ \times 0.005 \\ \hline 65 \end{array}$$

1.3 ← 1 decimal place; 0.005 ← 3 decimal places — The answer will have 1 + 3 = 4 decimal places.

We then place the decimal point in the result.

$$\begin{array}{r} 1.3 \\ \times\ 0.005 \\ \hline 0.0065 \end{array}$$

Add 2 placeholder zeros and position the decimal point so that the product has 4 decimal places.

Self Check 2
Multiply: (0.0002)7.2.

Answer 0.00144

Heating costs

CALCULATOR SNAPSHOT

When billing a household, a gas company converts the amount of natural gas used into units of heat energy called *therms*. The number of therms used by a household in one month and the cost per therm are shown below.

Customer charge . . . . 39 therms @ $0.72264

To find the total charges for the month, we multiply the number of therms by the cost per therm: 39 · 0.72264.

39 [×] .72264 [=] 28.18296

Rounding to the nearest cent, we see that the total charge is $28.18.

Self Check 3
Multiply: 178(2.7).

EXAMPLE 3 Multiply: 234(3.1).

Solution

```
   234   ← No decimal places  }
×  3.1   ← 1 decimal place    }  The answer will have 0 + 1 = 1 decimal place.
  23 4
 702
 725.4   Locate the decimal point so that the answer has 1 decimal place.
```

Answer 480.6

Multiplying decimals by powers of 10

The numbers 10, 100, and 1,000 are called *powers of 10,* because they are the results when we evaluate 10^1, 10^2, and 10^3, respectively. To develop a rule to determine the product when multiplying a decimal and a power of 10, we will multiply 8.675 by three different powers of 10.

Multiply: 8.675 · **10**

```
    8.675
×      10
     0000
    8675
   86.750
```

The answer is 86.75.

Multiply: 8.675 · **100**

```
    8.675
×     100
     0000
    0000
   8675
  867.500
```

The answer is 867.5

Multiply: 8.675 · **1,000**

```
    8.675
×    1000
     0000
    0000
   0000
  8675
 8675.000
```

The answer is 8,675.

We can make observations about the results.

- In each case, the answer contains the same digits as the factor 8.675.
- When we inspect the answers, the decimal point in the first factor 8.675 appears to be moved to the right by the multiplication process. The number of decimal places it moves depends on the power of 10 by which 8.675 is multiplied.

One zero in 10
8.675 · 10 = 86.75
It moves one place to the right.

Two zeros in 100
8.675 · 100 = 867.5
It moves two places to the right.

Three zeros in 1,000
8.675 · 1,000 = 8675.
It moves three places to the right.

These observations suggest the following rule.

Multiplying a decimal by a power of 10

To multiply a decimal by a power of 10, move the decimal point to the right the same number of places as there are zeros in the power of 10.

Self Check 4
Find each product:
a. 0.721 · 100
b. 6.08(1,000)

EXAMPLE 4 Find each product: **a.** 2.81 · 10 and **b.** 0.076 · 10,000.

Solution

a. 2.81 · 10 = 28.1   Since 10 has 1 zero, move the decimal point 1 place to the right.

b. $0.076 \cdot 10{,}000 = 0760.$ Since 10,000 has 4 zeros, move the decimal point 4 places to the right. Write a placeholder zero (shown in blue).

$= 760$

Answers **a.** 72.1, **b.** 6,080

EXAMPLE 5 Tachometers.

A tachometer indicates the engine speed of a vehicle, in revolutions per minute (rpm). What engine speed is indicated by the tachometer in Figure 5-6?

FIGURE 5-6

Solution The needle is pointing to 4.5. The notation "RPM × 1000" on the tachometer instructs us to multiply 4.5 by 1,000 to find the engine speed.

$4.5 \cdot 1{,}000 = 4500.$ Since 1,000 has 3 zeros, move the decimal point 3 places to the right. Write two placeholder zeros.

$= 4{,}500$

The engine speed is 4,500 rpm.

Overtime — THINK IT THROUGH

"Employees covered by the Fair Labor Standards Act must receive overtime pay for hours worked in excess of 40 in a workweek of at least 1.5 times their regular rates of pay." U.S. Department of Labor

The map of the United States shown below is divided into nine regions. The average hourly wage for private industry workers in each region is also listed in the legend below the map. Determine the average hourly wage for the region where you live. Then calculate the corresponding average hourly overtime wage for that region.

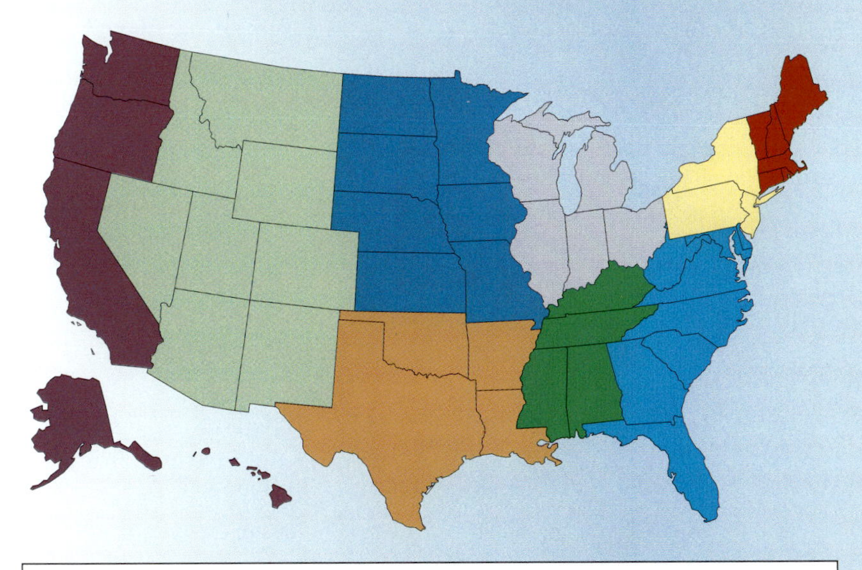

West North Central: $16.30
Mountain: $15.65
Pacific: $19.11
East South Central: $13.97
East North Central: $17.16
West South Central: $15.22
New England: $20.10
Middle Atlantic: $19.08
South Atlantic: $15.88

Multiplying signed decimals

Recall that the product of two numbers with like signs is positive, and the product of two numbers with unlike signs is negative.

Self Check 6
Multiply:
a. $6.6(-5.5)$
b. $(-44.968)(-100)$

Answers **a.** -36.3, **b.** $4{,}496.8$

EXAMPLE 6 Multiply: **a.** $-1.8(4.5)$ and **b.** $(-1{,}000)(-59.08)$.

Solution

a. Since the decimals have unlike signs, their product is negative.

$-1.8(4.5) = -8.1$ Multiply the absolute values, 1.8 and 4.5, to get 8.1. Make the result negative.

b. Since the decimals have like signs, their product is positive.

$(-1{,}000)(-59.08) = 59{,}080$ Multiply the absolute values, 1,000 and 59.08. Since 1,000 has 3 zeros, move the decimal point 3 places to the right. Write a placeholder zero.

Self Check 7
Evaluate:
a. $(-1.3)^2$
b. $(0.09)^2$

Answers **a.** 1.69, **b.** 0.0081

EXAMPLE 7 Evaluate: **a.** $(2.4)^2$ and **b.** $(-0.05)^2$.

Solution

a. $(2.4)^2 = (2.4)(2.4)$ Write 2.4 as a factor 2 times.
$= 5.76$ Perform the multiplication.

b. $(-0.05)^2 = (-0.05)(-0.05)$ Write -0.05 as a factor 2 times.
$= 0.0025$ Perform the multiplication. The product of two decimals with like signs is positive.

Order of operations

In the remaining examples, we use the rules for the order of operations to evaluate expressions involving decimals.

Self Check 8
Evaluate:
$-2|-4.4 + 5.6| + (-0.8)^2$.

Answer -1.76

EXAMPLE 8 Evaluate: $-(0.6)^2 + 5|-3.6 + 1.9|$.

Solution

$-(0.6)^2 + 5|-3.6 + 1.9| = -(0.6)^2 + 5|-1.7|$ Perform the addition within the absolute value symbols.
$= -(0.6)^2 + 5(1.7)$ Simplify: $|-1.7| = 1.7$.
$= -0.36 + 5(1.7)$ Find the power: $(0.6)^2 = 0.36$.
$= -0.36 + 8.5$ Perform the multiplication: $5(1.7) = 8.5$.
$= 8.14$ Perform the addition.

EXAMPLE 9 Evaluate $6.28r(h + r)$ for $h = 3.1$ and $r = 6$.

Solution

$6.28r(h + r) = 6.28(6)(3.1 + 6)$ Replace r with 6 and h with 3.1.

$= 6.28(6)(9.1)$ Perform the addition within the parentheses: $3.1 + 6 = 9.1$.

$= 37.68(9.1)$ Perform the multiplication: $6.28(6) = 37.68$.

$= 342.888$ Perform the multiplication.

Self Check 9

Evaluate $1.3pr^3$ for $p = 3.14$ and $r = 3$.

Answer 110.214

EXAMPLE 10 Weekly earnings.

A cashier's workweek is 40 hours. After his daily shift is over, he can work overtime at a rate 1.5 times his regular rate of \$7.50 per hour. How much money will he earn in a week if he works 6 hours of overtime?

Solution First, we need to find his overtime rate, which is 1.5 times his regular rate of \$7.50 per hour.

$1.5(7.50) = 11.25$

His overtime rate is \$11.25 per hour.

To find his total weekly earnings, we use the following fact.

The regular rate	times	40 hours	plus	the overtime rate	times	overtime hours worked	equals	his total earnings.

$7.50(40) + 11.25(6) = 300 + 67.50$ Perform the multiplications.

$= 367.50$ Perform the addition.

The cashier's earnings for the week are \$367.50.

Section 5.3 STUDY SET

VOCABULARY *Fill in the blanks.*

1. In the multiplication problem $2.89 \cdot 15.7$, the numbers 2.89 and 15.7 are called ________. The answer, 45.373, is called the ________.
2. Numbers such as 10, 100, and 1,000 are called ________ of 10.

CONCEPTS *Fill in the blanks.*

3. To multiply decimals, multiply them as if they were ________ numbers. The number of decimal places in the product is the same as the ________ of the decimal places in the factors.
4. To multiply a decimal by a power of 10, move the decimal point to the ________ the same number of decimal places as the number of ________ in the power of 10.
5. **a.** Multiply $\frac{3}{10}$ and $\frac{7}{100}$.

   b. Now write both fractions from part a as decimals. Multiply them in that form. Compare your results from parts a and b.
6. **a.** Multiply 0.11 and 0.3.

   b. Now write both decimals in part a as fractions. Multiply them in that form. Compare your results from parts a and b.

NOTATION

7. Suppose that the result of multiplying two decimals is 2.300. Write this result in simpler form.
8. When we move the decimal point to the right, does the decimal number get larger or smaller?

PRACTICE *Perform each multiplication.*

9. $(0.4)(0.2)$
10. $(0.2)(0.3)$
11. $(-0.5)(0.3)$
12. $(0.6)(-0.7)$
13. $(1.4)(0.7)$
14. $(2.1)(0.4)$
15. $(0.08)(0.9)$
16. $(0.003)(0.9)$
17. $(-5.6)(-2.2)$
18. $(-7.1)(-4.1)$
19. $(-4.9)(0.001)$
20. $(0.001)(-7.09)$
21. $(-0.35)(0.24)$
22. $(-0.85)(0.42)$
23. $(-2.13)(4.05)$
24. $(3.06)(-1.82)$
25. $16 \cdot 0.6$
26. $24 \cdot 0.8$
27. $-7(8.1)$
28. $-5(4.7)$
29. $0.04(306)$
30. $0.02(417)$
31. $60.61(-0.3)$
32. $-70.07 \cdot 0.6$
33. $-0.2(0.3)(-0.4)$
34. $-0.1(-2.2)(0.5)$
35. $5.5(10)(-0.3)$
36. $6.2(100)(-0.8)$
37. $4.2 \cdot 10$
38. $10 \cdot 7.1$
39. $67.164 \cdot 100$
40. $708.199 \cdot 100$
41. $-0.056(10)$
42. $-100(0.0897)$
43. $1{,}000(8.05)$
44. $23.7(1{,}000)$
45. $0.098(10{,}000)$
46. $3.63(10{,}000)$
47. $-0.2 \cdot 1{,}000$
48. $-1{,}000 \cdot 1.9$

Complete each table.

49.

Decimal	Its square
0.1	
0.2	
0.3	
0.4	
0.5	
0.6	
0.7	
0.8	
0.9	

50.

Decimal	Its cube
0.1	
0.2	
0.3	
0.4	
0.5	
0.6	
0.7	
0.8	
0.9	

Find each power.

51. $(1.2)^2$
52. $(2.3)^2$
53. $(-1.3)^2$
54. $(-2.5)^2$

Evaluate each expression.

55. $-4.6(23.4 - 19.6)$
56. $6.9(9.8 - 8.9)$
57. $(-0.2)^2 + 2(7.1)$
58. $(-6.3)(3) - (1.2)^2$
59. $(-0.7 - 0.5)(2.4 - 3.1)$
60. $(-8.1 - 7.8)(0.3 + 0.7)$
61. $(0.5 + 0.6)^2(-3.2)$
62. $(-5.1)(4.9 - 3.4)^2$
63. $|-2.6| \cdot |-7.2|$
64. $4|-3.1| + 5|-5.5|$
65. $|-2.6 - 6.7|^2$
66. $-3|-8.16 + 9.9|$

Evaluate each expression.

67. $3.14 + 2(d - t)$ for $d = 1.2$ and $t = -6.7$
68. $-8h^2 - rh$ for $r = 2.1$ and $h = -0.02$
69. $t + 0.5rt^2$ for $t = -0.4$ and $r = 100$
70. $1{,}000(x - y)(x + y)$ for $x = 9.8$ and $y = 1.3$
71. $10|a^2 - b^2|$ for $a = -1.1$ and $b = 2.2$
72. $t|r| + t|s|$ for $r = -0.021$, $s = -0.016$, and $t = 100$

APPLICATIONS

73. CONCERT SEATING Two types of tickets were sold for a concert. Floor seating cost \$12.50 a ticket, and balcony seats were \$15.75.
    a. Complete the following table and find the receipts from each type of ticket.
    b. Find the total receipts from the sale of both types of tickets.

Ticket type	Price	Number sold	Receipts
Floor		1,000	
Balcony		100	

74. CITY PLANNING In the city map on the next page, the streets form a grid. The lines are 0.35 mile apart. Find the distance of each trip.
    a. The airport to the Convention Center
    b. City Hall to the Convention Center
    c. The airport to City Hall

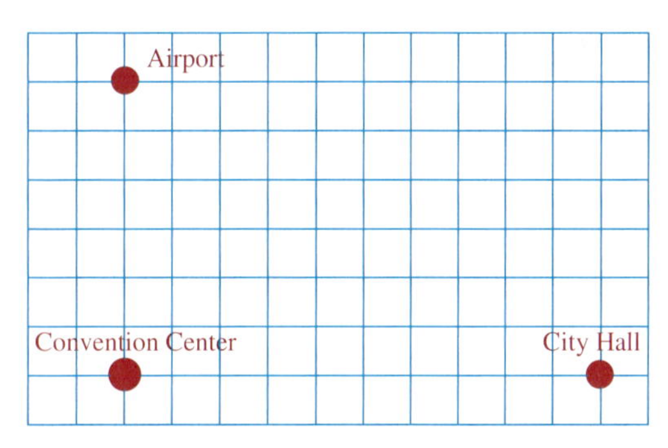

75. STORM DAMAGE After a rainstorm, the saturated ground under a hilltop house began to give way. A survey team noted that the house dropped 0.57 inch initially. In the next two weeks, the house fell 0.09 inch per week. How far did the house fall during this three-week period?

76. WATER USAGE In May, the water level of a reservoir reached its high mark for the year. During the summer months, as water usage increased, the level dropped. In the months of May and June, it fell 4.3 feet each month. In August, because of high temperatures, it fell another 8.7 feet. By September, how far below the year's high mark had the water level fallen?

77. WEIGHTLIFTING The barbell is evenly loaded with iron plates. How much plate weight is loaded on the barbell?

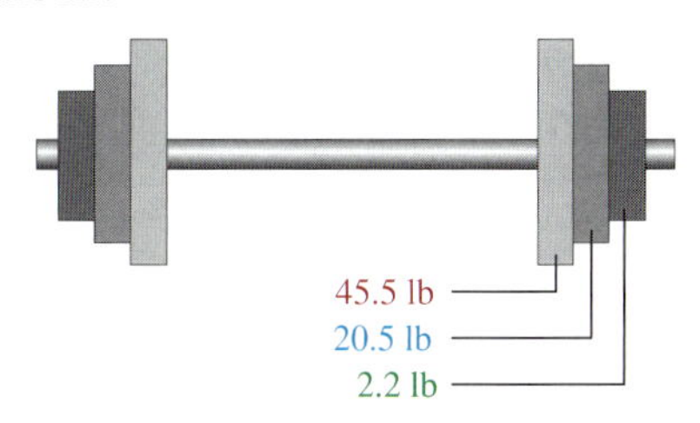

78. PLUMBING BILLS In the following illustration, an invoice for plumbing work is torn. What is the charge for the 4 hours of work? What is the total charge?

Carter Plumbing 100 W. Dalton Ave.		Invoice #210
Standard service charge 4 hr @ $40.55/hr		$25.75
Total		

79. BAKERY SUPPLIES A bakery buys various types of nuts as ingredients for cookies. Complete the table by filling in the cost of each purchase.

Type of nut	Price per pound	Pounds	Cost
Almonds	$5.95	16	
Walnuts	$4.95	25	
Peanuts	$3.85	x	

80. RETROFITS The illustration shows the width of the three columns of an existing freeway overpass. A computer analysis indicates that each column needs to be increased in width by a factor of 1.4 to ensure stability during an earthquake. According to the analysis, how wide should each of the columns be?

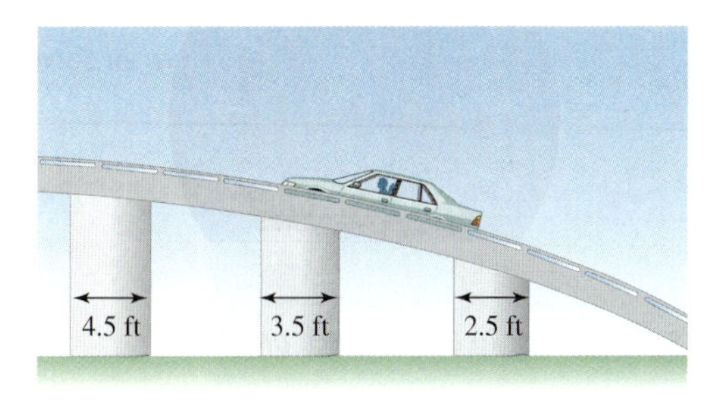

81. POOL CONSTRUCTION Long bricks, called *coping,* can be used to outline the edge of a swimming pool. How many meters of coping will be needed in the construction of the swimming pool shown?

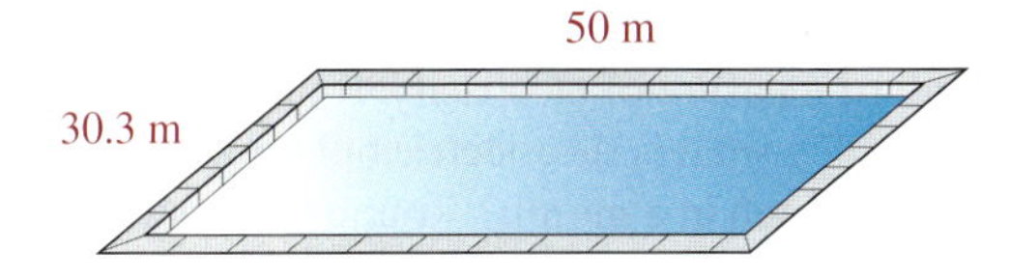

82. SOCCER A soccer goal measures 24 feet wide by 8 feet high. Major league soccer officials are proposing to increase its width by 1.5 feet and increase its height by 0.75 foot.

a. What is the area of the goal opening now?

b. What would it be if their proposal is adopted?

c. How much area would be added?

83. BIOLOGY DNA is found in cells. It is referred to as the genetic "blueprint." In humans, it determines such traits as eye color, hair color, and height. A model of DNA appears on the right. If Å = 0.000000004 inch, determine the three dimensions shown in the illustration.

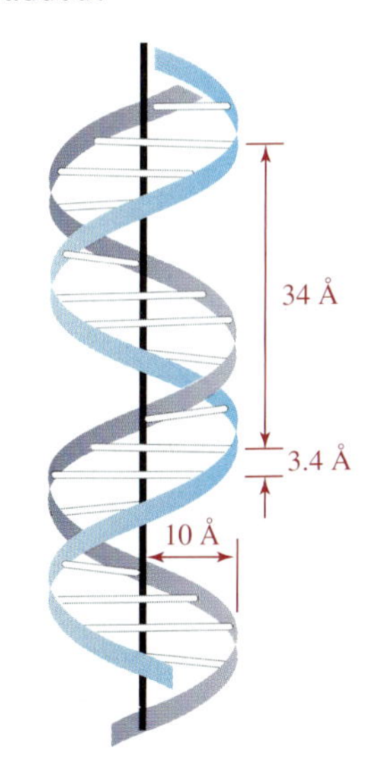

6. How many more cubic feet of storage do you get with the deluxe model compared to the economy model?

7. Three roommates are planning on purchasing the deluxe model and splitting the cost evenly. How much will each have to pay?

8. What is the energy cost per year to run the deluxe model?

9. If you make a $220 down payment on the standard model, how much of the cost is left to finance?

10. The economy model can be expected to last for 10 years. What would be the total energy cost over that period?

Estimate the answer to each problem. Does the result on the calculator display seem reasonable?

11. 25.9 + 345.1 + 0.09 | 347.78 |

12. 8,345.889 − 345.6 | 8000.289 |

13. 42,090.8 + 3,303.09 | 45393.89 |

14. 10.007 − 0.626 | 3.747 |

15. 9.8(8.8) | 86.24 |

16. $\frac{24.56}{2.2}$ | 1.116363636 |

17. 53 · 5.61 | 241.23 |

18. 89.11 ÷ 22.707 | 39.24340 |

5.5 Fractions and Decimals

- Writing fractions as equivalent decimals
- Repeating decimals
- Rounding repeating decimals
- Graphing fractions and decimals
- Problems involving fractions and decimals

In this section, we will further investigate the relationship between fractions and decimals.

Writing fractions as equivalent decimals

To write $\frac{5}{8}$ as a decimal, we use the fact that $\frac{5}{8}$ indicates the division 5 ÷ 8. We can convert $\frac{5}{8}$ to decimal form by performing the division.

```
     .625
  8)5.000    Write a decimal point and additional zeros to the right of 5.
    4 8
      20
      16
       40
       40
        0 ← The remainder is 0.
```

Thus, $\frac{5}{8} = 0.625$.

Writing a fraction as a decimal

To write a fraction as a decimal, divide the numerator of the fraction by its denominator.

EXAMPLE 1 Write $\frac{3}{4}$ as a decimal.

Solution We divide the numerator by the denominator.

$$\begin{array}{r} .75 \\ 4\overline{)3.00} \\ \underline{2\,8} \\ 20 \\ \underline{20} \\ 0 \end{array}$$

Write a decimal point and two zeros to the right of 3.

0 ← The remainder is 0.

Thus, $\frac{3}{4} = 0.75$.

Self Check 1

Write $\frac{3}{16}$ as a decimal.

Answer 0.1875

In Example 1, the division process ended because a remainder of zero was obtained. In this case, we call the quotient, 0.75, a **terminating decimal.**

Repeating decimals

Sometimes, when we are finding a decimal equivalent of a fraction, the division process never gives a remainder of zero. In this case, the result is a **repeating decimal.** Examples of repeating decimals are 0.4444. . . and 1.373737. . . . The three dots tell us that a block of digits repeats in the pattern shown. Repeating decimals can be written using a bar over the repeating block of digits. For example, 0.4444. . . can be written as $0.\overline{4}$, and 1.373737. . . can be written as $1.\overline{37}$.

! COMMENT When using an overbar to write a repeating decimal, use the fewest digits necessary to show the repeating block of digits.

$0.333\ldots = 0.\overline{333}$ $\qquad$ $6.7454545\ldots = 6.7\overline{454}$

$0.333\ldots = 0.\overline{3}$ $\qquad$ $6.7454545\ldots = 6.7\overline{45}$

EXAMPLE 2 Write $\frac{5}{12}$ as a decimal.

Solution We use division to find the decimal equivalent.

$$\begin{array}{r} .4166 \\ 12\overline{)5.0000} \\ \underline{4\,8} \\ 20 \\ \underline{12} \\ 80 \\ \underline{72} \\ 80 \\ \underline{72} \\ 8 \end{array}$$

Write a decimal point and four zeros to the right of 5.

It is apparent that 8 will continue to reappear as the remainder. Therefore, 6 will continue to reappear in the quotient. Since the repeating pattern is now clear, we can stop the division.

Thus, $\frac{5}{12} = 0.41\overline{6}$.

Self Check 2

Write $\frac{4}{11}$ as a decimal.

Answer $0.\overline{36}$

Every fraction can be written as either a terminating decimal or a repeating decimal. For this reason, the set of fractions (**rational numbers**) forms a subset of the set of decimals called the set of **real numbers.** The set of real numbers corresponds to *all* points on a number line.

Not all decimals are terminating or repeating decimals. For example,

0.2020020002 . . .

does not terminate, and it has no repeating block of digits. This decimal cannot be written as a fraction with an integer numerator and a nonzero integer denominator. Thus, it is not a rational number. It is an example from the set of **irrational numbers.**

Rounding repeating decimals

When a fraction is written in decimal form, the result is either a terminating or a repeating decimal. Repeating decimals are often rounded to a specified place value.

EXAMPLE 3 Write $\frac{1}{3}$ as a decimal and round to the nearest hundredth.

Solution First, we divide the numerator by the denominator to find the decimal equivalent of $\frac{1}{3}$.

$$\begin{array}{r} 0.333 \\ 3\overline{)1.000} \\ \underline{9} \\ 10 \\ \underline{9} \\ 10 \\ \underline{9} \\ 1 \end{array}$$

Write a decimal point and additional zeros to the right of 1.

We see that the division process never gives a remainder of zero. When we write $\frac{1}{3}$ in decimal form, the result is the repeating decimal $0.333\ldots = 0.\overline{3}$.

To find a decimal approximation of $\frac{1}{3}$ to the nearest hundredth, we proceed as follows.

0.333. . .

Round 0.333 to the nearest hundredth by examining the test digit in the thousandths column.

Since 3 is less than 5, we round down, and we have

$\frac{1}{3} \approx 0.33$

Read $\approx$ as "is approximately equal to."

Self Check 4
Write $\frac{7}{24}$ as a decimal and round to the nearest thousandth.

EXAMPLE 4 Write $\frac{2}{7}$ as a decimal and round to the nearest thousandth.

Solution

$$\begin{array}{r} .2857 \\ 7\overline{)2.0000} \\ \underline{1\,4} \\ 60 \\ \underline{56} \\ 40 \\ \underline{35} \\ 50 \\ \underline{49} \\ 1 \end{array}$$

Write a decimal point and additional zeros to the right of 2.

To round to the thousandths column, we must divide to the ten thousandths column.

0.2857

Round 0.2857 to the nearest thousandth by examining the test digit in the ten thousandths column.

Since 7 is greater than 5, we round up, and $\frac{2}{7} \approx 0.286$.

Answer 0.292

The fixed-point key

CALCULATOR SNAPSHOT

After performing a calculation, a scientific calculator can round the result to a given decimal place. This is done using the *fixed-point key.* As we did in Example 4, let's find the decimal equivalent of $\frac{2}{7}$ and round to the nearest thousandth. This time, we will use a calculator.

First, we set the calculator to round to the third decimal place (thousandths) by pressing FIX 3. Then we press

2 ÷ 7 = 0.286

Thus, $\frac{2}{7} \approx 0.286$. To round to the nearest tenth, we would fix 1; to round to the nearest hundredth, we would fix 2, and so on.

If your calculator does not have a fixed-point key, see the owner's manual.

EXAMPLE 5 Write $5\frac{3}{8}$ in decimal form.

Solution To write a mixed number in decimal form, recall that a mixed number is made up of a whole-number part and a fractional part. Since we can write $5\frac{3}{8}$ as $5 + \frac{3}{8}$, we need only consider how to write $\frac{3}{8}$ as a decimal.

$$\begin{array}{r} .375 \\ 8\overline{)3.000} \\ \underline{2\,4} \\ 60 \\ \underline{56} \\ 40 \\ \underline{40} \\ 0 \end{array}$$

Write a decimal point and three zeros to the right of 3.

Thus, $5\frac{3}{8} = 5 + \frac{3}{8} = 5 + 0.375 = 5.375$. We would obtain the same result if we changed $5\frac{3}{8}$ to the improper fraction $\frac{43}{8}$ and divided 43 by 8.

Self Check 5

Write $8\frac{19}{20}$ in decimal form.

Answer 8.95

Graphing fractions and decimals

The number line can be used to show the relationship between fractions and their respective decimal equivalents. Figure 5-8 shows some commonly used fractions that have terminating decimal equivalents. For example, we see from the graph that $\frac{13}{16} = 0.8125$.

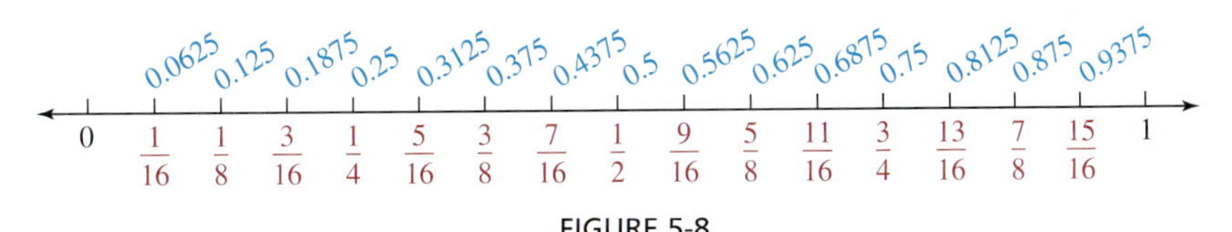

FIGURE 5-8

The number line in Figure 5-9 shows some commonly used fractions that have repeating decimal equivalents.

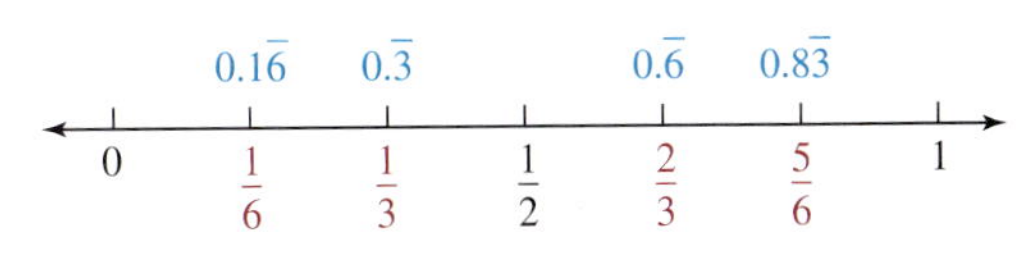

FIGURE 5-9

Problems involving fractions and decimals

Numerical expressions can contain both fractions and decimals. In the following examples, we show how different methods can be used to solve problems of this type.

Self Check 6
Evaluate by working in terms of fractions: $0.53 - \frac{1}{6}$.

Answer $\frac{109}{300}$

EXAMPLE 6 Evaluate $\frac{1}{3} + 0.27$ by working in terms of fractions.

Solution We write 0.27 as a fraction and add it to $\frac{1}{3}$.

$$\frac{1}{3} + 0.27 = \frac{1}{3} + \frac{27}{100}$$ Replace 0.27 with $\frac{27}{100}$.

$$= \frac{1 \cdot \mathbf{100}}{3 \cdot \mathbf{100}} + \frac{27 \cdot \mathbf{3}}{100 \cdot \mathbf{3}}$$ The LCD for $\frac{1}{3}$ and $\frac{27}{100}$ is 300. Express each fraction in terms of 300ths.

$$= \frac{100}{300} + \frac{81}{300}$$ Multiply in the numerators and in the denominators.

$$= \frac{181}{300}$$ Add the numerators and write the sum over the common denominator, 300.

Self Check 7
Evaluate by working in terms of decimals: $0.53 - \frac{1}{6}$.

Answer 0.36

EXAMPLE 7 Evaluate $\frac{1}{3} + 0.27$ by working in terms of decimals.

Solution We have seen that the decimal equivalent of $\frac{1}{3}$ is the repeating decimal 0.333. . . . To add $\frac{1}{3}$ to 0.27, we round 0.333. . . to the nearest hundredth: $\frac{1}{3} \approx 0.33$.

$$\frac{1}{3} + 0.27 \approx 0.33 + 0.27$$ Approximate $\frac{1}{3}$ with the decimal 0.33.

$$\approx 0.60$$ Perform the addition.

In the previous two examples, we evaluated $\frac{1}{3} + 0.27$ in different ways. In Example 6, we obtained the exact answer, $\frac{181}{300}$. In Example 7, we obtained an approximation, 0.60. It is apparent that the results are in agreement when we write $\frac{181}{300}$ in decimal form: $\frac{181}{300} = 0.60333. . . .$

Self Check 8
Perform the operations: $(-0.6)^2 + (2.3)\left(\frac{1}{8}\right)$.

EXAMPLE 8 Perform the operations: $\left(\frac{4}{5}\right)(1.35) + (0.5)^2$.

Solution It appears simplest to work in terms of decimals. We use division to find the decimal equivalent of $\frac{4}{5}$.

$$\begin{array}{r} .8 \\ 5\overline{)4.0} \\ \underline{4\,0} \\ 0 \end{array}$$ Write a decimal point and one zero to the right of the 4.

Now we use the rules for the order of operations to evaluate the given expression.

$\left(\frac{4}{5}\right)(1.35) + (0.5)^2 = (\mathbf{0.8})(1.35) + (0.5)^2$	Replace $\frac{4}{5}$ with its decimal equivalent, 0.8.
$= (0.8)(1.35) + 0.25$	Find the power: $(0.5)^2 = 0.25$.
$= 1.08 + 0.25$	Perform the multiplication: $(0.8)(1.35) = 1.08$.
$= 1.33$	Perform the addition.

Answer 0.6475

EXAMPLE 9 **Shopping.** During a trip to the grocery store, a shopper purchased $\frac{3}{4}$ pound of fruit, priced at \$0.88 a pound, and $\frac{1}{3}$ pound of fresh-ground coffee, selling for \$6.60 a pound. Find the total cost of these items.

Solution To find the cost of each item, we multiply the amount purchased by its unit price. Then we add the two individual costs to obtain the total cost.

Cost of fruit	plus	cost of coffee	equals	total cost.
$\left(\frac{3}{4}\right)(0.88)$	$+$	$\left(\frac{1}{3}\right)(6.60)$	$=$	total cost

Because 0.88 is divisible by 4 and 6.60 is divisible by 3, we can work with the decimals and fractions in this form; no conversion is necessary.

$\left(\frac{3}{4}\right)(0.88) + \left(\frac{1}{3}\right)(6.60) = \left(\frac{3}{4}\right)\left(\frac{0.88}{1}\right) + \left(\frac{1}{3}\right)\left(\frac{6.60}{1}\right)$	Express 0.88 as $\frac{0.88}{1}$ and 6.60 as $\frac{6.60}{1}$.
$= \frac{2.64}{4} + \frac{6.60}{3}$	Multiply the numerators and the denominators.
$= 0.66 + 2.20$	Perform each division.
$= 2.86$	Perform the addition.

The total cost of the items is \$2.86.

Section 5.5 STUDY SET

VOCABULARY *Fill in the blanks.*

1. The decimal form of the fraction $\frac{1}{3}$ is a ________ decimal, which is written $0.\overline{3}$ or 0.3333. . . .

2. The decimal form of the fraction $\frac{2}{5}$ is a ________ decimal, which is written 0.4.

3. The ________ equivalent of $\frac{1}{16}$ is 0.0625.

4. We read $\approx$ as "is ________ equal to."

CONCEPTS

5. **a.** What division is indicated by $\frac{7}{8}$?
   b. Fill in the blank: To write a fraction as a decimal, divide the ________ of the fraction by its denominator.

6. Insert the proper symbol $<$ or $>$ in the blank.
   a. $0.\overline{6}$ ☐ 0.7 **b.** $0.\overline{6}$ ☐ 0.6

7. When we round 0.272727. . . to the nearest hundredth, is the result larger or smaller than the original number?

8. Write each decimal in fraction form.

a. 0.7 **b.** 0.77

9. Graph: $1\frac{3}{4}$, -0.75, $0.\overline{6}$, and $-3.8\overline{3}$.

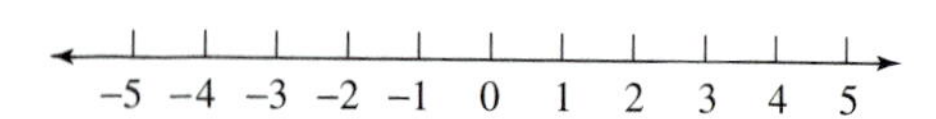

10. Graph: $2\frac{7}{8}$, -2.375, $0.\overline{3}$, and $4.1\overline{6}$.

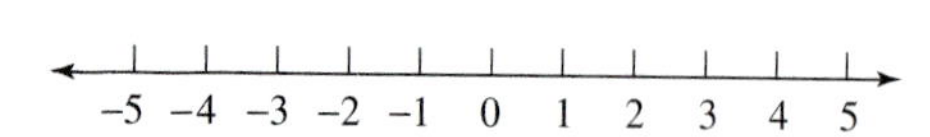

11. Determine whether each statement is true or false.

a. $\frac{1}{3} = 0.3$ **b.** $\frac{3}{4} = 0.75$

c. $20\frac{1}{2} = 20.5$ **d.** $\frac{1}{16} = 0.1\overline{6}$

12. When evaluating the expression $0.25 + \left(2.3 + \frac{2}{5}\right)^2$, would it be easier to work in terms of fractions or in terms of decimals?

NOTATION

13. Examine the color portion of the long division below.

a. Will the remainder ever be zero?

b. What can be deduced about the decimal equivalent of $\frac{5}{6}$?

$$\begin{array}{r} .833 \\ 6\overline{)5.000} \\ \underline{4\,8} \\ 20 \\ \underline{18} \\ 20 \end{array}$$

14. Write each repeating decimal using an overbar.

a. 0.888. . . **b.** 0.323232. . .

c. 0.56333. . . **d.** 0.8898989. . .

PRACTICE

Write each fraction in decimal form.

15. $\frac{1}{2}$ **16.** $\frac{1}{4}$

17. $-\frac{5}{8}$ **18.** $-\frac{3}{5}$

19. $\frac{9}{16}$ **20.** $\frac{3}{32}$

21. $-\frac{17}{32}$ **22.** $-\frac{15}{16}$

$\frac{11}{[illegible]}$ **24.** $\frac{19}{25}$

25. $\frac{31}{40}$ **26.** $\frac{17}{20}$

27. $-\frac{3}{200}$ **28.** $-\frac{21}{50}$

29. $\frac{1}{500}$ **30.** $\frac{1}{250}$

Write each fraction in decimal form. Use an overbar.

31. $\frac{2}{3}$ **32.** $\frac{7}{9}$

33. $\frac{5}{11}$ **34.** $\frac{4}{15}$

35. $-\frac{7}{12}$ **36.** $-\frac{17}{22}$

37. $\frac{1}{30}$ **38.** $\frac{1}{60}$

Write each fraction in decimal form. Round to the nearest hundredth.

39. $\frac{7}{30}$ **40.** $\frac{14}{15}$

41. $\frac{17}{45}$ **42.** $\frac{8}{9}$

Write each fraction in decimal form. Round to the nearest thousandth.

43. $\frac{5}{33}$ **44.** $\frac{5}{12}$

45. $\frac{10}{27}$ **46.** $\frac{17}{21}$

Write each fraction in decimal form. Round to the nearest hundredth.

47. $\frac{4}{3}$ **48.** $\frac{10}{9}$

49. $-\frac{34}{11}$ **50.** $-\frac{25}{12}$

Write each mixed number in decimal form. Round to the nearest hundredth when the result is a repeating decimal.

51. $3\frac{3}{4}$ **52.** $5\frac{4}{5}$

53. $-8\frac{2}{3}$ **54.** $-1\frac{7}{9}$

55. $12\frac{11}{16}$ **56.** $32\frac{1}{8}$

57. $203\frac{11}{15}$ **58.** $568\frac{23}{30}$

Fill in the correct symbol (< or >) to make a true statement. (Hint: Express each number as a decimal.)

59. $\frac{7}{8}$ ▢ 0.895

60. 4.56 ▢ $4\frac{2}{5}$

61. $-\frac{11}{20}$ ▢ $-0.\overline{4}$

62. $-9.\overline{09}$ ▢ $-9\frac{1}{11}$

Evaluate each expression. Work in terms of fractions.

63. $\frac{1}{9} + 0.3$

64. $\frac{2}{3} + 0.1$

65. $0.9 - \frac{7}{12}$

66. $0.99 - \frac{5}{6}$

67. $\frac{5}{11}(0.3)$

68. $(0.9)\left(\frac{1}{27}\right)$

69. $\frac{1}{3}\left(-\frac{1}{15}\right)(0.5)$

70. $(-0.4)\left(\frac{5}{18}\right)\left(-\frac{1}{3}\right)$

71. $\frac{1}{4}(0.25) + \frac{15}{16}$

72. $\frac{2}{5}(0.02) - (0.04)$

Evaluate each expression to the nearest hundredth.

73. $0.24 + \frac{1}{3}$

74. $0.02 + \frac{5}{6}$

75. $5.69 - \frac{5}{12}$

76. $3.19 - \frac{2}{3}$

77. $\frac{3}{4}(0.43) - \frac{1}{12}$

78. $-\frac{2}{5}(0.33) + 0.45$

Evaluate each expression. Work in terms of decimals.

79. $(3.5 + 6.7)\left(-\frac{1}{4}\right)$

80. $\left(-\frac{5}{8}\right)(5.3 - 3.9)$

81. $\left(\frac{1}{5}\right)^2(1.7)$

82. $(2.35)\left(\frac{2}{5}\right)^2$

83. $7.5 - (0.78)\left(\frac{1}{2}\right)$

84. $8.1 - \left(\frac{3}{4}\right)(0.12)$

85. $\frac{3}{8}(-3.2) + (4.5)\left(-\frac{1}{9}\right)$

86. $(-0.8)\left(\frac{1}{4}\right) + \left(\frac{1}{3}\right)(0.39)$

Evaluate each expression. Round to the nearest hundredth.

87. $\dfrac{3\frac{1}{5} + 2\frac{1}{2}}{5.69} + 3\frac{1}{4}$

88. $4\frac{2}{3} - \dfrac{2.7 - \frac{7}{8}}{0.12}$

Evaluate each algebraic expression.

89. $\frac{4}{3}pr^3$ for $p = 3.14$ and $r = 3$

90. $\frac{1}{3}pr^2h$ for $p = 3.14$, $r = 6$, and $h = 12$

Write each fraction in decimal form.

91. $\frac{23}{101}$

92. $\frac{1}{99}$

93. $\frac{2{,}046}{55}$

94. $-\frac{11}{128}$

APPLICATIONS

95. DRAFTING The architect's scale below has several measuring edges. The edge marked 16 divides each inch into 16 equal parts. Find the decimal form for each fractional part of 1 inch that is highlighted on the scale.

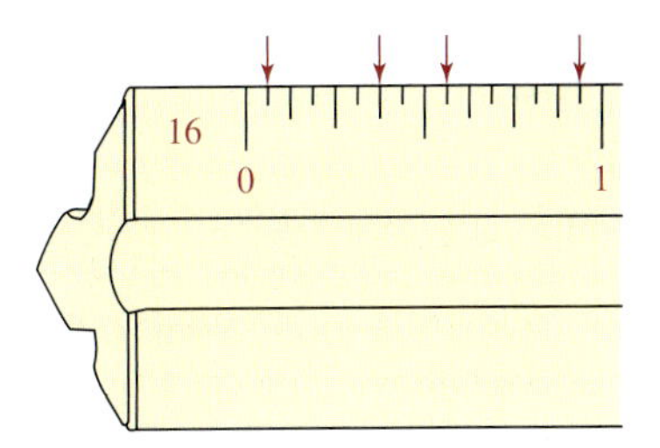

96. FREEWAY SIGNS The freeway sign below gives the number of miles to the next three exits. Convert the mileages to decimal notation.

BARRANCA AVE.	$\frac{3}{4}$ mi
210 FREEWAY	$2\frac{1}{4}$ mi
ADA ST.	$3\frac{1}{2}$ mi

97. GARDENING Two brands of replacement line for a lawn trimmer are labeled in different ways. On one package, the line's thickness is expressed as a decimal; on the other, as a fraction. Which line is thicker?

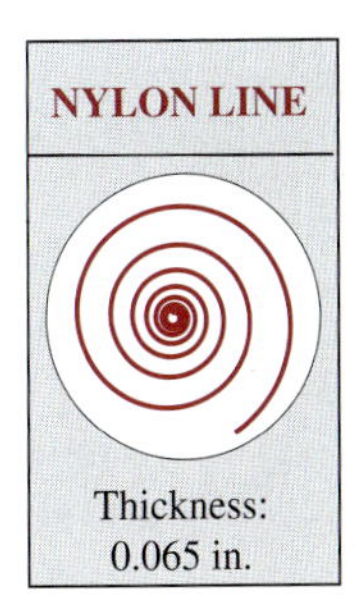

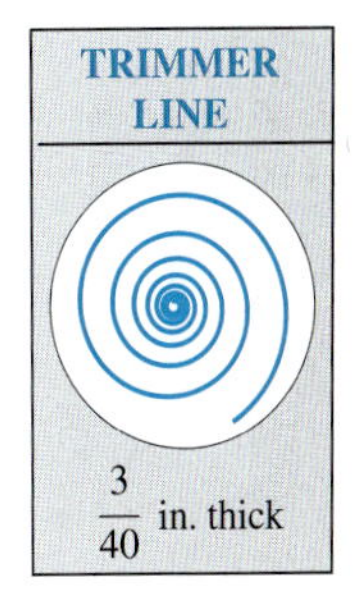

98. AUTO MECHANICS While doing a tuneup, a mechanic checks the gap on one of the spark plugs of a car to be sure it is firing correctly. The owner's manual states that the gap should be $\frac{2}{125}$ inch. The gauge the mechanic uses to check the gap is in decimal notation; it registers 0.025 inch. Is the spark plug gap too large or too small?

99. HORSE RACING In thoroughbred racing, the time a horse takes to run a given distance is measured using fifths of a second. For example, 55^2 (read "fifty-five and two") means $55\frac{2}{5}$ seconds. The illustration lists four split times for a horse. Express the times in decimal form.

Speedy Flight	Turfway Park, Ky	3-year-old	
17 May 97	$1\frac{1}{16}$ mile	:23[2] :23[4] :24[1] :32[3]	

100. GEOLOGY A geologist weighed a rock sample at the site where it was discovered and found it to weigh $17\frac{7}{8}$ lb. Later, a more accurate digital scale in the laboratory gave the weight as 17.671 lb. What is the difference in the two measurements?

101. WINDOW REPLACEMENTS The amount of sunlight that comes into a room depends on the area of the windows in the room. What is the area of the window shown below?

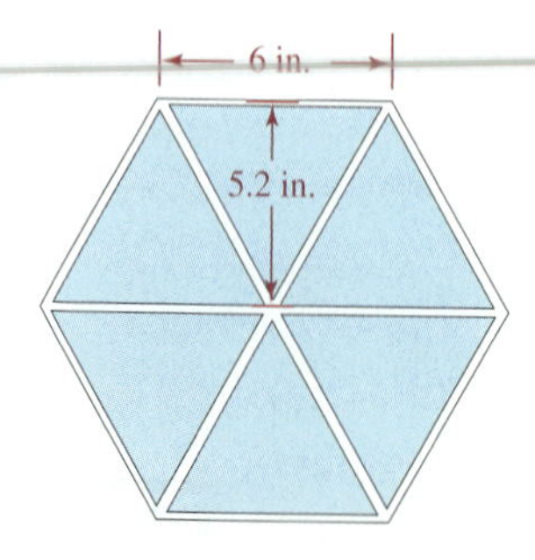

102. FIRE CONTAINMENT A command post asked each of three fire crews to estimate the length of the fire line they were fighting. Their reports came back in different forms, as indicated in the illustration. Find the perimeter of the fire.

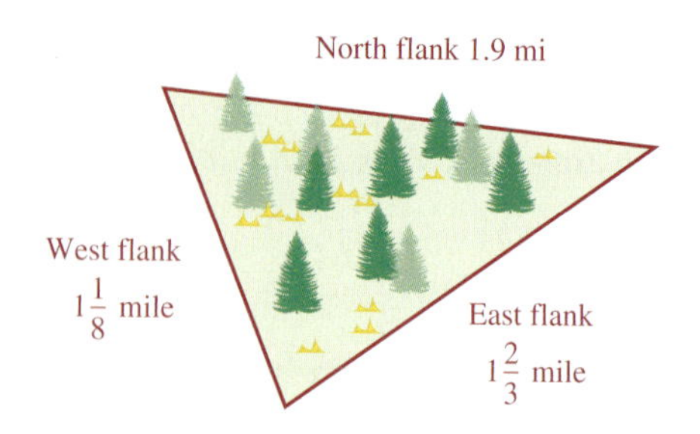

WRITING

103. Explain the procedure used to write a fraction in decimal form.

104. Compare and contrast the two numbers 0.5 and $0.\overline{5}$.

105. A student represented the repeating decimal 0.1333. . . as $0.\overline{1333}$. Is this correct? Explain why or why not.

106. Is 0.10100100010000 . . . a repeating decimal? Explain why or why not.

REVIEW

107. Add: $-2 + (-3) + 10 + (-6)$.

108. Evaluate: $-3 + 2[-3 + (2 - 7)]$.

109. Simplify: $3T - 4T + 2(-4t)$.

110. In the expression $6x^2 + 3x + 7$, what is the coefficient of the second term?

111. Simplify: $4x^2 + 2x^2$.

112. Two pieces of pipe have lengths x and $x + 10$. Which is the longer piece of pipe?

5.6 Solving Equations Containing Decimals

- Solving equations using the properties of equality
- Simplifying expressions to solve equations
- Problem solving with equations

We have studied how to add, subtract, multiply, and divide decimals. We will now use these skills to solve equations containing decimals. We will use the following strategy to solve these equations.

Strategy for solving equations

To simplify the equation:

1. Clear the equation of any fractions.
2. Use the distributive property to remove parentheses.
3. Combine like terms on either side of the equation.

To isolate the variable:

4. Apply the addition and subtraction properties of equality to get the variables on one side of the equation and the constant terms on the other.
5. Continue to combine like terms when necessary.
6. Undo the operations of multiplication and division to isolate the variable.

Solving equations using the properties of equality

Recall that the addition and subtraction properties of equality allow us to add the same number to or subtract the same number from both sides of an equation.

EXAMPLE 1 Solve each equation: **a.** $x + 3.5 = 7.8$ and **b.** $y - 1.23 = -4.52$.

Solution

a. To isolate x, we undo the addition of 3.5 by subtracting 3.5 from both sides of the equation.

$$x + 3.5 = 7.8$$

$$x + 3.5 - 3.5 = 7.8 - 3.5 \quad \text{Subtract 3.5 from both sides.}$$

$$x = 4.3 \quad \text{Simplify.}$$

b. To isolate y, we undo the subtraction of 1.23 by adding 1.23 to both sides of the equation.

$$y - 1.23 = -4.52$$

$$y - 1.23 + 1.23 = -4.52 + 1.23 \quad \text{Add 1.23 to both sides.}$$

$$y = -3.29 \quad \text{Simplify.}$$

Verify each result.

Self Check 1

Solve each equation:

a. $4.6 + t = 15.7$

b. $-1.24 = r - 0.04$

Answers **a.** 11.1, **b.** −1.2

The multiplication property of equality states that we can multiply both sides of an equation by the same nonzero number.

EXAMPLE 2 Solve: $\frac{m}{2} = -24.8$.

Solution To isolate m, we undo the division by 2 by multiplying both sides of the equation by 2.

$$\frac{m}{2} = -24.8$$

$$2\left(\frac{m}{2}\right) = 2(-24.8) \quad \text{Multiply both sides by 2.}$$

$$m = -49.6 \quad \text{Perform the multiplications.}$$

Check the result.

Self Check 2

Solve: $\frac{y}{3} = -13.11$.

Answer −39.33

The division property of equality states that we can divide both sides of an equation by the same nonzero number.

Self Check 3
Solve: $-22.32 = -3.1m$.

Answer 7.2

EXAMPLE 3 Solve: $-4.6x = -9.66$.

Solution To isolate x, we undo the multiplication by -4.6 by dividing by -4.6.

$$-4.6x = -9.66$$

$$\frac{-4.6x}{-4.6} = \frac{-9.66}{-4.6} \quad \text{Divide both sides by } -4.6.$$

$$x = 2.1 \quad \text{Perform the divisions.}$$

Check the result.

Sometimes, more than one property must be used to solve an equation. In the next example, we use the addition property of equality and the division property of equality.

Self Check 4
Solve: $-4.2h + 3.14 = 1.88$.

Answer 0.3

EXAMPLE 4 Solve: $8.1y - 6.04 = -13.33$.

Solution The left-hand side involves a multiplication and a subtraction. To solve the equation, we must undo these operations, but in the opposite order. We begin by undoing the subtraction.

$$8.1y - 6.04 = -13.33$$

$$8.1y - 6.04 + 6.04 = -13.33 + 6.04 \quad \text{To undo the subtraction of 6.04, add 6.04 to both sides.}$$

$$8.1y = -7.29 \quad \text{Simplify: } -13.33 + 6.04 = -7.29.$$

$$\frac{8.1y}{8.1} = \frac{-7.29}{8.1} \quad \text{To undo the multiplication of 8.1, divide both sides by 8.1.}$$

$$y = -0.9 \quad \text{Perform the divisions.}$$

Check:

$$8.1y - 6.04 = -13.33$$

$$8.1(-0.9) - 6.04 \stackrel{?}{=} -13.33 \quad \text{Substitute } -0.9 \text{ for } y.$$

$$-7.29 - 6.04 \stackrel{?}{=} -13.33 \quad \text{Perform the multiplication: } 8.1(-0.9) = -7.29.$$

$$-13.33 = -13.33 \quad \text{Perform the subtraction by adding the opposite: } -7.29 - 6.04 = -7.29 + (-6.04) = -13.33.$$

The solution is -0.9.

Simplifying expressions to solve equations

Recall that to *combine like terms* means to simplify the sum (or difference) of like terms.

Self Check 5
Simplify: $4.06a - 6.71 - 3.04a$.

Answer $1.02a - 6.71$

EXAMPLE 5 Simplify: $9.9b + 5.4 - 2.6b$.

Solution This expression involves three terms. We can combine the two that are like terms.

$$9.9b + 5.4 - 2.6b = 7.3b + 5.4 \quad \text{Subtract: } 9.9 - 2.6 = 7.3. \text{ Keep the variable } b.$$

Sometimes we must combine like terms in order to isolate the variable and solve an equation.

EXAMPLE 6 Solve: $-22.46 + 3.2t + 1.9t = 52$.

Solution First, we combine the like terms on the left-hand side.

$$-22.46 + \mathbf{3.2t} + \mathbf{1.9t} = 52$$

$$-22.46 + \mathbf{5.1t} = 52 \quad \text{Combine like terms: } 3.2t + 1.9t = 5.1t.$$

$$-22.46 + 5.1t + \mathbf{22.46} = 52 + \mathbf{22.46} \quad \text{To eliminate } -22.46 \text{ from the left-hand side, add 22.46 to both sides.}$$

$$5.1t = 74.46 \quad \text{Simplify.}$$

$$\frac{5.1t}{\mathbf{5.1}} = \frac{74.46}{\mathbf{5.1}} \quad \text{To undo the multiplication by 5.1, divide both sides by 5.1.}$$

$$t = 14.6 \quad \text{Perform the divisions.}$$

Self Check 6
Solve:
$-1.9 + 2.8x - 1.4x = 12.24$.

Answer 10.1

EXAMPLE 7 Solve: $0.2s - 3 = 0.7s + 1.5$.

Solution We isolate the variable terms on the right-hand side and isolate the constant terms on the left-hand side of the equation.

$$0.2s - 3 = 0.7s + 1.5$$

$$0.2s - 3 - \mathbf{0.2s} = 0.7s + 1.5 - \mathbf{0.2s} \quad \text{Eliminate } 0.2s \text{ from the left-hand side by subtracting } 0.2s \text{ from both sides.}$$

$$-3 = 0.5s + 1.5 \quad \text{Combine like terms.}$$

$$-3 - \mathbf{1.5} = 0.5s + 1.5 - \mathbf{1.5} \quad \text{To undo the addition of 1.5, subtract 1.5 from both sides.}$$

$$-3 + (-1.5) = 0.5s \quad \text{On the left, write the subtraction as addition of the opposite. On the right, simplify.}$$

$$-4.5 = 0.5s \quad \text{Perform the addition: } -3 + (-1.5) = -4.5.$$

$$\frac{-4.5}{\mathbf{0.5}} = \frac{0.5s}{\mathbf{0.5}} \quad \text{To undo the multiplication by 0.5, divide both sides by 0.5.}$$

$$-9 = s \quad \text{Perform the divisions.}$$

Self Check 7
Solve:
$6.1b - 5.5 = 5.2b + 5.3$.

Answer 12

EXAMPLE 8 Solve: $5(x + 1.3) = -9.9$.

Solution First, we remove the parentheses by applying the distributive property.

$$5(x + 1.3) = -9.9$$

$$5x + 6.5 = -9.9 \quad \text{Distribute the 5: } 5 \cdot 1.3 = 6.5.$$

$$5x + 6.5 - \mathbf{6.5} = -9.9 - \mathbf{6.5} \quad \text{To undo the addition of 6.5, subtract 6.5 from both sides.}$$

$$5x = -9.9 + (-6.5) \quad \text{On the right-hand side, add the opposite of 6.5.}$$

$$5x = -16.4 \quad \text{Perform the addition.}$$

$$\frac{5x}{\mathbf{5}} = \frac{-16.4}{\mathbf{5}} \quad \text{To undo the multiplication by 5, divide both sides by 5.}$$

$$x = -3.28 \quad \text{Perform the division.}$$

Self Check 8
Solve: $2(4.1 + c) = -19.4$.

Answer −13.8

Problem solving with equations

EXAMPLE 9 **Business expenses.** A business decides to rent a copy machine (Figure 5-10) instead of buying one. Under the rental agreement, the company is charged \$65 per month plus 2¢ for every copy made. If the business has budgeted \$125 for copier expenses each month, how many copies can be made before exceeding the budget?

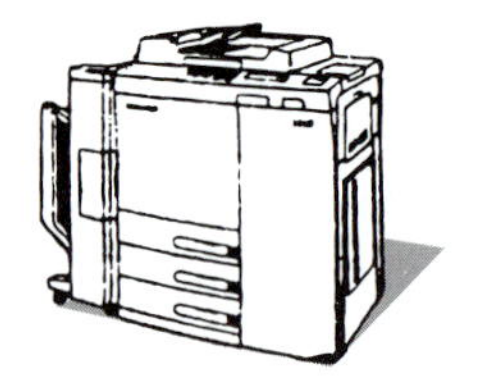

FIGURE 5-10

Analyze the problem

- The basic rental charge is \$65 a month.
- There is a 2¢ charge for each copy made.
- \$125 is budgeted for copier expenses each month.
- We must find the maximum number of copies that can be made each month.

Form an equation

Let x = the maximum number of copies that can be made. We can write the amount budgeted for copier expenses in two ways.

The basic fee	plus	the cost of the copies	is	the amount budgeted each month.

We can find the total cost of the copies by multiplying the cost per copy by the maximum number of copies that can be made. Notice that the costs are expressed in terms of dollars and cents. We need to work in terms of one unit, so we write 2¢ as \$0.02 and work in terms of dollars.

65	plus	0.02	·	the maximum number of copies made	is	125.
65	+	0.02	·	x	=	125

Solve the equation

$$65 + 0.02x = 125$$

$$65 + 0.02x - \mathbf{65} = 125 - \mathbf{65}$$ To undo the addition of 65, subtract 65 from both sides.

$$0.02x = 60$$ Simplify.

$$\frac{0.02x}{\mathbf{0.02}} = \frac{60}{\mathbf{0.02}}$$ To undo the multiplication by 0.02, divide both sides by 0.02.

$$x = 3{,}000$$ Perform the divisions.

State the conclusion

The business can make up to 3,000 copies each month without exceeding its budget.

Check the result

If we multiply the cost per copy and the maximum number of copies, we get $\$0.02 \cdot 3{,}000 = \$60$. Then we add the \$65 monthly fee: $\$60 + \$65 = \$125$. The answer checks.

Section 5.6 STUDY SET

VOCABULARY *Fill in the blanks.*

1. To ______ an equation, we isolate the variable on one side of the equals symbol.
2. $4.1(x + 3) = 4.1x + 4.1(3)$ is an example of the use of the __________ property.
3. In the term $5.65t$, the number 5.65 is called the __________.
4. A ________ is a letter that is used to stand for a number.

CONCEPTS

5. Show that 1.7 is a solution of $2.1x - 6.3 = -2.73$ by checking it.
6. Show that 0.04 is a solution of $\frac{y}{2} + 0.7 = 0.72$ by checking it.
7. For which problem below does the instruction *simplify* apply?

   $7.8x + 9.1 = 12.4$ or $7.8x + 9.1 + 12.4$

8. **a.** What operations are performed on the variable?

   $$\frac{m}{2.1} - 7.4 = 5.6$$

   b. In what order should the operations be undone to isolate the variable?
9. Write each amount of money as a dollar amount.
   a. 25 cents **c.** 250 cents
   b. 1 penny **d.** 99 cents
10. Why can't the expression $5.6A + 3.4a$ be simplified?
11. What algebraic concept is shown below?

    $3.1(6 - 0.3h)$

12. Rewrite each subtraction as addition of the opposite.
    a. $4.02 - (-1.7)$
    b. $y - (-0.6)$

NOTATION *Complete each solution to solve the equation.*

13. $$0.6s - 2.3 = -1.82$$
    $$0.6s - 2.3 + \square = -1.82 + \square$$
    $$\square = 0.48$$
    $$\frac{0.6s}{\square} = \frac{0.48}{\square}$$
    $$s = 0.8$$

14. $$\frac{x}{2} + 1 = -5.2$$
    $$\frac{x}{2} + 1 - \square = -5.2 - \square$$
    $$\square = -6.2$$
    $$2\left(\frac{x}{2}\right) = 2(\square)$$
    $$x = -12.4$$

PRACTICE *Combine like terms.*

15. $8.7x + 1.4x$
16. $45.1t + 38.6t$
17. $0.05h - 0.03h$
18. $67.89j - 54.73j$
19. $3.1r - 5.5r - 1.3r$
20. $3.8x - 6.5x - 2.4x$
21. $3.2 - 8.78x + 9.1$
22. $25.04 - 5.6w - 12.02$
23. $5.6x - 8.3 - 6.1x + 12.2$
24. $-17.3y - 8.01 + 12.2y - 4.4$
25. $0.05(100 - x) + 0.04x$
26. $0.06(1{,}000 - y) + 0.04y$

Solve each equation. Check the result.

27. $x + 8.1 = 9.8$
28. $6.75 + y = 8.99$
29. $7.08 = t - 0.03$
30. $14.1 = k - 13.1$
31. $-5.6 + h = -17.1$
32. $-0.05 + x = -1.25$
33. $7.75 = t - (-7.85)$
34. $3.33 = y - (-5.55)$
35. $2x = -8.72$
36. $3y = -12.63$
37. $-3.51 = -2.7x$
38. $-1.65 = -0.5f$
39. $\frac{x}{2.04} = -4$
40. $\frac{y}{2.22} = -6$
41. $\frac{-x}{5.1} = -4.4$
42. $\frac{-t}{8.1} = -3$
43. $\frac{1}{3}x = -7.06$
44. $\frac{1}{5}x = -3.02$
45. $\frac{x}{100} = 0.004$
46. $\frac{y}{1{,}000} = 0.0606$
47. $2x + 7.8 = 3.4$
48. $3x - 1.2 = -4.8$
49. $-0.8 = 5y + 9.2$
50. $-9.9 = 6t + 14.1$
51. $0.3x - 2.1 = 7.2$
52. $0.4a + 3.3 = -5.1$
53. $-1.5b + 2.7 = 1.2$
54. $-2.1x - 3.1 = 5.3$
55. $0.4a - 6 + 0.5a = -5.73$
56. $0.1t - 0.7t + 4 = 3.46$

57. $2(t - 4.3) + 1.2 = -6.2$
58. $3(y - 1.1) + 3.2 = 2.3$
59. $1.2x - 1.3 = 2.4x + 0.02$
60. $-4.4y - 1.3 = -5.1y - 5.08$
61. $53.7t - 10.1 = 46.3t + 4.7$
62. $37.1w + 12.2 = 16.8w + 93.4$
63. $2.1x - 4.6 = 7.3x - 11.36$
64. $4.1y + 5.7 = 6.4y + 0.87$
65. $0.06x + 0.09(100 - x) = 8.85$
66. $0.08(1{,}000 - x) + 0.6x = 72.72$

APPLICATIONS *Complete each solution.*

67. PETITION DRIVES On weekends, a college student works for a political organization, collecting signatures for a petition drive. Her pay is $15 a day plus 30 cents for each signature she obtains. How many signatures does she have to collect to make $60 a day?

Analyze the problem

- Her base pay is ____ dollars a day.
- She makes ____ cents for each signature.
- She wants to make ____ dollars a day.
- Find the number of ____ she needs to get.

Form an equation Let x = ______________________

We need to work in terms of the same units, so we write 30 cents as ____.

If we multiply the pay per signature by the number of signatures, we get the money she makes just from collecting signatures. Therefore, ____ = total amount (in dollars) made from collecting signatures.

We can express the money she earns in a day in two ways.

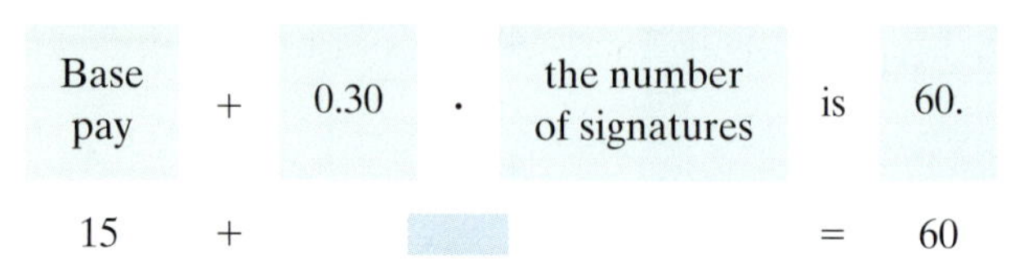

Solve the equation

15 + ____ = ____
____ = 45
$x = 150$

State the conclusion

____

Check the result

If she collects ____ signatures, she will make 0.30 · ____ = ____ dollars from signatures. If we add this to $15, we get $60. The answer checks.

68. HIGHWAY CONSTRUCTION A 12.8-mile highway is in its third and final year of construction. In the first year, 2.3 miles of the highway were completed. In the second year, 4.9 miles were finished. How many more miles of the highway need to be completed?

Analyze the problem

- The planned highway is ____ miles long.
- The 1st year, ____ miles were completed.
- The 2nd year, ____ miles were completed.
- Find the number of ____ yet to be completed.

A diagram will help us understand the problem.

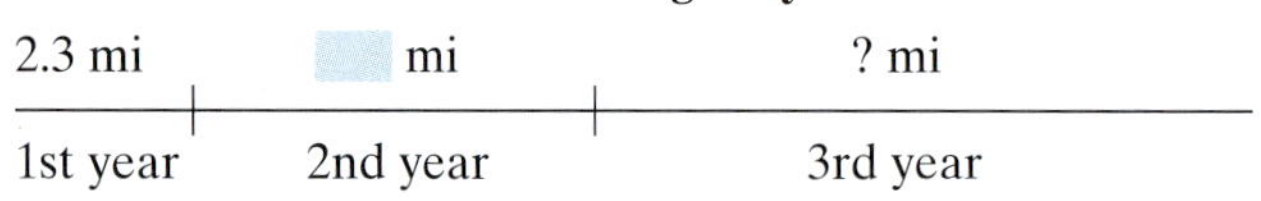

Form an equation

Let x = ______________________

We can express the length of the highway in two ways.

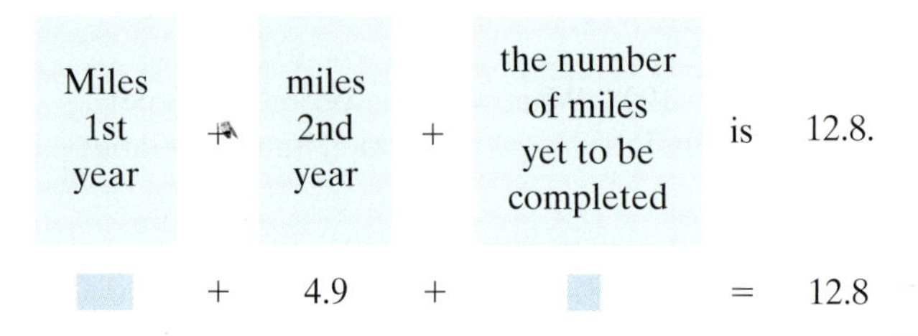

Solve the equation

2.3 + 4.9 + ____ = 12.8
____ + ____ = 12.8
x = ____

State the conclusion

____

Check the result

Add: ____ + ____ + ____ = 12.8. The answer checks.

Choose a variable to represent the unknown. Then write and solve an equation to answer the question.

69. DISASTER RELIEF After hurricane damage estimated at $27.9 million, a county looked to three sources for relief. Local agencies contributed $6.8 million toward the cleanup. A state emergency fund offered another $12.5 million. When applying for federal government help, how much should the county ask for?

70. TELETHONS Midway through a telethon, the donations had reached $16.7 million. How much more was donated in the second half of the program if the final total pledged was $30 million?

71. GRADE POINT AVERAGES After receiving her grades for the fall semester, a college student noticed that her overall GPA had dropped by 0.18. If her new GPA was 3.09, what was her GPA at the beginning of the fall semester?

72. MONTHLY PAYMENTS A food dehydrator offered on a home shopping channel can be purchased by making 3 equal monthly payments. If the price is \$113.25, how much is each monthly payment?

73. POINTS PER GAME As a senior, a college basketball player's scoring average was double that of her junior season. If she averaged 21.4 points a game as a senior, how many did she average as a junior?

74. NUTRITION One 3-ounce serving of broiled ground beef has 7 grams of saturated fat. This is 14 times the amount of saturated fat in 1 cup of cooked crab meat. How many grams of saturated fat are in 1 cup of cooked crab meat?

75. FUEL EFFICIENCY Each year, the Federal Highway Administration determines the number of vehicle-miles traveled in the country and divides it by the amount of fuel consumed to get an average miles per gallon (mpg). The illustration shows how the figure has changed over the years to reach a high of 16.7 mpg in 1998. What was the average miles per gallon in 1960?

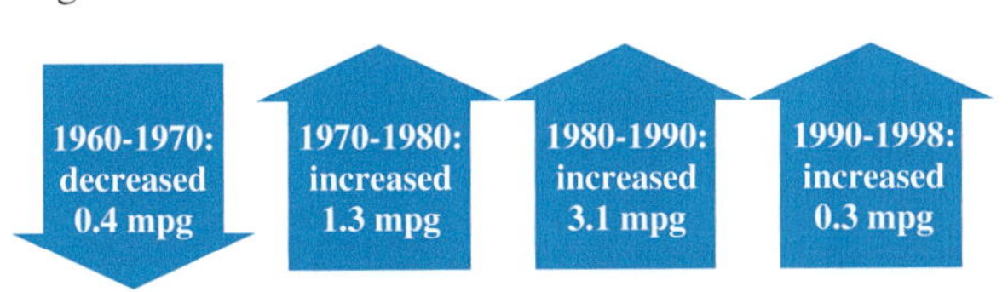

76. RATINGS REPORTS The illustration shows the prime-time television ratings for the week of January 3, 2000. If the Fox network ratings had been $\frac{1}{2}$ point higher, there would have been a three-way tie for second place. What prime-time rating did Fox have that week?

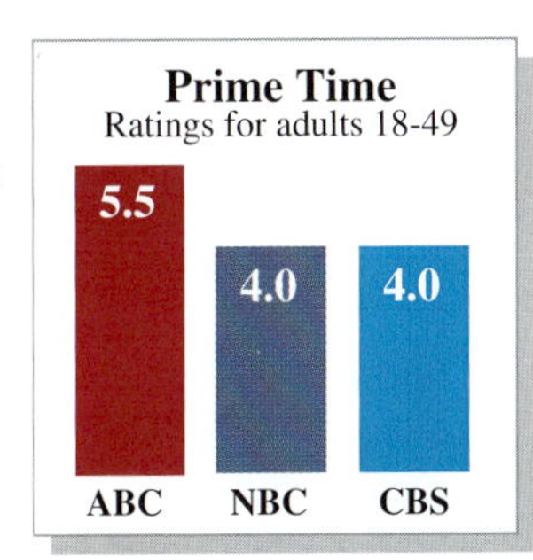

Source: Neilsen Media Research

77. CALLIGRAPHY A city honors its citizen of the year with a framed certificate. A calligrapher charges \$20 for the frame and then 15 cents a word for writing out the proclamation. If the city charter prohibits gifts in excess of \$50, what is the maximum number of words that can be printed on the award?

78. HELIUM BALLOONS The organizer of a jog-a-thon wants an archway of balloons constructed at the finish line of the race. A company charges a \$100 setup fee and then 8 cents for every balloon. How many balloons will be used if \$300 is spent for the decoration?

WRITING

79. Did you encounter any differences in solving equations containing decimals as compared to solving equations containing only integers? Explain your answer.

80. In the following case, why is it rather easy to apply the distributive property?

$$100(0.07x + 5.16)$$

REVIEW

81. Add: $-\frac{2}{3} + \frac{3}{4}$.

82. Add: $-\frac{2}{3} + \frac{1}{x}$.

83. Evaluate $x^3 - y^3$ for $x = -\frac{1}{2}$ and $y = -1$.

84. Multiply: $2\frac{1}{3} \cdot 4\frac{1}{2}$.

85. Evaluate: $\frac{-3 - 3}{-3 + 4}$.

86. Write a complex fraction using $-\frac{4}{5}$ and $\frac{1}{5}$. Then simplify it.

5.7 Square Roots

- Square roots
- Evaluating numerical expressions containing radicals
- Square roots of fractions and decimals
- Using a calculator to find square roots
- Approximating square roots

There are six basic operations of arithmetic. We have seen the relationships between addition and subtraction and between multiplication and division. In this section, we will explore the relationship between raising a number to a power and finding a root. Decimals will play an important role in this discussion.

Square roots

When we raise a number to the second power, we are squaring it, or finding its **square.**

The square of 6 is 36, because $6^2 = 36$.

The square of -6 is 36, because $(-6)^2 = 36$.

The **square root** of a given number is a number whose square is the given number. For example, the square roots of 36 are 6 and -6, because either number, when squared, yields 36. We can express this concept using symbols.

Square root

A number b is the **square root** of a if $b^2 = a$.

Self Check 1
Find the square roots of 64.

Answers 8 and -8

EXAMPLE 1 Find the square roots of 49.

Solution Ask yourself, What number was squared to obtain 49? The two answers are

$7^2 = 49$ and $(-7)^2 = 49$

Thus, 7 and -7 are the square roots of 49.

In Example 1, we saw that 49 has two square roots—one positive and one negative. The symbol $\sqrt{}$ is called a **radical symbol** and is used to indicate a positive square root.

When a number, called the **radicand,** is written under a radical symbol, we have a **radical expression.** Some examples of radical expressions are

$\sqrt{36}$ $\sqrt{100}$ $\sqrt{144}$ $\sqrt{81}$ In the radical expression $\sqrt{36}$, 36 is called the radicand.

To evaluate (or simplify) a radical expression, we need to find the positive square root of the radicand. For example, if we evaluate $\sqrt{36}$ (read as "the square root of 36"), the result is

$\sqrt{36} = 6$

because $6^2 = 36$. The negative square root of 36 is denoted $-\sqrt{36}$, and we have

$-\sqrt{36} = -6$ Read as "the negative square root of 36 is -6" or "the opposite of the square root of 36 is -6."

Self Check 2
Simplify each expression:
a. $\sqrt{144}$ and **b.** $-\sqrt{81}$.

Answers **a.** 12, **b.** -9

EXAMPLE 2 Simplify: **a.** $\sqrt{81}$ and **b.** $-\sqrt{100}$.

Solution

a. $\sqrt{81}$ means the positive square root of 81.

$\sqrt{81} = 9$, because $9^2 = 81$

b. $-\sqrt{100}$ means the opposite (or negative) of the square root of 100. Since $\sqrt{100} = 10$, we have

$-\sqrt{100} = -10$

COMMENT Radical expressions such as

$$\sqrt{-36} \qquad \sqrt{-100} \qquad \sqrt{-144} \qquad \sqrt{-81}$$

do not represent real numbers. This is because there are no real numbers that, when squared, give a negative number.

Be careful to note the difference between expressions such as $-\sqrt{36}$ and $\sqrt{-36}$. We have seen that $-\sqrt{36}$ is a real number: $-\sqrt{36} = -6$. On the other hand, $\sqrt{-36}$ is not a real number.

Evaluating numerical expressions containing radicals

Numerical expressions can contain radical expressions. When applying the rules for the order of operations, we treat a radical expression as we would a power.

EXAMPLE 3 Evaluate: **a.** $\sqrt{64} + \sqrt{9}$ and **b.** $-\sqrt{25} - \sqrt{4}$.

Solution

a. $\sqrt{64} + \sqrt{9} = 8 + 3$ Evaluate each radical expression first.
$= 11$ Perform the addition.

b. $-\sqrt{25} - \sqrt{4} = -5 - 2$ Evaluate each radical expression first.
$= -7$ Perform the subtraction.

Self Check 3
Evaluate:
a. $\sqrt{121} + \sqrt{1}$ and
b. $-\sqrt{9} - \sqrt{16}$.

Answers **a.** 12, **b.** −7

EXAMPLE 4 Evaluate: **a.** $6\sqrt{100}$ and **b.** $-5\sqrt{16} + 3\sqrt{9}$.

Solution

a. We note that $6\sqrt{100}$ means $6 \cdot \sqrt{100}$.

$6\sqrt{100} = 6(10)$ Simplify the radical first.
$= 60$ Perform the multiplication.

b. $-5\sqrt{16} + 3\sqrt{9} = -5(4) + 3(3)$ Simplify each radical first.
$= -20 + 9$ Perform the multiplications.
$= -11$ Perform the addition.

Self Check 4
Evaluate each expression:
a. $8\sqrt{64}$ and
b. $-6\sqrt{25} + 2\sqrt{36}$.

Answers **a.** 64, **b.** −18

Square roots of fractions and decimals

So far, we have found square roots of whole numbers. We can also find square roots of fractions and decimals.

EXAMPLE 5 Find each square root: **a.** $\sqrt{\frac{25}{64}}$ and **b.** $\sqrt{0.81}$.

Solution

a. $\sqrt{\frac{25}{64}} = \frac{5}{8}$, because $\left(\frac{5}{8}\right)^2 = \frac{25}{64}$.

b. $\sqrt{0.81} = 0.9$, because $(0.9)^2 = 0.81$.

Self Check 5
Find each square root:
a. $\sqrt{\frac{16}{49}}$ and **b.** $\sqrt{0.04}$.

Answers **a.** $\frac{4}{7}$, **b.** 0.2

Using a calculator to find square roots

We can also use a calculator to find square roots.

CALCULATOR SNAPSHOT

Finding a square root

We use the $\boxed{\sqrt{\ }}$ key (square root key) on a scientific calculator to find square roots. For example, to find $\sqrt{729}$, we enter these numbers and press these keys.

729 $\boxed{\sqrt{\ }}$ | 27

We have found that $\sqrt{729} = 27$. To check this result, we need to square 27. This can be done by entering the numbers 2 and 7 and pressing the $\boxed{x^2}$ key. We obtain 729. Thus, 27 is the square root of 729.

Approximating square roots

Numbers whose square roots are whole numbers are called **perfect squares.** The perfect squares that are less than or equal to 100 are

0, 1, 4, 9, 16, 25, 36, 49, 64, 81, 100

To find the square root of a number that is not a perfect square, we can use a calculator. For example, to find $\sqrt{17}$, we enter these numbers and press the square root key.

17 $\sqrt{\ }$

The display reads 4.123105626. This result is not exact, because $\sqrt{17}$ is a **nonterminating decimal** that never repeats. $\sqrt{17}$ is an *irrational number.* Together, the rational and the irrational numbers form the set of *real numbers.* Rounding to the nearest thousandth, we have

$$\sqrt{17} = 4.123$$

Self Check 6

Use a scientific calculator to find each square root. Round to the nearest hundredth.

a. $\sqrt{607.8}$

b. $\sqrt{0.076}$

Answers **a.** 24.65, **b.** 0.28

EXAMPLE 6 Use a scientific calculator to find each square root. Round to the nearest hundredth.

a. $\sqrt{373}$, **b.** $\sqrt{56.2}$, and **c.** $\sqrt{0.0045}$.

Solution

a. From the calculator, we get $\sqrt{373} \approx 19.31320792$. Rounded to the nearest hundredth, $\sqrt{373}$ is 19.31.

b. From the calculator, we get $\sqrt{56.2} \approx 7.496665926$. Rounded to the nearest hundredth, $\sqrt{56.2}$ is 7.50.

c. From the calculator, we get $\sqrt{0.0045} \approx 0.067082039$. Rounded to the nearest hundredth, $\sqrt{0.0045}$ is 0.07.

Section 5.7 STUDY SET

VOCABULARY *Fill in the blanks.*

1. When we find what number is squared to obtain a given number, we are finding the square ______ of the given number.
2. Whole numbers such as 25, 36, and 49 are called ________ squares because their square roots are whole numbers.
3. The symbol $\sqrt{\ }$ is called a ________ symbol. It indicates that we are to find a ________ square root.
4. The decimal number that represents $\sqrt{17}$ is a ______________ decimal — it never ends.
5. In $\sqrt{26}$, the number 26 is called the _________.
6. The symbol $\approx$ means ______________________.

CONCEPTS *Fill in the blanks.*

7. The square of 5 is $\square$, because $(5)^2 = \square$.
8. The square of $\frac{1}{4}$ is $\square$, because $\left(\frac{1}{4}\right)^2 = \square$.
9. $\sqrt{49} = 7$, because $\square = 49$.
10. $\sqrt{4} = 2$, because $\square = 4$.
11. $\sqrt{\frac{9}{16}} = \square$, because $\left(\frac{3}{4}\right)^2 = \frac{9}{16}$.
12. $\sqrt{0.16} = \square$, because $(0.4)^2 = 0.16$.
13. Without evaluating the following square roots, write them in order, from smallest to largest: $\sqrt{23}$, $\sqrt{11}$, $\sqrt{27}$, $\sqrt{6}$.
14. Without evaluating the following square roots, write them in order from smallest to largest: $-\sqrt{13}$, $-\sqrt{5}$, $-\sqrt{17}$, $-\sqrt{37}$.
15. Find each square root.
    a. $\sqrt{1}$ b. $\sqrt{0}$
16. Multiplication can be thought of as the opposite of division. What is the opposite of finding the square root of a number?

Use a calculator.

17. a. Use a calculator to approximate $\sqrt{6}$ to the nearest tenth.
    b. Square the result from part a.
    c. Find the difference between 6 and the answer to part b.
18. a. Use a calculator to approximate $\sqrt{6}$ to the nearest hundredth.
    b. Square the result from part a.
    c. Find the difference between the answer to part b and 6.
19. Graph: $\sqrt{9}$ and $-\sqrt{5}$.

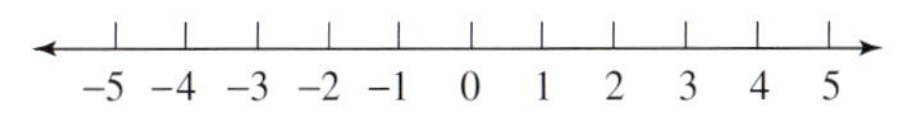

20. Graph: $-\sqrt{3}$ and $\sqrt{7}$.

−5 −4 −3 −2 −1 0 1 2 3 4 5

21. Between what two whole numbers would each square root be located when graphed on the number line?
    a. $\sqrt{19}$ b. $\sqrt{87}$
22. Between what two whole numbers would each square root be located when graphed on the number line?
    a. $\sqrt{33}$ b. $\sqrt{50}$

NOTATION *Complete each solution to evaluate the expression.*

23. $-\sqrt{49} + \sqrt{64} = \square + \square$
    $= 1$
24. $2\sqrt{100} - 5\sqrt{25} = 2(\square) - 5(\square)$
    $= \square - 25$
    $= -5$

PRACTICE *Evaluate each expression without using a calculator.*

25. $\sqrt{16}$
26. $\sqrt{64}$
27. $-\sqrt{121}$
28. $-\sqrt{144}$
29. $-\sqrt{0.49}$
30. $-\sqrt{0.64}$
31. $\sqrt{0.25}$
32. $\sqrt{0.36}$
33. $\sqrt{0.09}$
34. $\sqrt{0.01}$
35. $-\sqrt{\frac{1}{81}}$
36. $-\sqrt{\frac{1}{4}}$
37. $-\sqrt{\frac{16}{9}}$
38. $-\sqrt{\frac{64}{25}}$
39. $\sqrt{\frac{4}{25}}$
40. $\sqrt{\frac{36}{121}}$

41. $5\sqrt{36} + 1$

42. $2 + 6\sqrt{16}$

43. $-4\sqrt{36} + 2\sqrt{4}$

44. $-6\sqrt{81} + 5\sqrt{1}$

45. $\sqrt{\frac{1}{16}} - \sqrt{\frac{9}{25}}$

46. $\sqrt{\frac{25}{9}} - \sqrt{\frac{64}{81}}$

47. $5(\sqrt{49})(-2)$

48. $(-\sqrt{64})(-2)(3)$

49. $\sqrt{0.04} + 2.36$

50. $\sqrt{0.25} + 4.7$

51. $-3\sqrt{1.44}$

52. $-2\sqrt{1.21}$

Use a calculator to complete each square root table. Round to the nearest thousandth when an answer is not exact.

53.

Number	Square root
1	
2	
3	
4	
5	
6	
7	
8	
9	
10	

54.

Number	Square root
10	
20	
30	
40	
50	
60	
70	
80	
90	
100	

Use a calculator to evaluate each of the following.

55. $\sqrt{1{,}369}$

56. $\sqrt{841}$

57. $\sqrt{3{,}721}$

58. $\sqrt{5{,}625}$

Use a calculator to approximate each of the following to the nearest hundredth.

59. $\sqrt{15}$

60. $\sqrt{51}$

61. $\sqrt{66}$

62. $\sqrt{204}$

Use a calculator to approximate each of the following to the nearest thousandth.

63. $\sqrt{24.05}$

64. $\sqrt{70.69}$

65. $-\sqrt{11.1}$

66. $\sqrt{0.145}$

Use a calculator to evaluate each expression. If an answer is not exact, round to the nearest ten thousandth.

67. $\sqrt{24{,}000{,}201}$

68. $-\sqrt{4.012009}$

69. $-\sqrt{0.00111}$

70. $\sqrt{\frac{27}{44}}$

APPLICATIONS

Square roots have been used to express various lengths. Solve each problem by evaluating any square roots. You may need to use a calculator. Round to the nearest tenth, if necessary.

71. CARPENTRY Find the length of the slanted side of each roof truss.

a.

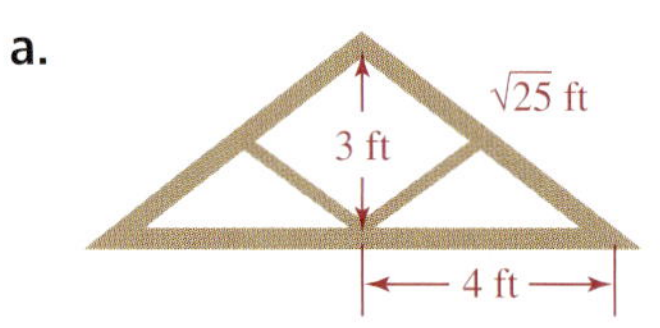

b.

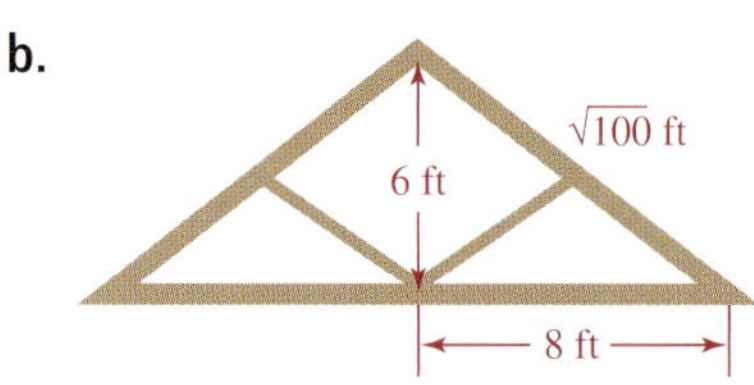

72. RADIO ANTENNAS How far from the base of the following antenna is each guy wire anchored to the ground? (The measurements are in feet.)

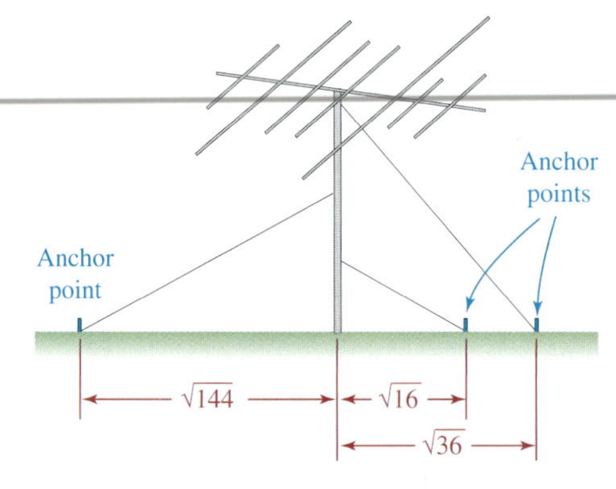

73. BASEBALL The illustration shows some dimensions of a major league baseball field. How far is it from home plate to second base?

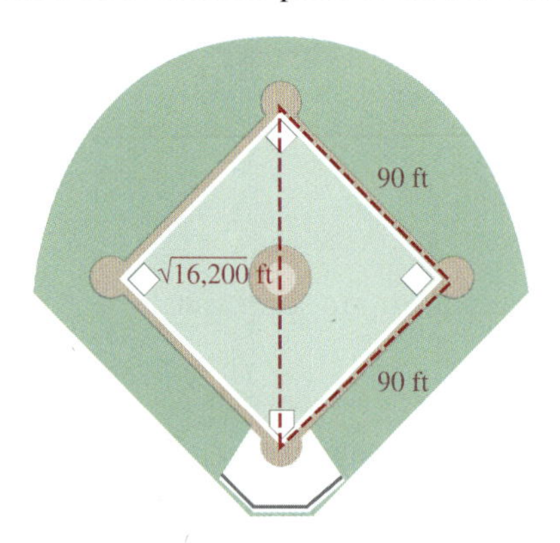

74. SURVEYING Use the imaginary triangles set up by a surveyor to find the length of each lake. (The measurements are in meters.)

a.

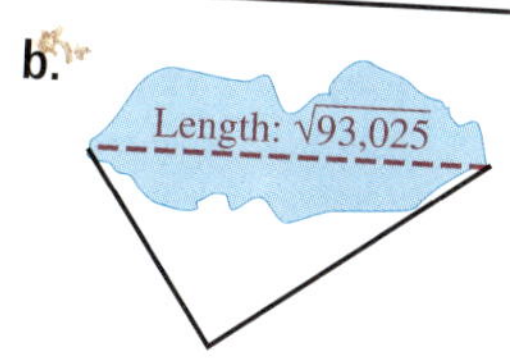

b.

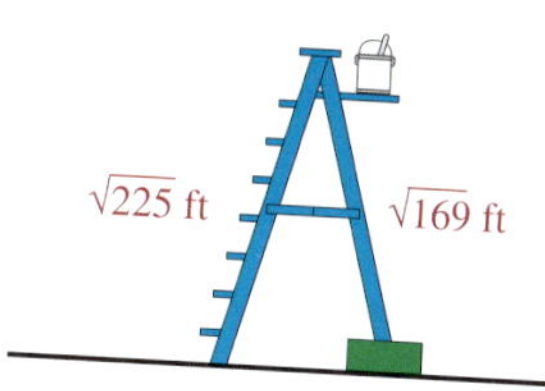

75. BIG-SCREEN TELEVISIONS The picture screen on a television set is measured diagonally. What size screen is shown?

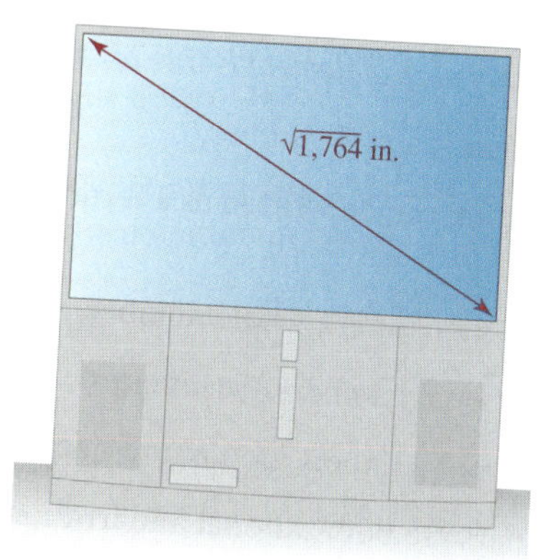

76. LADDERS A painter's ladder is shown. How long are the legs of the ladder?

$\sqrt{225}$ ft $\sqrt{169}$ ft

WRITING

77. When asked to find $\sqrt{16}$, a student answered 8. Explain his misunderstanding of square root.

78. Explain the difference between the square and the square root of a number.

79. What is a nonterminating decimal? Use an example in your explanation.

80. What do you think might be meant by the term *cube root?*

81. Explain why $\sqrt{-4}$ does not represent a real number.

82. Is there a difference between $-\sqrt{25}$ and $\sqrt{-25}$? Explain.

REVIEW

83. When solving the equation $2x - 5 = 11$, what operations must be undone in order to isolate the variable?

84. Simplify: $\dfrac{-\frac{2}{3}}{8}$.

85. Evaluate: $5(-2)^2 - \dfrac{16}{4}$.

86. Translate to mathematical symbols: four less than twice x.

87. List the set of whole numbers.

88. Determine whether y is used as a factor or as a term: $5x + y + 3x$.

89. Solve: $8 + \dfrac{a}{5} = 14$.

90. Insert the proper symbol, $<$ or $>$, in the blank to make a true statement: $-15 \ \square \ -14$.

SECTION 5.7 *Square Roots*

The number b is a *square root* of a if $b^2 = a$.

A *radical sign* $\sqrt{}$ is used to indicate a positive square root. The square root of a *perfect square* is a whole number.

94. Fill in the blanks.

a. The symbol $\sqrt{}$ is called a ______ symbol. **b.** $\sqrt{64} = 8$, because ______ $= 64$.

Evaluate each expression without using a calculator.

95. $\sqrt{49}$ **96.** $-\sqrt{16}$ **97.** $\sqrt{100}$ **98.** $\sqrt{0.09}$

99. $\sqrt{\frac{64}{25}}$ **100.** $\sqrt{0.81}$ **101.** $-\sqrt{\frac{1}{36}}$ **102.** $\sqrt{0}$

103. Between what two whole numbers would $\sqrt{83}$ be located when graphed on a number line?

A square root can be approximated using a calculator.

104. Use a calculator to approximate $\sqrt{11}$. Round to the nearest tenth. Now square the approximation. How close is it to 11?

105. Graph each square root on the number line: $\sqrt{3}$, $-\sqrt{2}$, and $\sqrt{0}$.

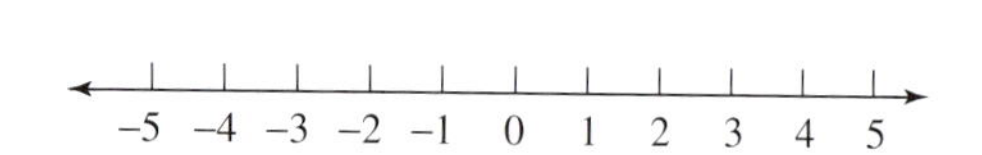

When evaluating an expression containing square roots, treat a radical as you would a power when applying the rules for the order of operations.

Evaluate each expression without using a calculator.

106. $-3\sqrt{100}$ **107.** $5\sqrt{0.25}$

108. $-3\sqrt{49} - \sqrt{36}$ **109.** $\sqrt{\frac{9}{100}} + \sqrt{1.44}$

110. Use a calculator to approximate $\sqrt{19}$ to the nearest hundredth.

CHAPTER 5 TEST

1. Express the amount of the square that is shaded, using a fraction and a decimal.

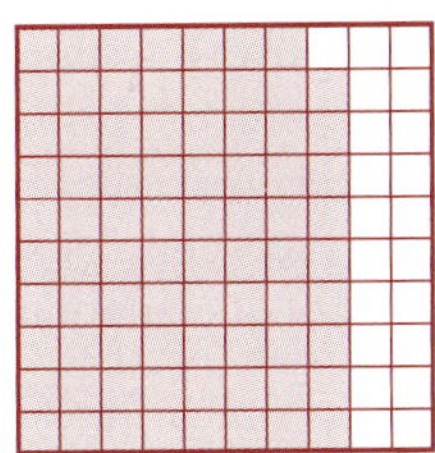

2. WATER PURITY A county health department sampled the pollution content of tap water in five cities, with the results shown. Rank the cities in order, from dirtiest tap water to cleanest.

City	Pollution, parts per million
Monroe	0.0909
Covington	0.0899
Paston	0.0901
Cadia	0.0890
Selway	0.1001

3. Write each decimal in words and then as a fraction or mixed number. **Do not simplify the fraction.**

   a. SKATEBOARDING Gary Hardwick of Carlsbad, California, set the skateboard speed record of 62.55 mph in 1998.

   b. MONEY A dime weighs 0.08013 ounce.

4. Round to the nearest thousandth: 33.0495.

5. SKATING RECEIPTS At an ice-skating complex, receipts on Friday were \$30.25 for indoor skating and \$62.25 for outdoor skating. On Saturday, the corresponding amounts were \$40.50 and \$75.75. Find the total receipts for the two days.

6. Perform each operation in your head.

   a. $567.909 \div 1{,}000$ b. $0.00458 \cdot 100$

7. EARTHQUAKE FAULT LINES After an earthquake, geologists found that the ground on the west side of the fault line had dropped 0.83 inch. The next week, a strong aftershock caused the same area to sink 0.19 inch deeper. How far did the ground on the west side of the fault drop because of the seismic activity?

Perform each operation.

8. $2 + 4.56 + 0.89 + 3.3$

9. $45.2 - 39.079$

10. $(0.32)^2$

11. $-6.7(-2.1)$

12. NEW YORK CITY Central Park, which lies in the middle of Manhattan, is the city's best-known park. If it is 2.5 miles long and 0.5 mile wide, what is its area?

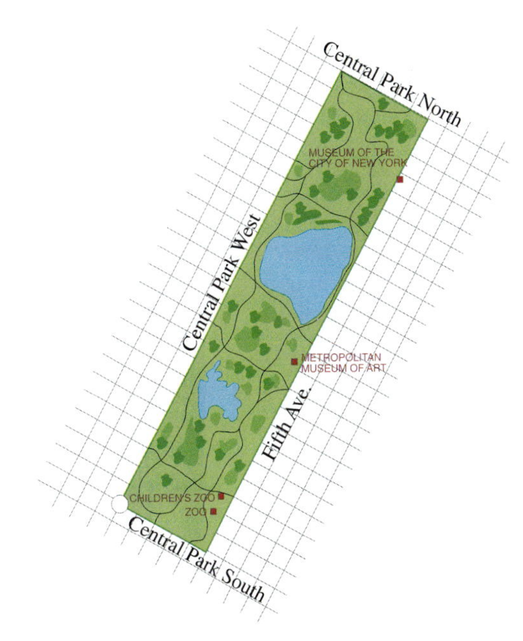

13. TELEPHONE BOOKS To print a telephone book, 565 sheets of paper were used. If the book is 2.3 inches thick, what is the thickness of each sheet of paper? (Round to the nearest thousandth of an inch.)

14. Evaluate: $4.1 - (3.2)(0.4)^2$.

15. Write each fraction as a decimal.

   a. $\frac{17}{50}$ b. $\frac{5}{12}$

16. Perform the division and round to the nearest hundredth: $\frac{12.146}{-5.3}$.

17. Divide: $11\overline{)13}$.

18. RATINGS The seven top-rated cable television programs for the week of February 8–14 are given below. What are the mean, median, mode, and range of the ratings? Round to the nearest tenth.

Show/day/time/network	Rating
1. "WCW Monday," Mon. 9 P.M., TNT	4.5
2. "WCW Monday," Mon. 10 P.M., TNT	4.4
3. "WCW Monday," Mon. 8 P.M., TNT	3.9
4. "WWF Special," Sat. 9 P.M., USA	3.6
5. "WWF Wrestling," Sun. 7 P.M., USA	3.1
6. "Dog Show," Tues. 8 P.M., USA	3.1
7. "WWF Special," Sat. 8 P.M., USA	2.9

19. STATISTICS The following graph has an asterisk * that refers readers to a note at the bottom. In your own words, complete the explanation of the term *median.*

Family Debt Grows

Median family indebtedness grew by 42% between 1995 and 1998. according to a Federal Reserve survey of consumer finances.

Median* amount of debt

1995 **$23,400**

1998 **$33,300**

*Median means that............

Source: *Los Angeles Times* (February 1, 2000)

20. Graph $\frac{3}{8}$ and $-\frac{4}{5}$ on the number line. Label each point using the decimal equivalent of the given fractions.

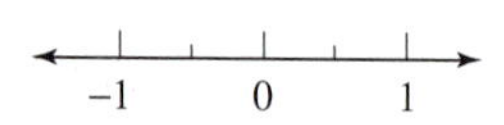

21. Find the exact answer: $\frac{2}{3} + 0.7$.

22. Simplify: $6.18s + 8.9 - 1.22s - 6.6$.

23. Simplify: $2.1(x - 3) + 3.1(x - 4)$.

Solve each equation.

24. $-2.4t = 16.8$

25. $-0.008 + x = 6$

26. $0.3x - 0.53 = 0.0225 + 1.6x$.

27. CHEMISTRY In a lab experiment, a chemist mixed three compounds together to form a mixture weighing 4.37 g. Later, she discovered that she had forgotten to record the weight of compound C in her notes. Find the weight of compound C used in the experiment.

	Weight
Compound A	1.86 g
Compound B	2.09 g
Compound C	?
Mixture total	4.37 g

28. WEDDING COSTS A printer charges a setup fee of $24 and then 95 cents for each wedding announcement printed (tax included). If a couple has budgeted $100 for printing costs, how many announcements can they have made?

29. Fill in the blank: $\sqrt{144} = 12$, because ▭ $= 144$.

30. Graph: $\sqrt{2}$ and $-\sqrt{5}$.

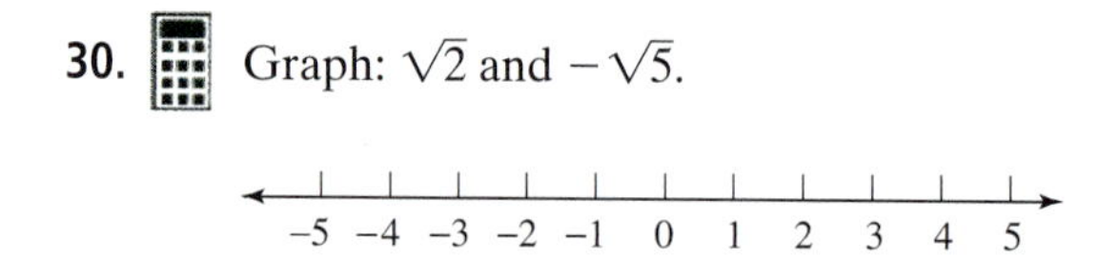

Evaluate each expression.

31. $-2\sqrt{25} + 3\sqrt{49}$

32. $\sqrt{\frac{1}{36}} - \sqrt{\frac{1}{25}}$

33. Insert the proper symbol < or > to make a true statement.

a. -6.78 ▭ -6.79

b. $\frac{3}{8}$ ▭ 0.3

c. $\sqrt{\frac{16}{81}}$ ▭ $\frac{16}{81}$

d. 0.45 ▭ $0.\overline{45}$

34. Find each square root.

a. $-\sqrt{0.04}$

b. $\sqrt{1.69}$

35. Although the decimal 3.2999 contains more digits than 3.3, it is smaller than 3.3. Explain why this is so.

36. What is a repeating decimal? Give an example.

CHAPTERS 1–5 CUMULATIVE REVIEW EXERCISES

1. THE EXECUTIVE BRANCH The annual salaries for the President and the Vice President of the United States are \$400,000 and \$203,000, respectively. How much more money does the President make than the Vice President during a four-year term?

2. Use the variables x, y, and z to write the associative property of addition.

3. Divide: $43\overline{)1,161}$.

4. How many thousands are there in one million?

5. Find the prime factorization of 220.

6. List the factors of 20, from least to greatest.

7. List the set of whole numbers.

8. Add: $-8 + (-5)$.

9. Fill in the blank to make the statement true: Subtraction is the same as ________ the opposite.

10. Complete the solution to evaluate the expression.

$$\begin{aligned}(-6)^2 - 2(5 - 4\cdot 2) &= (-6)^2 - 2(5 - \square)\\ &= (-6)^2 - 2(\square)\\ &= \square - 2(-3)\\ &= 36 - (\square)\\ &= 36 + \square\\ &= 42\end{aligned}$$

11. Consider the division statement $\frac{-15}{-5} = 3$. What is its related multiplication statement?

12. Find the power: $(-1)^5$.

13. Solve: $8 - 2d = -5 - 5$.

14. Solve: $0 = 6 + \frac{c}{-5}$.

15. Evaluate: $|-7(5)|$.

16. What is the opposite of -102?

17. A chain is x yards long. Express its length in feet.

18. CHECKING ACCOUNTS After a deposit of \$995, a student's checking account was still \$105 overdrawn. What was the balance in the account before the deposit?

19. See the illustration.

   a. Let k represent the length of the key. Write an algebraic expression that represents the length of the match.

   b. Let m represent the length of the match. Write an algebraic expression that represents the length of the key.

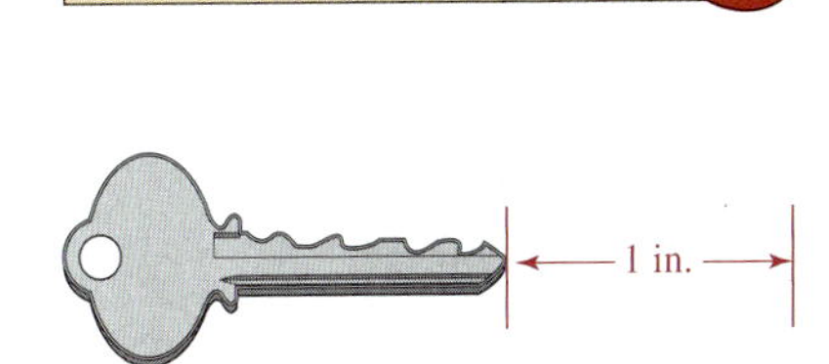

20. The expression $2(4x) + 2(5)$ is the result of an application of the distributive property. What was the original expression?

21. How many terms does the expression $6x^2 - 3x + 18$ have?

22. Simplify: $3w - 8w$.

23. Solve: $-(5x - 4) + 6(2x - 7) = -3$.

24. What fraction of the stripes in the flag are white?

25. Although the fractions listed below look different, they all represent the same value. What concept does this illustrate?

$$\frac{1}{2} = \frac{2}{4} = \frac{3}{6} = \frac{4}{8} = \frac{5}{10} = \frac{6}{12}$$

26. Simplify: $\frac{90x^2}{126x}$.

Perform the operations.

27. $\dfrac{3}{8} \cdot \dfrac{7}{16}$

28. $-\dfrac{15}{8y} \div \dfrac{10}{y^3}$

29. $\dfrac{4}{m} + \dfrac{2}{7}$

30. $-4\dfrac{1}{4}\left(-4\dfrac{1}{2}\right)$

31. $76\dfrac{1}{6} - 49\dfrac{7}{8}$

32. $\dfrac{\frac{5}{27}}{-\frac{5}{9}}$

33. Solve: $\dfrac{2}{3}y = -30$.

34. Solve: $\dfrac{d}{6} - \dfrac{2}{3} = \dfrac{d}{12}$.

35. KITES Find the area of the kite.

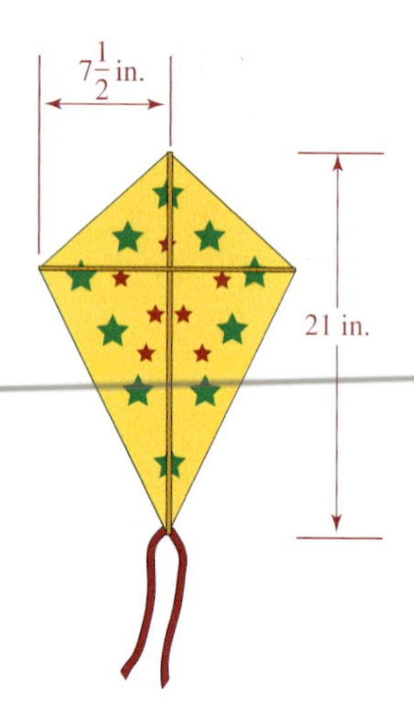

36. Graph: $\{-3\frac{1}{4}, 0.75, -1.5, -\frac{9}{8}, 3.8, \sqrt{4}\}$.

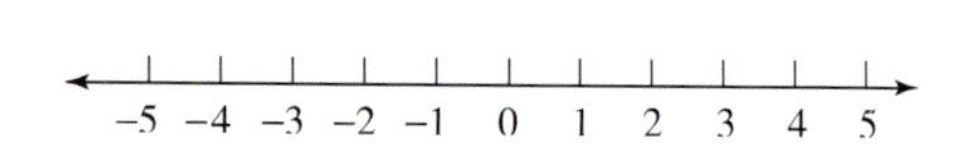

37. GLASS Some electronic and medical equipment uses glass that is only 0.00098 inch thick. Round this number to the nearest thousandth.

38. Place the proper symbol > or < in the box to make the statement true.

356.1978 ☐ 356.22

Perform the operations.

39. $-1.8(4.52)$

40. $\dfrac{-21.28}{-3.8}$

41. $56.012(100)$

42. $\dfrac{0.897}{10{,}000}$

43. Evaluate: $-9.1 - (-6.05 - 51)$.

44. WEEKLY SCHEDULES Refer to the illustration. Determine the number of hours during a week that an adult spends, on average, watching television.

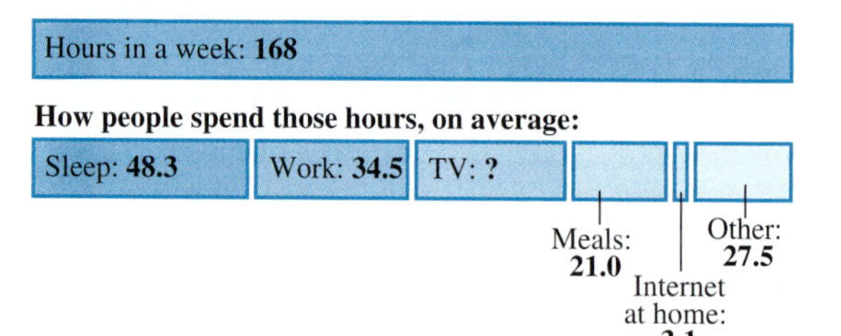

Source: National Sleep Foundation and the U.S. Bureau of Statistics

45. LITERATURE The novel *Fahrenheit 451,* by Ray Bradbury, is a story about censorship and book burning. Use the formula $C = \dfrac{5}{9}(F - 32)$ to convert 451° F to degrees Celsius. Round to the nearest tenth of a degree.

46. TEAM GPA The grade point averages of the players on a badminton team are listed below. Find the mean, median, and mode of the team's GPAs.

3.04	4.00	2.75	3.23	3.87	2.20
3.02	2.25	2.99	2.56	3.58	2.75

47. Write $\dfrac{5}{12}$ as a decimal. Use an overbar.

48. Solve: $0.2t - 3 = 0.7t + 1.5$.

49. CONCESSIONAIRES At a ballpark, a vendor is paid \$22 a game plus 35¢ for each bag of peanuts she sells. How many bags of peanuts must she sell to make \$50 a game?

50. Evaluate: $-4\sqrt{36} + 2\sqrt{81}$.

CHAPTER 6

Graphing, Exponents, and Polynomials

CORBIS

Many U.S. cities are laid out in rectangular grids. One of the best examples of this is Salt Lake City, Utah. The street system is a grid based on the four streets bordering Temple Square. Most streets run precisely north-south or east-west. Streets have names such as Second East, Third East, Fourth East, and so on. Each block has one hundred house numbers; the evens on one side, the odds on the other. Salt Lake City residents claim that the grid system does make finding things relatively simple.

To learn more about grids and how they are used in mathematics, visit *The Learning Equation* on the Internet at http://tle.brookscole.com. (The log-in instructions are in the Preface.) For Chapter 6, the online lesson is:

- *TLE* Lesson 12: The Coordinate Plane

Check Your Knowledge

1. The x and y axes divide the rectangular coordinate system into four ________.
2. The point represented by the ordered pair (0, 0) is called the ______.
3. The graph of a linear equation is a _____.
4. An ordered pair is a ________ of an equation in two variables if the numbers in the ordered pair satisfy the equation.

Determine whether the given ordered pair is a solution of the equation.

5. $(1, 1), x - 3y = -2$
6. $(6, -1), x - 2y = 7$
7. Graph and label the ordered pairs: $(2, 3), (3, -1), (-2, -3), (-3, 1)$, and $\left(0, \frac{3}{2}\right)$.
8. Give the coordinates of each point shown in the graph.

PROBLEM 7

(3, 2)
(−1, 1)
(−2, 0)
(2, −3)
(0, −4)

PROBLEM 8

9. **a.** Complete the table of solutions for $2x - y = 3$. Then graph the equation. Label three points on the graph.

$$2x - y = 3$$

x	y
0	
	0
2	

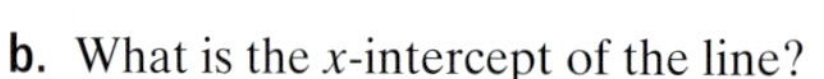

b. What is the x-intercept of the line?

c. What is the y-intercept of the line?

10. Graph: $x = -2$.

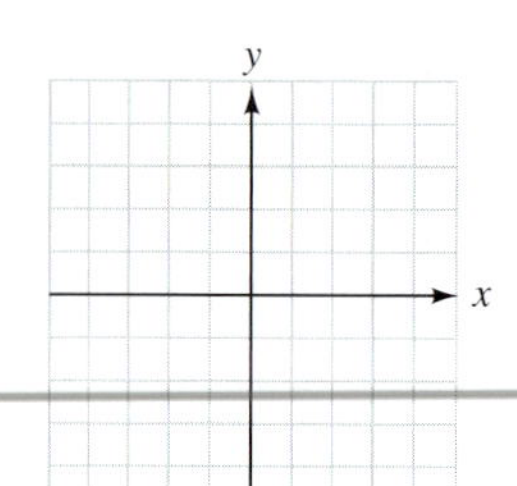

11. Graph: $y = 1$.

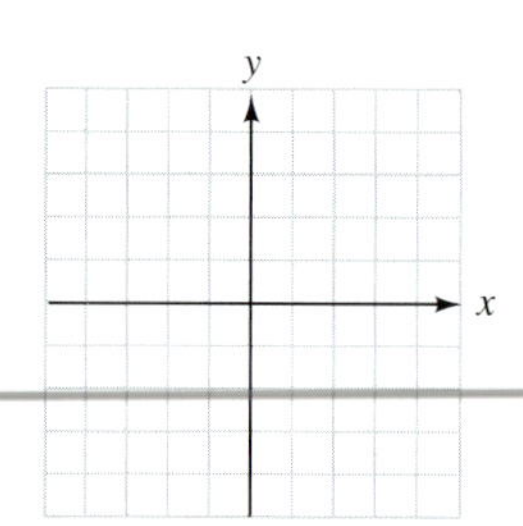

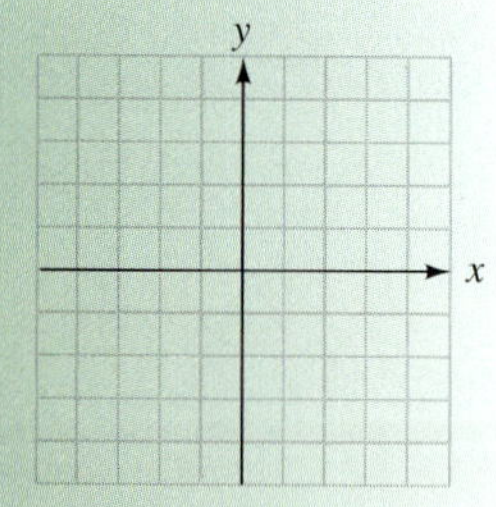

PROBLEM 9

12. Simplify.
    a. $x^2 \cdot x \cdot x \cdot x^4$ **b.** $2x(3x^5)$ **c.** $-2x^2y^3(-3xy^2)$
    d. $(x^2)^3$ **e.** $(3xy^3)^2$ **f.** $(x^3)^4(x^2)^3$
13. Consider $4x^2 - 9x$.
    a. Identify the polynomial as a monomial, a binomial, or a trinomial.
    b. What is the degree of the polynomial?
14. Evaluate $-16t^2 + 28t + 8$ for $t = -1$.
15. Graph: $y = x^2 + 1$.

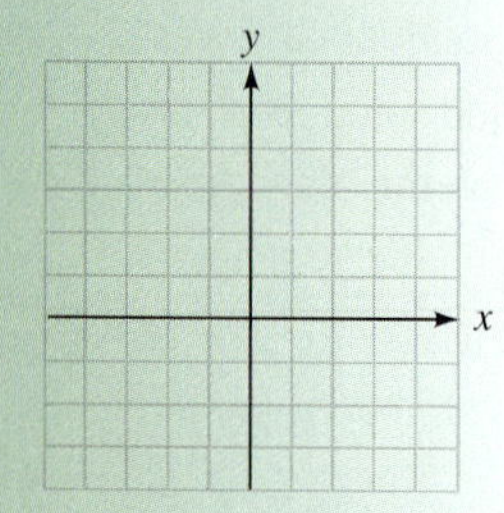

PROBLEM 15

Perform the operations.

16. $(x^2 - 3x + 4) + (x^3 - 2x^2 + x + 4)$
17. $(y^2 - 4y + 7) - (2y^2 + 2y - 3)$
18. $(-3xy^2)(-4xyz^2)$
19. $2x(x^2 + 3x - 5)$
20. $(2y - 7)(2y + 7)$
21. $(2x + 3)(4x - 7)$
22. $(2x - 3)(x^2 - x + 1)$

Study Skills Workshop

STUDYING FOR TESTS

Doing homework regularly is important, but tests require another type of strategy for preparation. Before you take a test, it is important that you have done your best to commit the important concepts to memory.

How Much Time Should I Devote to Preparing for a Test? The time needed to prepare for a test will vary according to how well you have understood and memorized the basic concepts. Plan to prepare at least four days in advance of your test and schedule this on your study calendar.

How Do I Prepare? Four days before the test Know exactly what material the test will cover. Then imagine that you could bring one $8\frac{1}{2}'' \times 11''$ sheet of paper to the test. What would you write on that sheet? Go through each section (and your reworked notes) to identify the key rules and definitions to include on your study sheet. Keep this paper with you all the time until the test, and review it whenever you can.

Three days before the test Go through your reworked notes and find the examples that your instructor gave you in class. Add any examples you might have missed to your paper and continue to look at it whenever you have time.

Two days before the test Use a chapter test in your textbook, or make up your own practice test by choosing a sampling of problems from your text or reworked notes. You can also take a quiz on chapter and section material online using the iLrn Web site rather than create your own test. This site is located at www.iLrn.com, and the quizzes are found in the "Tutorial" section. Try to include the same number of questions that your real test will have. Choose problems that have a solution that can be checked. Then, *with your book closed,* take a *timed* trial. Don't be upset if you can't do everything perfectly; it's only a trial. When you are done, check your answers. Be honest with yourself! Make a list of the topics that were difficult and add these to your study sheet.

One day before the test Get help, if necessary, with yesterday's problems from your instructor during office hours, from a tutor at the tutorial center, or from a classmate. Practice your trial test again, without books or notes. Go back over any problem you didn't get correct. Get plenty of rest the night before your test.

Test day Review your study sheet, if you have time. Focus on how well you have prepared and relax as much as possible. When taking your test, complete the problems that you are sure of first. Skip the problems that you don't understand right away, and return to them later. Be aware of time throughout the test so that you don't spend too long on any one problem.

ASSIGNMENT

1. Four days before your test, prepare your study sheet.
2. Three days before your test, go through all of the examples in your reworked notes.
3. Two days before your test, make a written practice test and time youreslf on it. Then check your answers.
4. The day before the test, fix the problems that you did incorrectly. Use your textbook, notes, classmate, tutor, or instructor for help. Review all of your practice problems until you can do them without any help. Gather the materials that you will need for your test and get enough rest.
5. On test day, arrive in class a few minutes early. Look over your study sheet if you have time. Relax as much as possible, knowing that you have prepared well.

It is said that a picture is worth 1,000 words. In algebra, pictures of equations are given by their graphs.

6.1 The Rectangular Coordinate System

- Equations containing two variables
- The rectangular coordinate system
- Graphing ordered pairs of real numbers
- Finding coordinates
- An application of the coordinate system

We have seen that business and statistical information is often presented in tables or graphs. In algebra, we also present information that way. For example, Figure 6-1 shows a table and a graph that are related to the equation $d = 4t$, a formula that gives the distance d (in miles) that a woman can walk in a time t (in hours) at a rate of 4 miles per hour.

To find the distance she can walk in 3 hours, we substitute 3 for t in the formula $d = 4t$ and simplify.

$$d = 4t$$
$$= 4(3) \quad \text{Substitute 3 for } t.$$
$$= 12 \quad \text{Perform the multiplication.}$$

In 3 hours, she can walk 12 miles. This result and others are shown in the table and graph.

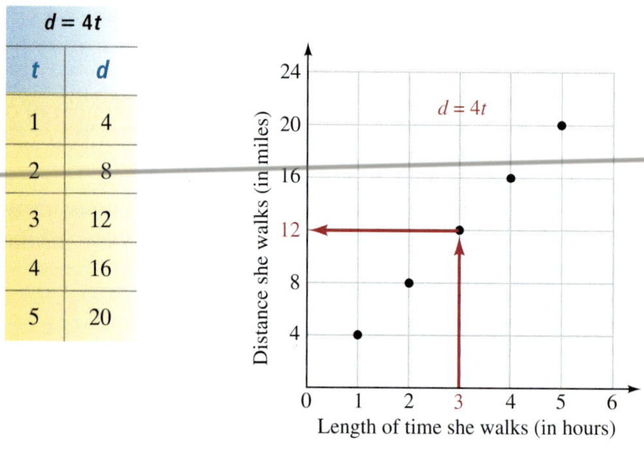

$d = 4t$

t	d
1	4
2	8
3	12
4	16
5	20

FIGURE 6-1

From either the table or the graph, we can see the following.

- In 1 hour, she can walk a distance of 4 miles.
- In 2 hours, she can walk a distance of 8 miles.
- In 3 hours, she can walk a distance of 12 miles.
- In 4 hours, she can walk a distance of 16 miles.
- In 5 hours, she can walk a distance of 20 miles.

In the first two sections of this chapter, we will discuss how to construct tables and graphs like those shown in Figure 6-1.

Equations containing two variables

So far, we have worked with equations containing one variable. For example, we solved equations such as $2x + 4 = 12$. We will now work with equations that contain two variables, such as $2x + y = 12$.

The equation

$$2x + y = 12$$

sets up a correspondence between x and y. The solutions of this equation are pairs of real numbers. For example, the pair $x = 3$ and $y = 6$ is a solution, because the equation is true when $x = 3$ and $y = 6$.

$$2x + y = 12$$
$$2(3) + 6 \stackrel{?}{=} 12 \quad \text{Substitute 3 for } x \text{ and 6 for } y.$$
$$6 + 6 \stackrel{?}{=} 12 \quad \text{Perform the multiplication: } 2(3) = 6.$$
$$12 = 12 \quad \text{Perform the addition.}$$

Since the pair $x = 3$ and $y = 6$ is a solution, we say that it **satisfies** the equation. The pair $x = 3$ and $y = 6$ can be written in the form (3, 6). When a pair is written in this form, the first number is the x-value and the second number is the y-value. Since order is important when writing a pair in (x, y) form, we call it an **ordered pair.**

! COMMENT Don't get confused by this new use of parentheses. The notation (3, 6) represents an ordered pair, whereas 3(6) indicates multiplication.

The ordered pair (2, 8) is also a solution of $2x + y = 12$, because it satisfies the equation.

$$2x + y = 12$$
$$2(2) + 8 \stackrel{?}{=} 12 \quad \text{Substitute 2 for } x \text{ and 8 for } y.$$
$$4 + 8 \stackrel{?}{=} 12 \quad \text{Perform the multiplication: } 2(2) = 4.$$
$$12 = 12 \quad \text{Perform the addition.}$$

EXAMPLE 1

Determine whether $(-1, 12)$ is a solution of $2x + y = 12$.

Solution For the ordered pair $(-1, 12)$, $x = -1$ and $y = 12$. We substitute -1 for x and 12 for y and see whether the equation is satisfied.

$$2x + y = 12$$
$$2(-1) + 12 \stackrel{?}{=} 12 \quad \text{Substitute } -1 \text{ for } x \text{ and 12 for } y.$$
$$-2 + 12 \stackrel{?}{=} 12 \quad \text{Perform the multiplication: } 2(-1) = -2.$$
$$10 = 12 \quad \text{Perform the addition.}$$

Since $10 = 12$ is false, the pair does not satisfy the given equation. The ordered pair $(-1, 12)$ is not a solution of $2x + y = 12$.

Self Check 1
Determine whether $(-2, 16)$ is a solution of $2x + y = 12$.

Answer yes

EXAMPLE 2

Complete the ordered pairs so that each one satisfies $4x + 2y = 2$:
a. (0,) and **b.** (, 2).

Solution

a. To complete the ordered pair (0,), we substitute 0 for x and solve for y.

$$4x + 2y = 2 \quad \text{This is the given equation.}$$
$$4(0) + 2y = 2 \quad \text{Substitute 0 for } x.$$
$$0 + 2y = 2 \quad \text{Multiply: } 4(0) = 0.$$
$$2y = 2 \quad \text{Add: } 0 + 2y = 2y.$$
$$y = 1 \quad \text{Divide both sides by 2.}$$

When $x = 0$, $y = 1$, and the ordered pair (0, 1) satisfies the equation.

Self Check 2
Complete the ordered pairs so that each one satisfies $2x - 7y = 14$:
a. (7,)
b. (, 1)

b. To complete the ordered pair (, 2), we substitute 2 for y and solve for x.

$$4x + 2y = 2 \quad \text{This is the given equation.}$$
$$4x + 2(2) = 2 \quad \text{Substitute 2 for } y.$$
$$4x + 4 = 2 \quad \text{Multiply: } 2(2) = 4.$$
$$4x = -2 \quad \text{Subtract 4 from both sides.}$$
$$x = -\frac{1}{2} \quad \text{Divide both sides by 4 and simplify: } \tfrac{-2}{4} = -\tfrac{1}{2}.$$

When $y = 2$, $x = -\frac{1}{2}$, and the ordered pair $(-\frac{1}{2}, 2)$ satisfies the equation.

Answers **a.** (7, 0), **b.** (10.5, 1)

The solutions of $4x + 2y = 2$ that we found in Example 2 can be listed in a **table of solutions.**

x	y	(x, y)
0	1	(0, 1)
$-\frac{1}{2}$	2	$(-\frac{1}{2}, 2)$

Self Check 3
Complete the table of solutions for $2x + 3y = 5$.

x	y	(x, y)
4		(4,)
	−5	(, −5)

EXAMPLE 3 Complete the table of solutions for $3x + 2y = 5$.

x	y	(x, y)
3		(3,)
	4	(, 4)

Solution If we substitute 3 for x in $3x + 2y = 5$ and solve for y, we get $y = -2$. We then enter −2 in the first row of the table. If we substitute 4 for y in $3x + 2y = 5$ and solve for x, we get $x = -1$. We then enter −1 in the second row of the table. The completed table is as follows.

x	y	(x, y)
3	**−2**	(3, **−2**)
−1	4	(**−1**, 4)

Answer

x	y	(x, y)
4	−1	(4, −1)
10	−5	(10, −5)

We have seen that solutions of an equation containing two variables are ordered pairs and that the ordered pairs can be listed in a table. We will now introduce a way to represent ordered pairs as points on a graph.

The rectangular coordinate system

Ordered pairs of real numbers can be displayed on a grid called a **rectangular coordinate system.** This system is also called the *Cartesian coordinate system* after its developer, René Descartes, a 17th-century French mathematician.

The rectangular coordinate system consists of two number lines, called the **x-axis** and the **y-axis,** as shown in Figure 6-2(a) on the next page. The two axes intersect at a point called the **origin,** which is the 0 point on each axis. The positive direction on the x-axis is to the right, and the positive direction on the y-axis is upward.

The two axes divide the coordinate system into four regions, called **quadrants,** which are numbered using Roman numerals, as shown in Figure 6-2(a). The axes are not considered to be in any quadrant.

! COMMENT If no scale is given on the x- and y-axes, we assume that the grid lines are one unit apart.

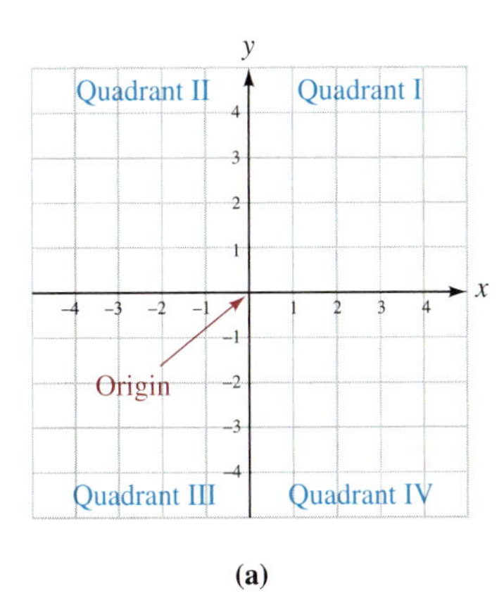

(a)

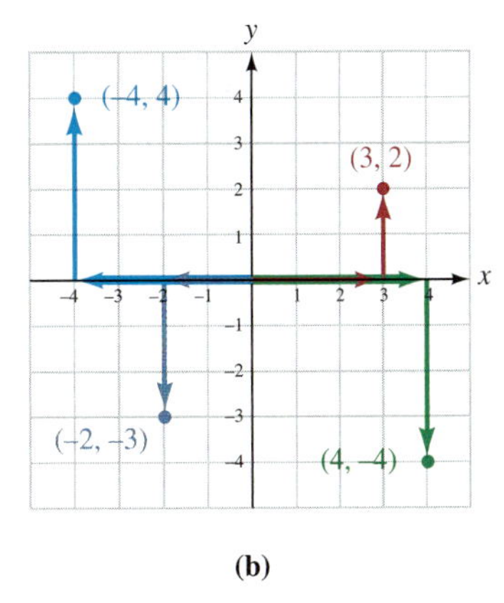

(b)

FIGURE 6-2

Graphing ordered pairs of real numbers

The process of locating an ordered pair on the rectangular coordinate system is called **graphing** or **plotting** the point. In Figure 6-2(b), we plot four ordered pairs.

- To plot the ordered pair (3, 2), we start at the origin and move 3 units to the right along the x-axis and then 2 units up in the y direction and draw a dot (shown in red). This locates the point with an ***x*-coordinate** of 3 and a ***y*-coordinate** of 2. The ordered pair (3, 2) gives the **coordinates of the point** and the point is the **graph** of the ordered pair (3, 2). This point lies in quadrant I.
- To plot the point with coordinates $(-4, 4)$, we start at the origin and move 4 units to the left and then 4 units up and draw a dot (shown in blue). This point lies in quadrant II.
- To plot the point with coordinates $(-2, -3)$, we start at the origin and move 2 units to the left and then 3 units down and draw a dot (shown in purple). This point lies in quadrant III.
- To plot the point with coordinates $(4, -4)$, we start at the origin and move 4 units to the right and then 4 units down and draw a dot (shown in green). This point lies in quadrant IV.

! COMMENT The order of the coordinates of a point is important. The point with coordinates $(-4, 4)$ is not the same as the point with coordinates $(4, -4)$.

EXAMPLE 4 Graph each ordered pair: $(0, 3)$, $(4, 0)$, $(-\frac{5}{2}, 1)$, and $(2, -1.5)$.

Solution Since no scale is indicated on the axes, we assume that they are scaled in units of 1. We note that $-\frac{5}{2} = -2\frac{1}{2}$ and $-1.5 = -1\frac{1}{2}$. The points are graphed in Figure 6-3 on the next page.

Self Check 4

Graph each ordered pair: $(-3, 0), (0, 4)\ (2.5, -4), (-\frac{7}{2}, -\frac{5}{2})$.

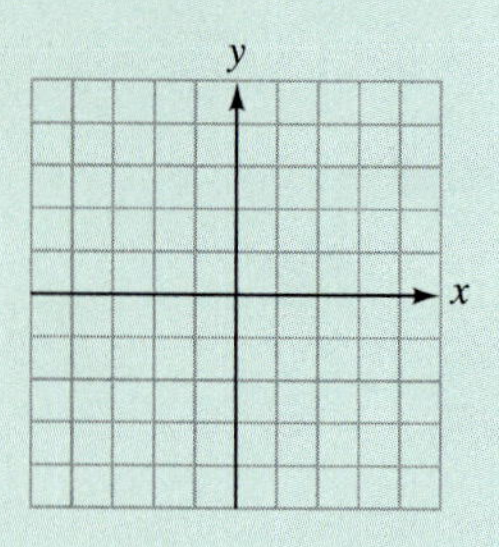

Answers

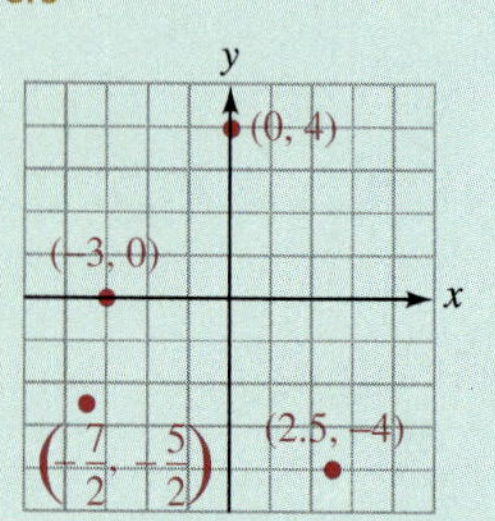

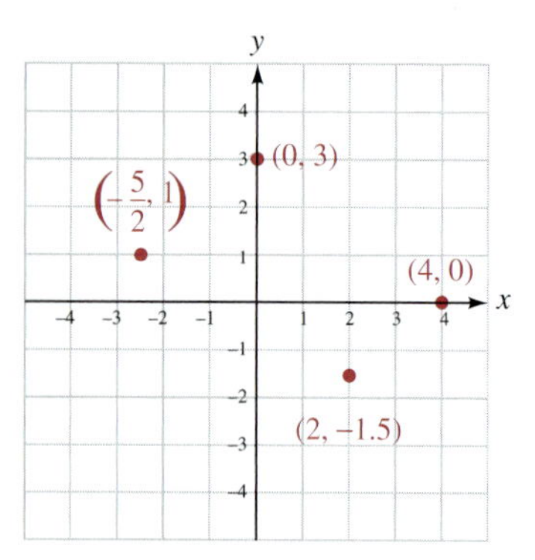

FIGURE 6-3

Finding coordinates

To find the coordinates of a point on the rectangular coordinate system, we must determine how far it is to the left or right of the y-axis and how far it is above or below the x-axis.

Self Check 5

Find the coordinates of each point in the graph below.

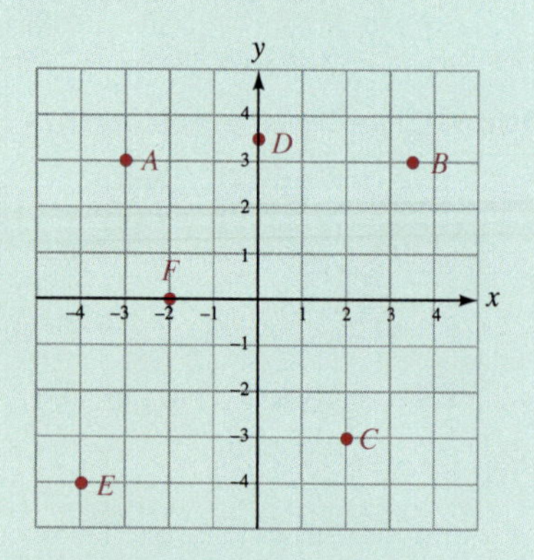

Answers $A(-3, 3), B(3.5, 3), C(2, -3), D(0, 3.5), E(-4, -4), F(-2, 0)$

EXAMPLE 5 Find the coordinates of each point in Figure 6-4.

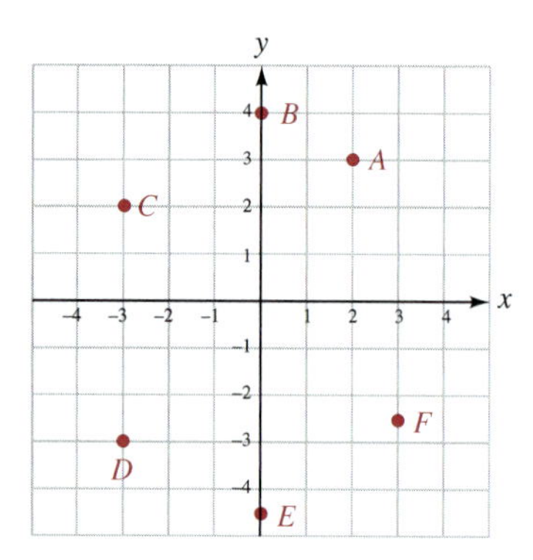

FIGURE 6-4

Solution Since point A is 2 units to the right of the y-axis and 3 units above the x-axis, its coordinates are (2, 3). The coordinates of the other points are found in the same way: $B(0, 4), C(-3, 2), D(-3, -3), E(0, -4.5), F(3, -2.5)$.

An application of the coordinate system

Many cities are laid out in a rectangular grid. For example, on the east side of Rockford, Illinois, all streets run north and south, and all avenues run east and west, as shown in Figure 6-5 on the next page.

In this rectangular grid system, it is very easy to find an address. For example,

- Don Smith lives at the corner of Fourth Street and Third Avenue.
- Mia Vang lives at the corner of Seventh Street and Fifth Avenue.
- The grocery store is at the corner of Second Street and Sixth Avenue.

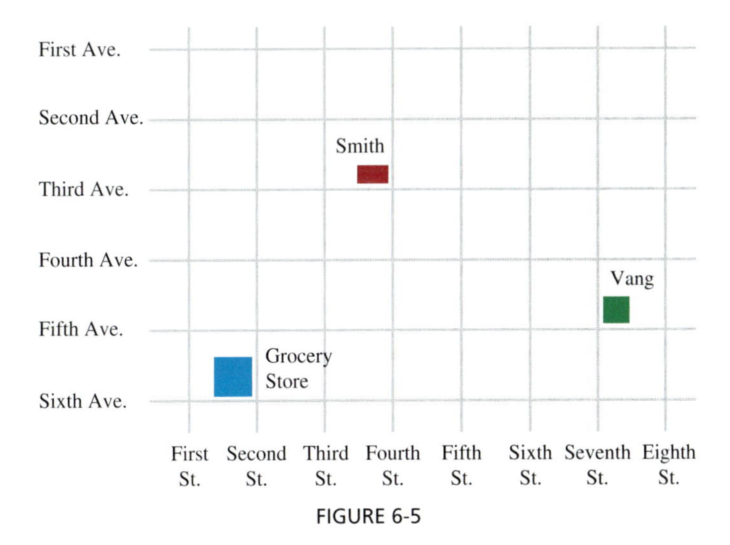

FIGURE 6-5

Population Shift

THINK IT THROUGH

"The interior west states clearly won the national growth sweepstakes in the 1990's. Five western states, Nevada, Arizona, Utah, Colorado, and Idaho, plus Georgia, stand out as growing twice as fast as the national average."

Center of the American West

In the illustration below, data from the 2000 census were used to draw a north–south line so that half of the nation's population lived east and half lived west of it, and an east–west line was drawn so that half of the nation's population lived north and half lived south of it. The point of intersection of the lines occurs in northeast Daviess County, Indiana. It could be thought of as the "center" of the U.S. population in 2000. If the 2000 census recorded the population of the United States to be 285,230,516, how many people lived in each quadrant created by the lines in the illustration?

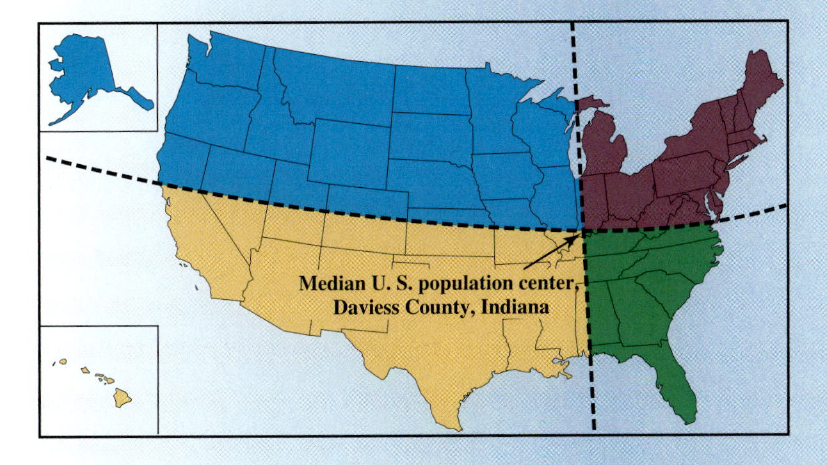

Section 6.1 STUDY SET

VOCABULARY *Fill in the blanks.*

1. The pair (2, 5) is called an ________ pair.
2. The equation $x + y = 4$ contains two ________. The ________ of this equation are pairs of numbers.
3. Since the ordered pair (1, 3) satisfies $x + y = 4$, we say that (1, 3) is a ________ of the equation.
4. The rectangular coordinate ________ is shown. It consists of two ________ lines. Label the x-axis and the y-axis.
5. The rectangular coordinate system is sometimes called the ________ coordinate system.

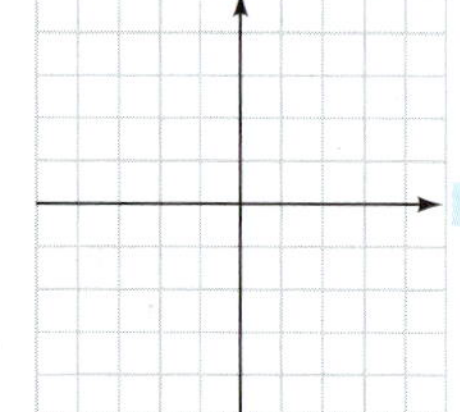

6. The x- and y-axes divide the rectangular coordinate system into four regions called ________.
7. The point where the x- and y-axes cross is called the ________.
8. In the ordered pair $(-2, 4)$, -2 is the x-________, and 4 is the y-coordinate.

CONCEPTS

9. BURNING CALORIES The table shows the number of calories a 140-pound woman would burn doing light activities such as office work, cleaning house, or playing golf. Create a graph using the data.

Minutes of activity	Calories burned
1	4
2	8
3	12
4	16

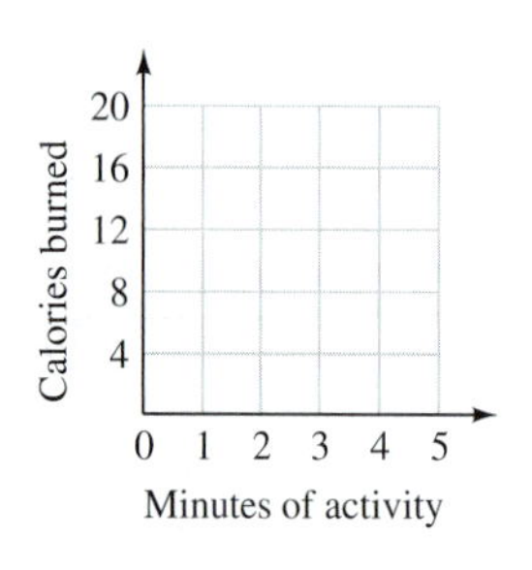

10. CONCRETE The graph shows the number of shovels of sand that should be used for a given number of shovels of cement when mixing concrete for a walkway. Create a table using the data.

Parts cement	Parts sand

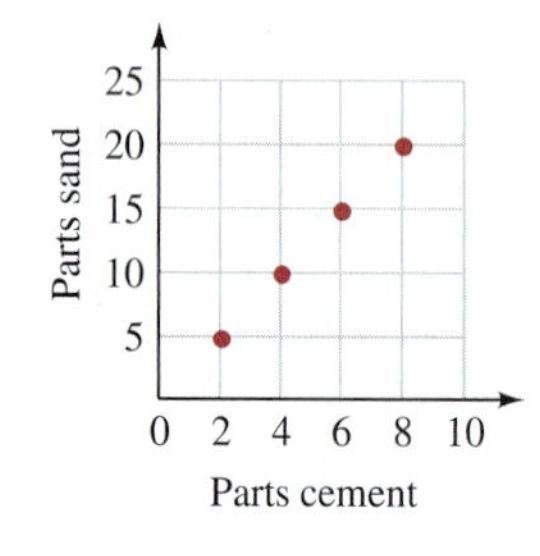

11. Is (2, 3) a solution of $2x + 3y = 14$?
12. Is (4, 1) a solution of $3x - 2y = 10$?

Fill in the blanks.

13. To plot the point with coordinates $(3, -4)$, we start at the ________ and move 3 units to the ________ and then move 4 units ________.
14. To plot the point with coordinates $(-2, 3)$, we start at the ________ and move 2 units to the ________ and then move 3 units ________.

NOTATION *Complete each solution.*

15. For the equation $4x + 3y = 14$, find the value of y when $x = 2$.

$$\begin{aligned} 4x + 3y &= 14 \\ 4(\square) + 3y &= 14 \\ \square + 3y &= 14 \\ 3y &= \square \\ y &= 2 \end{aligned}$$

16. For the equation $4x - 3y = 12$, find the value of x when $y = 6$.

$$\begin{aligned} 4x - 3y &= 12 \\ 4x - 3(\square) &= 12 \\ 4x - \square &= 12 \\ 4x &= \square \\ x &= 7.5 \end{aligned}$$

17. Is the point $(3, -\frac{5}{2})$ the same as $(3, -2.5)$?
18. List the Roman numerals from 1 to 4. How are they used in this section?

PRACTICE *Complete each statement.*

19. $3x + y = 12$
    a. If $x = 0$, then $y = \square$.
    b. If $y = 0$, then $x = \square$.
    c. If $x = 2$, then $y = \square$.
20. $4x + 3y = 24$
    a. If $x = 0$, then $y = \square$.
    b. If $y = 0$, then $x = \square$.
    c. If $y = 2$, then $x = \square$.

Complete the ordered pairs so that they are solutions of the equation.

21. $2x + y = 8$: a. (0,), b. (, 0), and c. (, 2)
22. $x - 3y = 5$: a. (2,), b. (, 3), and c. (26,)

Complete each table of solutions.

23. $5x - 4y = 20$

x	y	(x, y)
0		(0,)
	0	(, 0)
	5	(, 5)

24. $7x - y = 21$

x	y	(x, y)
0		(0,)
	0	(, 0)
2		(2,)

Graph each point on the coordinate grid.

25. $(1, 3), (-2, 4), (-3, -2), (3, -2)$

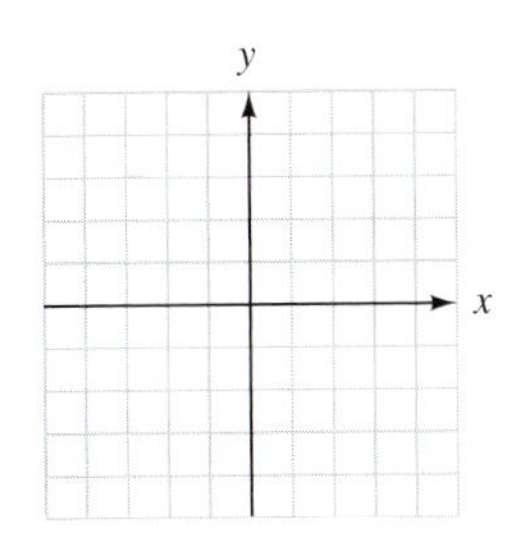

26. $(-\frac{3}{2}, 2), (3, -\frac{5}{2}), (0, -3), (3, 0)$

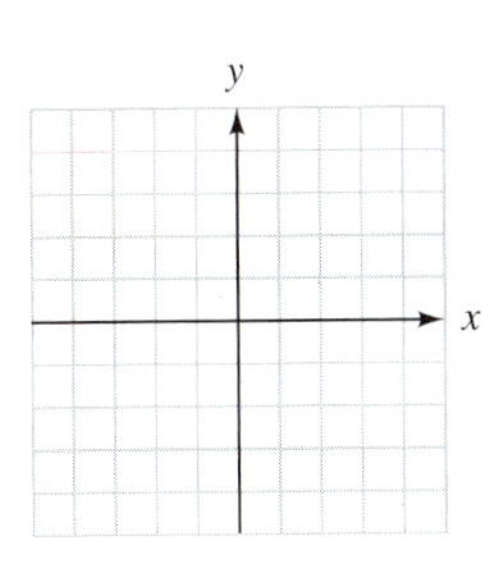

27. $(-4, -3), (1.5, 1.5), (-3.5, 0), (0, 3.5)$

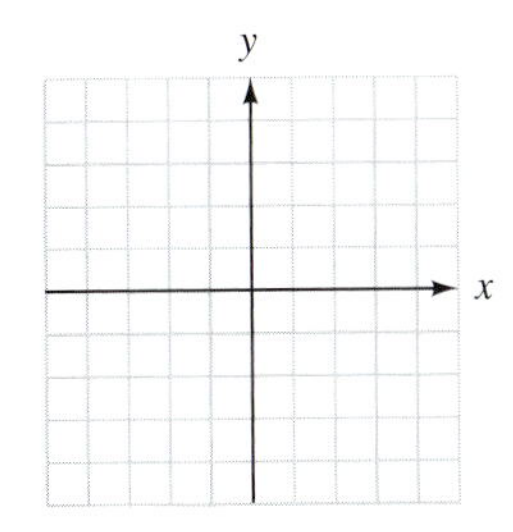

28. $(0, 0), (-\frac{1}{2}, \frac{5}{2}), (5, -5), (-5, 5)$

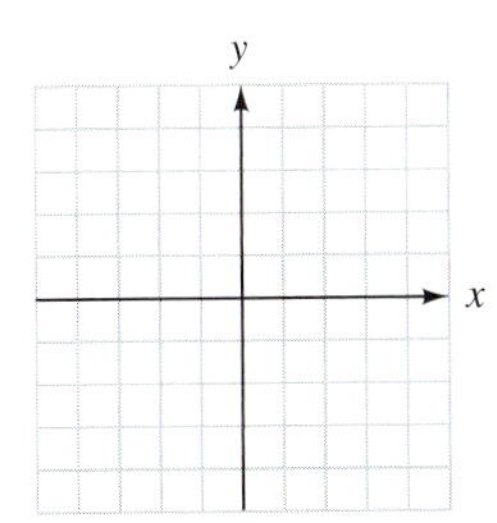

Find the coordinates of each point shown in the graph.

29.

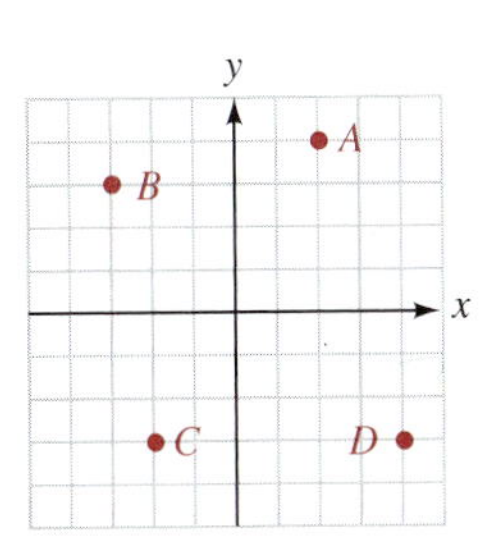

30.

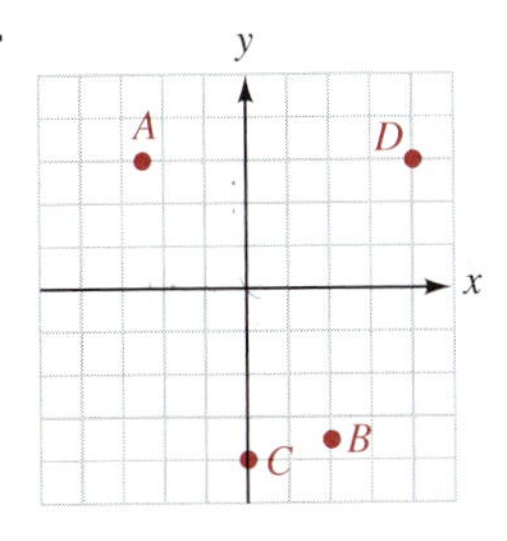

31.

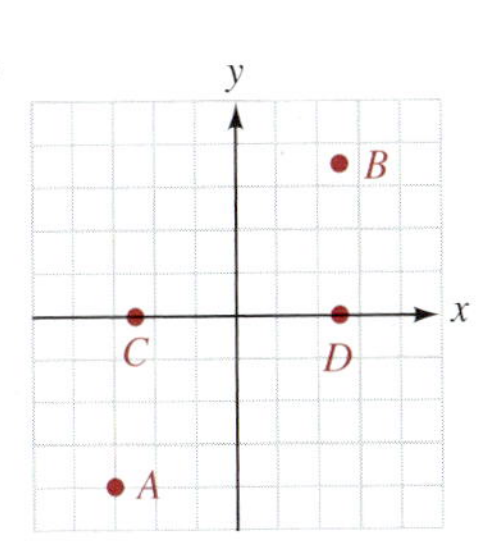

32.

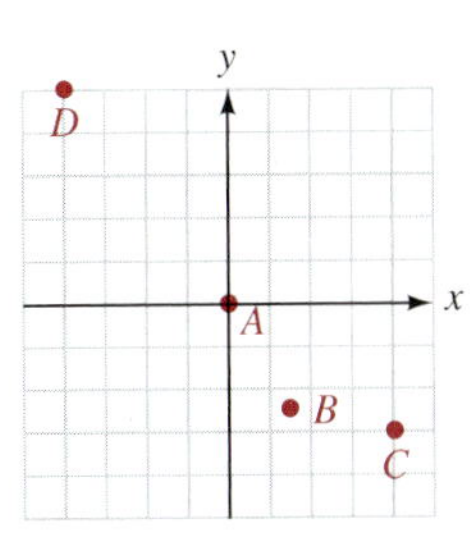

APPLICATIONS

33. MAPS Road maps usually have a coordinate system to help locate cities. Use the following map to locate Rockford, Mount Carroll, Harvard, and the intersection of State Highway 251 and U.S. Highway 30. Express each answer in the form (number, letter).

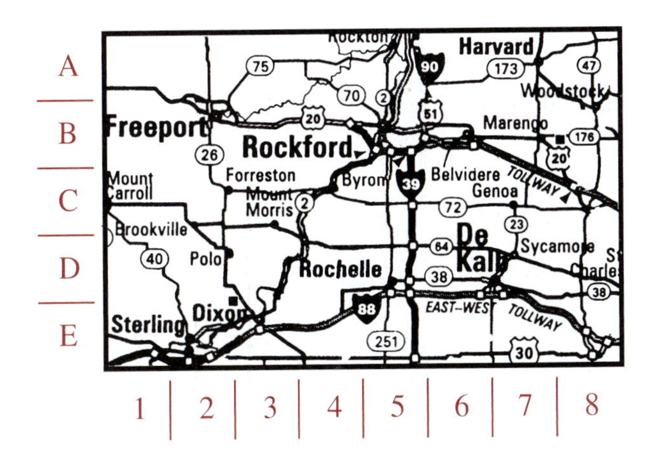

34. BATTLESHIPS In the game Battleship, the player uses coordinates to drop depth charges from a battleship to hit a hidden submarine. Find the coordinates that should be used to make three hits on the submarine shown.

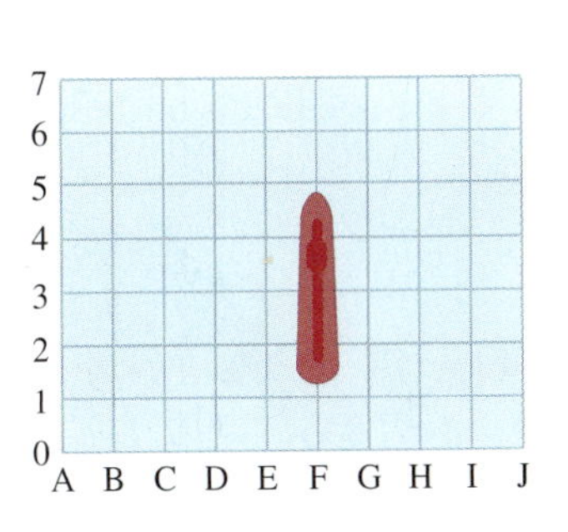

35. EARTHQUAKES The map shown on the next page shows the area where damage was caused by an earthquake.

a. Find the coordinates of the epicenter (the source of the quake).

b. Was damage done at the point (4, 5)?

c. Was damage done at the point (−1, −4)?

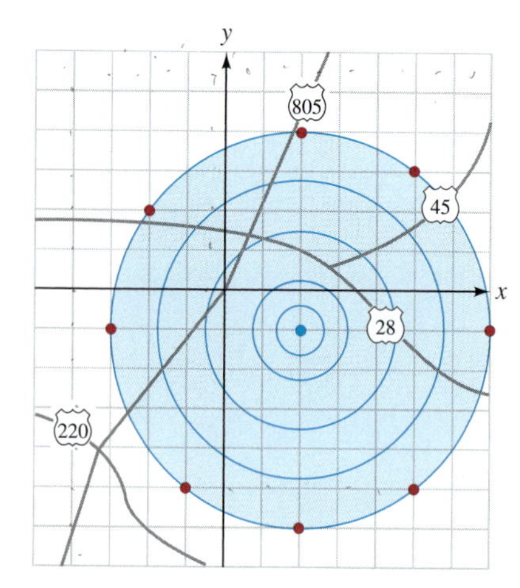

36. AUTOMATION A robot can be programmed to make welds on a car chassis. To do this, an imaginary coordinate system is superimposed on the side of the car. Using the commands Up, Down, Left, and Right, write a set of instructions for the robot arm to move from its initial position to weld the points A, B, C, and D, in that order.

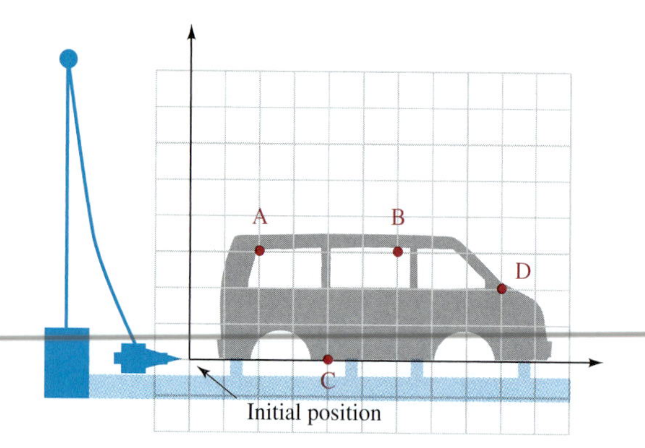

37. THE GLOBE A coordinate system that is used to locate places on the surface of the Earth uses a series of curved lines running north and south and east and west, as shown. List the cities in order, beginning with the one that is farthest east on this map.

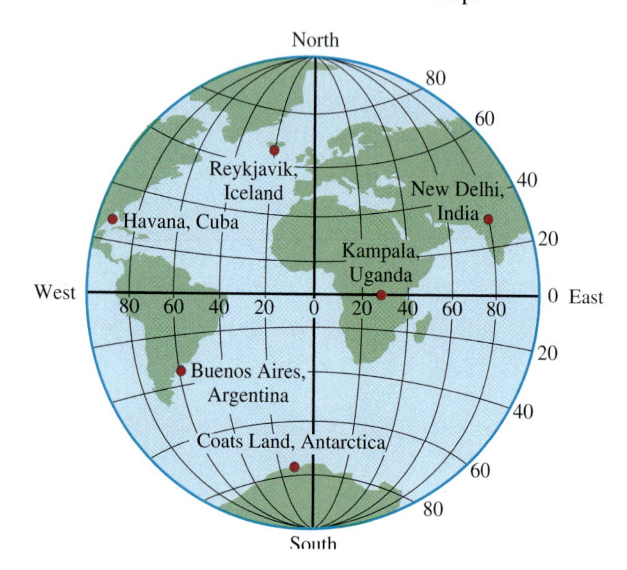

38. BLOOD TRANSFUSIONS The shaded boxes in the illustration indicate the compatibility of the major blood groups, AB, A, B, and O. Red cells of the donor are mixed with serum of the recipient to test for clumping. List the compatible blood groups as ordered pairs of the form (donor, recipient).

Serum of recipient / group	Red cells of donor AB	A	B	O
AB				
A				
B				
O				

39. COOKING Use the information from the table to complete the graph in the illustration for 2, 4, 6, 8, and 10 servings of instant mashed potatoes.

Number of servings	2	4	6	8	10
Flakes (cups)	$\frac{2}{3}$	$1\frac{1}{3}$	2	$2\frac{2}{3}$	$3\frac{1}{3}$

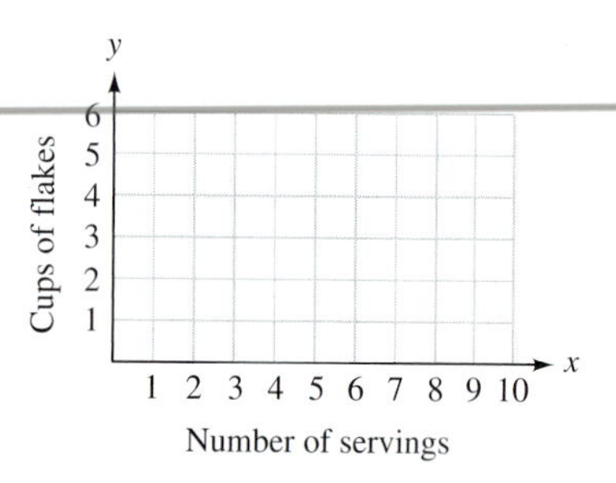

40. DICE The red point in Illustration (a) represents one of the 36 possible outcomes when two fair dice are rolled a single time. Draw the appropriate number of dots on the top face of each die in Illustration (b) to illustrate this outcome.

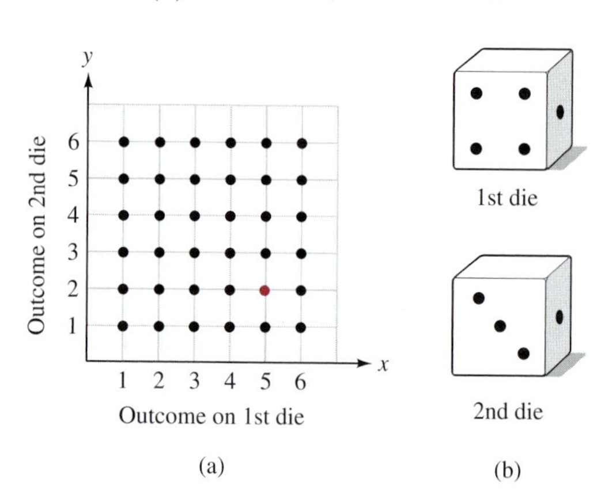

(a) (b)

WRITING

41. Explain why the point with coordinates $(-4, 4)$ is not the same as the point with coordinates $(4, -4)$.

42. Explain the difference between $3(6)$, $(3, 6)$, and $(3 + 6)$.

43. Explain how to plot the point with coordinates of $(4, -3)$.

44. Explain why the coordinates of the origin are $(0, 0)$.

REVIEW *Evaluate each expression.*

45. $(-8 - 5) - 3$

46. $-1 - [5 - (-3)]$

47. $(-4)^2 - 3^2$

48. $-5 - \frac{24}{6} - 8(-3)$

Solve each equation and check the result.

49. $\frac{x}{3} + 3 = 10$

50. $-3x - 4 = 8$

51. $5 - (7 - x) = -5$

52. $2(y + 6) - 4 = 2$

Use a scientific calculator to find each value.

53. $(4^2)^4$

54. $(3^4)^3$

6.2 Graphing Linear Equations

- Making a table of solutions
- Graphing linear equations
- Using intercepts to graph linear equations
- Graphing equations that are solved for y
- Graphing equations of the form $y = b$ and $x = a$

In the previous section, we saw that solutions of equations containing the variables x and y were ordered pairs of real numbers (x, y). We also saw that ordered pairs can be graphed on a rectangular coordinate system. In this section, we will use these skills to see how plotting points can give the graph of an equation.

Making a table of solutions

To find solutions of equations in x and y, we can pick numbers at random, substitute them for x, and find the corresponding values of y. For example, to find an ordered pair that satisfies the equation $2x - 3y = 6$, we can let $x = 0$ and solve for y.

$$2x - 3y = 6$$

$$2(0) - 3y = 6 \quad \text{Substitute 0 for } x.$$

$$0 - 3y = 6 \quad \text{Perform the multiplication: } 2(0) = 0.$$

$$-3y = 6 \quad \text{Perform the subtraction: } 0 - 3y = -3y.$$

$$\frac{-3y}{-3} = \frac{6}{-3} \quad \text{To undo the multiplication by } -3 \text{, divide both sides by } -3.$$

$$y = -2 \quad \text{Perform the division: } \tfrac{6}{-3} = -2.$$

We have found that the ordered pair $(0, -2)$ is a solution of $2x - 3y = 6$. As we find solutions, we will list them in a table of solutions shown below.

x	y	(x, y)
0	-2	$(0, -2)$

To find another solution of the equation, we let $x = -3$, and solve for y.

$$2x - 3y = 6$$

$$2(-3) - 3y = 6 \quad \text{Substitute } -3 \text{ for } x.$$

$$-6 - 3y = 6 \quad \text{Perform the multiplication: } 2(-3) = -6.$$

$$-3y = 12 \quad \text{To eliminate the } -6 \text{ from the left-hand side, add 6 to both sides.}$$

$$\frac{-3y}{-3} = \frac{12}{-3} \quad \text{To undo the multiplication by } -3\text{, divide both sides by } -3.$$

$$y = -4 \quad \text{Perform the division: } \tfrac{12}{-3} = -4.$$

x	y	(x, y)
0	−2	(0, −2)
−3	**−4**	**(−3, −4)**

A second solution is $(-3, -4)$, and we list it in the table.

We can also find solutions of $2x - 3y = 6$ by picking a number for y and finding the corresponding value of x. For example, we can let $y = 0$ and solve for x.

$$2x - 3y = 6$$

$$2x - 3(0) = 6 \quad \text{Substitute 0 for } y.$$

$$2x - 0 = 6 \quad \text{Perform the multiplication: } 3(0) = 0.$$

$$2x = 6 \quad \text{Perform the subtraction: } 2x - 0 = 2x.$$

$$\frac{2x}{2} = \frac{6}{2} \quad \text{To undo the multiplication by 2, divide both sides by 2.}$$

$$x = 3 \quad \text{Perform the division: } \tfrac{6}{2} = 3.$$

x	y	(x, y)
0	−2	(0, −2)
−3	−4	(−3, −4)
3	**0**	**(3, 0)**

A third solution is $(3, 0)$, which we also add to the table.

If we pick $y = 2$, we have

$$2x - 3y = 6$$

$$2x - 3(2) = 6 \quad \text{Substitute 2 for } y.$$

$$2x - 6 = 6 \quad \text{Perform the multiplication: } 3(2) = 6.$$

$$2x = 12 \quad \text{To undo the subtraction of 6, add 6 to both sides.}$$

$$\frac{2x}{2} = \frac{12}{2} \quad \text{To undo the multiplication by 2, divide both sides by 2.}$$

$$x = 6 \quad \text{Perform the division: } \tfrac{12}{2} = 6.$$

x	y	(x, y)
0	−2	(0, −2)
−3	−4	(−3, −4)
3	0	(3, 0)
6	**2**	**(6, 2)**

A fourth solution is $(6, 2)$. We enter it in the table.

Since any choice of x will give a corresponding value of y, and any choice of y will give a corresponding value of x, it is apparent that $2x - 3y = 6$ has *infinitely many solutions.* We have found four of them: $(0, -2)$, $(-3, -4)$, $(3, 0)$, and $(6, 2)$.

Graphing linear equations

To graph $2x - 3y = 6$, we plot the ordered pairs that we have just found. See Figure 6-6(a). In Figure 6-6(b), we see that the four points lie on a line.

In Figure 6-6(c), we draw a straight line through the points, because the graph of any solution of $2x - 3y = 6$ will lie on this line. The arrowheads show that the line continues forever in both directions. The line is a picture of all of the solutions of $2x - 3y = 6$, and it is called a **graph** of the equation. Every point on the line is a

2x − 3y = 6		
x	y	(x, y)
0	−2	(0, −2)
−3	−4	(−3 −4)
3	0	(3, 0)
6	2	(6, 2)

(a)

solution of the equation, and every solution of the equation is on the line. Equations whose graphs are straight lines, such as $2x - 3y = 6$, are called **linear equations.**

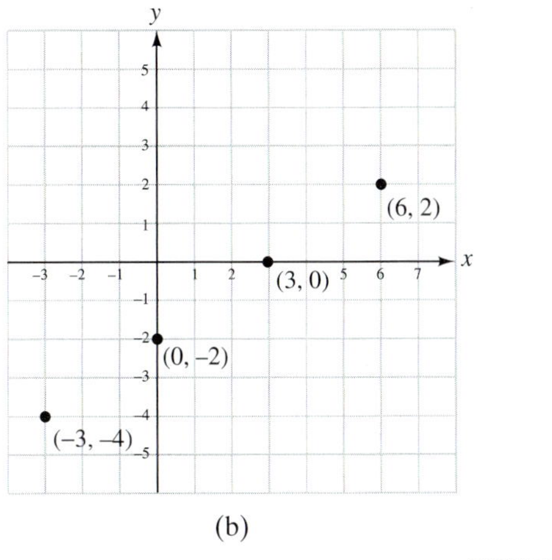

(b)

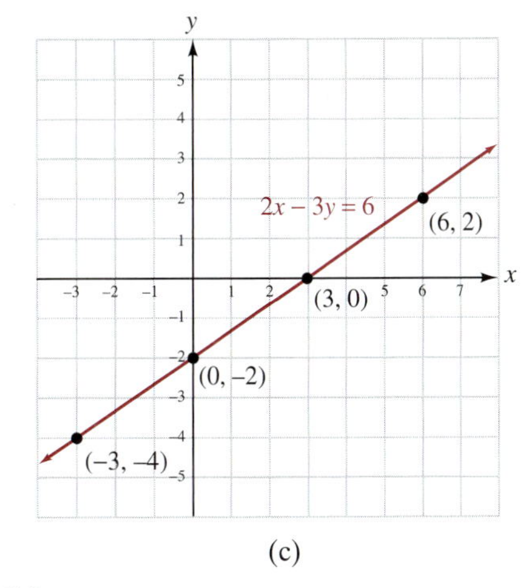

(c)

FIGURE 6-6

When we graphed $2x - 3y = 6$, we did more work than necessary. Since two points determine a line, only two points were necessary to graph the equation. However, it is always a good idea to plot a third point as a check. If the three points do not lie on a straight line, then at least one of them is in error. To graph a linear equation, we will follow these steps.

Strategy for graphing linear equations in x and y

1. Find two pairs (x, y) that satisfy the equation by picking arbitrary numbers for x (or y) and solving for the corresponding values of y (or x). A third point acts as a check.
2. Plot each resulting pair (x, y) on a rectangular coordinate system. If they do not lie on a straight line, check your calculations.
3. Use a straightedge to draw the line passing through the three points. Draw arrowheads to indicate that the line continues forever in both directions.

Using intercepts to graph linear equations

In Figure 6-6(c), the line crosses the y-axis at the point with coordinates $(0, -2)$ and crosses the x-axis at the point with coordinates $(3, 0)$. These points have special names.

y- and x-intercepts

The point where the graph of a linear equation crosses the y-axis is called the **y-intercept.** To find the coordinates of the y-intercept, let $x = 0$ and solve for y.

The point where the graph of a linear equation crosses the x-axis is called the **x-intercept.** To find the coordinates of the x-intercept, let $y = 0$ and solve for x.

EXAMPLE 1 Graph $3x + 2y = 6$ by plotting the x- and y-intercepts and a third point on the line.

Solution To find the coordinates of the x-intercept, we substitute 0 for y in $3x + 2y = 6$ and find that $x = 2$. The x-intercept is $(2, 0)$.

Self Check 1
Graph: $2x + 3y = 6$.

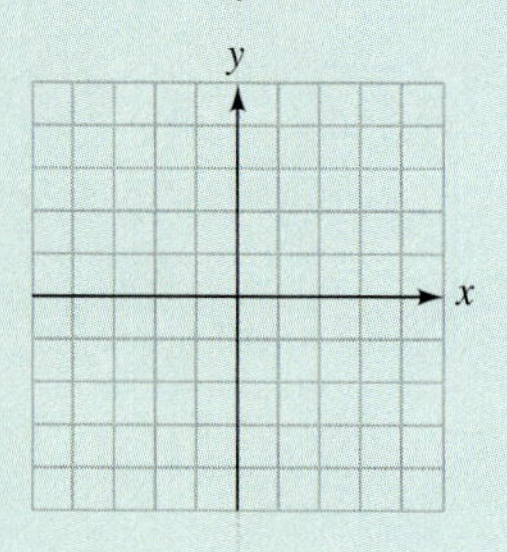

Answer

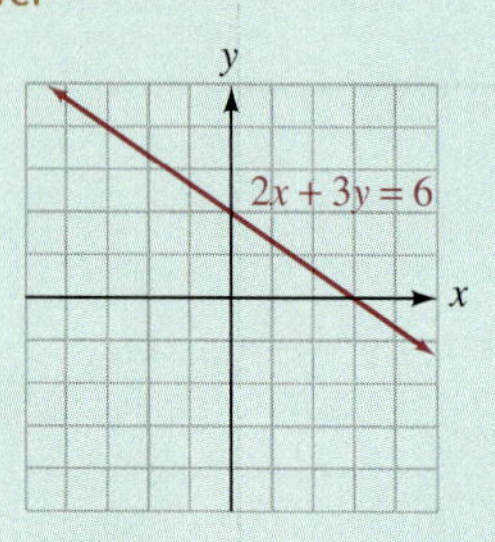

To find the coordinates of the y-intercept, we substitute 0 for x in $3x + 2y = 6$ and find that $y = 3$. The y-intercept is $(0, 3)$.

To find a third point on the line, we let $x = 1$. Then $y = \frac{3}{2}$.

We list the ordered pairs $(2, 0)$, $(0, 3)$, and $(1, \frac{3}{2})$ in a table. To get the graph, we plot each point and join them with a line, as shown in Figure 6-7.

	$3x + 2y = 6$		
	x	y	(x, y)
x-intercept ⟶	2	0	$(2, 0)$
y-intercept ⟶	0	3	$(0, 3)$
	1	$\frac{3}{2}$	$(1, \frac{3}{2})$

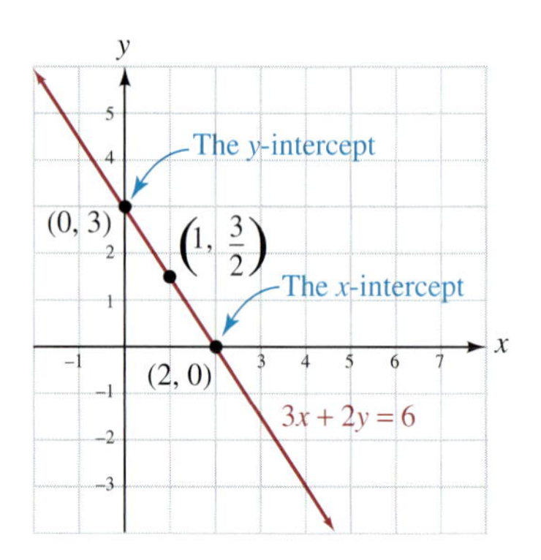

FIGURE 6-7

Graphing equations that are solved for y

We have graphed the equation $3x + 2y = 6$ and gotten a line. Equations of a line can occur in other forms, such as $y = 2x - 1$. Since this equation is solved for y, the value of y depends on the value chosen for x. For this reason, we call y the **dependent variable** and x the **independent variable.**

To graph $y = 2x - 1$, we will find three ordered pairs (x, y) that satisfy the equation. Since x is the independent variable, it is easiest to pick three values for x and solve for y. If we let $x = 0$, we will find the coordinates of the y-intercept. We may pick any three values for x, but it is a good idea to make $x = 0$ one of those choices, because it makes the computations easy.

$y = 2x - 1$
$= 2(\mathbf{0}) - 1$ Substitute 0 for x.
$= 0 - 1$ Perform the multiplication: $2(0) = 0$.
$= -1$ Perform the subtraction.

x	y	(x, y)
0	**−1**	**(0, −1)**

The y-intercept is $(0, -1)$. If we let $x = 1$, we have

$y = 2x - 1$
$= 2(\mathbf{1}) - 1$ Substitute 1 for x.
$= 2 - 1$ Perform the multiplication: $2(1) = 2$.
$= 1$ Perform the subtraction.

x	y	(x, y)
0	−1	(0, −1)
1	**1**	**(1, 1)**

The ordered pair $(1, 1)$ satisfies the equation. If we let $x = -2$, we have

$y = 2x - 1$
$= 2(\mathbf{-2}) - 1$ Substitute -2 for x.
$= -4 - 1$ Perform the multiplication: $2(-2) = -4$.
$= -5$ Perform the subtraction.

x	y	(x, y)
0	−1	(0, −1)
1	1	(1, 1)
−2	**−5**	**(−2, −5)**

The ordered pair $(-2, -5)$ satisfies the equation. We copy the final table of solutions in Figure 6-8, plot the points, and draw a straight line through them. This line is the graph of the equation.

$y = 2x - 1$		
x	y	(x, y)
0	−1	(0, −1)
1	1	(1, 1)
−2	−5	(−2, −5)

We can pick any three values for x.

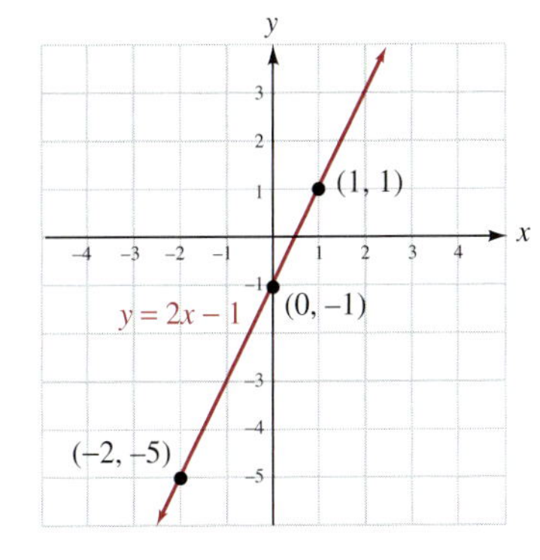

FIGURE 6-8

EXAMPLE 2 Graph: $y = \frac{1}{2}x + 1$.

Solution We pick three values for x and then calculate each corresponding value of y. If we let $x = 0$, then

$$
\begin{aligned}
y &= \frac{1}{2}(\mathbf{0}) + 1 && \text{Substitute 0 for } x. \\
&= 0 + 1 && \text{Perform the multiplication.} \\
&= 1 && \text{The point } (0, 1) \text{ lies on the graph. It is the } y\text{-intercept.}
\end{aligned}
$$

Since each value of x will be multiplied by $\frac{1}{2}$, we choose values for x that can be divided by 2 to make the multiplication easy. If we let $x = 4$, then

$$
\begin{aligned}
y &= \frac{1}{2}(\mathbf{4}) + 1 && \text{Substitute 4 for } x. \\
&= \frac{4}{2} + 1 && \text{Multiply: } \frac{1}{2}(4) = \frac{1}{2}\left(\frac{4}{1}\right) = \frac{4}{2}. \\
&= 2 + 1 && \text{Divide: } \frac{4}{2} = 2. \\
&= 3 && \text{The point } (4, 3) \text{ lies on the graph.}
\end{aligned}
$$

If we let $x = -4$, then

$$
\begin{aligned}
y &= \frac{1}{2}(\mathbf{-4}) + 1 && \text{Substitute } -4 \text{ for } x. \\
&= \frac{-4}{2} + 1 && \frac{1}{2}(-4) = \frac{1}{2}\left(\frac{-4}{1}\right) = \frac{-4}{2}. \\
&= -2 + 1 && \text{Perform the division: } \frac{-4}{2} = -2. \\
&= -1 && \text{The point } (-4, -1) \text{ lies on the graph.}
\end{aligned}
$$

We list these coordinates in a table of solutions, plot each point, and draw a straight line through them, as shown in Figure 6-9 on the next page.

Self Check 2
Graph: $y = 3x - 2$.

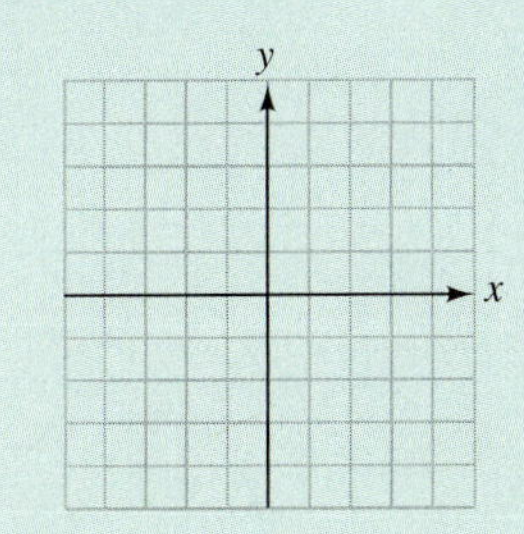

Answer

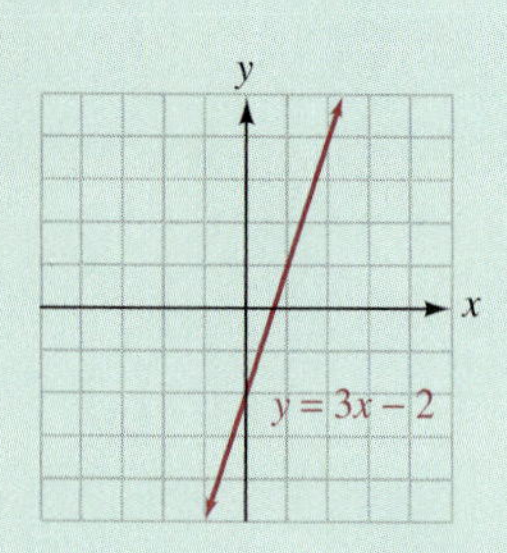

$y = \frac{1}{2}x + 1$		
x	y	(x, y)
0	1	(0, 1)
4	3	(4, 3)
−4	−1	(−4, −1)

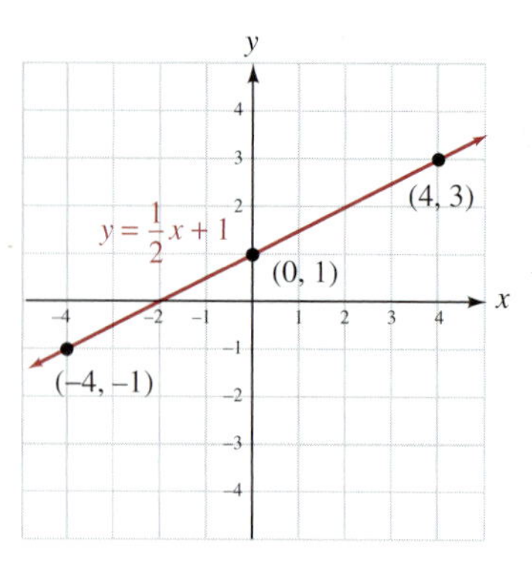

FIGURE 6-9

Self Check 3
Graph: $y = -25x + 50$.

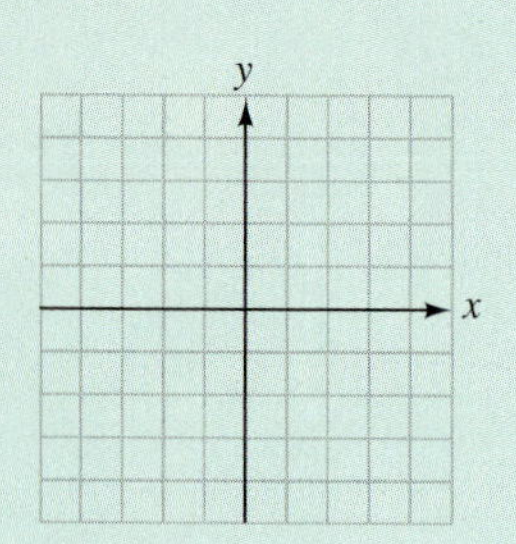

Answer (scaling may vary)

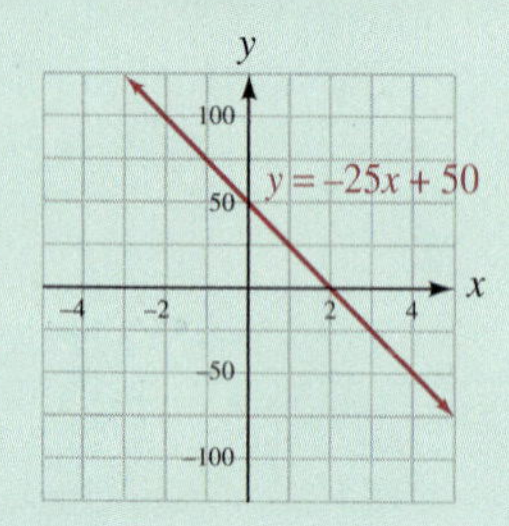

EXAMPLE 3 Graph: $y = 20x$.

Solution We begin by randomly selecting three values for x: −2, 0, and 2. If $x = -2$, we can calculate the corresponding value of y by substituting −2 for x in $y = 20x$.

$$y = 20x$$
$$= 20(-2) \quad \text{Substitute } -2 \text{ for } x.$$
$$= -40 \quad \text{Perform the multiplication.}$$

We see that $x = -2$ and $y = -40$ is a solution of $y = 20x$. In a similar manner, we find the corresponding values for y when x is 0 and 2 and enter them in the table in Figure 6-10.

Because of the sizes of the y-coordinates of the points (−2, −40) and (2, 40), we must adjust the scale on the y-axis. (If we used grid lines 1 unit apart, the graph would be very large.)

One way to accommodate these points is to scale the y-axis in units of 5, 10, or 20. If we choose divisions of 20 units, plot the three solutions from the table, and draw a line through them, we get the graph shown in Figure 6-10.

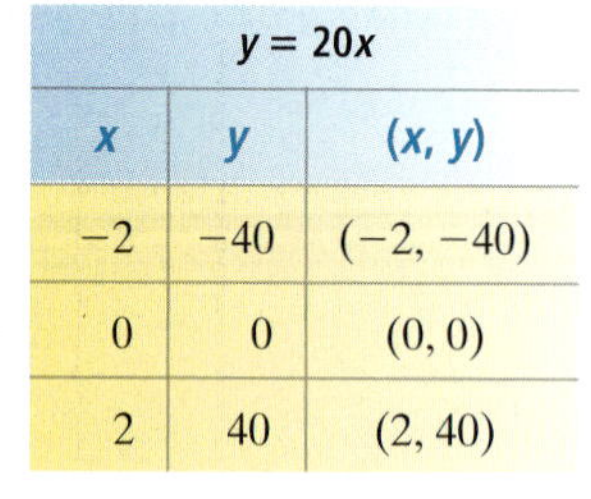

$y = 20x$		
x	y	(x, y)
−2	−40	(−2, −40)
0	0	(0, 0)
2	40	(2, 40)

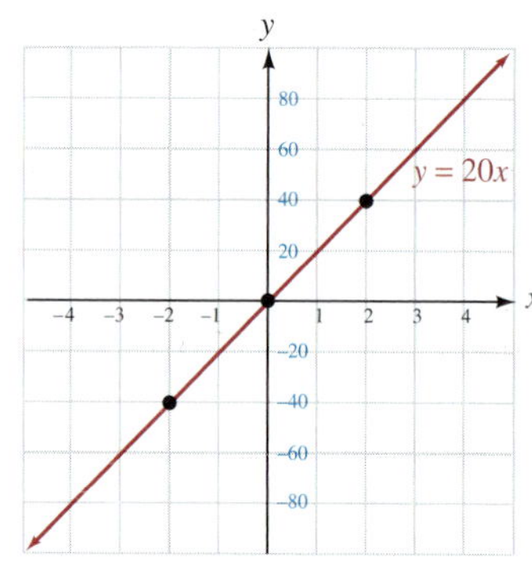

FIGURE 6-10

Graphing equations of the form $y = b$ and $x = a$

When a linear equation contains only one variable, such as $y = 4$ or $x = -2$, its graph is either a horizontal or a vertical line.

EXAMPLE 4 Graph: $y = 4$.

Solution We can write this equation as $0x + y = 4$. Since the coefficient of x is 0, the numbers assigned to x have no effect on y. The value of y is always 4. For example, if $x = 5$, then $y = 4$.

$$0x + y = 4$$
$$0(5) + y = 4 \quad \text{Substitute 5 for } x.$$
$$0 + y = 4 \quad \text{Multiply: } 0(5) = 0.$$
$$y = 4 \quad \text{Add: } 0 + y = y.$$

Similarly, if $x = 2$, then $y = 4$, and if $x = -2$, then $y = 4$. To graph the equation, we plot the points and draw the line, as shown in Figure 6-11.

$y = 4$		
x	y	(x, y)
5	4	(5, 4)
2	4	(2, 4)
−2	4	(−2, 4)

All y-values are 4.

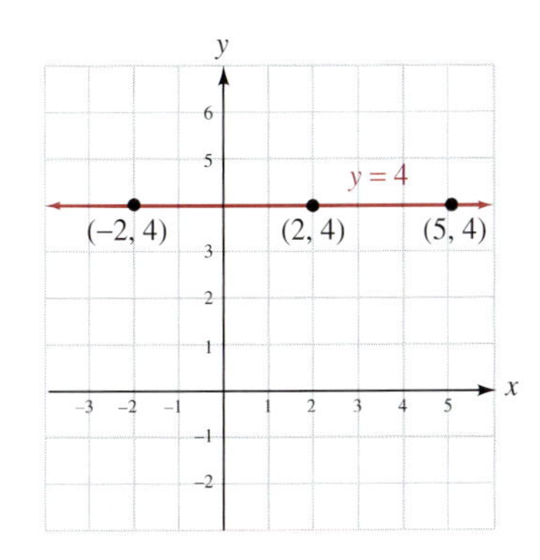

FIGURE 6-11

Self Check 4
Graph: $y = -2$.

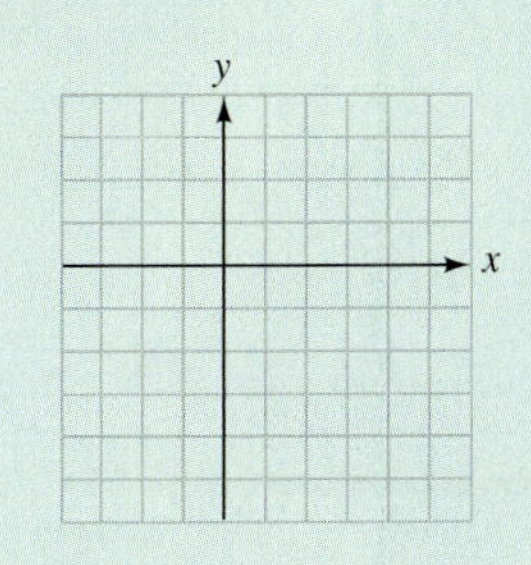

Answer

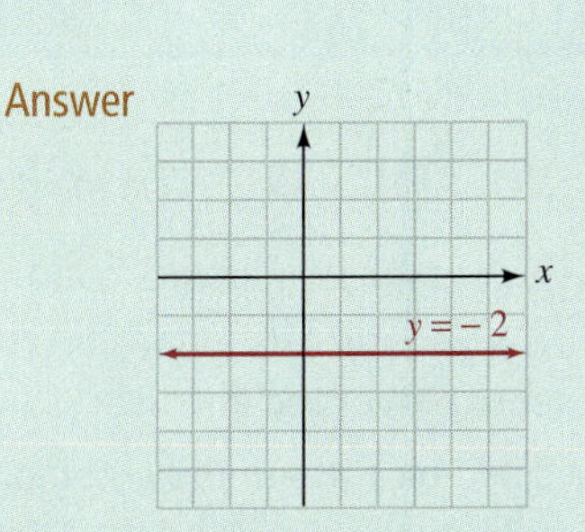

EXAMPLE 5 Graph: $x = -2$.

Solution We can write this equation as $x + 0y = -2$. Since the coefficient of y is 0, the numbers assigned to y have no effect on x. The value of x is always −2. For example, if $y = 1$, then $x = -2$.

$$x + 0y = -2$$
$$x + 0(1) = -2 \quad \text{Substitute 1 for } y.$$
$$x + 0 = -2 \quad \text{Multiply: } 0(1) = 0.$$
$$x = -2 \quad \text{Add: } x + 0 = x.$$

Similarly, if $y = 2$, then $x = -2$, and if $y = -2$, then $x = -2$. To graph the equation, we plot the points and draw the line, as shown in Figure 6-12.

$x = -2$		
x	y	(x, y)
−2	1	(−2, 1)
−2	2	(−2, 2)
−2	−2	(−2, −2)

All x-values are −2.

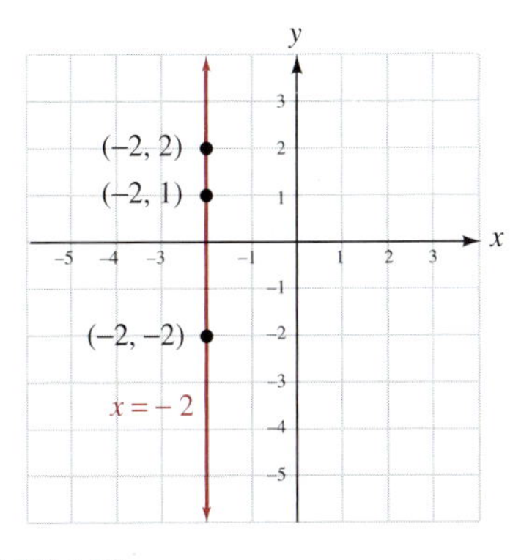

FIGURE 6-12

Self Check 5
Graph: $x = 3$.

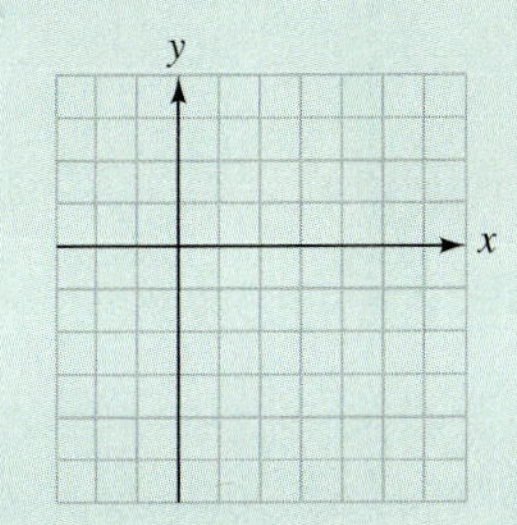

Answer

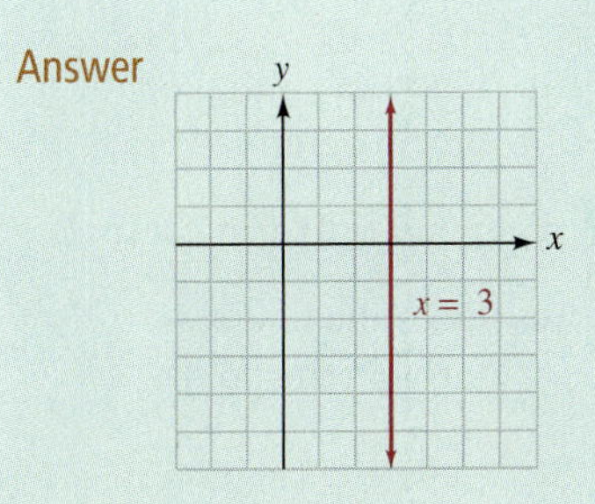

The results of Example 4 and Self Check 4 suggest that the graphs of equations of the form $y = b$ are horizontal lines. The results of Example 5 and Self Check 5 suggest that the graphs of equations of the form $x = a$ are vertical lines. We now summarize these observations.

Equations of horizontal and vertical lines

The equation $y = b$ represents a horizontal line that intersects the y-axis at the point $(0, b)$. If $b = 0$, the line is the x-axis.

The equation $x = a$ represents a vertical line that intersects the x-axis at the point $(a, 0)$. If $a = 0$, the line is the y-axis.

Section 6.2 STUDY SET

VOCABULARY *Fill in the blanks.*

1. The graph of a linear equation is a ______.
2. If a point lies on a line, its coordinates ________ the equation of the line.
3. The point where the graph of a linear equation crosses the x-axis is called the ___________.
4. The y-intercept is the point where the graph of a linear equation crosses the _______.
5. In the equation $y = 7x + 2$, x is called the ____________ variable.
6. In the equation $y = 7x + 2$, y is called the __________ variable.

CONCEPTS *Fill in the blanks.*

7. The graph of the equation $y = 3$ is a __________ line.
8. The graph of the equation $x = -2$ is a _________ line.
9. **a.** Find the y-intercept of the line graphed.

   b. What is its x-intercept?

   c. Does the line pass through the point $(4, 3)$?

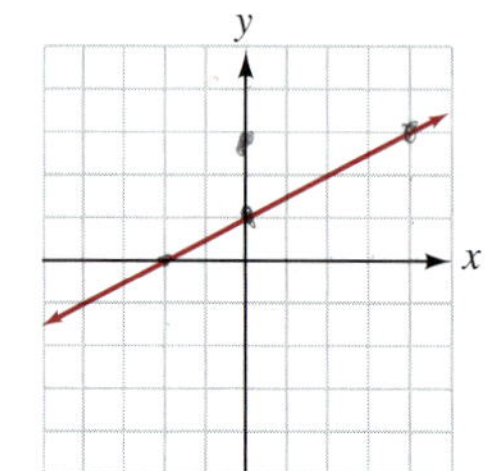

10. **a.** Find the y-intercept of the line graphed.

    b. What is its x-intercept?

    c. Does the line pass through the point $(1, -1)$?

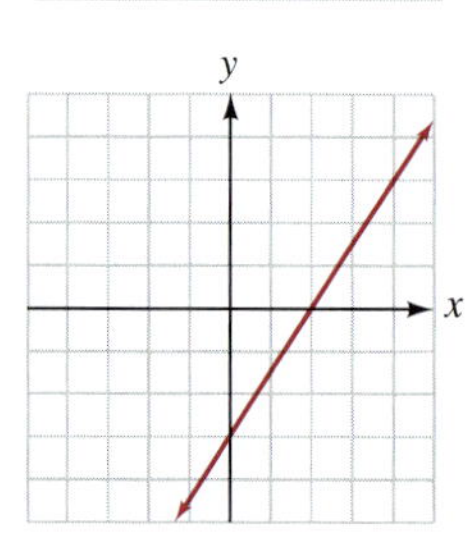

11. What is wrong with the graph of $x - y = 3$ shown below?

x	y	(x, y)
0	−3	(0, −3)
3	0	(3, 0)
1	−2	(1, −2)

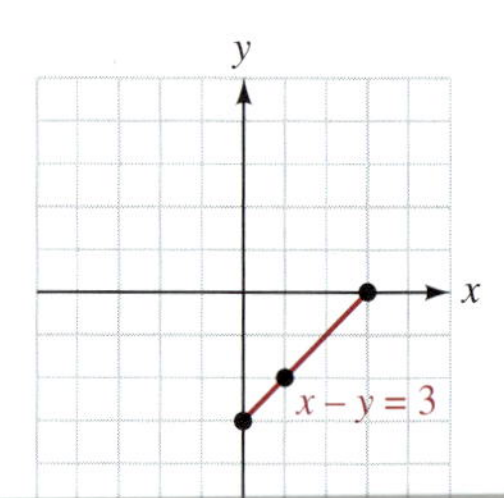

12. To graph $y = -x + 1$, a student constructed a table of solutions and plotted the ordered pairs as shown. Instead of drawing a curve through the points, what should he have done?

$y = -x + 1$		
x	y	(x, y)
−3	−2	(−3, −2)
0	1	(0, 1)
2	−1	(2, −1)

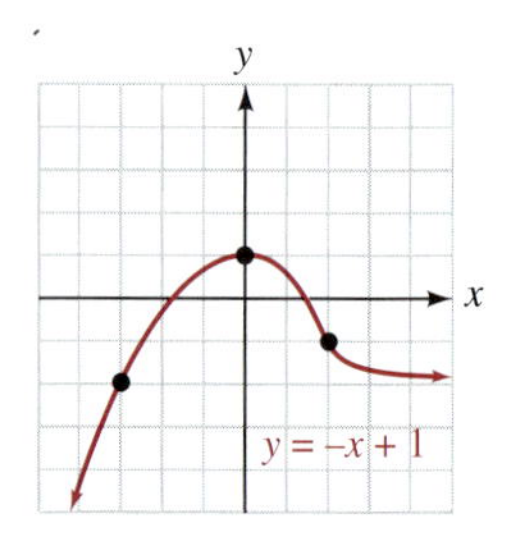

13. The graph of a linear equation is shown. Determine six solutions of the equation from the graph.

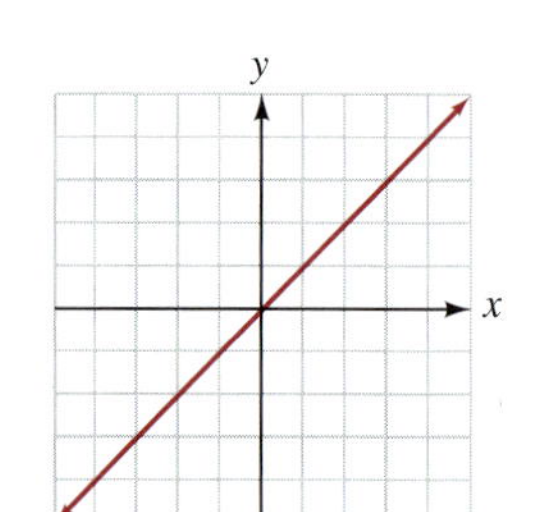

14. The graph of a linear equation is shown. What three points were apparently plotted to obtain the graph? Show them in the table of solutions.

x	y	(x, y)

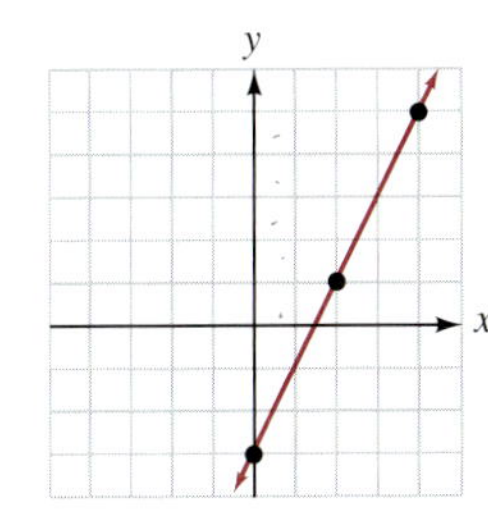

NOTATION *Complete each solution.*

15. Consider the equation $2x - 4y = 8$. Find y if $x = 3$.

$$2x - 4y = 8$$
$$2(\ \ ) - 4y = 8$$
$$\ \ - 4y = 8$$
$$-4y = \ \ $$
$$y = -\frac{1}{2}$$

16. Consider the equation $y = \frac{2}{3}x - 1$. Find y when $x = -3$.

$$y = \frac{2}{3}x - 1$$
$$y = \frac{2}{3}(\ \ ) - 1$$
$$y = \ \ - 1$$
$$y = -3$$

PRACTICE *Complete each table of solutions.*

17. $2x - 5y = 10$

x	y	(x, y)
5		
−5		
10		

18. $3x + 4y = 18$

x	y	(x, y)
	0	
	3	
	6	

19. $y = 2x - 3$

x	y	(x, y)
3		
−4		
6		

20. $y = -3x + 4$

x	y	(x, y)
4		
−3		

Find the coordinates of the y- and x-intercepts of the graph of each equation.

21. $x + y = 5$

22. $x - y = 2$

23. $4x + 5y = 20$

24. $3x - 5y = 15$

Complete the table and graph the equation.

25. $x + y = 5$ (see Exercise 21)

x	y
0	
	0
2	

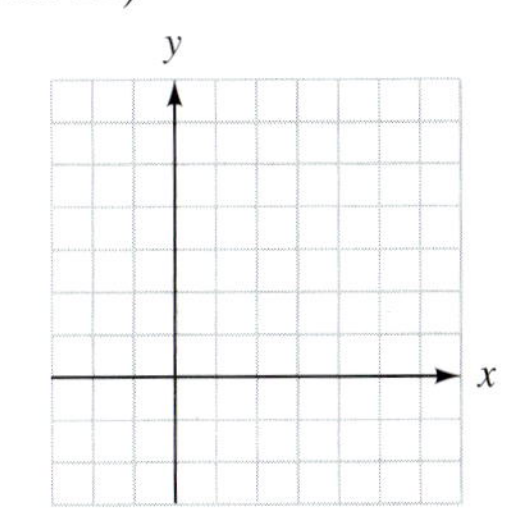

26. $x - y = 2$ (see Exercise 22)

x	y
0	
	0
−2	

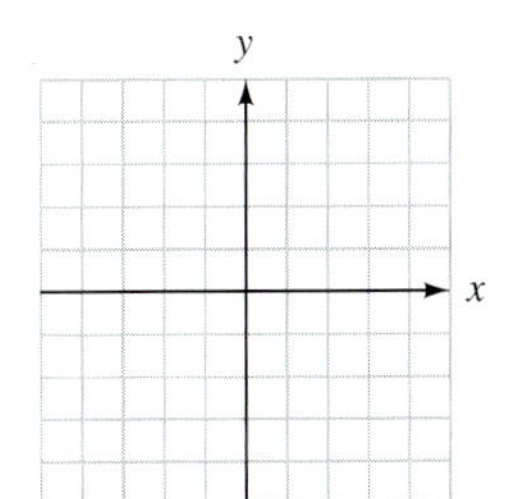

27. $4x + 5y = 20$ (see Exercise 23)

x	y
0	
	0
1	

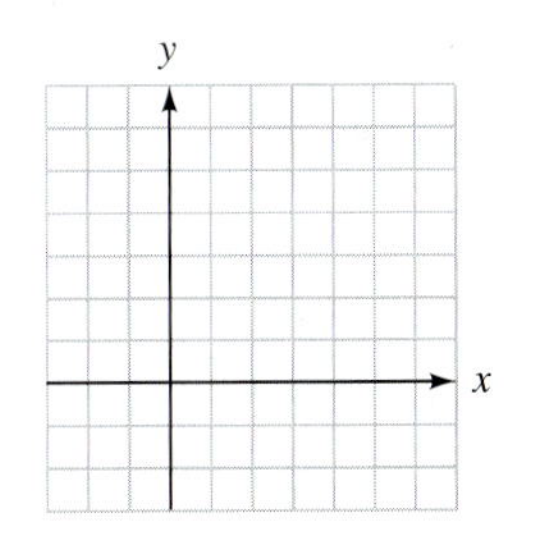

28. $3x - 5y = 15$ (see Exercise 24)

x	y
0	
	0
3	

29. $x - 2y = -4$

x	y
0	
	0
4	

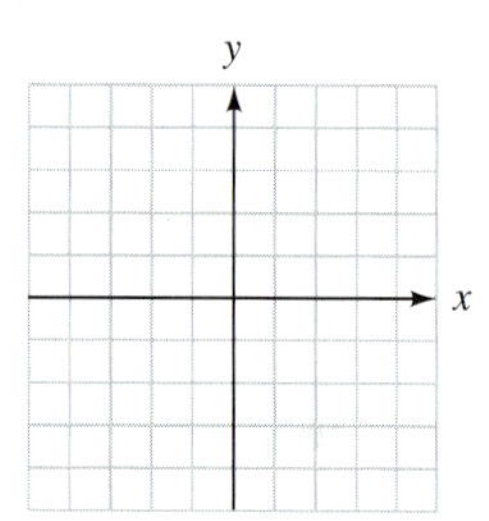

30. $3x + y = -3$

x	y
0	
	0
1	

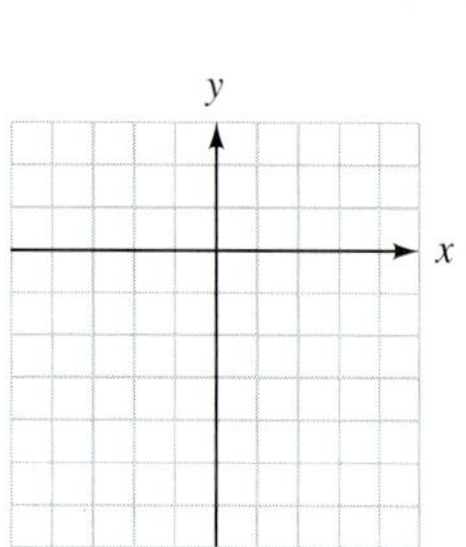

Graph each equation.

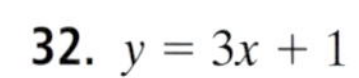

31. $y = 2x - 5$

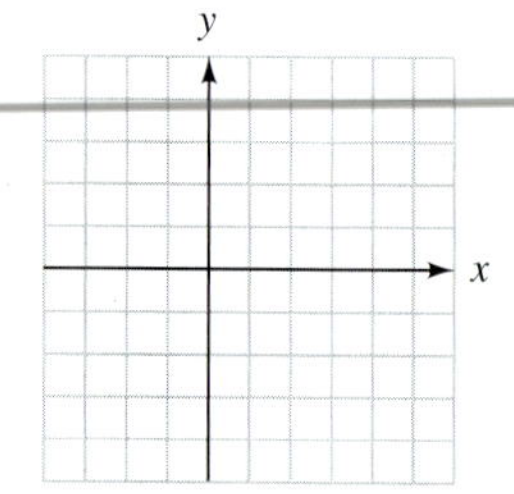

32. $y = 3x + 1$

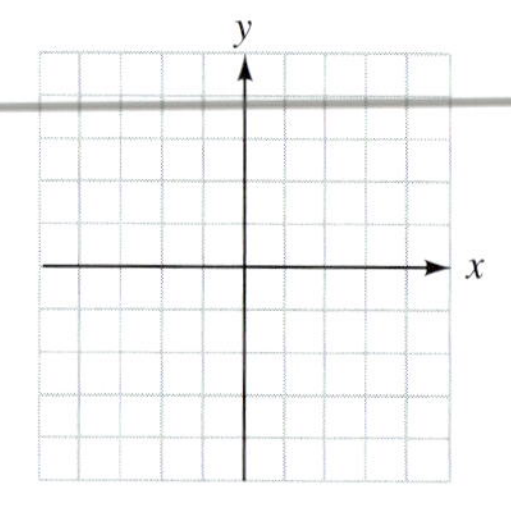

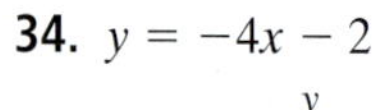

33. $y = -3x + 2$

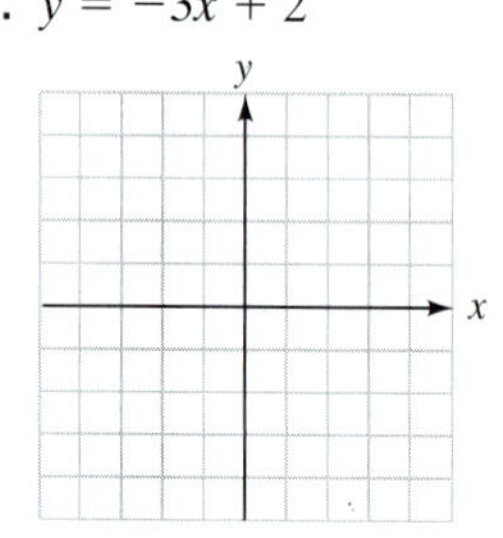

34. $y = -4x - 2$

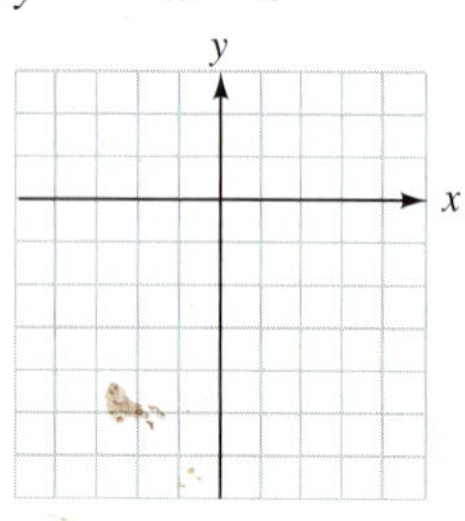

35. $y = -\frac{1}{2}x + 1$

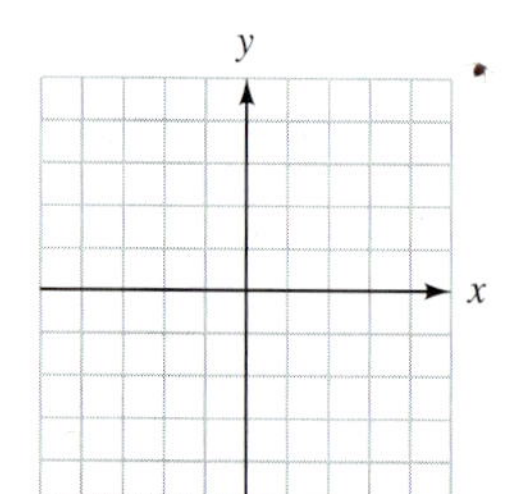

36. $y = \frac{2}{3}x - 2$

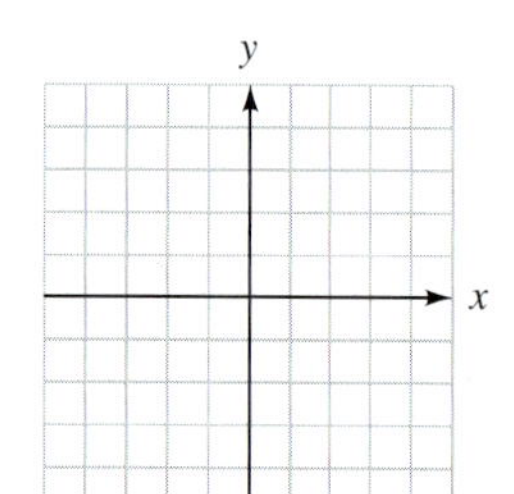

37. $y = 2x$

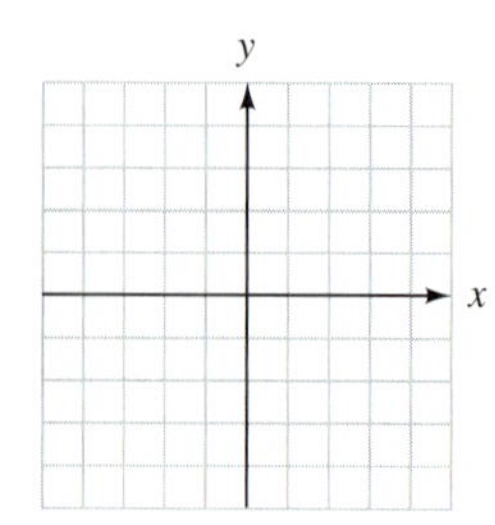

38. $y = -2x$

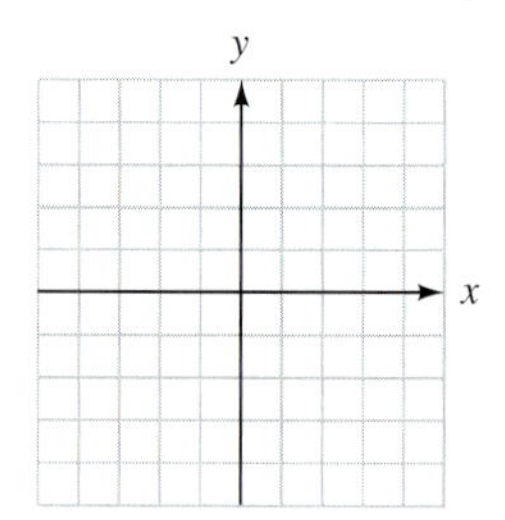

39. $y = \frac{x}{3}$

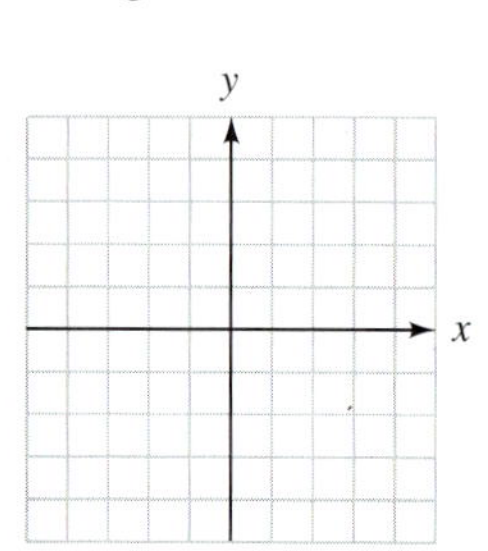

40. $y = x$

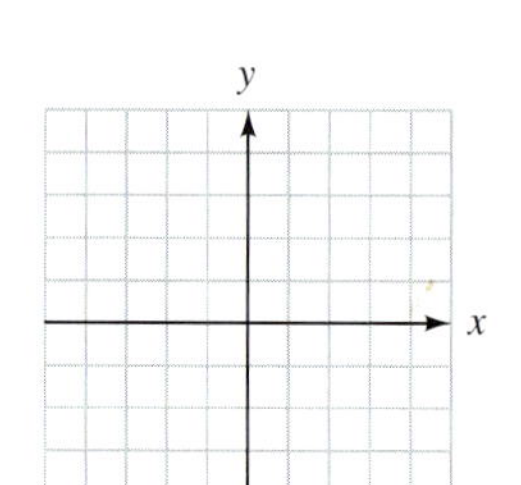

41. $y = 5$

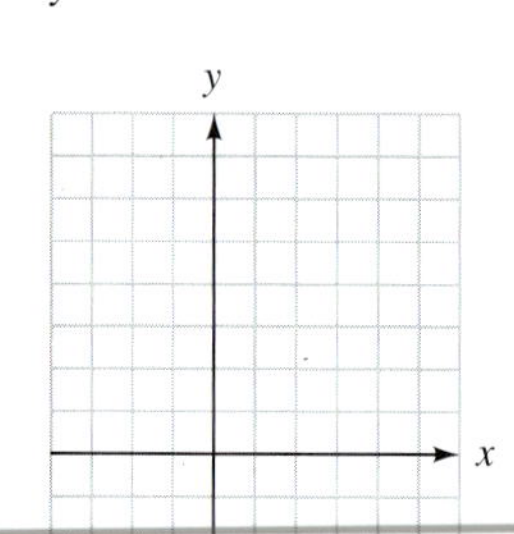

42. $x = -2$

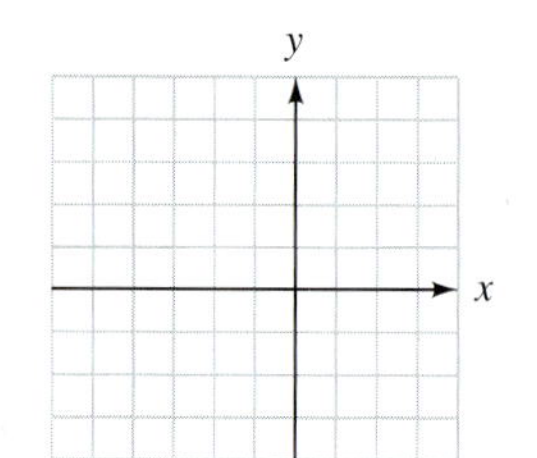

43. $x = 4$

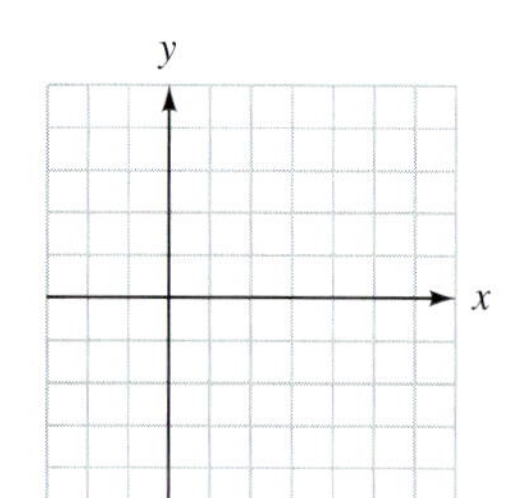

44. $y = -3$

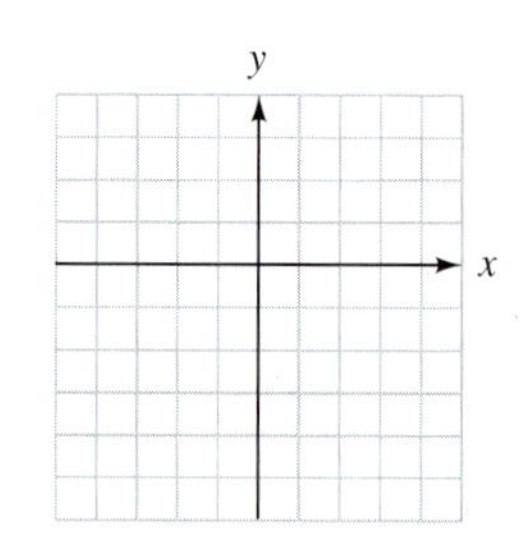

Make a table showing three solutions of each equation. Determine an appropriate scale for the y-axis and graph the equation.

45. $y = 100x$

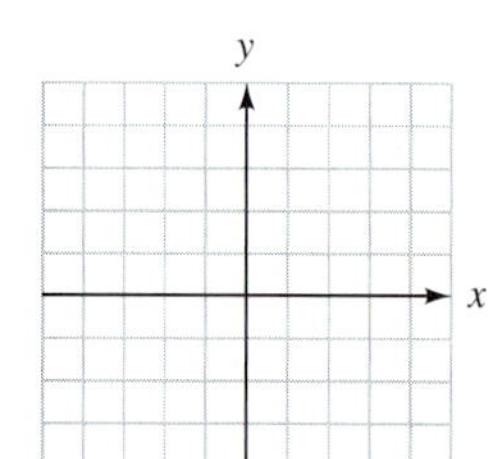

46. $y = -30x$

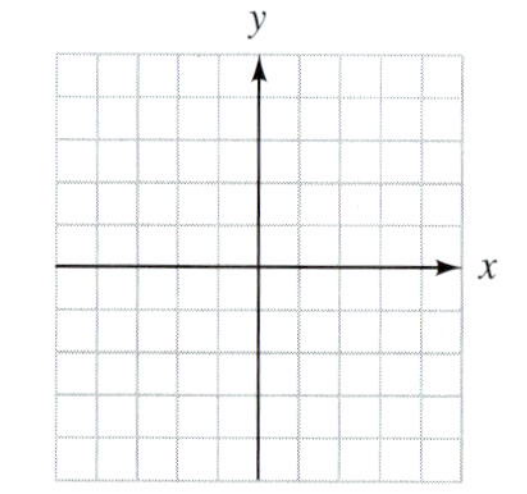

47. $y = -50x - 25$

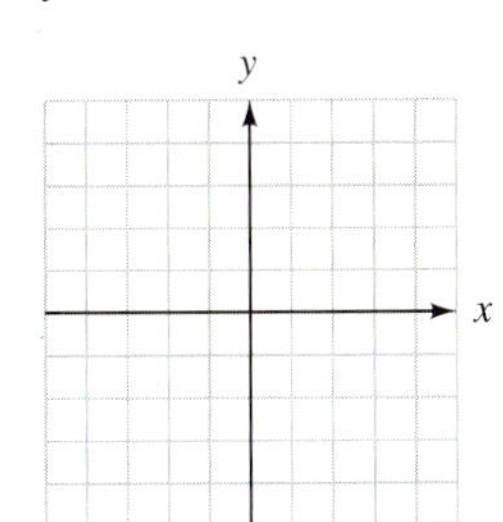

48. $y = 200x - 400$

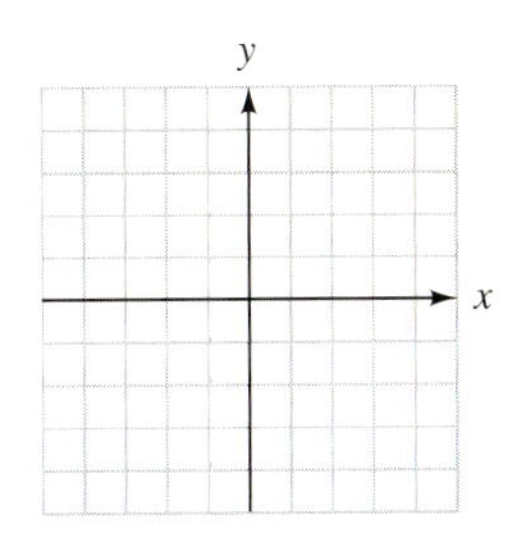

APPLICATIONS

49. HOURLY WAGES The following table gives the amount y (in dollars) that a student can earn by working x hours. Plot the ordered pairs, draw a line connecting the points, and estimate how much the student will earn in 3 hours.

x	y
2	15
4	30
6	45

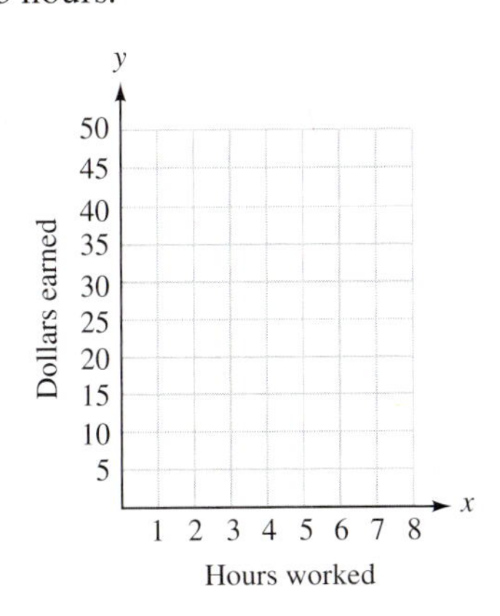

50. VALUE OF A CAR The following table shows the value y (in thousands of dollars) of a car that is x years old. Plot the ordered pairs, draw a line connecting the points, and estimate the value of the car when it is 7 years old.

x	y
3	7
4	5.5
5	4

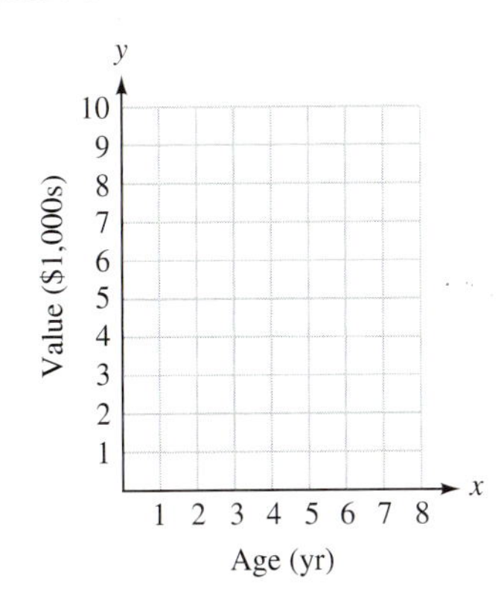

51. DISTANCE, RATE, AND TIME The formula $d = 2t$ gives the distance d (in miles) that a child can walk in a time t (in hours) at the rate of 2 mph. Complete the table of solutions, and then graph the equation to get a picture of the relationship between distance and time. (*Hint:* Plot t on the horizontal axis and d on the vertical axis.)

$d = 2t$	
t	d
1	
2	
3	
4	
5	

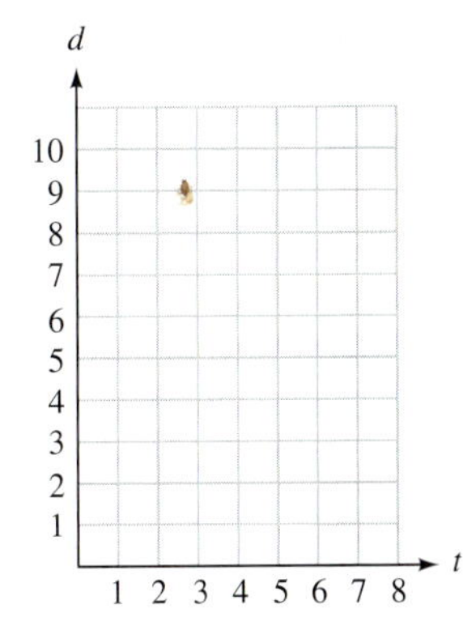

52. INVESTMENTS If \$100 is invested in a savings account paying 6% per year simple interest, the amount A in the account over a period of time t is given by the formula $A = 6t + 100$. Complete the table of solutions, and then graph this equation to get a picture of how the account grows over a period of time. (*Hint:* Plot t on the horizontal axis and A on the vertical axis.)

$A = 6t + 100$	
t	A
1	
2	
3	
4	
5	
6	

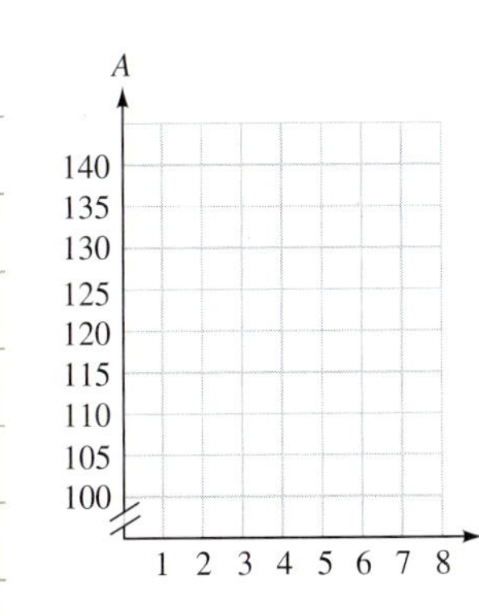

The symbol ⫽ is used to indicate a break in the labeling of the vertical axis.

53. AIR TRAFFIC CONTROL The equations describing the paths of two airplanes are $y = -\frac{1}{2}x + 3$ and $3y = 2x + 2$. Each equation is graphed on the radar screen. If they are flying at the same altitude, is there a possibility of a midair collision? If so, where?

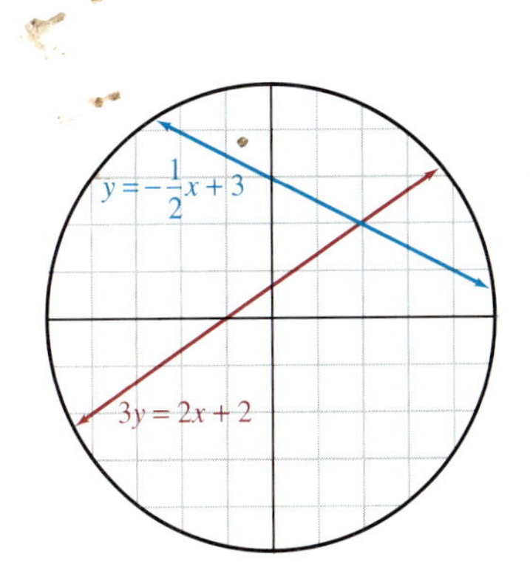

b. $(y^2)(y^4) = y^{2+4}$ Since the bases are the same, add the exponents and keep the common base, which is y.

$= y^6$ Perform the addition: $2 + 4 = 6$.

c. $x^2x^4x^9 = x^{2+4+9}$ Since the bases are the same, add the exponents and keep the common base, which is x.

$= x^{15}$ Perform the addition: $2 + 4 + 9 = 15$.

Answers **a.** 5^9, **b.** m^5, **c.** b^{13}

! COMMENT We cannot use the product rule for exponents to simplify an expression such as $x^4 + x^3$, because it is not a product. Nor can we use it to simplify $x^4 \cdot y^3$, because the bases are not the same.

The product rule for exponents can be used to simplify more complicated algebraic expressions involving multiplication.

Self Check 3
Simplify each product:
a. $4m \cdot 6m^5$
b. $8r^3(-5r^2)$

EXAMPLE 3 Simplify each product: **a.** $3a(5a^2)$ and **b.** $-2t^2 \cdot 6t^6$.

Solution

a. $3a(5a^2) = (3 \cdot 5)(a \cdot a^2)$ Use the commutative and associative properties to change the order and regroup the factors.

$= (3 \cdot 5)(a^1 \cdot a^2)$ Recall that $a = a^1$.

$= 15a^{1+2}$ Perform the multiplication: $3 \cdot 5 = 15$. Then add the exponents and keep the common base, which is a.

$= 15a^3$ Perform the addition: $1 + 2 = 3$.

b. $-2t^2 \cdot 6t^6 = (-2 \cdot 6)(t^2 \cdot t^6)$ Change the order of the factors and regroup them.

$= -12t^{2+6}$ Perform the multiplication: $-2 \cdot 6 = -12$. Then add the exponents and keep the common base.

$= -12t^8$ Perform the addition: $2 + 6 = 8$.

Answers **a.** $24m^6$, **b.** $-40r^5$

Exponential expressions often contain more than one variable.

Self Check 4
Simplify the following:
a. $c^3d^2 \cdot cd^5$
b. $-7a^2b^3(8a^4b^5)$

EXAMPLE 4 Simplify the following: **a.** $n^2m \cdot n^8m^3$ and **b.** $4xy^2(-3x^2y^3)$.

Solution

a. $n^2m \cdot n^8m^3 = (n^2 \cdot n^8)(m \cdot m^3)$ Change the order and group the factors with like bases.

$= n^{2+8} \cdot m^{1+3}$ Add the exponents of the like bases. Recall that $m = m^1$.

$= n^{10}m^4$ Perform the additions.

b. $4xy^2(-3x^2y^3) = [4(-3)](x \cdot x^2)(y^2 \cdot y^3)$ Group the factors with like bases.

$= -12 \cdot x^{1+2} \cdot y^{2+3}$ Perform the multiplication. Add the exponents of the like bases.

$= -12x^3y^5$ Perform the additions.

Answers **a.** c^4d^7, **b.** $-56a^6b^8$

Pin Code Choices

THINK IT THROUGH

"According to a Student Monitor LLC survey, ATM debit card ownership among college students has almost doubled from 30 percent to 57 percent in the past four years." *BYU Newsletter*, Oct 2002

In 2002, there were 13.9 billion ATM transactions in the United States. On average, that's more than 38 million a day! Before each transaction, the card owner is required to enter his or her PIN (personal identification number). When an ATM card is issued, many financial institutions have the applicant select a four-digit PIN. Write the number of choices as an exponential expression, and then evaluate it.

The power rule for exponents

To develop the power rule for exponents, we consider the expression $(x^2)^5$. Notice that the base, x^2, is raised to a power. Therefore, we are working with a power of a power. We will again rely on the definition of an exponent to find a rule for simplifying this exponential expression.

$(x^2)^5 = x^2 \cdot x^2 \cdot x^2 \cdot x^2 \cdot x^2$ The exponent 5 tells us to write the base x^2 five times.

$= x^{2+2+2+2+2}$ Since the bases are alike, add the exponents and keep the common base.

$= x^{10}$ Perform the addition: $2 + 2 + 2 + 2 + 2 = 10$.

Notice that the exponent of the result is the *product* of the exponents in $(x^2)^5$.

Product of the exponents

$$(x^2)^5 = x^{2\cdot 5} = x^{10}$$

This observation suggests the following rule.

The power rule for exponents

For any number x and any positive integers m and n,

$$(x^m)^n = x^{m\cdot n} \qquad \text{or, more simply,} \qquad (x^m)^n = x^{mn}$$

To raise an exponential expression to a power, keep the base and multiply the exponents.

EXAMPLE 5 Simplify each expression: **a.** $(2^3)^7$ and **b.** $(b^5)^3$.

Solution

a. $(2^3)^7 = 2^{3\cdot 7}$ Use the power rule for exponents by keeping the base and multiplying the exponents.

$= 2^{21}$ Perform the multiplication: $3 \cdot 7 = 21$.

b. $(b^5)^3 = b^{5\cdot 3}$ Keep the base and multiply the exponents.

$= b^{15}$ Perform the multiplication: $5 \cdot 3 = 15$.

Self Check 5

Simplify each expression:

a. $(4^2)^6$

b. $(y^6)^4$

Answers **a.** 4^{12}, **b.** y^{24}

In some cases, when simplifying algebraic expressions involving exponents, we must apply two rules of exponents.

Self Check 6
Simplify:
a. $(x^4)^2(x^3)^3$
b. $(x^4x^2)^3$

Answers **a.** x^{17}, **b.** x^{18}

EXAMPLE 6 Simplify: **a.** $(n^3)^4(n^2)^5$ and **b.** $(n^2n^3)^5$.

Solution

a. $(n^3)^4(n^2)^5 = n^{3\cdot4}\cdot n^{2\cdot5}$ Keep each base and multiply their exponents.

$= n^{12}\cdot n^{10}$ Perform the multiplications: $3\cdot4 = 12$ and $2\cdot5 = 10$.

$= n^{12+10}$ Since the bases are alike, keep the base and add the exponents.

$= n^{22}$ Perform the addition: $12 + 10 = 22$.

b. $(n^2n^3)^5 = (n^{2+3})^5$ Work within the parentheses first. Since the bases are alike, keep the base and add the exponents.

$= (n^5)^5$ Perform the addition: $2 + 3 = 5$.

$= n^{5\cdot5}$ Keep the base and multiply the exponents.

$= n^{25}$ Perform the multiplication: $5\cdot5 = 25$.

The power rule for products

The exponential expression $(2x)^4$ has an exponent of 4 and a base of $2x$. The base $2x$ is a product, since $2x = 2\cdot x$. Therefore, $(2x)^4$ is a power of a product. To find a rule to simplify it, we will again use the definition of exponent.

$(2x)^4 = 2x\cdot2x\cdot2x\cdot2x$ Write the base, $2x$, as a factor 4 times.

$= (2\cdot2\cdot2\cdot2)(x\cdot x\cdot x\cdot x)$ Apply the commutative and associative properties of multiplication to change the order and group like factors.

$= 2^4x^4$ The factor 2 and the factor x are both repeated 4 times. Apply the definition of an exponent.

The result has factors of 2 and x. In the original problem, they were within the parentheses. Each is now raised to the fourth power.

Each factor within the parentheses ends up being raised to the 4th power.

$$(2x)^4 = 2^4x^4$$

This observation suggests the following rule.

The power rule for products

For any numbers x and y, and any positive integer m,

$$(xy)^m = x^my^m$$

To raise a product to a power, raise each factor of the product to that power.

Self Check 7
Simplify each expression:
a. $(10c)^2$
b. $(5rs)^3$

EXAMPLE 7 Simplify each expression: **a.** $(8a)^2$ and **b.** $(2bx)^3$.

Solution

a. $(8a)^2 = 8^2a^2$ To raise $8a$ to the 2nd power, raise each factor of the product to the 2nd power.

$= 64a^2$ Find the power: $8^2 = 64$.

b. $(2bx)^3 = 2^3b^3x^3$ — To raise $2bx$ to the 3rd power, raise each factor of the product to the 3rd power.

$= 8b^3x^3$ — Find the power: $2^3 = 8$.

Answers **a.** $100c^2$, **b.** $125r^3s^3$

EXAMPLE 8 Simplify each expression: **a.** $(10a^2)^3$ and **b.** $(3c^5d^3)^4$.

Solution

a. $(10a^2)^3 = 10^3(a^2)^3$ — To raise $10a^2$ to the 3rd power, raise each factor of the product, 10 and a^2, to the 3rd power.

$= 10^3a^{2\cdot 3}$ — To raise a^2 to a power, keep the base and multiply the exponents.

$= 10^3a^6$ — Perform the multiplication: $2 \cdot 3 = 6$.

$= 1{,}000a^6$ — Find the power: $10^3 = 1{,}000$.

b. $(3c^5d^3)^4 = 3^4(c^5)^4(d^3)^4$ — To raise $3c^5d^3$ to the 4th power, raise each factor of the product, 3, c^5, and d^3, to the 4th power.

$= 3^4c^{5\cdot 4}d^{3\cdot 4}$ — To raise c^5 and d^3 to powers, keep the bases and multiply their exponents.

$= 3^4c^{20}d^{12}$ — Perform the multiplications: $5 \cdot 4 = 20$ and $3 \cdot 4 = 12$.

$= 81c^{20}d^{12}$ — Find the power: $3^4 = 81$.

Self Check 8
Simplify each expression:
a. $(3n^2)^3$
b. $(6h^2s^9)^2$

Answers **a.** $27n^6$, **b.** $36h^4s^{18}$

EXAMPLE 9 Simplify: $(2a^2)^2(4a^3)^3$.

Solution

$(2a^2)^2(4a^3)^3 = 2^2(a^2)^2 \cdot 4^3(a^3)^3$ — To raise $2a^2$ and $4a^3$ to powers, raise the factors of each product to the appropriate power.

$= 2^2a^{2\cdot 2} \cdot 4^3 \cdot a^{3\cdot 3}$ — To raise a^2 and a^3 to powers, keep the bases and multiply the exponents.

$= 2^2a^4 \cdot 4^3a^9$ — Perform the multiplications: $2 \cdot 2 = 4$ and $3 \cdot 3 = 9$.

$= (2^2 \cdot 4^3)(a^4 \cdot a^9)$ — Change the order of the factors and group like bases.

$= (2^2 \cdot 4^3)(a^{4+9})$ — To multiply $a^4 \cdot a^9$, keep the base and add the exponents.

$= (2^2 \cdot 4^3)a^{13}$ — Perform the addition: $4 + 9 = 13$.

$= (4 \cdot 64)a^{13}$ — Find the powers: $2^2 = 4$ and $4^3 = 64$.

$= 256a^{13}$ — Perform the multiplication: $4 \cdot 64 = 256$.

Self Check 9
Simplify: $(4y^3)^2(3y^4)^3$.

Answer $432y^{18}$

Section 6.3 STUDY SET

VOCABULARY *Fill in the blanks.*

1. In x^n, x is called the ______ and n is called the ______.
2. x^2 is the second ______ of x, or we can read it as "x ______."
3. $x^m \cdot x^n$ is the product of two exponential expressions with ______ bases.
4. $(x^m)^n$ is a power of a ______.
5. $(2x)^n$ is a ______ raised to a power.
6. In x^{m+n}, $m + n$ is the ______ of m and n.

CONCEPTS

7. Represent each repeated multiplication using exponents.
   a. $x \cdot x \cdot x \cdot x \cdot x \cdot x \cdot x$
   b. $x \cdot x \cdot y \cdot y \cdot y$
   c. $3 \cdot 3 \cdot 3 \cdot 3 \cdot a \cdot a \cdot b \cdot b \cdot b$
8. Write each exponential expression as repeated multiplication.
   a. a^3b^5
   b. $(x^2)^3$
   c. $(2a)^3$
9. Write a product of two exponential expressions with like variable bases. Then simplify it using a rule of exponents.
10. Write a power of a product and then simplify it using a rule of exponents.
11. Write a power of a power and then simplify it using a rule of exponents.
12. What algebraic property allows us to change the order of the factors of a multiplication?
13. Complete each rule for exponents.
    a. $x^m x^n = \square$
    b. $(x^m)^n = \square$
    c. $(ax)^n = \square$
14. In each case, explain how the expression has been improperly simplified.
    a. $2^3 \cdot 2^4 = 2^{12}$
    b. $3^3 \cdot 3^4 = 9^7$
    c. $(2^3)^4 = 2^7$
15. Write each expression without an exponent.
    a. 2^1 b. $(-10)^1$ c. x^1
16. Find each power.
    a. 2^3 b. 4^3 c. 5^3
17. Simplify each expression, if possible.
    a. $x \cdot x$ and $x + x$
    b. $x \cdot x^2$ and $x + x^2$
    c. $x^2 \cdot x^2$ and $x^2 + x^2$
18. Simplify each expression, if possible.
    a. $a \cdot a$ and $a - a$
    b. $2a \cdot a$ and $2a - a$
    c. $2a \cdot 3a$ and $2a - 3a$
19. Simplify each expression, if possible.
    a. $4x \cdot x$ and $4x + x$
    b. $4x \cdot 3x$ and $4x + 3x$
    c. $4x^2 \cdot 3x$ and $4x^2 + 3x$
20. Simplify each expression, if possible.
    a. $ab(-2ab)$ and $ab - 2ab$
    b. $-2ab(3ab^2)$ and $-2ab + 3ab^2$
    c. $-2ab^2(-3ab^2)$ and $-2ab^2 - 3ab^2$
21. Evaluate the exponential expression x^{m+n} for $x = 3$, $m = 2$, and $n = 1$.
22. Evaluate the exponential expression $(x^m)^n$ for $x = 2$, $m = 3$, and $n = 2$.

NOTATION *Complete each solution.*

23. $x^5 \cdot x^7 = x^{\square + \square}$
    $= x^{12}$
24. $(x^5)^4 = x^{\square \cdot \square}$
    $= x^{20}$
25. $(2x^4)(8x^3) = (2 \cdot 8)(\square \cdot \square)$
    $= 16x^{\square + \square}$
    $= 16x^7$
26. $(2x^2)^3 = 2^3(x^2)^3$
    $= 2^3x^{\square \cdot \square}$
    $= 2^3x^{\square}$
    $= 8x^6$

PRACTICE *Simplify each product.*

27. $x^2 \cdot x^3$
28. $t^4 \cdot t^3$
29. x^3x^7
30. y^2y^5
31. $f^5(f^8)$
32. $g^6(g^2)$
33. $n^{24} \cdot n^8$
34. $m^9 \cdot m^{61}$
35. $l^4 \cdot l^5 \cdot l$
36. $w^4 \cdot w \cdot w^3$
37. $x^6(x^3)x^2$
38. $y^5(y^2)(y^3)$
39. $2^4 \cdot 2^8$
40. $3^4 \cdot 3^2$
41. $5^6(5^2)$
42. $(8^3)(8^4)$
43. $2x^2 \cdot 4x$
44. $5y \cdot 6y^3$
45. $5t \cdot t^9$
46. $f^4 \cdot 3f$
47. $-6x^3(4x^2)$
48. $-7y^5(5y^3)$
49. $-x \cdot x^3$
50. $8x^6(-x)$
51. $6y(2y^3)3y^4$
52. $2d(5d^4)(d^2)$
53. $-2t^3(-4t^2)(-5t^5)$
54. $-7k^5(-3k^3)(-2k^9)$
55. $xy^2 \cdot x^2y$
56. $s^2t \cdot st$

57. $b^3 \cdot c^2 \cdot b^5 \cdot c^6$

58. $h^3 \cdot f^3 \cdot f^2 \cdot h^4$

59. $x^4y(xy)$

60. $(ab)(ab^2)$

61. $a^2b \cdot b^3a^2$

62. $w^2y \cdot yw^4$

63. $x^5y \cdot y^6$

64. $a^7 \cdot b^2a^4$

65. $3x^2y^3 \cdot 6xy$

66. $25a^3b \cdot 2ab^5$

67. $xy^2 \cdot 16x^3$

68. $mn^4 \cdot 8n^3$

69. $-6f^2t(4f^4t^3)$

70. $(-5a^2b^2)(5a^3b^6)$

71. $ab \cdot ba \cdot a^2b$

72. $xy \cdot y^2x \cdot x^2y$

73. $-4x^2y(-3x^2y^2)$

74. $-2rt^4(-5r^2t^2)$

Simplify each expression.

75. $(x^2)^4$

76. $(y^6)^3$

77. $(m^{50})^{10}$

78. $(n^{25})^4$

79. $(2a)^3$

80. $(3x)^3$

81. $(xy)^4$

82. $(ab)^8$

83. $(3s^2)^3$

84. $(5f^6)^2$

85. $(2s^2t^3)^2$

86. $(4h^5y^6)^2$

87. $(x^2)^3(x^4)^2$

88. $(a^5)^2(a^3)^3$

89. $(c^5)^3 \cdot (c^3)^5$

90. $(y^2)^8 \cdot (y^8)^2$

91. $(2a^4)^2(3a^3)^2$

92. $(5x^3)^2(2x^4)^3$

93. $(3a^3)^3(2a^2)^3$

94. $(6t^5)^2(2t^2)^2$

95. $(x^2x^3)^{12}$

96. $(a^3a^3)^3$

97. $(2b^4b)^5$

98. $(3y^2y^5)^3$

WRITING

99. Explain the difference between x^2 and $2x$.

100. Explain why the rules of exponents do not apply to $x^2 + x^3$.

101. One of the rules of exponents is that the power of a product is the product of the powers. Use a specific example to explain this rule.

102. To find the result when *multiplying* two exponential expressions with like bases, we must *add* the exponents. Explain why this is so.

REVIEW

103. JEWELRY A lot of what we refer to as gold jewelry is actually made of a combination of gold and another metal. For example, 18-karat gold is $\frac{18}{24}$ gold by weight. Simplify this fraction.

104. When evaluated, what is the sign of $(-13)^5$?

105. Divide: $\frac{-25}{-5}$.

106. How much did the temperature change if it went from $-4°$ to $-17°$?

107. Evaluate: $2\left(\frac{12}{-3}\right) + 3(5)$.

108. Solve: $-4 - 6 = x + 1$.

109. Solve: $-x = -12$.

110. Divide: $\frac{0}{10}$.

6.4 Introduction to Polynomials

- Polynomials
- Classifying polynomials
- Degree of a polynomial
- Evaluating polynomials
- Graphing equations involving polynomials

We have graphed the equations $y = 2x - 1$ and $y = \frac{1}{2}x + 1$. The expressions $2x - 1$ and $\frac{1}{2}x + 1$ are examples of algebraic expressions called *polynomials.* In this section, we will define polynomials, classify them into groups, and show how to evaluate them. Finally, we will show how to graph equations involving polynomials.

Polynomials

Recall that an **algebraic term,** or simply a **term,** is a number or a product of a number and one or more variables, which may be raised to powers. Some examples of terms are

$$17, \quad 5x, \quad 6t^2, \quad \text{and} \quad -8z^3$$

The coefficients of these terms are 17, 5, 6, and -8, respectively.

Polynomials

A **polynomial** is a term or a sum of terms in which all variables have whole-number exponents.

Some examples of polynomials are

$$0, \quad 8y^2, \quad 2x + 1, \quad 4y^2 - 2y + 3, \quad \text{and} \quad 7a^3 + 2a^2 - a - 1$$

The polynomial $8y^2$ has one term. The polynomial $2x + 1$ has two terms, $2x$ and 1. Since $4y^2 - 2y + 3$ can be written as $4y^2 + (-2y) + 3$, it is the sum of three terms, $4y^2$, $-2y$, and 3.

Classifying polynomials

We classify some polynomials by the number of terms they contain. A polynomial with one term is called a **monomial.** A polynomial with two terms is called a **binomial.** A polynomial with three terms is called a **trinomial.** Some examples of these polynomials are shown in Table 6-1.

Monomials	Binomials	Trinomials
$5x^2$	$2x - 1$	$5t^2 + 4t + 3$
$-6x$	$18a^2 - 4a$	$27x^3 - 6x + 2$
29	$-27z^4 + 7z^2$	$32r^2 + 7r - 12$

TABLE 6-1

EXAMPLE 1 Classify each polynomial as a monomial, a binomial, or a trinomial: **a.** $3x + 4$, **b.** $3x^2 + 4x - 12$, and **c.** $25x^3$.

Solution

a. Since $3x + 4$ has two terms, it is a binomial.

b. Since $3x^2 + 4x - 12$ has three terms, it is a trinomial.

c. Since $25x^3$ has one term, it is a monomial.

Self Check 1

Classify each polynomial as a monomial, a binomial, or a trinomial:

a. $5x$

b. $8x^2 + 7$

c. $x^2 - 2x - 1$

Answers **a.** monomial, **b.** binomial, **c.** trinomial

Degree of a polynomial

The monomial $7x^3$ is called a **monomial of third degree** or a **monomial of degree 3,** because the variable occurs three times as a factor.

- $5x^2$ is a monomial of degree 2. Because the variable occurs two times as a factor: $x^2 = x \cdot x$.
- $-8x^4$ is a monomial of degree 4. Because the variable occurs four times as a factor: $x^4 = x \cdot x \cdot x \cdot x$.
- $\frac{1}{2}x^5$ is a monomial of degree 5. Because the variable occurs five times as a factor: $x^5 = x \cdot x \cdot x \cdot x \cdot x$.

We define the degree of a polynomial by considering the degrees of each of its terms.

Degree of a polynomial

The **degree of a polynomial** is the same as the degree of its term with largest degree.

For example,

- $x^2 + 5x$ is a binomial of degree 2, because the degree of its term with largest degree (x^2) is 2.
- $4y^3 + 2y - 7$ is a trinomial of degree 3, because the degree of its term with largest degree ($4y^3$) is 3.
- $\frac{1}{2}z + 3z^4 - 2z^2$ is a trinomial of degree 4, because the degree of its term with largest degree ($3z^4$) is 4.

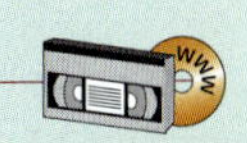

EXAMPLE 2 Find the degree of each polynomial: **a.** $-2x + 4$, **b.** $5t^3 + t^4 - 7$, and **c.** $3 - 9z + 6z^2 - z^3$.

Solution

a. Since $-2x$ can be written as $-2x^1$, the degree of the term with largest degree is 1. Thus, the degree of the polynomial is 1.

b. In $5t^3 + t^4 - 7$, the degree of the term with largest degree (t^4) is 4. Thus, the degree of the polynomial is 4.

c. In $3 - 9z + 6z^2 - z^3$, the degree of the term with largest degree ($-z^3$) is 3. Thus, the degree of the polynomial is 3.

Self Check 2
Find the degree of each polynomial:
a. $3p^3$
b. $17r^4 + 2r^8 - r$
c. $-2g^5 - 7g^6 + 12g^7$

Answers **a.** 3, **b.** 8, **c.** 7

Evaluating polynomials

When a number is substituted for the variable in a polynomial, the polynomial takes on a numerical value. Finding this value is called **evaluating the polynomial.**

EXAMPLE 3 Evaluate each polynomial for $x = 3$: **a.** $3x - 2$ and **b.** $-2x^2 + x - 3$.

Solution

a.
$$\begin{aligned} 3x - 2 &= 3(3) - 2 && \text{Substitute 3 for } x. \\ &= 9 - 2 && \text{Multiply: } 3(3) = 9. \\ &= 7 && \text{Subtract: } 9 - 2 = 7. \end{aligned}$$

b.
$$\begin{aligned} -2x^2 + x - 3 &= -2(3)^2 + 3 - 3 && \text{Substitute 3 for } x. \\ &= -2(9) + 3 - 3 && \text{Square 3: } 3 \cdot 3 = 9. \\ &= -18 + 3 - 3 && \text{Multiply: } -2(9) = 18. \\ &= -15 - 3 && \text{Add: } -18 + 3 = -15. \\ &= -18 && \text{Subtract: } -15 - 3 = -18. \end{aligned}$$

Self Check 3
Evaluate each polynomial for $x = -1$:
a. $-2x^2 - 4$
b. $3x^2 - 4x + 1$

Answers **a.** -6, **b.** 8

Self Check 4
Find the height of the object in 2 seconds.

Answer 0 ft

EXAMPLE 4 **Height of an object.** The polynomial $-16t^2 + 28t + 8$ gives the height (in feet) of an object t seconds after it has been thrown straight upward. Find the height of the object in 1 second.

Solution To find the height at 1 second, we evaluate the polynomial at $t = 1$.

$$
\begin{aligned}
-16t^2 + 28t + 8 &= -16(\mathbf{1})^2 + 28(\mathbf{1}) + 8 && \text{Substitute 1 for } t. \\
&= -16(1) + 28(1) + 8 && \text{Square 1: } 1 \cdot 1 = 1. \\
&= -16 + 28 + 8 && \text{Multiply: } -16(1) = -16 \text{ and } 28(1) = 28. \\
&= 12 + 8 && \text{Add: } -16 + 28 = 12. \\
&= 20 && \text{Add: } 12 + 8 = 20.
\end{aligned}
$$

At 1 second, the object is 20 feet above the ground.

Graphing equations involving polynomials

In the previous section, we graphed the linear equations

$y = 2x - 1$ and $y = \frac{1}{2}x + 1$ — The polynomials $2x - 1$ and $\frac{1}{2}x + 1$ are both of degree 1.

We will now graph some equations involving second-degree polynomials. The method we will use is like that used to graph linear equations earlier in this chapter.

Self Check 5
Graph: $y = 3x^2$.

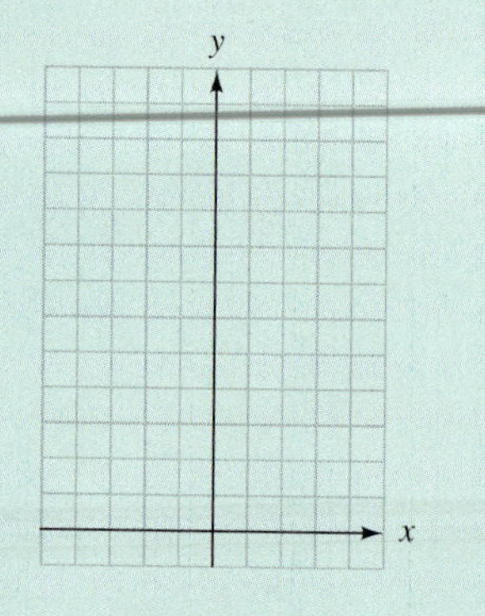

Answer

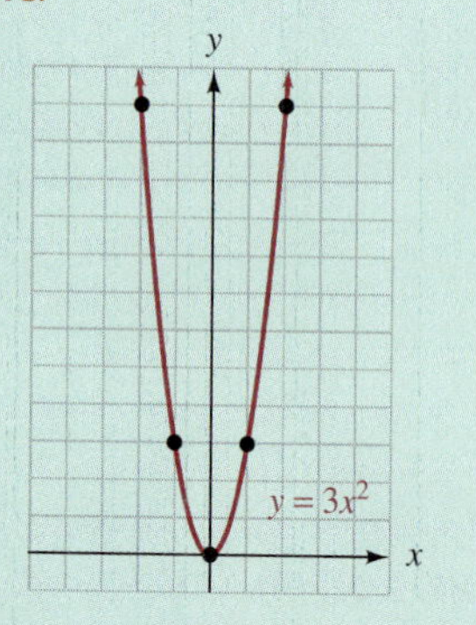

EXAMPLE 5 Graph: $y = 2x^2$.

Solution To make a table of solutions for $y = 2x^2$, we choose numbers for x and find the corresponding values of y. If $x = -2$, we get

$$
\begin{aligned}
y &= 2\mathbf{x}^2 \\
&= 2(\mathbf{-2})^2 && \text{Substitute } -2 \text{ for } x. \\
&= 2(4) && \text{Square } -2 \text{ first: } (-2)^2 = -2(-2) = 4. \\
&= 8 && \text{Perform the multiplication.}
\end{aligned}
$$

Thus, $x = -2$ and $y = 8$ is a solution of $y = 2x^2$. In a similar way, we find the corresponding values for y when x is -1, 0, 1, and 2 and enter them in the table in Figure 6-13. If we plot the ordered pairs and join the points with a smooth curve, we get a U-shaped figure, called a **parabola.**

$y = 2x^2$		
x	y	(x, y)
-2	8	$(-2, 8)$
-1	2	$(-1, 2)$
0	0	$(0, 0)$
1	2	$(1, 2)$
2	8	$(2, 8)$

↑ We can choose any values for x.

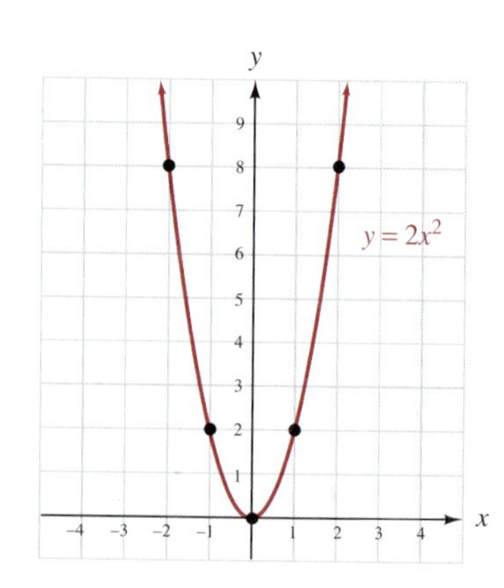

FIGURE 6-13

! COMMENT When an equation is not linear, its graph is not a straight line. To graph nonlinear equations, we must usually plot many points to recognize the shape of the curve.

EXAMPLE 6 Graph: $y = -x^2 + 3$.

Solution We make a table of values by substituting numbers for x and finding the corresponding values of y. For example, if we substitute -3 for x, we get

$$\begin{aligned} y &= -x^2 + 3 \\ &= -(-3)^2 + 3 && \text{Don't forget to write the } - \text{ sign shown in blue.} \\ &= -9 + 3 && \text{Square } -3\text{: } (-3)(-3) = 9. \\ &= -6 && \text{Add: } -9 + 3 = -6. \end{aligned}$$

The coordinates of this point and others are shown in the table in Figure 6-14. To get the graph, we plot each of these points and join them with a smooth curve. The graph is a parabola opening downward.

$y = -x^2 + 3$

x	y	(x, y)
−3	−6	(−3, −6)
−2	−1	(−2, −1)
−1	2	(−1, 2)
0	3	(0, 3)
1	2	(1, 2)
2	−1	(2, −1)
3	−6	(3, −6)

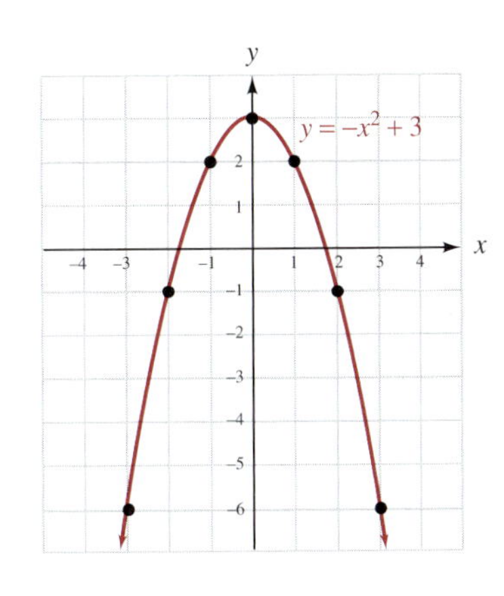

FIGURE 6-14

Self Check 6

Graph: $y = x^2 - 1$

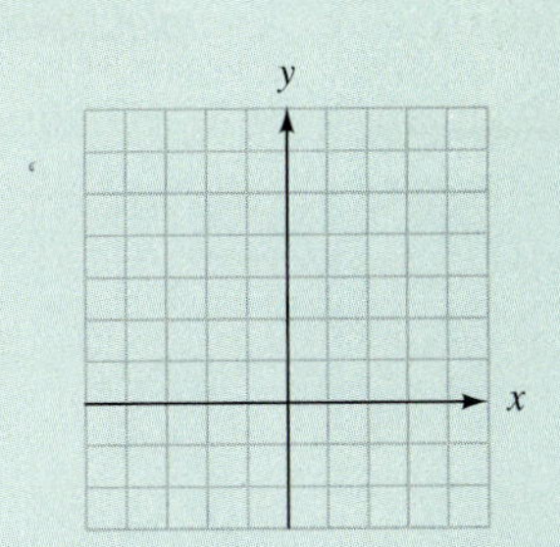

Answer

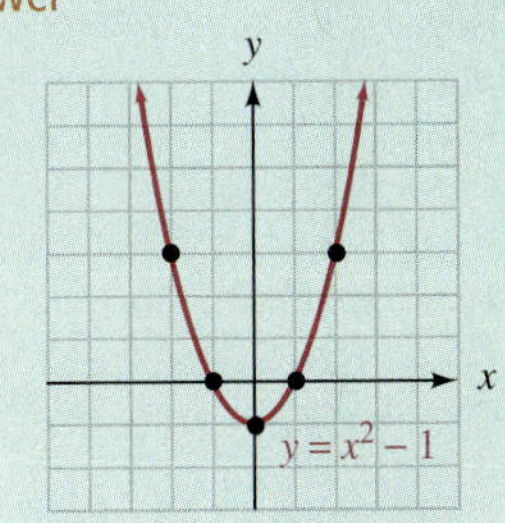

Section 6.4 STUDY SET

VOCABULARY *Fill in the blanks.*

1. A polynomial with one term, such as $2x^4$, is called a __________.
2. A polynomial with three terms, such as $x^2 - 3x + 1$, is called a __________.
3. A polynomial with two terms, such as $x^2 - 25$, is called a __________.
4. The degree of a polynomial is the same as the ________ of its term with largest degree.

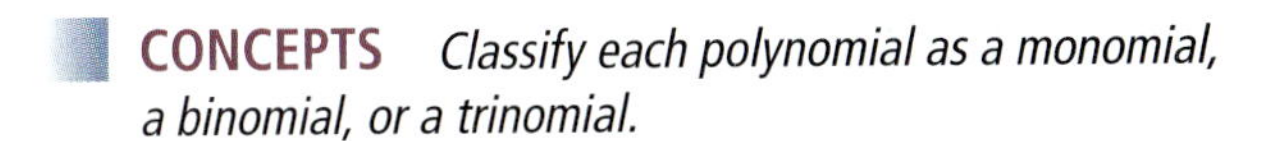

CONCEPTS *Classify each polynomial as a monomial, a binomial, or a trinomial.*

5. $3x^2 - 4$
6. $5t^2 - t + 1$
7. $17e^4$
8. $x^2 + x + 7$
9. $25u^2$
10. $x^2 - 9$
11. $q^5 + q^2 + 1$
12. $4d^3 - 3d^2$

Find the degree of each polynomial.

13. $5x^3$

14. $3t^5 + 3t^2$

15. $2x^2 - 3x + 2$

16. $\frac{1}{2}p^4 - p^2$

17. $2m$

18. $7q - 5$

19. $25w^6 + 5w^7$

20. $p^6 - p^8$

NOTATION *Complete each solution.*

21. Evaluate $3a^2 + 2a - 7$ for $a = 2$.

$$\begin{aligned} 3a^2 + 2a - 7 &= 3(\underline{\quad})^2 + 2(\underline{\quad}) - 7 \\ &= 3(\underline{\quad}) + \underline{\quad} - 7 \\ &= 12 + 4 - 7 \\ &= \underline{\quad} - 7 \\ &= 9 \end{aligned}$$

22. Evaluate $-q^2 - 3q + 2$ for $q = -1$.

$$\begin{aligned} -q^2 - 3q + 2 &= -(\underline{\quad})^2 - 3(\underline{\quad}) + 2 \\ &= -(\underline{\quad}) - 3(-1) + 2 \\ &= -1 + \underline{\quad} + 2 \\ &= \underline{\quad} + 2 \\ &= 4 \end{aligned}$$

PRACTICE *Evaluate each polynomial for the given value.*

23. $3x + 4$ for $x = 3$

24. $\frac{1}{2}x - 3$ for $x = -6$

25. $2x^2 + 4$ for $x = -1$

26. $-\frac{1}{2}x^2 - 1$ for $x = 2$

27. $0.5t^3 - 1$ for $t = 4$

28. $0.75a^2 + 2.5a + 2$ for $a = 0$

29. $\frac{2}{3}b^2 - b + 1$ for $b = 3$

30. $3n^2 - n + 2$ for $n = 2$

31. $-2s^2 - 2s + 1$ for $s = -1$

32. $-4r^2 - 3r - 1$ for $r = -2$

Graph each equation.

33. $y = x^2$

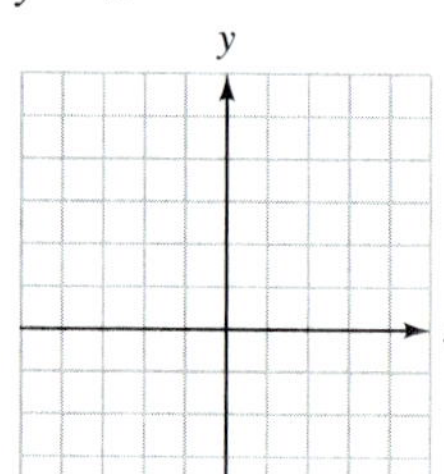

34. $y = -x^2$

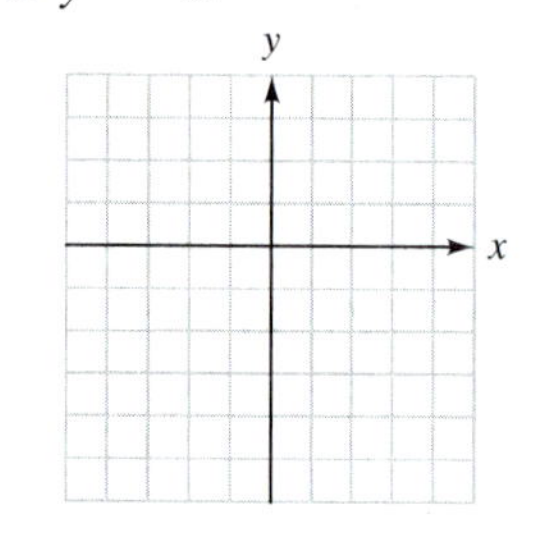

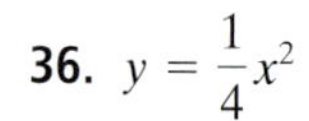

35. $y = \frac{1}{2}x^2$

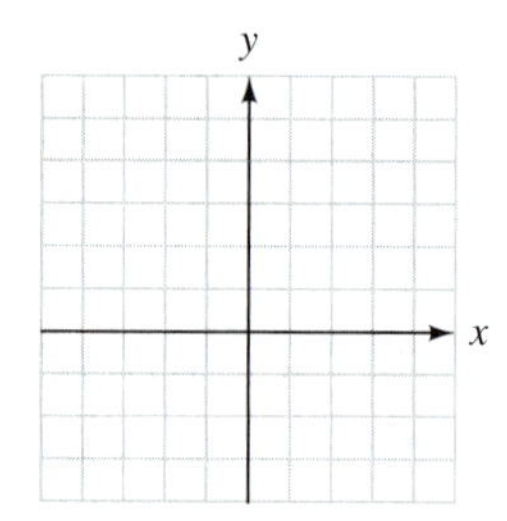

36. $y = \frac{1}{4}x^2$

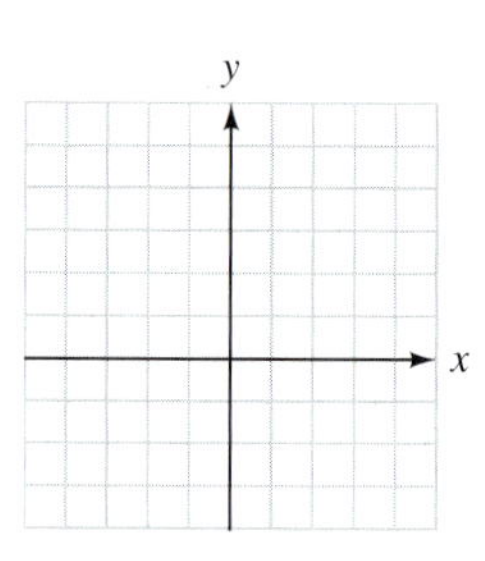

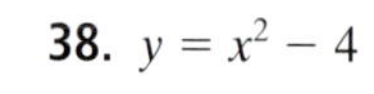

37. $y = -x^2 + 1$

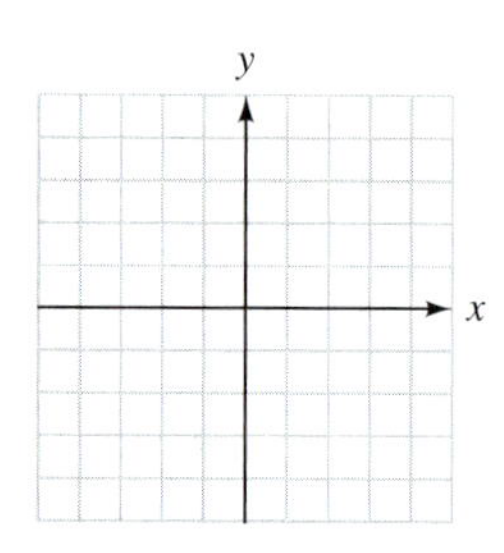

38. $y = x^2 - 4$

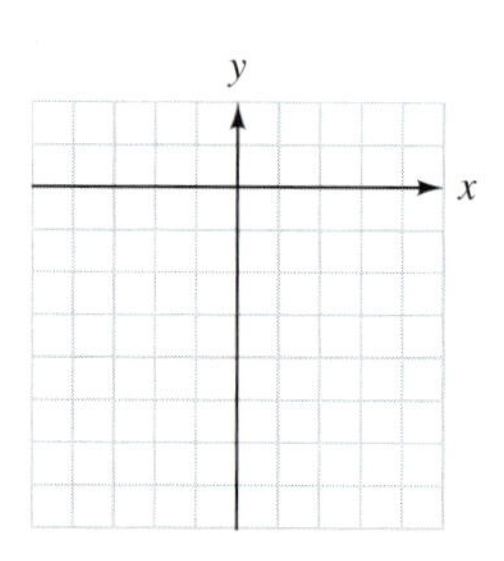

39. $y = 2x^2 - 3$

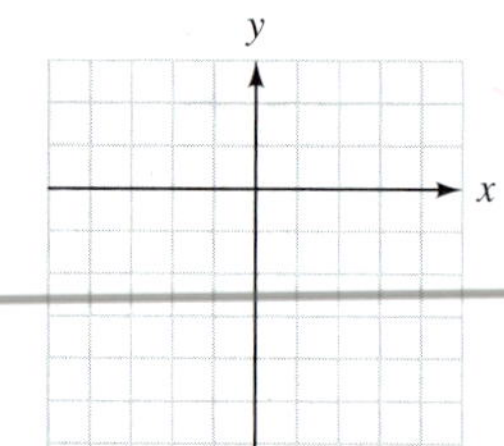

40. $y = -2x^2 + 2$

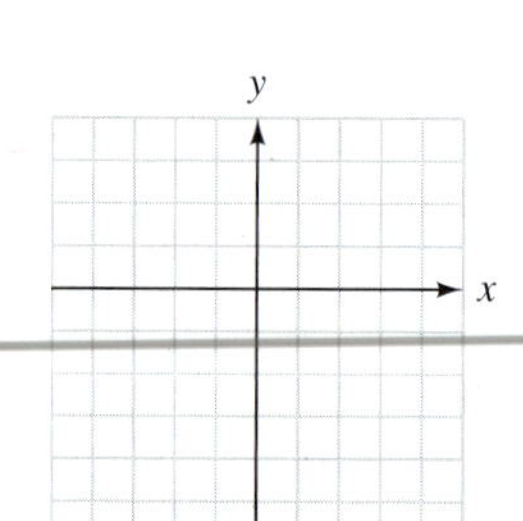

APPLICATIONS *The height h (in feet) of a ball shot straight up with an initial velocity of 64 feet per second is given by the equation $h = -16t^2 + 64t$. Find the height of the ball after the given number of seconds.*

41. 0 second

42. 1 second

43. 2 seconds

44. 4 seconds

The number of feet that a car travels before stopping depends on the driver's reaction time and the braking distance. (See the illustration on the next page.) For one driver, the stopping distance d is given by the equation $d = 0.04v^2 + 0.9v$, where v is the velocity of the car. Find the stopping distance for each of the following speeds.

45. 30 mph

46. 50 mph

47. 60 mph

48. 70 mph

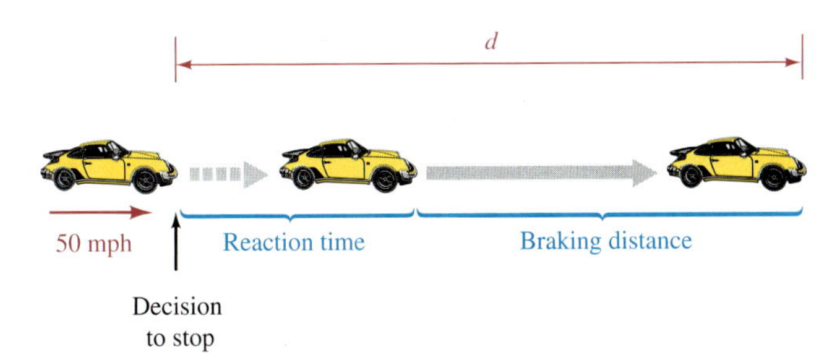

49. BRIDGES The cable of the suspension bridge below hangs in the shape of a parabola. Use information from the graph to complete the table.

x	0	2	4	−2	−4
y					

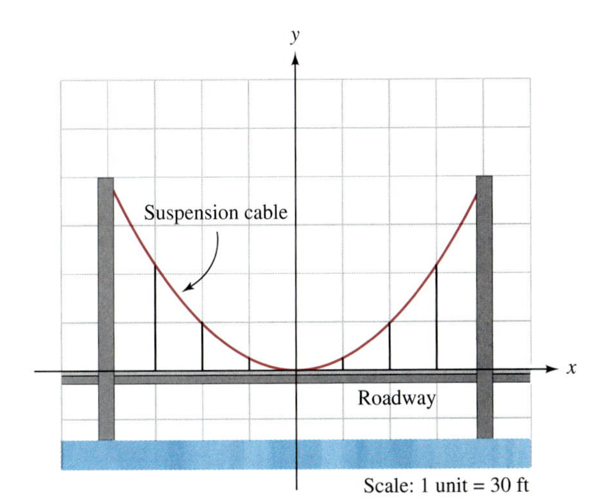

50. FIRE BOATS A stream of water shot from a high-pressure hose on a fire boat travels in the shape of a parabola. Use information from the graph to complete the table.

x	0	1	2	3	4
y					

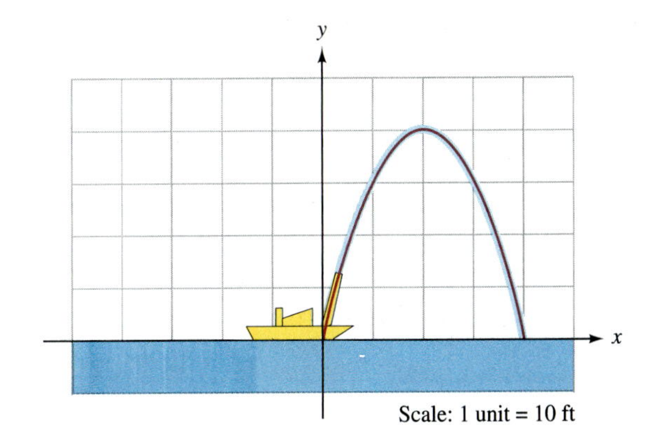

WRITING

51. Explain how to find the degree of the polynomial. $2x^3 + 5x^5 - 7x$.

52. Explain how to evaluate the polynomial. $-2x^2 - 3$ for $x = 5$.

53. Explain how to graph $y = x^2 - 3$.

54. Graph $y = \frac{1}{2}x^2$, $y = x^2$, and $y = 2x^2$ and explain what happens when the coefficient of x^2 gets larger.

REVIEW *Perform the operations.*

55. $\frac{2}{3} + \frac{4}{3}$

56. $\frac{1}{2} + \frac{2}{3}$

57. $\frac{36}{7} - \frac{23}{7}$

58. $\frac{5}{14} - \frac{4}{21}$

59. $\frac{5}{12} \cdot \frac{18}{5}$

60. $\frac{23}{25} \div \frac{46}{5}$

Solve each equation.

61. $x - 4 = 12$

62. $4z = 108$

63. $2(x - 3) = 6$

64. $3(a - 5) = 4(a + 9)$

6.5 Adding and Subtracting Polynomials

- Adding polynomials
- Subtracting polynomials

Polynomials can be added, subtracted, and multiplied just like numbers in arithmetic. In this section, we will show how to find sums and differences of polynomials.

Adding polynomials

Recall that like terms have exactly the same variables and the same exponents. For example, the monomials

$3z^2$ and $-2z^2$ are like terms. Both have the same variable (z) with the same exponent (2).

However, the monomials

$7b^2$ and $8a^2$ are not like terms. They have different variables.

$32p^2$ and $25p^3$ are not like terms. The exponents of p are different.

Also recall that we can combine like terms by adding their coefficients and keeping the same variables and exponents. For example,

$$2y + 5y = (2 + 5)y \qquad \text{and} \qquad -3x^2 + 7x^2 = (-3 + 7)x^2$$
$$= 7y \qquad\qquad\qquad\qquad = 4x^2$$

Thus, *to add monomials that are like terms, we add the coefficients and keep the same variables and exponents.*

Self Check 1
Add: $7y^3 + 12y^3$.

Answer $19y^3$

EXAMPLE 1 Add: $5x^3 + 7x^3$.

Solution Since the monomials are like terms, we add the coefficients and keep the variables and exponents.

$$5x^3 + 7x^3 = 12x^3$$

Self Check 2
Add: $3.2m^3 + 4.5m^3 + 7.2m^3$.

Answer $14.9m^3$

EXAMPLE 2 Add: $\frac{3}{2}t^2 + \frac{5}{2}t^2 + \frac{7}{2}t^2$.

Solution Since the monomials are like terms, we add the coefficients and keep the variables and exponents.

$$\frac{3}{2}t^2 + \frac{5}{2}t^2 + \frac{7}{2}t^2 = \left(\frac{3}{2} + \frac{5}{2} + \frac{7}{2}\right)t^2$$

$$= \frac{15}{2}t^2$$ To add the fractions, add the numerators and keep the denominator: $3 + 5 + 7 = 15$.

To add two polynomials, we write a + sign between them and combine like terms.

Self Check 3
Add: $(5y - 2) + (-3y + 7)$.

Answer $2y + 5$

EXAMPLE 3 Add: $2x + 3$ and $7x - 1$.

Solution

$(2x + 3) + (7x - 1)$ — Write a + sign between the binomials.

$= (2x + 7x) + (3 - 1)$ — Use the associative and commutative properties to group like terms together.

$= 9x + 2$ — Combine like terms.

The binomials in Example 3 can be added by writing the polynomials so that like terms are in columns.

$$\begin{array}{r} 2x + 3 \\ + \underline{7x - 1} \\ 9x + 2 \end{array}$$ Add the like terms, one column at a time.

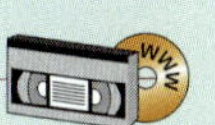

EXAMPLE 4 Add: $(5x^2 - 2x + 4) + (3x^2 - 5)$.

Solution

$$(5x^2 - 2x + 4) + (3x^2 - 5)$$
$$= (5x^2 + 3x^2) + (-2x) + (4 - 5)$$ Use the associative and commutative properties to group like terms together.
$$= 8x^2 - 2x - 1$$ Combine like terms.

Self Check 4
Add:
$(2b^2 - 4b) + (b^2 + 3b - 1)$.

Answer $3b^2 - b - 1$

The polynomials in Example 4 can be added by writing the polynomials so that like terms are in columns.

$$\begin{array}{r} 5x^2 - 2x + 4 \\ + \underline{3x^2 \qquad - 5} \\ 8x^2 - 2x - 1 \end{array}$$ Add the like terms, one column at a time.

EXAMPLE 5 Add: $(3.7x^2 + 4x - 2) + (7.4x^2 - 5x + 3)$.

Solution

$$(3.7x^2 + 4x - 2) + (7.4x^2 - 5x + 3)$$
$$= (3.7x^2 + 7.4x^2) + (4x - 5x) + (-2 + 3)$$ Use the associative and commutative properties to group like terms together.
$$= 11.1x^2 - x + 1$$ Combine like terms.

Self Check 5
Add:
$(s^2 + 1.2s - 5) + (3s^2 - 2.5s + 4)$.

Answer $4s^2 - 1.3s - 1$

The trinomials in Example 5 can be added by writing them so that like terms are in columns.

$$\begin{array}{r} 3.7x^2 + 4x - 2 \\ + \underline{7.4x^2 - 5x + 3} \\ 11.1x^2 - \quad x + 1 \end{array}$$ Add the like terms, one column at a time.

Subtracting polynomials

To subtract one monomial from another, we add the opposite of the monomial that is to be subtracted. In symbols, $x - y = x + (-y)$.

EXAMPLE 6 Subtract: $8x^2 - 3x^2$.

Solution

$$8x^2 - 3x^2 = 8x^2 + (-3x^2)$$ Add the opposite of $3x^2$.
$$= 5x^2$$ Add the coefficients and keep the same variable and exponent.

Self Check 6
Subtract: $6y^3 - 9y^3$.

Answer $-3y^3$

To subtract polynomials, we also add the opposite. For example, to subtract $3n^2 - 4n + 2$ from $5n^2 + 2n - 3$, we proceed as follows.

$(5n^2 + 2n - 3) - (3n^2 - 4n + 2)$

$= (5n^2 + 2n - 3) + [-(3n^2 - 4n + 2)]$ Add the opposite.

$= (5n^2 + 2n - 3) + (-3n^2 + 4n - 2)$ Use the distributive property to change signs.

$= (5n^2 - 3n^2) + (2n + 4n) + (-3 - 2)$ Use the associative and commutative properties to group like terms together.

$= 2n^2 + 6n - 5$ Combine like terms.

These polynomials can be subtracted by writing them so that like terms are in columns.

$$\begin{array}{r} 5n^2 + 2n - 3 \\ -\underline{(3n^2 - 4n + 2)} \end{array} \longrightarrow \begin{array}{r} 5n^2 + 2n - 3 \\ + \underline{-3n^2 + 4n - 2} \\ 2n^2 + 6n - 5 \end{array}$$ Change signs and add.

Self Check 7
Subtract:
$(3.3a - 5) - (7.8a + 2)$.

Answer $-4.5a - 7$

EXAMPLE 7 Subtract: $(3x - 4.2) - (5x + 7.2)$.

Solution

$(3x - 4.2) - (5x + 7.2)$

$= (3x - 4.2) + [-(5x + 7.2)]$ Add the opposite.

$= (3x - 4.2) + (-5x - 7.2)$ Use the distributive property to remove parentheses.

$= (3x - 5x) + (-4.2 - 7.2)$ Use the associative and commutative properties to group like terms together.

$= -2x - 11.4$ Combine like terms.

The binomials in Example 7 can be subtracted by writing them so that like terms are in columns.

$$\begin{array}{r} 3x - 4.2 \\ -\underline{(5x + 7.2)} \end{array} \longrightarrow \begin{array}{r} 3x - 4.2 \\ + \underline{-5x - 7.2} \\ -2x - 11.4 \end{array}$$ Change signs and add.

Self Check 8
Substract:
$(5y^2 - 4y + 2) - (3y^2 + 2y - 1)$.

Answer $2y^2 - 6y + 3$

EXAMPLE 8 Subtract: $(3x^2 - 4x - 6) - (2x^2 - 6x + 12)$.

Solution

$(3x^2 - 4x - 6) - (2x^2 - 6x + 12)$

$= (3x^2 - 4x - 6) + [-(2x^2 - 6x + 12)]$ Add the opposite.

$= (3x^2 - 4x - 6) + (-2x^2 + 6x - 12)$ Use the distributive property to remove parentheses.

$= (3x^2 - 2x^2) + (-4x + 6x) + (-6 - 12)$ Use the associative and commutative properties to group like terms together.

$= x^2 + 2x - 18$ Combine like terms.

The trinomials in Example 8 can be subtracted by writing them so that like terms are in columns.

$$\begin{array}{r} 3x^2 - 4x - 6 \\ -\underline{(2x^2 - 6x + 12)} \end{array} \longrightarrow \begin{array}{r} 3x^2 - 4x - 6 \\ + \underline{-2x^2 + 6x - 12} \\ x^2 + 2x - 18 \end{array}$$ Change signs and add.

Section 6.5 STUDY SET

VOCABULARY *Fill in the blanks.*

1. If two algebraic terms have exactly the same variables and exponents, they are called ______ terms.
2. $3x^3$ and $3x^2$ are ________ terms.

CONCEPTS *Fill in the blanks.*

3. To add two monomials, we add the __________ and keep the same __________ and exponents.
4. To subtract one monomial from another, we add the __________ of the monomial that is to be subtracted.

Determine whether the monomials are like terms. If they are, combine them.

5. $3y, 4y$
6. $3x^2, 5x^2$
7. $3x, 3y$
8. $3x^2, 6x$
9. $3x^3, 4x^3, 6x^3$
10. $-2y^4, -6y^4, 10y^4$
11. $-5x^2, 13x^2, 7x^2$
12. $23, 12x, 25x^2$

NOTATION *Complete each solution.*

13. $(3x^2 + 2x - 5) + (2x^2 - 7x)$
$= (3x^2 + \square) + (2x - \square) + (-5)$
$= \square + (-5x) - 5$
$= 5x^2 - 5x - 5$

14. $(3x^2 + 2x - 5) - (2x^2 - 7x)$
$= (3x^2 + 2x - 5) + [-(\square - 7x)]$
$= (3x^2 + 2x - 5) + (\square)$
$= (\square) + (2x + 7x) + (-5)$
$= x^2 + 9x - 5$

PRACTICE *Add the polynomials.*

15. $4y + 5y$
16. $-2x + 3x$
17. $-8t^2 - 4t^2$
18. $15x^2 + 10x^2$
19. $3s^2 + 4s^2 + 7s^2$
20. $-2a^3 + 7a^3 - 3a^3$
21. $(3x + 7) + (4x - 3)$
22. $(2y - 3) + (4y + 7)$
23. $(2x^2 + 3) + (5x^2 - 10)$
24. $(-4a^2 + 1) + (5a^2 - 1)$
25. $(5x^3 - 4.2x) + (7x^3 - 10.7x)$
26. $(-4.3a^3 + 25a) + (5.8a^3 - 10a)$
27. $(3x^2 + 2x - 4) + (5x^2 - 17)$
28. $(5a^2 - 2a) + (-2a^2 + 3a + 4)$
29. $(7y^2 + 5y) + (y^2 - y - 2)$
30. $(4p^2 - 4p + 5) + (6p - 2)$
31. $(3x^2 - 3x - 2) + (3x^2 + 4x - 3)$
32. $(4c^2 + 3c - 2) + (3c^2 + 4c + 2)$
33. $(3n^2 - 5.8n + 7) + (-n^2 + 5.8n - 2)$
34. $(-3t^2 - t + 3.4) + (3t^2 + 2t - 1.8)$

35. $\begin{array}{r} 3x^2 + 4x + 5 \\ + \underline{2x^2 - 3x + 6} \end{array}$

36. $\begin{array}{r} 2x^2 - 3x + 5 \\ + \underline{-4x^2 - x - 7} \end{array}$

37. $\begin{array}{r} -3x^2 \qquad - 7 \\ + \underline{-4x^2 - 5x + 6} \end{array}$

38. $\begin{array}{r} 4x^2 - 4x + 9 \\ + \underline{\qquad 9x - 3} \end{array}$

39. $\begin{array}{r} -3x^2 + 4x + 25.4 \\ + \underline{5x^2 - 3x - 12.5} \end{array}$

40. $\begin{array}{r} -6x^3 - 4.2x^2 + 7 \\ + \underline{-7x^3 + 9.7x^2 - 21} \end{array}$

Subtract the polynomials.

41. $32u^3 - 16u^3$
42. $25y^2 - 7y^2$
43. $18x^5 - 11x^5$
44. $17x^6 - 22x^6$
45. $(4.5a + 3.7) - (2.9a - 4.3)$
46. $(5.1b - 7.6) - (3.3b + 5.9)$
47. $(-8x^2 - 4) - (11x^2 + 1)$
48. $(5x^3 - 8) - (2x^3 + 5)$
49. $(3x^2 - 2x - 1) - (-4x^2 + 4)$
50. $(7a^2 + 5a) - (5a^2 - 2a + 3)$
51. $(3.7y^2 - 5) - (2y^2 - 3.1y + 4)$
52. $(t^2 - 4.5t + 5) - (2t^2 - 3.1t - 1)$
53. $(2b^2 + 3b - 5) - (2b^2 - 4b - 9)$
54. $(3a^2 - 2a + 4) - (a^2 - 3a + 7)$
55. $(5p^2 - p + 7.1) - (4p^2 + p + 7.1)$
56. $(m^2 - m - 5) - (m^2 + 5.5m - 7.5)$

57. $\begin{array}{r} 3x^2 + 4x - 5 \\ \underline{-(-2x^2 - 2x + 3)} \end{array}$

58. $\begin{array}{r} 3y^2 - 4y + 7 \\ \underline{-(6y^2 - 6y - 13)} \end{array}$

59. $\begin{array}{r} -2x^2 - 4x + 12 \\ \underline{-(10x^2 + 9x - 24)} \end{array}$

60. $\begin{array}{r} 25x^3 - 45x^2 + 31x \\ -(12x^3 + 27x^2 - 17x) \\ \hline \end{array}$

61. $\begin{array}{r} 4x^3 - 3x + 10 \\ -(5x^3 - 4x - 4) \\ \hline \end{array}$

62. $\begin{array}{r} 3x^3 + 4x^2 + 12 \\ -(-4x^3 + 6x^2 - 3) \\ \hline \end{array}$

APPLICATIONS

Consider the following information: If a house is purchased for $85,000 and is expected to appreciate $700 per year, its value y after x years is given by the equation $y = 700x + 85{,}000$.

63. VALUE OF A HOUSE Find the expected value of the house after 10 years.

64. VALUE OF A HOUSE A second house is purchased for $102,000 and is expected to appreciate $900 per year. Find an equation that will give the value y of the house after x years.

65. VALUE OF A HOUSE Find the value of the house discussed in Exercise 64 after 12 years.

66. VALUE OF TWO HOUSES Find a single polynomial equation that will give the combined value y of both houses after x years.

67. VALUE OF TWO HOUSES In two ways, find the value of the two houses after 15 years.
   a. By substituting into the polynomial equations $y = 700x + 85{,}000$ and $y = 900x + 102{,}000$ and adding
   b. By substituting into the result of Exercise 66

68. VALUE OF TWO HOUSES In two ways, find the value of the two houses after 25 years.
   a. By substituting into the polynomial equations $y = 700x + 85{,}000$ and $y = 900x + 102{,}000$ and adding
   b. By substituting into the result of Exercise 66

Consider the following information. A young couple bought two cars, one for $8,500 and the other for $10,200. The first car is expected to depreciate $800 per year and the second car $1,100 per year.

69. VALUE OF A CAR Write an equation that will give the value y of the first car after x years.

70. VALUE OF A CAR Write an equation that will give the value y of the second car after x years.

71. VALUE OF TWO CARS Find a single equation that will give the value y of both cars after x years.

72. VALUE OF TWO CARS In two ways, find the value of the two cars after 6 years.

WRITING

73. What are *like terms?*

74. Explain how to add two polynomials.

75. Explain how to subtract two polynomials.

76. When two binomials are added, is the result always a binomial? Explain.

REVIEW

77. BASKETBALL SHOES Use the following information to find how much lighter the Kevin Garnett shoe is than the Michael Jordan shoe.

Nike Air Garnett III	**Air Jordan XV**
Synthetic fade mesh and leather. Sizes: $6\frac{1}{2}$–18. Weight: 13.8 oz	Full-grain leather upper with woven leather pattern. Sizes: $6\frac{1}{2}$–18. Weight: 14.6 oz

78. AEROBICS The number of calories burned when doing step aerobics depends on the step height. Use the following table to find how many more calories are burned during a 10-minute workout using an 8-inch step instead of a 4-inch step.

Step height (in.)	Calories burned per minute
4	4.5
6	5.5
8	6.4
10	7.2

Source: *Reebok Instructor News* (Vol. 4, No. 3, 1991)

79. THE PANAMA CANAL A ship entering the Panama Canal from the Atlantic Ocean is lifted up 85 feet to Lake Gatun by the Gatun Lock system. Then the ship is lowered 31 feet by the Pedro Miguel Lock. By how much must the ship be lowered by the Miraflores Lock system for it to reach the Pacific Ocean water level?

80. CANAL LOCKS What is the combined length of the system of locks in the Panama Canal shown below? Express your answer as a mixed number and as a decimal, rounded to the nearest tenth.

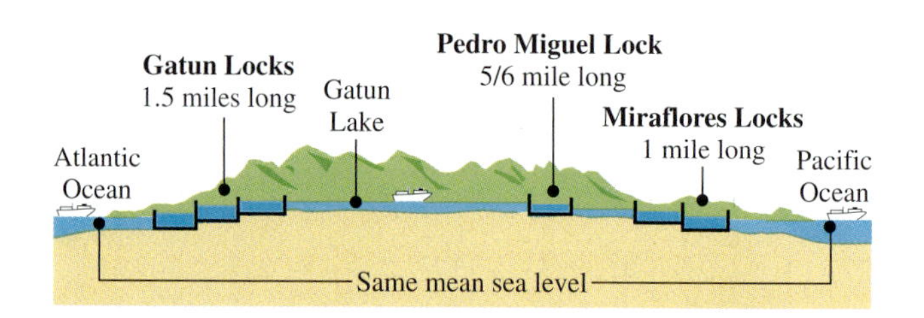

6.6 Multiplying Polynomials

- Multiplying monomials • Multiplying a polynomial by a monomial
- Multiplying a binomial by a binomial • Multiplying a polynomial by a binomial

In this section, we will discuss how to multiply polynomials.

Multiplying monomials

To multiply $4x^2$ by $-2x^3$, we use the commutative and associative properties of multiplication to group the numerical factors and the variable factors and multiply.

$$\begin{aligned} 4x^2(-2x^3) &= 4(-2)x^2x^3 \\ &= -8x^5 \end{aligned}$$

This example suggests the following rule.

Multiplying two monomials

To multiply two monomials, multiply the numerical factors and then multiply the variable factors.

EXAMPLE 1 Multiply: **a.** $3y \cdot 6y$ and **b.** $-3x^5(2x^5)$.

Solution

a. $3y \cdot 6y = (3 \cdot 6)(y \cdot y)$ Multiply the numerical factors and multiply the variables.

$= 18y^2$ Multiply: $3 \cdot 6 = 18$ and $y \cdot y = y^2$.

b. $-3x^5(2x^5) = (-3 \cdot 2)(x^5 \cdot x^5)$ Multiply the numerical factors and multiply the variables.

$= -6x^{10}$ Multiply: $-3 \cdot 2 = -6$ and $x^5 \cdot x^5 = x^{10}$.

Self Check 1
Multiply: $-7a^3 \cdot 2a^5$.

Answer $-14a^8$

Multiplying a polynomial by a monomial

To find the product of a polynomial and a monomial, we use the distributive property. To multiply $x + 4$ by $3x$, for example, we proceed as follows.

$\mathbf{3x}(x + 4) = \mathbf{3x}(x) + \mathbf{3x}(4)$ Use the distributive property.

$= 3x^2 + 12x$ Multiply the monomials: $3x(x) = 3x^2$ and $3x(4) = 12x$.

The results of this example suggest the following rule.

Multiplying polynomials by monomials

To multiply a polynomial by a monomial, use the distributive property to remove parentheses and simplify.

Self Check 2
Multiply:
a. $3y(5y^3 - 4y)$
b. $5x(3x^2 - 2x + 3)$

Answers **a.** $15y^4 - 12y^2$, **b.** $15x^3 - 10x^2 + 15x$

EXAMPLE 2 Multiply: **a.** $2a^2(3a^2 - 4a)$ and **b.** $2x(3x^2 + 2x - 3)$.

Solution

a. $2a^2(3a^2 - 4a) = 2a^2(3a^2) - 2a^2(4a)$ Use the distributive property.
$= 6a^4 - 8a^3$ Multiply: $2a^2(3a^2) = 6a^4$ and $2a^2(4a) = 8a^3$.

b. $2x(3x^2 + 2x - 3) = 2x(3x^2) + 2x(2x) - 2x(3)$ Use the distributive property.
$= 6x^3 + 4x^2 - 6x$ Multiply: $2x(3x^2) = 6x^3$, $2x(2x) = 4x^2$, and $2x(3) = 6x$.

Multiplying a binomial by a binomial

To multiply two binomials, we must use the distributive property more than once. For example, to multiply $2x + 3$ by $3x - 5$, we proceed as follows.

$(3x - 5)(2x + 3) = (3x - 5)2x + (3x - 5)3$ Distribute the factor of $3x - 5$ over the two terms within $(2x + 3)$.
$= 2x(3x - 5) + 3(3x - 5)$ Use the commutative property of multiplication.
$= 2x(3x) - 2x(5) + 3(3x) - 3(5)$ Use the distributive property twice.
$= 6x^2 - 10x + 9x - 15$ Perform the multiplications.
$= 6x^2 - x - 15$ Combine like terms: $-10x + 9x = -x$.

The results of this example suggest the following rule.

Multiplying a binomial by a binomial

To multiply two binomials, multiply each term of one binomial by each term of the other binomial and combine like terms.

Self Check 3
Multiply: $(3x - 2)(2x + 3)$.

Answer $6x^2 + 5x - 6$

EXAMPLE 3 Multiply: $(2x - 4)(3x + 5)$.

Solution

$(2x - 4)(3x + 5) = (2x - 4)3x + (2x - 4)5$ Each term within $(3x + 5)$ is multiplied by $2x - 4$.
$= 3x(2x - 4) + 5(2x - 4)$ Use the commutative property of multiplication.
$= 3x(2x) - 3x(4) + 5(2x) - 5(4)$ Use the distributive property twice.
$= 6x^2 - 12x + 10x - 20$ Perform the multiplications.
$= 6x^2 - 2x - 20$ Combine like terms: $-12x + 10x = -2x$.

EXAMPLE 4 Find each square: $(5x - 4)^2$.

Solution In the expression $(5x - 4)^2$, the binomial $5x - 4$ is the base and 2 is the exponent.

$(5x - 4)^2 = (5x - 4)(5x - 4)$	Write $5x - 4$ as a factor two times.
$= (5x - 4)5x - (5x - 4)4$	Distribute the factor of $5x - 4$ over each term within $(5x - 4)$.
$= 5x(5x - 4) - 4(5x - 4)$	Change the order of the factors.
$= 5x(5x) - 5x(4) - 4(5x) + 4(4)$	Distribute the multiplication by $5x$. Distribute the multiplication by 4.
$= 25x^2 - 20x - 20x + 16$	Perform the multiplications.
$= 25x^2 - 40x + 16$	Simplify: $-20x - 20x = -40x$.

Self Check 4
Find each square: $(5x + 4)^2$.

Answer $25x^2 + 40x + 16$

! COMMENT A common error when squaring a binomial is to square only its first and second terms. For example, it is incorrect to write

$$(5x - 4)^2 = (5x)^2 - (4)^2 = 25x^2 - 16$$

Multiplying a polynomial by a binomial

We must also use the distributive property more than once to multiply a polynomial by a binomial. For example, to multiply $3x^2 + 3x - 5$ by $2x + 3$, we proceed as follows.

$$\begin{aligned}(2x + 3)(3x^2 + 3x - 5) &= (2x + 3)3x^2 + (2x + 3)3x - (2x + 3)5 \\ &= 3x^2(2x + 3) + 3x(2x + 3) - 5(2x + 3) \\ &= 6x^3 + 9x^2 + 6x^2 + 9x - 10x - 15 \\ &= 6x^3 + 15x^2 - x - 15\end{aligned}$$

EXAMPLE 5 Multiply: $(3a + 1)(3a^2 + 2a + 2)$.

Solution

$$\begin{aligned}&(3a + 1)(3a^2 + 2a + 2) \\ &\quad = (3a + 1)3a^2 + (3a + 1)2a + (3a + 1)2 \\ &\quad = 3a^2(3a + 1) + 2a(3a + 1) + 2(3a + 1) \\ &\quad = 3a^2(3a) + 3a^2(1) + 2a(3a) + 2a(1) + 2(3a) + 2(1) \\ &\quad = 9a^3 + 3a^2 + 6a^2 + 2a + 6a + 2 \\ &\quad = 9a^3 + 9a^2 + 8a + 2\end{aligned}$$

Self Check 5
Multiply: $(x - 2)(3x^2 + 4x + 1)$.

Answer $3x^3 - 2x^2 - 7x - 2$

We can use a column format to multiply polynomials. To do so, we multiply each term in the top polynomial by each term in the bottom polynomial. To make the addition easy, we will keep like terms in columns. As an example, we can multiply $2x - 4$ by $3x + 2$ as follows.

$$\begin{array}{rr} & 2x - 4 \\ & \times\ 3x + 2 \\ 3x(2x-4) \rightarrow & 6x^2 - 12x \\ 2(2x-4) \rightarrow & + \ 4x - 8 \\ & 6x^2 - \ 8x - 8 \end{array}$$

Self Check 6
Multiply $3y^2 - 5y + 4$ by $4y - 3$.

Answer $12y^3 - 29y^2 + 31y - 12$

EXAMPLE 6 Multiply $3a^2 - 4a + 7$ by $2a + 5$.

Solution

$$\begin{array}{rrrr} 3a^2 & - \ 4a & + 7 & \\ \times & 2a & + 5 & \\ 6a^3 - & 8a^2 & + 14a & \\ & + 15a^2 & - 20a & + 35 \\ 6a^3 + & 7a^2 & - \ 6a & + 35 \end{array}$$

Section 6.6 STUDY SET

VOCABULARY *Fill in the blanks.*

1. A polynomial with exactly one term is called a ________.
2. A polynomial with exactly two terms is called a ________.
3. A polynomial with exactly ▭ terms is called a trinomial.
4. $a(b + c) = ab + ac$ illustrates the ________ property.

CONCEPTS *Fill in the blanks.*

5. To multiply two monomials, multiply the ________ factors and then multiply the variable ________.
6. To multiply a polynomial by a monomial, use the ________ property to remove parentheses and simplify.
7. To multiply two binomials, multiply each ______ of one binomial by each term of the other binomial and combine ______ terms.
8. To multiply a polynomial by a binomial, we must use the distributive ________ more than once.

NOTATION *Complete each solution.*

9. $3x(2x - 5) = 3x(\ \square\ ) - 3x(\ \square\ )$
   $= 6x^2 - 15x$
10. $(3x + 1)(2x^2 - 3x - 2)$
   $= (3x + 1)\square - (3x + 1)\square - (3x + 1)\square$
   $= 2x^2(3x + 1) - 3x(3x + 1) - 2(3x + 1)$
   $= 2x^2(\ \square\ ) + 2x^2(1) - 3x(3x) - 3x(1)$
   $- 2(3x) - 2(\ \square\ )$
   $= 6x^3 + 2x^2 - \square - 3x - \square - 2$
   $= 6x^3 - 7x^2 - 9x - 2$

PRACTICE *Find each product.*

11. $(3x^2)(4x^3)$
12. $(-2a^3)(3a^2)$
13. $(3b^2)(-2b)$
14. $(3y)(-y^4)$
15. $(-2x^2)(3x^3)$
16. $(-7x^3)(-3x^3)$
17. $\left(-\frac{2}{3}y^5\right)\left(\frac{3}{4}y^2\right)$
18. $\left(\frac{2}{5}r^4\right)\left(\frac{3}{5}r^2\right)$
19. $3(x + 4)$
20. $-3(a - 2)$
21. $-4(t + 7)$
22. $6(s^2 - 3)$
23. $3x(x - 2)$
24. $4y(y + 5)$
25. $-2x^2(3x^2 - x)$
26. $4b^3(2b^2 - 2b)$
27. $2x(3x^2 + 4x - 7)$
28. $3y(2y^2 - 7y - 8)$
29. $-p(2p^2 - 3p + 2)$
30. $-2t(t^2 - t + 1)$
31. $3q^2(q^2 - 2q + 7)$
32. $4v^3(-2v^2 + 3v - 1)$
33. $(a + 4)(a + 5)$
34. $(y - 3)(y + 5)$
35. $(3x - 2)(x + 4)$
36. $(t + 4)(2t - 3)$
37. $(2a + 4)(3a - 5)$
38. $(2b - 1)(3b + 4)$

Square each binomial.

39. $(2x + 3)^2$
40. $(2y + 5)^2$
41. $(2x - 3)^2$
42. $(2y - 5)^2$
43. $(5t + 1)^2$
44. $(5t - 1)^2$

Multiply the polynomials.

45. $(2x + 1)(3x^2 - 2x + 1)$
46. $(x + 2)(2x^2 + x - 3)$
47. $(x - 1)(x^2 + x + 1)$
48. $(x + 2)(x^2 - 2x + 4)$
49. $(x + 2)(x^2 - 3x + 1)$
50. $(x + 3)(x^2 + 3x + 2)$

Find each product.

51. $\begin{array}{r} 4x + 3 \\ \underline{x + 2} \end{array}$
52. $\begin{array}{r} 5r + 6 \\ \underline{2r - 1} \end{array}$
53. $\begin{array}{r} 4x - 2 \\ \underline{3x + 5} \end{array}$
54. $\begin{array}{r} 6r + 5 \\ \underline{2r - 3} \end{array}$
55. $\begin{array}{r} x^2 - x + 1 \\ \underline{x + 1} \end{array}$
56. $\begin{array}{r} 4x^2 - 2x + 1 \\ \underline{2x + 1} \end{array}$

APPLICATIONS

57. GEOMETRY Express the area of the rectangle below.

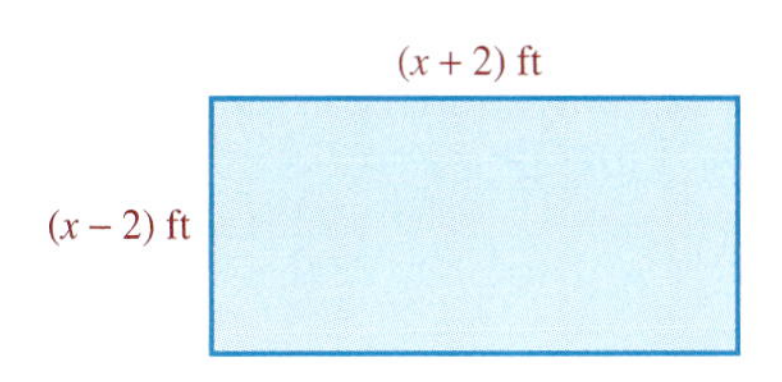

58. SAILING The height h of the triangular sail below is $4x$ feet, and the base b is $(3x - 2)$ feet. Express the area of the sail. (*Hint:* The area of a triangle is given by the formula $A = \frac{1}{2}bh$.)

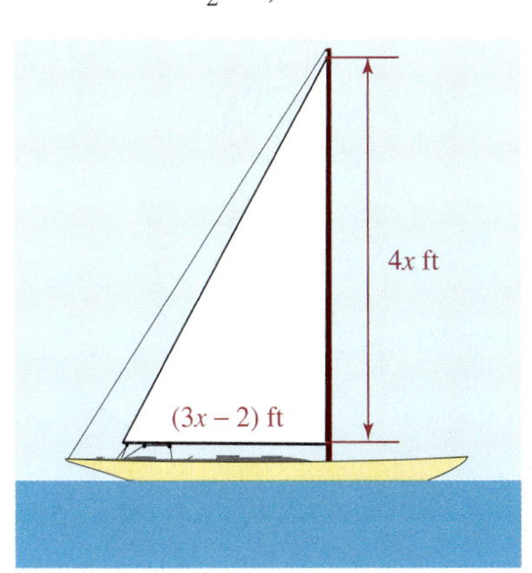

59. ECONOMICS The revenue R received from selling clock radios is the product of their price and the number that are sold. If the price of each radio is given by the formula $-\frac{x}{100} + 30$ and x is the number sold, find a formula that gives the amount of revenue received.

60. ECONOMICS If the pricing formula given in Problem 59 changes to $-\frac{x}{100} + 40$, find the formula for revenue received.

WRITING

61. Explain how to multiply two binomials.
62. Explain how to find $(2x + 1)^2$.
63. Explain why $(x + 1)^2 \neq x^2 + 1^2$. (Read $\neq$ as "is not equal to.")
64. If two terms are to be added, they have to be like terms. If two terms are to be multiplied, must they be like terms? Explain.

REVIEW

65. THE EARTH It takes 23 hours, 56 minutes, and 4.091 seconds for the Earth to rotate around its axis once. Write 4.091 in words.

66. TAKE-OUT FOODS The following sticker shows the amount and the price per pound of some spaghetti salad that was purchased at a delicatessen. Find the total price of the salad.

67. What is $\frac{7}{64}$ in decimal form?
68. Solve: $1.7x + 1.24 = -1.4x - 0.62$.
69. Add: $56.09 + 78 + 0.567$.
70. Subtract: $-679.4 - (-599.89)$.
71. Evaluate: $\sqrt{16} + \sqrt{36}$.
72. Divide: $103.6 \div 0.56$.

KEY CONCEPT

Graphing

Ordered pairs of real numbers can be graphed on a *rectangular coordinate system.*

Refer to the illustration below.

1. Label the x- and y-axes.
2. Find the coordinates of point P.
3. Plot the points $(-3, 2)$, $(-3, -2)$, and $(3, -2)$.
4. What are the coordinates of the origin?
5. Label each of the quadrants. In what quadrant do the points have a negative x-coordinate and a positive y-coordinate?
6. Graph the points: $(4, 0)$, $(0, 4)$, $(-4, 0)$, and $(0, -4)$.

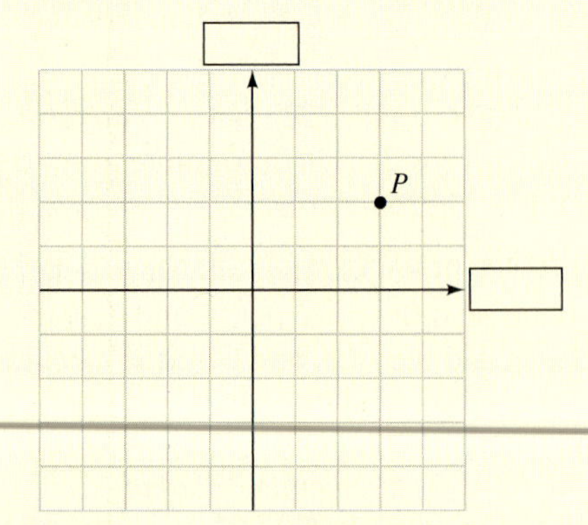

To graph linear equations on a rectangular coordinate system, we must find the coordinates of several points that satisfy the equation.

7. Graph: $2x - 4y = 8$.

Step 1: Complete the table of solutions.

Step 2: Plot the points listed in the table of solutions.

Step 3: Draw a straight line through the points.

x	y	(x, y)
0		
	0	
2		

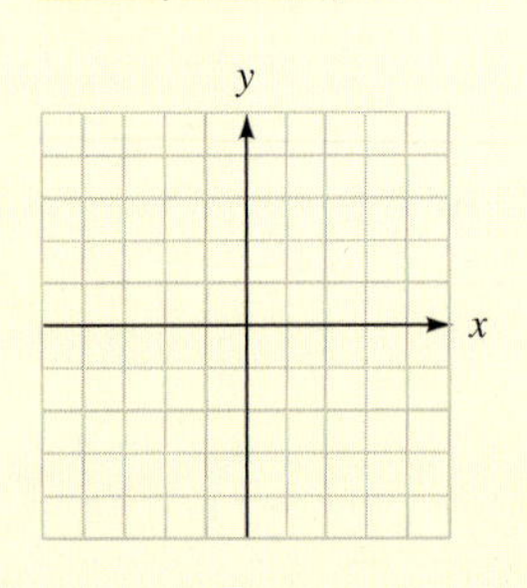

8. Graph: $y = -2x + 1$.

Step 1: Complete the table of solutions.

Step 2: Plot the points listed in the table of solutions.

Step 3: Draw a straight line through the points.

x	y	(x, y)
−2		
0		
2		

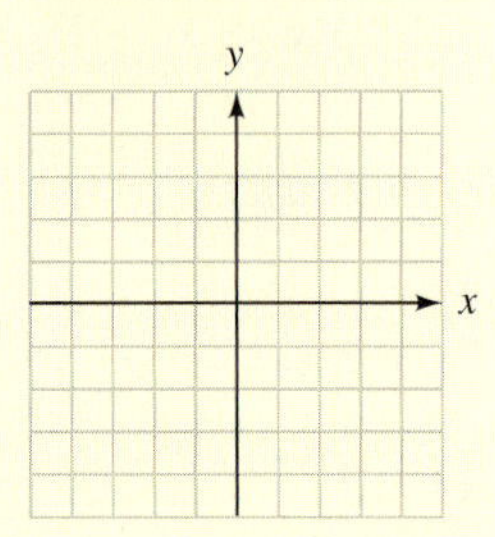

The right-hand side of the equation $y = x^2 + 1$ is a second- degree polynomial. Equations of this type are graphed in the same way as we graph linear equations.

9. Graph: $y = x^2 + 1$.

Step 1: Complete the table of solutions.

Step 2: Plot the points listed in the table of solutions.

Step 3: Join the points with a smooth curve.

x	y	(x, y)
−2		
−1		
0		
1		
2		

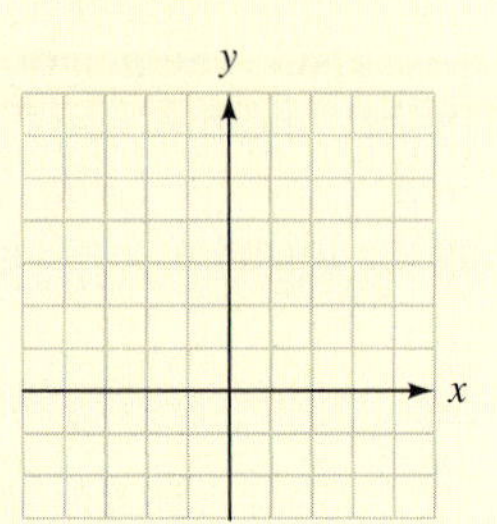

ACCENT ON TEAMWORK

SECTION 6.1

CAMPUS MAPS Get a map of your college campus and use a black marker to draw a rectangular coordinate system on the map. The size of the grid you use will depend on the size of the map. Determine what school landmark should serve as the origin of the coordinate system. List the coordinates of important locations on your campus.

SECTION 6.2

TRANSLATIONS On the same rectangular coordinate system, sketch the graphs of

$$y = x$$
$$y = x + 2$$
$$y = x - 2$$

How are the graphs similar? How are they different?

SECTION 6.3

RULES FOR EXPONENTS Have one student in your group write the three rules for exponents introduced in Section 6.3 on separate 3×5 cards. Have another student write a word description of the rules on separate cards. Finally, have a third student write an example of the use of the rules on separate cards.

When the three sets of cards are completed, put them together, shuffle them, and then work together as a group to match the symbolic description, the word description, and the example for each of the rules.

SECTION 6.4

POLYNOMIALS Write a polynomial using the variable x that meets the following conditions:

- It has degree 3.
- It has three terms.
- It does not have an x^2 term.
- The coefficients of the first and second terms are opposites, and the coefficient of the last term is twice the coefficient of the first term.
- When it is evaluated for $x = 0$, the result is 6.

SECTION 6.5

ADDING AND SUBTRACTING POLYNOMIALS IN VERTICAL FORM Fill in the blanks.

a. $$\begin{array}{r} \square x^2 - x + 6 \\ + \; 2x^2 + \square x - 8 \\ \hline 7x^2 + 6x - \square \end{array}$$

b. $$\begin{array}{r} 12x^2 - 3x - \square \\ - \; 6x^2 + \square x - 7 \\ \hline \square x^2 - 7x - 4 \end{array}$$

SECTION 6.6

MULTIPLYING POLYNOMIALS To multiply $(2x + 1)(3x - 5)$ using a table, we enter the terms $2x$ and $+1$ of the binomial $2x + 1$ in the leftmost column, as shown. We enter the terms $3x$ and -5 of the binomial $3x - 5$ in the top row as shown.

Multiply	$3x$	-5
$2x$		
$+1$		

We then multiply each term in the leftmost column by each term in the top row and enter each result in the proper box. To begin, multiply $2x$ and $3x$. The result is shown in red in the illustration below. Then we multiply $2x$ and -5. The result is shown in blue. Next, we multiply $+1$ and $3x$. The result is shown in green. Finally, we multiply $+1$ and -3. The result is shown in purple.

Multiply	$3x$	-5
$2x$	$6x^2$	$-10x$
$+1$	$+3x$	-5

To complete the process, we combine the like terms along the diagonal and write the final result as shown below.

Multiply	$3x$	-5
$2x$	$6x^2$	$-10x$
$+1$	$+3x$	-5

$+3x - 10x = -7x$

$$(2x + 1)(3x - 5) = 6x^2 - 7x - 5$$

Use a table to find each product.

a. $(4x + 3)(5x - 1)$
b. $(3x - 7)(4x - 3)$
c. $(9x + 2)(8x + 1)$
d. $(6x + 5)(6x - 5)$
e. $(2x^2 + x - 4)(4x - 7)$

CHAPTER REVIEW

SECTION 6.1 *The Rectangular Coordinate System*

CONCEPTS

A solution of an *equation in two variables* is an ordered pair of real numbers.

REVIEW EXERCISES

1. Determine whether $(2, -3)$ is a solution of $2x + 5y = -11$.

2. Determine whether $(-3, 2)$ is a solution of $3x - 5y = 19$.

3. Complete the solutions of the equation $3x - 4y = 12$. $(0,\quad)$ and $(-4,\quad)$

4. Complete the table of solutions for $y = -3x - 2$.

x	*y*	(*x*, *y*)
1		(1,)
3		(3,)
−2		(−2,)

Ordered pairs of real numbers can be graphed on a *rectangular coordinate system.*

The *x-axis* and the *y-axis* divide the coordinate system into four regions, called *quadrants.*

5. Graph the points with coordinates $(2, 3)$, $(-3, 4)$, $(-1.5, -3)$, and $(\frac{7}{2}, -1)$.

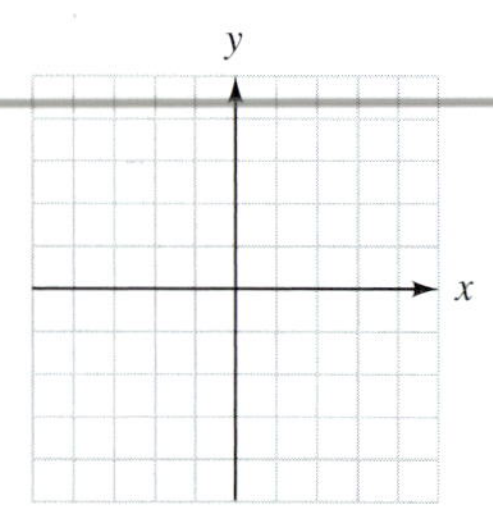

6. Give the coordinates of each point shown in the illustration below.

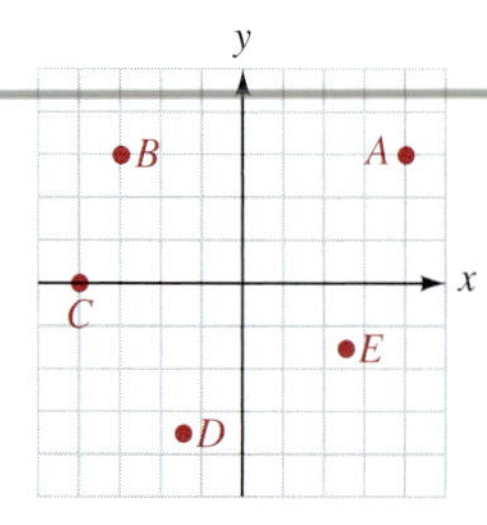

7. In what quadrant does the point $(-3, -4)$ lie?

Coordinate systems have many applications in the real world.

8. Your ticket at the theater is for seat B-10. Locate your seat on the diagram.

SECTION 6.2 *Graphing Linear Equations*

Equations whose graphs are straight lines are called *linear equations.*

The points where a graph crosses the x- and y-axes, respectively, are called the *x-intercept* and the *y-intercept.*

We can graph a linear equation by constructing a table of solutions, plotting the points, and drawing a straight line through the points.

In the equation $y = \frac{1}{2}x - 1$, x is the *independent variable* and y is the *dependent variable.*

The graph of any equation of the form $y = b$ is a *horizontal line.*

The graph of any equation of the form $x = a$ is a *vertical line.*

9. Complete the table and then graph $3x - y = 5$. Label the x-intercept and the y-intercept.

x	y
0	
	0
1	−2

10. Make a table showing three solutions of $y = \frac{1}{2}x - 1$. Then graph it.

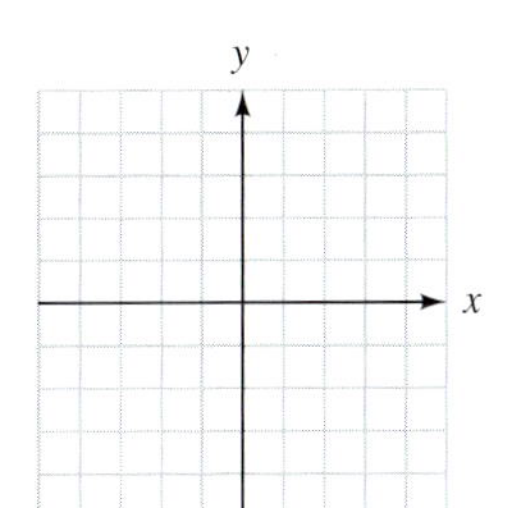

11. Graph $y = 2$.

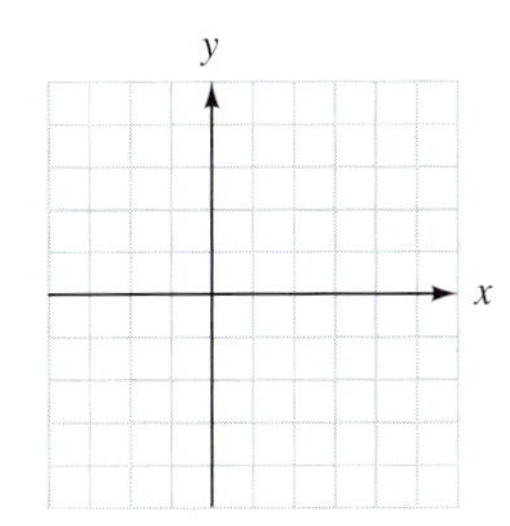

12. Graph $x = 1$.

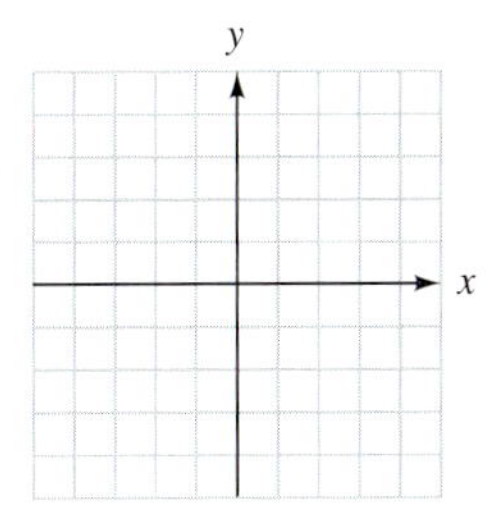

13. The line graphed is a picture of all the solutions of $y = x + 1$. Explain.

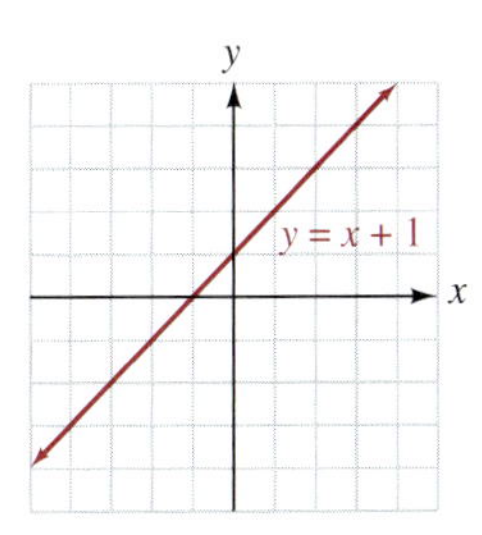

SECTION 6.3 *Multiplication Rules for Exponents*

Exponents represent repeated multiplication.

The *product rule for exponents:*

$$x^m x^n = x^{m+n}$$

14. What repeated multiplication does $(4h)^3$ represent?

15. Write this expression using exponents: $5 \cdot 5 \cdot d \cdot d \cdot d \cdot m \cdot m \cdot m \cdot m$

Simplify each expression.

16. h^6h^4

17. $t^3(t^5)$

18. $w^2 \cdot w \cdot w^4$

19. $4^7 \cdot 4^5$

Simplify each product.

20. $2b^2 \cdot 4b^5$

21. $-6x^3(4x)$

22. $-2f^2(-4f)(3f^4)$

23. $-ab \cdot b \cdot a$

24. $xy^4 \cdot xy^2$

25. $(mn)(mn)$

26. $3z^3 \cdot 9m^3z^4$

27. $-5cd(4c^2d^5)$

The *power rule for exponents:*

$$(x^m)^n = x^{m \cdot n}$$

The *power rule for products:*

$$(xy)^m = x^m y^m$$

Simplify each expression.

28. $(v^3)^4$

29. $(3y)^3$

30. $(5t^4)^2$

31. $(2a^4b^5)^3$

Simplify each expression.

32. $(c^4)^5(c^2)^3$

33. $(3s^2)^3(2s^3)^2$

34. $(c^4c^3)^2$

35. $(2xx^2)^3$

SECTION 6.4 Introduction to Polynomials

A *monomial* is a polynomial with one term. A *binomial* is a polynomial with two terms. A *trinomial* is a polynomial with three terms.

Classify each polynomial as a monomial, a binomial, or a trinomial.

36. $3x^2 + 4x - 5$

37. $3t^2$

38. $2x^2 - 1$

The *degree* of a polynomial is the same as the degree of its term with largest degree.

Give the degree of each polynomial.

39. $3x^2 + 2x^3$

40. $3t^4 - 4t^2 - 3$

41. $3q^2 - 4q^5$

A polynomial has a *numerical value* for specific values of its variable.

42. Evaluate $3x^2 - 2x - 1$ for $x = 2$.

43. Evaluate $2t^2 + t - 2$ for $t = -3$.

We can graph *second-degree polynomials* by constructing a table of solutions, plotting the points, and joining them with a smooth curve.

44. Graph: $y = x^2 - 3$.

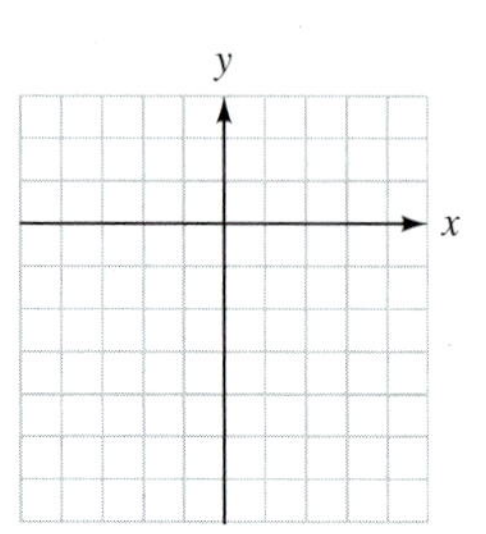

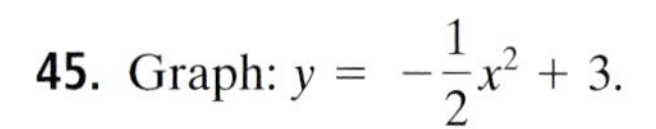

45. Graph: $y = -\frac{1}{2}x^2 + 3$.

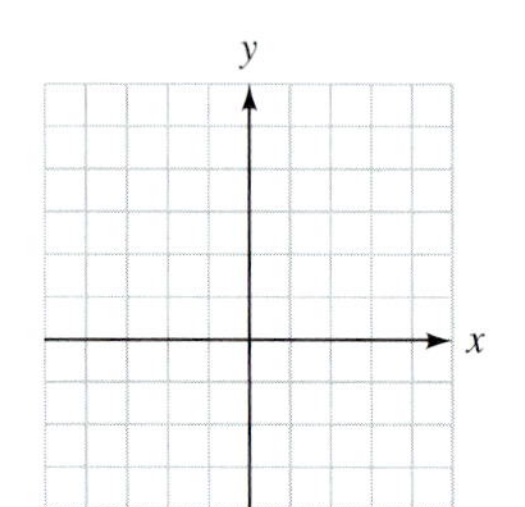

SECTION 6.5 Adding and Subtracting Polynomials

To *add monomials,* add the coefficients and keep the same variables and exponents.

Add the monomials.

46. $3x^3 + 2x^3$

47. $\frac{1}{2}p^2 + \frac{5}{2}p^2 + \frac{7}{2}p^2$

To *add two polynomials,* write + between them and combine like terms.

Add the polynomials.

48. $(3x - 1) + (6x + 5)$

49. $(3x^2 - 2x + 4) + (-x^2 - 1)$

Add the polynomials.

50. $\begin{array}{r} 5x - 2 \\ + \underline{3x + 5} \end{array}$

51. $\begin{array}{r} 3x^2 - 2x + 7 \\ + \underline{-5x^2 + 3x - 5} \end{array}$

Subtract the monomials.

52. $16p^3 - 9p^3$

53. $4y^2 - 9y^2$

To *subtract two polynomials,* add the opposite of the polynomial that is to be subtracted.

Subtract the polynomials.

54. $(2.5x + 4) - (1.4x + 12)$

55. $(3z^2 - z + 4) - (2z^2 + 3z - 2)$

Subtract the polynomials.

56. $\begin{array}{r} 5x - 2 \\ -\underline{(3x + 5)} \end{array}$

57. $\begin{array}{r} 3x^2 - 2x + 7 \\ -\underline{(-5x^2 + 3x - 5)} \end{array}$

SECTION 6.6 *Multiplying Polynomials*

To *multiply two monomials,* multiply the numerical factors and multiply the variable factors.

Multiply the monomials.

58. $3x^2 \cdot 5x^3$

59. $(3z^2)(-2z^2)$

To *multiply a polynomial by a monomial,* use the distributive property to remove parentheses and combine like terms.

To *multiply two binomials,* multiply each term of one binomial by each term of the other binomial and combine like terms.

To *multiply two polynomials,* multiply each term of one polynomial by each term of the other polynomial and combine like terms.

Multiply.

60. $2x^2(3x + 2)$

61. $-5t^3(7t^2 - 6t - 2)$

62. $(2x - 1)(3x + 2)$

63. $(5t + 4)(7t - 6)$

64. $\begin{array}{r} 5x - 2 \\ \times \underline{3x + 5} \end{array}$

65. $\begin{array}{r} 3x + 2 \\ \times \underline{5x - 5} \end{array}$

66. $(3x + 2)(2x^2 - x + 1)$

67. $(2r - 3)(3r^2 + 2r - 3)$

68. $\begin{array}{r} 5x^2 - 2x + 3 \\ \times \underline{\quad 3x + 5} \end{array}$

69. $\begin{array}{r} 3x^2 - 2x - 1 \\ \times \underline{\quad 5x - 2} \end{array}$

70. Explain why $(x + 2)^2 \neq x^2 + 4$.

CHAPTER 6 TEST

1. Determine whether $(-1, 2)$ is a solution of $4x + 5y = 6$.

2. Determine whether $(3, -2)$ is a solution of $3x - 2y = -13$.

3. Complete the ordered pairs so that each one satisfies the equation $x - 2y = 4$.

$(0,\quad), (\;, 0)$ and $(2,\quad)$

4. Complete the table of solutions for $y = \frac{1}{3}x + 1$.

x	y	(x, y)
0		
3		
−3		

5. PANTS SALE See the illustration. List the pant sizes that are not available as ordered pairs of the form (waist, length).

*These pants in your size or they're free. Guaranteed!**

Stonewash jeans, $31.99-$39.99

*Our size guarantee is good only for the following sizes:

Length	Waist	30	31	32	33	34	36	38
	30	X	X	X	X	X	X	
	32		X	X	X	X	X	X
	34			X	X	X	X	X

6. Graph the ordered pairs:

$(4, 2), (-1, 3), (-2, 0)$, and $(4, -3)$

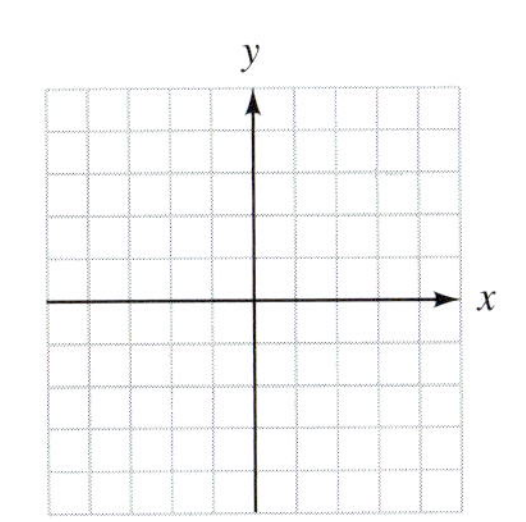

7. Give the coordinates of each point on the graph.

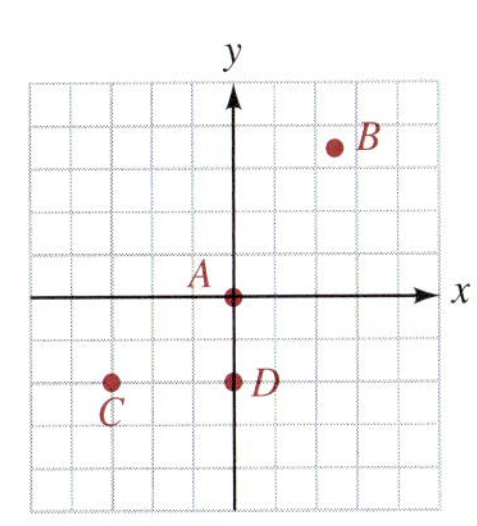

8. Graph: $2x - y = 4$.

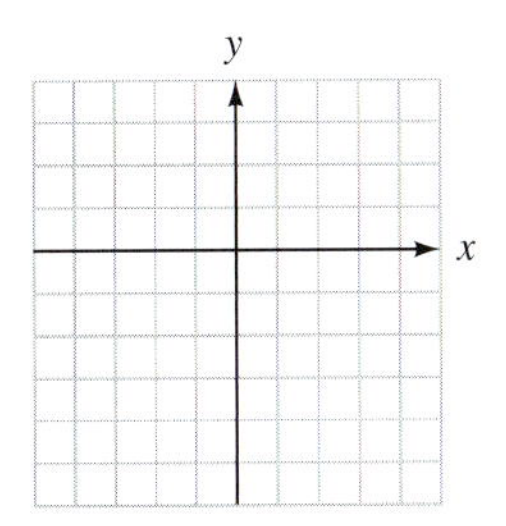

9. a. What is the x-intercept of the line graphed in Problem 8?

b. What is the y-intercept of the line graphed in Problem 8?

10. Graph: $y = -\frac{3}{2}x - 1$.

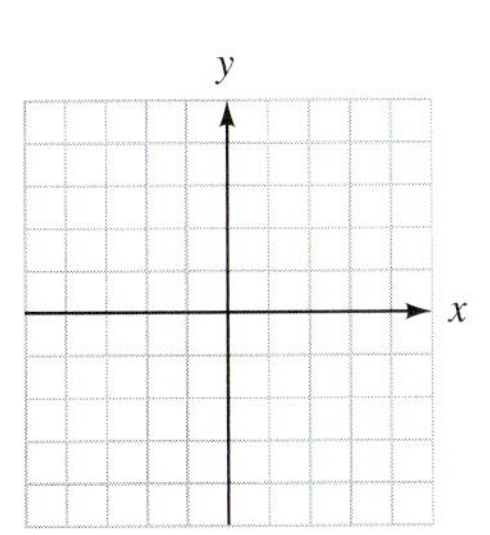

11. Graph: $y = -2$.

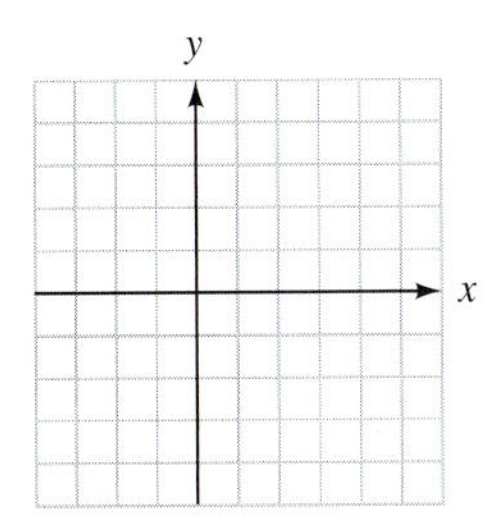

12. Graph: $x = 3$.

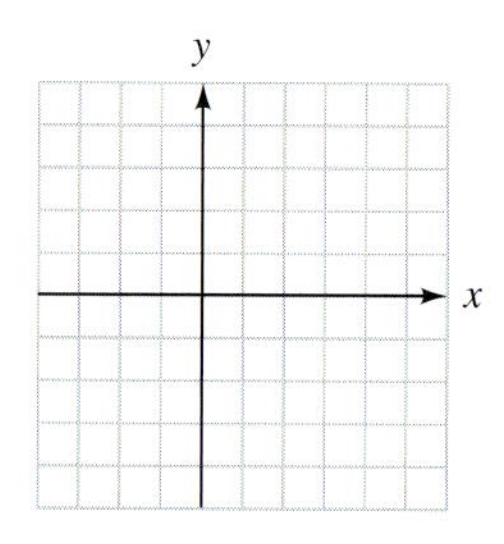

13. Simplify each expression.

a. h^2h^4 **b.** $-7x^3(4x^2)$

c. $b^2 \cdot b \cdot b^5$ **d.** $-3g^2k^3(-8g^3k^{10})$

14. Simplify each expression.

a. $(f^3)^5$ **b.** $(2a^2b)^2$

c. $(x^2)^3(x^3)^3$ **d.** $(x^2x^3)^3$

Classify each polynomial as a monomial, a binomial, or a trinomial.

15. $5x^2 + 4x$

16. $-3x^2 - 2x + 3$

Give the degree of each polynomial.

17. $3t^4 - 2t^3 + 5t^6 - t$

18. $7q^7 + 5q^5 - 8q^2$

Evaluate each polynomial at the given value.

19. $3x^2 - 2x + 4$ for $x = 3$

20. $-2r^2 - r + 3$ for $r = -1$

Graph each equation.

21. $y = 2x^2$

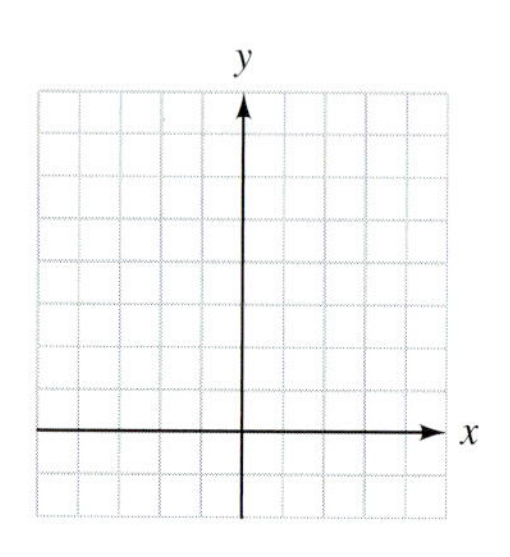

22. $y = -x^2 + 4$

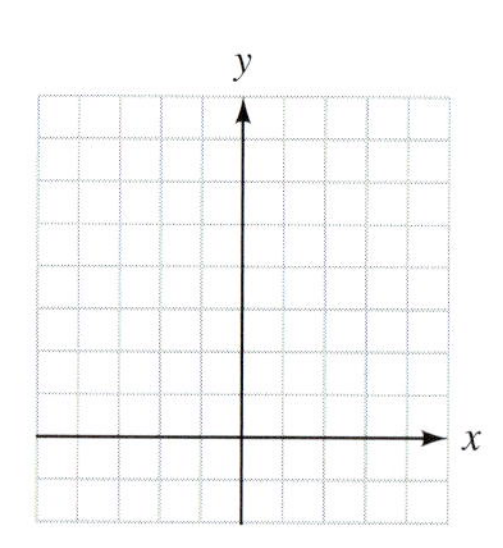

23. Add: $(3x^2 + 2x) + (2x^2 - 5x + 4)$.

24. Add:
$$\begin{array}{r} 4x^2 - 5x + 5 \\ \underline{+\ 3x^2 + 7x - 7} \end{array}$$

25. Subtract: $(2.1p^2 - 2p - 2) - (3.3p^2 - 5p - 2)$.

26. Subtract:
$$\begin{array}{r} 3d^2 - 3d + 7.2 \\ \underline{-(-5d^2 + 6d - 5.3)} \end{array}$$

Find each product.

27. $(-2x^3)(4x^2)$

28. $3y^2(y^2 - 2y + 3)$

29. $(2x - 5)(3x + 4)$

30. $(2x - 3)(x^2 - 2x + 4)$

31. Are the points with coordinates $(1, -2)$ and $(-2, 1)$ the same? Explain.

32. Explain what is meant when we say that the equation $x + y = 8$ has infinitely many solutions.

CHAPTERS 1–6 CUMULATIVE REVIEW EXERCISES

Consider the number 6,245,867.

1. Round to the nearest thousand.

2. Round to the nearest million.

Find the perimeter of each figure.

3. A rectangle that is 8 meters long and 3 meters wide

4. A square with sides that are 13 inches long

5. PARKING The dimensions of various types of rectangular parking spaces are given in the table. Complete the table.

Type	Length (ft)	Width (ft)	Area (ft^2)
Standard space	20	9	
Standard space adjacent to a wall	20	10	
Parallel space	25	10	
Compact space	17	8	

6. HEALTH A person's blood pressure is a combination of two measurements, and it is normally written as a fraction of the form $\frac{\text{systolic}}{\text{diastolic}}$. (The fraction is not simplified.) Study the graph below, and then complete this sentence: In healthy persons, blood pressure increases from about $\frac{\quad}{\quad}$ in infants, to about $\frac{\quad}{\quad}$ at age 30, to about $\frac{\quad}{\quad}$ at age 40 and over.

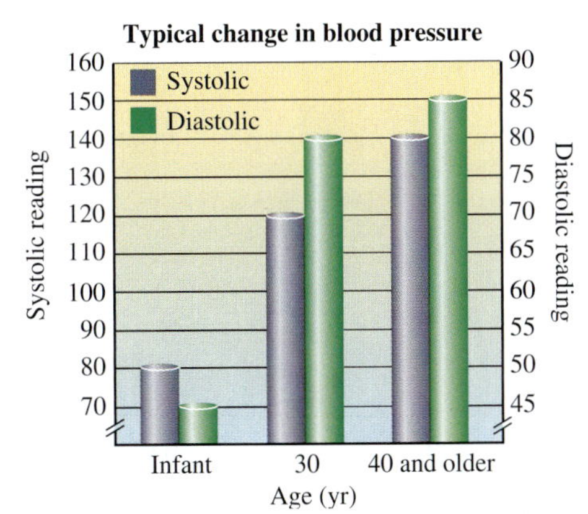

Source: *Microsoft Encarta 98 Encyclopedia*

Find the prime factorization of each number.

7. 120

8. 525

9. LAKE TAHOE Because of the growth of algae, a scientific study concluded that the legendary clarity of Lake Tahoe, which straddles the California/Nevada border, could be doomed. Study the data below. How has the visibility changed since the 1960s?

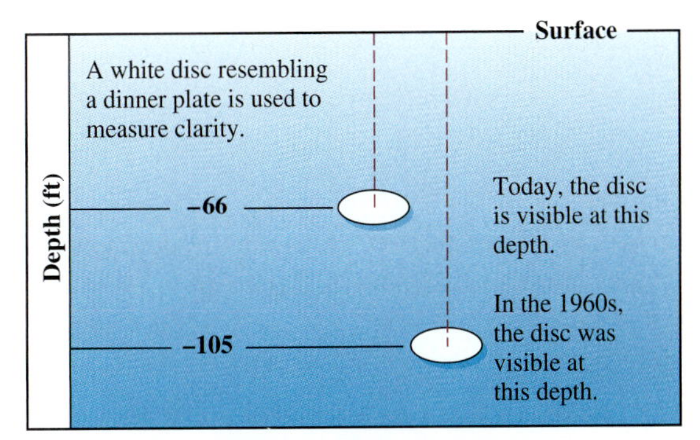

Source: *Los Angeles Times* (February 16, 2000)

10. Evaluate $\frac{6x + x^3}{|x|}$ for $x = -2$.

11. Evaluate: $12 - 2[1 - (-8 + 2)]$.

12. Evaluate: -3^2.

Combine like terms.

13. $5x - 11x$

14. $-4(x - 3y) + 5x - 2y$

Solve each equation and check the result.

15. $4x + 3 = 11$

16. $2z + 12 = 6z - 4$

17. $\frac{t}{3} + 2 = -4$

18. $2y + 7 = 2 - (4y + 7)$

Perform the operations and simplify.

19. $\frac{5}{10b^3} \cdot 2b^2$

20. $-4\frac{1}{4} \div 4\frac{1}{2}$

21. $34\frac{1}{9} - 13\frac{5}{6}$

22. $\frac{5}{m} - \frac{n}{5}$

Solve each equation and check the result.

23. $\frac{7}{8}t = -28$

24. $\frac{4}{5}x = \frac{3}{4}x + \frac{1}{2}$

25. PAPER SHREDDERS A paper shredder cuts paper into $\frac{1}{4}$-inch-wide strips. If an $8\frac{1}{2} \times 11$ in. piece of notebook paper is fed into the shredder as shown, into how many strips will it be shredded?

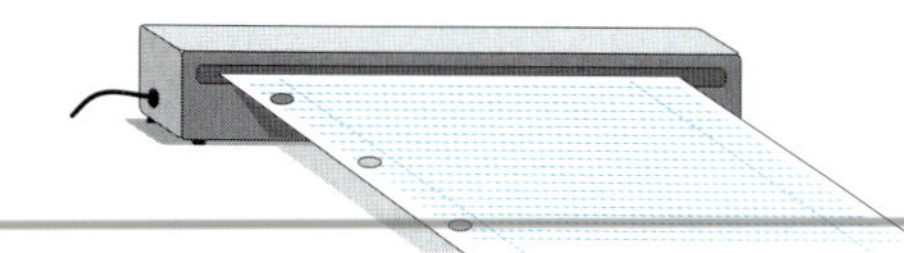

26. Explain why dividing a number by $\frac{1}{4}$ is the same as multiplying it by 4.

27. Round 57.574 to the nearest hundredth.

28. Add: 29.703 + 321.35.

29. Subtract: 287.23 − 179.97.

30. Multiply: 7.89×0.27.

31. Divide: $3.8\overline{)17.746}$.

32. Write $\frac{35}{99}$ as a decimal.

Write each number as a decimal. Round to the nearest tenth, if necessary.

33. $5\frac{5}{8}$

34. $-4\frac{7}{9}$

Solve each equation and check the result.

35. $3.2x = 74.46 - 1.9x$

36. $-5.2x = 108 - 6.1x$

37. $-2(x - 2.1) = -2.4$

38. $\frac{1}{5}x - 2.5 = -17.2$

39. EARTHQUAKES Listed below are the magnitudes of the 15 largest earthquakes in the United States, according to the U.S. Geological Survey. What are the mean, median, mode, and range of the listed magnitudes? (Round to the nearest tenth.)

9.2	1964	Alaska	7.9	1812	Missouri
8.8	1957	Alaska	7.9	1857	California
8.7	1965	Alaska	7.9	1868	Hawaii
8.3	1938	Alaska	7.9	1900	Alaska
8.3	1958	Alaska	7.9	1987	Alaska
8.2	1899	Alaska	7.8	1872	California
8.2	1899	Alaska	7.8	1892	California
8.0	1986	Alaska			

40. DRIVING What will be the reading on the odometer if a motorist travels at the rate shown on the speedometer for 3.5 hours?

41. PETITION DRIVES A worker for a political organization is to collect signatures for a petition drive. Her pay is \$20 plus 5¢ per signature. How many signatures must she get to earn \$60?

42. CONCERT TICKETS Seven-eighths of the total number of tickets sold for a concert were ordered by mail. The remaining 200 tickets were purchased at the concert hall box office. How many tickets were sold for the concert?

Simplify each expression.

43. $\sqrt{121}$

44. $\sqrt{\frac{81}{4}}$

45. $\sqrt{0.25}$

46. $3\sqrt{144} - \sqrt{49}$

47. Is $(-2, 3)$ a solution of $4x - 5y = -23$?

48. Graph: $3x - 4y = 12$.

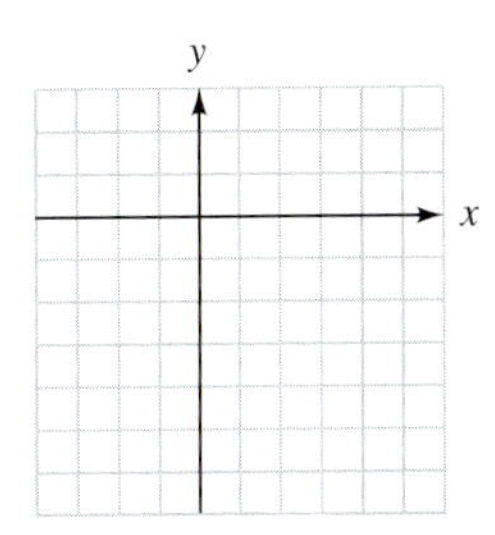

Simplify each product.

49. p^4p^5

50. $(2q^2)(-5q^6)$

51. $(p^3q^2)(p^3q^4)$

52. $(3a^2)^3(-a^3)^3$

Perform the operations.

53. $(3x^2 - 5x) - (2x^2 + x - 3)$

54. $(2x + 3)(3x - 1)$

CHAPTER 7

Percent

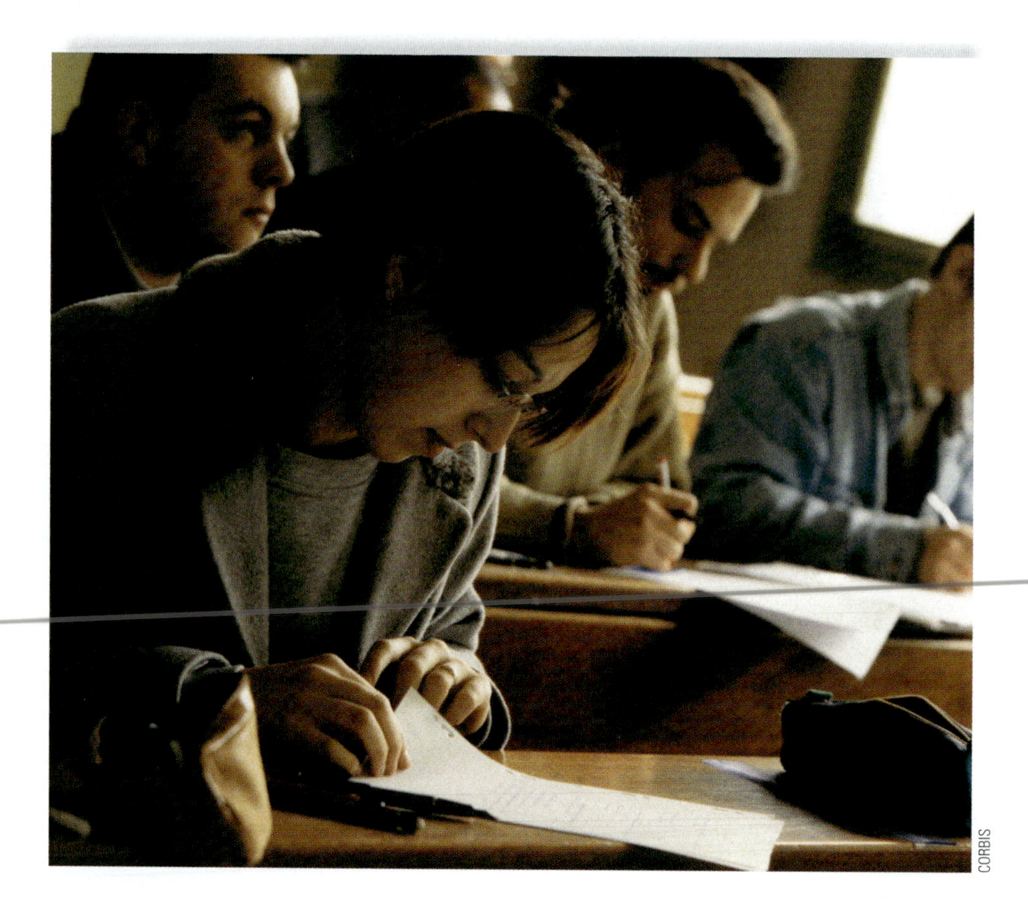

CORBIS

Percents are commonly used to present numerical information. The word percent comes from the Latin phrase *per centum*, which means parts per one hundred. Many instructors use a percent grading scale when evaluating their students' work. For example, if a student correctly answers 85 out of 100 true/false questions on a history exam, the student's grade on the exam can be expressed as 85%.

To learn more about percent and its many applications, visit The Learning Equation on the Internet at http://tle.brookscole.com. (The log-in instructions are in the Preface.) For Chapter 7, the online lesson is:

- *TLE* Lesson 13: Percent

Check Your Knowledge

1. Percent means parts per ________.
2. In the statement "40 is 50% of 80," 40 is the ________, 50% is the ________, and 80 is the ________.
3. The difference between the original price and the sale price of an item is called the ________.
4. In banking, the original amount of money borrowed or deposited is known as the ________.
5. ________ interest is interest paid on the principal and the accumulated interest. Interest computed on only the original principal is called ________ interest.
6. Change each fraction to a decimal and to a percent.

   a. $\frac{3}{4}$ b. $\frac{5}{8}$ c. $\frac{29}{20}$

7. Change each decimal to a percent and to a fraction.

   a. 0.35 b. 3.98 c. 0.105

8. Change each percent to a decimal and to a fraction.

   a. 25% b. 200% c. 0.5%

9. If a glass is 65% full, what percent of the glass is empty?
10. Find the exact percent equivalent for each fraction.

    a. $\frac{3}{8}$ b. $\frac{5}{6}$

11. Change $\frac{2}{3}$ to a percent.

    a. Give the exact percent.

    b. Round your answer to part a to the nearest tenth of a percent.

12. What is 65% of 500?
13. What percent of 200 is 34?
14. 13 is 25% of what number?
15. What percent of 50 is 125?
16. A pen is normally priced at $14.95. The sale price is 20% off the normal price. What are the amount of the discount and the sale price?
17. A CD player is on sale for $24.95, which is $5.00 off the regular price. Find the regular price and the discount rate. Round to the nearest one percent.
18. If the sales tax rate is 8.25%, what is the price of an item when the sales tax amount is $2.47? Round to the nearest cent.
19. Find the simple interest on a $2,000.00 savings account for one year if the interest rate is 2.3%.
20. Find the account balance after two years on a $1,500.00 investment with earnings of 5.6% compounded quarterly.
21. Only 3 of 46 parking spaces in a parking lot were not taken. What percent of the parking lot spaces were filled? Round to the nearest one percent.
22. If Sheila borrows $1,000 with simple annual interest at 18%, what will be the payoff amount if she pays off the loan after 3 months?
23. If Leslie wants to tip 15% on a $35.40 meal, what will the total cost be, including the tip?
24. A quiz has 15 questions. Assuming that the questions are weighted equally, how many questions must George answer correctly to score at least 85%?

Study Skills Workshop

HELP OUTSIDE OF CLASS

Have you ever had the experience of understanding everything your instructor is saying in class, only to go home, try a homework problem, and be completely stumped? This is a common complaint among math students. The key to success is to take care of these problems before you go on to tackle new material. Below are some suggestions for finding help outside of your classroom.

Instructor Office Hours. Your instructor may hold office hours during the week to help students with questions. Usually these hours are listed in your syllabus, and you do not need to make appointments to see your instructor at these times. Remember to bring a list of questions that are giving you trouble and try to pinpoint exactly where in the process you are getting stuck. This will help your instructor answer your questions efficiently and effectively.

Tutorial Centers. Many colleges have tutorial centers where students can meet with a tutor, either one-on-one or in a group of students taking the same class. Tutorial centers usually offer their services for free and have regular hours of operation. When you visit your tutor, bring your list of questions that detail where in the process you're having difficulty.

Math Labs. Some colleges have math labs or learning centers where students can drop in at their convenience to have their math questions answered or where they can hang out and work on their homework. If this is available at your college, try to organize your study calendar to spend some time there doing your homework.

Study Groups. Study groups are groups of classmates who meet outside of class to discuss homework problems or study for tests. Study groups work best when they are relatively small (no more than four members), when they meet regularly, and when they follow these guidelines:

- Members should have attempted all homework problems before meeting.
- No one person in the group should be responsible for doing all of the work or all of the explaining.
- The group should meet in a place where members can spread out and talk, not in a quiet area of the library.
- Members should practice verbalizing and explaining processes and concepts to others in the study group. The best way to really learn a topic is by teaching it to someone else.

ASSIGNMENT

1. List your instructor's office hours and location. Next, pay a visit to your instructor during his or her office this week, even if you don't have any homework questions.
2. List the hours that your college's tutorial center is open, the location of the center, and how to make an appointment with a tutor.
3. Find out whether your college has a math lab or learning center. If so, list its hours of operation, location, and rules.
4. Find at least two other students who can meet for a study group. Plan to meet two days before your next homework assignment is due and follow the guidelines given above. After your group has met, evaluate how well it worked. Is there anything you might do to make it better the next time?

Percents are based on the number 100. They offer us a standardized way to measure and describe many situations in our daily lives.

7.1 Percents, Decimals, and Fractions

- The meaning of percent
- Changing a percent to a fraction
- Changing a percent to a decimal
- Changing a decimal to a percent
- Changing a fraction to a percent

Percents are a popular way to present numeric information. Stores use them to advertise discounts, manufacturers use them to describe the content of their products, and banks use them to list interest rates for loans and savings accounts. Newspapers are full of statistics presented in percent form. In this section, we introduce percent and show how fractions, decimals, and percents are interrelated.

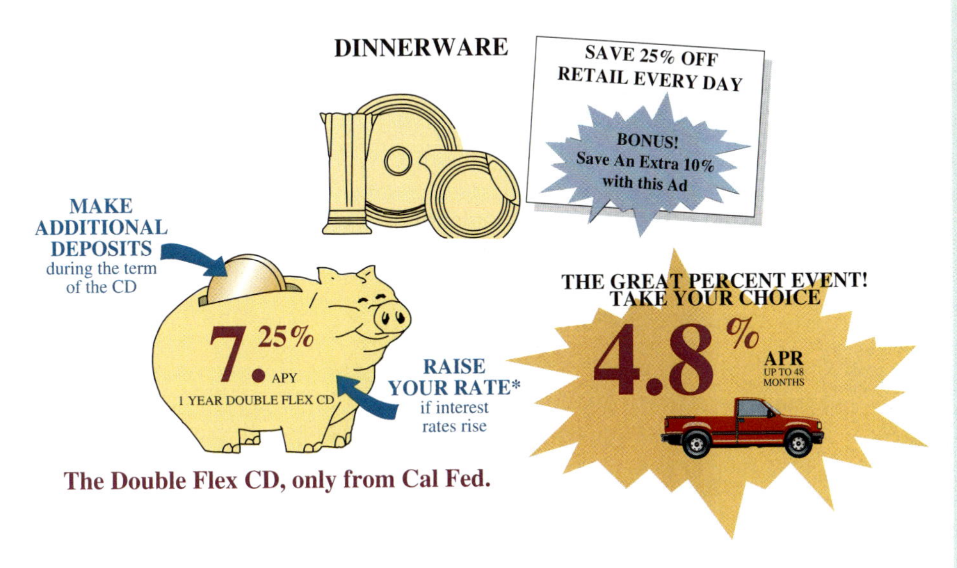

The meaning of percent

A percent tells us the number of parts per 100. You can think of a percent as the *numerator* of a fraction that has a denominator of 100.

Percent

Percent means parts per one hundred.

In Figure 7-1, 93 out of 100 equal-sized squares are shaded. Thus, $\frac{93}{100}$ or 93 percent of the figure is shaded. The word *percent* can be written using the symbol %, so 93% of Figure 7-1 is shaded.

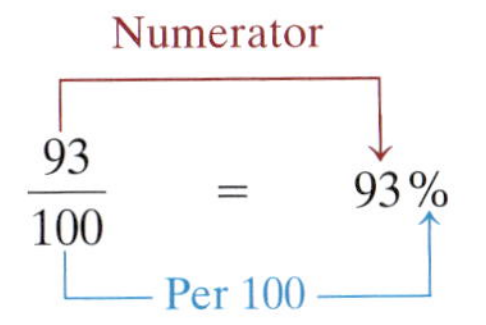

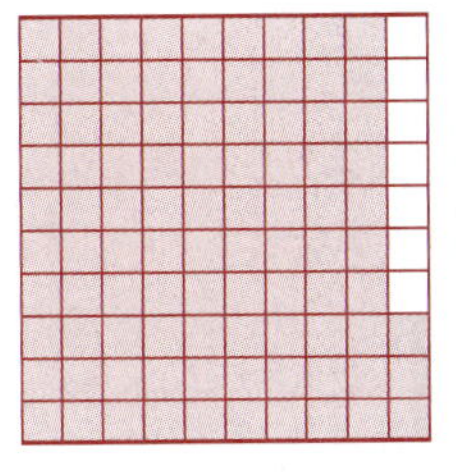

FIGURE 7-1

If the entire grid in Figure 7-1 had been shaded, we would say that 100 out of the 100 squares, or 100%, was shaded. Using this fact, we can determine what percent of the figure is *not* shaded by subtracting the percent of the figure that is shaded from 100%.

$$100\% - 93\% = 7\%$$

So 7% of Figure 7-1 is not shaded.

Changing a percent to a fraction

To change a percent to an equivalent fraction, we use the definition of percent.

Changing a percent to a fraction

To change a percent to a fraction, drop the % symbol and write the given number over 100. Then simplify the fraction, if possible.

Self Check 1

An average watermelon is 92% water. Write this percent as a fraction.

Answer $\frac{23}{25}$

EXAMPLE 1 **Earth.** The chemical makeup of the Earth's atmosphere is 78% nitrogen, 21% oxygen, and 1% other gases. Write each percent as a fraction.

Solution We begin with nitrogen.

$78\% = \frac{78}{100}$ Use the definition of percent: 78% means 78 parts per one hundred. This fraction can be simplified.

$= \frac{\overset{1}{\cancel{2}} \cdot 39}{\underset{1}{\cancel{2}} \cdot 50}$ Factor 78 as $2 \cdot 39$ and 100 as $2 \cdot 50$. Divide out the common factor of 2.

$= \frac{39}{50}$ Perform the multiplication in the numerator and in the denominator.

Nitrogen makes up $\frac{78}{100}$, or $\frac{39}{50}$, of the Earth's atmosphere.

Oxygen makes up 21% or $\frac{21}{100}$ of the Earth's atmosphere. Other gases make up 1% or $\frac{1}{100}$ of the atmosphere.

Self Check 2

In 2002, 13.3% of the U.S. labor force belonged to a union. Write this percent as a fraction.

Answer $\frac{133}{1,000}$

EXAMPLE 2 **Unions.** In 2003, 12.9% of the U.S. labor force belonged to a union. Write this percent as a fraction.

Solution

$12.9\% = \frac{12.9}{100}$ Drop the % symbol and write 12.9 over 100.

$= \frac{12.9 \cdot \mathbf{10}}{100 \cdot \mathbf{10}}$ To obtain a whole number in the numerator, multiply by 10. This will move the decimal point 1 place to the right. Multiply the denominator by 10 as well.

$= \frac{129}{1,000}$ Perform the multiplication in the numerator and in the denominator.

Thus, $12.9\% = \frac{129}{1,000}$. This means that 129 out of every 1,000 workers in the U.S. labor force belonged to a union in 2003.

EXAMPLE 3 Write $66\frac{2}{3}\%$ as a fraction.

Solution

$66\frac{2}{3}\% = \frac{66\frac{2}{3}}{100}$ Drop the % symbol and write $66\frac{2}{3}$ over 100.

$= 66\frac{2}{3} \div 100$ The fraction bar indicates division.

$= \frac{200}{3} \cdot \frac{1}{100}$ Change $66\frac{2}{3}$ to an improper fraction and then multiply by the reciprocal of 100.

$= \frac{2 \cdot 100 \cdot 1}{3 \cdot 100}$ Multiply the numerators and the denominators. Factor 200 as $2 \cdot 100$.

$= \frac{2 \cdot \overset{1}{\cancel{100}} \cdot 1}{3 \cdot \underset{1}{\cancel{100}}}$ Divide out the common factor of 100.

$= \frac{2}{3}$ Multiply in the numerator. Multiply in the denominator.

Self Check 3

Write $83\frac{1}{3}\%$ as a fraction.

Answer $\frac{5}{6}$

Changing a percent to a decimal

To write a percent as a decimal, recall that a percent can be written as a fraction with denominator 100, and that a denominator of 100 indicates division by 100.

Consider 14.25%, which means 14.25 parts per 100.

$14.25\% = \frac{14.25}{100}$ Use the definition of percent: write 14.25 over 100.

$= 14.25 \div 100$ The fraction bar indicates division.

$= 0.14\,25$ To divide a decimal by 100, move the decimal point 2 places to the left.

$= 0.1425$

This example suggests the following procedure.

> **Changing a percent to a decimal**
>
> To change a percent to a decimal, drop the % symbol and divide by 100 by moving the decimal point 2 places to the left.

EXAMPLE 4

The music industry. Figure 7-2 shows that the compact disc has become the format of choice among most consumers. What percent of all music sold is produced on CDs? Write the percent as a decimal.

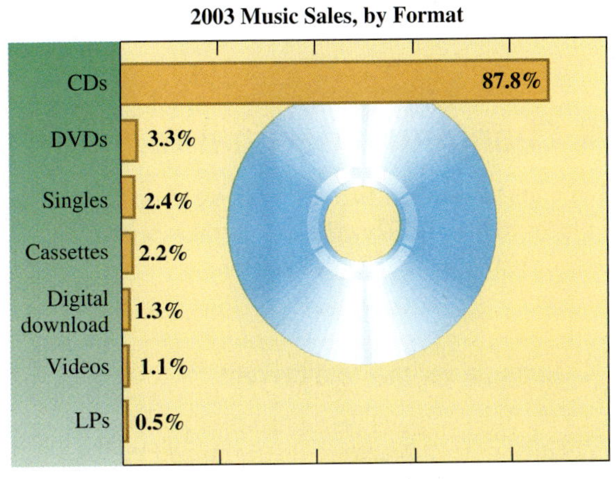

Source: *The Recording Industry Association of America*

FIGURE 7-2

Self Check 4

What percent of all music sold is produced on LPs (long-playing vinyl record albums)? Write the percent as a decimal.

Solution From the graph, we see that 87.8% of all music sold is produced on CDs. To write 87.8% as a decimal, we proceed as follows.

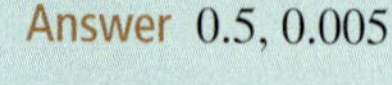

Answer 0.5, 0.005

$87.8\% = .87\,8$	Drop the % symbol and divide by 100 by moving the decimal point 2 places to the left.
$= 0.878$	Write a 0 to the left of the decimal point.

Self Check 5

Write 600% as a decimal.

Answer 6

EXAMPLE 5 Write 310% as a decimal.

Solution The whole number 310 has an understood decimal point to the right of 0.

$310\% = 310.0\%$	Write a decimal point and a 0 to the right of 310.
$= 3.10\,0$	Drop the % symbol and divide by 100 by moving the decimal point 2 places to the left.
$= 3.100$	
$= 3.1$	Drop the unnecessary zeros to the right of the 1.

Self Check 6

Write $15\frac{3}{4}\%$ as a decimal.

Answer 0.1575

EXAMPLE 6 **States.** The population of Oregon is approximately $1\frac{1}{4}\%$ of the population of the United States. Write this percent as a decimal.

Solution To change a percent to a decimal, we drop the % symbol and divide by 100 by moving the decimal point 2 places to the left. In this case, however, there is no decimal point in $1\frac{1}{4}\%$ to move. Since $1\frac{1}{4} = 1 + \frac{1}{4}$, and since the decimal equivalent of $\frac{1}{4}$ is 0.25, we can write $1\frac{1}{4}\%$ in an equivalent form as 1.25%.

$1\frac{1}{4}\% = 1.25\%$	Write $1\frac{1}{4}$ as 1.25.
$= 0.01\,25$	Drop the % symbol and divide by 100 by moving the decimal point 2 places to the left.
$= 0.0125$	

Changing a decimal to a percent

To change a percent to a decimal, we drop the % symbol and move the decimal point 2 places to the left. To write a decimal as a percent, we move the decimal point 2 places to the right and insert a % symbol.

> **Changing a decimal to a percent**
>
> To change a decimal to a percent, multiply the decimal by 100 by moving the decimal point 2 places to the right, and then insert a % symbol.

EXAMPLE 7 **Geography.** Land areas make up 0.291 of the Earth's surface. Write this decimal as a percent.

Solution

$0.291 = 0\,29.1\%$ Multiply the decimal by 100 by moving the decimal point 2 places to the right, and then insert a % symbol.

$= 29.1\%$

Self Check 7
Write 0.5343 as a percent.

Answer 53.43%

Changing a fraction to a percent

We will use a two-step process to change a fraction to a percent. First, we write the fraction as a decimal. Then we change that decimal to a percent.

Fraction  Decimal  Percent

Changing a fraction to a percent

To change a fraction to a percent:

1. Write the fraction as a decimal by dividing its numerator by its denominator.
2. Multiply the decimal by 100 by moving the decimal point 2 places to the right.
3. Insert a % symbol.

EXAMPLE 8 **Television.** The highest-rated television show of all time was the episode of "M*A*S*H" that aired on February 28, 1983. Surveys found that three out of every five American households watched this show. Express the rating as a percent.

Solution 3 out of 5 we can express as $\frac{3}{5}$. We need to change this fraction to a decimal.

$$\begin{array}{r} 0.6 \\ 5\overline{)3.0} \\ \underline{3\,0} \\ 0 \end{array}$$

Write 3 as 3.0 and then divide the numerator by the denominator.

$\frac{3}{5} = 0.6$ The result is a terminating decimal.

$0.6 = 0\,60.\%$ Write a placeholder 0 to the right of the 6. Multiply the decimal by 100 by moving the decimal point 2 places to the right, and then insert a % symbol.

$= 60\%$

So 60% of American households watched the episode of "M*A*S*H."

Self Check 8
Write $\frac{7}{8}$ as a percent.

Answer 87.5%

In Example 8, the result of the division was a terminating decimal. Sometimes when we change a fraction to a decimal, the result of the division is a repeating decimal.

EXAMPLE 9 Write $\frac{5}{6}$ as a percent.

Solution The first step is to change $\frac{5}{6}$ to a decimal.

$$6\overline{)5.0000}$$ with quotient 0.8333: 4 8, 20, 18, 20, 18, 20

Write 5 as 5.0000. Divide the numerator by the denominator.

$\frac{5}{6} = 0.8333...$ The result is a repeating decimal.

$= 0\,83.33...\%$ Change 0.8333... to a percent. Multiply the decimal by 100 by moving the decimal point 2 places to the right, and then insert a % symbol.

$= 83.33...\%$ 83.333... is a repeating decimal.

We must now decide whether we want an approximation or an exact answer. For an approximation, we can round 83.333. . .% to a specific place value. For an exact answer, we can represent the repeating part of the decimal using an equivalent fraction.

Approximation

$\frac{5}{6} = 83.33...\%$

$\approx 83.3\%$ Round to the nearest tenth.

$\frac{5}{6} \approx 83.3\%$

Exact answer

$\frac{5}{6} = 83.3333...\%$

$= 83\frac{1}{3}\%$ Use the fraction $\frac{1}{3}$ to represent .333. . . .

$\frac{5}{6} = 83\frac{1}{3}\%$

Some percents occur so frequently that it is useful to memorize their fractional and decimal equivalents.

Percent	Decimal	Fraction
1%	0.01	$\frac{1}{100}$
10%	0.1	$\frac{1}{10}$
20%	0.2	$\frac{1}{5}$
25%	0.25	$\frac{1}{4}$

Percent	Decimal	Fraction
$33\frac{1}{3}\%$	0.3333. . .	$\frac{1}{3}$
50%	0.5	$\frac{1}{2}$
$66\frac{2}{3}\%$	0.6666. . .	$\frac{2}{3}$
75%	0.75	$\frac{3}{4}$

Section 7.1 STUDY SET

VOCABULARY *Fill in the blanks.*

1. ________ means parts per one hundred.
2. When we change a fraction to a decimal, the result is either a __________ or a repeating decimal.

CONCEPTS *Fill in the blanks.*

3. To write a percent as a fraction, drop the % symbol and write the given number over ______. Then ________ the fraction, if possible.

4. To change a percent to a decimal, drop the % symbol and divide by 100 by moving the decimal point 2 places to the ______.

5. To change a decimal to a percent, multiply the decimal by 100 by moving the decimal point 2 places to the ______, and then insert a % symbol.

6. To write a fraction as a percent, first write the fraction as a ________. Then multiply the decimal by 100 by moving the decimal point 2 places to the ______, and insert a % symbol.

NOTATION

7. **a.** See the illustration. Express the amount of the figure that is shaded as a decimal, a percent, and a fraction.

b. What percent of the figure is not shaded?

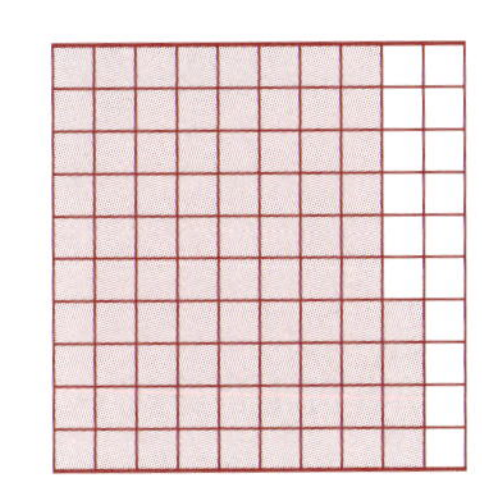

8. In the illustration below, each set of 100 squares represents 100%. What percent is shaded?

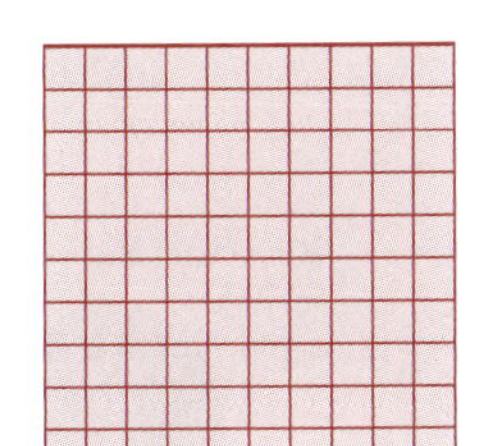

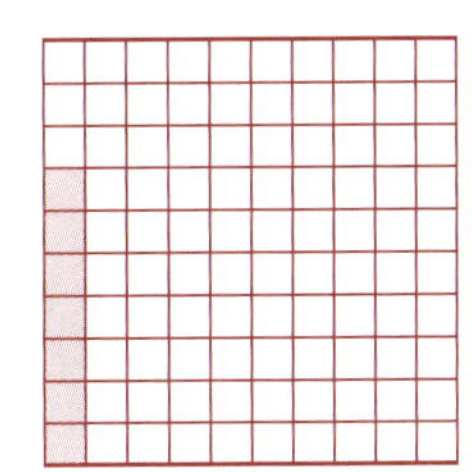

PRACTICE *Change each percent to a fraction. Simplify when necessary.*

9. 17% **10.** 31%

11. 5% **12.** 4%

13. 60% **14.** 40%

15. 125% **16.** 210%

17. $\frac{2}{3}\%$ **18.** $\frac{1}{5}\%$

19. $5\frac{1}{4}\%$ **20.** $6\frac{3}{4}\%$

21. 0.6% **22.** 0.5%

23. 1.9% **24.** 2.3%

Change each percent to a decimal.

25. 19% **26.** 83%

27. 6% **28.** 2%

29. 40.8% **30.** 34.2%

31. 250% **32.** 600%

33. 0.79% **34.** 0.01%

35. $\frac{1}{4}\%$ **36.** $8\frac{1}{5}\%$

Change each decimal to a percent.

37. 0.93 **38.** 0.44

39. 0.612 **40.** 0.727

41. 0.0314 **42.** 0.0021

43. 8.43 **44.** 7.03

45. 50 **46.** 3

47. 9.1 **48.** 8.7

Change each fraction to a percent.

49. $\frac{17}{100}$ **50.** $\frac{29}{100}$

51. $\frac{4}{25}$ **52.** $\frac{47}{50}$

53. $\frac{2}{5}$ **54.** $\frac{21}{50}$

55. $\frac{21}{20}$ **56.** $\frac{33}{20}$

57. $\frac{5}{8}$ **58.** $\frac{3}{8}$

59. $\frac{3}{16}$ **60.** $\frac{1}{32}$

Find the exact equivalent percent for each fraction.

61. $\frac{2}{3}$ **62.** $\frac{1}{6}$

63. $\frac{1}{12}$ **64.** $\frac{4}{3}$

Express each of the given fractions as a percent. Round to the nearest hundredth.

65. $\frac{1}{9}$ **66.** $\frac{2}{3}$

67. $\frac{5}{9}$ **68.** $\frac{7}{3}$

APPLICATIONS

69. U.N. SECURITY COUNCIL The United Nations has 191 members. The United States, Russia, the United Kingdom, France, and China, along with 10 other nations, make up the Security Council.

a. What fraction of the members of the United Nations belong to the Security Council?

b. Write your answer to part a in percent form. (Round to the nearest one percent.)

70. ECONOMIC FORECASTS One economic indicator of the national economy is the number of orders placed by manufacturers. One month, the number of orders rose one-fourth of one percent.

a. Write this using a % symbol.

b. Express it as a fraction.

c. Express it as a decimal.

71. PIANO KEYS Of the 88 keys on a piano, 36 are black.

a. What fraction of the keys are black?

b. What percent of the keys are black? (Round to the nearest one percent.)

72. INTEREST RATES Write as a decimal the interest rate associated with each of these accounts.

a. Home loan: 7.75%

b. Savings account: 5%

c. Credit card: 14.25%

73. THE HUMAN SPINE The human spine consists of a group of bones (vertebrae).

a. What fraction of the vertebrae are lumbar?

b. What percent of the vertebrae are lumbar? (Round to the nearest one percent.)

c. What percent of the vertebrae are cervical? (Round to the nearest one percent.)

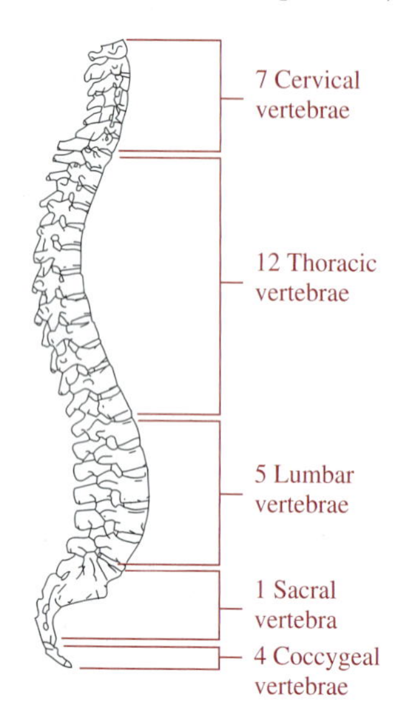

74. REGIONS OF THE COUNTRY The continental United States is divided into seven regions.

a. What percent of the 50 states are in the Rocky Mountain region?

b. What percent of the 50 states are in the Midwestern region?

c. What percent of the 50 states are not located in any of the seven regions shown here?

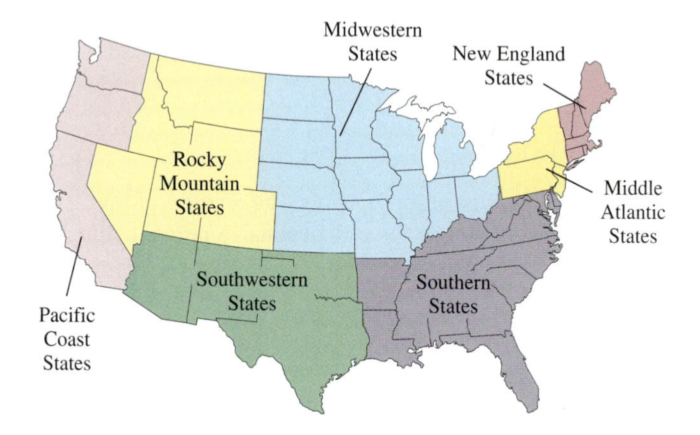

75. STEEP GRADE Sometimes, signs are used to warn truckers when they are approaching a steep grade on the highway. For a 5% grade, how many feet does the road rise over a 100-foot run?

5% Grade Ahead

?

100 ft

76. COMPANY LOGOS In the illustration, what part of the company's logo is red? Express your answer as a percent, a fraction, and a decimal. **Do not round.**

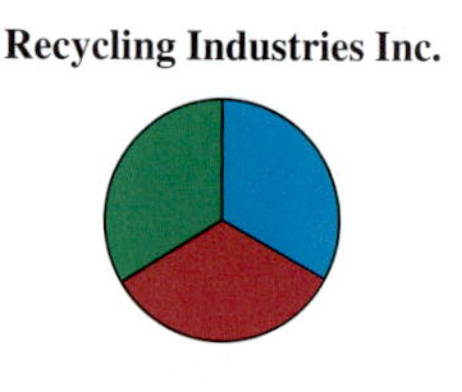

77. SOAP Ivory soap claims to be $99\frac{44}{100}$% pure. Write this percent as a decimal.

78. DRUNK DRIVING In most states, it is illegal to drive with a blood alcohol concentration of 0.08% or more. Change this percent to a fraction. **Do not simplify.** Explain what the numerator and the denominator of the fraction represent.

79. BASKETBALL In the following table, we see that Chicago has won 60 of 67, or $\frac{60}{67}$ of its games. In what form is the team's winning percentage presented in the newspaper? Express it as a percent.

Eastern conference			
Team	W	L	Pct.
Chicago	60	7	.896

80. WON-LOST RECORDS In sports, when a team wins as many as it loses, it is said to be playing "500 ball." Examine the following table and explain the significance of the number 500.

Eastern conference			
Team	W	L	Pct.
Orlando	33	33	.500

81. HUMAN SKIN The illustration shows roughly what percent each section of the body represents of the total skin area. Determine the missing percent, and then complete the bar graph below.

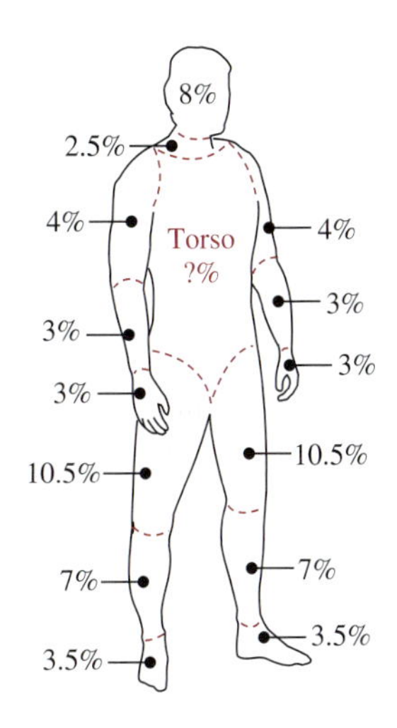

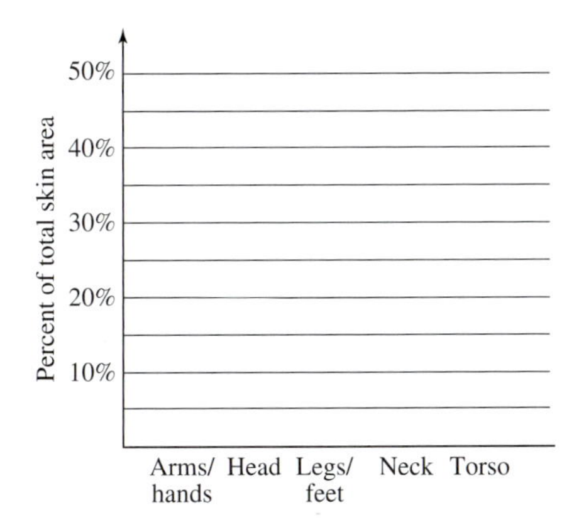

82. RAP MUSIC The following table shows what percent rap/hip-hop music sales were of total U.S. dollar sales of recorded music for the years 1997–2003. On the illustration in the next column, construct a line graph using the given data.

1997	1998	1999	2000	2001	2002	2003
10.1%	9.7%	10.8%	12.9%	11.4%	13.8%	13.3%

Source: *The Recording Industry Association of America*

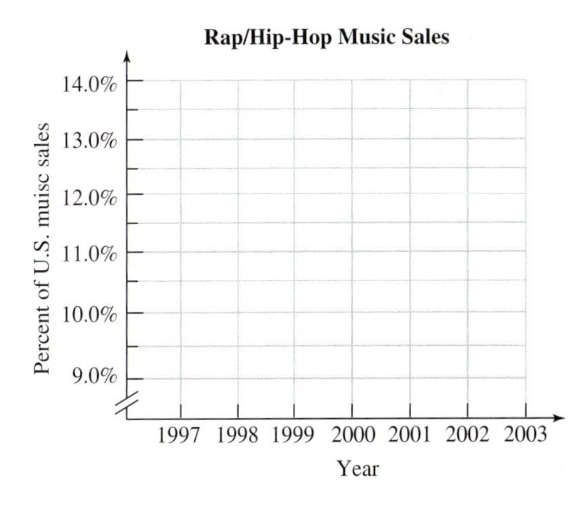

83. CHARITIES A 1998 fact sheet released by the American Red Cross stated, "For the past three fiscal years, an average of 92 cents of every dollar spent by the Red Cross went to programs and services to help those in need." What percent of the money spent by the Red Cross went to programs and services?

84. TAXES Santa Anita Thoroughbred Racetrack in Arcadia, California, has to pay a one-third of 1% tax on all the money wagered at the track. Write the percent as a fraction.

A calculator will be helpful to solve these problems.

85. BIRTHDAYS If the day of your birthday represents $\frac{1}{365}$ of a year, what percent of the year is it? Round to the nearest hundredth of a percent.

86. POPULATION As a fraction, each resident of the United States represents approximately $\frac{1}{295{,}000{,}000}$ of the population. Express this as a percent. Round to one nonzero digit.

WRITING

87. If you were writing advertising, which form do you think would attract more customers: "25% off" or "$\frac{1}{4}$ off"? Explain.

88. Many coaches ask their players to give a 110% effort during practices and games. What do you think this means? Is it possible?

89. Explain how to change a fraction to a percent.

90. Explain how an amusement park could have an attendance that is 103% of capacity.

91. CHAMPIONS Muhammad Ali won 92% of his professional boxing matches. Does that mean he had exactly 100 fights and won 92 of them? Explain.

92. CALCULATORS To change the fraction $\frac{15}{16}$ to a percent, a student used a calculator to divide 15 by 16. The display is shown below.

0.9375

Now what keys should the student press to change this decimal to a percent?

REVIEW

93. Solve: $-\frac{2}{3}x = -6$.

94. Add: $\frac{1}{3} + \frac{1}{4} + \frac{1}{2}$.

95. Complete the ordered pairs so that each one satisfies $y = 2x + 3$: (2,), (4,), (0,).

96. Write an algebraic expression that represents the area of a square with a side that is x feet long.

97. Multiply: $(x + 1)(x + 2)$.

98. Subtract: $41 - 10.287$.

7.2 Solving Percent Problems

- Percent problems • Finding the amount • Finding the percent • Finding the base
- Restating the problem • An alternative approach: the percent formula • Circle graphs

Percent problems occur in three forms. In this section, we will study a single procedure that can be used to solve all three types. It involves the equation-solving skills that we studied earlier.

Percent problems

The articles on the front page of the newspaper in Figure 7-3 suggest three types of percent problems.

- In the labor article, if we want to know how many union members voted to accept the new offer, we would ask:

  What number is 84% of 500?

- In the article on drinking water, if we want to know what percent of the wells are safe, we would ask:

  38 is what percent of 40?

- In the article on new appointees, if we want to know how many examiners are on the State Board, we would ask:

  6 is 75% of what number?

DAILY NEWS

Circulation | Monday, March 23 | 50 cents

Transit Strike Averted!

Labor: 84% of 500-member union votes to accept new offer

Drinking Water
38 of 40 Wells Declared Safe

New Appointees
These six area residents now make up 75% of the State Board of Examiners

FIGURE 7-3

These percent problems have features in common.

- Each problem contains the word *is*. Here, *is* can be translated to an = symbol.
- Each of the problems contains a phrase such as *what number* or *what percent*. In other words, there is an unknown quantity that can be represented by a variable.
- Each problem contains the word *of*. In this context, *of* means multiply.

These observations suggest that each of the percent problems can be translated into an equation. The equation, called a **percent equation,** will contain a variable, and the operation of multiplication will be involved.

Finding the amount

To solve the labor union problem, we translate the words into an equation and then solve it.

What number	is	84%	of	500?	
↓	↓	↓	↓	↓	
x	$=$	84%	$\cdot$	500	Translate to mathematical symbols.

$x = 0.84 \cdot 500$ Change 84% to a decimal: 84% = 0.84.

$= 420$ Perform the multiplication.

We have found that 420 is 84% of 500. That is, 420 union members voted to accept the new offer.

! COMMENT When solving percent equations, always write the percent as a decimal or a fraction before performing any calculations. For example, in the previous problem, we wrote 84% as 0.84 before multiplying by 500.

Percent problems involve a comparison of numbers or quantities. In the statement "420 is 84% of 500," the number 420 is called the **amount,** 84% is the **percent,** and 500 is called the **base.** Think of the base as the standard of comparison—it represents the whole of some quantity. The amount is a part of the base, but it can exceed the base when the percent is more than 100%. The percent, of course, has the % symbol.

EXAMPLE 1 What number is 160% of 15.8?

Solution First, we translate the words into an equation.

What number	is	160%	of	15.8?	
↓		↓		↓	
x	$=$	160%	$\cdot$	15.8	x is the amount, 160% is the percent, and 15.8 is the base.

Then we solve the equation.

$x = 1.6 \cdot 15.8$ Change 160% to a decimal: 160% = 1.6.

$= 25.28$ Perform the multiplication.

Thus, 25.28 is 160% of 15.8.

Self Check 1
What number is 240% of 80?

Answer 192

Finding the percent

In the drinking water problem, we must find the percent. Once again, we translate the words of the problem into an equation and solve it.

38	is	what percent	of	40?	
↓		↓		↓	
38	=	x	·	40	38 is the amount, x is the percent, and 40 is the base.

$38 = 40x$ Use the commutative property of multiplication to write $x \cdot 40$ as $40x$.

$\frac{38}{40} = \frac{40x}{40}$ To undo the multiplication by 40, divide both sides by 40.

$0.95 = x$ Perform the divisions.

$x = 0.95$ Since $0.95 = x$, $x = 0.95$.

$x = 95\%$ To change a decimal to a percent, multiply the decimal by 100 by moving the decimal point 2 places to the right, and then insert a % symbol.

Thus, 38 is 95% of 40. That is, 95% of the wells referred to in the article were declared safe.

Self Check 2

9 is what percent of 16?

Answer 56.25%

EXAMPLE 2 14 is what percent of 32?

Solution First, we translate the words into an equation.

14	is	what percent	of	32?	
↓		↓		↓	
14	=	x	·	32	14 is the amount, x is the percent, and 32 is the base.

Then we solve the equation.

$14 = 32x$ Rewrite the right-hand side: $x \cdot 32 = 32x$.

$\frac{14}{32} = \frac{32x}{32}$ To undo the multiplication by 32, divide both sides by 32.

$0.4375 = x$ Perform the divisions.

$43.75\% = x$ Change 0.4375 to a percent. Multiply the decimal by 100 by moving the decimal point 2 places to the right, and then insert a % symbol.

Thus, 14 is 43.75% of 32.

CALCULATOR SNAPSHOT Cost of an air bag

An air bag is estimated to add an additional $500 to the cost of a car. What percent of the $16,295 sticker price is the cost of the air bag?

First, we translate the words into an equation.

What percent	of	the $16,295 sticker price	is	the cost of the air bag?	
↓		↓		↓	
x	·	16,295	=	500	500 is the amount, x is the percent, and 16,295 is the base.

Then we solve the equation.

$16{,}295x = 500$ — $x \cdot 16{,}295 = 16{,}295x$.

$\frac{16{,}295x}{\mathbf{16{,}295}} = \frac{500}{\mathbf{16{,}295}}$ — To undo the multiplication by 16,295, divide both sides by 16,295.

$x = \frac{500}{16{,}295}$

To perform the division using a calculator, enter these numbers and press these keys.

500 [÷] 16295 [=] — 0.030684259

This display gives the answer in decimal form. To change it to a percent, we multiply the result by 100 and insert a % symbol. This moves the decimal point 2 places to the right. If we round to the nearest tenth of a percent, the cost of the air bag is about 3.1% of the sticker price.

Finding the base

In the problem about the State Board of Examiners, we must find the base. As before, we translate the words of the problem into an equation and solve it.

6	is	75%	of	what number?
↓		↓		↓
6	=	75%	·	x

6 is the amount, 75% is the percent, and x is the base.

$6 = 0.75x$ — Change 75% to 0.75.

$\frac{6}{\mathbf{0.75}} = \frac{0.75x}{\mathbf{0.75}}$ — To undo the multiplication by 0.75, divide both sides by 0.75.

$8 = x$ — Perform the divisions.

Thus, 6 is 75% of 8. That is, there are 8 examiners on the State Board.

EXAMPLE 3 31.5 is $33\frac{1}{3}$% of what number?

Solution

31.5	is	$33\frac{1}{3}$%	of	what number?
↓		↓		↓
31.5	=	$33\frac{1}{3}$%	·	x

31.5 is the amount, $33\frac{1}{3}$% is the percent, and x is the base.

In this case the computations can be made easier by changing the percent to a fraction instead of to a decimal. We write $33\frac{1}{3}$% as a fraction and proceed as follows.

$31.5 = \frac{1}{3} \cdot x$ — $33\frac{1}{3}\% = \frac{1}{3}$.

$\mathbf{3} \cdot 31.5 = \mathbf{3} \cdot \frac{1}{3}x$ — To isolate x on the right-hand side, multiply both sides by 3.

$94.5 = x$ — Perform the multiplications: $3 \cdot 31.5 = 94.5$ and $3 \cdot \frac{1}{3} = 1$.

Thus, 31.5 is $33\frac{1}{3}$% of 94.5.

Self Check 3

150 is $66\frac{2}{3}$% of what number?

Answer 225

Restating the problem

Not all percent problems are presented in the form we have been studying. In Example 4, we must examine the given information carefully so that we can restate the problem in the familiar form.

EXAMPLE 4 **Housing.** In an apartment complex, 110 of the units are currently being rented. This represents an 88% occupancy rate. How many units are there in the complex?

Solution An occupancy rate of 88% means that 88% of the units are occupied. We restate the problem in the form we have been studying.

110	is	88%	of	what number?
↓		↓		↓
110	=	88%	·	x

110 is the amount, 88% is the percent, and x is the base.

Now we solve the equation.

$110 = 0.88x$ Change 88% to a decimal: 88% = 0.88.

$\frac{110}{\mathbf{0.88}} = \frac{0.88x}{\mathbf{0.88}}$ To undo the multiplication by 0.88, divide both sides by 0.88.

$125 = x$ Perform the divisions.

The complex has 125 units.

An alternative approach: the percent formula

In any percent problem, the relationship between the amount, the percent, and the base is as follows: *Amount is percent of base.* This relationship is shown in the **percent formula.**

The percent formula

Amount = percent · base

The percent formula can be used as an alternative way to solve percent problems. With this method, we need to identify the *amount* (the part that is compared to the whole), the *percent* (indicated by the % symbol or the word *percent*), and the *base* (the whole of some quantity, usually following the word *of*).

Self Check 5
What number is 240% of 80?

EXAMPLE 5 What number is 160% of 15.8?

Solution In this example, the percent is 160% and the base is 15.8, the number following the word *of.* We can let A stand for the amount and use the percent formula.

Amount	=	percent	·	base
↓		↓		↓
A	=	160%	·	15.8

Substitute 160% for the percent and 15.8 for the base.

The statement $A = 160\% \cdot 15.8$ is an equation, with the amount A being the unknown. We can find the unknown amount by multiplication.

$A = 1.6 \cdot 15.8$ Change 160% to a decimal: 160% = 1.6.

$= 25.28$ Perform the multiplication.

Thus, 25.28 is 160% of 15.8. Note that we got the same result in Example 1.

Answer 192

EXAMPLE 6 14 is what percent of 32?

Solution In this example, 14 is the amount and 32 is the base. Once again, we use the percent formula and let p stand for the percent.

Amount	=	percent	·	base
↓		↓		↓
14	=	p	·	32

Substitute 14 for the amount and 32 for the base.

The statement $14 = p \cdot 32$ is an equation, with the percent p being the unknown. We can find the unknown percent by division.

$14 = p \cdot 32$ This is the equation to solve.

$14 = 32p$ Rewrite the right-hand side: $p \cdot 32 = 32p$.

$\frac{14}{\mathbf{32}} = \frac{32p}{\mathbf{32}}$ To undo the multiplication by 32, divide both sides by 32.

$0.4375 = p$ Perform the divisions: $\frac{14}{32} = 0.4375$.

$p = 43.75\%$ To change the decimal to a percent, multiply the decimal by 100 by moving the decimal point 2 places to the right, and then insert a % symbol.

Thus, 14 is 43.75% of 32. Note that we got the same result in Example 2.

Self Check 6

9 is what percent of 16?

Answer 56.25%

EXAMPLE 7 31.5 is $33\frac{1}{3}\%$ of what number?

Solution In this example, 31.5 is the amount and $33\frac{1}{3}\%$ is the percent. To find the base (which we will call b), we form an equation using the percent formula.

Amount	=	percent	·	base
↓		↓		↓
31.5	=	$33\frac{1}{3}\%$	·	b

Substitute 31.5 for the amount and $33\frac{1}{3}\%$ for the percent.

The statement $31.5 = 33\frac{1}{3}\% \cdot b$ is an equation, with the base b being the unknown. We can find the unknown base by multiplication.

Self Check 7

150 is $66\frac{2}{3}\%$ of what number?

$$31.5 = 33\frac{1}{3}\% \cdot b \quad \text{This is the equation to solve.}$$

$$31.5 = \frac{1}{3}b \quad 33\frac{1}{3}\% = \frac{33\frac{1}{3}}{100} = \frac{1}{3}.$$

$$3 \cdot 31.5 = 3 \cdot \frac{1}{3}b \quad \text{To isolate } b \text{ on the right-hand side, multiply both sides by 3.}$$

$$94.5 = b \quad \text{Perform the multiplication: } 31.5 \cdot 3 = 94.5.$$

Answer 225

Thus, 31.5 is $33\frac{1}{3}\%$ of 94.5. Note that we got the same result in Example 3.

Circle graphs

Percents are used with **circle graphs,** or **pie charts,** as a way of presenting data for comparison. In Figure 7-4, the entire circle represents the total amount of electricity generated in the United States in 2003. The pie-shaped pieces of the graph show the relative sizes of the energy sources used to generate the electricity. For example, we see that the greatest amount of electricity (51%) was generated from coal. Note that if we add the percents from all categories (51% + 3% + 7% + 16% + 20% + 3%), the sum is 100%.

The 100 tick marks equally spaced around the circle serve as a visual aid when constructing a circle graph. For example, to represent hydropower as 7%, a line was drawn from the center of the circle to a tick mark. Then we counted off 7 ticks and drew a second line from the center to that tick to complete the pie-shaped wedge.

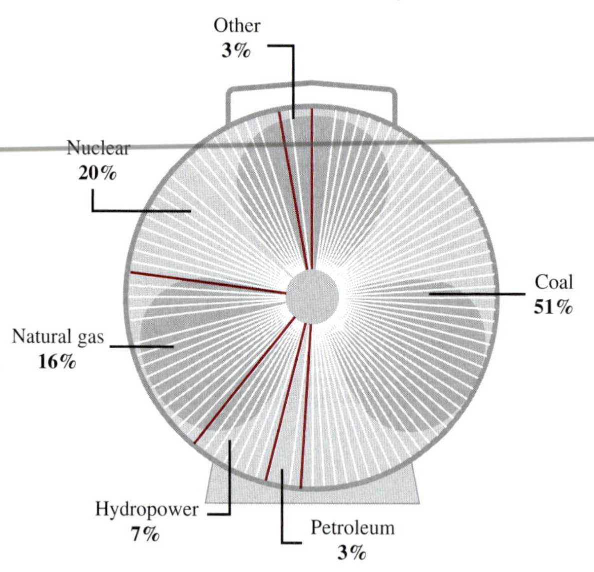

FIGURE 7-4

EXAMPLE 8 Presidential elections.

Results from the 2004 presidential election are shown in Figure 7-5 on the right. Use the information to find the number of states won by George W. Bush.

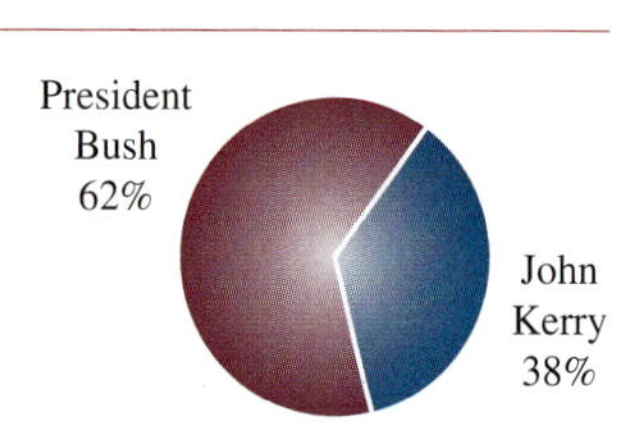

2004 Presidential Election
States won by each candidate

FIGURE 7-5

Solution The circle graph shows that George W. Bush was victorious in 62% of the 50 states. Here, the percent is 62% and the base is 50. We use the percent formula and solve for the amount.

Amount	=	percent	·	base	
A	=	62%	·	50	Substitute 62% for the percent and 50 for the base.

$A = 0.62 \cdot 50$ Change 62% to a decimal: 62% = 0.62. This is the equation to solve.

$A = 31$ Perform the multiplication.

George W. Bush won 31 states in the 2004 presidential election.

Community College Students

THINK IT THROUGH

"Community Colleges are centers of educational opportunity. More than 100 years ago, this unique, American invention put publicly funded higher education at close-to-home facilities and initiated a practice of welcoming all who desire to learn, regardless of wealth, heritage or previous academic experience. Today, the community college of continues the process of making higher education available to a maximum number of people at 1,166 public and independent community colleges." The American Association of Community Colleges (AACC)

Over 33,500 students responded to the 2002 Community College Survey of Student Engagement. Several of the results are shown below. Study each circle graph and then complete its legend.

Enrollment in Community Colleges

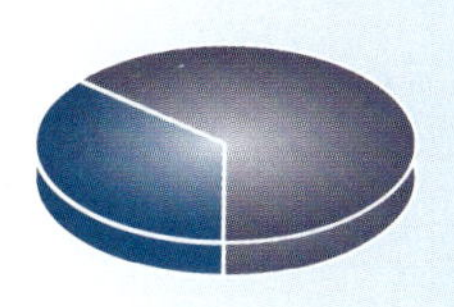

- 64% are enrolled in college part time.
- ?

How Much Reading Are Community College Students Doing?

- 31% of full-time students read four or fewer assigned textbooks, manuals, or books during the current school year
- ?

Community College Students Who Discussed Ideas with Instructors outside of Class

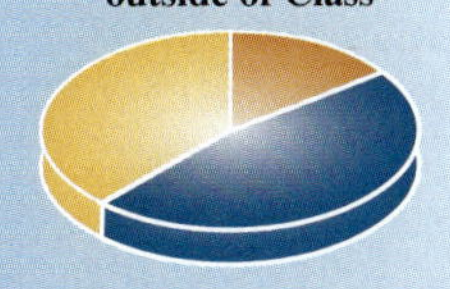

- 15% often or very often
- 47% never
- ?

Section 7.2 STUDY SET

VOCABULARY *Fill in the blanks.*

1. In a circle ______, pie-shaped wedges are used to show the division of a whole quantity into its component parts.

2. In the statement "45 is 90% of 50," 45 is the ______, 90% is the ______, and 50 is the ______.

CONCEPTS *Translate each sentence into a percent equation.* ***Do not solve the equation.***

3. What number is 10% of 50?

4. 16 is 55% of what number?

5. 48 is what percent of 47?

6. 12 is what percent of 20?

7. When we compute with percents, the percent must be changed to a decimal or a fraction. Change each percent to a decimal.

   a. 12% b. 5.6%

   c. 125% d. $\frac{1}{4}\%$

8. When we compute with percents, the percent must be changed to a decimal or a fraction. Change each percent to a fraction.

   a. $33\frac{1}{3}\%$ b. $66\frac{2}{3}\%$

   c. $16\frac{2}{3}\%$ d. $83\frac{1}{3}\%$

9. Without doing the calculation, determine whether 120% of 55 is more than 55 or less than 55.

10. Without doing the calculation, determine whether 12% of 55 is more than 55 or less than 55.

11. Solve each of the following problems in your head.

   a. What is 100% of 25?

   b. What percent of 132 is 132?

   c. What number is 87% of 100?

12. To solve the problem

   15 is what percent of 75?

   a student wrote a percent equation, solved it, and obtained $x = 0.2$. For her answer, the student wrote

   15 is 0.2% of 75.

   Explain her error.

13. E-MAIL The circle graph shows the types of e-mail messages a typical Internet user receives. *Spam* is the name given "junk" e-mail that is sent to a large number of people to promote products or services. What percent of e-mail messages is spam?

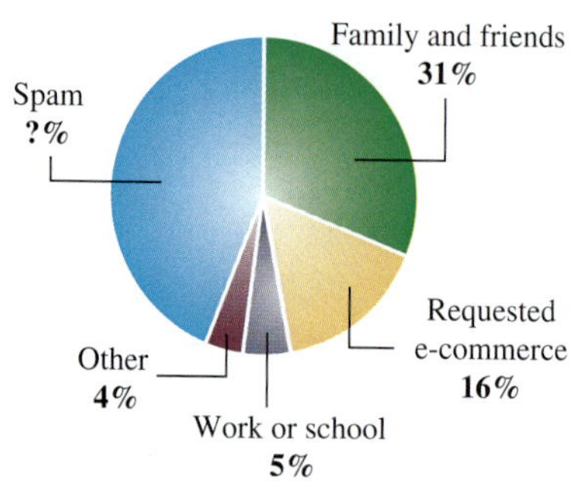

Source: *USA Today*, October 31, 2003

14. HOUSING In the last quarter of 2000, approximately 105.5 million housing units in the United States were occupied. Use the data in the circle graph in the next column to determine what percent were owner-occupied.

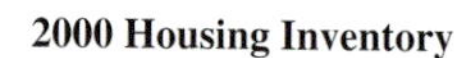

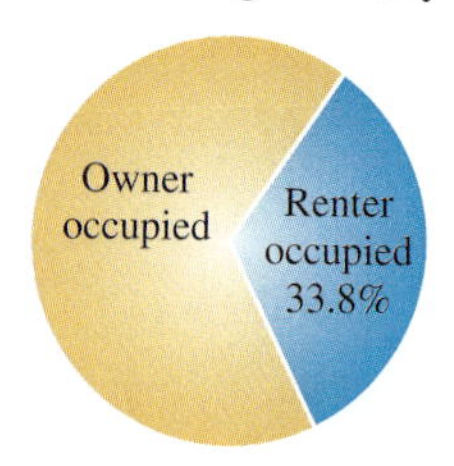

Source: *The U.S. Census Bureau*

NOTATION

15. How is each of the following words or phrases translated in this section?

   a. of

   b. is

   c. what number

16. a. Write the repeating decimal shown in the calculator display as a percent. Use an overbar.

   0.456666666

   b. Round your answer to part a to the nearest hundredth of a percent.

   c. Write your answer to part a using a fraction.

PRACTICE *Solve each problem by solving a percent equation.*

17. What number is 36% of 250?
18. What number is 82% of 300?
19. 16 is what percent of 20?
20. 13 is what percent of 25?
21. 7.8 is 12% of what number?
22. 39.6 is 44% of what number?
23. What number is 0.8% of 12?
24. What number is 5.6% of 4,040?
25. 0.5 is what percent of 40,000?
26. 0.3 is what percent of 15?
27. 3.3 is 7.5% of what number?
28. 8.4 is 20% of what number?
29. Find $7\frac{1}{4}\%$ of 600.
30. Find $1\frac{3}{4}\%$ of 800.
31. 102% of 105 is what number?
32. 210% of 66 is what number?
33. $33\frac{1}{3}\%$ of what number is 33?

34. $66\frac{2}{3}\%$ of what number is 28?

35. $9\frac{1}{2}\%$ of what number is 5.7?

36. $\frac{1}{2}\%$ of what number is 5,000?

37. What percent of 8,000 is 2,500?

38. What percent of 3,200 is 1,400?

Use a circle graph to illustrate the given data. A circle divided into 100 sections is provided to aid in the graphing process.

39. ENERGY Complete the graph to show what percent of the total U.S. energy produced was provided by each source in 2003.

Renewable	12%
Nuclear	11%
Coal	31%
Natural gas	29%
Petroleum	17%

Source: *Energy Information Administration*

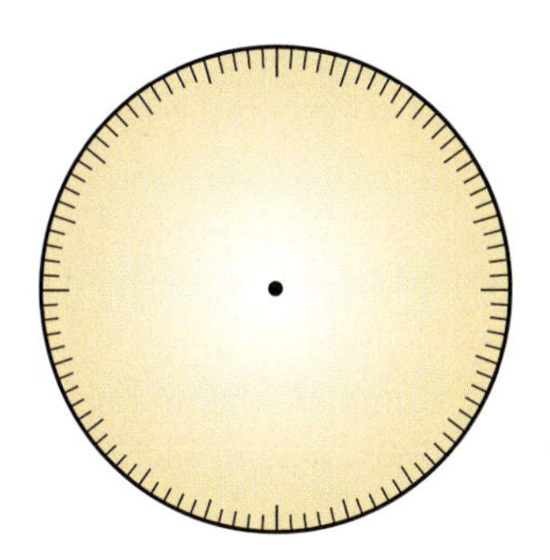

Source: Energy Information Administration

40. GREENHOUSE EFFECT Complete the graph to show what percent of the total U.S. emissions from human activities in 2002 came from each greenhouse gas.

Carbon dioxide	83%
Nitrous oxide	6%
Methane	9%
PFCs, HFCs	2%

Source: *The World Almanac 2005*

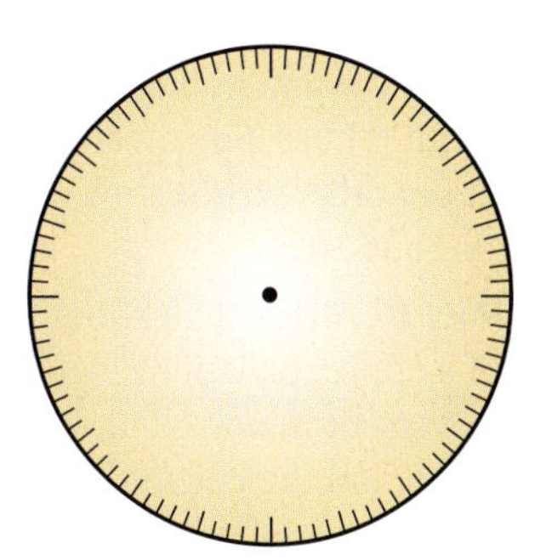

APPLICATIONS

41. CHILD CARE After the first day of registration, 84 children had been enrolled in a new day care center. That represented 70% of the available slots. What was the maximum number of children the center could enroll?

42. RACING PROGRAMS One month before a stock car race, the sale of ads for the official race program was slow. Only 12 pages, or just 60% of the available pages, had been sold. What was the total number of pages devoted to advertising in the program?

43. GOVERNMENT SPENDING The illustration shows the breakdown of federal spending for fiscal year 2003. If the total spending was approximately \$1,800 billion, how many dollars were spent on Social Security, Medicare, and other retirement programs?

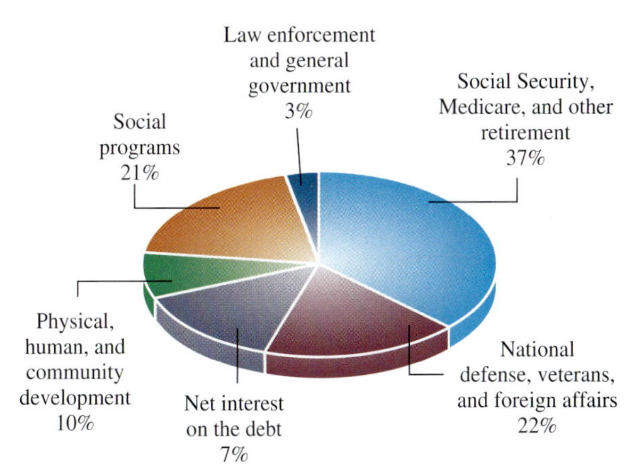

Source: 2004 Federal Income Tax Form 1040

44. GOVERNMENT INCOME Complete the table by finding what percent of total federal government income each source provided in 2003. Round to the nearest percent. Then complete the circle graph on the next page.

Total income, fiscal year 2003: \$2,200 billion		
Source of income	**Amount**	**Percent of total**
Social Security, Medicare, unemployment taxes	\$726 billion	
Personal income taxes	\$814 billion	
Corporate income taxes	\$132 billion	
Excise, estate, customs taxes	\$154 billion	
Borrowing to cover deficit	\$374 billion	

Source: 2003 Federal Income Tax Form

2003 Federal Income Sources

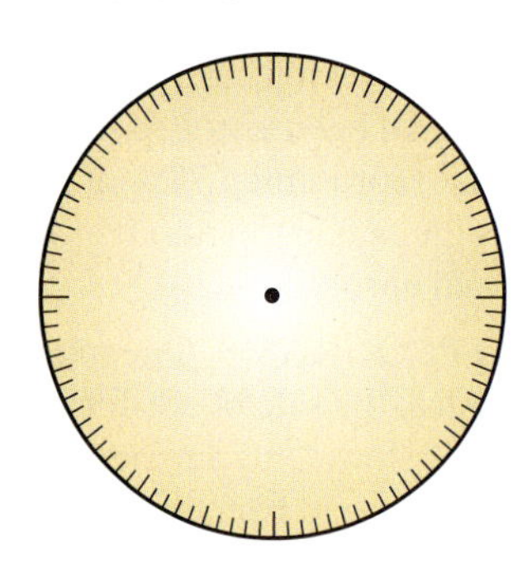

45. THE INTERNET The message at the bottom of the computer screen indicates that 24% of the 50K bytes of information that the user has decided to view have been downloaded to her computer. How many more bytes of information must be downloaded? (50K stands for 50,000.)

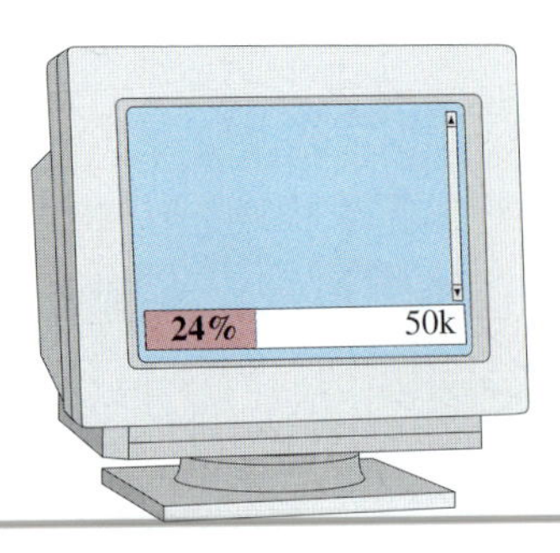

46. REBATES A long-distance telephone company offered its customers a rebate of 20% of the cost of all long-distance calls made in the month of July. One customer's calls are listed in the table. What amount will this customer receive in the form of a rebate?

Date	Time	Place called	Min.	Amount
Jul 4	3:48 P.M.	Denver	47	$3.80
Jul 9	12:00 P.M.	Detroit	68	$7.50
Jul 20	8:59 A.M.	San Diego	70	$9.45

47. PRODUCT PROMOTIONS To promote sales, a free 6-ounce bottle of shampoo is packaged with every large bottle. Use the information on the package to find how many ounces of shampoo the large bottle contains.

48. NUTRITION FACTS The nutrition label on a package of corn chips is shown (g stands for grams and mg for milligrams).

a. How many milligrams of sodium are in one serving of chips?

b. According to the label, what percent of the daily value of sodium is this?

c. What daily value of sodium intake is deemed healthy?

Nutrition Facts
Serving Size: 1 oz. (28g/About 29 chips)
Servings Per Container: About 11

Amount Per Serving	
Calories 160	Calories from Fat 90
	% Daily Value
Total fat 10g	**15%**
Saturated fat 1.5 g	**7%**
Cholesterol 0mg	**0%**
Sodium 240mg	**12%**
Total carbohydrate 15g	**5%**
Dietary fiber 1g	**4%**
Sugars less than 1g	
Protein 2g	

49. DRIVER'S LICENSES On the written part of his driving test, a man answered 28 out of 40 questions correctly. If 70% correct is passing, did he pass the test?

50. THE ALPHABET What percent of the English alphabet do the vowels a, e, i, o, and u make up? (Round to the nearest one percent.)

51. MIXTURES Complete the table to find the number of gallons of sulfuric acid in each of two storage tanks.

	Gallons of solution in tank	% sulfuric acid	Gallons of sulfuric acid in tank
Tank 1	60	50%	
Tank 2	40	30%	

52. CUSTOMER GUARANTEES To assure its customers of low prices, the Home Club offers a "10% Plus" guarantee. If the customer finds the same item selling for less somewhere else, he or she receives the difference in price, plus 10% of the difference. A woman bought miniblinds at the Home Club for $120 but later saw the same blinds on sale for $98 at another store. How much can she expect to be reimbursed?

53. MAKING COPIES The zoom key on the control panel of a copier programs is to print a magnified or reduced copy of the original document. If the zoom is set at 180% and the original document contains type that is 1.5 inches tall, what will be the height of the type on the copy?

54. MAKING COPIES The zoom setting for a copier is entered as a decimal: 0.98. Express it as a percent and find the resulting type size on the copy if the original has type 2 inches in height.

55. INSURANCE The cost to repair a car after a collision was $4,000. The automobile insurance policy paid the entire bill except for a $200 deductible, which the driver paid. What percent of the cost did he pay?

56. FLOOR SPACE A house has 1,200 square feet on the first floor and 800 square feet on the second floor. What percent of the square footage of the house is on the first floor?

57. MAJORITIES In Los Angeles City Council races, if no candidate receives more than 50% of the vote, a runoff election is held between the first- and second-place finishers. From the election results in the table, determine whether there must be a runoff election for District 10.

City council	District 10
Nate Holden	8,501
Madison T. Shockley	3,614
Scott Suh	2,630
Marsha Brown	2,432

58. PORTS In 2002, the busiest port in the United States was the Port of South Louisiana, which handled 216,396,497 tons of goods. Of that amount, 124,908,067 tons were domestic goods and 91,488,430 tons were foreign. What percent of the total was domestic? Round to the nearest tenth of a percent.

WRITING

59. Explain the relationship in a percent problem between the amount, the percent, and the base.

60. Write a real-life situation that could be described by "9 is what percent of 20?"

61. Explain why 150% of a number is more than the number.

62. Explain why "Find 9% of 100" is an easy problem to solve.

REVIEW

63. Add: $2.78 + 6 + 9.09 + 0.3$.

64. Evaluate: $\sqrt{64} + 3\sqrt{9}$.

65. On a number line, which number is closer to 5: 4.9 or 5.001?

66. Is the x-axis horizontal or vertical?

67. Multiply: $34.5464 \cdot 1{,}000$.

68. Find the power: $(0.2)^3$.

69. Solve: $0.4x + 1.2 = -7.8$.

70. If $d = 4t$, find d when $t = 25$.

7.3 Applications of Percent

• Taxes • Commissions • Percent of increase or decrease • Discounts

In this section, we discuss applications of percent. Three of them (taxes, commissions, and discounts) are directly related to purchasing. A solid understanding of these concepts will make you a better consumer. The fourth application uses percent to describe increases or decreases of such things as unemployment and grocery store sales.

Taxes

The sales receipt in Figure 7-6 on the next page gives a detailed account of what items were purchased, how many of each were purchased, and the price of each item.

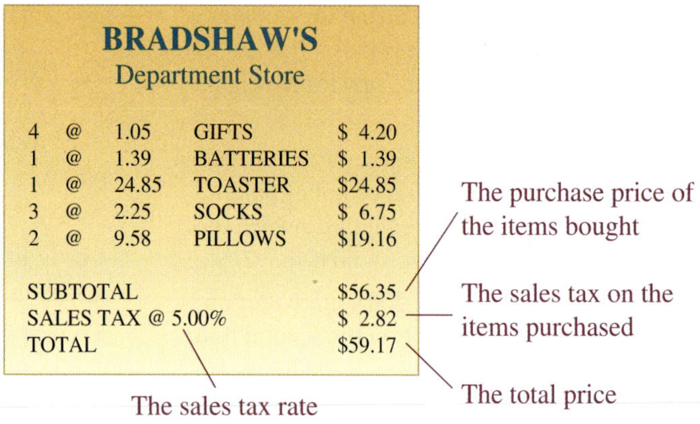

		BRADSHAW'S		
		Department Store		
4	@	1.05	GIFTS	$ 4.20
1	@	1.39	BATTERIES	$ 1.39
1	@	24.85	TOASTER	$24.85
3	@	2.25	SOCKS	$ 6.75
2	@	9.58	PILLOWS	$19.16
SUBTOTAL				$56.35
SALES TAX @ 5.00%				$ 2.82
TOTAL				$59.17

The purchase price of the items bought

The sales tax on the items purchased

The total price

The sales tax rate

FIGURE 7-6

The receipt shows that the $56.35 purchase price (labeled *subtotal*) was taxed at a **rate** of 5%. Sales tax of $2.82 was charged. The sales tax was then added to the subtotal to get the total price of $59.17.

Finding the total price

Total price = purchase price + sales tax

In Example 1, we verify that the amount of sales tax shown on the receipt in Figure 7-6 is correct.

Self Check 1
What would the sales tax be if the $56.35 purchase were made in Texas, which has a 6.25% state sales tax?

Answer $3.52

EXAMPLE 1 **Sales tax.** Find the sales tax on a purchase of $56.35 if the sales tax rate is 5%.

Solution First we write the problem so that we can translate it into an equation. The rate is 5%. We are to find the amount of the tax.

What number	is	5%	of	56.35?
x	$=$	5%	$\cdot$	56.35

$x = 0.05 \cdot 56.35$ Change 5% to a decimal: 5% = 0.05.

$= 2.8175$ Perform the multiplication.

Rounding to the nearest cent (hundredths), we find that the sales tax would be $2.82. The sales receipt in Figure 7-6 is correct.

In addition to sales tax, we pay many other types of taxes in our daily lives. Income tax, gasoline tax, and Social Security tax are just a few.

Self Check 2
A tax of $5,250 had to be paid on an inheritance of $15,000. What is the inheritance tax rate?

EXAMPLE 2 **Withholding tax.** A waitress found that $11.04 was deducted from her weekly gross earnings of $240 for federal income tax. What withholding tax rate was used?

Solution First, we write the problem in a form that can be translated into an equation. We need to find the tax rate.

11.04	is	what percent	of	240?
11.04	=	x	$\cdot$	240

$11.04 = 240x$ Rewrite the right-hand side: $x \cdot 240 = 240x$.

$\frac{11.04}{\mathbf{240}} = \frac{240x}{\mathbf{240}}$ To undo the multiplication by 240, divide both sides by 240.

$0.046 = x$ Perform the divisions.

$4.6\% = x$ Change 0.046 to a percent.

The withholding tax rate was 4.6%.

Answer 35%

Commissions

Instead of working for a salary or getting paid at an hourly rate, many salespeople are paid on **commission.** They earn an amount based on the goods or services they sell.

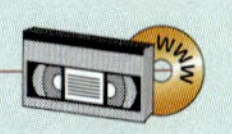

EXAMPLE 3 **Appliance sales.** The commission rate for a salesperson at an appliance store is 16.5%. Find his commission from the sale of a refrigerator costing $499.95.

Solution We write the problem so that it can be translated into an equation. We are to find the amount of the commission.

What number	is	16.5%	of	499.95?
x	=	16.5%	$\cdot$	499.95

$x = 0.165 \cdot 499.95$ Change 16.5% to a decimal: $16.5\% = 0.165$.

$= 82.49175$ Use a calculator to perform the multiplication.

Rounding to the nearest cent (hundredth), we find that the commission is $82.49.

Self Check 3
An insurance salesperson receives a 4.1% commission on each $120 premium paid by a client. What is the amount of the commission on this premium?

Answer $4.92

Percent of increase or decrease

Percents can be used to describe how a quantity has changed. For example, consider Figure 7-7, which compares the number of hours of work it took the average U.S. worker to earn enough to buy a dishwasher in 1950 and 1998.

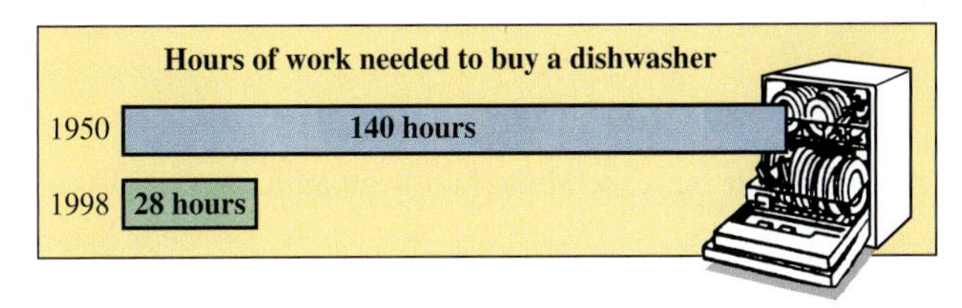

Source: Federal Reserve Bank of Dallas

FIGURE 7-7

From the figure, we see that the number of hours an average American had to work in order to buy a dishwasher has decreased over the years. To describe this decrease using a percent, we first subtract to find the amount of the decrease.

$140 - 28 = 112$ Subtract the hours of work needed in 1998 from the hours of work needed in 1950.

Next, we find what percent of the original number of hours of work needed in 1950 this difference represents.

112	is	what percent	of	140?
112	=	x	$\cdot$	140

$112 = 140x$ Rewrite the right-hand side: $x \cdot 140 = 140x$.

$\frac{112}{140} = \frac{140x}{140}$ To undo the multiplication by 140, divide both sides by 140.

$0.8 = x$ Perform the divisions.

$80\% = x$ Change 0.8 to a percent.

From 1950 to 1998, there was an 80% decrease in the number of hours it took the average U.S. worker to earn enough to buy a dishwasher.

Finding the percent of increase or decrease

To find the percent of increase or decrease:

1. Subtract the smaller number from the larger to find the amount of increase or decrease.
2. Find what percent the difference is of the original amount.

Self Check 4

In one school district, the number of home-schooled children increased from 15 to 150 in 4 years. Find the percent of increase.

EXAMPLE 4 JFK.

A 1996 auction included an oak rocking chair used by President John F. Kennedy in the Oval Office. The chair, originally valued at $5,000, sold for $453,500. Find the percent of increase in the value of the rocking chair.

Solution First, we find the amount of increase.

$453{,}500 - 5{,}000 = 448{,}500$ Subtract the original value from the price paid at auction.

The rocking chair increased in value by $448,500. Next, we find what percent of the original value the increase represents.

448,500	is	what percent	of	5,000?
448,500	=	x	$\cdot$	5,000

$448{,}500 = 5{,}000x$ Rewrite the right-hand side: $x \cdot 5{,}000 = 5{,}000x$.

$\frac{448{,}500}{5{,}000} = \frac{5{,}000x}{5{,}000}$ To undo the multiplication by 5,000, divide both sides by 5,000.

$89.7 = x$ Perform the divisions.

$8{,}970\% = x$ Change 89.7 to a percent.

The Kennedy rocking chair increased in value by an amazing 8,970%.

Answer 900%

EXAMPLE 5

Population decline. Norfolk, Virginia, experienced a decrease in population over the ten-year period from 1990 to 2000. Use the information in Figure 7-8 to determine the population of Norfolk in 2000.

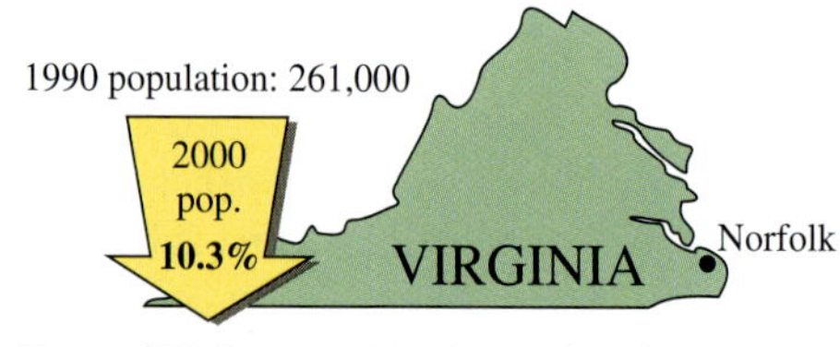

Source: U.S. Bureau of the Census (1999)

FIGURE 7-8

Solution In 1990, the population was 261,000. We are told that the number fell, and we need to find out by how much. To do so, we solve the following percent formula.

Amount	=	percent	·	base
A	=	10.3%	·	261,000

$A = 0.103 \cdot 261{,}000$ Change 10.3% to a decimal: 10.3% = 0.103.

$A = 26{,}883$ Perform the multiplication.

From 1990 to 2000, the population decreased by 26,883. To find the city's population in 2000, we subtract the decrease from the population in 1990.

$261{,}000 - 26{,}883 = 234{,}117$

In 2000, the population of Norfolk, Virginia, was 234,117.

We can solve this problem in another way. If the population of Norfolk decreased by 10.3%, then the population in 2000 was $100\% - 10.3\% = 89.7\%$ of the population in 1990. Using this approach, we can find the 2000 population directly by solving the following percent formula.

Amount	=	percent	·	base
A	=	89.7%	·	261,000

$A = 0.897 \cdot 261{,}000$ Change 89.7% to a decimal: 89.7% = 0.897.

$A = 234{,}117$ Perform the multiplication.

As before, we see that the population of Norfolk in 2000 was 234,117.

Fastest Growing Occupations

THINK IT THROUGH

"With the Baby Boomer generation aging, more medical professionals will be required to manage the health needs of the elderly."

Jonathan Stanewick, Compensation Analyst, Salary.com, 2004

The table below shows predictions by the U.S. Department of Labor, Bureau of Labor Statistics, of the fastest growing occupations for the years 2002–2012. Find the percent increase in the number of jobs for each occupation. Which occupation has the greatest percent increase?

Occupation	Number of jobs in 2002	Predicted number of jobs in 2012	Education or training required
Home health aide	580,000	859,000	On-the-job training or AA degree
Medical assistant	365,000	579,000	On-the-job training or AA degree
Physician assistant	63,000	94,000	Bachelor's or Master's degree
Social service assistant	305,000	454,000	On-the-job training or AA degree
Systems, data analyst	186,000	292,000	Bachelor's degree

Discounts

The difference between the original price and the sale price of an item is called the **discount.** If the discount is expressed as a percent of the selling price, it is called the **rate of discount.** We will use the information in the advertisement shown in Figure 7-9 to discuss how to find a discount and how to find a discount rate.

FIGURE 7-9

Self Check 6
Sunglasses, regularly selling for \$15.40, are discounted 15%. Find the sale price.

Answer \$13.09

EXAMPLE 6 **Shoe sales.** Find the amount of the discount on the pair of men's basketball shoes shown in Figure 7-9. Then find the sale price.

Solution To find the discount, we find 25% of the regular price, \$59.80.

What number	is	25%	of	59.80?
x	=	25%	·	59.80

$x = 0.25 \cdot 59.80$ Change 25% to a decimal: 25% = 0.25.
$= 14.95$ Perform the multiplication.

The discount is \$14.95. To find the sale price, we subtract the amount of the discount from the regular price.

$59.80 - 14.95 = 44.85$

The sale price of the men's basketball shoes is \$44.85.

In Example 6, we used the following formula to find the sale price.

Finding the sale price

Sale price = original price − discount

Self Check 7
An early-bird special at a restaurant offers a \$10.99 prime rib dinner for only \$7.95 if it is ordered before 6 P.M. Find the rate of discount. Round to the nearest one percent.

EXAMPLE 7 **Discounts.** What is the rate of discount on the ladies' aerobic shoes advertised in Figure 7-9?

Solution We can think of this as a percent-of-decrease problem. We first compute the amount of the discount. This decrease in price is found using subtraction.

$39.99 - 21.99 = 18$

The shoes are discounted \$18. Now we find what percent of the original price the discount is.

18	is	what percent	of	39.99?
18	=	x	·	39.99

$$18 = 39.99x \qquad x \cdot 39.99 = 39.99x.$$

$$\frac{18}{\mathbf{39.99}} = \frac{39.99x}{\mathbf{39.99}} \qquad \text{To undo the multiplication by 39.99, divide both sides by 39.99.}$$

$$0.450113 \approx x \qquad \text{Perform the division.}$$

$$45.0113\% \approx x \qquad \text{Change 0.450113 to a percent.}$$

Rounded to the nearest one percent, the discount rate is 45%.

Answer 28%

Section 7.3 STUDY SET

VOCABULARY *Fill in the blanks.*

1. Some salespeople are paid on ____________. It is based on a percent of the total dollar amount of the goods or services they sell.
2. When we use percent to describe how a quantity has increased when compared to its original value, we are finding the percent of __________.
3. The difference between the original price and the sale price of an item is called the __________.
4. The ______ of a sales tax is expressed as a percent.

CONCEPTS

5. Fill in the blanks: To find the percent decrease, __________ the smaller number from the larger number to find the amount of decrease. Then find what percent that difference is of the __________ amount.
6. NEWSPAPERS The table below shows how the circulations of two daily newspapers changed from 1997 to 2003.

CIRCULATION		
	Miami Herald	**USA Today**
1997	356,803	1,629,665
2003	315,850	2,154,539

Source: *The World Almanac 2005*

   a. What was the *amount of decrease* of the *Miami Herald*'s circulation?

   b. What was the *amount of increase* of *USA Today*'s circulation?

APPLICATIONS *Solve each problem. If a percent answer is not exact, round to the nearest one percent.*

7. SALES TAXES The state sales tax rate in Utah is 4.75%. Find the sales tax on a dining room set that sells for $900.
8. SALES TAXES Find the sales tax on a pair of jeans costing $40 if they are purchased in Arkansas, which has a sales tax rate of 4.625%.
9. ROOM TAXES After checking out of a hotel, a man noticed that the hotel bill included an additional charge labeled *room tax.* If the price of the room was $129 plus a room tax of $10.32, find the room tax rate.
10. EXCISE TAXES While examining her monthly telephone bill, a woman noticed an additional charge of $1.24 labeled *federal excise tax.* If the basic service charges for that billing period were $42, what is the federal excise tax rate?
11. SALES RECEIPTS Complete the following sales receipt by finding the subtotal, the sales tax, and the total.

NURSERY CENTER
Your one-stop garden supply

3 @ 2.99	PLANTING MIX	$ 8.97
1 @ 9.87	GROUND COVER	$ 9.87
2 @ 14.25	SHRUBS	$28.50
SUBTOTAL		$
SALES TAX @ 6.00%		$
TOTAL		$

12. SALES RECEIPTS Complete the following sales receipt by finding the prices, the subtotal, the sales tax, and the total.

McCOY'S FURNITURE

1 @ 450.00	SOFA	$
2 @ 90.00	END TABLES	$
1 @ 350.00	LOVE SEAT	$
SUBTOTAL		$
SALES TAX @ 4.20%		$
TOTAL		$

13. SALES TAX In order to raise more revenue, some states raise the sales tax rate. How much additional money will be collected on the sale of a $15,000 car if the sales tax rate is raised 1%?

14. FOREIGN TRAVEL Value-added tax is a consumer tax imposed on goods and services. Currently, there are VAT systems in place all around the world. (The United States is one of the few industrialized nations not using a value-added tax system.) Complete the table by determining the VAT a traveler would pay in each country on a dinner costing $20.95.

Country	VAT rate	Tax on a $20.95 dinner
Canada	7%	
Germany	16%	
England	17.5%	
Sweden	25%	

15. PAYCHECKS Use the information on the paycheck stub to find the tax rate for the federal withholding, worker's compensation, Medicare, and Social Security taxes that were deducted from the gross pay.

6286244	
Issue date: 03-27-05	
GROSS PAY	$360.00
TAXES	
FED. TAX	$ 28.80
WORK. COMP.	$ 4.32
MEDICARE	$ 5.22
SOCIAL SECURITY	$ 22.32
NET PAY	$299.34

16. GASOLINE TAXES In one state, a gallon of unleaded gasoline sells for $1.89. This price includes federal and state taxes that total approximately $0.54. Therefore, the price of a gallon of gasoline, before taxes, is about $1.35. What is the tax rate on gasoline?

17. OVERTIME Factory management wants to reduce the number of overtime hours by 25%. If the total number of overtime hours is 480 this month, what is the target number of overtime hours for next month?

18. COST-OF-LIVING INCREASES If a woman making $32,000 a year receives a cost-of-living increase of 2.4%, how much is her raise? What is her new salary?

19. REDUCED CALORIES A company advertised its new, improved chips as having 36% fewer calories per serving than the original style. How many calories are in a serving of the new chips if a serving of the original style contained 150 calories?

20. POLICE FORCE A police department plans to increase its 80-person force by 5%. How many additional officers will be hired? What will be the new size of the department?

21. ENDANGERED SPECIES The illustration shows the total number of endangered and threatened plant and animal species for each of the years 1993–1999, as determined by the U.S. Fish and Wildlife Service. When was there a decline in the total? To the nearest percent, find the percent of decrease in the total for that period.

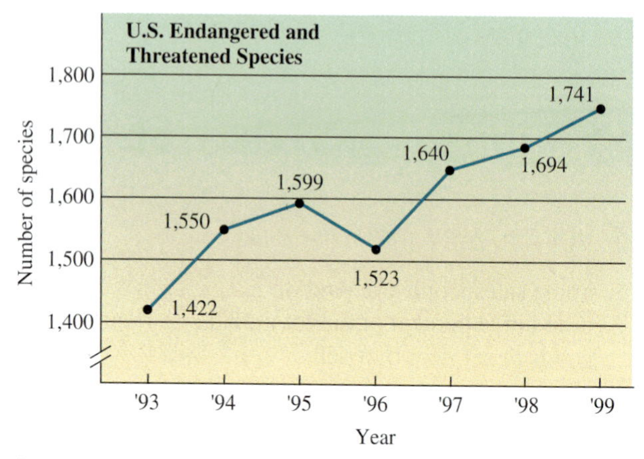

Source: *The New York Times 2000 Almanac*

22. CROP DAMAGE After flooding damaged much of the crop, the cost of a head of lettuce jumped from $0.99 to $2.20. What percent of increase is this?

23. CAR INSURANCE A student paid a car insurance premium of $400 every three months. Then the premium dropped to $360, because she qualified for a good-student discount. What was the percent of decrease in the premium?

24. BUS PASSES To increase the number of riders, a bus company reduced the price of a monthly pass from $112 to $98. What was the percent of decrease?

25. LAKE SHORELINES Because of a heavy spring runoff, the shoreline of a lake increased from 5.8 miles to 7.6 miles. What was the percent of increase in the shoreline?

26. BASEBALL The illustration shows the path of a baseball hit 110 mph, with a launch angle of 35 degrees, at sea level and at Coors Field, home of the Colorado Rockies. What is the percent of increase in the distance the ball travels at Coors Field?

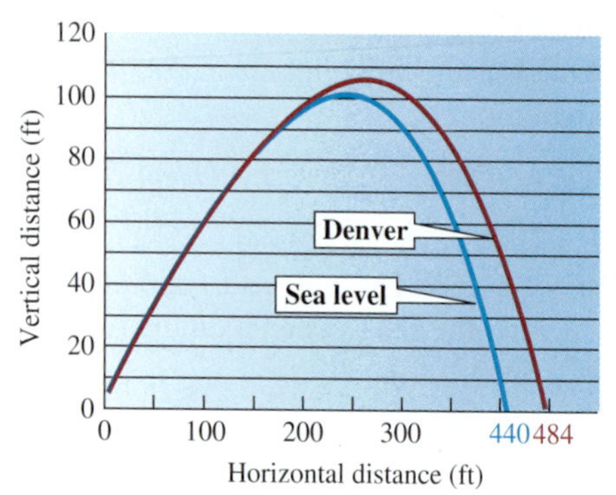

Source: *Los Angeles Times*

27. EARTH MOVING The illustration shows the typical soil volume change during earth moving. (One cubic yard of soil fits in a cube that is 1 yard long, 1 yard wide, and 1 yard high.)

a. Find the percent of increase in the soil volume as it goes through step 1 of the process.

b. Find the percent of decrease in the soil volume as it goes through step 2 of the process.

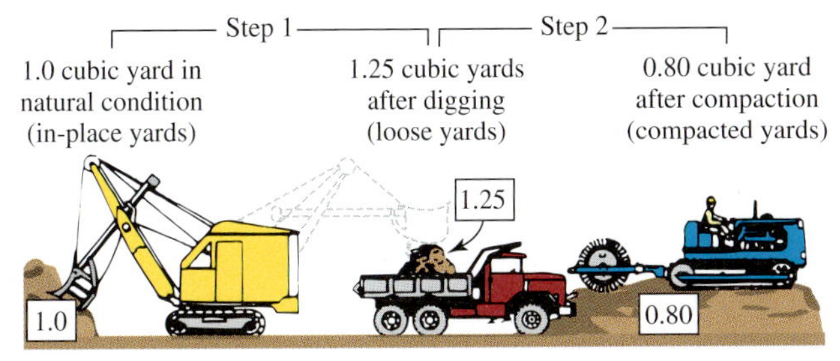

Source: U.S. Department of the Army

28. PARKING The management of a mall has decided to increase the parking area. The plans are shown. What will be the percent of increase in the parking area once the project is completed?

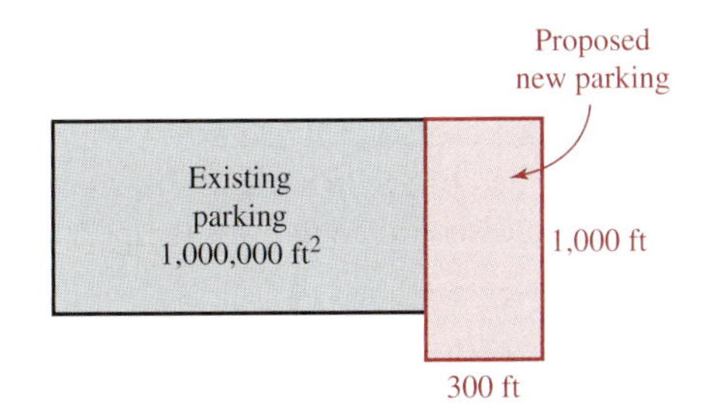

29. REAL ESTATE After selling a house for $198,500, a real estate agent split the 6% commission with another agent. How much did each person receive?

30. MEDICAL SUPPLIES A salesperson for a medical supplies company is paid a commission of 9% for orders under $8,000. For orders exceeding $8,000, she receives an additional 2% in commission on the total amount. What is her commission on a sale of $14,600?

31. SPORTS AGENTS A sports agent charges her clients a fee to represent them during contract negotiations. The fee is based on a percent of the contract amount. If the agent earned $37,500 when her client signed a $2,500,000 professional football contract, what rate did she charge for her services?

32. ART GALLERIES An art gallery displays paintings for artists and receives a commission from the artist when a painting is sold. What is the commission rate if a gallery received $135.30 when a painting was sold for $820?

33. CONCERT PARKING A concert promoter gets $33\frac{1}{3}\%$ of the revenue the arena receives from its parking concession the night of the performance. How much can the promoter make if 6,000 cars are anticipated and parking costs $6 a car?

34. KITCHENWARE A homemaker invited her neighbors to a kitchenware party to show off cookware and utensils. As party hostess, she received 12% of the total sales. How much was purchased if she received $41.76 for hosting the party?

35. WATCHES Find the regular price and the rate of discount for the watch that is on sale.

36. STEREOS Find the regular price and the rate of discount for the stereo system that is on sale.

37. RINGS What does a ring regularly sell for if it has been discounted 20% and is on sale for $149.99? (*Hint:* The ring is selling for 80% of its regular price.)

38. BLINDS What do vinyl blinds regularly sell for if they have been discounted 55% and are on sale for $49.50? (*Hint:* The blinds are selling for 45% of their regular price.)

39. DVRs What are the sale price and the discount rate for a digital video recorder with remote that regularly sells for $399.97 and is being discounted $50?

40. CAMCORDERS What are the sale price and the discount rate for a camcorder that regularly sells for $559.97 and is being discounted $80?

41. REBATES Find the discount, the discount rate, and the reduced price for a case of motor oil if a shopper receives the manufacturer's rebate mentioned in the following ad.

42. COUPONS Find the discount, the discount rate, and the reduced price for a box of cereal that normally sells for $3.29 if a shopper presents the following coupon at a store that doubles the value of the coupon.

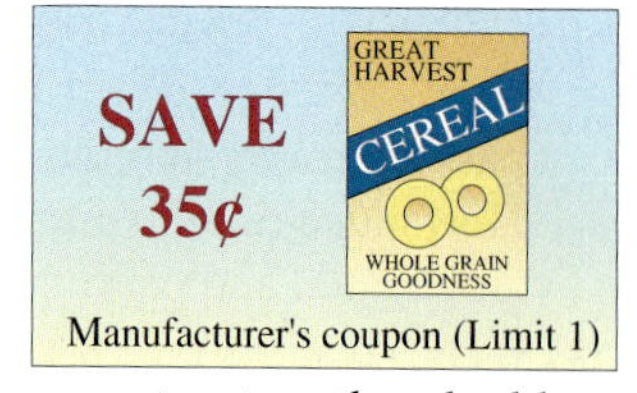

43. SHOPPING Determine the Home Shopping Network price of the ring described below if it sells it for 55% off of the retail price. Ignore shipping and handling costs.

Item 169-117
2.75 lb ctw
10K Blue Topaz Ring
6, 7, 8, 9, 10
Retail value $170
HSN Price
$??.??
S&H $5.95

44. INFOMERCIALS The host of a TV infomercial says that the suggested retail price of a rotisserie grill is $249.95 and that it is now offered "for just 4 easy payments of only $39.95." What is the discount, and what is the discount rate?

WRITING

45. List the pros and cons of working on commission.

46. In Example 6, explain why you get the correct answer for the sale price by finding 75% of the regular price.

47. Explain the difference between a tax and a tax rate.

48. Explain how to find the sale price of an item if you know the regular price and the discount rate.

REVIEW

49. Multiply: $-5(-5)(-2)$.

50. Solve: $4(x + 6) = 12$.

51. Evaluate the expression $2a^2(a + b)$ for $a = -2$ and $b = -1$.

52. Let x represent the height of an elm tree. A pine tree is 15 feet taller than the elm. Write an algebraic expression that represents the height of the pine tree.

53. How many eggs in d dozen?

54. Evaluate: $-4 - (-7)$.

55. Evaluate: $|-5 - 8|$.

56. A store clerk earns x dollars an hour. How much will she earn in a 40-hour week?

57. Graph each point: $(-3, 4)$, $(4, 3.5)$, $(-2, -\frac{5}{2})$, $(0, -4)$, $(\frac{3}{2}, 0)$, and $(3, -4)$.

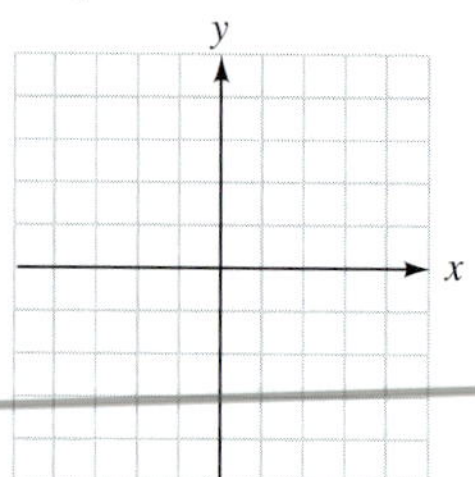

58. Complete the table of solutions for $y = 2x$.

x	y	(x, y)
−2		
0		
3		

Estimation

We will now discuss some estimation methods that can be used when working with percent. To begin, we consider a way to find 10% of a number quickly. Recall that 10% of a number is found by multiplying the number by 10% or 0.1. When multiplying a number by 0.1, we simply move the decimal point 1 place to the left to find the result.

EXAMPLE 1 **10% of a number.** Find 10% of 234.

Solution To find 10% of 234, move the decimal point 1 place to the left.

234 = 23.4 0

Thus, 10% of 234 is 23.4, or approximately 23.

To find 15% of a number, first find 10% of the number. Then find half of that to obtain the other 5%. Finally, add the two results.

EXAMPLE 2 **Estimating 15% of a number.** Estimate 15% of 78.

Solution

10% of 78 is 7.8, or about 8. ⟶ 8
Add half of 8 to get the other 5%. ⟶ + 4
12

Thus, 15% of 78 is approximately 12.

To find 20% of a number, first find 10% of it and then double that result. A similar procedure can be used when working with any multiple of 10%.

EXAMPLE 3 **Estimating 20% of a number.** Estimate 20% of 3,234.15.

Solution 10% of 3,234.15 is 323.415 or about 323. To find 20%, double that.

Thus, 20% of 3,234.15 is approximately 646.

EXAMPLE 4 **1% of a number.** Find 1% of 0.8.

Solution To find 1% of a number, multiply it by 0.01, because 1% = 0.01. When multiplying a number by 0.01, simply move the decimal point 2 places to the left to find the result.

0.8 = .00 8

= 0.008

Thus, 1% of 0.8 is 0.008.

EXAMPLE 5 **50% of a number.** Find 50% of 2,800,000,000.

Solution To find 50% of a number means to find $\frac{1}{2}$ of that number. To find one-half of a number, simply divide it by 2. Thus, 50% of 2,800,000,000 is 2,800,000,000 ÷ 2 = 1,400,000,000.

To find 25% of a number, first find 50% of it and then divide that result by 2.

EXAMPLE 6 Estimating 25% of a number. Estimate 25% of 16,813.

Solution 16,813 is about 16,800. Half of that is 8,400. Thus, 50% of 16,813 is approximately 8,400.

To estimate 25% of 16,813, divide 8,400 by 2. Thus, 25% of 16,813 is approximately 4,200.

100% of a number is the number itself. To find 200% of a number, double the number.

EXAMPLE 7 Estimating 200% of a number. Estimate 200% of 65.198.

Solution 65.198 is about 65. To find 200% of 65, double it. Thus, 200% of 65.198 is approximately $65 \cdot 2$ or 130.

STUDY SET *Estimate each answer.*

1. COLLEGE COURSES 20% of the 815 students attending a small college were enrolled in a science course. How many students is this?
2. SPECIAL OFFERS In the grocery store, a 65-ounce bottle of window cleaner was marked "25% free." How many ounces are free?
3. DISCOUNTS By how much is the price of a VCR discounted if the regular price of $196.88 is reduced by 30%?
4. TIPPING A restaurant tip is normally 15% of the cost of the meal. Find the tip on a dinner costing $38.64.
5. FIRE DAMAGE An insurance company paid 50% of the $107,809 it cost to rebuild a home that was destroyed by fire. How much did the insurance company pay?
6. SAFETY INSPECTIONS Of the 2,580 vehicles inspected at a safety checkpoint, 10% had code violations. How many cars had code violations?
7. WEIGHTLIFTING A 158-pound weightlifter can bench press 200% of his body weight. How many pounds can he bench press?
8. TESTING On a 120-question true/false test, 5% of a student's answers were wrong. How many questions did she miss?
9. TRAFFIC STUDIES According to an electronic traffic monitor, 20% of the 650 motorists who passed it were speeding. How many of these motorists were speeding?
10. SELLING HOMES A homeowner has been told she will recoup 70% of her $5,000 investment if she paints her home before selling it. What is the potential payback if she paints her home?

Approximate the percent and then estimate each answer.

11. NO-SHOWS The attendance at a seminar was only 31% of what the organizers had anticipated. If 68 people were expected, how many actually attended the seminar?
12. "A" STUDENTS Of the 900 students in a school, 16% were on the principal's honor roll. How many students were on the honor roll?
13. INTERNET SURVEY The illustration shows an online survey question. How many people voted yes?

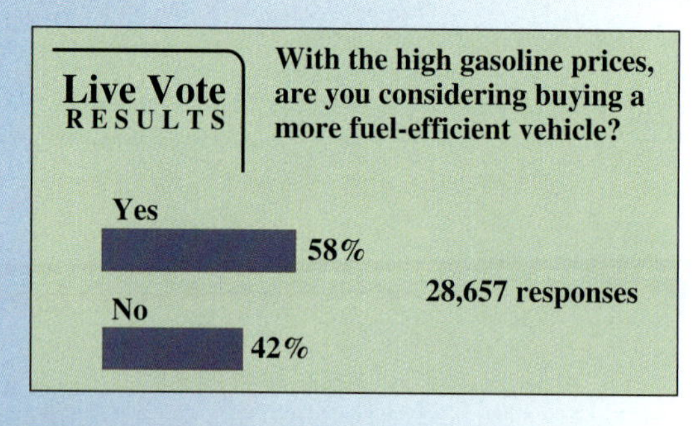

14. MEDICARE The Medicare payroll tax rate is 1.45%. How much Medicare tax will be deducted from a paycheck of $596?
15. VOTING On election day, 48% of the 6,200 workers at the polls were volunteers. How many volunteers helped with the election?
16. BUDGETS Each department at a college was asked to cut its budget by 21%. By how much money should the mathematics department budget be reduced if it is currently $4,515?

7.4 Interest

- Simple interest
- Compound interest

When money is borrowed, the lender expects to be paid back the amount of the loan plus an additional charge for the use of the money. The additional charge is called **interest.** When money is deposited in a bank, the depositor is paid for the use of the money. The money the deposit earns is also called interest. In general, interest is money that is paid for the use of money.

Simple interest

Interest is calculated in one of two ways: either as **simple interest** or as **compound interest.** We will begin by discussing simple interest. First, we need to introduce some key terms associated with borrowing or lending money.

Principal: the amount of money that is invested, deposited, or borrowed.

Interest rate: a percent that is used to calculate the amount of interest to be paid. It is usually expressed as an annual (yearly) rate.

Time: the length of time (usually in years) that the money is invested, deposited, or borrowed.

The amount of interest to be paid depends on the principal, the rate, and the time. That is why all three are usually mentioned in advertisements for bank accounts, investments, and loans. (See Figure 7-10.)

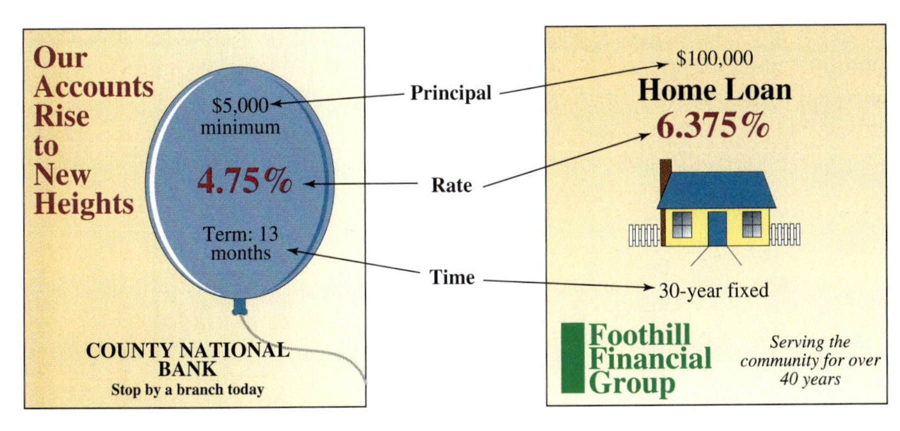

FIGURE 7-10

Simple interest is interest earned on the original principal. It is found by using a formula.

Simple interest formula

Interest = principal · rate · time

or

$I = Prt$

where the rate r is expressed as an annual rate and the time t is expressed in years.

Self Check 1

If \$4,200 is invested for 2 years at a rate of 4% annual interest, how much interest is earned?

Answer \$336

EXAMPLE 1 \$3,000 is invested for 1 year at a rate of 5%. How much interest is earned?

Solution We will use the formula $I = Prt$ to calculate the interest earned. The principal is \$3,000, the interest rate is 5% (or 0.05), and the time is 1 year.

$P = 3{,}000 \qquad r = 5\% = 0.05 \qquad t = 1 \text{ year}$

$I = Prt$ Write the interest formula.

$= 3{,}000 \cdot 0.05 \cdot 1$ Substitute the values for P, r, and t.

$= 150$ Perform the multiplication.

The interest earned in 1 year is \$150.

The information given in this problem and the result can be presented in a table.

Principal	Rate	Time	Interest earned
\$3,000	5%	1 year	\$150

When we use the formula $I = Prt$, the time must be expressed in years. If the time is given in days or months, we rewrite it as a fractional part of a year. For example, a 30-day investment lasts $\frac{30}{365}$ of a year, since there are 365 days in a year. For a 6-month loan, we express the time as $\frac{6}{12}$ or $\frac{1}{2}$ of a year, since there are 12 months in a year.

Self Check 2

How much must be repaid if \$3,200 is borrowed at a rate of 15% for 120 days?

Answer \$3,357.81

EXAMPLE 2 To start a carpet-cleaning business, a couple borrows \$5,500 to purchase equipment and supplies. If the loan has a 14% interest rate, how much must they repay at the end of the 90-day period?

Solution First, we find the amount of interest paid on the loan. We must rewrite the time (90-day period) as a fractional part of a 365-day year.

$P = 5{,}500 \qquad r = 14\% = 0.14 \qquad t = \frac{90}{365}$

$I = Prt$ Write the interest formula.

$= 5{,}500 \cdot 0.14 \cdot \frac{90}{365}$ Substitute the values for P, r, and t.

$= \frac{5{,}500}{1} \cdot \frac{0.14}{1} \cdot \frac{90}{365}$ Write 5,500 and 0.14 as fractions.

$= \frac{69{,}300}{365}$ Use a calculator to multiply the numerators. Multiply the denominators.

≈ 189.86 Use a calculator to perform the division. Round to the nearest cent.

The interest on the loan is \$189.86. To find how much they must pay back, we add the principal and the interest.

$5{,}500 + 189.86 = 5{,}689.86$

The couple must pay back \$5,689.86 at the end of 90 days.

Compound interest

Most savings accounts pay **compound interest** rather than simple interest. Compound interest is interest paid on accumulated interest. To illustrate this concept, suppose that $2,000 is deposited in a savings account at a rate of 5% for 1 year. We can use the formula $I = Prt$ to calculate the interest earned at the end of 1 year.

$$I = Prt$$
$$= 2{,}000 \cdot 0.05 \cdot 1 \quad \text{Substitute for } P, r, \text{ and } t.$$
$$= 100 \quad \text{Perform the multiplication.}$$

Interest of $100 was earned. At the end of the first year, the account contains the interest ($100) plus the original principal ($2,000), for a balance of $2,100.

Suppose that the money remains in the savings account for another year at the same interest rate. For the second year, interest will be paid on a principal of $2,100. That is, during the second year, we earn *interest on the interest* as well as on the original $2,000 principal. Using $I = Prt$, we can find the interest earned in the second year.

$$I = Prt$$
$$= 2{,}100 \cdot 0.05 \cdot 1 \quad \text{Substitute for } P, r, \text{ and } t.$$
$$= 105 \quad \text{Perform the multiplication.}$$

In the second year, $105 of interest is earned. The account now contains that interest plus the $2,100 principal, for a total of $2,205.

As Figure 7-11 shows, we calculated the simple interest two times to find the compound interest.

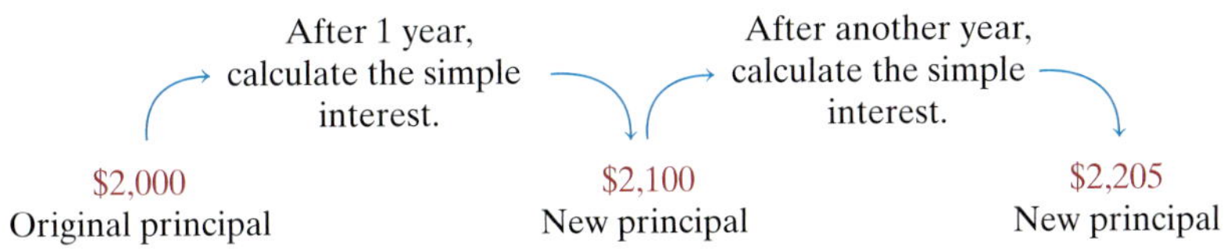

After 1 year, calculate the simple interest.
After another year, calculate the simple interest.

$2,000 Original principal → $2,100 New principal → $2,205 New principal

FIGURE 7-11

If we compute the *simple interest* on $2,000, at 5% for 2 years, the interest earned is $I = 2{,}000 \cdot 0.05 \cdot 2 = 200$. Thus, the account balance would be $2,200. Comparing the balances, the account earning compound interest will contain $5 more than the account earning simple interest.

In the previous example, the interest was calculated at the end of each year, or **annually.** When compounding, we can compute the interest in other time increments, such as **semiannually** (twice a year), **quarterly** (four times a year), or even **daily.**

EXAMPLE 3 Compound interest. As a gift for her newborn granddaughter, a grandmother opens a $1,000 savings account in the baby's name. The interest rate is 4.2%, compounded quarterly. Find the amount of money the child will have in the bank on her first birthday.

Solution If the interest is compounded quarterly, the interest will be computed four times in one year. To find the amount of interest $1,000 will earn in the first quarter of the year, we use the simple interest formula, where t is $\frac{1}{4}$ of a year.

Interest earned in the first quarter

$$P = 1{,}000 \qquad r = 4.2\% = 0.042 \qquad t = \frac{1}{4}$$

$$I = 1{,}000 \cdot 0.042 \cdot \frac{1}{4}$$
$$= \$10.50$$

The interest earned in the first quarter is $10.50. This now becomes part of the principal for the second quarter.

$$\$1{,}000 + \$10.50 = \$1{,}010.50$$

To find the amount of interest \$1,010.50 will earn in the second quarter of the year, we use the simple interest formula, where t is again $\frac{1}{4}$ of a year.

$$P = 1{,}010.50 \qquad r = 0.042 \qquad t = \frac{1}{4}$$

$$I = 1{,}010.50 \cdot 0.042 \cdot \frac{1}{4}$$

$$\approx \$10.61 \quad \text{(Rounded)}$$

The interest earned in the second quarter is \$10.61. This becomes part of the principal for the third quarter.

\$1,010.50 + \$10.61 = \$1,021.11

To find the interest \$1,021.11 will earn in the third quarter of the year, we proceed as follows.

$$P = 1{,}021.11 \qquad r = 0.042 \qquad t = \frac{1}{4}$$

$$I = 1{,}021.11 \cdot 0.042 \cdot \frac{1}{4}$$

$$\approx \$10.72 \quad \text{(Rounded)}$$

The interest earned in the third quarter is \$10.72. This now becomes part of the principal for the fourth quarter.

\$1,021.11 + \$10.72 = \$1,031.83

To find the interest \$1,031.83 will earn in the fourth quarter, we again use the simple interest formula.

$$P = 1{,}031.83 \qquad r = 0.042 \qquad t = \frac{1}{4}$$

$$I = 1{,}031.83 \cdot 0.042 \cdot \frac{1}{4}$$

$$\approx \$10.83 \quad \text{(Rounded)}$$

The interest earned in the fourth quarter is \$10.83. Adding this to the existing principal, we get

\$1,031.83 + \$10.83 = \$1,042.66

The amount that has accumulated in the account after four quarters, or 1 year, is \$1,042.66.

Computing compound interest by hand is tedious. The **compound interest formula** can be used to find the total amount of money that an account will contain at the end of the term.

Compound interest formula

The total amount A in an account can be found using the formula

$$A = P\left(1 + \frac{r}{n}\right)^{nt}$$

where P is the principal, r is the annual interest rate expressed as a decimal, t is the length of time in years, and n is the number of compoundings in one year.

A calculator is often helpful in solving compound interest problems.

Compound interest

CALCULATOR SNAPSHOT

A businessman invests \$9,250 at 7.6% interest, to be compounded monthly. To find what the investment will be worth in 3 years, we use the compound interest formula with the following values.

$$P = \$9{,}250, \quad r = 7.6\% = 0.076, \quad t = 3 \text{ years}, \quad n = 12 \text{ times a year (monthly)}$$

We apply the compound interest formula.

$$A = P\left(1 + \frac{r}{n}\right)^{nt}$$ Write the compound interest formula.

$$= 9{,}250\left(1 + \frac{0.076}{12}\right)^{12(3)}$$ Substitute the values of P, r, t, and n.

$$= 9{,}250\left(1 + \frac{0.076}{12}\right)^{36}$$ Simplify the exponent: $12(3) = 36$.

To evaluate the expression on the right-hand side of the equation, we enter these numbers and press these keys.

9250 [×] [(] 1 [+] .076 [÷] 12 [)] [y^x] 36 [=]   11610.43875

Rounded to the nearest cent, the amount in the account after 3 years will be \$11,610.44.

If your calculator does not have parenthesis keys, calculate the sum within the parentheses first. Then find the power. Finally, multiply by 9,250.

EXAMPLE 4 A man deposited \$50,000 in a long-term account at 6.8% interest, compounded daily. How much money will he be able to withdraw in 7 years if the principal is to remain in the bank?

Solution "Compounded daily" means that compounding will be done 365 times in a year.

$$P = \$50{,}000 \qquad r = 6.8\% = 0.068 \qquad t = 7 \text{ years} \qquad n = 365 \text{ times a year}$$

$$A = P\left(1 + \frac{r}{n}\right)^{nt}$$ Write the compound interest formula.

$$= 50{,}000\left(1 + \frac{0.068}{365}\right)^{365(7)}$$ Substitute the values of P, r, t, and n.

$$= 50{,}000\left(1 + \frac{0.068}{365}\right)^{2{,}555}$$ Perform the multiplication: $365(7) = 2{,}555$.

$$\approx 80{,}477.58$$ Use a calculator. Round to the nearest cent.

The account will contain \$80,477.58 at the end of 7 years. To find the amount the man can withdraw, we subtract.

$$80{,}477.58 - 50{,}000 = 30{,}477.58$$

The man can withdraw \$30,477.58 without having to touch the \$50,000 principal.

Self Check 4
Find the amount of interest \$25,000 will earn in 10 years if it is deposited in an account at 5.99% interest, compounded daily.

Answer \$20,505.20

Section 7.4 STUDY SET

VOCABULARY *Fill in the blanks.*

1. In banking, the original amount of money borrowed or deposited is known as the ________.
2. Borrowers pay ________ to lenders for the use of their money.
3. The percent that is used to calculate the amount of interest to be paid is called the ________ rate.
4. ________ interest is interest paid on accumulated interest.
5. Interest computed on only the original principal is called ________ interest.
6. Percent means parts per ________.

CONCEPTS

7. When we do calculations with percents, they must be changed to decimals or fractions. Change each percent to a decimal.

   a. 7% **b.** 9.8% **c.** $6\frac{1}{4}\%$

8. Express each of the following as a fraction of a year. Simplify the fraction.

   a. 6 months **b.** 90 days
   c. 120 days **d.** 1 month

9. Complete the table by finding the simple interest earned.

Principal	Rate	Time	Interest earned
$10,000	6%	3 years	

10. Determine how many times a year the interest on a savings account is calculated if the interest is compounded

    a. semiannually **b.** quarterly
    c. daily **d.** monthly

11. **a.** What concept studied in this section is illustrated by the following diagram?
    b. What was the original principal?
    c. How many times was the interest found?
    d. How much interest was earned on the first compounding?
    e. For how long was the money invested?

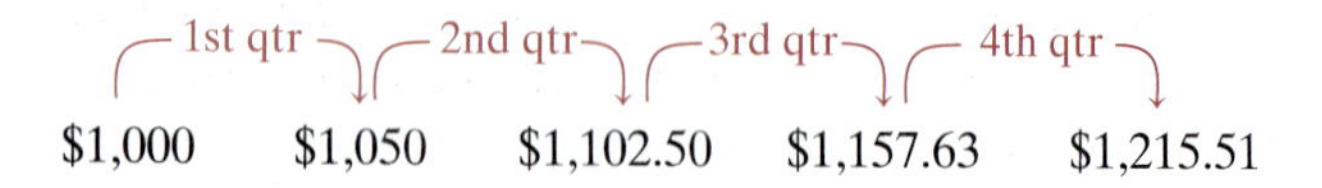

12. $3,000 is deposited in a savings account that earns 10% interest compounded annually. Complete the series of calculations shown in the illustration to find how much money will be in the account at the end of 2 years.

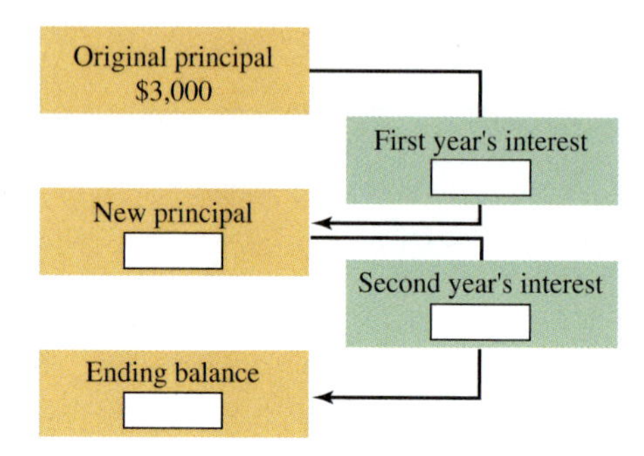

NOTATION

13. In the formula $I = Prt$, what operations are indicated by Prt?
14. In the formula $A = P\left(1 + \frac{r}{n}\right)^{nt}$, how many operations must be performed to find A?

APPLICATIONS *In the following problems, use simple interest.*

15. RETIREMENT INCOME A retiree invests $5,000 in a savings plan that pays 6% per year. What will the account balance be at the end of the first year?
16. INVESTMENTS A developer promised a return of 8% annual interest on an investment of $15,000 in her company. How much could an investor expect to make in the first year?
17. REMODELING A homeowner borrows $8,000 to pay for a kitchen remodeling project. The terms of the loan are 9.2% annual interest and repayment in 2 years. How much interest will be paid on the loan?
18. CREDIT UNIONS A farmer borrowed $7,000 from a credit union. The money was loaned at 8.8% annual interest for 18 months. How much money did the credit union charge him for the use of the money?
19. MEETING PAYROLLS In order to meet end-of-the-month payroll obligations, a small business had to borrow $4,200 for 30 days. How much did the business have to repay if the interest rate was 18%?

20. CAR LOANS To purchase a car, a man takes out a loan for \$2,000. If the interest rate is 9% per year, how much interest will he have to pay at the end of the 120-day loan period?

21. SAVINGS ACCOUNTS Find the interest earned on \$10,000 at $7\frac{1}{4}\%$ for 2 years. Use the table to organize your work.

P	*r*	*t*	*I*

22. TUITION A student borrows \$300 from an educational fund to pay for books for spring semester. If the loan is for 45 days at $3\frac{1}{2}\%$ annual interest, what will the student owe at the end of the loan period?

23. LOAN APPLICATIONS Complete the loan application below.

Loan Application Worksheet

1. Amount of loan (principal) _\$1,200.00_
2. Length of loan (time) _2 YEARS_
3. Annual percentage rate _8%_
4. Interest charged ______
5. Total amount to be repaid ______
6. Check method of repayment:
   ☐ 1 lump sum ☑ monthly payments

   Borrower agrees to pay _24_ equal payments of ______ to repay loan.

24. LOAN APPLICATIONS Complete the loan application below.

Loan Application Worksheet

1. Amount of loan (principal) _\$810.00_
2. Length of loan (time) _9 mos._
3. Annual percentage rate _12%_
4. Interest charged ______
5. Total amount to be repaid ______
6. Check method of repayment:
   ☐ 1 lump sum ☑ monthly payments

   Borrower agrees to pay _9_ equal payments of ______ to repay loan.

25. LOW-INTEREST LOANS An underdeveloped country receives a low-interest loan from a bank to finance the construction of a water treatment plant. What must the country pay back at the end of 2 years if the loan is for \$18 million at 2.3%?

26. CITY REDEVELOPMENT A city is awarded a low-interest loan to help renovate the downtown business district. The \$40-million loan, at 1.75%, must be repaid in $2\frac{1}{2}$ years. How much interest will the city have to pay?

A calculator will be helpful in solving these problems.

27. COMPOUNDING ANNUALLY If \$600 is invested in an account that earns 8%, compounded annually, what will the account balance be after 3 years?

28. COMPOUNDING SEMIANNUALLY If \$600 is invested in an account that earns annual interest of 8%, compounded semiannually, what will the account balance be at the end of 3 years?

29. COLLEGE FUNDS A ninth-grade student opens a savings account that locks her money in for 4 years at an annual rate of 6%, compounded daily. If the initial deposit is \$1,000, how much money will be in the account when she begins college in four years?

30. CERTIFICATES OF DEPOSIT A 3-year certificate of deposit pays an annual rate of 5%, compounded daily. The maximum allowable deposit is \$90,000. What is the most interest a depositor can earn from the CD?

31. TAX REFUNDS A couple deposits an income tax refund check of \$545 in an account paying an annual rate of 4.6%, compounded daily. What will the size of the account be at the end of 1 year?

32. INHERITANCES After receiving an inheritance of \$11,000, a man deposits the money in an account paying an annual rate of 7.2%, compounded daily. How much money will be in the account at the end of 1 year?

33. LOTTERIES Suppose you won \$500,000 in the lottery and deposited the money in a savings account that paid an annual rate of 6% interest, compounded daily. How much interest would you earn each year?

34. CASH GIFTS After receiving a \$250,000 cash gift, a university decides to deposit the money in an account paying an annual rate of 5.88%, compounded quarterly. How much money will the account contain in 5 years?

WRITING

35. What is the difference between simple and compound interest?

36. Explain: *Interest is the amount of money paid for the use of money.*

37. On some accounts, banks charge a penalty if the depositor withdraws the money before the end of the term. Why would a bank do this?

38. Explain why it is better for a depositor to open a savings account that pays 5% interest, compounded daily, than one that pays 5% interest, compounded monthly.

REVIEW

39. Evaluate: $\sqrt{\frac{1}{4}}$.

40. Find the power: $\left(\frac{1}{4}\right)^2$.

41. Is the point $(2, -3)$ on the line $y = 2x - 10$?

42. Subtract: $32u^3 - 22u^3$.

43. Solve: $\frac{2}{3}x = -2$.

44. Divide: $-12\frac{1}{2} \div 5$.

45. How many terms does the polynomial $2x^2 - 3x + 5$ contain?

46. Evaluate: $(0.2)^2 - (0.3)^2$.

47. In which quadrant does the point $(-2, -2)$ lie?

48. Multiply: $2x^2(3x + 2)$.

KEY CONCEPT

Equivalent Expressions

Equivalent expressions do not look the same, but they represent the same amount. One of the major objectives of this course is for you to develop a knack for recognizing situations where an equivalent expression should be written.

Write an equivalent expression for each quantity and describe the concept you applied.

1. $\frac{10}{24}$
2. $3 \cdot x \cdot x \cdot x$
3. $2x + 3x$
4. $-3 - (-8)$
5. $x^3 \cdot x^2$
6. Write 0.125 as a percent.
7. Write $\frac{2}{3}$ as a percent.
8. $\frac{4}{5} - \frac{1}{5}$
9. $4x^2 + 1 - 2x^2$
10. $\frac{\frac{1}{5}}{\frac{3}{4}}$
11. $\frac{6}{6}$
12. Write 5.1% as a decimal.
13. $(-5)(-6)$
14. $\frac{x}{1}$
15. $\frac{2x}{2}$
16. $-x + x$
17. $2 + 3 \cdot 5$
18. $a + 0$
19. $2(x + 5)$
20. $|-4|$
21. $\frac{2}{3} \cdot \frac{3}{2}$
22. $\sqrt{49}$

ACCENT ON TEAMWORK

SECTION 7.1

M & M'S Give each member of your group a bag of M & M's candies.

a. Determine what percent of the total number of M & M's in your bag are yellow. Do the same for each of the other colors. Enter the results in the table. (Round to the nearest one percent.)

M & M's color	Percent
Yellow	
Brown	
Green	
Red	
Blue	

b. Present the data in the table using the circle graph. Compare your graph to the graphs made by the other members of your group. Do the colors occur in the same percentages in each of the bags?

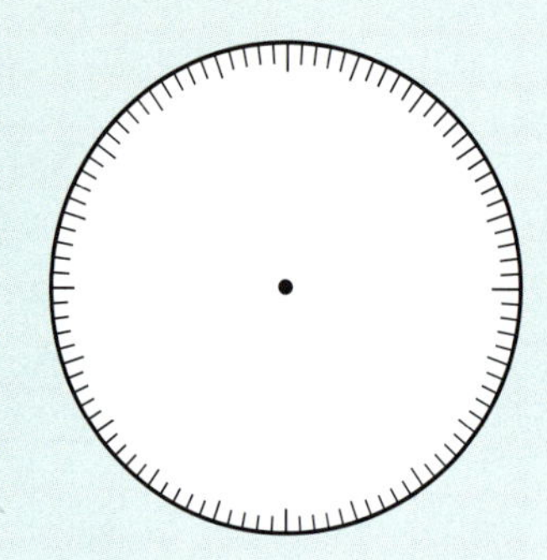

SECTION 7.2

NUTRITION Have each person in your group bring in a nutrition label like the one shown and write the name of the food product on the back. Have the members of the group exchange labels. With the label that you receive, determine what percent of the total calories come from fat.

The USDA recommends that no more than 30% of a person's daily calories should come from fat. Which products exceed the recommendation?

Nutrition Facts
Serving Size 1 meal

Amount Per Serving

Calories 560	Calories from Fat 190
	% Daily Value
Total fat 21g	**32%**
Saturated fat 9 g	**43%**
Cholesterol 60mg	**20%**
Sodium 2110mg	**88%**
Total carbohydrate 67g	**22%**
Dietary fiber 7g	**29%**
Sugars less than 25g	
Protein 27g	

SECTION 7.3

ENROLLMENTS From your school's admissions office, get the enrollment figures for the last ten years. Calculate the percent of increase (or decrease) in enrollment for each of the following periods.

- Ten years ago to the present
- Five years ago to the present
- One year ago to the present

NEWSPAPER ADS Have each person in your group find a newspaper advertisement for some item that is on sale. The ad should include only two of the four details listed below.

- The regular price
- The sale price
- The discount
- The discount rate

For example, the following ad gives the regular price and the sale price, but it doesn't give the discount or the discount rate.

Have the members of your group exchange ads. Determine the two missing details on the ad that you receive. In your group, which item had the highest discount rate?

SECTION 7.4

INTEREST RATES Recall that interest is money that the borrower pays to the lender for the use of the money. The amount of interest that the borrower must pay depends on the interest rate charged by the lender.

Have members of your group call banks, savings and loans, credit unions, and other financial services to get the lending rates for various types of loans. (See the yellow pages of the phone book.)

Find out what rate is charged by credit cards such as VISA, department stores, and gasoline companies. List the interest rates in order, from greatest to least, and present your findings to the class.

CHAPTER REVIEW

SECTION 7.1 Percents, Decimals, and Fractions

CONCEPTS

Percent means parts per one hundred.

To change a percent to a fraction, drop the % symbol and put the given number over 100.

To change a percent to a decimal, drop the % symbol and divide by 100 by moving the decimal point 2 places to the left.

To change a decimal to a percent, multiply the decimal by 100 by moving the decimal point 2 places to the right, and then insert a % symbol.

To change a fraction to a percent, write the fraction as a decimal by dividing its numerator by its denominator. Multiply the decimal by 100 by moving the decimal point 2 places to the right, and then insert a % symbol.

REVIEW EXERCISES

Express the amount of each figure that is shaded as a percent, as a decimal, and as a fraction. Each set of squares represents 100%.

1. 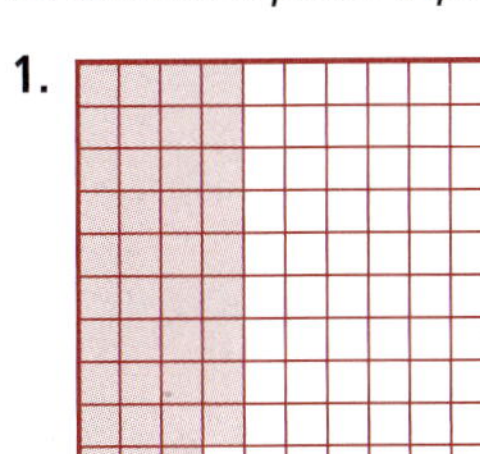

2. 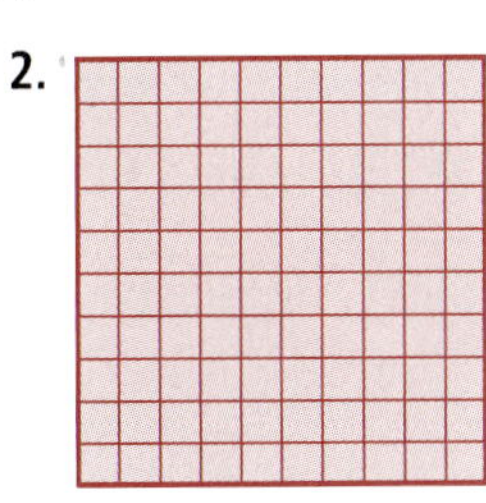 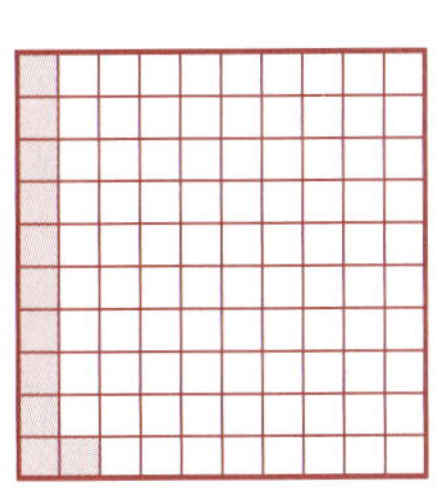

3. In Problem 1, what percent of the figure is not shaded?

Change each percent to a fraction.

4. 15% **5.** 120% **6.** $9\frac{1}{4}\%$ **7.** 0.1%

Change each percent to a decimal.

8. 27% **9.** 8% **10.** 155% **11.** $1\frac{4}{5}\%$

Change each decimal to a percent.

12. 0.83 **13.** 0.625 **14.** 0.051 **15.** 6

Change each fraction to a percent.

16. $\frac{1}{2}$ **17.** $\frac{4}{5}$ **18.** $\frac{7}{8}$ **19.** $\frac{1}{16}$

Find the exact percent equivalent for each fraction.

20. $\frac{1}{3}$ **21.** $\frac{5}{6}$

Change each fraction to a percent. Round to the nearest hundredth of a percent.

22. $\frac{5}{9}$ **23.** $\frac{8}{3}$

24. BILL OF RIGHTS There are 27 amendments to the Constitution of the United States. The first ten are known as the Bill of Rights. What percent of the amendments were adopted after the Bill of Rights? (Round to the nearest one percent.)

25. Express one-tenth of one percent as a fraction.

SECTION 7.2 *Solving Percent Problems*

The percent formula:

Amount = percent · base

We can translate a percent problem from words into an equation. A *variable* is used to stand for the unknown number; *is* can be translated to an = sign; and *of* means multiply.

26. Identify the amount, the base, and the percent in the statement "15 is $33\frac{1}{3}$% of 45."

27. Translate the given sentence into a percent equation.

What number is 32% of 96?

Solve each percent problem.

28. What number is 40% of 500?

29. 16% of what number is 20?

30. 1.4 is what percent of 80?

31. $66\frac{2}{3}$% of 3,150 is what number?

32. Find 220% of 55.

33. What is 0.05% of 60,000?

34. RACING The nitro–methane fuel mixture used to power some experimental cars is 96% nitro and 4% methane. How many gallons of each fuel component are needed to fill a 15-gallon fuel tank?

35. HOME SALES After the first day on the market, 51 homes in a new subdivision had already sold. This was 75% of the total number of homes available. How many homes were originally for sale?

36. HURRICANES In a mobile home park, 96 of the 110 trailers were either damaged or destroyed by hurricane winds. What percent is this? (Round to the nearest one percent.)

37. TIPPING The cost of dinner for a family of five at a restaurant was \$36.20. Find the amount of the tip if it should be 15% of the cost of dinner.

A *circle graph* is a way of presenting data for comparison. The sizes of the segments of the circle indicate the percents of the whole represented by each category.

38. AIR POLLUTION Complete the circle graph to show the given data.

Sources of carbon monoxide air pollution	
Transportation vehicles	63%
Fuel combustion in homes, offices, electrical plants	12%
Industrial processes	8%
Solid-waste disposal	3%
Miscellaneous	14%

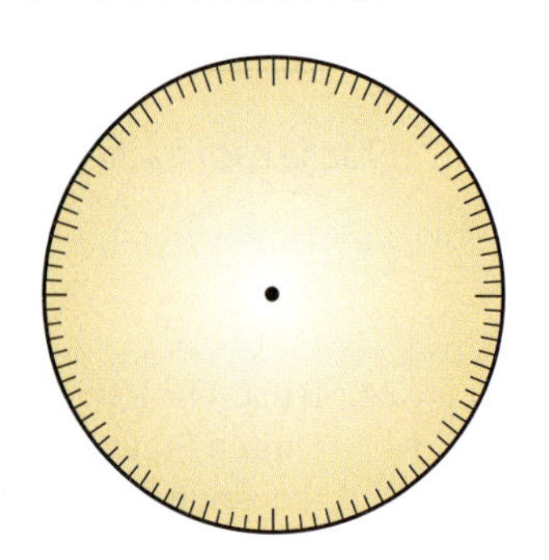

39. EARTH'S SURFACE The surface of the Earth is approximately 196,800,000 square miles. Use the graph to determine the number of square miles of the Earth's surface that are covered with water.

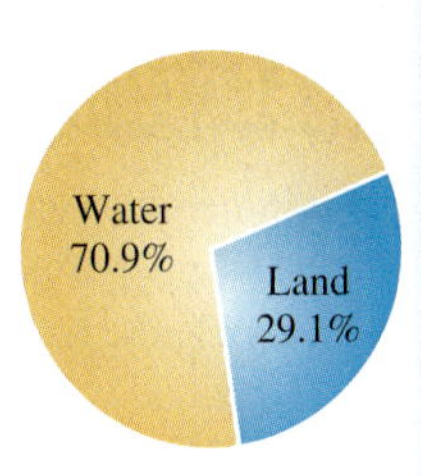

SECTION 7.3 Applications of Percent

To find the total price of an item:

Total price = purchase price + sales tax

40. SALES RECEIPT Complete the sales receipt.

CAMERA CENTER	
35mm Canon Camera	$59.99
SUBTOTAL	$59.99
SALES TAX @ 5.5%	
TOTAL	

41. TAX RATES Find the sales tax rate if the sales tax is $492 on the purchase of an automobile priced at $12,300.

Commission is based on a percent of the total dollar amount of the goods or services sold.

42. COMMISSIONS If the commission rate is 6%, find the commission earned by an appliance salesperson who sells a washing machine for $369.97 and a dryer for $299.97.

To find *percent of increase or decrease:*

1. Subtract the smaller number from the larger to find the amount of increase or decrease.
2. Find what percent the difference is of the original amount.

43. Fill in the blank: Always find the percent of increase or decrease of a quantity with respect to the ________ amount.

44. TROOP SIZE The size of a peacekeeping force was increased from 10,000 to 12,500 troops. What percent of increase is this?

45. GAS MILEAGE Experimenting with a new brand of gasoline in her truck, a woman found that the gas mileage fell from 18.8 to 17.0 miles per gallon. What percent of decrease is this? (Round to the nearest tenth of a percent.)

The difference between the original price and the sale price of an item is called the *discount.*

To find the *sale price:*

Sale price = original price − discount

46. TOOL CHESTS Use the information in the ad to find the discount, the original price, and the discount rate on the tool chest.

SECTION 7.4 *Interest*

Simple interest is interest earned on the original principal and is found using the formula

$$I = Prt$$

where P is the principal, r is the annual interest rate, and t is the length of time in years.

47. Find the interest earned on $6,000 invested at 8% per year for 2 years. Use the following chart to organize your work.

P	r	t	I

48. CODE VIOLATIONS A business was ordered to correct safety code violations in a production plant. To pay for the needed corrections, the company borrowed $10,000 at 12.5% for 90 days. Find the total amount that had to be paid after 90 days.

Compound interest is interest earned on interest.

49. MONTHLY PAYMENTS A couple borrows $1,500 for 1 year at $7\frac{3}{4}\%$ and decides to repay the loan by making 12 equal monthly payments. How much will each monthly payment be?

The compound interest formula:

$$A = P\left(1 + \frac{r}{n}\right)^{nt}$$

where A is the amount in the account, P is the principal, r is the annual interest rate, n is the number of compoundings in one year, and t is the length of time in years.

50. Find the amount of money that will be in a savings account at the end of 1 year if $2,000 is the initial deposit and the annual interest rate of 7% is compounded semiannually. (*Hint:* Find the simple interest twice.)

51. Find the amount that will be in a savings account at the end of 3 years if a deposit of $5,000 earns interest at an annual rate of $6\frac{1}{2}\%$, compounded daily.

52. CASH GRANTS Each year a cash grant is given to a deserving college student. The grant consists of the interest earned that year on a $500,000 savings account. What is the cash award for the year if the money is invested at an annual rate of 8.3%, compounded daily?

CHAPTER 7 TEST

1. Express the amount of the figure that is shaded as a percent, as a fraction, and as a decimal.

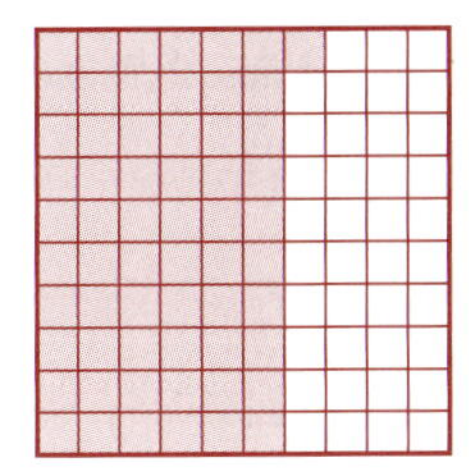

2. In the illustration, each set of 100 squares represents 100%. Express as a percent the amount of the figure that is shaded. Then express that percent as a fraction and as a decimal.

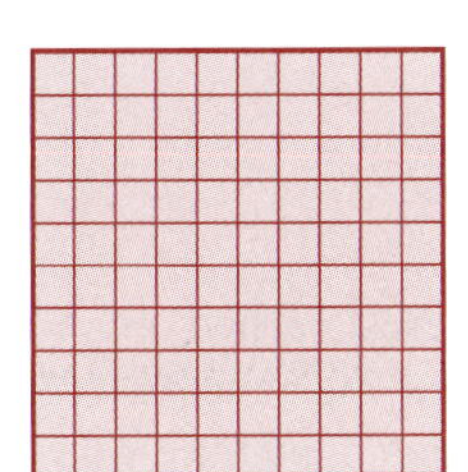 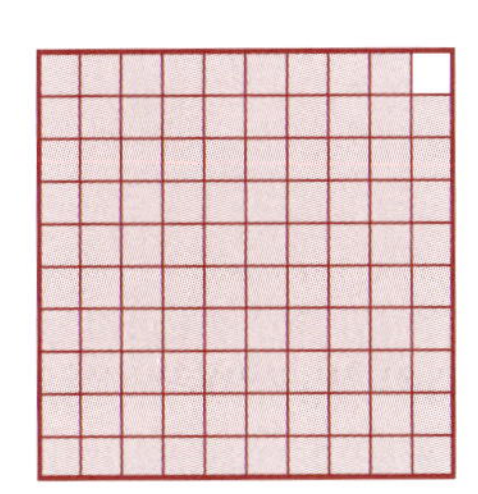

3. Change each percent to a decimal.

   a. 67% b. 12.3% c. $9\frac{3}{4}\%$

4. Change each fraction to a percent.

   a. $\frac{1}{4}$ b. $\frac{5}{8}$ c. $\frac{3}{25}$

5. Change each decimal to a percent.

   a. 0.19 b. 3.47 c. 0.005

6. Change each percent to a fraction.

   a. 55% b. 0.01% c. 125%

7. Change $\frac{7}{30}$ to a percent. Round to the nearest hundredth of a percent.

8. WEATHER REPORTS A weather reporter states that there is a 40% chance of rain. What are the chances that it will not rain?

9. Find the exact percent equivalent for the fraction $\frac{2}{3}$.

10. Find the exact percent equivalent for the fraction $\frac{1}{4}$.

11. SHRINKAGE Refer to the label on a pair of jeans.

    a. How much length will be lost due to shrinkage?

    b. What will be the resulting length?

WAIST	INSEAM
33	34

Expect shrinkage of approximately **3%** in length after the jeans are washed.

12. 65 is what percent of 1,000?

13. TIPPING Find the amount of a 15% tip on a meal costing $25.40.

14. FUGITIVES As of October 2004, 450 of the 479 fugitives who have appeared on the FBI's Ten Most Wanted list have been apprehended. What percent is this? Round to the nearest tenth of a percent.

15. SWIMMING WORKOUTS A swimmer was able to complete 18 laps before a shoulder injury forced him to stop. This was only 20% of a typical workout. How many laps does he normally complete during a workout?

16. COLLEGE EMPLOYEES The 700 employees at a community college fall into three major categories, as shown. How many employees are in administration?

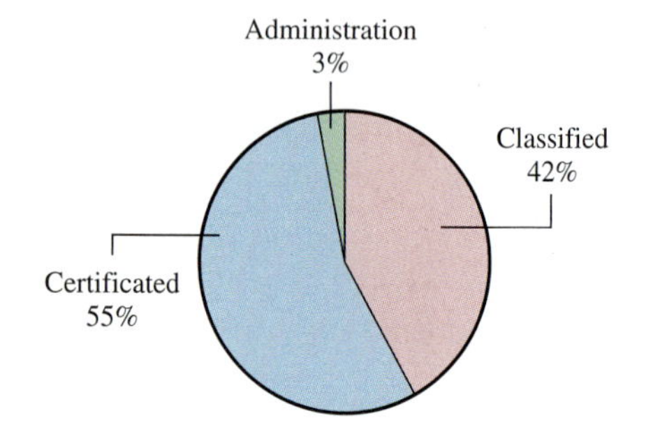

17. Find 24% of 600.

18. HAIRCUTS The illustration shows the number of minutes it took the average U.S. worker to earn enough to pay for a man's haircut in 1950 and 1998. Find the percent of decrease, to the nearest one percent.

19. INSURANCE An insurance salesperson receives a 4% commission on the annual premium of any policy she sells. Find her commission on a homeowner's policy if the premium is $898.

20. COST-OF-LIVING INCREASES A teacher earning $40,000 just received a cost-of-living increase of 3.6%. What is the teacher's new salary?

21. CAR WAX SALE A car waxing kit, regularly priced at $14.95, is on sale for $3 off. What are the sale price, the discount, and the rate?

22. POPULATION INCREASES After a new freeway was completed, the population of a city it passed through increased from 12,808 to 15,565 in two years. What percent of increase is this? (Round to the nearest one percent.)

23. Find the simple interest on a loan of $3,000 at 5% per year for 1 year.

24. Find the amount of interest earned on an investment of $24,000 paying an annual rate of 6.4% interest, compounded daily for 3 years.

25. POLITICAL ADS Explain what is unclear about the ad shown below.

26. In Section 7.4, we discussed *interest.* What is interest?

CHAPTERS 1–7 CUMULATIVE REVIEW EXERCISES

1. SHAQUILLE O'NEAL Use the data in the table to complete the line graph that shows the growth of Shaquille O'Neal, the Los Angeles Lakers' center.

Age (yr)	4	6	8	10	12	16	21	28
Weight (lb)	56	82	108	139	192	265	302	315

Based on data from *Los Angeles Times*

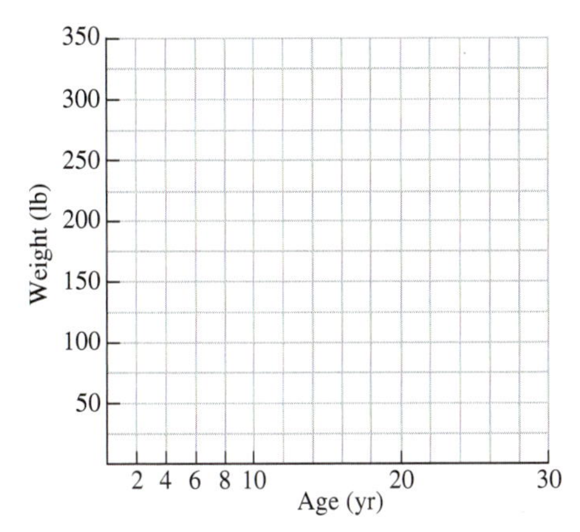

2. State the commutative property of multiplication.

3. **a.** Find the factors of 40.

   b. Find the prime factorization of 40.

4. AUTO INSURANCE See the premium comparison below. What is the average (mean) six-month insurance premium for the companies listed?

Allstate	**\$2,672**	**Mercury**	**\$1,370**
Auto Club	**\$1,680**	**State Farm**	**\$2,737**
Farmers	**\$2,485**	**20th Century**	**\$1,692**

Criteria: Six-month premium. Husband, 45, drives a 1995 Explorer, 12,000 annual miles. Wife, 43, drives a 1996 Dodge Caravan, 12,000 annual miles. Son, 17, is an occasional operator. All have clean driving records.

5. PAINTING A square tarp has sides 8 feet long. When it is laid out on a floor, how much area will it cover?

6. Evaluate: $-12 - (-5)$.

7. Evaluate: $12 - 2[-8 - 2^4(-1)]$.

8. Evaluate: $|-55|$.

9. Solve: $6 = 2 - 2x$.

10. Translate into mathematical symbols: 16 less than twice the total t.

11. FRUIT STORAGE Use the formula $C = \frac{5(F - 32)}{9}$ to complete the label on the box of bananas shown.

12. Solve: $-(3x - 3) = 6(2x - 7)$.

13. SPELLING What fraction of the letters in the word *Mississippi* are vowels?

14. Simplify: $\frac{10y}{15y}$.

Perform the operations.

15. $-\frac{16a}{35} \cdot \frac{25}{48a^2}$

16. $4\frac{2}{5} \div 11$

17. $\frac{4}{m} + \frac{2}{7}$

18. $34\frac{1}{9} - 13\frac{5}{6}$

19. Solve: $\frac{5}{6}y = -25$.

20. Solve: $\frac{y}{6} = \frac{y}{12} + \frac{2}{3}$.

Perform the operations.

21. $78.1 - 7.81$

22. $2.13(-4.05)$

23. $0.752(1{,}000)$

24. $\frac{241.86}{2.9}$

25. Evaluate $\frac{a - b}{0.5b - 0.4a}$ for $a = 3.6$ and $b = -1.5$. Round to the nearest hundredth.

26. Round 452.0298 to the nearest thousandth.

27. Write $\frac{11}{15}$ as a decimal. Use an overbar.

28. Solve: $\frac{y}{2.22} = -5$.

29. Evaluate: $3\sqrt{81} - 8\sqrt{49}$.

30. LABOR COSTS On the repair bill shown, one line cannot be read. How many hours of labor did it take to repair the car?

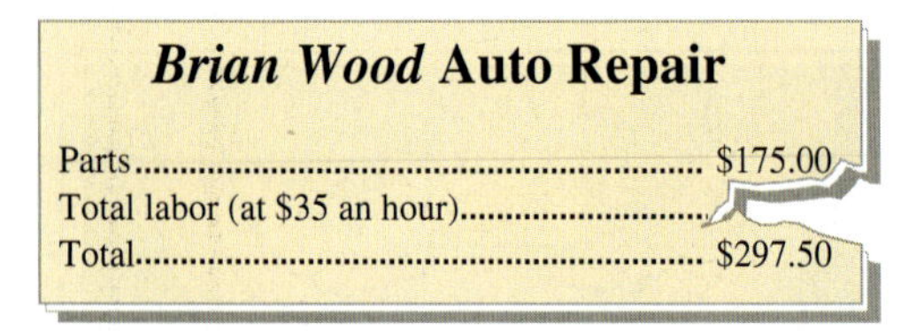
***Brian Wood* Auto Repair**

Parts $175.00
Total labor (at $35 an hour) ....
Total $297.50

Perform the operations.

31. $(m^2 - m - 5) - (3m^2 + 2m - 8)$

32. $(3x - 2)(x + 4)$

33. $(2y - 5)^2$

Simplify each expression.

34. $y^2 \cdot y^5$

35. $(h^5)^4$

36. $(2a^3b^6)^3$

37. $-7g^5(8g^4)$

38. Graph the points: $(-1, 3)$, $(0, 1.5)$, $(-4, -4)$, $(2, \frac{7}{2})$, and $(4, 0)$.

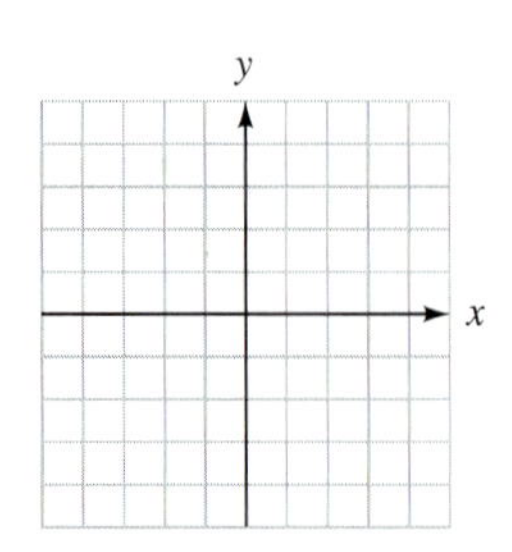

39. Graph: $3x - 3y = 9$.

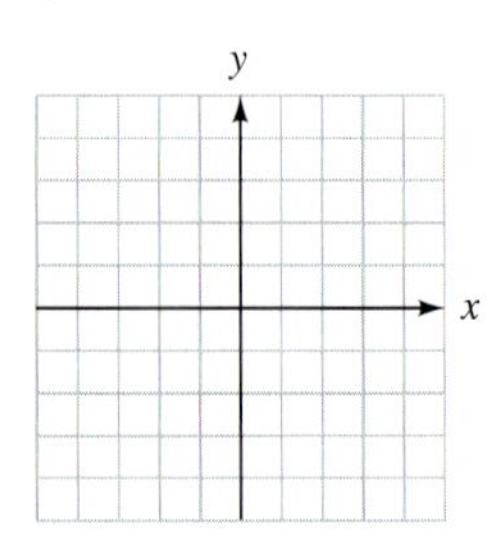

40. Graph: $y = -x - 1$.

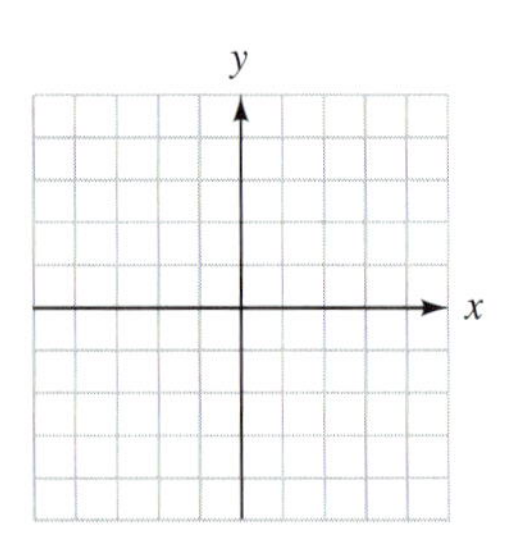

41. Complete the table.

Percent	Decimal	Fraction
	0.29	
47.3%		
		$\frac{7}{8}$

42. 16% of what number is 20?

43. TIPPING Complete the sales draft below if a 15% tip, rounded up to the next dollar, is to be left for the waiter.

STEAK STAMPEDE
Bloomington, MN
Server #12\ AT

VISA	67463777288
NAME	DALTON/ LIZ
AMOUNT	$75.18
GRATUITY $	____________
TOTAL $	____________

44. GENEALOGY Through an extensive computer search, a genealogist determined that worldwide, 180 out of every 10 million people had his last name. What percent is this?

45. SAVINGS ACCOUNT Find the simple interest earned on $10,000 at $7\frac{1}{4}\%$ for 2 years.

CHAPTER 8

Ratio, Proportion, and Measurement

CORBIS

TLE We are all familiar with the basic American units of measurement: feet, pounds, and gallons. However, most of the countries in the world use a different system of measurement called the *metric system*. In the metric system, the basic unit of length is the meter, the basic unit of weight is the gram, and the basic unit of capacity is the liter.

The metric system was invented by French scientists in the late 18th century. Until then, units of measure differed from country to country and even village to village. The objective was to create worldwide standard units of measurement based on the decimal system rather than fractions.

To learn more about the metric system, visit The Learning Equation on the Internet at http://tle.brookscole.com. (The log-in instructions are in the Preface.) For Chapter 8, the online lesson is:

- *TLE* Lesson 14: Measurement

Check Your Knowledge

1. A ______ is a quotient of two numbers or a quotient of two quantities with the same units.
2. A __________ is a statement that two ratios or rates are equal.
3. Inches, feet, and miles are examples of American units of ________. Meters, grams, and liters are units of measurement in the ________ system.
4. In the American system, temperatures are measured in degrees __________. In the metric system, temperatures are measured in degrees ________.

Reduce ratios to lowest terms.

5. What is the ratio of 14 pounds to 10 pounds?
6. What is the ratio of 8 inches to 3 feet?
7. David drives a hybrid gas–electric automobile. Recently he drove 517 miles using 11 gallons of gasoline. How many miles per gallon (mpg) did he get?
8. Specialty coffee is for sale in 12-ounce bags for \$8.95 and in 2-pound bags for \$23.95. Which is the better buy?
9. Chelsea threw 20 darts and hit the target 16 times. Find her ratio of hits to misses?
10. Determine whether each statement is a proportion.

    a. $\frac{33}{44} = \frac{39}{52}$ b. $\frac{3.5}{2.7} = \frac{10.5}{8.3}$
11. Are the numbers 17, 27 and 125, 199 proportional?

Solve each proportion.

12. $\frac{12}{x} = \frac{3}{17}$
13. $-\frac{3}{8} = \frac{15}{x}$
14. $\frac{13.2}{4} = \frac{x}{7.8}$
15. $\frac{x}{0.04} = \frac{-0.35}{0.07}$
16. Convert 5 meters to centimeters.
17. How many yards are in 29 feet?
18. How many feet are in 5 miles?
19. How many inches are in 7 yards?
20. Convert 100 meters to yards (1 m $\approx$ 1.0936 yd). Round to one decimal place.
21. Convert 5 feet 9 inches to centimeters (1 in. $\approx$ 2.54 cm).
22. To estimate the size of the coyote population in a state park, rangers trapped, tagged, and released 50 coyotes. Later, a random sample of 100 coyotes included only 2 tagged animals. Estimate the size of the coyote population.
23. One-half inch on a map corresponds to 25 miles. Two cities are 175 miles apart. How many inches separate the two cities on the map?
24. Convert −40° Fahrenheit to Celsius. $\left(\textit{Hint: } C = \frac{5}{9}(F - 32)\right)$

Study Skills Workshop

GETTING READY FOR THE FINAL EXAM

Final exams are stressful for many students; most math teachers give comprehensive finals (covering material from the beginning to the end of the course), and the sheer number of topics to study may be overwhelming. Relax! If you have been following the guidelines presented in these workshops, a good deal of your work is already done. Preparing for a final exam is much like preparing for other tests, but it may require a little more time. Planning for this at least a week in advance will give you plenty of time and will reduce the amount of anxiety you may feel.

Study Sheets. In the chapter 4 workshop, you learned to prepare study sheets (or index cards, or audiotapes) for each test. You can use these to begin your final exam preparation. Review each study sheet and make a note of only those things that you did not remember. Try to reduce all of your study materials to a list that will fit on one sheet of paper. This will be your final exam study sheet. Carry it with you all week, and review it whenever you have the opportunity.

Returned Tests. Go over all of your returned tests and correct any mistakes you made. Note any topic you may have omitted from your final exam study sheet and add items if necessary. Make a list of any test problems that you still are unsure about.

Practice Final. Just like the practice tests you made, find problems for which you can find a solution (examples in your text, odd homework problems, and examples from your notes are all good sources of these types of problems). Make sure you get a good representation of problems from all of the chapters you studied. Make a practice final that has approximately the same number of questions as will be on your real final exam.

Help from Others. It's a good idea to compare notes with other students in your class; you might even swap practice finals. If you have problems from old tests that you can't answer, see either a tutor or your instructor during office hours.

Final Exam Day. Make sure you have all of the materials you will need for your final exam. Be sure you are rested, and plan to arrive early. Make sure you know the correct time for your final exam, because finals are often given at times that are different from your class meeting time. Always work on problems that you are sure of first and save the most difficult ones for last.

ASSIGNMENT

1. Make a calendar a week before your final for studying for this exam and all other exams that you will have to take. Plan to study at least two hours each day.
2. Prepare a study sheet for your final exam.
3. Make a practice final. Find at least one other student in your class with whom you can swap practice tests. At least two days before your final, take a practice final under conditions that are as close as possible to those you will have in your real exam (closed book, timed, etc.).
4. See your tutor or instructor with any unanswered questions at least one day before the final.
5. The night before your final, make sure you have all of the materials that you will need. Get a good night's sleep.
6. On the day of the final, go over problems that you feel confident about. Do not try to learn new material on the day of the final if you feel stressed.

Rates

When we compare two different kinds of quantities, we call the comparison a **rate,** and we can write it as a fraction. For example, on the label of the can of paint in Figure 8-2, we see that one quart of paint is needed for every 200 square feet to be painted. Writing this as a rate, we have

$$\frac{1 \text{ quart}}{200 \text{ square feet}}$$ Read as "1 quart per 200 square feet."

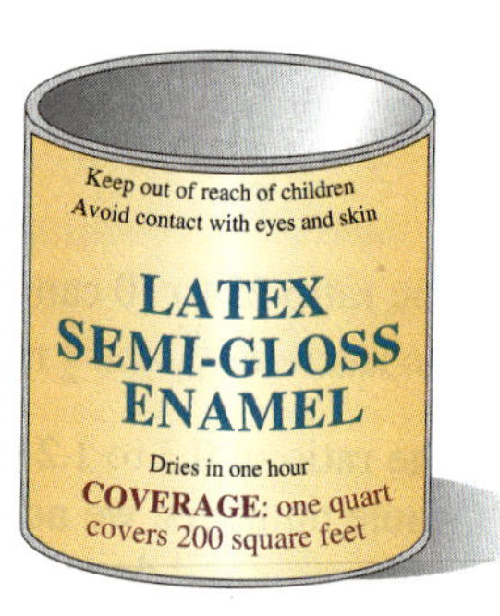

FIGURE 8-2

When writing a rate, always include the units.

Rates

A **rate** is a quotient of two quantities with different units.

Self Check 4
The fastest-growing flowering plant on record grew 12 feet in 14 days. What was its rate of growth over this period?

Answer $\frac{6 \text{ feet}}{7 \text{ days}}$

EXAMPLE 4 **Snowfall.** According to the *Guinness Book of World Records,* a total of 78 inches of snow fell at Mile 47 Camp, Cooper River Division, Arkansas, in a 24-hour period in 1963. What was the rate of snowfall?

Solution We begin by comparing the amount of snow, 78 inches, to the elapsed time, 24 hours. Then we simplify the fraction.

$$\frac{78 \text{ inches}}{24 \text{ hours}} = \frac{\overset{1}{\cancel{6}} \cdot 13 \text{ inches}}{4 \cdot \underset{1}{\cancel{6}} \text{ hours}}$$ Factor 78 and 24. Then divide out the common factor 6.

The snow fell at a rate of 13 inches per 4 hours: $\frac{13 \text{ inches}}{4 \text{ hours}}$.

THINK IT THROUGH **Student-to-Instructor Ratio**

"A more personal classroom atmosphere can sometimes be an easier adjustment for college freshmen. They are less likely to feel like a number, a feeling that can sometimes impact students' first semester grades."
From *The Importance of Class* Size by Stephen Pemberton

The data below come from a nationwide study of mathematics programs at two-year colleges. Determine which course has the lowest student-to-instructor ratio. (Assume that there is one instructor per section.)

	Prealgebra	Elementary Algebra	Intermediate Algebra
Students enrolled	81,903	268,152	243,828
Number of sections	3,561	11,173	9,378

Source: Conference Board of the Mathematical Science, 2000 CBMS Survey of Undergraduate Programs

! COMMENT Unlike in the fraction $\frac{a}{b}$, b can be zero in a rate. For example, the rate of women to men on the 1999 U.S. Women's World Cup soccer team is expressed as $\frac{20 \text{ women}}{0 \text{ men}}$. Such applications are rare, however.

Unit rates

A **unit rate** is a rate in which the denominator is 1. To illustrate the concept of a unit rate, suppose a driver makes the 354-mile trip from Pittsburgh to Indianapolis in 6 hours. Then the motorist's rate (or more specifically, rate of speed) is given by

$$\frac{354 \text{ miles}}{6 \text{ hours}} = \frac{\overset{1}{\cancel{6}} \cdot 59 \text{ miles}}{\underset{1}{\cancel{6}} \cdot 1 \text{ hours}} = \frac{59 \text{ miles}}{1 \text{ hour}}$$ Factor 354 as 6 · 59 and divide out the common factor of 6.

We can also find the unit rate by dividing 354 by 6.

```
   59
6)354
  30
   54
   54
    0
```

The unit rate $\frac{59 \text{ miles}}{1 \text{ hour}}$ can be expressed in any of the following forms:

$$59 \frac{\text{miles}}{\text{hour}}, \quad 59 \text{ miles per hour}, \quad 59 \text{ miles/hour}, \quad \text{or} \quad 59 \text{ mph}$$

EXAMPLE 5 A student earns \$152 for working 16 hours in a bookstore. Find his hourly rate of pay.

Solution We can write the rate of pay as

$$\text{Rate of pay} = \frac{\$152}{16 \text{ hr}}$$ Compare the amount of money earned to the number of hours worked.

To find the rate of pay for 1 hour of work, we divide 152 by 16.

```
     9.5
16)152.0
   144
     8 0     Write a decimal point and a zero to the right of 2.
     8 0
       0
```

The unit rate of pay is $\frac{\$9.50}{1 \text{ hour}}$, which can be written as \$9.50 per hour.

Self Check 5
Joan earns \$436 per 40-hour week managing a dress shop. Find her hourly rate of pay.

Answer \$10.90 per hour

EXAMPLE 6 **Energy consumption.** One household used 795 kilowatt-hours (kwh) of electricity during a 30-day period. Find the rate of energy consumption in kilowatt-hours per day.

Self Check 6
To heat a house for 30 days, a furnace burned 69 therms of natural gas. Find the rate of gas consumption in therms per day.

Answer 2.3 therms per day

Solution We can write the rate of energy consumption as

$$\text{Rate of energy consumption} = \frac{795 \text{ kwh}}{30 \text{ days}}$$

To find the unit rate, we divide 795 by 30.

$$\text{Unit rate of energy consumption} = \frac{26.5 \text{ kwh}}{1 \text{ day}}$$

The rate of energy consumption was 26.5 kilowatt-hours per day.

CALCULATOR SNAPSHOT

Computing gas mileage

A man drove from Houston to St. Louis — a total of 775 miles. Along the way, he stopped for gas three times, pumping 10.5, 11.3, and 8.75 gallons of gas. He started with the tank half full and ended with the tank half full. To find how many miles he got per gallon (mpg), we need to compare the total distance to the total number of gallons of gas consumed.

$$\frac{775 \text{ miles}}{(10.5 + 11.3 + 8.75) \text{ gallons}}$$

We can simplify this rate by entering these numbers and pressing these keys on a scientific calculator.

775 [÷] [(] 10.5 [+] 11.3 [+] 8.75 [)] [=]   25.368249

To the nearest hundredth, he got 25.37 mpg.

Unit costs

If a store sells 5 pounds of coffee for $18.75, a consumer might want to know what the coffee costs per pound. When we find the cost of 1 pound of the coffee, we are finding a *unit cost.* To find the unit cost of an item, we begin by comparing its cost to its quantity.

$$\frac{\$18.75}{5 \text{ pounds}}$$

Then we divide the cost by the number of items.

$$5\overline{)18.75} = 3.75$$

The unit cost of the coffee is $3.75 per pound.

Self Check 7
A fast-food restaurant sells a 12-ounce cola for 72¢ and a 16-ounce cola for 99¢. Which is the better buy?

EXAMPLE 7 **Comparison shopping.** Olives come packaged in a 10-ounce jar, which sells for $2.49, or in a 6-ounce jar, which sells for $1.53. (See Figure 8-3.) Which is the better buy?

Solution To find the better buy, we must find each unit cost.

FIGURE 8-3

10-oz jar:

$$\frac{\$2.49}{10 \text{ oz}} = \frac{249¢}{10 \text{ oz}}$$ Change \$2.49 to 249 cents.

$$= 24.9¢ \text{ per oz}$$ Divide 249 by 10.

6-oz jar:

$$\frac{\$1.53}{6 \text{ oz}} = \frac{153¢}{6 \text{ oz}}$$ Change \$1.53 to 153 cents.

$$= 25.5¢ \text{ per oz}$$ Perform the division.

One ounce for 24.9¢ is a better buy than one ounce for 25.5¢. The unit cost is lower when olives are packaged in 10-ounce jars, so that is the better buy.

Answer the 12-oz cola

Section 8.1 STUDY SET

VOCABULARY *Fill in the blanks.*

1. A ______ is a quotient of two numbers or a quotient of two quantities with the same units.
2. A quotient of two quantities with different units is called a ______.
3. When the price of candy is advertised as \$1.75 per pound, we are told its unit ______.
4. A ______ rate is a rate in which the denominator is 1.

CONCEPTS

5. To write the ratio $\frac{15}{24}$ in lowest terms, we divide out any common factors of the numerator and denominator. What common factor do they have?
6. Complete the solution. Write the ratio $\frac{14}{21}$ in lowest terms.

$$\frac{14}{21} = \frac{2 \cdot 7}{\square \cdot \square} = \frac{2 \cdot \overset{1}{\cancel{7}}}{\square \cdot \underset{1}{\cancel{7}}} = \frac{\square}{\square}$$

7. Consider the ratio $\frac{0.5}{0.6}$. By what number should we multiply numerator and denominator to make this a ratio of whole numbers?
8. What should be done to write the ratio $\frac{15 \text{ inches}}{22 \text{ inches}}$ in simplest form?
9. Since a ratio is a comparison of quantities with the same units, how should the ratio $\frac{11 \text{ minutes}}{1 \text{ hour}}$ be rewritten?
10. a. Consider the rate $\frac{\$248}{16 \text{ hours}}$. How can we find the unit rate (\$ per hour)?

    b. Consider the rate $\frac{\$7.95}{3 \text{ pairs}}$. How can we find the unit cost of a pair of socks?

NOTATION

11. Refer to the illustration. Write the ratio of the flag's length to its width using a fraction, using the word *to*, and using a colon.

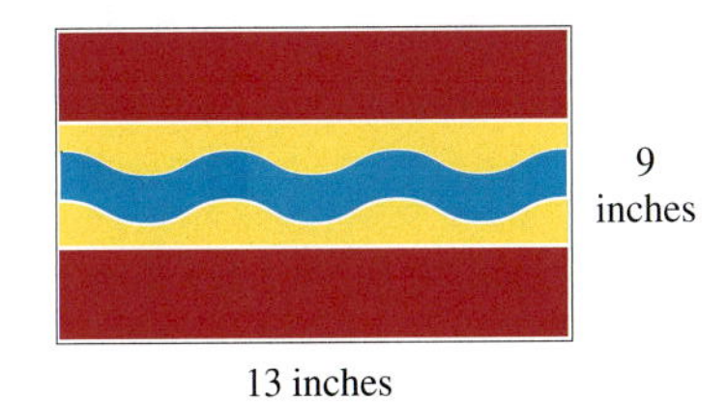

12. The rate $\frac{55 \text{ miles}}{1 \text{ hour}}$ can be expressed as
    - 55 ______ ______ ______ in three words
    - 55 ______ ______ in two words
    - 55 ___ ___ ___ in three letters

PRACTICE *Write each ratio as a fraction in simplest form.*

13. 5 to 7
14. 3 to 5
15. 17 to 34
16. 19 to 38
17. 22:33
18. 14:21
19. 1.5:2.4
20. 0.9:0.6
21. 7 to 24.5
22. 0.65 to 0.15
23. 4 ounces to 12 ounces
24. 3 inches to 15 inches
25. 12 minutes to 1 hour
26. 8 ounces to 1 pound

27. 3 days to 1 week

28. 4 inches to 2 yards

29. 18 months to 2 years

30. 8 feet to 4 yards

Refer to the monthly budget shown below. Give each ratio in lowest terms.

31. Find the total amount of the budget.

32. Find the ratio of the amount budgeted for rent to the total budget.

33. Find the ratio of the amount budgeted for food to the total budget.

34. Find the ratio of the amount budgeted for the phone to the total budget.

Item	Amount
Rent	$800
Food	$600
Gas and electric	$180
Phone	$100
Entertainment	$120

Refer to the list of tax deductions shown below. Give each ratio in lowest terms.

35. Find the total amount of the deductions.

36. Find the ratio of the real estate tax deduction to the total deductions.

37. Find the ratio of the charitable contributions to the total deductions.

38. Find the ratio of the mortgage interest deduction to the union dues deduction.

Item	Amount
Medical expenses	$875
Real estate taxes	$1,250
Charitable contributions	$1,750
Mortgage interest	$4,375
Union dues	$500

Write each rate as a fraction in simplest form.

39. 64 feet in 6 seconds

40. 45 applications for 18 openings

41. 84 made out of 100 attempts

42. 75 days on 20 gallons of water

43. 3,000 students over a 16-year career

44. 16 right compared to 34 wrong

45. 18 beats every 12 measures

46. 1.5 inches as a result of 30 turns

Write each phrase as a unit rate.

47. 60 revolutions in 5 minutes

48. 14 trips every 2 months

49. 12 errors in 8 hours

50. $50,000 paid over 10 years

51. 245 presents for 35 children

52. 108 occurrences in a 12-month period

53. 4,000,000 people living in 12,500 square miles

54. 117.6 pounds of pressure on 8 square inches

Find each unit cost.

55. $3.50 for 50 feet

56. 150 barrels cost $4,950.

57. 65 ounces sell for 78 cents.

58. They charged $48 for 15 minutes.

59. Four of us donated a total of $272.

60. For 7 dozen, you will pay $10.15.

61. $4 billion over a 5-month span

62. 7,020 pesos will buy six stickers.

APPLICATIONS

63. ART HISTORY Leonardo da Vinci drew the human figure shown within a square. (All four sides of a square are the same length.) What is the ratio of the length of the man's outstretched arms to his height?

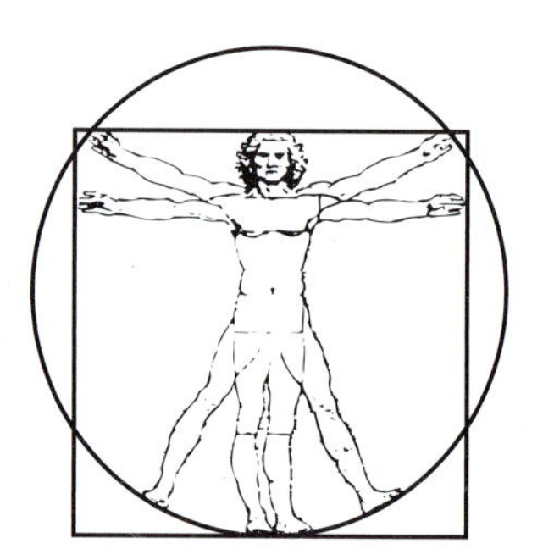

64. FLAGS The flag shown is composed of squares. (All four sides of a square are the same length.) Find the ratio of the width of the flag to its length.

65. GEAR RATIOS Refer to the illustration. Find the ratio of the number of teeth of the larger gear to the number of teeth of the smaller gear.

66. BANKRUPTCIES After declaring bankruptcy, a company could reimburse its creditors only 5¢ on the dollar. Write this as a ratio in lowest terms.

67. COOKING A recipe from *Easy Living* magazine is shown. Write the ratio of sugar to milk as a fraction. **Do not simplify the ratio.**

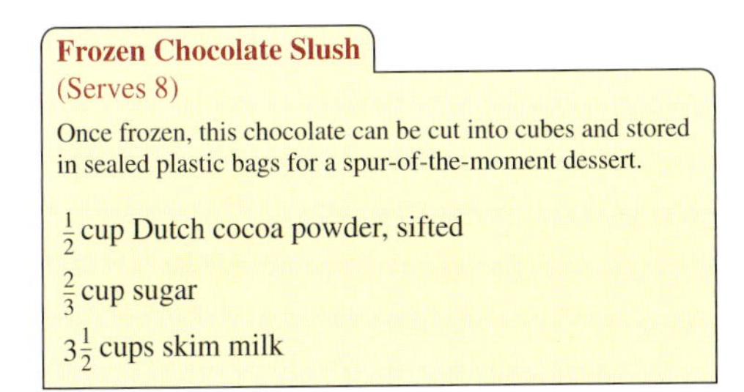

Frozen Chocolate Slush
(Serves 8)

Once frozen, this chocolate can be cut into cubes and stored in sealed plastic bags for a spur-of-the-moment dessert.

$\frac{1}{2}$ cup Dutch cocoa powder, sifted

$\frac{2}{3}$ cup sugar

$3\frac{1}{2}$ cups skim milk

68. HEARING From the graph, determine the ratio of hearing loss in males as compared to females.

Hearing loss, by gender

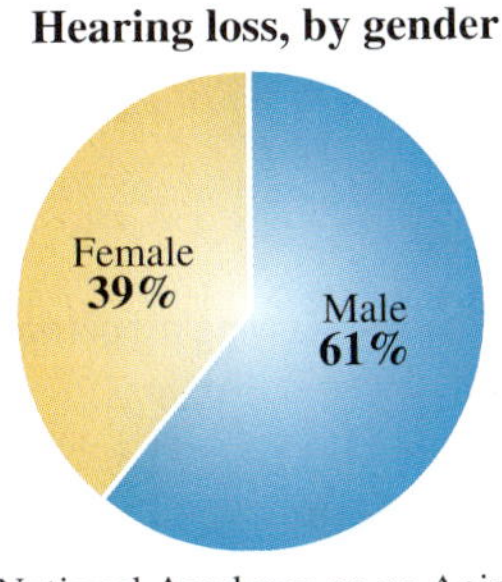

Source: National Academy on an Aging Society

69. SOFTBALL Lisa Fernandez led the U.S. women's softball team in winning a gold medal at the 2004 Olympic games. Her hitting statistics are shown below. What was her rate of hits (H) to at-bats (AB) during the Olympic competition?

	BA	AB	H	R	2B	3B	HR	RBI
Fernandez	.545	22	12	3	3	0	1	8

70. TYPING A secretary typed a document containing 330 words in 5 minutes. How many words per minute did he type?

71. CPR A paramedic performed 125 compressions to 50 breaths on an adult with no pulse. What compressions-to-breaths rate did the paramedic use?

72. INTERNET SALES A web site determined that it had 112,500 hits in one month. Of those visiting the site, 4,500 made purchases. How many did not make a purchase? What was the browser/buyers unit rate for the web site that month?

73. AIRLINE COMPLAINTS An airline had 3.29 complaints for every 1,000 passengers. Write this rate as a fraction of whole numbers.

74. FINGERNAILS On average, fingernails grow 0.02 inch per week. Write this rate using whole numbers.

75. FACULTY-STUDENT RATIOS At a college, there are 125 faculty members and 2,000 students. Find the rate of faculty to students. (This is often referred to as the faculty-to-student *ratio,* even though the units are different.)

76. PARKING METERS A parking meter requires 25¢ for 20 minutes of parking. What is the unit cost?

77. UNIT COSTS A driver pumped 17 gallons of gasoline into his tank at a cost of $32.13. Find the unit cost of the gasoline.

78. UNIT COSTS A 50-pound bag of grass seed costs $222.50. Find the unit cost of grass seed.

79. UNIT COSTS A 12-ounce can of cranberry juice sells for 84¢. Give the unit cost in cents per ounce.

80. UNIT COSTS A 24-ounce package of green beans sells for $1.29. Give the unit cost in cents per ounce.

81. COMPARISON SHOPPING A 6-ounce can of orange juice sells for 89¢, and an 8-ounce can sells for $1.19. Which is the better buy?

82. COMPARISON SHOPPING A 30-pound bag of fertilizer costs $12.25, and an 80-pound bag costs $30.25. Which is the better buy?

83. COMPARISON SHOPPING A certain brand of cold and sinus medication is sold in 20-tablet boxes for \$4.29 and in 50-tablet boxes for \$9.59. Which is the better buy?

84. COMPARISON SHOPPING Which tire shown is the better buy?

85. COMPARING SPEEDS A car travels 345 miles in 6 hours, and a truck travels 376 miles in 6.2 hours. Which vehicle is going faster?

86. READING SPEEDS One seventh-grader read a 54-page book in 40 minutes. Another read an 80-page book in 62 minutes. If the books were equally difficult, which student read faster?

87. EMPTYING TANKS An 11,880-gallon tank can be emptied in 27 minutes. Find the rate of flow in gallons per minute.

88. RATE OF PAY Ricardo worked for 27 hours to help insulate a hockey arena. For his work, he received \$337.50. Find his hourly rate of pay.

89. AUTO TRAVEL A car's odometer reads 34,746 at the beginning of a trip. Five hours later, it reads 35,071. How far has the car traveled? What is the average rate of speed?

90. RATES OF SPEED An airplane travels from Chicago to San Francisco, a distance of 1,883 miles, in 3.5 hours. Find the average rate of speed of the plane.

91. GAS MILEAGE One car went 1,235 miles on 51.3 gallons of gasoline, and another went 1,456 miles on 55.78 gallons. Which car got the better gas mileage?

92. ELECTRICITY RATES In one community, a bill for 575 kilowatt-hours of electricity is \$38.81. In a second community, a bill for 831 kwh is \$58.10. In which community is electricity cheaper?

WRITING

93. Are the ratios 3 to 1 and 1 to 3 the same? Explain why or why not.

94. Give three examples of ratios (or rates) that you have encountered in the past week.

95. How will the topics studied in this section make you a better shopper?

96. What is a unit rate? Give some examples.

REVIEW *Perform each operation.*

97. $3.05 + 17.17 + 25.317$

98. $3.5\overline{)157.85}$

99. $13.2 + 25.07 \cdot 7.16$

100. $\frac{4}{3} - \frac{1}{4}$

101. $5 - 3\frac{1}{4}$

102. Complete the table of solutions for $3x - 5y = 15$ and graph the equation.

x	y
0	
	0
3	

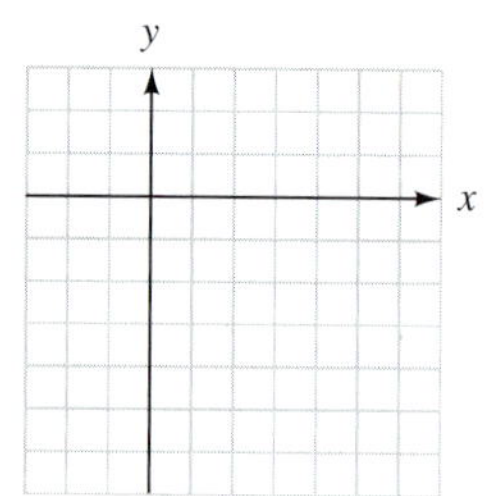

8.2 Proportions

- Proportions
- Means and extremes of a proportion
- Solving proportions
- Writing proportions to solve problems

A ladder can be dangerous if used improperly. A safety pamphlet states, "When setting up an extension ladder, use the *4-to-1 rule*—For every 4 feet of ladder height, position the legs of the ladder 1 foot away from the base of the wall." The 4-to-1 rule for ladders can be expressed using a ratio.

$$\frac{4 \text{ feet}}{1 \text{ foot}} = \frac{4 \cancel{\text{feet}}}{1 \cancel{\text{foot}}} = \frac{4}{1}$$

In Figure 8-4, the 4-to-1 rule was used to position the legs of a ladder properly, 3 feet from the base of a 12-foot-high wall. We can write a ratio comparing the ladder's height to its distance from the wall.

$$\frac{12 \text{ feet}}{3 \text{ feet}} = \frac{12 \cancel{\text{feet}}}{3 \cancel{\text{feet}}} = \frac{12}{3}$$

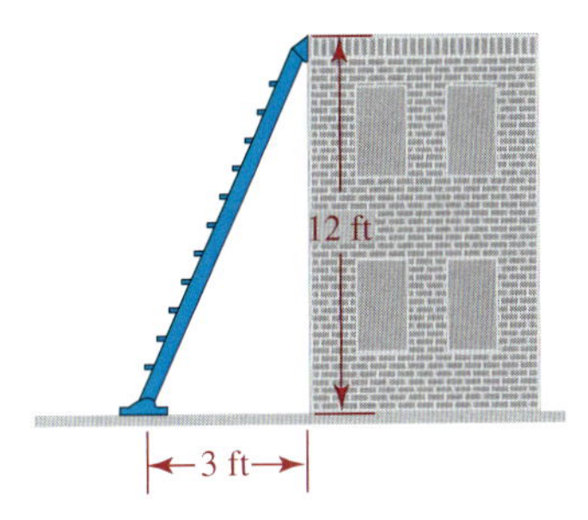

FIGURE 8-4

Since this ratio satisfies the 4-to-1 rule, the two ratios $\frac{4}{1}$ and $\frac{12}{3}$ must be equal. Therefore, we have

$$\frac{4}{1} = \frac{12}{3}$$

Such equations, which show that two ratios are equal, are called *proportions.* In this section, we will introduce the concept of proportion, and we will use proportions to solve many different types of problems.

Proportions

Proportions

A **proportion** is a statement that two ratios (or rates) are equal.

Some examples of proportions are

$$\frac{1}{2} = \frac{3}{6}, \quad \frac{3 \text{ waiters}}{7 \text{ tables}} = \frac{9 \text{ waiters}}{21 \text{ tables}}, \quad \text{and} \quad \frac{a}{b} = \frac{c}{d}$$

- The proportion $\frac{1}{2} = \frac{3}{6}$ can be read as "1 is to 2 as 3 is to 6."
- The proportion $\frac{3 \text{ waiters}}{7 \text{ tables}} = \frac{9 \text{ waiters}}{21 \text{ tables}}$ can be read as "3 waiters are to 7 tables as 9 waiters are to 21 tables."
- The proportion $\frac{a}{b} = \frac{c}{d}$ can be read as "*a* is to *b* as *c* is to *d*."

The terms of the proportion $\frac{a}{b} = \frac{c}{d}$ are numbered as follows.

First term ⟶ a, Second term ⟶ b, Third term ⟵ c, Fourth term ⟵ d

$$\frac{a}{b} = \frac{c}{d}$$

Means and extremes of a proportion

In any proportion, the first and fourth terms are called the **extremes.** The second and third terms are called the **means.**

In the proportion $\frac{1}{2} = \frac{3}{6}$, 1 and 6 are the **extremes** and 2 and 3 are the **means.**

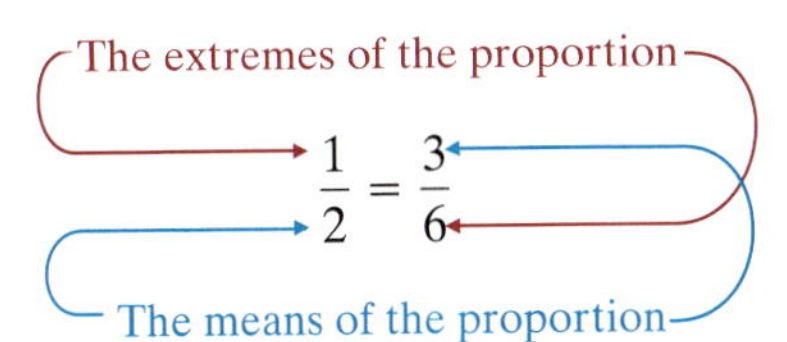

In this proportion, the product of the extremes is equal to the product of the means.

$1 \cdot 6 = 6$ and $2 \cdot 3 = 6$

This example illustrates a fundamental property of proportions.

Fundamental property of proportions

In any proportion, the product of the extremes is equal to the product of the means.

In the proportion $\frac{a}{b} = \frac{c}{d}$, a and d are the extremes and b and c are the means. We can show that the product of the extremes (ad) is equal to the product of the means (bc) by multiplying both sides of the proportion by bd and observing that $ad = bc$.

$$\frac{a}{b} = \frac{c}{d}$$

$$\frac{bd}{1} \cdot \frac{a}{b} = \frac{bd}{1} \cdot \frac{c}{d}$$ To eliminate the fractions, multiply both sides by $\frac{bd}{1}$.

$$\frac{abd}{b} = \frac{bcd}{d}$$ Multiply the numerators and multiply the denominators.

$$ad = bc$$ Divide out the common factors: $\frac{b}{b} = 1$ and $\frac{d}{d} = 1$.

Since $ad = bc$, the product of the extremes equals the product of the means.

To determine whether an equation is a proportion, we can check to see whether the product of the extremes is equal to the product of the means.

Self Check 1

Determine whether the equation is a proportion:

$$\frac{6}{13} = \frac{18}{39}$$

Answer yes

EXAMPLE 1 Determine whether each equation is a proportion: **a.** $\frac{3}{7} = \frac{9}{21}$ and **b.** $\frac{8}{3} = \frac{13}{5}$.

Solution In each case, we check to see whether the product of the extremes is equal to the product of the means.

a. The product of the extremes is $3 \cdot 21 = 63$. The product of the means is $7 \cdot 9 = 63$. Since the products are equal, the equation is a proportion: $\frac{3}{7} = \frac{9}{21}$.

$3 \cdot 21 = 63$ $7 \cdot 9 = 63$

$$\frac{3}{7} = \frac{9}{21}$$

The product of the extremes and the product of the means are also known as **cross products.**

b. The product of the extremes is $8 \cdot 5 = 40$. The product of the means is $3 \cdot 13 = 39$. Since the cross products are not equal, the equation is not a proportion: $\frac{8}{3} \neq \frac{13}{5}$.

$8 \cdot 5 = 40$ $3 \cdot 13 = 39$

$$\frac{8}{3} = \frac{13}{5}$$

When two pairs of numbers such as 2, 3 and 8, 12 form a proportion, we say that they are **proportional.** To show that 2, 3 and 8, 12 are proportional, we check to see whether the equation

$$\frac{2}{3} = \frac{8}{12}$$

is a proportion. To do so, we find the product of the extremes and the product of the means:

$$2 \cdot 12 = 24 \qquad 3 \cdot 8 = 24$$

Since the cross products are equal, the equation is a proportion, and the numbers are proportional.

EXAMPLE 2 Determine whether 3, 7 and 36, 91 are proportional.

Solution We check to see whether $\frac{3}{7} = \frac{36}{91}$ is a proportion by finding two products.

$3 \cdot 91 = 273$ The product of the extremes

$7 \cdot 36 = 252$ The product of the means

Since the cross products are not equal, the numbers are not proportional.

Self Check 2
Determine whether 6, 11 and 54, 99 are proportional.

Answer yes

Solving proportions

Suppose we know three terms in the following proportion

$$\frac{?}{5} = \frac{24}{20}$$

To find the missing term, we represent it with x, multiply the extremes and multiply the means, set them equal, and solve for x.

$$\frac{x}{5} = \frac{24}{20}$$

$20x = 5 \cdot 24$ In a proportion, the product of the extremes is equal to the product of the means.

$20x = 120$ Perform the multiplication: $5 \cdot 24 = 120$.

$\frac{20x}{\mathbf{20}} = \frac{120}{\mathbf{20}}$ To undo the multiplication by 20, divide both sides by 20.

$x = 6$ Perform the divisions.

To check this result, we substitute 6 for x in $\frac{x}{5} = \frac{24}{20}$ and find the cross products.

Check: $\frac{6}{5} \stackrel{?}{=} \frac{24}{20}$ $\qquad 6 \cdot 20 = \mathbf{120}$

$\qquad 5 \cdot 24 = \mathbf{120}$

Since the cross products are equal, x is 6.

Self Check 3

Solve the proportion $\frac{15}{x} = \frac{20}{32}$ for x. Check the result.

Answer 24

EXAMPLE 3 Solve the proportion $\frac{12}{18} = \frac{3}{x}$ for x.

Solution

$$\frac{12}{18} = \frac{3}{x}$$

$12 \cdot x = 18 \cdot 3$ — In a proportion, the product of the extremes equals the product of the means.

$12x = 54$ — Multiply: $18 \cdot 3 = 54$.

$\frac{12x}{\mathbf{12}} = \frac{54}{\mathbf{12}}$ — To undo the multiplication by 12, divide both sides by 12.

$x = \frac{9}{2}$ — Simplify: $\frac{54}{12} = \frac{\overset{1}{\cancel{6}} \cdot 9}{2 \cdot \underset{1}{\cancel{6}}} = \frac{9}{2}$.

Thus, x is $\frac{9}{2}$. Check this result in the proportion.

Self Check 4

Find the second term of the proportion $\frac{6.7}{x} = \frac{33.5}{38}$.

Answer 7.6

EXAMPLE 4 Find the third term of the proportion $\frac{3.5}{7.2} = \frac{x}{15.84}$.

Solution

$$\frac{3.5}{7.2} = \frac{x}{15.84}$$

$3.5(15.84) = 7.2x$ — In a proportion, the product of the extremes equals the product of the means.

$55.44 = 7.2x$ — Multiply: $3.5(15.84) = 55.44$.

$\frac{55.44}{\mathbf{7.2}} = \frac{7.2x}{\mathbf{7.2}}$ — To undo the multiplication by 7.2, divide both sides by 7.2.

$7.7 = x$ — Perform the divisions.

The third term is 7.7. Check the result in the proportion.

CALCULATOR SNAPSHOT Solving proportions with a calculator

To solve the proportion in Example 4 with a calculator, we can proceed as follows.

$$\frac{3.5}{7.2} = \frac{x}{15.84}$$

$\frac{3.5(15.84)}{7.2} = x$ — Multiply both sides by 15.84 to isolate x.

We can find x by entering these numbers and pressing these keys on a scientific calculator.

3.5 [×] 15.84 [÷] 7.2 [=] 7.7

Thus, $x = 7.7$.

EXAMPLE 5 Solve the proportion: $\frac{2a+1}{4} = \frac{10}{8}$.

Solution

$$\frac{2a+1}{4} = \frac{10}{8}$$

$8(2a + 1) = 40$ In a proportion, the product of the extremes equals the product of the means.

$16a + 8 = 40$ Distribute the multiplication by 8.

$16a + 8 - 8 = 40 - 8$ To undo the addition of 8, subtract 8 from both sides.

$16a = 32$ Simplify: $8 - 8 = 0$ and $40 - 8 = 32$.

$\frac{16a}{16} = \frac{32}{16}$ To undo the multiplication by 16, divide both sides by 16.

$a = 2$ Perform the divisions.

Thus, the solution is 2. Check this result in the proportion.

Self Check 5
Solve the proportion:
$\frac{3m-1}{2} = \frac{12.5}{5}$.

Answer 2

Writing proportions to solve problems

We can use proportions to solve many real-world problems. If we are given a ratio (or rate) comparing two quantities, the words of the problem can be translated to a proportion, and we can solve it to find the unknown.

EXAMPLE 6 **Grocery shopping.** If 5 apples cost \$1.15, how much will 16 apples cost?

Solution Let c represent the cost of 16 apples. If we compare the number of apples to their cost, we know that the two rates are equal.

5 apples is to \$1.15 as 16 apples is to \$$c$.

5 apples → $\frac{5}{1.15} = \frac{16}{c}$ ← 16 apples
Cost of 5 apples → ← Cost of 16 apples

To find the cost of 16 apples, we solve the proportion for c.

$5 \cdot c = 1.15(16)$ In a proportion, the product of the extremes is equal to the product of the means.

$5c = 18.4$ Perform the multiplication: $1.15(16) = 18.4$.

$\frac{5c}{5} = \frac{18.4}{5}$ To undo the multiplication by 5, divide both sides by 5.

$c = 3.68$ Perform the divisions.

Sixteen apples will cost \$3.68. To check the result, we substitute 3.68 for c in the proportion and find the cross products.

Check: $\frac{5}{1.15} \stackrel{?}{=} \frac{16}{3.68}$ $5 \cdot 3.68 = \mathbf{18.4}$
$1.15 \cdot 16 = \mathbf{18.4}$

The cross products are equal. The result 3.68 checks.

Self Check 6
If 9 tickets to a concert cost \$112.50, how much will 15 tickets cost?

Answer \$187.50

In Example 6, we could have compared the cost of the apples to the number of apples: \$1.15 is to 5 apples as \$$c$ is to 16 apples. This would have led to the proportion

$$\begin{array}{r} \text{Cost of 5 apples} \rightarrow \\ \text{5 apples} \rightarrow \end{array} \frac{1.15}{5} = \frac{c}{16} \begin{array}{l} \leftarrow \text{Cost of 16 apples} \\ \leftarrow \text{16 apples} \end{array}$$

If we solve this proportion for c, we obtain the same result: $c = 3.68$.

! COMMENT When we solve problems using proportions, it is a good practice to make sure that the units of the numerators are the same and the units of the denominators are the same. For Example 6, it would be incorrect to write

$$\begin{array}{r} \text{Cost of 5 apples} \rightarrow \\ \text{5 apples} \rightarrow \end{array} \frac{1.15}{5} = \frac{16}{c} \begin{array}{l} \leftarrow \text{16 apples} \\ \leftarrow \text{Cost of 16 apples} \end{array}$$

EXAMPLE 7 **Scale drawings.** A **scale** is a ratio (or rate) that compares the size of a model, drawing, or map to the size of an actual object. The airplane in Figure 8-5 is drawn using a scale of 1 inch: 6 feet. This means that 1 inch on the drawing is actually 6 feet on the plane. The distance from wing tip to wing tip (the wingspan) on the drawing is 5 inches. Find the actual wingspan of the plane.

FIGURE 8-5

Solution Let w represent the actual wingspan of the plane. Since 1 inch corresponds to 6 feet as 5 inches corresponds to w feet, we can write the proportion.

$$\begin{array}{r} \text{Measure on drawing} \rightarrow \\ \text{Measure on plane} \rightarrow \end{array} \frac{1}{6} = \frac{5}{w} \begin{array}{l} \leftarrow \text{Measure on drawing} \\ \leftarrow \text{Measure on plane} \end{array}$$

$1 \cdot w = 6 \cdot 5$ In a proportion, the product of the extremes is equal to the product of the means.

$w = 30$ Perform the multiplications.

The actual wingspan of the plane is 30 feet. Check the result by finding the cross products.

EXAMPLE 8 **Baking.** A recipe for rhubarb cake calls for $1\frac{1}{4}$ cups of sugar for every $2\frac{1}{2}$ cups of flour. How many cups of flour are needed if the baker intends to use 3 cups of sugar?

Solution Let f represent the number of cups of flour to be mixed with the sugar. The ratios of the cups of sugar to the cups of flour are equal. We have $1\frac{1}{4}$ cups sugar is to $2\frac{1}{2}$ cups flour as 3 cups sugar is to f cups flour. We can write the proportion.

$$\frac{1\frac{1}{4}}{2\frac{1}{2}} = \frac{3}{f}$$

Cups sugar → $1\frac{1}{4}$; 3 ← Cups sugar

Cups flour → $2\frac{1}{2}$; f ← Cups flour

$\frac{1.25}{2.5} = \frac{3}{f}$ Change the fractions to decimals: $1\frac{1}{4} = 1.25$ and $2\frac{1}{2} = 2.5$.

$1.25f = 2.5 \cdot 3$ In a proportion, the product of the extremes is equal to the product of the means.

$1.25f = 7.5$ Perform the multiplication: $2.5 \cdot 3 = 7.5$.

$\frac{1.25f}{\mathbf{1.25}} = \frac{7.5}{\mathbf{1.25}}$ To undo the multiplication by 1.25, divide both sides by 1.25.

$f = 6$ Perform the divisions.

The baker should use 6 cups of flour.

Self Check 8

How many cups of sugar will be needed to make several rhubarb cakes that will require a total of 25 cups of flour?

Answer 12.5 cups

Section 8.2 STUDY SET

VOCABULARY *Fill in the blanks.*

1. A __________ is a statement that two ratios or rates are equal.
2. In $\frac{1}{2} = \frac{5}{10}$, the terms 1 and 10 are called the __________ of the proportion. The terms 2 and 5 are called the ________ of the proportion.
3. Consider the proportion $\frac{3}{4} = \frac{15}{20}$. The two ______ products are $3 \cdot 20 = 60$ and $8 \cdot 15 = 60$.
4. When two pairs of numbers form a proportion, we say that the numbers are __________.

CONCEPTS *Fill in the blanks.*

5. The equation $\frac{a}{b} = \frac{c}{d}$ will be a proportion if the product ___ is equal to the product ___.
6. ___ $\cdot 10 =$ ___ $2 \cdot$ ___ $=$ ___

$$\frac{9}{2} = \frac{45}{10}$$

7. Write each statement as a proportion.
   a. 5 is to 8 as 15 is to 24.
   b. 3 teacher's aides are to 25 children as 12 teacher's aides are to 100 children.
8. Consider the proportion $\frac{3}{4} = \frac{15}{20}$. What are the two cross products?
9. For every 15 feet of chain link fencing, 4 support posts are used. How many support posts will be needed for 300 feet of chain link fence? Which of the following proportions could be used to solve this problem?

   i. $\frac{15}{4} = \frac{300}{x}$ ii. $\frac{15}{4} = \frac{x}{300}$

   iii. $\frac{4}{15} = \frac{300}{x}$ iv. $\frac{4}{15} = \frac{x}{300}$

10. Write a problem that could be solved using the following proportion.

Ounces of cashews → 4, 10 ← Ounces of cashews
Calories → 639, x ← Calories

$$\frac{4}{639} = \frac{10}{x}$$

NOTATION *Complete each solution to solve for x.*

11. $\frac{12}{18} = \frac{x}{24}$

$12 \cdot 24 = \square$

$\square = 18x$

$\frac{288}{\square} = \frac{18x}{\square}$

$16 = x$

12. $\frac{14}{x} = \frac{49}{17.5}$

$14 \cdot \square = 49x$

$\square = 49x$

$\frac{245}{\square} = \frac{49x}{\square}$

$5 = x$

PRACTICE *Determine whether each statement is a proportion.*

13. $\frac{9}{7} = \frac{81}{70}$

14. $\frac{5}{2} = \frac{20}{8}$

15. $\frac{7}{3} = \frac{14}{6}$

16. $\frac{13}{19} = \frac{65}{95}$

17. $\frac{9}{19} = \frac{38}{80}$

18. $\frac{40}{29} = \frac{29}{22}$

19. $\frac{10.4}{3.6} = \frac{41.6}{14.4}$

20. $\frac{13.23}{3.45} = \frac{39.96}{11.35}$

21. $\frac{\frac{2}{3}}{\frac{5}{8}} = \frac{\frac{4}{5}}{\frac{9}{16}}$

22. $\frac{\frac{3}{2}}{\frac{8}{9}} = \frac{\frac{1}{4}}{\frac{4}{27}}$

23. $\frac{4\frac{1}{6}}{\frac{12}{7}} = \frac{2\frac{3}{16}}{\frac{9}{10}}$

24. $\frac{2\frac{1}{2}}{\frac{4}{5}} = \frac{3\frac{3}{4}}{\frac{9}{10}}$

Solve for the variable in each proportion. **Check each result.**

25. $\frac{2}{3} = \frac{x}{6}$

26. $\frac{3}{6} = \frac{x}{8}$

27. $\frac{5}{10} = \frac{3}{c}$

28. $\frac{7}{14} = \frac{2}{x}$

29. $\frac{6}{x} = \frac{8}{4}$

30. $\frac{4}{x} = \frac{2}{8}$

31. $\frac{x}{8} = \frac{9}{2}$

32. $\frac{x}{2} = \frac{18}{6}$

33. $\frac{x+1}{5} = \frac{3}{15}$

34. $\frac{x-1}{7} = \frac{2}{21}$

35. $\frac{x+3}{12} = \frac{-7}{6}$

36. $\frac{x+7}{-4} = \frac{1}{4}$

37. $\frac{4-x}{13} = \frac{11}{26}$

38. $\frac{5-x}{17} = \frac{13}{34}$

39. $\frac{2x+1}{18} = \frac{14}{3}$

40. $\frac{2x-1}{18} = \frac{9}{54}$

41. $\frac{4{,}000}{x} = \frac{3.2}{2.8}$

42. $\frac{0.4}{1.6} = \frac{96.7}{x}$

43. $\frac{\frac{1}{2}}{\frac{1}{5}} = \frac{x}{2\frac{1}{4}}$

44. $\frac{x}{4\frac{1}{10}} = \frac{3\frac{3}{4}}{1\frac{7}{8}}$

APPLICATIONS *Set up and solve a proportion.*

45. SCHOOL LUNCHES A manager of a school cafeteria orders 750 pudding cups. What will the order cost if she purchases them wholesale, 6 cups for $1.75?

46. CLOTHES SHOPPING As part of a spring clearance, a men's store put dress shirts on sale, 2 for $25.98. How much will a businessman pay if he buys five shirts?

47. GARDENING Three packets of garden seeds sell for 98¢. A Girl Scout troop leader needs to purchase three dozen packets. What will they cost?

48. COOKING A recipe for spaghetti sauce requires four 16-ounce bottles of ketchup to make 2 gallons of sauce. How many bottles of ketchup are needed to make 10 gallons of sauce?

49. BUSINESS PERFORMANCE The bar graph shows the yearly costs incurred and the revenue received by a business. How do the ratios of costs to revenue for 2003 and 2004 compare?

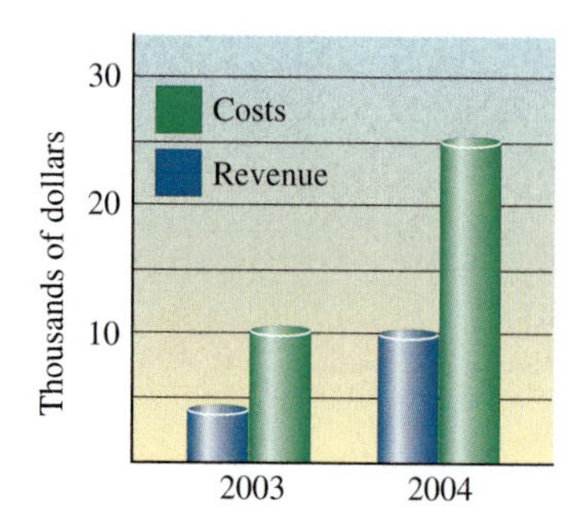

50. RAMP Write a ratio of the rise to the run for each ramp shown on the next page. Set the ratios equal. Is the resulting proportion true? Is one ramp steeper than the other?

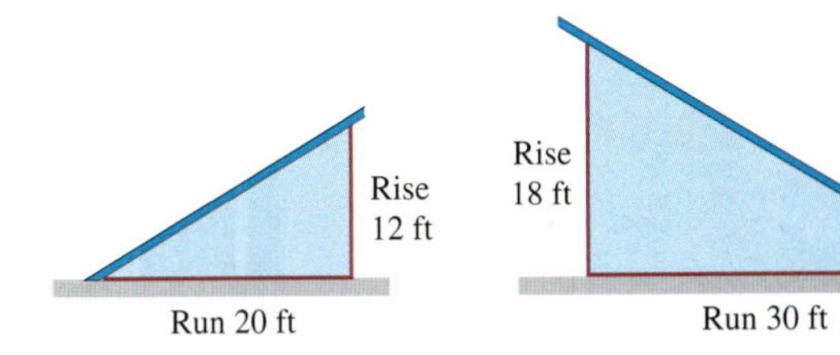

51. MIXING PERFUMES A perfume is to be mixed in the ratio of 3 drops of pure essence to 7 drops of alcohol. How many drops of pure essence should be mixed with 56 drops of alcohol?

52. MAKING COLOGNE A cologne can be made by mixing 2 drops of pure essence with 5 drops of distilled water. How much water should be used with 15 drops of pure essence?

53. LAB WORK In a red blood cell count, a drop of the patient's diluted blood is placed on a grid like that shown. Instead of counting each and every red blood cell in the 25-square grid, a technician just counts the number of cells in the five highlighted squares. Then he or she uses a proportion to estimate the total red blood cell count. If there are 195 red blood cells in the blue squares, about how many red blood cells would there be in the entire grid?

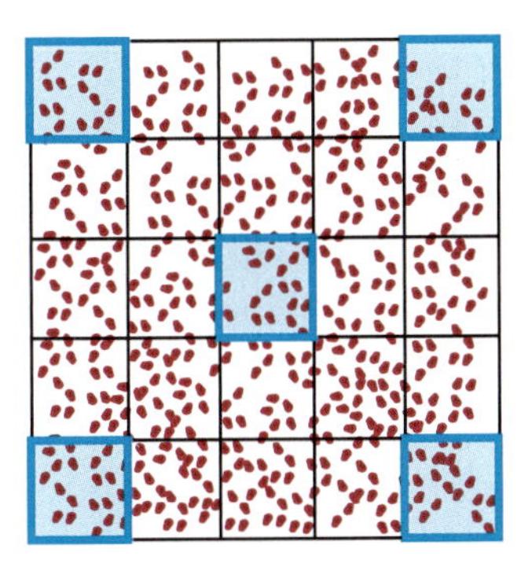

54. DOSAGE The proper dose of a certain medication for a 30-pound child is shown. At this rate, what would be the dose for a 45-pound child?

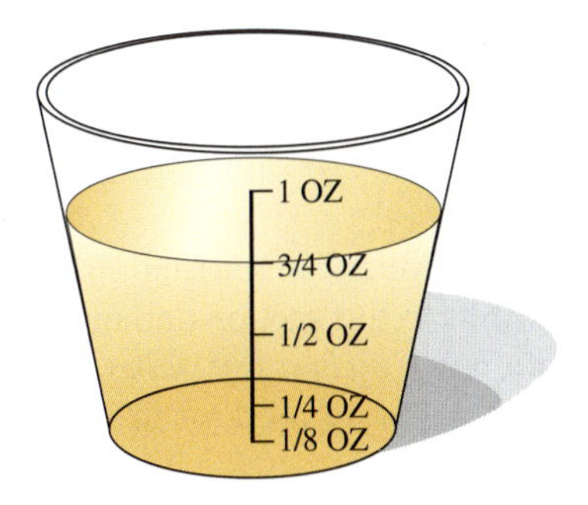

55. MAKING COOKIES A recipe for chocolate chip cookies calls for $1\frac{1}{4}$ cups of flour and 1 cup of sugar. The recipe will make $3\frac{1}{2}$ dozen cookies. How many cups of flour will be needed to make 12 dozen cookies?

56. MAKING BROWNIES A recipe for brownies calls for 4 eggs and $1\frac{1}{2}$ cups of flour. If the recipe makes 15 brownies, how many cups of flour will be needed to make 130 brownies?

57. COMPUTER SPEED Using the *Mathematica 3.0* program, a Dell Dimension XPS R350 (Pentium II) computer can perform a set of 15 calculations in 2.85 seconds. How long will it take the computer to perform 100 such calculations?

58. QUALITY CONTROL Out of a sample of 500 men's shirts, 17 were rejected because of crooked collars. How many crooked collars would you expect to find in a run of 15,000 shirts?

59. FUEL CONSUMPTION A "high mobility multipurpose wheeled vehicle" is better known as a Hummer. Under normal conditions, a Hummer can travel 325 miles on a full tank (25 gallons) of diesel. How far can it travel on its auxiliary tank, which holds 17 gallons of diesel?

60. ANNIVERSARY GIFTS A florist sells a dozen long-stemmed red roses for \$57.99. In honor of their 16th wedding anniversary, a man wants to buy 16 roses for his wife. What will the roses cost?

61. PAYCHECKS Sanchez earns \$412 for a 40-hour week. If he missed 10 hours of work last week, how much did he get paid?

62. STAFFING A school board has determined that there should be 3 teachers for every 50 students. Complete the table by filling in the number of teachers needed at each school.

	Glenwood High	Goddard Junior High	Sellers Elementary
Enrollment	2,700	1,900	850
Teachers			

63. BLUEPRINTS The scale for the blueprint shown on the next page tells the reader that a $\frac{1}{4}$-inch length $(\frac{1}{4}'')$ on the drawing corresponds to an actual size of 1 foot (1′0″). Suppose the length of the kitchen is $2\frac{1}{2}$ inches on the blueprint. How long is the actual kitchen?

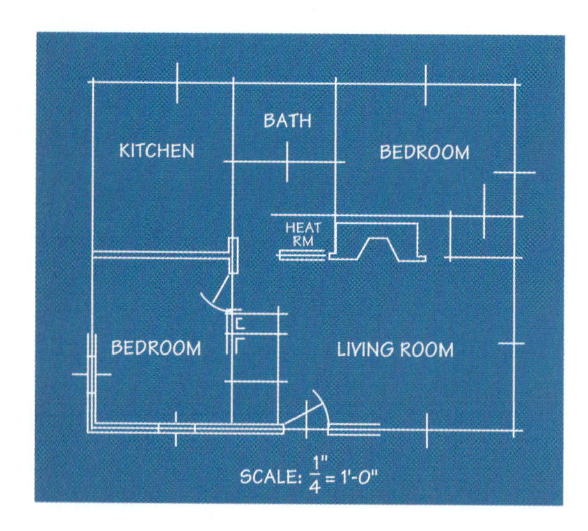

64. DRAFTING In a scale drawing, a 280-foot antenna tower is drawn 7 inches high. The building next to it is drawn 2 inches high. How tall is the actual building?

65. MODEL RAILROADING An HO-scale model railroad engine is 9 inches long. If HO scale is 87 feet to 1 foot, how long is a real engine?

66. MODEL RAILROADING An N-scale model railroad caboose is 4 inches long. If N scale is 169 feet to 1 foot, how long is a real caboose?

67. MINIATURES The ratio in the illustration indicates that 1 inch on the model carousel is equivalent to 160 inches on the actual carousel. How wide should the model be if the actual carousel is 35 feet wide?

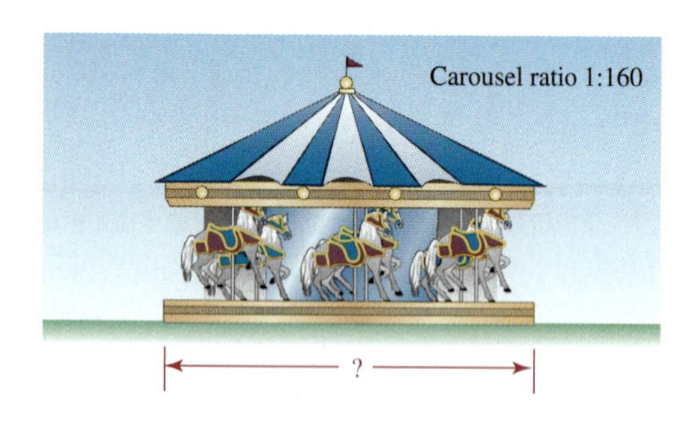

68. MIXING FUELS The instructions on a can of oil intended to be added to lawn mower gasoline read as shown. Are these instructions correct? (*Hint:* There are 128 ounces in 1 gallon.)

Recommended	Gasoline	Oil
50 to 1	6 gal	16 oz

WRITING

69. Explain the difference between a ratio and a proportion.

70. Explain how to determine whether $\frac{3.2}{3.7} = \frac{5.44}{6.29}$ is a true proportion.

71. DOLLHOUSES The following paragraph is from a book about dollhouses. What concept from this section is mentioned?

> Today, the internationally recognized scale for dollhouses and miniatures is 1 in. = 1 ft. This is small enough to be defined as a miniature, yet not too small for all details of decoration and furniture to be seen clearly.

72. Write a problem about a situation you encounter in your daily life that could be solved by using a proportion.

REVIEW

73. Change $\frac{9}{10}$ to a percent.

74. Change $\frac{7}{8}$ to a percent.

75. Change $33\frac{1}{3}\%$ to a fraction.

76. Multiply: $(2x - 1)(3x + 2)$.

77. Find $\frac{1}{2}\%$ of 520.

78. SHOPPING Bill purchased a shirt on sale for \$17.50. Find the original cost of the shirt if it was marked down 30%.

8.3 American Units of Measurement

- American units of length • Converting units of length • American units of weight
- American units of capacity • Units of time

Two common systems of measurement are the American (or English) system and the metric system. We will discuss American units in this section and metric units in the next section. Some common American units are *inches, feet, miles, ounces, pounds, tons, cups, pints, quarts,* and *gallons.* These units are used when measuring length, weight, and capacity.

- A newborn baby is 20 inches long.
- The distance from St. Louis to Memphis is 285 miles.
- First-class postage for a letter that weighs less than 1 ounce is 37¢.
- The largest pumpkin ever grown weighed 1,092 pounds.
- Milk is sold in quart and gallon containers.

American units of length

A ruler is one of the most common devices used for measuring distances or lengths. Figure 8-6 shows only a portion of a ruler; most rulers are 12 inches (1 foot) long. Since 12 inches = 1 foot, a ruler is divided into 12 equal distances of 1 inch. Each inch is divided into halves of an inch, quarters of an inch, eighths of an inch, and sixteenths of an inch. Four distances are measured using the ruler shown in Figure 8-6.

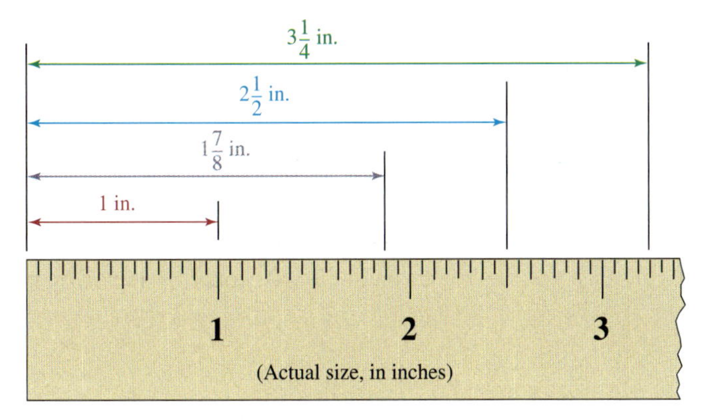

FIGURE 8-6

Each point on a ruler, like each point on a number line, has a number associated with it. This number is the distance between the point and 0.

EXAMPLE 1 To the nearest $\frac{1}{4}$ inch, find the length of the nail in Figure 8-7.

Solution We place the end of the ruler by one end of the nail and note that the other end of the nail is closer to the $2\frac{1}{2}$-inch mark than to the $2\frac{1}{4}$-inch mark on the ruler. To the nearest quarter-inch, the nail is $2\frac{1}{2}$ inches long.

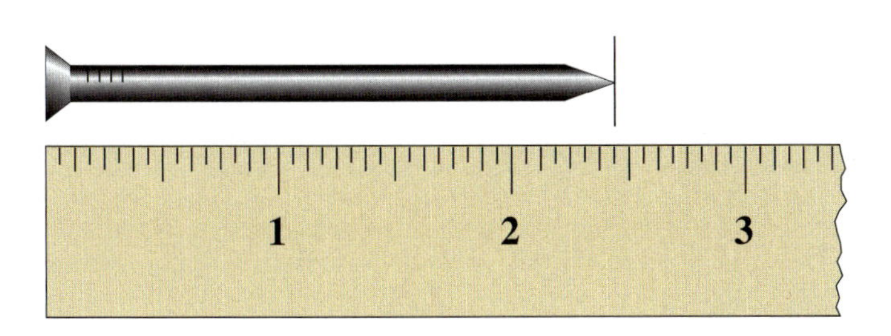

FIGURE 8-7

Self Check 1

To the nearest $\frac{1}{4}$ inch, find the width of the circle below.

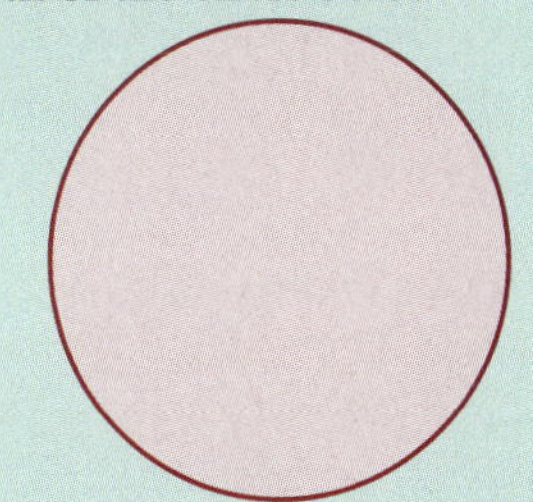

Answer $1\frac{1}{4}$ in.

Self Check 2

To the nearest $\frac{1}{8}$ inch, find the length of the jumbo paper clip below.

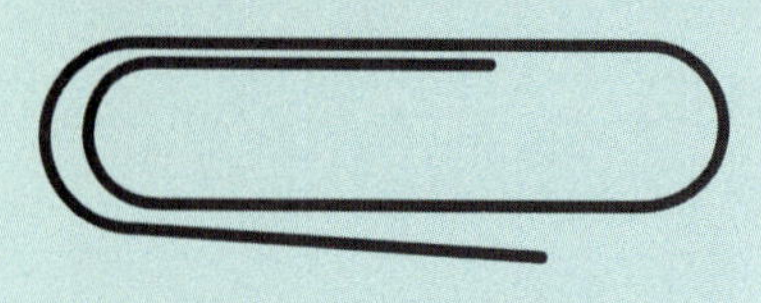

Answer $1\frac{7}{8}$ in.

EXAMPLE 2 To the nearest $\frac{1}{8}$ inch, find the length of the paper clip in Figure 8-8.

Solution We place the end of the ruler by one end of the paper clip and note that the other end is closer to the $1\frac{3}{8}$-inch mark than to the $1\frac{1}{2}$-inch mark on the ruler. To the nearest eighth of an inch, the paper clip is $1\frac{3}{8}$ inches long.

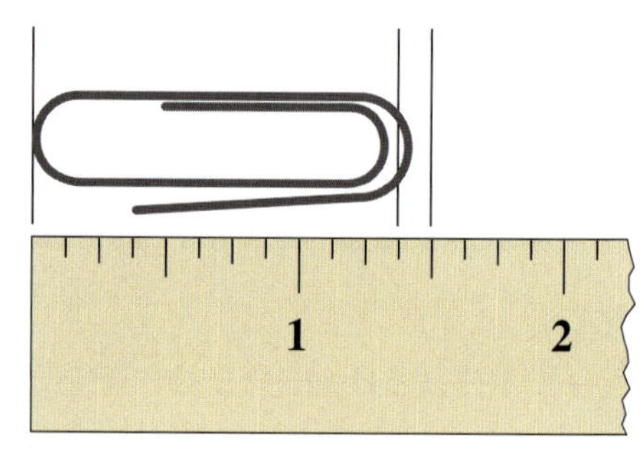

FIGURE 8-8

Converting units of length

American units of length are related in the following ways.

American units of length

1 foot (ft) = 12 inches (in.) 1 yard (yd) = 36 inches

1 yard = 3 feet 1 mile (mi) = 5,280 feet

To convert from one unit to another, we use *unit conversion factors.* To find the unit conversion factor between yards and feet, we begin with this fact:

$$3 \text{ ft} = 1 \text{ yd}$$

If we divide both sides of this equation by 1 yard, we get

$$\frac{3 \text{ ft}}{\mathbf{1 \text{ yd}}} = \frac{1 \text{ yd}}{\mathbf{1 \text{ yd}}}$$

$$\frac{3 \text{ ft}}{1 \text{ yd}} = 1 \quad \text{A number divided by itself is 1: } \frac{1 \text{ yd}}{1 \text{ yd}} = 1.$$

The fraction $\frac{3 \text{ ft}}{1 \text{ yd}}$ is called a **unit conversion factor,** because its value is 1. It can be read as "3 feet per yard." Since this fraction is equal to 1, multiplying a length by this fraction does not change its measure; it only changes the *units* of measure.

To convert units of length, we use the following unit conversion factors.

To convert from	Use the unit conversion factor	To convert from	Use the unit conversion factor
feet to inches	$\frac{12 \text{ in.}}{1 \text{ ft}}$	inches to feet	$\frac{1 \text{ ft}}{12 \text{ in.}}$
yards to feet	$\frac{3 \text{ ft}}{1 \text{ yd}}$	feet to yards	$\frac{1 \text{ yd}}{3 \text{ ft}}$
yards to inches	$\frac{36 \text{ in.}}{1 \text{ yd}}$	inches to yards	$\frac{1 \text{ yd}}{36 \text{ in.}}$
miles to feet	$\frac{5{,}280 \text{ ft}}{1 \text{ mi}}$	feet to miles	$\frac{1 \text{ mi}}{5{,}280 \text{ ft}}$

EXAMPLE 3 Convert 7 yards to feet.

Solution To convert from yards to feet, we must use a unit conversion factor that relates feet to yards. Since there are 3 feet per yard, we multiply 7 yards by the unit conversion factor $\frac{3\text{ ft}}{1\text{ yd}}$ to get

$$7\text{ yd} = \frac{7\text{ yd}}{1} \cdot \frac{\mathbf{3\text{ ft}}}{\mathbf{1\text{ yd}}}$$ Write 7 yd as a fraction: $7\text{ yd} = \frac{7\text{ yd}}{1}$. Then multiply by 1: $\frac{3\text{ ft}}{1\text{ yd}} = 1$.

$$= \frac{7\,\overset{1}{\cancel{\text{yd}}}}{1} \cdot \frac{3\text{ ft}}{1\,\underset{1}{\cancel{\text{yd}}}}$$ The units of yards divide out.

$$= 7 \cdot 3\text{ ft}$$

$$= 21\text{ ft}$$ Multiply: $7 \cdot 3 = 21$.

Seven yards is equal to 21 feet.

Self Check 3
Convert 9 yards to feet.

Answer 27 ft

Notice that in Example 3, we eliminated the units of yards and introduced the units of feet by multiplying by the appropriate unit conversion factor. In general, a unit conversion factor is a fraction with the following form.

$$\frac{\text{Unit we want to introduce}}{\text{Unit we want to eliminate}} \begin{array}{l} \leftarrow \text{Numerator} \\ \leftarrow \text{Denominator} \end{array}$$

EXAMPLE 4 Convert $1\frac{3}{4}$ feet to inches.

Solution To convert from feet to inches, we must use a unit conversion factor that relates inches to feet. Since there are 12 inches per foot, we multiply $1\frac{3}{4}$ feet by the unit conversion factor $\frac{12\text{ in.}}{1\text{ ft}}$ to get

$$1\frac{3}{4}\text{ ft} = \frac{7}{4}\text{ ft} \cdot \frac{\mathbf{12\text{ in.}}}{\mathbf{1\text{ ft}}}$$ Write $1\frac{3}{4}$ as an improper fraction: $1\frac{3}{4} = \frac{7}{4}$. Multiply by 1: $\frac{12\text{ in.}}{1\text{ ft}} = 1$.

$$= \frac{7}{4}\,\overset{1}{\cancel{\text{ft}}} \cdot \frac{12\text{ in.}}{1\,\underset{1}{\cancel{\text{ft}}}}$$ The units of feet divide out.

$$= \frac{7 \cdot 12}{4 \cdot 1}\text{ in.}$$ Multiply the fractions.

$$= 21\text{ in.}$$ Simplify: $\frac{7 \cdot 12}{4 \cdot 1} = \frac{7 \cdot 3 \cdot \overset{1}{\cancel{4}}}{\underset{1}{\cancel{4}} \cdot 1} = 7 \cdot 3 = 21$.

$1\frac{3}{4}$ feet is equal to 21 inches.

Self Check 4
Convert 1.5 feet to inches.

Answer 18 in.

Sometimes we must use two unit conversion factors in combination to eliminate the given units while introducing the desired units.

CALCULATOR SNAPSHOT

Finding the length of a football field in miles

A football field (including the end zones) is 120 yards long. To find this distance in miles, we set up the problem so that the units of yards divide out and leave us with units of miles. Since there are 3 feet per yard and 5,280 feet per mile, we multiply 120 yards by $\frac{3\text{ ft}}{1\text{ yd}}$ and $\frac{1\text{ mi}}{5{,}280\text{ ft}}$.

$$120\text{ yd} = 120\text{ yd} \cdot \frac{3\text{ ft}}{1\text{ yd}} \cdot \frac{1\text{ mi}}{5{,}280\text{ ft}}$$

Use two unit conversion factors: $\frac{3\text{ ft}}{1\text{ yd}} = 1$ and $\frac{1\text{ mi}}{5{,}280\text{ ft}} = 1$.

$$= \frac{120\text{ }\overset{1}{\cancel{\text{yd}}}}{1} \cdot \frac{3\text{ }\overset{1}{\cancel{\text{ft}}}}{1\text{ }\underset{1}{\cancel{\text{yd}}}} \cdot \frac{1\text{ mi}}{5{,}280\text{ }\underset{1}{\cancel{\text{ft}}}}$$

Divide out the units of yards and feet.

$$= \frac{120 \cdot 3}{5{,}280}\text{ mi}$$

Multiply the fractions.

We can do this arithmetic using a scientific calculator by entering these numbers and pressing these keys.

120 [×] 3 [÷] 5280 [=]   0.0681818

To the nearest hundredth, a football field is 0.07 mile long.

American units of weight

American units of weight are related in the following ways.

American units of weight

1 pound (lb) = 16 ounces (oz)   1 ton = 2,000 pounds

To convert units of weight, we use the following unit conversion factors.

To convert from	Use the unit conversion factor	To convert from	Use the unit conversion factor
pounds to ounces	$\frac{16\text{ oz}}{1\text{ lb}}$	ounces to pounds	$\frac{1\text{ lb}}{16\text{ oz}}$
tons to pounds	$\frac{2{,}000\text{ lb}}{1\text{ ton}}$	pounds to tons	$\frac{1\text{ ton}}{2{,}000\text{ lb}}$

Self Check 5
Convert 60 ounces to pounds.

EXAMPLE 5 Convert 40 ounces to pounds.

Solution Since there is 1 pound per 16 ounces, we multiply 40 ounces by the unit conversion factor $\frac{1\text{ lb}}{16\text{ oz}}$ to get

$$40\text{ oz} = \frac{40\text{ oz}}{1} \cdot \frac{\mathbf{1\text{ lb}}}{\mathbf{16\text{ oz}}}$$

Write 40 oz as a fraction: 40 oz = $\frac{40\text{ oz}}{1}$. Then multiply by 1: $\frac{1\text{ lb}}{16\text{ oz}} = 1$.

$$= \frac{40\text{ }\overset{1}{\cancel{\text{oz}}}}{1} \cdot \frac{1\text{ lb}}{16\text{ }\underset{1}{\cancel{\text{oz}}}}$$

The units of ounces divide out.

$$= \frac{40}{16}\text{ lb}$$

Multiply the fractions.

There are two ways to complete the solution. First, we can divide out the common factors of the numerator and denominator and then write the result as a mixed number.

$$\frac{40}{16}\text{ lb} = \frac{5\cdot\overset{1}{\cancel{8}}}{2\cdot\underset{1}{\cancel{8}}}\text{ lb} = \frac{5}{2}\text{ lb} = 2\frac{1}{2}\text{ lb}$$

A second approach is to divide the numerator by the denominator and express the result as a decimal.

$$\frac{40}{16}\text{ lb} = 2.5\text{ lb} \qquad \text{Perform the division: } 40 \div 16 = 2.5.$$

Forty ounces is equal to $2\frac{1}{2}$ lb (or 2.5 lb).

Answer $3\frac{3}{4}$ lb = 3.75 lb

EXAMPLE 6 Convert 25 pounds to ounces.

Solution Since there are 16 ounces per pound, we multiply 25 pounds by the unit conversion factor $\frac{16\text{ oz}}{1\text{ lb}}$ to get

$$\begin{aligned} 25\text{ lb} &= \frac{25\text{ lb}}{1}\cdot\frac{\mathbf{16\text{ oz}}}{\mathbf{1\text{ lb}}} && \text{Multiply by 1: } \frac{16\text{ oz}}{1\text{ lb}} = 1. \\ &= \frac{25\,\overset{1}{\cancel{\text{lb}}}}{1}\cdot\frac{16\text{ oz}}{1\,\underset{1}{\cancel{\text{lb}}}} && \text{The units of pounds divide out.} \\ &= 25\cdot 16\text{ oz} \\ &= 400\text{ oz} && \text{Multiply: } 25\cdot 16 = 400. \end{aligned}$$

Twenty-five pounds is equal to 400 ounces.

Self Check 6
Convert 60 pounds to ounces.

Answer 960 oz

Finding the weight of a car in pounds

CALCULATOR SNAPSHOT

A BMW 323Ci convertible weighs 1.78 tons. To find its weight in pounds, we set up the problem so that the units of tons divide out and leave us with pounds. Since there are 2,000 pounds per ton, we multiply by $\frac{2{,}000\text{ lb}}{1\text{ ton}}$.

$$\begin{aligned} 1.78\text{ tons} &= \frac{1.78\text{ tons}}{1}\cdot\frac{2{,}000\text{ lb}}{1\text{ ton}} && \text{Multiply by 1: } \frac{2{,}000\text{ lb}}{1\text{ ton}} = 1. \\ &= \frac{1.78\,\overset{1}{\cancel{\text{tons}}}}{1}\cdot\frac{2{,}000\text{ lb}}{1\,\underset{1}{\cancel{\text{ton}}}} && \text{Divide out the units of tons.} \\ &= 1.78\cdot 2{,}000\text{ lb} \end{aligned}$$

We can do this multiplication using a scientific calculator by entering these numbers and pressing these keys.

1.78 [×] 2000 [=] 3560

The convertible weighs 3,560 pounds.

American units of capacity

American units of capacity are related as follows.

American units of capacity

1 cup (c) = 8 fluid ounces (fl oz)
1 pint (pt) = 2 cups (c)
1 quart (qt) = 2 pints (pt)
1 gallon (gal) = 4 quarts (qt)

To convert units of capacity, we use the following unit conversion factors.

To convert from	Use the unit conversion factor	To convert from	Use the unit conversion factor
cups to ounces	$\frac{8 \text{ fl oz}}{1 \text{ c}}$	ounces to cups	$\frac{1 \text{ c}}{8 \text{ fl oz}}$
pints to cups	$\frac{2 \text{ c}}{1 \text{ pt}}$	cups to pints	$\frac{1 \text{ pt}}{2 \text{ c}}$
quarts to pints	$\frac{2 \text{ pt}}{1 \text{ qt}}$	pints to quarts	$\frac{1 \text{ qt}}{2 \text{ pt}}$
gallons to quarts	$\frac{4 \text{ qt}}{1 \text{ gal}}$	quarts to gallons	$\frac{1 \text{ gal}}{4 \text{ qt}}$

Self Check 7
How many pints are in 1 gallon?

EXAMPLE 7 If a recipe calls for 3 pints of milk, how many fluid ounces of milk should be used?

Solution Since there are 2 cups per pint and 8 fluid ounces per cup, we multiply 3 pints by unit conversion factors of $\frac{2 \text{ c}}{1 \text{ pt}}$ and $\frac{8 \text{ fl oz}}{1 \text{ c}}$.

$$3 \text{ pt} = \frac{3 \text{ pt}}{1} \cdot \frac{2 \text{ c}}{1 \text{ pt}} \cdot \frac{8 \text{ fl oz}}{1 \text{ c}}$$ Use two unit conversion factors: $\frac{2 \text{ c}}{1 \text{ pt}} = 1$ and $\frac{8 \text{ fl oz}}{1 \text{ c}} = 1$.

$$= \frac{3 \cancel{\text{pt}}}{1} \cdot \frac{2 \cancel{\text{c}}}{1 \cancel{\text{pt}}} \cdot \frac{8 \text{ fl oz}}{1 \cancel{\text{c}}}$$ Divide out the units of pints and cups.

$$= 3 \cdot 2 \cdot 8 \text{ fl oz}$$

$$= 48 \text{ fl oz}$$

Since 3 pints is equal to 48 fluid ounces, 48 fluid ounces of milk should be used.

Answer 8 pt

Units of time

Units of time are related in the following ways.

Units of time

1 minute (min) = 60 seconds (sec) 1 hour (hr) = 60 minutes
1 day = 24 hours

To convert units of time, we use the following unit conversion factors.

To convert from	Use the unit conversion factor	To convert from	Use the unit conversion factor
minutes to seconds	$\frac{60 \text{ sec}}{1 \text{ min}}$	seconds to minutes	$\frac{1 \text{ min}}{60 \text{ sec}}$
hours to minutes	$\frac{60 \text{ min}}{1 \text{ hr}}$	minutes to hours	$\frac{1 \text{ hr}}{60 \text{ min}}$
days to hours	$\frac{24 \text{ hr}}{1 \text{ day}}$	hours to days	$\frac{1 \text{ day}}{24 \text{ hr}}$

EXAMPLE 8 **Astronomy.** A lunar eclipse occurs when the Earth is between the sun and the moon in such a way that the Earth's shadow darkens the moon. (See Figure 8-10, which is not to scale.) A total lunar eclipse can last as long as 105 minutes. How many hours is this?

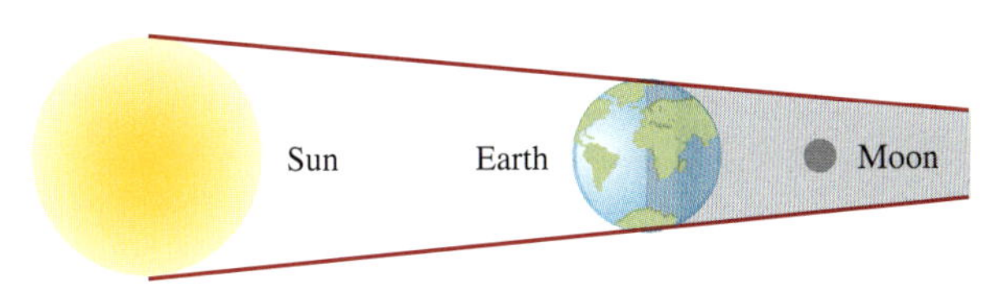

FIGURE 8-10

Solution Since there is 1 hour for every 60 minutes, we multiply 105 by the unit conversion factor $\frac{1\text{ hr}}{60\text{ min}}$ to get

$$105\text{ min} = \frac{105\text{ min}}{1} \cdot \frac{\mathbf{1\ hr}}{\mathbf{60\ min}} \quad \text{Multiply by 1: } \frac{1\text{ hr}}{60\text{ min}} = 1$$

$$= \frac{105\,\cancel{\text{min}}}{1} \cdot \frac{1\text{ hr}}{60\,\cancel{\text{min}}} \quad \text{The units of minutes divide out.}$$

$$= \frac{105}{60}\text{ hr} \quad \text{Multiply the fractions.}$$

$$= \frac{\cancel{3} \cdot \cancel{5} \cdot 7}{2 \cdot 2 \cdot \cancel{3} \cdot \cancel{5}}\text{ hr} \quad \text{Prime factor 105 and 60. Then divide out common factors of the numerator and denominator.}$$

$$= \frac{7}{4}\text{ hr}$$

$$= 1\tfrac{3}{4}\text{ hr} \quad \text{Write } \tfrac{7}{4} \text{ as a mixed number.}$$

A total lunar eclipse can last as long as $1\frac{3}{4}$ hours.

Self Check 8
A solar eclipse (eclipse of the sun) can last as long as 450 seconds. How many minutes is this?

Answer $7\frac{1}{2}$ min

Section 8.3 STUDY SET

VOCABULARY *Fill in the blanks.*

1. Inches, feet, and miles are examples of American units of ________.
2. A ruler is used for measuring ________.
3. The value of any unit conversion factor is ☐.
4. Ounces, pounds, and tons are examples of American units of ________.
5. Some examples of American units of ________ are cups, pints, quarts, and gallons.
6. Some units of ______ are seconds, hours, and days.

CONCEPTS *Fill in the blanks.*

7. 12 in. = ☐ ft
8. ☐ ft = 1 yd
9. 1 mi = ☐ ft
10. 1 yd = ☐ in.
11. ☐ ounces = 1 pound
12. ☐ pounds = 1 ton
13. 1 cup = ☐ fluid ounces
14. 1 pint = ☐ cups

15. 2 pints = ☐ quart

16. 4 quarts = ☐ gallon

17. 1 day = ☐ hours

18. 2 hours = ☐ minutes

19. Determine which measurements the arrows point to on the ruler.

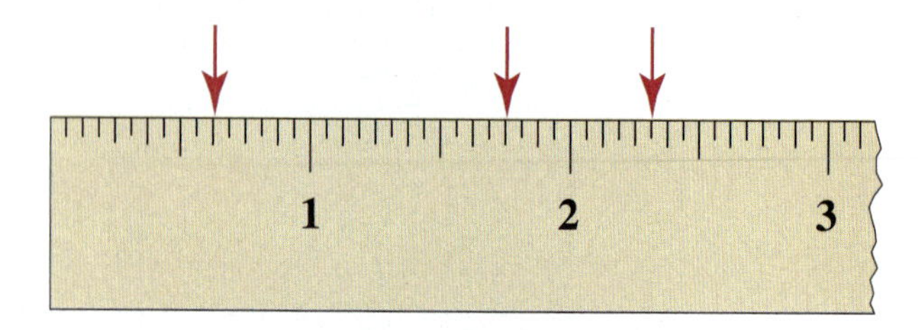

20. Determine which measurements the arrows point to on the ruler below, to the nearest $\frac{1}{8}$ inch.

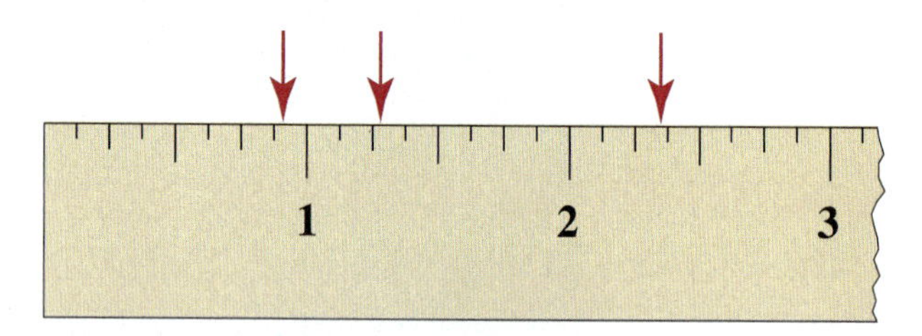

21. Write a unit conversion factor to convert the following.

a. Pounds to tons

b. Quarts to pints

22. Write the two unit conversion factors to convert the following.

a. Inches to yards

b. Days to minutes

23. Match each item with its proper measurement.

a. Length of the U.S. coastline
b. Height of a Barbie doll
c. Span of the Golden Gate Bridge
d. Width of a football field

i. $11\frac{1}{2}$ in.
ii. 4,200 ft
iii. 53.5 yd
iv. 12,383 mi

24. Match each item with its proper measurement.

a. Weight of the men's shot put used in track and field
b. Weight of an African elephant
c. Amount of gold that is worth $500

i. $1\frac{1}{2}$ oz
ii. 16 lb
iii. 7.2 tons

25. Match each item with its proper measurement.

a. Amount of blood in an adult
b. Size of the Exxon Valdez oil spill in 1989
c. Amount of nail polish in a bottle
d. Amount of flour to make 3 dozen cookies

i. $\frac{1}{2}$ fluid oz
ii. 2 cups
iii. 5 qt
iv. 10,080,000 gal

26. Match each item with its proper measurement.

a. Length of first U.S. manned space flight
b. A leap year
c. Time difference between New York and Fairbanks, Alaska
d. Length of Wright Brothers' first flight

i. 12 sec
ii. 15 min
iii. 4 hr
iv. 366 days

NOTATION *Complete each solution.*

27. Convert 12 yards to inches.

$$12 \text{ yd} = 12 \text{ yd} \cdot \frac{\square \text{ in.}}{1 \text{ yd}}$$
$$= 12 \cdot \square \text{ in.}$$
$$= 432 \text{ in.}$$

28. Convert 1 ton to ounces.

$$1 \text{ ton} = 1 \text{ ton} \cdot \frac{\square \text{ lb}}{1 \text{ ton}} \cdot \frac{\square \text{ oz}}{1 \text{ lb}}$$
$$= 1 \cdot 2{,}000 \cdot 16 \text{ oz}$$
$$= \square \text{ oz}$$

29. Convert 12 pints to gallons.

$$12 \text{ pt} = 12 \text{ pt} \cdot \frac{1 \text{ qt}}{\square \text{ pt}} \cdot \frac{1 \text{ gal}}{\square \text{ qt}}$$
$$= \square \cdot \frac{1}{2} \cdot \frac{1}{4} \text{ gal}$$
$$= 1.5 \text{ gal}$$

30. Convert 37,440 minutes to days.

$$37{,}440 \text{ min} = 37{,}440 \text{ min} \cdot \frac{1 \text{ hr}}{\square \text{ min}} \cdot \frac{1 \text{ day}}{\square \text{ hr}}$$
$$= \frac{\square}{60 \cdot 24} \text{ days}$$
$$= 26 \text{ days}$$

PRACTICE *Use a ruler with a scale in inches to measure each object to the nearest $\frac{1}{8}$ inch.*

31. The width of a dollar bill

32. The length of a dollar bill

33. The length (top to bottom) of this page

34. The length of the word supercalifragilisticexpialidocious.

Perform each conversion.

35. 4 feet to inches

36. 7 feet to inches

37. $3\frac{1}{2}$ feet to inches

38. $2\frac{2}{3}$ feet to inches

39. 24 inches to feet

40. 54 inches to feet

41. 8 yards to inches

42. 288 inches to yards

43. 90 inches to yards

44. 12 yards to inches

45. 56 inches to feet

46. 44 inches to feet

47. 5 yards to feet

48. 21 feet to yards

49. 7 feet to yards

50. $4\frac{2}{3}$ yards to feet

51. 15,840 feet to miles

52. 2 miles to feet

53. $\frac{1}{2}$ mile to feet

54. 1,320 feet to miles

55. 80 ounces to pounds

56. 8 pounds to ounces

57. 7,000 pounds to tons

58. 2.5 tons to ounces

59. 12.4 tons to pounds

60. 48,000 ounces to tons

61. 3 quarts to pints

62. 20 quarts to gallons

63. 16 pints to gallons

64. 3 gal to fluid ounces

65. 32 fluid ounces to pints

66. 2 quarts to fluid ounces

67. 240 minutes to hours

68. 2,400 seconds to hours

69. 7,200 minutes to days

70. 691,200 seconds to days

APPLICATIONS

71. THE GREAT PYRAMID The Great Pyramid in Egypt is about 450 feet high. Express this distance in yards.

72. THE WRIGHT BROTHERS In 1903, Orville Wright made the world's first sustained flight. It lasted 12 seconds, and the plane traveled 120 feet. Express the length of the flight in yards.

73. THE GREAT SPHINX The Great Sphinx of Egypt is 240 feet long. Express this in inches.

74. HOOVER DAM The Hoover Dam in Nevada is 726 feet high. Express this distance in inches.

75. THE SEARS TOWER The Sears Tower in Chicago has 110 stories and is 1,454 feet tall. To the nearest hundredth, express this height in miles.

76. NFL RECORDS Emmit Smith, the former Dallas Cowboys and Arizona Cardinals running back, holds the National Football League record for yards rushing in a career: 18,355. How many miles is this? Round to the nearest tenth of a mile.

77. NFL RECORDS When Dan Marino of the Miami Dolphins retired, it was noted that Marino's career passing total was nearly 35 miles! How many yards is this?

78. LEWIS AND CLARK The trail traveled by the Lewis and Clark expedition is shown. When the expedition reached the Pacific Ocean, Clark estimated that they had traveled 4,162 miles. (It was later determined that his guess was within 40 miles of the actual distance.) Express Clark's estimate of the distance in terms of feet.

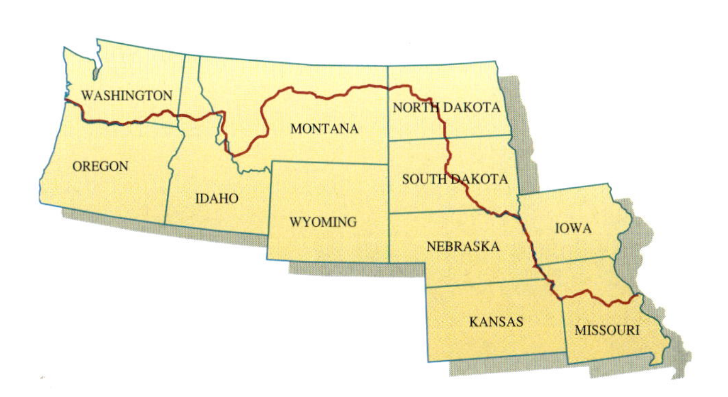

79. WEIGHT OF WATER One gallon of water weighs about 8 pounds. Express this weight in ounces.

80. WEIGHT OF A BABY A newborn baby weighed 136 ounces. Express this weight in pounds.

81. HIPPOS An adult hippopotamus can weigh as much as 9,900 pounds. Express this weight in tons.

82. ELEPHANTS An adult elephant can consume as much as 495 pounds of grass and leaves in one day. How many ounces is this?

83. BUYING PAINT A painter estimates that he will need 17 gallons of paint for a job. To take advantage of a closeout sale on quart cans, he decides to buy the paint in quarts. How many cans will he need to buy?

84. CATERING How many cups of apple cider can be dispensed from a 10-gallon container of cider?

85. SCHOOL LUNCHES Each student attending Eagle River Elementary School receives one pint of milk for lunch each day. If 575 students attend the school, how many gallons of milk are used each day?

86. RADIATORS The radiator capacity of a piece of earth-moving equipment is 39 quarts. If the radiator is drained and new coolant put in, how many gallons of new coolant will be used?

87. CAMPING How many ounces of camping stove fuel will fit in the container shown?

88. HIKING A college student walks 11 miles in 155 minutes. To the nearest tenth, how many hours does he walk?

89. SPACE TRAVEL The astronauts of the Apollo 8 mission, which was launched on December 21, 1968, were in space for 147 hours. How many days did the mission take?

90. AMELIA EARHART In 1935, Amelia Earhart became the first woman to fly across the Atlantic Ocean alone, establishing a new record for the crossing: 13 hours and 30 minutes. How many minutes is this?

WRITING

91. Explain how to find the unit conversion factor that will convert feet to inches.

92. Explain how to find the unit conversion factor that will convert pints to gallons.

REVIEW *Round each number as indicated.*

93. 3,673.263; nearest hundred

94. 3,673.263; nearest ten

95. 3,673.263; nearest hundredth

96. 3,673.263; nearest tenth

97. 0.100602; nearest thousandth

98. 0.100602; nearest hundredth

99. 0.09999; nearest tenth

100. 0.09999; nearest one

8.4 Metric Units of Measurement

- Metric units of length
- Converting units of length
- Metric units of mass
- Metric units of capacity
- Cubic centimeters

The metric system is the system of measurement used by most countries in the world. All countries, including the United States, use it for scientific purposes. The metric system, like our decimal numeration system, is based on the number 10. For this reason, converting from one metric unit to another is easier than with the American system.

Metric units of length

The basic metric unit of length is the **meter** (m). One meter is approximately 39 inches, slightly more than 1 yard. Figure 8-11 shows the relative sizes of a yardstick and a meterstick.

1 yard:
36 inches

1 meter:
about 39 inches

FIGURE 8-11

Larger and smaller units are created by prefixes in front of this basic unit, *meter.*

deka means tens
hecto means hundreds
kilo means thousands
deci means tenths
centi means hundredths
milli means thousandths

Metric units of length

1 dekameter (dam) = 10 meters.
1 dam is a little less than 11 yards.

1 hectometer (hm) = 100 meters.
1 hm is about 1 football field long, plus one end zone.

1 kilometer (km) = 1,000 meters.
1 km is about $\frac{3}{5}$ mile.

1 decimeter (dm) = $\frac{1}{10}$ of 1 meter.
1 dm is about the length of your palm.

1 centimeter (cm) = $\frac{1}{100}$ of 1 meter.
1 cm is about as wide as the nail of your little finger.

1 millimeter (mm) = $\frac{1}{1,000}$ of 1 meter.
1 mm is about the thickness of a dime.

TABLE 8-1

Figure 8-12 shows a portion of a metric ruler, scaled in centimeters, and a ruler scaled in inches. The rulers are used to measure several lengths.

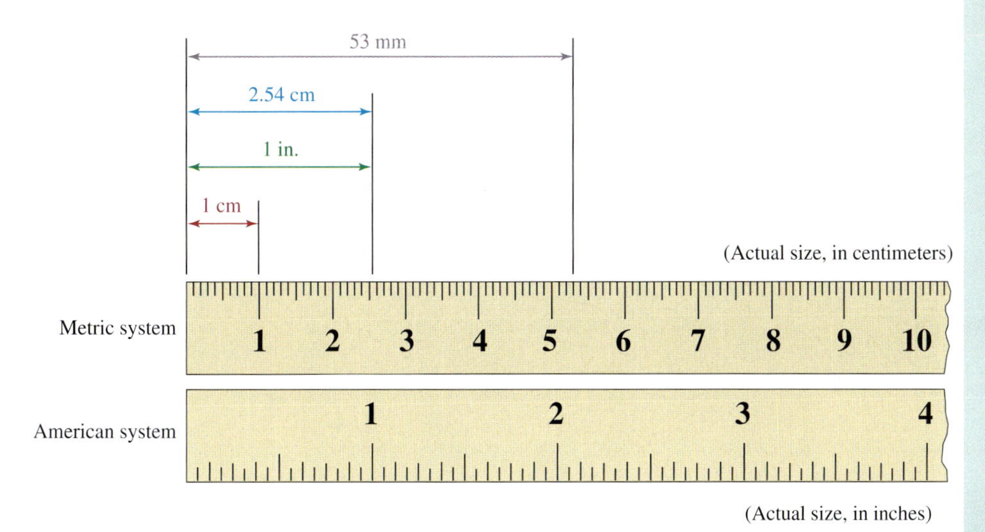

FIGURE 8-12

EXAMPLE 1 To the nearest centimeter, find the length of the nail in Figure 8-13.

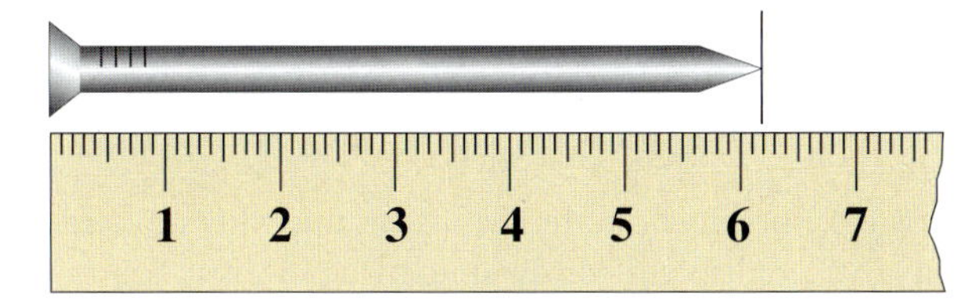

FIGURE 8-13

Solution We place the end of the ruler by one end of the nail and note that the other end of the nail is closer to the 6-cm mark than to the 7-cm mark on the ruler. To the nearest centimeter, the nail is 6 cm long.

Self Check 1
To the nearest centimeter, find the width of the circle below.

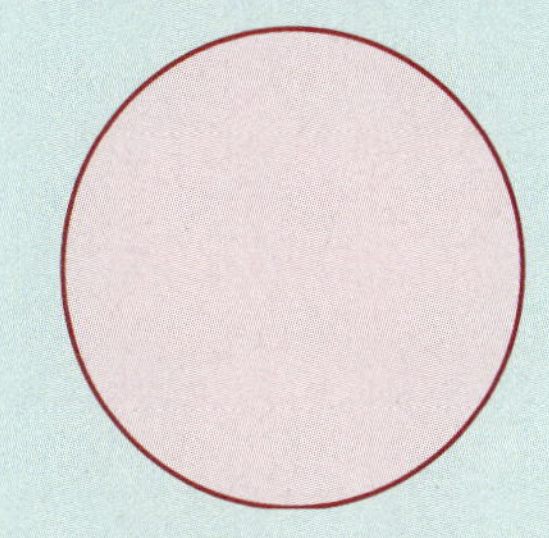

Answer 3 cm

Self Check 2
To the nearest millimeter, find the length of the jumbo paper clip below.

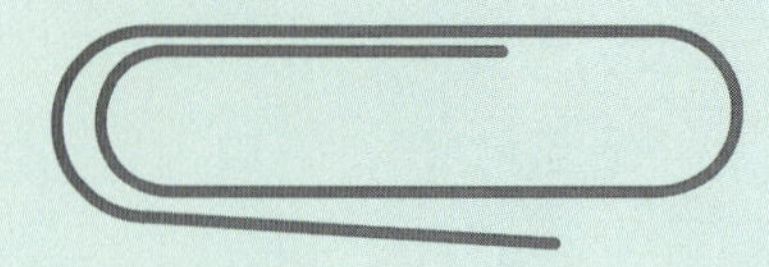

Answer 47 mm

EXAMPLE 2 To the nearest millimeter, find the length of the paper clip in Figure 8-14.

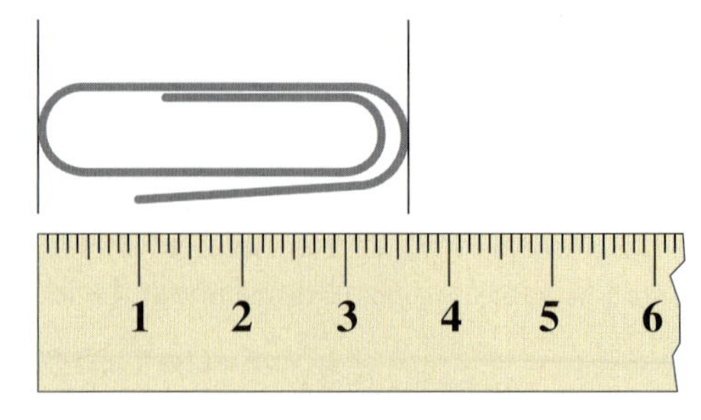

FIGURE 8-14

Solution On the ruler, each centimeter has been divided into 10 millimeters. We place the end of the ruler by one end of the paper clip and note that the other end is closer to the 36-mm mark than to the 37-mm mark on the ruler. To the nearest millimeter, the paper clip is 36 mm long.

Converting units of length

Metric units of length are related as shown in Table 8-2.

Metric units of length

1 kilometer (km) = 1,000 meters	or	1 meter = $\frac{1}{1,000}$ kilometer
1 hectometer (hm) = 100 meters	or	1 meter = $\frac{1}{100}$ hectometer
1 dekameter (dam) = 10 meters	or	1 meter = $\frac{1}{10}$ dekameter
1 decimeter (dm) = $\frac{1}{10}$ meter	or	1 meter = 10 decimeters
1 centimeter (cm) = $\frac{1}{100}$ meter	or	1 meter = 100 centimeters
1 millimeter (mm) = $\frac{1}{1,000}$ meter	or	1 meter = 1,000 millimeters

TABLE 8-2

We can use the information in the table to write unit conversion factors that can be used to convert metric units of length. For example, in the table we see that

1 meter = 100 centimeters

From this fact, we can write two unit conversion factors.

$$\frac{1\text{ m}}{100\text{ cm}} = 1 \quad \text{and} \quad \frac{100\text{ cm}}{1\text{ m}} = 1$$

To obtain the first unit conversion factor, divide both sides of the equation 1 m = 100 cm by 100 cm. To obtain the second unit conversion factor, divide both sides by 1 m.

One advantage of the metric system is that multiplying or dividing by a unit conversion factor involves multiplying or dividing by a power of 10.

Self Check 3
Convert 860 centimeters to meters.

EXAMPLE 3 Convert 350 centimeters to meters.

Solution Since there is 1 meter per 100 centimeters, we multiply 350 centimeters by the unit conversion factor $\frac{1\text{ m}}{100\text{ cm}}$ to get

$$350 \text{ cm} = \frac{350 \text{ cm}}{1} \cdot \frac{\mathbf{1\ m}}{\mathbf{100\ cm}}$$ Multiply by 1: $\frac{1 \text{ m}}{100 \text{ cm}} = 1$.

$$= \frac{350 \not{\text{cm}}}{1} \cdot \frac{1 \text{ m}}{100 \not{\text{cm}}}$$ The units of centimeters divide out.

$$= \frac{350}{100} \text{ m}$$

$$= 3.5 \text{ m}$$ Divide by 100 by moving the decimal point 2 places to the left.

Thus, 350 centimeters = 3.5 meters.

Answer 8.6 m

In Example 3, we converted 350 centimeters to meters using a unit conversion factor. We can also make this conversion by recognizing that all units of length in the metric system are powers of 10 of a meter. Converting from one unit to another is as easy as multiplying by the correct power of 10 or simply by moving a decimal point the correct number of places to the right or left. For example, in the chart below, we see that to convert from centimeters to meters, we move 2 places to the left.

km hm dam **m** dm **cm** mm

To go from centimeters to meters, we must move 2 places to the left.

If we write 350 centimeters as 350.0 centimeters, we can convert to meters by moving the decimal point 2 places to the left.

350.0 centimeters = 3.50 0 meters = 3.5 meters

With the unit conversion factor method or the chart method, we get 350 cm = 3.5 m.

! COMMENT When using a chart to help make a metric conversion, be sure to list the units from largest to smallest when reading from left to right.

EXAMPLE 4 Convert 2.4 meters to millimeters.

Solution Since there are 1,000 millimeters per meter, we multiply 2.4 meters by the unit conversion factor $\frac{1{,}000 \text{ mm}}{1 \text{ m}}$ to get

$$2.4 \text{ m} = \frac{2.4 \text{ m}}{1} \cdot \frac{\mathbf{1{,}000\ mm}}{\mathbf{1\ m}}$$ Multiply by 1: $\frac{1{,}000 \text{ mm}}{1 \text{ m}} = 1$.

$$= \frac{2.4 \not{\text{m}}}{1} \cdot \frac{1{,}000 \text{ mm}}{1 \not{\text{m}}}$$ The units of meters divide out.

$$= 2.4 \cdot 1{,}000 \text{ mm}$$

$$= 2{,}400 \text{ mm}$$ Multiply by 1,000 by moving the decimal point 3 places to the right.

Thus, 2.4 meters = 2,400 millimeters.

Self Check 4
Convert 5.3 meters to millimeters.

We can also make this conversion using a chart.

km hm dam **m** dm cm **mm**

From the chart, we see that we should move the decimal point 3 places to the right to convert from meters to millimeters.

2.4 meters = 2 400. millimeters = 2,400 millimeters

Answer 5,300 mm

Self Check 5
Convert 5.15 kilometers to centimeters.

EXAMPLE 5 Convert 3.2 kilometers to centimeters.

Solution To convert to centimeters, we set up the problem so kilometers divide out and leave us with units of centimeters. Since there are 1,000 meters per kilometer and 100 centimeters per meter, we multiply 3.2 kilometers by $\frac{1{,}000\text{ m}}{1\text{ km}}$ and $\frac{100\text{ cm}}{1\text{ m}}$.

$$3.2\text{ km} = \frac{3.2\ \cancel{\text{km}}}{1} \cdot \frac{1{,}000\ \cancel{\text{m}}}{1\ \cancel{\text{km}}} \cdot \frac{100\text{ cm}}{1\ \cancel{\text{m}}}$$ The units of kilometers and meters divide out.

$$= 3.2 \cdot 1{,}000 \cdot 100\text{ cm}$$

$$= 320{,}000\text{ cm}$$ Multiply by 1,000 and 100 by moving the decimal point 5 places to the right.

Thus, 3.2 kilometers = 320,000 centimeters.

Using a chart, we see that the decimal point should be moved 5 places to the right to convert kilometers to centimeters.

km hm dam m dm **cm** mm

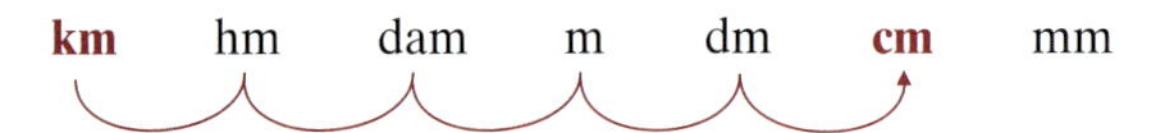

3.2 kilometers = 3 20000. centimeters = 320,000 centimeters

Answer 515,000 cm

Metric units of mass

The **mass** of an object is a measure of the amount of material in the object. When an object is moved about in space, its mass does not change. One basic unit of mass in the metric system is the **gram** (g). A gram is defined to be the mass of water contained in a cube having sides 1 centimeter long. (See Figure 8-15.)

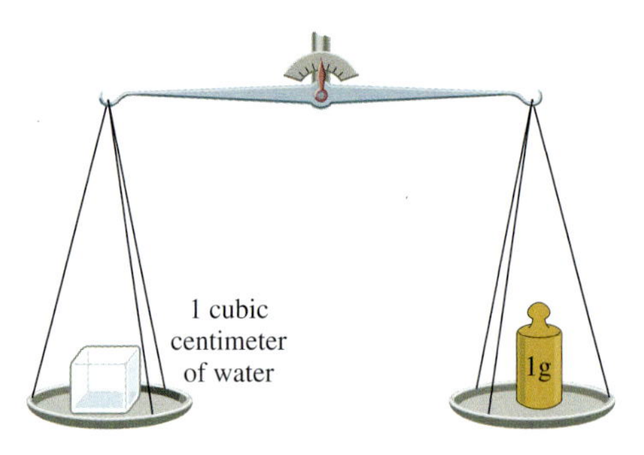

FIGURE 8-15

The **weight** of an object is determined by the Earth's gravitational pull on the object. Since gravitational pull on an object decreases as the object gets farther from Earth, the object weighs less as it gets farther from the Earth's surface. This is why astronauts experience weightlessness in space. However, since most of us remain near the Earth's surface, we will use the words *mass* and *weight* interchangeably. Thus, a mass of 30 grams is said to weigh 30 grams.

Metric units of mass are related as shown in Table 8-3.

Metric units of mass

1 kilogram (kg) = 1,000 grams	or	1 gram = $\frac{1}{1,000}$ kilogram
1 hectogram (hg) = 100 grams	or	1 gram = $\frac{1}{100}$ hectogram
1 dekagram (dag) = 10 grams	or	1 gram = $\frac{1}{10}$ dekagram
1 decigram (dg) = $\frac{1}{10}$ gram	or	1 gram = 10 decigrams
1 centigram (cg) = $\frac{1}{100}$ gram	or	1 gram = 100 centigrams
1 milligram (mg) = $\frac{1}{1,000}$ gram	or	1 gram = 1,000 milligrams

TABLE 8-3

Here are some examples of these units of mass.

- An average bowling ball weighs about 6 kilograms.
- A raisin weighs about 1 gram.
- A certain vitamin tablet contains 450 milligrams of calcium.

We can use the information in Table 8-3 to write unit conversion factors that can be used to convert metric units of mass. For example, in the table we see that

1 kilogram = 1,000 grams

From this fact, we can write two unit conversion factors.

$$\frac{1 \text{ kg}}{1{,}000 \text{ gm}} = 1 \quad \text{and} \quad \frac{1{,}000 \text{ gm}}{1 \text{ kg}} = 1$$

To obtain the first unit conversion factor, divide both sides of the equation 1 kg = 1,000 g by 1,000 g. To obtain the second unit conversion factor, divide both sides by 1 kg.

EXAMPLE 6 Convert 7.2 kilograms to grams.

Solution To convert to grams, we set up the problem so that the units of kilograms divide out and leave us with the units of grams. Since there are 1,000 grams per 1 kilogram, we multiply 7.2 kilograms by $\frac{1,000 \text{ g}}{1 \text{ kg}}$.

$$7.2 \text{ kg} = \frac{7.2 \not{\text{kg}}}{1} \cdot \frac{1{,}000 \text{ g}}{1 \not{\text{kg}}}$$ Divide out the units of kilograms.

$$= 7.2 \cdot 1{,}000 \text{ g}$$

$$= 7{,}200 \text{ g}$$ Perform the multiplication by moving the decimal point 3 places to the right.

Thus, 7.2 kilograms = 7,200 grams.

To use a chart to make the conversion, we list the metric units of weight from the largest (kilograms) to the smallest (milligrams).

kg hg dag **g** dg cg mg

From the chart, we see that we must move the decimal point 3 places to the right to change kilograms to grams.

7.2 kilograms = 7 200. grams = 7,200 grams

Self Check 6
Convert 5 kilograms to grams.

Answer 5,000 g

Self Check 7

One brand name for Verapamil is Isoptin. If a bottle of Isoptin contains 90 tablets, each containing 200 mg of active ingredient, how many centigrams of active ingredient are in the bottle?

Answer 1,800 cg

EXAMPLE 7 **Medications.** A bottle of Verapamil, a drug taken for high blood pressure, contains 30 tablets. If each tablet contains 180 mg of active ingredient, how many centigrams of active ingredient are in the bottle?

Solution Since there are 30 tablets and each one contains 180 mg of active ingredient, there are

$$30 \cdot 180 \text{ mg} = 5{,}400 \text{ mg}$$

of active ingredient in the bottle.

To convert milligrams to centigrams, we multiply 5,400 milligrams by $\frac{1 \text{ g}}{1{,}000 \text{ mg}}$ and $\frac{100 \text{ cg}}{1 \text{ g}}$ to get

$$5{,}400 \text{ mg} = \frac{5{,}400 \text{ mg}}{1} \cdot \frac{1 \text{ g}}{1{,}000 \text{ mg}} \cdot \frac{100 \text{ cg}}{1 \text{ g}}$$ Divide out the units of milligrams and grams.

$$= \frac{5{,}400 \cdot 100}{1{,}000} \text{ cg}$$ Multiply the fractions.

$$= 540 \text{ cg}$$ Simplify

There are 540 centigrams of active ingredient in the bottle.

Using a chart, we see that we must move the decimal point 1 place to the left to convert from milligrams to centigrams.

kg hg dag g dg **cg** **mg**

5,400 milligrams = 540.0 centigrams = 540 centigrams

Metric units of capacity

In the metric system, one basic unit of capacity is the **liter** (L), which is defined to be the capacity of a cube with sides 10 centimeters long. (See Figure 8-16.) A liter of liquid is slightly more than 1 quart.

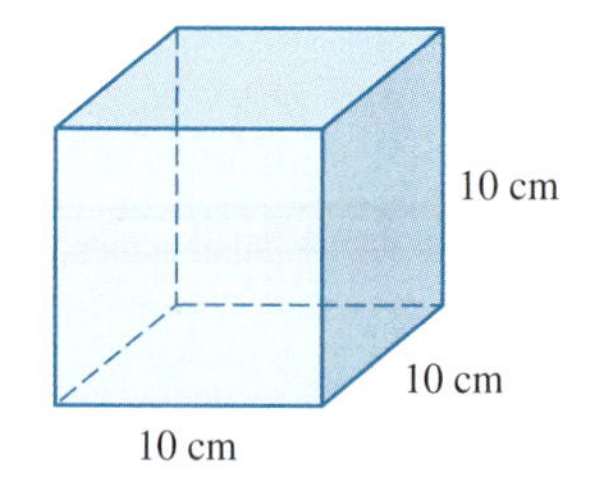

FIGURE 8-16

Metric units of capacity are related as shown in Table 8-4.

Metric units of capacity

1 kiloliter (kL) = 1,000 liters	or	1 liter = $\frac{1}{1{,}000}$ kiloliter
1 hectoliter (hL) = 100 liters	or	1 liter = $\frac{1}{100}$ hectoliter
1 dekaliter (daL) = 10 liters	or	1 liter = $\frac{1}{10}$ dekaliter
1 deciliter (dL) = $\frac{1}{10}$ liter	or	1 liter = 10 deciliters
1 centiliter (cL) = $\frac{1}{100}$ liter	or	1 liter = 100 centiliters
1 milliliter (mL) = $\frac{1}{1{,}000}$ liter	or	1 liter = 1,000 milliliters

TABLE 8-4

Here are some examples of these units of capacity.

- Soft drinks are sold in 2-liter plastic bottles.
- The fuel tank of a certain minivan can hold about 75 liters of gasoline.
- Chemists use glass cylinders, scaled in milliliters, to measure liquids.

We can use the information in Table 8-4 to write unit conversion factors that can be used to convert metric units of capacity. For example, in the table we see that

1 liter = 100 centiliters

From this fact, we can write two unit conversion factors.

$$\frac{1\text{ L}}{100\text{ cL}} = 1 \quad \text{and} \quad \frac{100\text{ cL}}{1\text{ L}} = 1$$

EXAMPLE 8 Soft drinks. How many centiliters are in three 2-liter bottles of cola?

Solution Three 2-liter bottles of cola contain 6 liters of cola. To convert to centiliters, we set up the problem so that liters divide out and leave us with centiliters. Since there are 100 centiliters per 1 liter, we multiply 6 liters by the unit conversion factor $\frac{100\text{ cL}}{1\text{ L}}$.

$$6\text{ L} = 6\text{ L} \cdot \frac{\mathbf{100\text{ cL}}}{\mathbf{1\text{ L}}} \qquad \text{Multiply by 1: } \frac{100\text{ cL}}{1\text{ L}} = 1.$$

$$= \frac{6\,\overset{1}{\cancel{\text{L}}}}{1} \cdot \frac{100\text{ cL}}{1\,\underset{1}{\cancel{\text{L}}}} \qquad \text{The units of liters divide out.}$$

$$= 6 \cdot 100\text{ cL}$$

$$= 600\text{ cL}$$

Thus, there are 600 centiliters in three 2-liter bottles of cola.

To make this conversion using a chart, we list the metric units of capacity in order from largest (kiloliter) to smallest (milliliter).

kL hL daL **L** dL **cL** mL

From the chart, we see that we should move the decimal point 2 places to the right to convert from liters to centiliters.

6 liters = 6 00. centiliters = 600 centiliters

Self Check 8

How many milliliters are in two 2-liter bottles of cola?

Answer 4,000 mL

Cubic centimeters

Another metric unit of capacity is the **cubic centimeter,** which is represented by the notation cm^3 or, more simply, cc. One milliliter and one cubic centimeter represent the same capacity.

$1\text{ mL} = 1\text{ cm}^3 = 1\text{ cc}$

The units of cubic centimeters are used frequently in medicine. For example, when a nurse administers an injection containing 5 cc of medication, the dose could also be expressed in milliliters.

5 cc = 5 mL

When a doctor orders that a patient be put on 1,000 cc of dextrose solution, the request could be expressed in these ways.

1,000 cc = 1,000 mL = 1 liter

Section 8.4 STUDY SET

VOCABULARY *Fill in the blanks.*

1. *Deka* means ______.
2. *Hecto* means __________.
3. *Kilo* means __________.
4. *Deci* means _______.
5. *Centi* means ___________.
6. *Milli* means ___________.
7. Meters, grams, and liters are units of measurement in the _______ system.
8. The _______ of an object is determined by the Earth's gravitational pull on the object.

CONCEPTS

9. To the nearest centimeter, determine which measurements the arrows point to on the ruler below.

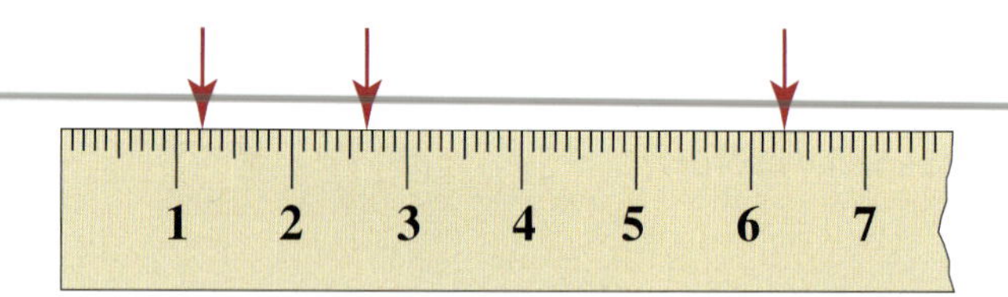

10. To the nearest millimeter, determine which measurements the arrows point to on the ruler below.

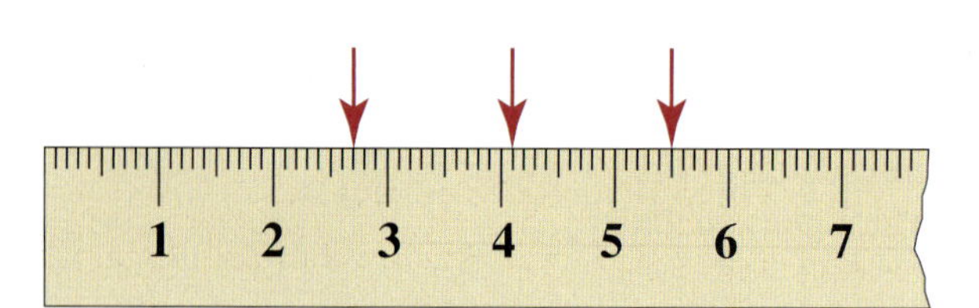

11. Write a unit conversion factor to convert the following.
    a. Meters to kilometers
    b. Grams to centigrams
    c. Liters to milliliters

12. Use the chart to determine how many decimal places and in which direction to move the decimal point when converting the following.
    a. Kilometers to centimeters

       km hm dam m dm cm mm

    b. Milligrams to grams

       kg hg dag g dg cg mg

    c. Hectoliters to centiliters

       kL hL daL L dL cL mL

13. Match each item with its proper measurement.
    a. Thickness of a phone book
    b. Length of the Amazon River
    c. Height of a soccer goal

    i. 6,275 km
    ii. 2 m
    iii. 6 cm

14. Match each item with its proper measurement.
    a. Weight of a giraffe
    b. Weight of a paper clip
    c. Active ingredient in an aspirin tablet

    i. 800 kg
    ii. 1 g
    iii. 325 mg

15. Match each item with its proper measurement.
    a. Amount of blood in an adult
    b. Cola in an aluminum can
    c. Kuwait's daily production of crude oil

    i. 290,000 kL
    ii. 6 L
    iii. 355 mL

16. Of the following objects on the next page, which can be used to measure the following?
    a. Millimeters
    b. Milligrams
    c. Milliliters

Balance

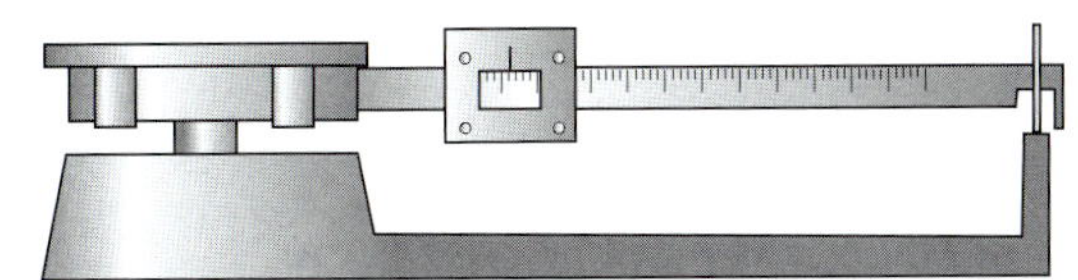

Beaker

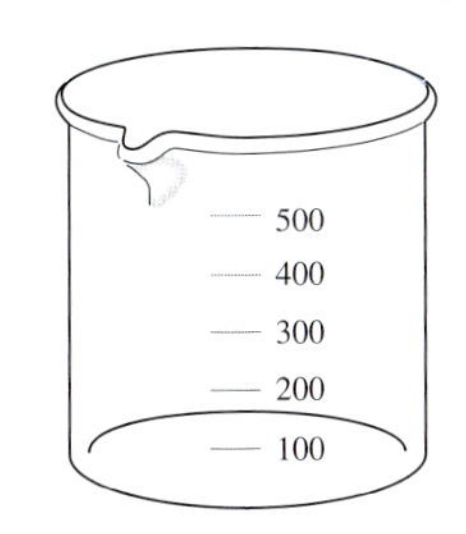

Micrometer

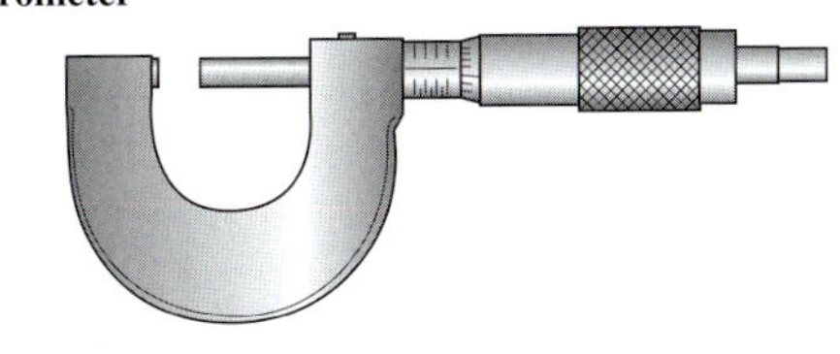

Fill in the blanks.

17. 1 dekameter = ____ meters

18. 1 decimeter = ____ meter

19. 1 centimeter = ____ meter

20. 1 kilometer = ____ meters

21. 1 millimeter = ____ meter

22. 1 hectometer = ____ meters

23. 1 gram = ____ milligrams

24. 100 centigrams = ____ gram

25. 1 kilogram = ____ grams

26. 1 milliliter = ____ cubic centimeter

27. 1 liter = ____ cubic centimeters

28. 1 kiloliter = ____ liters

29. 1 centiliter = ____ liter

30. 1 milliliter = ____ liter

31. 100 liters = ____ hectoliter

32. 10 deciliters = ____ liter

NOTATION *Complete each solution.*

33. Convert 20 centimeters to meters.

$$\begin{aligned} 20 \text{ cm} &= 20 \text{ cm} \cdot \frac{\square \text{ m}}{100 \text{ cm}} \\ &= \frac{20}{\square} \text{ m} \\ &= 0.2 \text{ m} \end{aligned}$$

34. Convert 300 centigrams to grams.

$$\begin{aligned} 300 \text{ cg} &= 300 \text{ cg} \cdot \frac{\square \text{ g}}{100 \text{ cg}} \\ &= \frac{\square}{100} \text{ g} \\ &= 3 \text{ g} \end{aligned}$$

35. Convert 2 kilometers to decimeters.

$$\begin{aligned} 2 \text{ km} &= 2 \text{ km} \cdot \frac{\square \text{ m}}{1 \text{ km}} \cdot \frac{10 \text{ dm}}{\square \text{ m}} \\ &= 2 \cdot \square \cdot 10 \text{ dm} \\ &= 20{,}000 \text{ dm} \end{aligned}$$

36. Convert 3 deciliters to milliliters.

$$\begin{aligned} 3 \text{ dL} &= 3 \text{ dL} \cdot \frac{1 \text{ L}}{\square \text{ dL}} \cdot \frac{\square \text{ mL}}{1 \text{ L}} \\ &= \frac{\square \cdot 1{,}000}{10} \text{ mL} \\ &= 300 \text{ mL} \end{aligned}$$

PRACTICE *Use a metric ruler to measure each object to the nearest millimeter.*

37. The length of a dollar bill

38. The width of a dollar bill

Use a metric ruler to measure each object to the nearest centimeter.

39. The length (top to bottom) of this page

40. The length of the word antidisestablishmentarianism

Convert each measurement between the given metric units.

41. 3 m = ____ cm

42. 5 m = ____ cm

43. 5.7 m = ____ cm

44. 7.36 km = ____ dam

45. 0.31 dm = ____ cm

46. 73.2 m = ____ dm

47. 76.8 hm = ________ mm

48. 165.7 km = _______ m

49. 4.72 cm = _____ dm

50. 0.593 cm = ________ dam

51. 453.2 cm = ______ m

52. 675.3 cm = ______ m

53. 0.325 dm = ______ m

54. 0.0034 mm = ________ m

55. 3.75 cm = _____ mm

56. 0.074 cm = _____ mm

57. 0.125 m = _____ mm

58. 134 m = _____ hm

59. 675 dam = _______ cm

60. 0.00777 cm = __________ dam

61. 638.3 m = ______ hm

62. 6.77 cm = _______ m

63. 6.3 mm = _____ cm

64. 6.77 mm = ______ cm

65. 695 dm = _____ m

66. 6,789 cm = ______ dm

67. 5,689 m = ______ km

68. 0.0579 km = _______ mm

69. 576.2 mm = ______ dm

70. 65.78 km = ______ dam

71. 6.45 dm = ________ km

72. 6.57 cm = _____ mm

73. 658.23 m = _______ km

74. 0.0068 hm = _______ km

75. 3 g = ______ mg

76. 5 g = _____ cg

77. 2 kg = ______ g

78. 4,000 g = ___ kg

79. 1,000 kg = ________ g

80. 2 kg = _______ cg

81. 500 mg = ____ g

82. 500 mg = ____ cg

83. 3 kL = ______ L

84. 500 mL = ____ L

85. 500 cL = ______ mL

86. 400 L = ___ hL

87. 10 mL = ___ cc

88. 2,000 cc = ___ L

APPLICATIONS

89. SPEED SKATING American Eric Heiden won an unprecedented five gold medals by capturing the men's 500 m, 1,000 m, 1,500 m, 5,000 m, and 10,000 m races at the 1980 Winter Olympic Games in Lake Placid, New York. Convert each race length to kilometers.

90. THE SUEZ CANAL The 163-km-long Suez Canal, connects the Mediterranean Sea with the Red Sea. It provides a shortcut for ships operating between European and American ports. Convert the length of the Suez Canal to meters.

91. HEALTH CARE Blood pressure is measured by a *sphygmomanometer.* The measurement is read at two points and is expressed, for example, as 120/80. This indicates a *systolic* pressure of 120 millimeters of mercury and a *diastolic* pressure of 80 millimeters of mercury. Convert each measurement to centimeters of mercury.

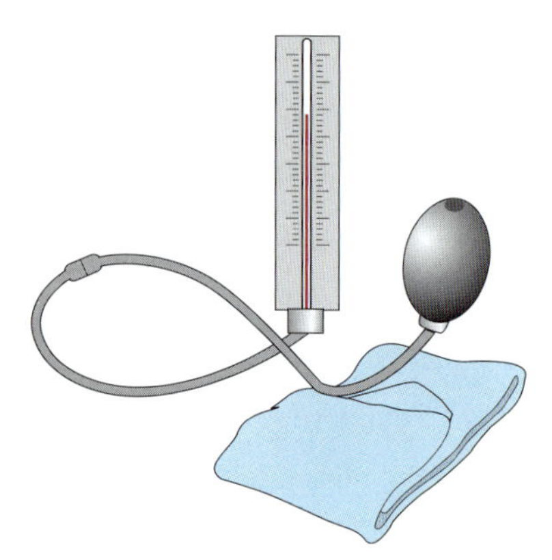

92. THE HANCOCK CENTER The John Hancock Center in Chicago has 100 stories and is 343 meters high. Give this height in hectometers.

93. WEIGHT OF A BABY A baby weighs 4 kilograms. Give this weight in centigrams.

94. JEWELRY A gold chain weighs 1,500 milligrams. Give this weight in grams.

95. CONTAINERS How many deciliters of root beer are in two 2-liter bottles?

96. BOTTLING How many liters of wine are in a 750-mL bottle?

97. BUYING OLIVES The net weight of a bottle of olives is 284 grams. Find the smallest number of bottles that must be purchased to have at least 1 kilogram of olives.

98. BUYING COFFEE A can of Cafe Vienna has a net weight of 133 grams. Find the smallest number of cans that must be packaged to have at least 1 metric ton of coffee. (*Hint:* 1 metric ton = 1,000 kg.)

99. MEDICINE A bottle of hydrochlorothiazine contains 60 tablets. If each tablet contains 50 milligrams of active ingredient, how many grams of active ingredient are in the bottle?

100. INJECTIONS The illustration shows a 3 cc syringe. Express its capacity in milliliters.

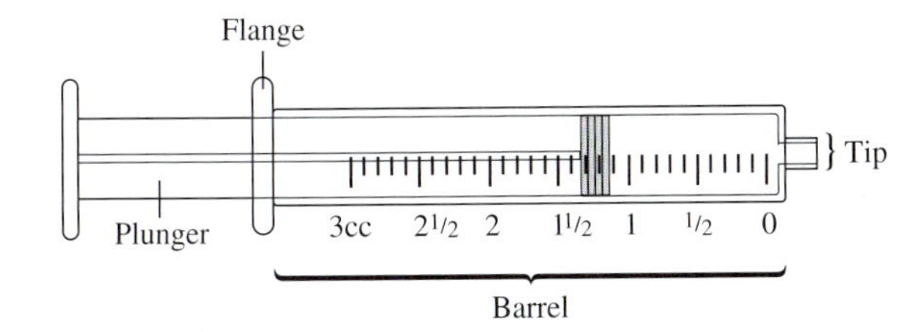

WRITING

101. To change 3.452 kilometers to meters, we can move the decimal point in 3.452 three places to the right to get 3,452 meters. Explain why.

102. To change 7,532 grams to kilograms, we can move the decimal point in 7,532 three places to the left to get 7.532 kilograms. Explain why.

103. A centimeter is one-hundredth of a meter. Make a list of other words that begin with the prefix *centi* or *cent* and write a definition for each.

104. List the advantages of the metric system of measurement as compared to the American system. There have been several attempts to bring the metric system into general use in the United States. Why do you think these efforts have been unsuccessful?

REVIEW

105. Find 7% of \$342.72.

106. Add: $(3x - 1) + (6x + 5)$.

107. \$32.16 is 8% of what amount?

108. Divide: $3\frac{1}{7} \div 2\frac{1}{2}$.

109. Simplify: $3\frac{1}{7} + 2\frac{1}{2} \cdot 3\frac{1}{3}$.

110. Solve: $\frac{x}{5} - 3 = -3$.

8.5 Converting Between American and Metric Units

- Converting between American and metric units
- Comparing American and metric units of temperature

It is often necessary to convert between American units and metric units. For example, we must convert units to answer the following questions.

- Which is higher: Pikes Peak (elevation 14,110 feet) or the Matterhorn (elevation 4,478 meters)?
- Does a 2-pound tub of butter weigh more than a 1-kilogram tub?
- Is a quart of soda pop more or less than a liter of soda pop?

In this section, we will discuss how to answer such questions.

Converting between American and metric units

We can convert between American and metric units of length using the table below.

Equivalent lengths	
American to metric	**Metric to American**
1 in. ≈ 2.54 cm	1 cm ≈ 0.3937 in.
1 ft ≈ 0.3048 m	1 m ≈ 3.2808 ft
1 yd ≈ 0.9144 m	1 m ≈ 1.0936 yd
1 mi ≈ 1.6093 km	1 km ≈ 0.6214 mi

Self Check 1
Refer to Figure 8-17. Find the inseam length, to the nearest inch.

Answer 30 in.

EXAMPLE 1 **Clothing sizes.** Figure 8-17 shows a label sewn into some pants for sale in the United States. Express the waist size to the nearest inch.

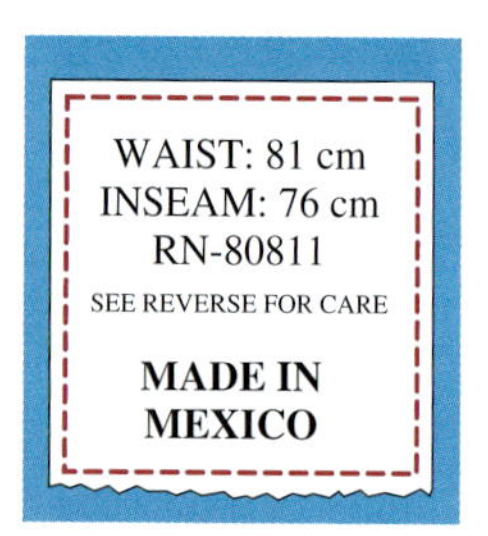

FIGURE 8-17

Solution We need to convert from metric to American units. From the table, we see that there is 0.3937 inch in 1 centimeter. To make the conversion, we substitute 0.3937 inch for 1 centimeter.

$$
\begin{aligned}
81 \text{ centimeters} &= 81(\textbf{centimeters}) \\
&\approx 81(\textbf{0.3937 in.}) && \text{Substitute 0.3937 inch for 1 centimeter.} \\
&\approx 31.8897 \text{ in.} && \text{Perform the multiplication.}
\end{aligned}
$$

To the nearest inch, the waist size is 32 inches.

Self Check 2
Which is longer: a 500-meter race or a 550-yard race?

Answer the 550-yard race

EXAMPLE 2 **Mountain elevations.** Pikes Peak, one of the most famous peaks in the Rocky Mountains, has an elevation of 14,110 feet. The Matterhorn, in the Swiss Alps, rises to an elevation of 4,478 meters. Which mountain is higher?

Solution To make a comparison, we must express the elevations in the same units. We will convert the elevation of Pikes Peak, which is given in feet, to meters.

$$
\begin{aligned}
14{,}110 \text{ feet} &= 14{,}110(\textbf{feet}) \\
&\approx 14{,}110(\textbf{0.3048 m}) && \text{Substitute 0.3048 meter for 1 foot.} \\
&\approx 4{,}300.728 \text{ m} && \text{Perform the multiplication.}
\end{aligned}
$$

Since the elevation of Pikes Peak is about 4,301 meters, we can conclude that the Matterhorn, with an elevation of 4,478 meters, is higher.

We can convert between American units of weight and metric units of mass by using the accompanying table.

Equivalent weights and masses	
American to metric	**Metric to American**
1 oz ≈ 28.35 g	1 g ≈ 0.035 oz
1 lb ≈ 0.454 kg	1 kg ≈ 2.2 lb

EXAMPLE 3 Change 50 pounds to grams.

Solution

$$\begin{aligned} 50 \text{ lb} &= 50(\mathbf{1\ lb}) \\ &= 50(\mathbf{16\ oz}) && \text{Substitute 16 ounces for 1 pound.} \\ &= 50(16)(\mathbf{1\ oz}) \\ &\approx 50(16)(\mathbf{28.35\ g}) && \text{Substitute 28.35 grams for 1 ounce.} \\ &\approx 22{,}680 \text{ g} && \text{Perform the multiplication.} \end{aligned}$$

Thus, 50 pounds is approximately 22,680 grams.

Self Check 3
Change 20 kilograms to pounds.

Answer 44 lb

EXAMPLE 4 Does a 2-pound tub of butter weigh more than a 1-kilogram tub?

Solution To decide which contains more butter, we can change 2 pounds to kilograms.

$$\begin{aligned} 2 \text{ lb} &= 2(\mathbf{1\ lb}) \\ &\approx 2(\mathbf{0.454\ kg}) && \text{Substitute 0.454 kilogram for 1 pound.} \\ &\approx 0.908 \text{ kg} && \text{Perform the multiplication.} \end{aligned}$$

Since a 2-pound tub weighs approximately 0.908 kilogram, the 1-kilogram tub weighs more.

Self Check 4
Who weighs more: a person who weighs 165 pounds or one who weighs 76 kilograms?

Answer the person who weighs 76 kg

We can convert between American and metric units of capacity by using the accompanying table.

Equivalent capacities	
American to metric	**Metric to American**
1 fl oz ≈ 0.030 L	1 L ≈ 33.8 fl oz
1 pt ≈ 0.473 L	1 L ≈ 2.1 pt
1 qt ≈ 0.946 L	1 L ≈ 1.06 qt
1 gal ≈ 3.785 L	1 L ≈ 0.264 gal

EXAMPLE 5 **Soft drinks.** A bottle of 7-UP contains 750 milliliters. Convert this measure to quarts.

Solution We convert milliliters to liters and then liters to quarts.

Self Check 5
A student bought a 355-mL can of cola. How many ounces of cola does the can contain?

$$
\begin{aligned}
750 \text{ mL} &= 750 \text{ mL} \cdot \frac{1 \text{ L}}{1{,}000 \text{ mL}} && \text{Use a unit conversion factor: } \frac{1 \text{ L}}{1{,}000 \text{ mL}} = 1. \\
&= \frac{750}{1{,}000} \text{ L} && \text{The units of mL divide out.} \\
&= \frac{3}{4} \textbf{ L} && \text{Simplify the fraction: } \frac{750}{1{,}000} = \frac{3 \cdot \cancel{250}^{1}}{4 \cdot \cancel{250}_{1}} = \frac{3}{4}. \\
&\approx \frac{3}{4}(\textbf{1.06 qt}) && \text{Substitute 1.06 quarts for 1 liter.} \\
&\approx 0.795 \text{ qt} && \text{Perform the multiplication.}
\end{aligned}
$$

The bottle contains approximately 0.795 quart.

Answer 12 oz

From the table of equivalent capacities, we see that 1 liter is equal to 1.06 quarts. Thus, 1 liter of soda pop is more than 1 quart of soda pop.

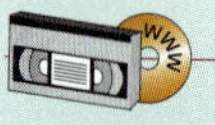

Self Check 6

Thirty-four fluid ounces of aged vinegar costs \$3.49. A 1-liter bottle of the same vinegar costs \$3.17. Which is the better buy?

EXAMPLE 6 **Comparison shopping.** A 2-quart bottle of soda pop is priced at \$1.89, and a 1-liter bottle is priced at 97¢. Which is the better buy?

Solution We can convert 2 quarts to liters and find the price per liter of the 2-quart bottle.

$$
\begin{aligned}
2 \text{ qt} &= 2(\textbf{1 qt}) \\
&\approx 2(\textbf{0.946 L}) && \text{Substitute 0.946 liter for 1 quart.} \\
&\approx 1.892 \text{ L} && \text{Perform the multiplication.}
\end{aligned}
$$

Thus, the 2-quart bottle contains approximately 1.892 liters. To find the price per liter of the 2-quart bottle, we divide $\frac{\$1.89}{1.892}$ to get

$$\frac{\$1.89}{1.892} \approx \$0.998942918$$

Since the price per liter of the 2-quart bottle is a little more than 99¢, the 1-liter bottle priced at 97¢ is the better buy.

Answer the 1-liter bottle

Comparing American and metric units of temperature

In the American system, we measure temperature using **degrees Fahrenheit** (°F). In the metric system, we measure temperature using **degrees Celsius** (°C). These two scales are shown on the thermometers in Figure 8-18 on the next page. From the figure, we can see that

- 212° F = 100° C Water boils.
- 32° F = 0° C Water freezes.
- 5° F = −15° C A cold winter day
- 95° F = 35° C A hot summer day

As we have seen, there is a formula that enables us to convert from degrees Fahrenheit to degrees Celsius. There is also a formula to convert from degrees Celsius to degrees Fahrenheit.

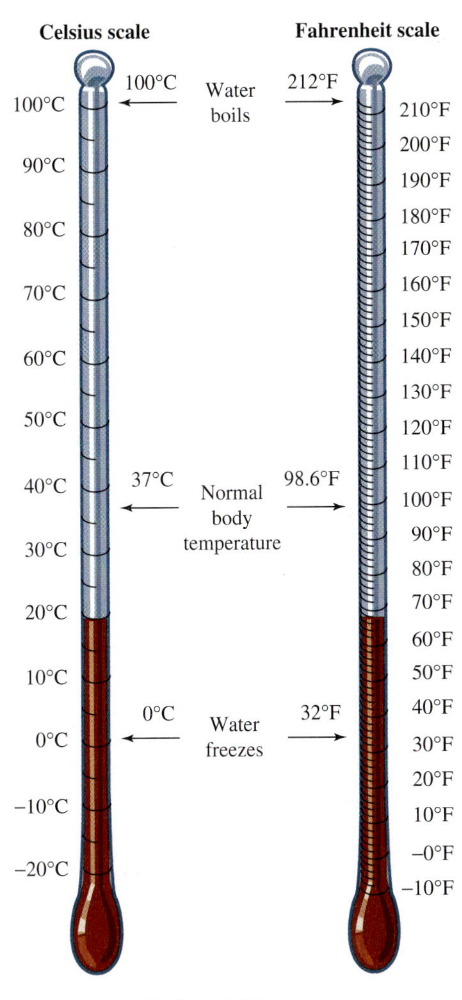

FIGURE 8-18

Conversion formulas for temperature

If F is the temperature in degrees Fahrenheit and C is the corresponding temperature in degrees Celsius, then

$$C = \frac{5}{9}(F - 32) \quad \text{and} \quad F = \frac{9}{5}C + 32$$

An alternative form of the formula $C = \frac{5}{9}(F - 32)$ is obtained by distributing the multiplication by 5 in the numerator to get $C = \frac{5F - 160}{9}$.

EXAMPLE 7 **Bathing.** Warm bath water is 90° F. Find the equivalent temperature in degrees Celsius.

Solution We substitute 90 for F in the formula $C = \frac{5}{9}(F - 32)$ and simplify.

Self Check 7

Hot coffee is 110° F. To the nearest tenth of a degree, express this temperature in degrees Celsius.

$$
\begin{aligned}
C &= \frac{5}{9}(F - 32) \\
&= \frac{5}{9}(90 - 32) && \text{Substitute 90 for } F. \\
&= \frac{5}{9}(58) && \text{Subtract: } 90 - 32 = 58. \\
&= 32.222\ldots && \text{Perform the arithmetic.}
\end{aligned}
$$

To the nearest tenth of a degree, the equivalent temperature is 32.2° C.

Answer 43.3° C

Self Check 8
To see whether a baby has a fever, her mother takes her temperature with a Celsius thermometer. If the reading is 38.8° C, does the baby have a fever? (*Hint:* Normal body temperature is 98.6° F.)

Answer yes

EXAMPLE 8 **Dishwashers.** A dishwasher manufacturer recommends that dishes be rinsed in hot water with a temperature of 60° C. Express this temperature in degrees Fahrenheit.

Solution We substitute 60 for C in the formula $F = \frac{9}{5}C + 32$ and simplify.

$$
\begin{aligned}
F &= \frac{9}{5}C + 32 \\
&= \frac{9}{5}(60) + 32 && \text{Substitute 60 for } C. \\
&= \frac{540}{5} + 32 && \text{Multiply: } \frac{9}{5}(60) = \frac{540}{5}. \\
&= 108 + 32 && \text{Perform the division.} \\
&= 140 && \text{Perform the addition.}
\end{aligned}
$$

The manufacturer recommends that dishes be rinsed in 140° F water.

THINK IT THROUGH Studying in Other Countries

"Over the past decade, the number of U.S. students studying abroad has more than doubled." From *The Open Doors 2003 Report*

In 2001/2002, a record number of 160,920 college students received credit for study abroad. Since students traveling to other countries are almost certain to come into contact with the metric system of measurement, they need to have a basic understanding of metric units.

Suppose a student studying overseas needs to purchase the following school supplies. For each item in red, circle the equivalent metric units of length, weight, or capacity.

1. $8\frac{1}{2}$ in. × 11 in. notebook paper:

216 meters × 279 meters　　216 centimeters × 279 centimeters　　216 millimeters × 279 millimeters

2. A backpack that can hold 20 pounds of books:

9 kilograms　　9 grams　　9 milligrams

3. $\frac{3}{4}$ fluid ounce bottle of Liquid Paper correction fluid:

22.5 hectoliters　　2.5 liters　　22.5 milliliters

Section 8.5 STUDY SET

VOCABULARY *Fill in the blanks.*

1. In the American system, temperatures are measured in degrees ________. In the metric system, temperatures are measured in degrees ________.
2. Inches and centimeters are used to measure ________. Gallons and liters are used to measure ________.

CONCEPTS

3. Which is longer?
   a. A yard or a meter?
   b. A foot or a meter?
   c. An inch or a centimeter?
   d. A mile or a kilometer?
4. Which is heavier?
   a. An ounce or a gram?
   b. A pound or a kilogram?
5. Which is the greater unit of capacity?
   a. A pint or a liter?
   b. A quart or a liter?
   c. A gallon or a liter?
6. a. What formula is used for changing degrees Celsius to degrees Fahrenheit?
   b. What formula is used for changing degrees Fahrenheit to degrees Celsius?

NOTATION *Complete each solution.*

7. Change 4,500 feet to kilometers.

$$4{,}500 \text{ ft} = 4{,}500(\quad \text{m})$$
$$= \quad \text{m}$$
$$= 1.3716 \text{ km}$$

8. Change 3 kilograms to ounces.

$$3 \text{ kg} = 3(\quad \text{lb})$$
$$= 3(2.2)(\quad \text{oz})$$
$$= 105.6 \text{ oz}$$

9. Change 8 liters to gallons.

$$8 \text{ L} = 8(\quad \text{gal})$$
$$= 2.112 \text{ gal}$$

10. Change 70°C to degrees Fahrenheit.

$$F = \frac{9}{5}C + 32$$
$$= \frac{9}{5}(\quad) + 32$$
$$= \quad + 32$$
$$= 158° \text{ F}$$

PRACTICE *Make each conversion. Since most conversions are approximate, answers will vary slightly depending on the method used.*

11. 3 ft = ☐ cm
12. 7.5 yd = ☐ m
13. 3.75 m = ☐ in.
14. 2.4 km = ☐ mi
15. 12 km = ☐ ft
16. 3,212 cm = ☐ ft
17. 5,000 in. = ☐ m
18. 25 mi = ☐ km
19. 37 oz = ☐ kg
20. 10 lb = ☐ kg
21. 25 lb = ☐ g
22. 7.5 oz = ☐ g
23. 0.5 kg = ☐ oz
24. 35 g = ☐ lb
25. 17 g = ☐ oz
26. 100 kg = ☐ lb
27. 3 fl oz = ☐ L
28. 2.5 pt = ☐ L
29. 7.2 L = ☐ fl oz
30. 5 L = ☐ qt
31. 0.75 qt = ☐ mL
32. 3 pt = ☐ mL
33. 500 mL = ☐ qt
34. 2,000 mL = ☐ gal
35. 50° F = ☐ C
36. 67.7° F = ☐ C
37. 50° C = ☐ F
38. 36.2° C = ☐ F
39. −10° C = ☐ F
40. −22.5° C = ☐ F
41. −5° F = ☐ C
42. −10° F = ☐ C

APPLICATIONS *Since most conversions are approximate, answers will vary slightly depending on the method used.*

43. THE MIDDLE EAST The distance between Jerusalem and Bethlehem is 8 kilometers. To the nearest mile, give this distance in miles.
44. THE DEAD SEA The Dead Sea is 80 kilometers long. To the nearest mile, give this distance in miles.
45. CHEETAHS A cheetah can run 112 kilometers per hour. How fast is this in mph?
46. LIONS A lion can run 50 mph. How fast is this in kilometers per hour?

47. MOUNT WASHINGTON The highest peak of the White Mountains of New Hampshire is Mount Washington, at 6,288 feet. To the nearest tenth, give this height in kilometers.

48. TRACK AND FIELD Track meets are held on an oval track such as the one shown. One lap around the track is usually 400 meters. However, some older tracks in the United States are 440-yard ovals. Are these two types of tracks the same length? If not, which is longer?

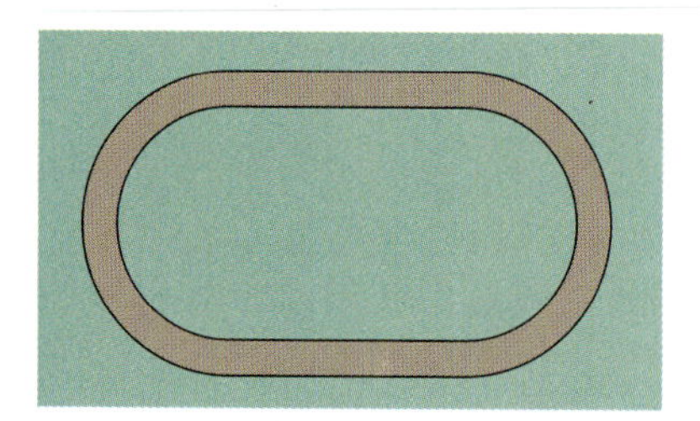

49. HAIR GROWTH When hair is short, its rate of growth averages about $\frac{3}{4}$ inch per month. How many centimeters is this a month?

50. WHALES An adult male killer whale can weigh as much as 12,000 pounds and be as long as 25 feet. Change these measurements to kilograms and meters.

51. WEIGHTLIFTING The table lists the personal best bench press records for two of the world's best powerlifters. Change each metric weight to pounds. Round to the nearest pound.

Name	Hometown	Bench press
Liz Willet	Ferndale, Washington	187 kg
Brian Siders	Charleston, W. Virginia	338 kg

52. WORDS OF WISDOM Refer to the wall hanging. Convert the first metric weight to ounces and the second to pounds. What famous saying results?

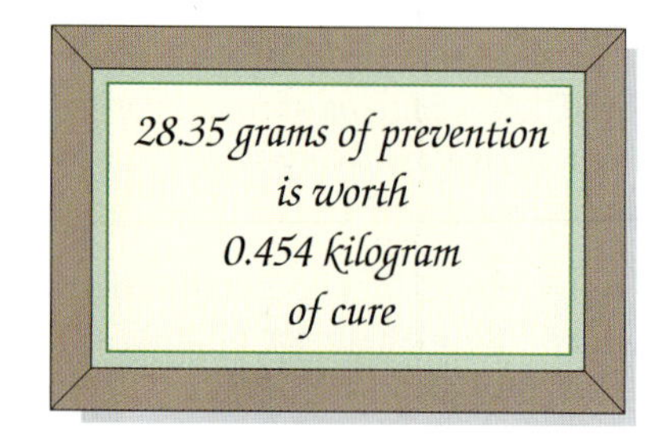

53. OUNCES AND FLUID OUNCES

a. There are 310 calories in 8 ounces of broiled chicken. Convert 8 ounces to grams.

b. There are 112 calories in a glass of fresh Valencia orange juice that holds 8 fluid ounces. Convert 8 fluid ounces to liters.

54. TRACK AND FIELD A shot put weighs 7.264 kilograms. Give this weight in pounds.

55. POSTAL REGULATIONS You can mail a package weighing up to 70 pounds via priority mail. Can you mail a package that weighs 32 kilograms by priority mail?

56. HEALTHY EATING Refer to the nutrition label for a packet of oatmeal shown. Change each circled weight to ounces.

Nutrition Facts
Serving Size: 1 Packet (46g)
Servings Per Container: 10

Amount Per Serving

Calories 170	Calories from Fat 20
	% Daily Value
Total fat 2g	**3%**
Saturated fat 0.5g	**2%**
Polyunsaturated Fat 0.5g	
Monounsaturated Fat 1g	
Cholesterol 0mg	**0%**
Sodium 250mg	**10%**
Total carbohydrate 35g	**12%**
Dietary fiber 3g	**12%**
Soluble Fiber 1g	
Sugars 16g	
Protein 4g	

57. COMPARISON SHOPPING Which is the better buy: 3 quarts of root beer for \$4.50 or 2 liters of root beer for \$3.60?

58. COMPARISON SHOPPING Which is the better buy: 3 gallons of antifreeze for \$10.35 or 12 liters of anti-freeze for \$10.50?

59. HOT SPRINGS The thermal springs in Hot Springs National Park in central Arkansas emit water as warm as 143° F. Change this temperature to degrees Celsius.

60. COOKING MEAT Meats must be cooked at high enough temperatures to kill harmful bacteria. According to the USDA and the FDA, the internal temperature for cooked roasts and steaks should be at least 145° F, and whole poultry should be 180° F. Convert these temperatures to degrees Celsius. Round up to the nearest degree.

61. TAKING A SHOWER When you take a shower, which water temperature would you choose: 15° C, 28° C, or 50° C?

62. DRINKING WATER To get a cold drink of water, which temperature would you choose: −2° C, 10° C, or 25° C?

63. SNOWY WEATHER At which temperatures might it snow: −5° C, 0° C, or 10° C?

64. AIR CONDITIONING At which outside temperature would you be likely to run the air conditioner: 15° C, 20° C, or 30° C?

WRITING

65. Explain how to change kilometers to miles.

66. Explain how to change 50° C to degrees Fahrenheit.

67. The United States is the only industrialized country in the world that does not officially use the metric system. Some people claim this is costing American businesses money. Do you think so? Why?

68. What is meant by the phrase *a table of equivalent measures?*

REVIEW *Combine like terms.*

69. $6y + 7 - y - 3$

70. $4x - 3y + 7 + 4x - 2 - 2y$

71. $-3(x - 4) - 2(2x + 6)$

72. $7(4 - t) + 2(3t - 8)$

Simplify each expression.

73. $x \cdot x \cdot x$

74. $a^3 \cdot a^5$

75. $3b(5b)$

76. $(x^2)^5$

KEY CONCEPT

Proportions

A **proportion** is a statement that two ratios or rates are equal.

Fill in the blanks as we set up a proportion to solve a problem.

1. TEACHER'S AIDES For every 15 children on the playground, a child care center is required to have 2 teacher's aides supervising. How many teacher's aides will be needed to supervise 75 children?

Step 1: Let x = the number of

________________________________.

If we compare the number of children to the number of teacher's aides, we know that the two rates must be equal.

15 children are to $\square$ aides as $\square$ children are to $\square$ aides.

Expressing this as a proportion, we have

$$\text{Number of children} \rightarrow \frac{15}{\square} = \frac{75}{\square} \leftarrow \text{Number of children}$$
$$\text{(denominators: Number of aides} \rightarrow \square, \ \square \leftarrow \text{Number of aides)}$$

In the proportion $\frac{15}{2} = \frac{75}{x}$, 15 and x are the *extremes* and 2 and 75 are the *means.* After setting up the proportion, we solve it using the fact that the product of the extremes is equal to the product of the means.

Step 2: Solve for x: $\frac{15}{2} = \frac{75}{x}$.

$\square \cdot x = 2 \cdot \square$	The product of the extremes equals the product of the means.
$15x = \square$	Perform the multiplication.
$\frac{15x}{\square} = \frac{150}{\square}$	Divide both sides by 15.
$x = \square$	Simplify.

To supervise 75 children, 10 teacher's aides are needed.

Step 3: To check the result, we substitute 10 for x in $\frac{15}{2} = \frac{75}{x}$ and find the cross products.

$$15 \cdot 10 = \mathbf{150} \qquad 2 \cdot 75 = \mathbf{150}$$
$$\frac{15}{2} \stackrel{?}{=} \frac{75}{10}$$

Since the cross products are equal, 10 is the solution.

Set up and solve each problem using a proportion.

2. PARKING A city code requires that companies provide 10 parking spaces for every 12 employees. How many spaces will be needed if a company employs 450 people?

3. MOTION PICTURES Every 2 seconds, 3 feet of motion picture film pass through the projector. How many feet of film are in a movie that runs for 120 minutes?

4. BEAUTY SUPPLIES A 0.5-fluid-ounce bottle of nail polish costs \$4.50. What would be the cost of one gallon of the nail polish? (*Hint:* 1 gallon = 128 fluid ounces)

ACCENT ON TEAMWORK

SECTION 8.1

ART HISTORY The illustration shows a drawing made by Leonardo Da Vinci (1452–1519) of a human figure within a square. We see that the man's height and the span of his outstretched arms are the same; that is, their ratio is 1 to 1. Use a tape measure to determine this ratio for each member of your group. Work in terms of inches. Are any ratios exactly 1 to 1?

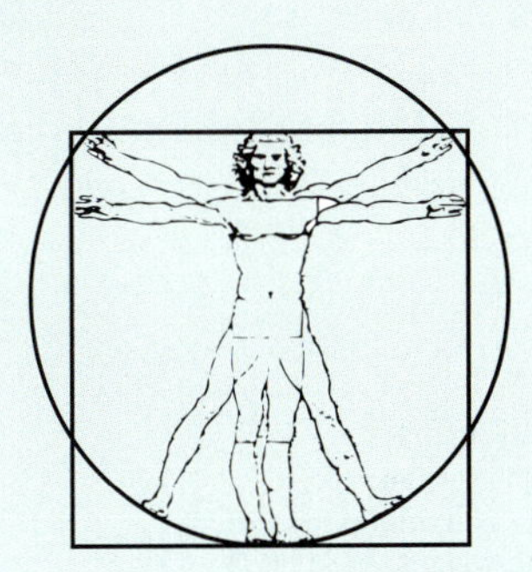

SECTION 8.2

ENLARGEMENTS Duplicating or enlarging a picture can be done by the *grid-transfer* method. To enlarge the picture of a cook shown in Illustration (a), begin by copying the markings in the square in the lower right-hand corner onto its corresponding square in the enlargement grid in Illustration (b). One by one, copy the contents of each square of the original picture to its counterpart in the enlargement.

SECTION 8.3

DISNEY CLASSICS The movie *20,000 Leagues Under the Sea* is a science fiction thriller about Captain Nemo and the crew of the submarine Nautilus. Use the fact that 1 mile $= \frac{1}{3}$ league to express a depth of 20,000 leagues in feet.

SECTION 8.4

METRIC MISHAP In 1999, NASA lost the $125-million Mars Climate Orbiter because a Lockheed Martin engineering team used American units of measurement, while NASA's team used the metric system. Use the Internet to research this incident and make a report to your class.

SECTION 8.5

TRUTH IN LABELING Have each member of your group bring in two items — one whose product label indicates capacity and another that indicates weight. For example, a bottle of shampoo could contain 15 fluid ounces (444 mL) or a can of soup could weigh 1 pound 3 ounces (539 g). Exchange your items with another person in your group. Check the accuracy of each label by converting from American units to metric units.

(a)

The original picture is drawn on a grid of $\frac{1}{4}$-inch squares. The enlargement is drawn using a grid of $\frac{1}{2}$-inch squares. By how much was the original picture enlarged?

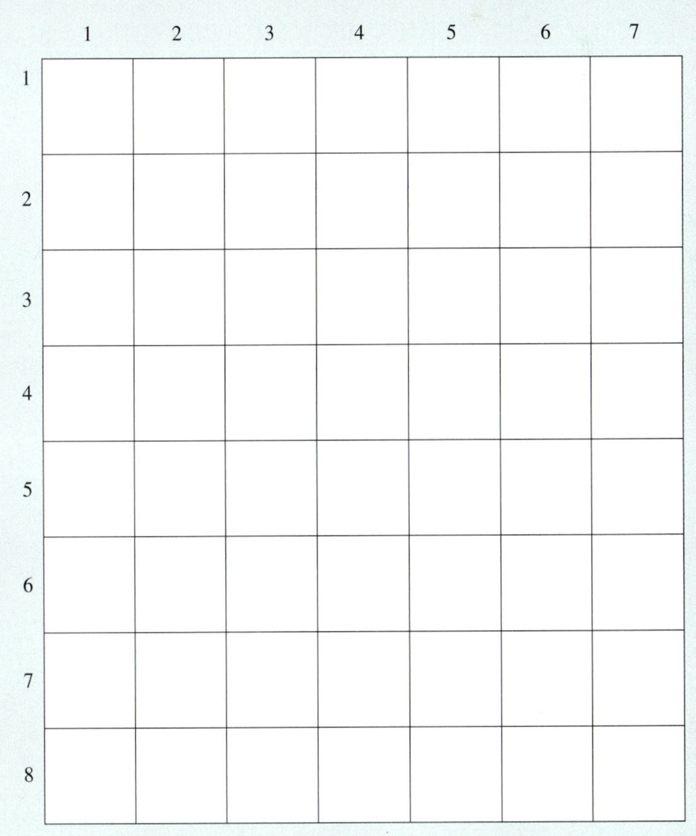

(b)

CHAPTER REVIEW

SECTION 8.1 Ratios

CONCEPTS

A *ratio* is a quotient of two numbers or a quotient of two quantities that have the same units.

REVIEW EXERCISES

Express each phrase as a ratio in lowest terms.

1. The ratio of 4 inches to 12 inches
2. The ratio of 8 ounces to 2 pounds
3. 21 : 14
4. 24 to 36
5. AIRCRAFT Specifications for a Boeing B-52 Stratofortress are given in the illustration. What is the ratio of the airplane's wingspan to its length?

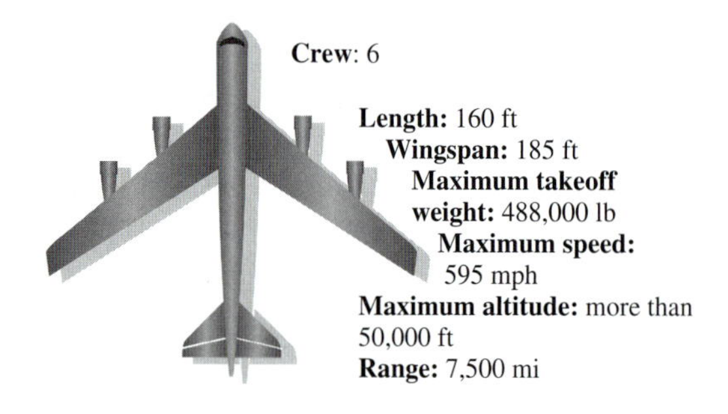

A *rate* is a comparison of two quantities with different units.

6. PAY SCALE Find the hourly rate of pay for a student who earned $333.25 for working 43 hours.

A *unit cost* is a comparison of the cost of an item to its quantity.

7. COMPARISON SHOPPING Mixed nuts come packaged in a 12-ounce can, which sells for $4.95, or an 8-ounce can, which sells for $3.25. Which is the better buy?

SECTION 8.2 Proportions

A *proportion* is a statement that two ratios (or rates) are equal.

Consider the proportion $\frac{5}{15} = \frac{25}{75}$.

8. Which term is the fourth term?
9. Which term is the second term?

In any proportion, the product of the *extremes* is equal to the product of the *means.*

Determine whether each statement is a proportion.

10. $\frac{15}{29} = \frac{105}{204}$
11. $\frac{17}{7} = \frac{204}{84}$

When two pairs of numbers a, b and c, d form a proportion, we say that the numbers are *proportional.*

Determine whether the numbers are proportional.

12. 5, 9 and 20, 36
13. 7, 13 and 29, 54

If three terms of a proportion are known, we can solve for the missing term.

Solve each proportion for the variable.

14. $\frac{12}{18} = \frac{3}{x}$

15. $\frac{4}{b} = \frac{2}{8}$

16. $\frac{c + 1}{5} = \frac{1}{5}$

17. $\frac{3p + 15}{2} = \frac{5}{3}$

18. PICKUP TRUCKS A Dodge Ram pickup truck can go 35 miles on 2 gallons of gas. How far can it go on 11 gallons?

19. QUALITY CONTROL In a manufacturing process, 12 parts out of 66 were found to be defective. How many defective parts will be expected in a run of 1,650 parts?

20. SCALE DRAWING The illustration shows an architect's drawing of a kitchen using a scale of $\frac{1}{8}$ inch to 1 foot ($\frac{1}{8}$″:1′0″). On the drawing, the length of the kitchen is $1\frac{1}{2}$ inches. How long is the actual kitchen? (The symbol ″ means inch and ′ means foot.)

ELEVATION B-B

SCALE: $\frac{1"}{8}$ = 1'0"

SECTION 8.3 *American Units of Measurement*

Common American units of length are *inches, feet, yards,* and *miles.*

12 in. = 1 ft

3 ft = 1 yd

36 in. = 1 yd

5,280 ft. = 1 mi

21. Use a ruler to measure the length of the computer mouse in the illustration, to the nearest quarter of an inch.

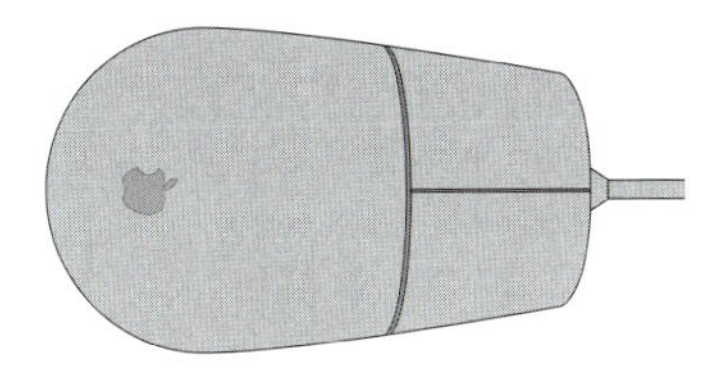

Unit conversion factors have a value of 1.

22. Write two unit conversion factors using the fact that 1 mile = 5,280 ft.

Multiplying a measurement by a unit conversion factor does not change the measure; it only changes the units of the measure.

Make each conversion.

23. 5 yards to feet

24. 6 yards to inches

25. 66 inches to feet

26. 25.5 feet to inches

27. 9,240 feet to miles

28. 1 mile to yards

Common American units of weight are *ounces, pounds,* and *tons.*

16 oz = 1 lb

2,000 lb = 1 ton

Make each conversion.

29. 32 ounces to pounds

30. 17.2 pounds to ounces

31. 3 tons to ounces

32. 4,500 pounds to tons

Common American units of capacity are *fluid ounces, cups, pints, quarts,* and *gallons.*

1 c = 8 fl oz

1 pt = 2 c

1 qt = 2 pt

1 gal = 4 qt

Make each conversion.

33. 5 pints to fluid ounces

34. 8 cups to gallons

35. 17 quarts to cups

36. 176 fluid ounces to quarts

37. 5 gallons to pints

38. 3.5 gallons to cups

Units of time are *seconds, minutes, hours,* and *days.*

1 min = 60 sec

1 hr = 60 min

1 day = 24 hr

Make each conversion.

39. 20 minutes to seconds

40. 900 seconds to minutes

41. 200 hours to days

42. 6 hours to minutes

43. 4.5 days to hours

44. 1 day to seconds

45. SKYSCRAPERS The Sears Tower in Chicago is 1,454 feet high. Express this distance in yards.

46. BOTTLING A magnum is a 2-quart bottle of wine. How many magnums are needed to hold 50 gallons of wine?

SECTION 8.4 Metric Units of Measurement

Common metric units of length are *millimeter, centimeter, decimeter, meter, dekameter, hectometer,* and *kilometer.*

1 mm = $\frac{1}{1{,}000}$ m

1 cm = $\frac{1}{100}$ m

1 dm = $\frac{1}{10}$ m

1 dam = 10 m

1 hm = 100 m

1 km = 1,000 m

47. Use a metric ruler to measure the length of the computer mouse in the illustration, to the nearest centimeter.

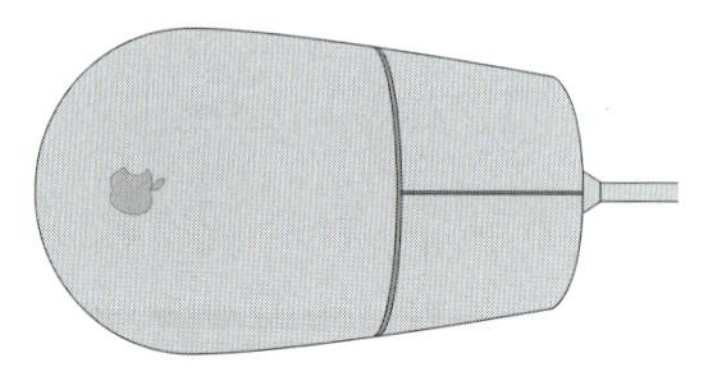

48. Write two unit conversion factors using the fact that 1 km = 1,000 m.

Make each conversion.

49. 475 centimeters to meters

50. 8 meters to millimeters

51. 3 dekameters to kilometers

52. 2 hectometers to decimeters

53. 5 kilometers to hectometers

54. 2,500 meters to hectometers

Common metric units of mass are *milligrams, centigrams, grams, kilograms,* and *metric tons.*

1 mg = $\frac{1}{1{,}000}$ g

1 cg = $\frac{1}{100}$ g

1 g = $\frac{1}{1{,}000}$ kg

Make each conversion.

55. 7 centigrams to milligrams

56. 800 centigrams to grams

57. 5,425 grams to kilograms

58. 5,425 grams to milligrams

59. 7,500 milligrams to grams

60. 5,000 centigrams to kilograms

61. PAIN RELIEVER A bottle of Extra Strength Tylenol contains 100 caplets of 500 milligrams each. How many grams of Tylenol are in the bottle?

Common metric units of capacity are *milliliters, centiliters, deciliters, liters, hectoliters,* and *kiloliters.*

$1 \text{ mL} = \frac{1}{1{,}000} \text{ L}$

$1 \text{ cL} = \frac{1}{100} \text{ L}$

$1 \text{ dL} = \frac{1}{10} \text{ L}$

$1 \text{ L} = 1{,}000 \text{ cc}$

$1 \text{ hL} = 100 \text{ L}$

$1 \text{ kL} = 1{,}000 \text{ L}$

Make each conversion.

62. 150 centiliters to liters

63. 3,250 liters to kiloliters

64. 1 hectoliter to deciliters

65. 400 milliliters to centiliters

66. 2 kiloliters to hectoliters

67. 4 deciliters to milliliters

68. SURGERY A dextrose solution is being administered to a patient intravenously using the apparatus shown. How many milliliters of solution does the IV bag hold?

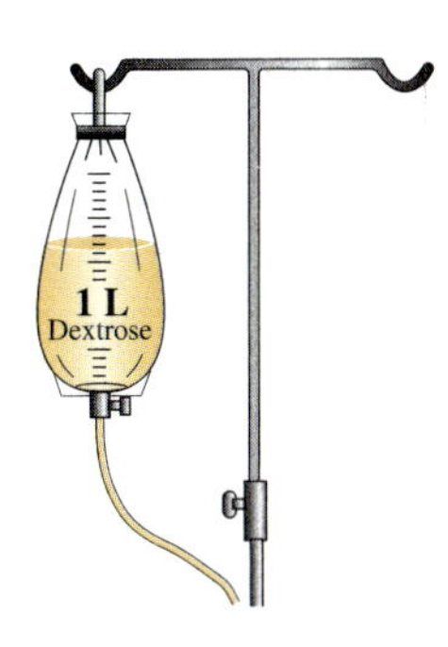

SECTION 8.5 Converting Between American and Metric Units

We can convert between American and metric units using the following:

$1 \text{ in.} \approx 2.54 \text{ cm}$
$1 \text{ ft} \approx 0.3048 \text{ m}$
$1 \text{ yd} \approx 0.9144 \text{ m}$
$1 \text{ mi} \approx 1.6093 \text{ km}$
$1 \text{ cm} \approx 0.3937 \text{ in.}$
$1 \text{ m} \approx 3.2808 \text{ ft}$
$1 \text{ m} \approx 1.0936 \text{ yd}$
$1 \text{ km} \approx 0.6214 \text{ mi}$

$1 \text{ oz} \approx 28.35 \text{ g}$
$1 \text{ lb} \approx 0.454 \text{ kg}$
$1 \text{ g} \approx 0.035 \text{ oz}$
$1 \text{ kg} \approx 2.2 \text{ lb}$

$1 \text{ fl oz} \approx 0.030 \text{ L}$
$1 \text{ pt} \approx 0.473 \text{ L}$
$1 \text{ qt} \approx 0.946 \text{ L}$
$1 \text{ gal} \approx 3.785 \text{ L}$
$1 \text{ L} \approx 33.8 \text{ fl oz}$
$1 \text{ L} \approx 2.1 \text{ pt}$
$1 \text{ L} \approx 1.06 \text{ qt}$
$1 \text{ L} \approx 0.264 \text{ gal}$

Two units used to measure temperature are degrees Fahrenheit and degrees Celsius.

$$C = \frac{5}{9}(F - 32)$$

$$F = \frac{9}{5}C + 32$$

69. SWIMMING Olympic-size swimming pools are 50 meters long. Express this distance in feet.

70. HIGH-RISE BUILDINGS The Sears Tower is 443 meters high, and the Empire State Building is 1,250 feet high. Which building is taller?

71. WESTERN SETTLERS The Oregon Trail was an overland route pioneers used in the 1840s through the 1870s to reach the Oregon Territory. It stretched 1,930 miles from Independence, Missouri, to Oregon City, Oregon. Find this distance to the nearest kilometer.

72. AIR JORDAN Michael Jordan is 6 feet, 6 inches tall. Express his height in centimeters.

Make each conversion.

73. 30 ounces to grams

74. 15 kilograms to pounds

75. 25 pounds to grams (Round to the nearest thousand.)

76. 2,000 pounds to kilograms (Round to the nearest ten.)

77. POLAR BEARS At birth, polar bear cubs weigh less than human babies — about 910 grams. Convert this to pounds.

78. BOTTLED WATER LaCroix® bottled water can be purchased in bottles containing 17 fluid ounces. Mountain Valley® water can be purchased in half-liter bottles. Which bottle contains more water?

79. COMPARISON SHOPPING One gallon of bleach costs \$1.39. A 5-liter economy bottle costs \$1.80. Which is the better buy?

80. Change 77° F to degrees Celsius.

81. Which temperature of water would you like to swim in: 10° C, 30° C, 50° C, or 70° C?

CHAPTER 8 TEST

Write each phrase as a ratio in lowest terms.

1. The ratio of 6 feet to 8 feet

2. The ratio of 8 ounces to 3 pounds

3. COMPARISON SHOPPING Two pounds of coffee can be purchased for \$3.38, and a 5-pound can can be purchased for \$8.50. Which is the better buy?

4. UTILITY COSTS A household used 675 kilowatt-hours of electricity during a 30-day month. Find the rate of electricity consumption in kilowatt-hours per day.

5. CHECKERS What is the ratio of the number of red squares to the number of black squares for the checker-board shown? Express your answer in three ways: as a fraction, using a colon, and using the word *to*.

Determine whether each statement is a proportion.

6. $\frac{25}{33} = \frac{350}{460}$

7. $\frac{2.2}{3.5} = \frac{1.76}{2.8}$

8. Are the numbers 7, 15 and 245, 525 proportional?

Solve each proportion.

9. $\frac{x}{3} = \frac{35}{7}$

10. $\frac{15.3}{x} = \frac{3}{12.4}$

11. $\frac{2x + 3}{5} = \frac{5}{1}$

12. $\frac{3}{2z - 1} = \frac{3}{5}$

13. SHOPPING If 13 ounces of tea costs \$2.79, how much would you expect to pay for 16 ounces?

14. COOKING A recipe calls for $\frac{2}{3}$ cup of sugar and 2 cups of flour. How much sugar should be used with 5 cups of flour?

15. Convert 180 inches to feet.

16. TOOLS If a 25-foot tape measure is completely extended, how many yards does it stretch?

17. Convert 10 pounds to ounces.

18. A car weighs 1.6 tons. Find its weight in pounds.

19. How many fluid ounces are in a 1-gallon carton of milk?

20. LITERATURE An excellent work of early science fiction is the book *Around the World in 80 Days* by Jules Verne (1828–1905). Convert 80 days to minutes.

21. A quart and a liter of fruit punch are shown. Which is the 1-liter carton?

22. The figures below show the relative lengths of a yardstick and a meterstick. Which one represents the meterstick?

23. An ounce and a gram are placed on the balance shown. On which side is the gram?

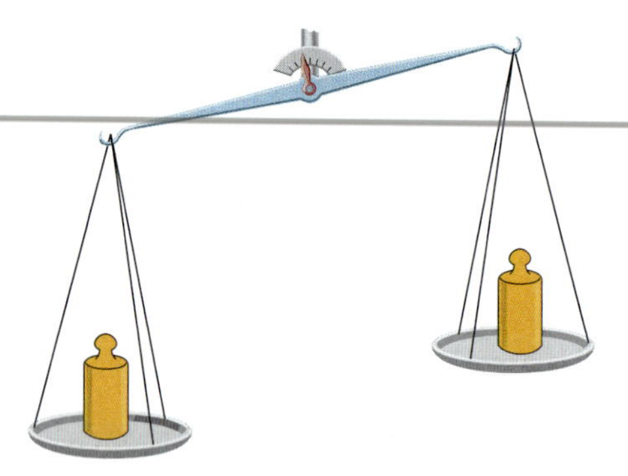

24. SPEED SKATING American Bonnie Blair won gold medals in the women's 500-meter speed skating competitions three times in the Winter Olympics. Convert the race length to kilometers.

25. How many centimeters are in 5 meters?

26. Convert 8,000 centigrams to kilograms.

27. Convert 70 liters to milliliters.

28. PRESCRIPTIONS A bottle contains 50 tablets, each containing 150 mg of medicine. How many grams of medicine does the bottle contain?

29. Which is the longer distance: a 100-yard race or an 80-meter race?

30. Which person is heavier: Jim, who weighs 160 pounds, or Ricardo, who weighs 72 kilograms?

31. COMPARISON SHOPPING A 2-quart bottle of soda pop costs \$1.73, and a 1-liter bottle costs 89¢. Which is the better buy? (*Hint:* 1 quart = 0.946 liter.)

32. COOKING MEAT The USDA recommends that turkey be cooked to a temperature of 83° C. Change this to degrees Fahrenheit. To be safe, round up to the next degree. (*Hint:* $F = \frac{9}{5}C + 32$.)

33. What is a scale drawing? Give an example.

34. Explain the benefits of the metric system of measurement as compared to the American system.

CHAPTERS 1–8 CUMULATIVE REVIEW EXERCISES

1. Write 64,502 in expanded notation.

2. Divide: $37\overline{)743}$.

3. ENLISTMENTS The table shows how the U.S. Army fared in reaching its recruiting goals in 2004. Complete the table. Use a negative number to denote a shortfall of enlistees.

	Goal	Enlistments	Outcome
Active	77,000	77,587	
Reserve	21,000	21,278	
National Guard	56,000	49,210	

Source: U.S. Army

4. Evaluate each expression, if possible.

   a. $0 + (-8)$ b. $\frac{-8}{0}$

   c. $0 - |-8|$ d. $\frac{0}{-8}$

   e. $0 - (-8)$ f. $0(-8)$

5. GOLF Tiger Woods won the 100th U.S. Open in June 2000 by the largest margin in the history of that tournament. If he shot 12 under par (-12) and the second-place finisher, Miguel Angel Jimenez, shot 3 over par ($+3$), what was his margin of victory?

6. Evaluate: -3^2 and $(-3)^2$.

7. Evaluate: $2 + 3[5(-6) - (1 - 10)]$.

8. What are the coefficients of the first term and the second term of $x^2 + 16x - 1$?

9. How many minutes are in h hours?

10. Simplify: $3x - 5 - 2x - 2$.

11. Solve: $7 + 2x = 2 - (4x + 7)$.

12. Give the formula that relates distance, rate, and time.

13. PHONE BOOKS A driver left a warehouse in the morning with 500 new telephone books loaded on his truck. His delivery route consisted of office buildings, each of which was to receive 5 of the books. The driver returned at the end of the day with 105 books on the truck. To how many office buildings did he deliver?

14. What is the formula for the area of a triangle?

15. Simplify: $\frac{16}{20}$.

16. Express $\frac{9}{10}$ as an equivalent fraction with a denominator of $60t$.

17. Divide: $-\frac{7}{8h} \div \frac{7}{8}$.

18. Find $\frac{1}{2}$ of $\frac{1}{2}$?

19. MOTORS What is the difference in horsepower (hp) between the two motors shown?

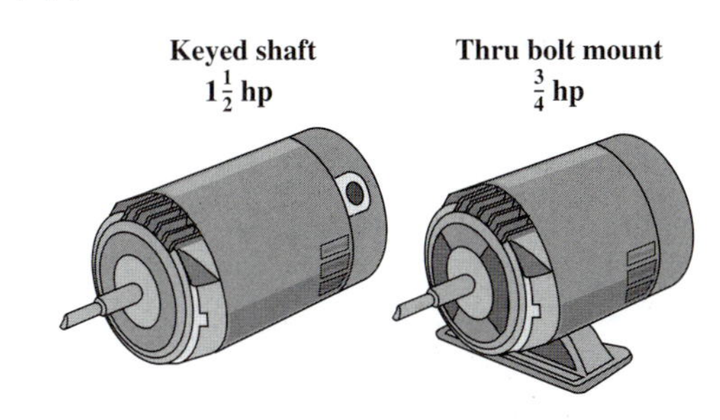

20. Solve: $\frac{5}{8}y = \frac{1}{10}$.

21. Solve: $\frac{2}{5}y + 1 = \frac{1}{3} + y$.

22. GLOBAL WARMING The average temperature in the United States during March through May 2000 was a record-setting 55.5° F. That was 0.4 degree warmer than the previous record, set in 1910. What was the previous spring temperature record?

23. Evaluate the expression $\frac{6.7 - x^2 + 1.6}{-x^3}$ for $x = -0.3$. Round to the nearest hundredth.

24. Write $\frac{1}{12}$ as a decimal.

25. Solve: $6(y - 1.1) + 3.2 = -1 + 3y$.

26. Evaluate: $3\sqrt{25} + 4\sqrt{4}$.

27. Plot the points: $(-4, -3)$, $(1.5, 1.5)$, $(-3, 0)$, $(0, 3\frac{1}{2})$.

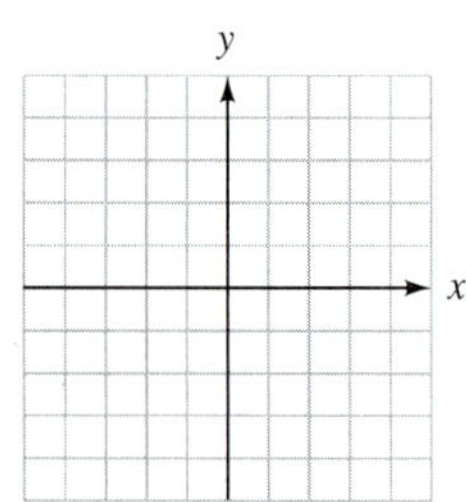

28. Graph each equation.

a. $x - 2y = -4$ **b.** $y = 2x^2$

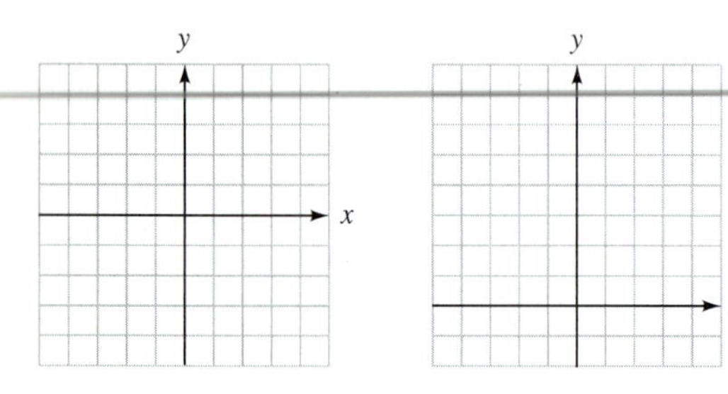

29. Subtract: $(5x^2 - 8x + 1) - (3x^2 - 2x + 3)$.

30. Multiply: $(3x + 2)(2x - 5)$.

31. Simplify each expression.

a. $s^6 \cdot s^7$

b. $(s^6)^7$

c. $(3a^2b^4)^3$

d. $-w^5(8w^3)$

32. 16 is what percent of 24?

33. What is the formula for simple interest?

34. Complete the table.

Percent	Decimal	Fraction
	0.99	
1.3%		
		$\frac{5}{16}$

35. GUITARS What are the regular price and the rate of discount for the guitar shown?

36. Express the phrase "3 inches to 15 inches" as a ratio in lowest terms.

37. SURVIVAL GUIDES

a. A person can go without food for about 40 days. How many hours is this?

b. A person can go without water for about 3 days. How many minutes is that?

c. A person can go without breathing oxygen for about 8 minutes. How many seconds is that?

38. Convert 40 ounces to pounds.

39. Convert 2.4 meters to millimeters.

40. Convert 320 grams to kilograms.

41. a. Which holds more: a 2-liter bottle or a 1-gallon bottle?

b. Which is longer: a meterstick or a yardstick?

42. BUILDING MATERIALS Which is the better buy: a 94-pound bag of cement for \$4.48 or a 45-kilogram bag of cement for \$4.56?

CHAPTER 9

Introduction to Geometry

Getty Images

TLE Many people enjoy tackling home-improvement projects themselves. These projects run more smoothly and are more cost efficient if carefully planned. This planning often requires some mathematics, and in particular, geometry. For example, to purchase the correct amount of materials to fence a yard, you need to know how to calculate *perimeter*. If you are painting a bedroom, you need to calculate the total *area* of the wall surfaces to determine the number of gallons of paint to buy.

To learn more about perimeter and area, visit The Learning Equation on the Internet at http://tle.brookscole.com. (The log-in instructions are in the Preface.) For Chapter 9, the online lesson is:

- *TLE* Lesson 15: Perimeter and Area

Check Your Knowledge

1. ________ lines do not intersect. If two lines intersect and form right angles, they are ________.
2. A polygon with four sides is called a ________. A ________ is a polygon with three sides.
3. A triangle that includes a 90° angle is called a ______ triangle. The longest side of such a triangle is the ________.
4. Triangles that are the same size and the same shape are called ________ triangles. If two triangles are the same shape but not necessarily the same size, they are called ______ triangles.
5. The distance around a polygon is called the ________. The measure of the surface enclosed by a polygon is called its ______.
6. A segment drawn from the center of a circle to a point on the circle is called a ______. The distance around a circle is called its ________. A ________ is a chord that passes through the center of a circle.
7. The space contained within a geometric solid is called its ________.
8. Match the descriptions with the angles.

   a. 37° ____ I. right angle

   b. 90° ____ II. straight angle

   c. 125° ____ III. acute angle

   d. 180° ____ IV. obtuse angle

9. In the notation $\angle ABC$, which point is the vertex of the angle?
10. Refer to the illustration on the left. Find z.

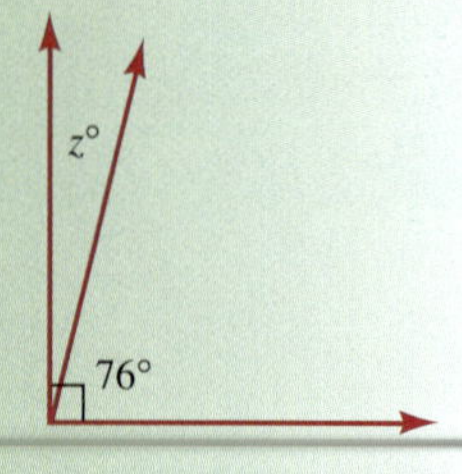

PROBLEM 10

11. Refer to the illustration on the left. Find y.

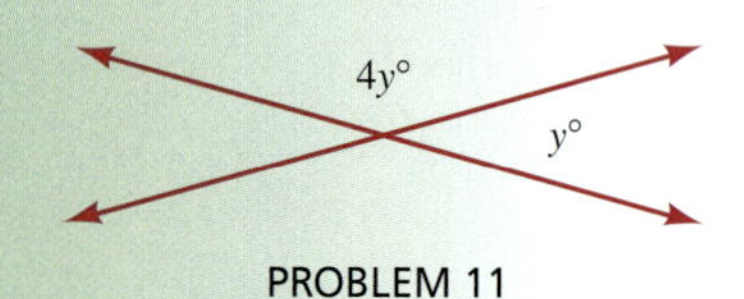

PROBLEM 11

12. Find the supplement of an angle measuring 119°.
13. Refer to the illustration, in which lines ℓ_1 and ℓ_2 are parallel.

    a. Find m(∠1). **b.** Find m(∠2).

    c. Find m(∠3). **d.** Find m(∠4).

    e. Find x.

14. If the measure of one angle of a triangle is 45° and the measure of another is 55°, what is the measure of the third angle?
15. A rectangle is 12 ft long and 5 ft wide.

    a. Find length of the diagonal of the rectangle.

    b. Find the area of the rectangle.

16. Find the area of a triangle with base 4.5 in. and height 7.8 in.
17. Find the area of a circle if the diameter is 10 ft.
18. Find the volume of a sphere that is 4 ft in diameter.
19. Find the volume of a rectangular solid with length 5 ft, width 4 ft, and height 10 ft.
20. Find the volume of a cylinder 8 ft in diameter and 10 ft tall.

Study Skills Workshop

PREPARING FOR YOUR NEXT MATH COURSE

According to Dr. Benjamin Bloom, 50% of your overall achievement in your next match course will be based on the math knowledge you have coming into that course [B. Bloom, *Human Characteristics and School Learning* (New York: McGraw-Hill Book Company, 1976)]. So, if you want to be successful in your next math course, it's a good idea to do some preparation in advance. How much you will have to do depends on how long a time period there is between your math courses, how well you remember material, and how well you did in your previous math course. It is always a good idea to take your next math course as soon as possible after this course is finished. The exception to this is that you should not take a math course during an accelerated term (like summer semesters at many colleges), unless you are very confident in your math abilities. It is a good idea to keep your textbook from this course until you finish your next course; resist the urge to make a couple of dollars by selling your book at textbook "buy-backs" on your campus. The textbook that you used this term is a good resource between terms and throughout your next course.

Here are some activities that will help you prepare for your next course:

- *Review old tests.* Try to work problems from old tests without peeking at the answers. If you get stuck, you can look — but then repeat the problem until you can do it without looking at the solution.
- *Use packaged software or online programs.* Using tutorials on the CD-ROM that came packaged with your text or on the iLrn Web site can help you refresh concepts.
- *Review chapter tests from your old textbook.* Try taking the tests at the end of each chapter in your textbook from this course. Check your answers and review the book for help in correcting the problems that you did incorrectly.
- *Try to get your new text before class starts.* If you have time, reading ahead in your new textbook is a good way to get a head start in your new course.
- *Consider getting a tutor.* If you didn't do as well as you would have liked in this course, consider hiring a tutor in the time between terms to learn some of the concepts that you missed in this course.

Geometry comes from the Greek words geo *(meaning earth) and* metron *(meaning measure).*

9.1 Some Basic Definitions

- Points, lines, and planes • Angles • Adjacent and vertical angles
- Complementary and supplementary angles

In this chapter, we will study two-dimensional geometric figures such as rectangles and circles. It is often necessary to find the perimeter or area of one of these figures. For example, to find the amount of fencing that is needed to enclose a circular garden, we must find the perimeter of a circle (called its *circumference*). To find the amount of paint needed to paint a room, we must find the area of its four rectangular walls.

We will also study three-dimensional figures such as cylinders and spheres. To find the amount of space enclosed within these figures, we must find their volumes.

Points, lines, and planes

Geometry is based on three words: **point, line,** and **plane.** Although we will make no attempt to define these words formally, we can think of a point as a geometric figure that has position but no length, width, or depth. Points are always labeled with capital letters. Point A is shown in Figure 9-1(a).

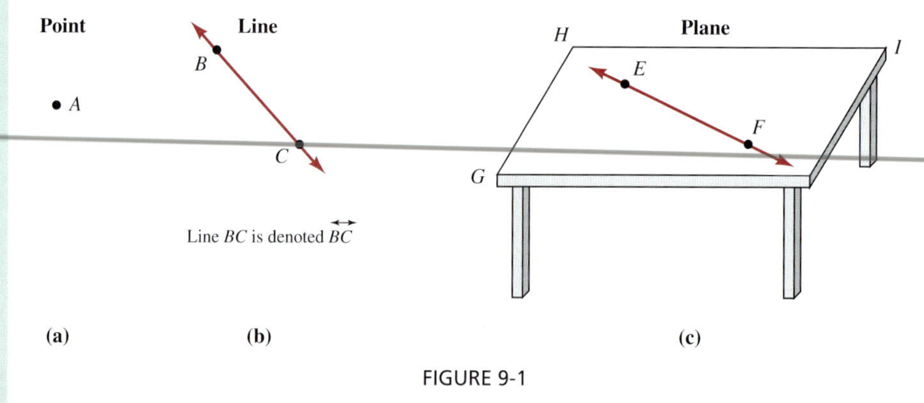

FIGURE 9-1

A line is infinitely long but has no width or depth. Figure 9-1(b) shows line BC, passing through points B and C. A plane is a flat surface, like a table top, that has length and width but no depth. In Figure 9-1(c), line EF lies in the plane GHI.

As Figure 9-1(b) illustrates, points B and C determine exactly one line, the line BC. In Figure 9-1(c), the points E and F determine exactly one line, the line EF. In general, any two points will determine exactly one line.

Other geometric figures can be created by using parts or combinations of points, lines, and planes.

Line segment

The **line segment** AB, denoted as $\overline{AB}$, is the part of a line that consists of points A and B and all points in between. (See Figure 9-2.) Points A and B are the **endpoints** of the segment.

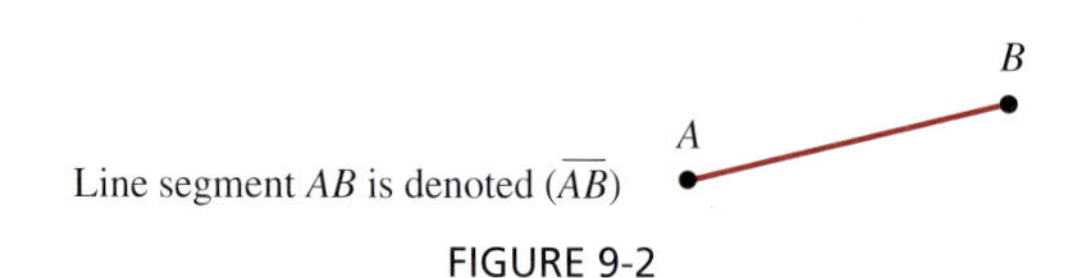

FIGURE 9-2

Every line segment has a **midpoint,** which divides the segment into two parts of equal length. In Figure 9-3, M is the midpoint of segment AB, because the measure of $\overline{AM}$, denoted as $m(\overline{AM})$, is equal to the measure of $\overline{MB}$, denoted as $m(\overline{MB})$.

$$m(\overline{AM}) = 4 - 1$$
$$= 3$$

and

$$m(\overline{MB}) = 7 - 4$$
$$= 3$$

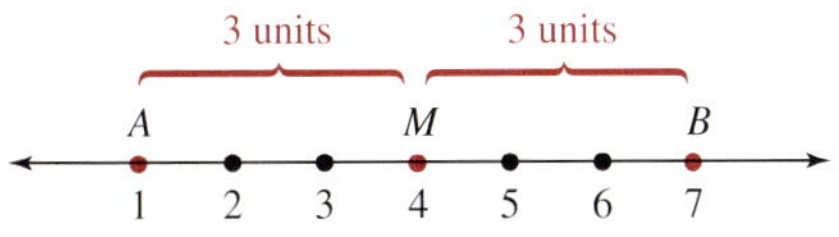

3 units 3 units

A M B

1 2 3 4 5 6 7

FIGURE 9.3

Since the measure of both segments is 3 units, $m(\overline{AM}) = m(\overline{MB})$.

When two line segments have the same measure, we say that they are **congruent.** Since $m(\overline{AM}) = m(\overline{MB})$, we can write

$\overline{AM} \cong \overline{MB}$ Read $\cong$ as "is congruent to."

Another geometric figure is the *ray,* as shown in Figure 9-4.

Ray

A **ray** is the part of a line that begins at some point (say, A) and continues forever in one direction. Point A is the **endpoint** of the ray.

Ray AB is denoted as $\overrightarrow{AB}$. The endpoint of the ray is always listed first.

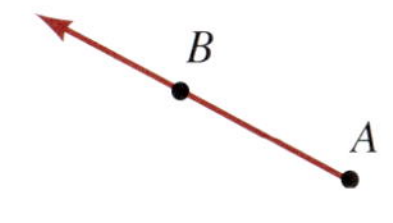

FIGURE 9-4

Angles

Angle

An **angle** is a figure formed by two rays with a common endpoint. The common endpoint is called the **vertex,** and the rays are called **sides.**

The angle in Figure 9-5 can be denoted as

$\angle BAC$, $\angle CAB$, $\angle A$, or $\angle 1$ The symbol $\angle$ means angle.

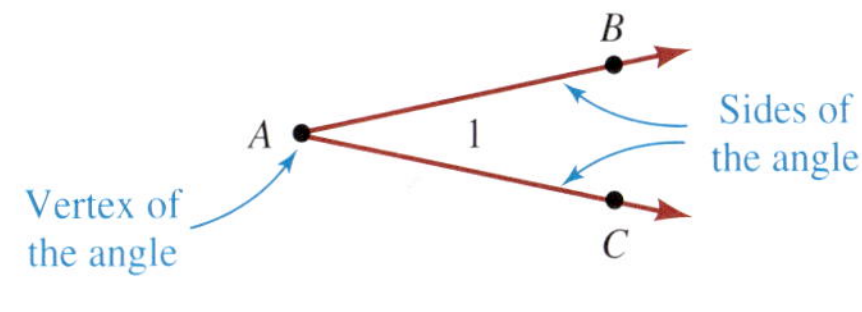

FIGURE 9-5

! COMMENT When using three letters to name an angle, be sure the letter name of the vertex is the middle letter.

Self Check 3
In Figure 9-11, find
a. m(∠2)
b. m(∠4)

EXAMPLE 3 In Figure 9-11, find **a.** m(∠1) and **b.** m(∠3).

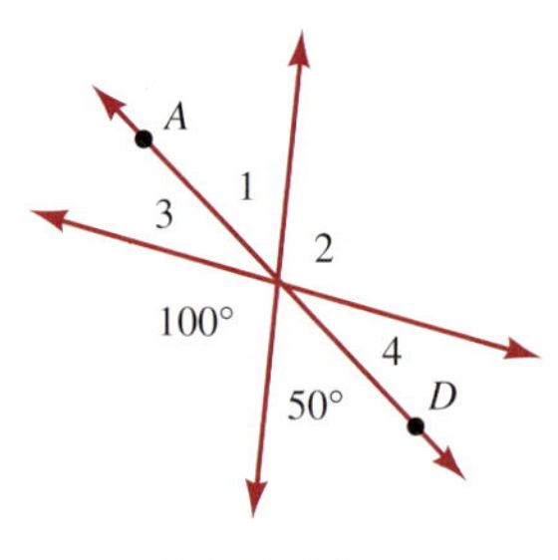

FIGURE 9-11

Solution

a. The 50° angle and ∠1 are vertical angles. Since vertical angles are congruent, m(∠1) = 50°.

b. Since AD is a line, the sum of the measures of ∠3, the 100° angle, and the 50° angle is 180°. If m(∠3) = x, we have

$$x + 100 + 50 = 180$$
$$x + 150 = 180 \quad \text{Perform the addition: } 100 + 50 = 150.$$
$$x = 30 \quad \text{Subtract 150 from both sides.}$$

Thus, m(∠3) = 30°.

Answers **a.** 100°, **b.** 30°

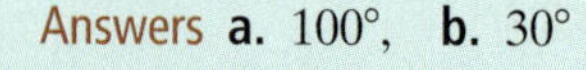

Self Check 4
In the figure below, find y.

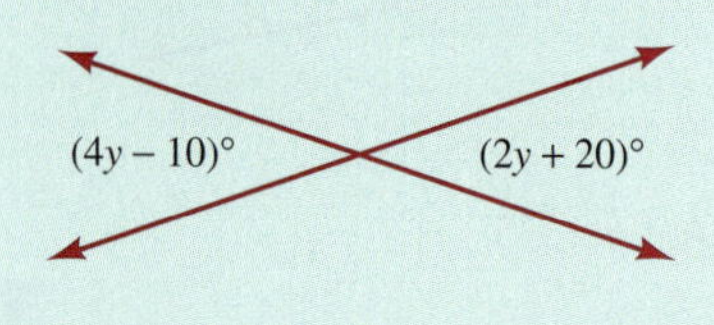

EXAMPLE 4 In Figure 9-12, **a.** find x, and **b.** find the measure of each vertical angle.

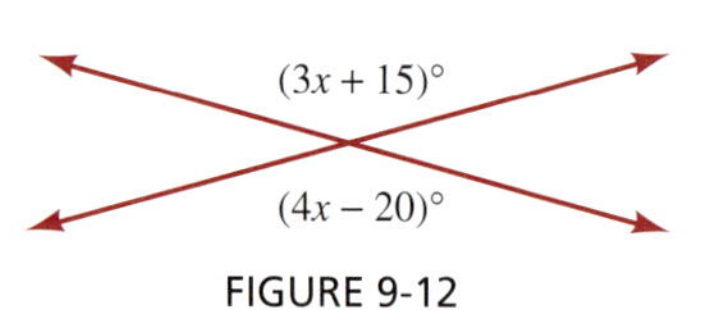

FIGURE 9-12

Solution

a. Since the angles are vertical angles, they have equal measures.

$$4x - 20 = 3x + 15$$
$$x - 20 = 15 \quad \text{To eliminate } 3x \text{ from the right-hand side, subtract } 3x \text{ from both sides.}$$
$$x = 35 \quad \text{To undo the subtraction of 20, add 20 to both sides.}$$

Thus, x is 35.

b. To find the measure of the vertical angles, we substitute 35 for x in either expression that represents their measure.

$$3x + 15 = 3(35) + 15$$
$$= 120$$

The measure of each vertical angle is 120°.

Answer 15

Complementary and supplementary angles

Complementary and supplementary angles

Two angles are **complementary angles** when the sum of their measures is 90°.

Two angles are **supplementary angles** when the sum of their measures is 180°.

EXAMPLE 5

a. Angles with measures of 60° and 30° are complementary angles, because the sum of their measures is 90°. Each angle is the complement of the other.

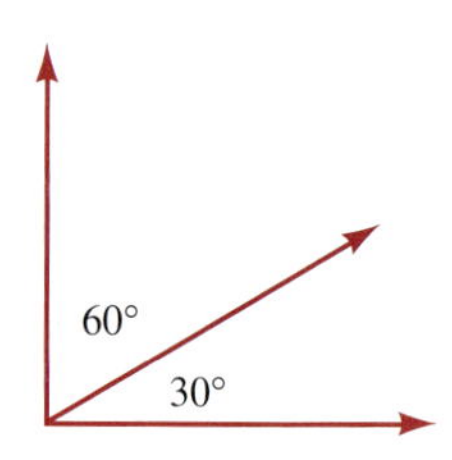

b. Angles of 130° and 50° are supplementary, because the sum of their measures is 180°. Each angle is the supplement of the other.

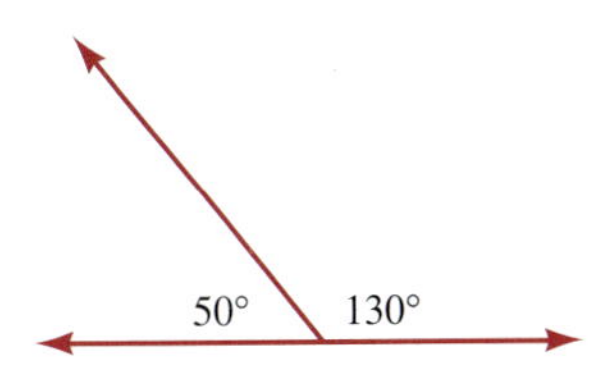

! COMMENT The definition of supplementary angles requires that the sum of *two* angles be 180°. Three angles of 40°, 60°, and 80° are not supplementary even though their sum is 180°.

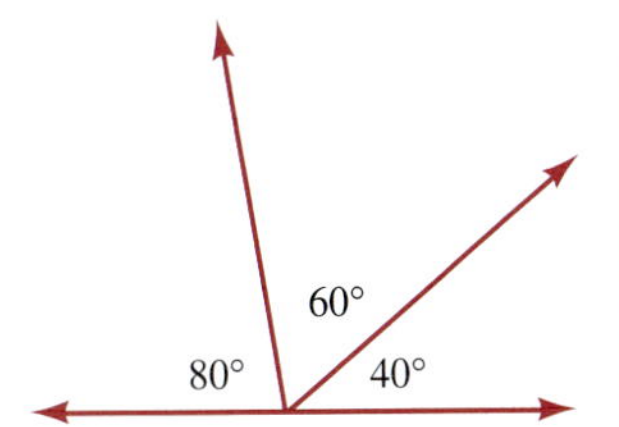

EXAMPLE 6

a. Find the complement of a 35° angle.

b. Find the supplement of a 105° angle.

Solution

a. See Figure 9-13. Let x represent the complement of the 35° angle. Since the angles are complementary, we have

$x + 35 = 90$ The sum of the angles' measures must be 90°

$x = 55$ To undo the addition of 35, subtract 35 from both sides.

The complement of 35° is 55°.

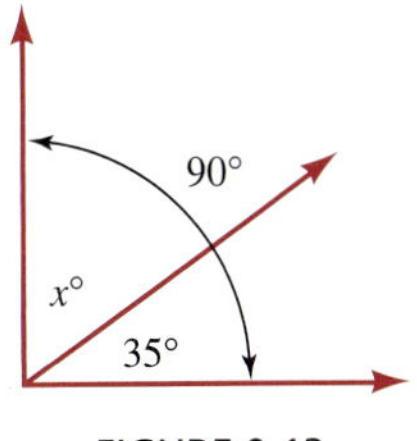

FIGURE 9-13

b. See Figure 9-14. Let y represent the supplement of the 105° angle. Since the angles are supplementary, we have

$y + 105 = 180$ The sum of the angles' measures must be 180°.

$y = 75$ To undo the addition of 105, subtract 105 from both sides.

The supplement of 105° is 75°.

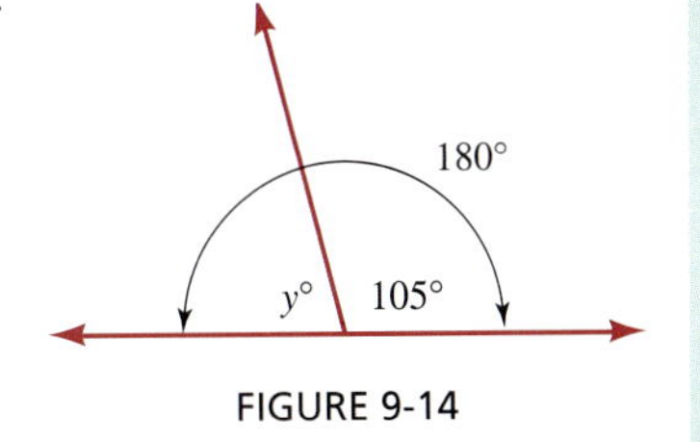

FIGURE 9-14

Self Check 6

a. Find the complement of a 50° angle.

b. Find the supplement of a 50° angle.

Answers a. 40°, **b.** 130°

Section 9.1 STUDY SET

VOCABULARY *Fill in the blanks.*

1. A line ________ has two endpoints.
2. Two points ________ at most one line.
3. A ________ divides a line segment into two parts of equal length.
4. An angle is measured in ________.
5. A ________ is used to measure angles.
6. An ________ angle is less than 90°.
7. A ________ angle measures 90°.
8. An ________ angle is greater than 90° but less than 180°.
9. The measure of a straight angle is ____.
10. Adjacent angles have the same vertex and are ________.
11. The sum of two ________ angles is 180°.
12. The sum of two complementary angles is ____.

CONCEPTS

Refer to the illustration and determine whether each statement is true. If a statement is false, explain why.

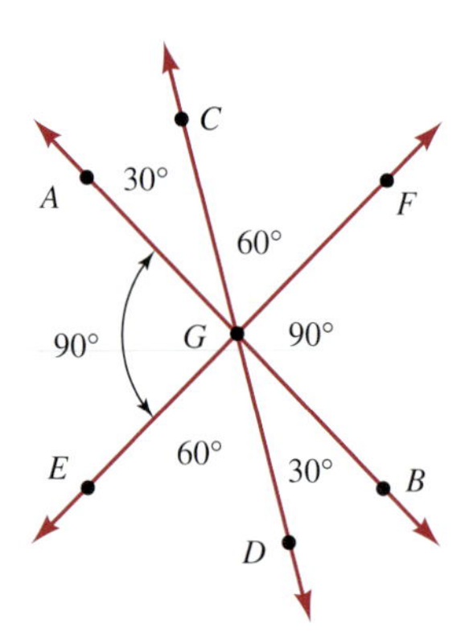

13. $\overrightarrow{GF}$ has point G as its endpoint.

14. $\overline{AG}$ has no endpoints.

15. Line CD has three endpoints.

16. Point D is the vertex of $\angle DGB$.

17. $m(\angle AGC) = m(\angle BGD)$

18. $\angle AGF \cong \angle BGE$

19. $\angle FGB \cong \angle EGA$

20. $\angle AGC$ and $\angle CGF$ are adjacent angles.

Refer to the illustration above and determine whether each angle is an acute angle, a right angle, an obtuse angle, or a straight angle.

21. $\angle AGC$

22. $\angle EGA$

23. $\angle FGD$

24. $\angle BGA$

25. $\angle BGE$

26. $\angle AGD$

27. $\angle DGC$

28. $\angle DGB$

Refer to the illustration below and determine whether each statement is true. If a statement is false, explain why.

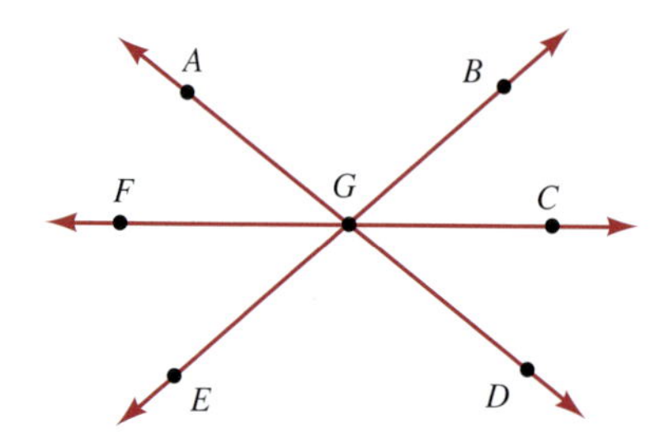

29. $\angle AGF$ and $\angle DGC$ are vertical angles.

30. $\angle FGE$ and $\angle BGA$ are vertical angles.

31. $m(\angle AGB) = m(\angle BGC)$

32. $\angle AGC \cong \angle DGF$

Refer to the illustration below and determine whether the angles are congruent.

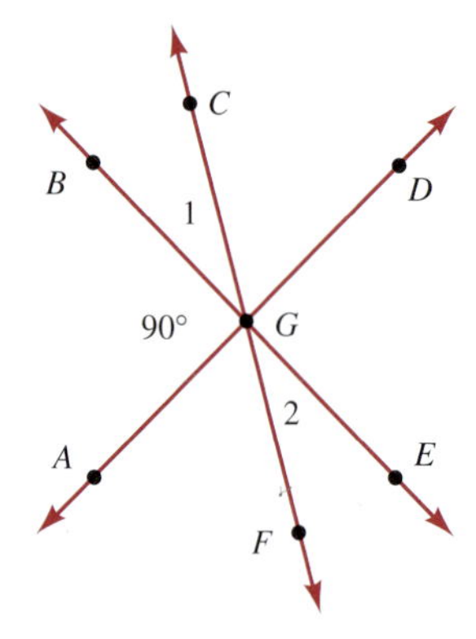

33. $\angle 1$ and $\angle 2$

34. $\angle FGB$ and $\angle CGE$

35. $\angle AGB$ and $\angle DGE$

36. $\angle CGD$ and $\angle CGB$

37. $\angle AGF$ and $\angle FGE$

38. $\angle AGB$ and $\angle BGD$

Refer to the illustration above and determine whether each statement is true.

39. $\angle 1$ and $\angle CGD$ are adjacent angles.

40. $\angle 2$ and $\angle 1$ are adjacent angles.

41. $\angle FGA$ and $\angle AGC$ are supplementary.

42. $\angle AGB$ and $\angle BGC$ are complementary.

43. $\angle AGF$ and $\angle 2$ are complementary.

44. $\angle AGB$ and $\angle EGD$ are supplementary.

45. $\angle EGD$ and $\angle DGB$ are supplementary.

46. $\angle DGC$ and $\angle AGF$ are complementary.

NOTATION

Fill in the blanks.

47. The symbol $\angle$ means ______.

48. The symbol $\overline{AB}$ is read as "____________ AB."

49. The symbol $\overrightarrow{AB}$ is read as "______ AB."

50. The symbol ____ is read as "is congruent to."

PRACTICE

Refer to the illustration and find the length of each line segment.

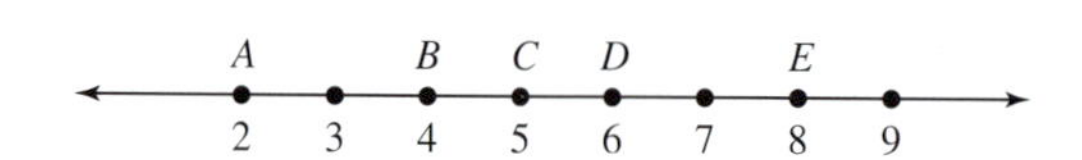

51. $\overline{AC}$

52. $\overline{BE}$

53. $\overline{CE}$

54. $\overline{BD}$

55. $\overline{CD}$

56. $\overline{DE}$

Refer to the illustration on the previous page and find each midpoint.

57. Find the midpoint of $\overline{AD}$.

58. Find the midpoint of $\overline{BE}$.

Use a protractor to measure each angle.

59. **60.**

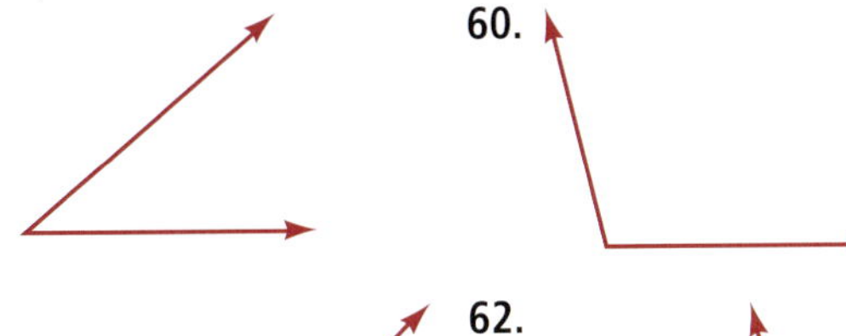

61. **62.**

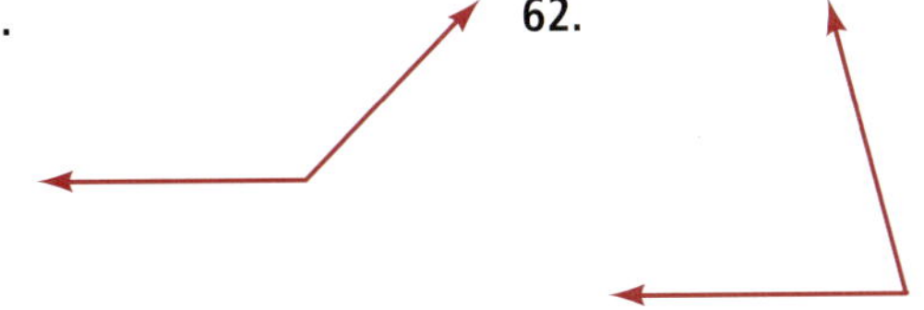

Find x.

63.

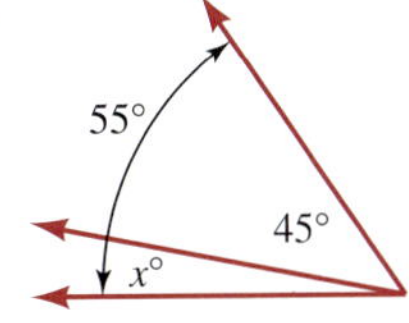

64.

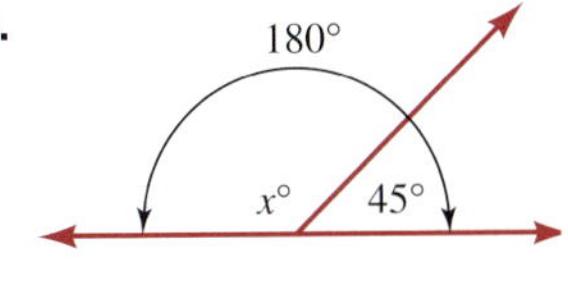

65.

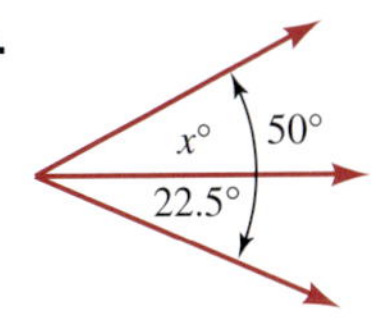

66.

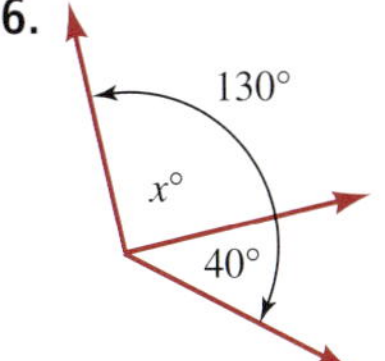

67.

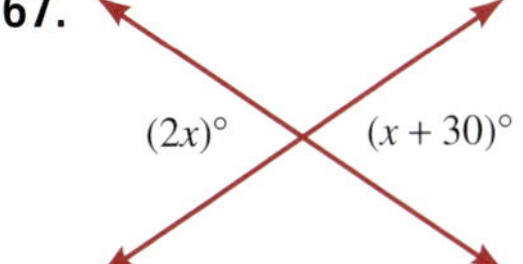

68.

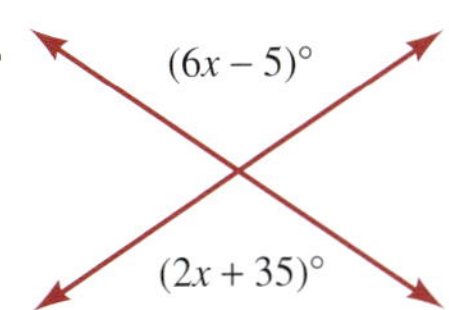

69.

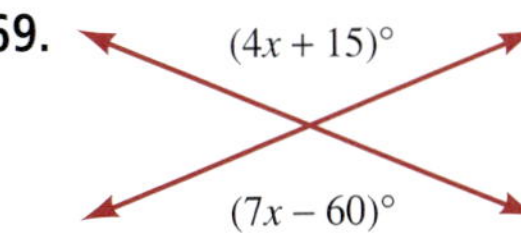

70.

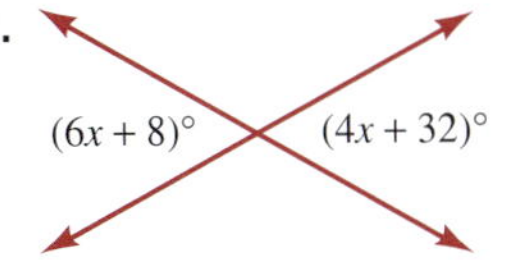

Let x represent the unknown angle measure. Draw a diagram, write an appropriate equation, and solve it for x.

71. Find the complement of a 30° angle.

72. Find the supplement of a 30° angle.

73. Find the supplement of a 105° angle.

74. Find the complement of a 75° angle.

Refer to the illustration, in which $m(\angle 1) = 50°$. *Find the measure of each angle or sum of angles.*

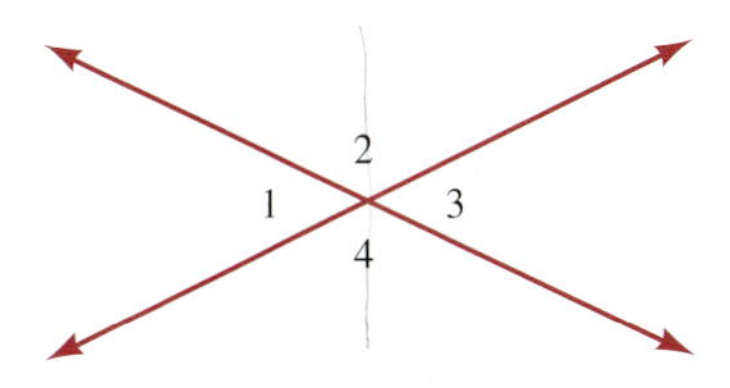

75. $\angle 4$

76. $\angle 3$

77. $m(\angle 1) + m(\angle 2) + m(\angle 3)$

78. $m(\angle 2) + m(\angle 4)$

Refer to the illustration, in which $m(\angle 1) + m(\angle 3) + m(\angle 4) = 180°$, $\angle 3 \cong \angle 4$, *and* $\angle 4 \cong \angle 5$. *Find the measure of each angle.*

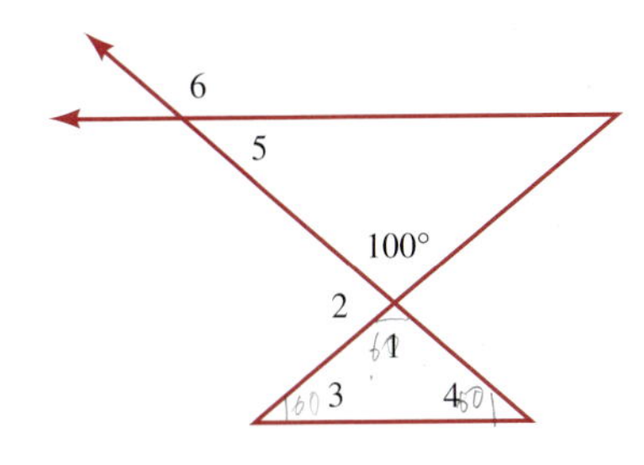

79. $\angle 1$

80. $\angle 2$

81. $\angle 3$

82. $\angle 6$

APPLICATIONS

83. BASEBALL Use the following definition to draw the strike zone for the player shown.

> The strike zone is that area over home plate the upper limit of which is a horizontal line at the midpoint between the top of the shoulders and the top of the uniform pants and the lower level is a line at the hollow beneath the kneecap.

84. PHYSICS The illustration shows a 15-pound block that is suspended with two ropes, one of which is horizontal. Classify each numbered angle in the illustration as acute, obtuse, or right.

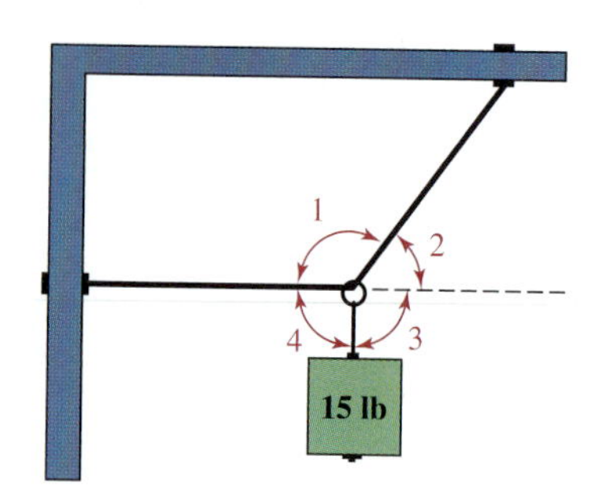

85. SYNTHESIZERS In the illustration, find x and y.

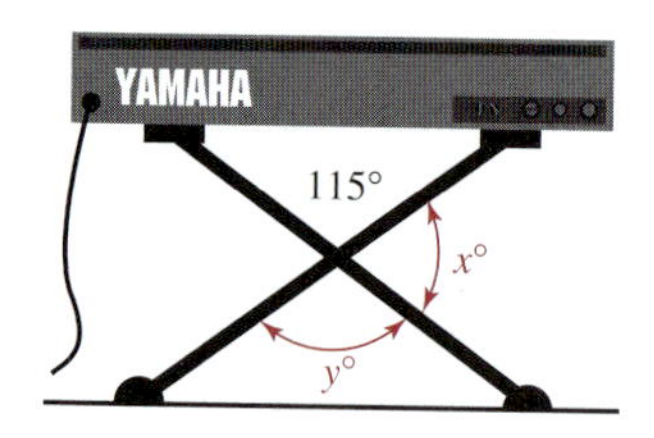

86. AVIATION In the illustration, how many degrees from the horizontal position are the wings of the airplane?

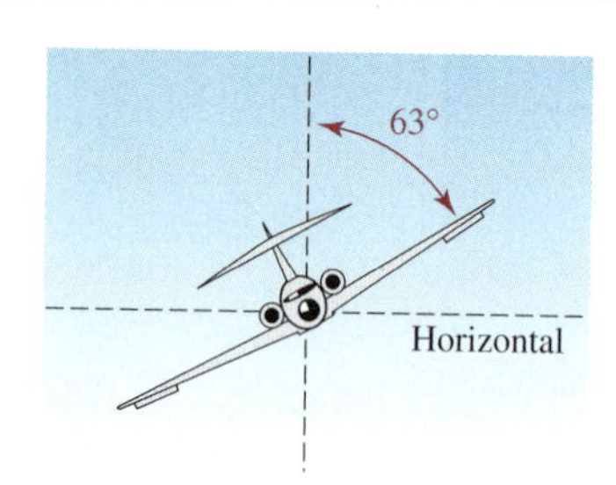

87. GARDENING What angle does the handle of the lawn mower make with the ground?

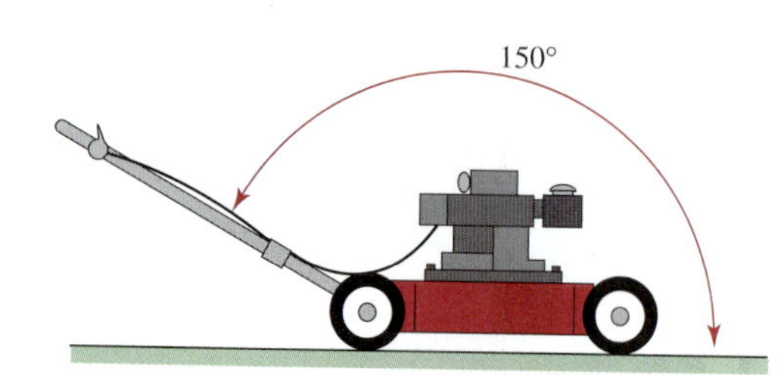

88. MUSICAL INSTRUMENTS Suppose you are a beginning band teacher describing the correct posture needed to play various instruments. Use the diagrams in the next column to approximate the angle measure at which each instrument should be held in relation to the student's body: **a.** flute, **b.** clarinet, **c.** trumpet.

WRITING

89. PHRASES Explain what you think each of these phrases means. How is geometry involved?

a. The president did a complete 180-degree flip on the subject of a tax cut.

b. The rollerblader did a "360" as she jumped off the ramp.

90. In the statements below, the ° symbol is used in two different ways. Explain the difference.

85° F and $m(\angle A) = 85°$

91. What is a protractor?

92. Explain the difference between a ray and a line segment.

93. Explain why an angle measuring 105° cannot have a complement.

94. Explain why an angle measuring 210° cannot have a supplement.

REVIEW

95. Find the power: 2^4.

96. Add: $\frac{1}{2} + \frac{2}{3} + \frac{3}{4}$.

97. Subtract: $\frac{3}{4} - \frac{1}{8} - \frac{1}{3}$.

98. Multiply: $\frac{5}{8} \cdot \frac{2}{15} \cdot \frac{6}{5}$.

99. Graph: $y = 2x - 5$.

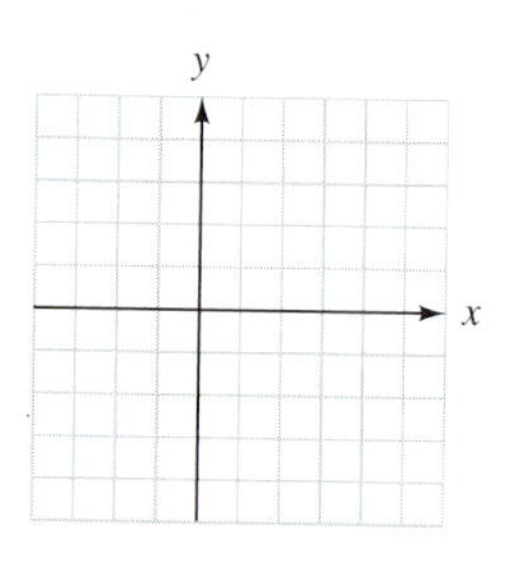

100. Find 7% of 7.

101. Solve the proportion: $\frac{x + 1}{18} = \frac{12.5}{45}$.

102. Convert 120 yards to feet.

9.2 Parallel and Perpendicular Lines

- Parallel and perpendicular lines • Transversals and angles • Properties of parallel lines

In this section, we will consider *parallel* and *perpendicular* lines. Since parallel lines are always the same distance apart, the railroad tracks shown in Figure 9-15(a) illustrate one application of parallel lines. Figure 9-15(b) shows one of the events of men's gymnastics, the parallel bars. Since perpendicular lines meet and form right angles, the monument and the ground shown in Figure 9-15(c) illustrate one application of perpendicular lines.

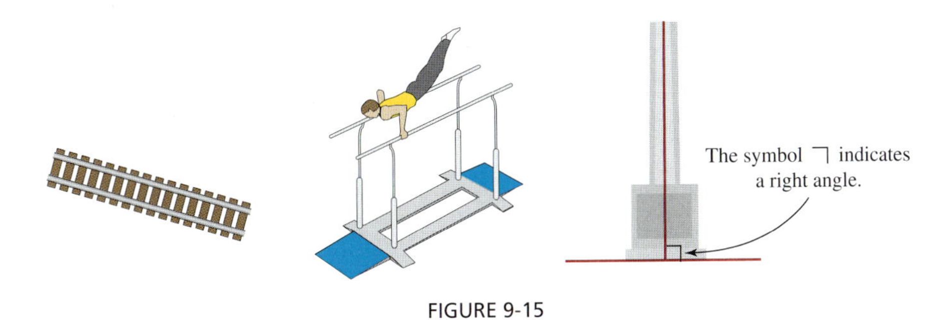

FIGURE 9-15

Parallel and perpendicular lines

If two lines lie in the same plane, they are called **coplanar.** Two coplanar lines that do not intersect are called **parallel lines.** See Figure 9-16(a).

Parallel lines

Parallel lines are coplanar lines that do not intersect.

If lines l_1 (read as "l sub 1") and l_2 (read as "l sub 2") are parallel, we can write $l_1 \parallel l_2$, where the symbol $\parallel$ is read as "is parallel to."

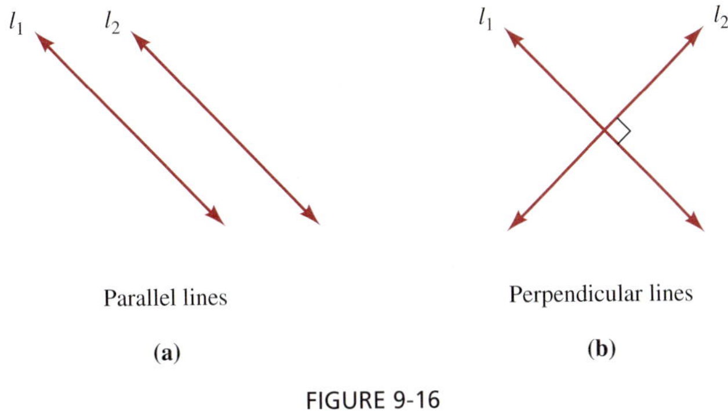

FIGURE 9-16

Perpendicular Lines

Perpendicular lines are lines that intersect and form right angles.

In Figure 9-16(b), $l_1 \perp l_2$, where the symbol $\perp$ is read as "is perpendicular to."

Transversals and angles

A line that intersects two or more coplanar lines is called a **transversal.** For example, line l_1 in Figure 9-17 is a transversal intersecting lines l_2, l_3, and l_4.

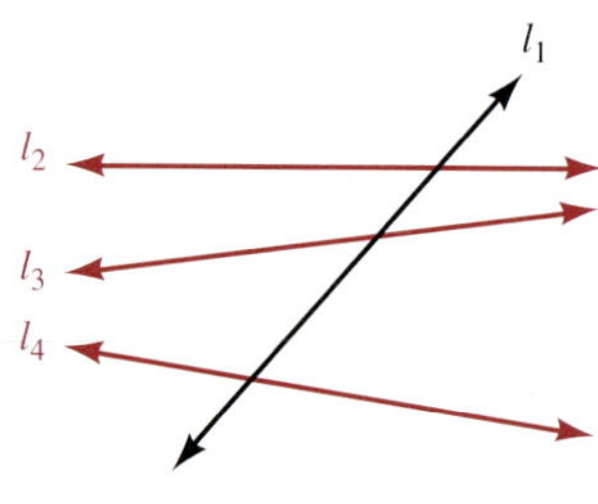

FIGURE 9-17

When two lines are cut by a transversal, the following types of angles are formed.

Alternate interior angles:

$\angle 4$ and $\angle 5$

$\angle 3$ and $\angle 6$

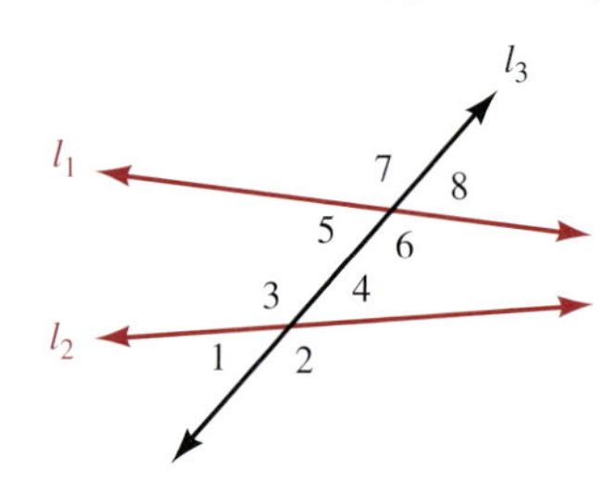

Corresponding angles:

$\angle 1$ and $\angle 5$

$\angle 3$ and $\angle 7$

$\angle 2$ and $\angle 6$

$\angle 4$ and $\angle 8$

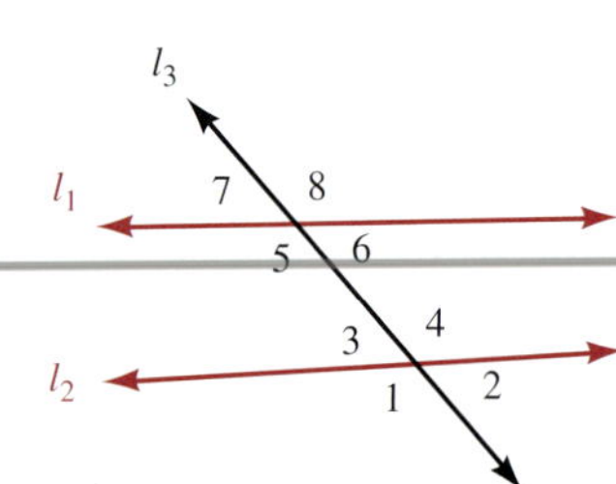

Interior angles:

$\angle 3$, $\angle 4$, $\angle 5$, and $\angle 6$

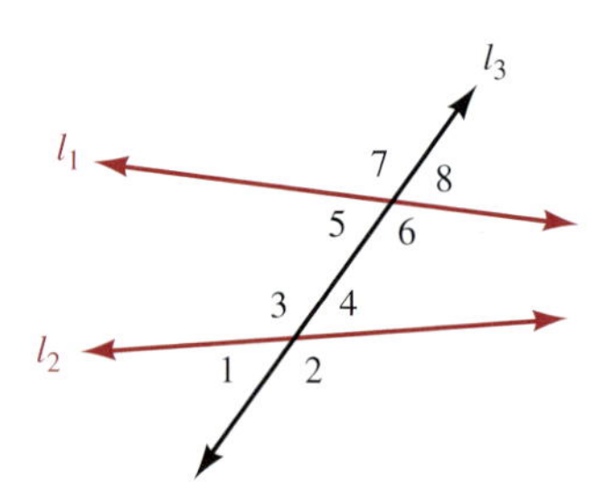

EXAMPLE 1 In Figure 9-18, identify **a.** all pairs of alternate interior angles, **b.** all pairs of corresponding angles, and **c.** all interior angles.

Solution

a. Pairs of alternate interior angles are

$\angle 3$ and $\angle 5$, $\angle 4$ and $\angle 6$

b. Pairs of corresponding angles are

$\angle 1$ and $\angle 5$, $\angle 4$ and $\angle 8$, $\angle 2$ and $\angle 6$, $\angle 3$ and $\angle 7$

c. Interior angles are

$\angle 3$, $\angle 4$, $\angle 5$, and $\angle 6$

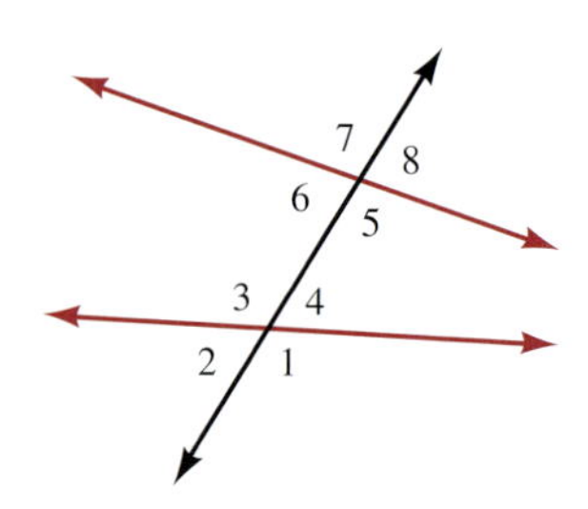

FIGURE 9-18

Properties of parallel lines

1. If two parallel lines are cut by a transversal, alternate interior angles are congruent. (See Figure 9-19.) If $l_1 \parallel l_2$, then $\angle 2 \cong \angle 4$ and $\angle 1 \cong \angle 3$.

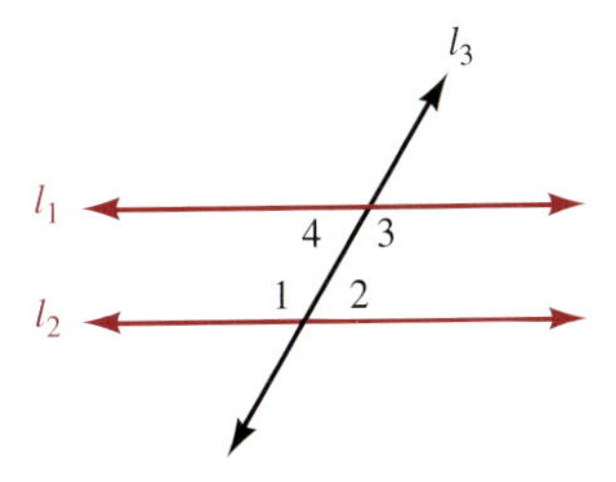

FIGURE 9-19

2. If two parallel lines are cut by a transversal, corresponding angles are congruent. (See Figure 9-20.) If $l_1 \parallel l_2$, then $\angle 1 \cong \angle 5$, $\angle 3 \cong \angle 7$, $\angle 2 \cong \angle 6$, and $\angle 4 \cong \angle 8$.
3. If two parallel lines are cut by a transversal, interior angles on the same side of the transversal are supplementary. (See Figure 9-21.) If $l_1 \parallel l_2$, then $\angle 1$ is supplementary to $\angle 2$ and $\angle 4$ is supplementary to $\angle 3$.

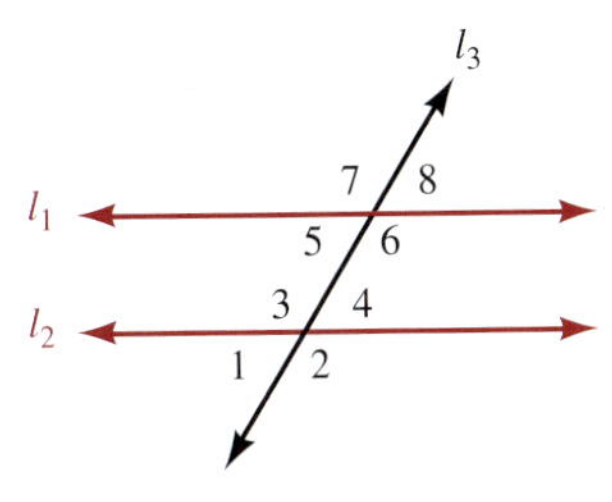

FIGURE 9-20

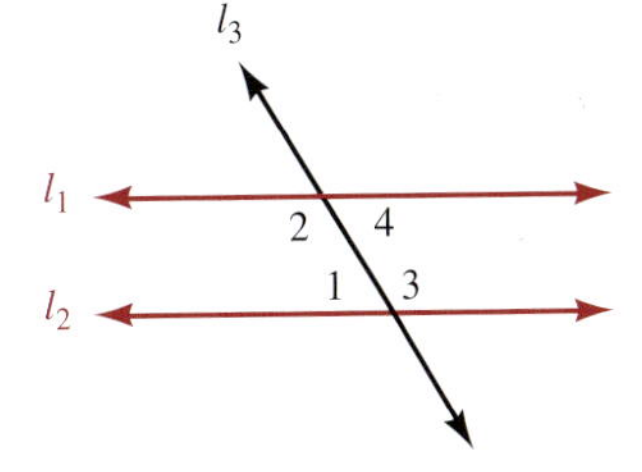

FIGURE 9-21

4. If a transversal is perpendicular to one of two parallel lines, it is also perpendicular to the other line. (See Figure 9-22.) If $l_1 \parallel l_2$ and $l_3 \perp l_1$, then $l_3 \perp l_2$.
5. If two lines are parallel to a third line, they are parallel to each other. (See Figure 9-23.) If $l_1 \parallel l_2$ and $l_1 \parallel l_3$, then $l_2 \parallel l_3$.

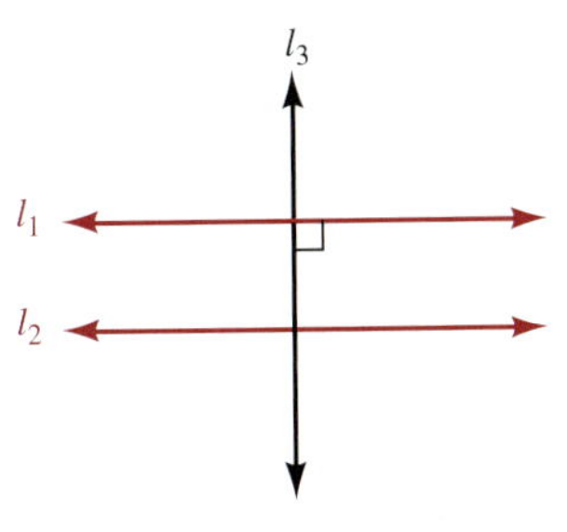

FIGURE 9-22

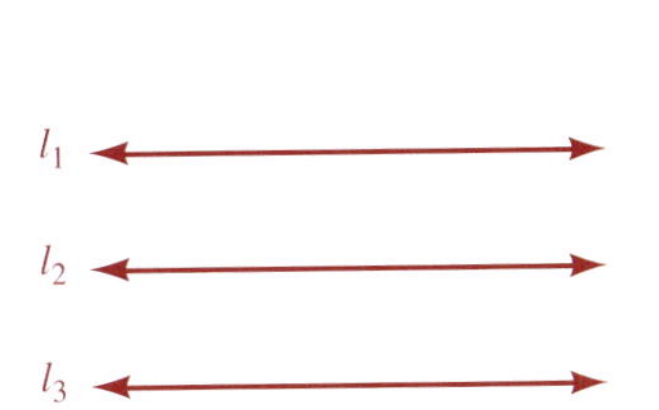

FIGURE 9-23

EXAMPLE 2 See Figure 9-24. If $l_1 \parallel l_2$ and $m(\angle 3) = 120°$, find the measures of the other angles.

Solution

$m(\angle 1) = 60°$ $\angle 3$ and $\angle 1$ are supplementary.

$m(\angle 2) = 120°$ Vertical angles are congruent: $m(\angle 2) = m(\angle 3)$.

$m(\angle 4) = 60°$ Vertical angles are congruent: $m(\angle 4) = m(\angle 1)$.

$m(\angle 5) = 60°$ If two parallel lines are cut by a transversal, alternate interior angles are congruent: $m(\angle 5) = m(\angle 4)$.

$m(\angle 6) = 120°$ If two parallel lines are cut by a transversal, alternate interior angles are congruent: $m(\angle 6) = m(\angle 3)$.

$m(\angle 7) = 120°$ Vertical angles are congruent: $m(\angle 7) = m(\angle 6)$.

$m(\angle 8) = 60°$ Vertical angles are congruent: $m(\angle 8) = m(\angle 5)$.

FIGURE 9-24

Self Check 2

If $l_1 \parallel l_2$ and $m(\angle 8) = 50°$, find the measures of the other angles. (See Figure 9-24.)

Answers $m(\angle 5) = 50°$, $m(\angle 7) = 130°$, $m(\angle 6) = 130°$, $m(\angle 3) = 130°$, $m(\angle 4) = 50°$, $m(\angle 1) = 50°$, $m(\angle 2) = 130°$

EXAMPLE 3 See Figure 9-25. If $\overline{AB} \parallel \overline{DE}$, which pairs of angles are congruent?

Solution Since $\overline{AB} \parallel \overline{DE}$, corresponding angles are congruent. So we have

$$\angle A \cong \angle 1 \quad \text{and} \quad \angle B \cong \angle 2$$

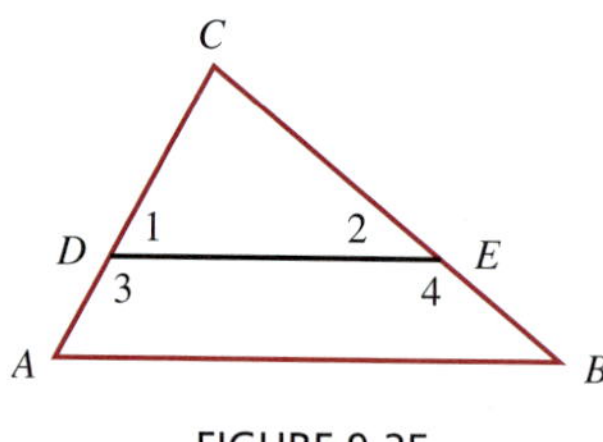

FIGURE 9-25

Self Check 4
In the figure below, $l_1 \parallel l_2$. Find y.

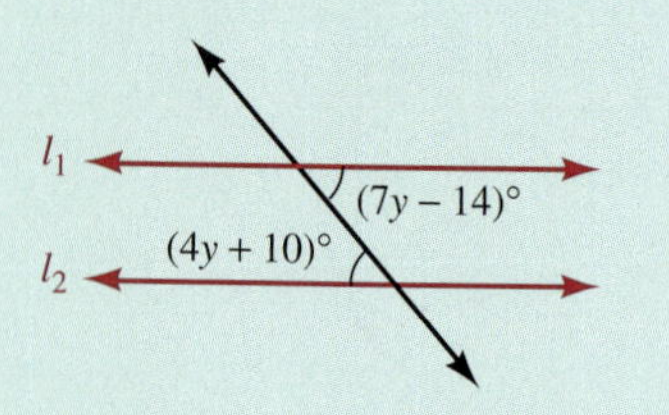

Answer 8

EXAMPLE 4 In Figure 9-26, $l_1 \parallel l_2$. Find x.

Solution The angles involving x are corresponding angles. Since $l_1 \parallel l_2$, all pairs of corresponding angles are congruent.

$9x - 15 = 6x + 30$	The angle measures are equal.
$3x - 15 = 30$	Subtract $6x$ from both sides.
$3x = 45$	To undo the subtraction of 15, add 15 to both sides.
$x = 15$	To undo the multiplication by 3, divide both sides by 3.

Thus, x is 15.

FIGURE 9-26

EXAMPLE 5 In Figure 9-27, $l_1 \parallel l_2$. Find x.

Solution Since the angles are interior angles on the same side of the transversal, they are supplementary.

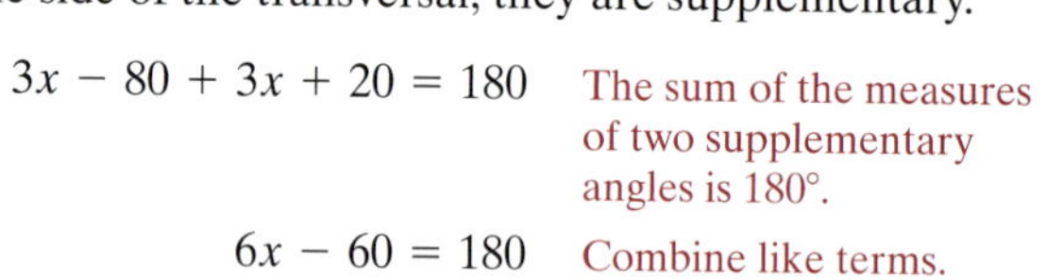

$3x - 80 + 3x + 20 = 180$	The sum of the measures of two supplementary angles is 180°.
$6x - 60 = 180$	Combine like terms.
$6x = 240$	To undo the subtraction of 60, add 60 to both sides.
$x = 40$	To undo the multiplication by 6, divide both sides by 6.

Thus, x is 40.

FIGURE 9-27

Section 9.2 STUDY SET

VOCABULARY *Fill in the blanks.*

1. Two lines in the same plane are ________.
2. ________ lines do not intersect.
3. If two lines intersect and form a right angle, they are ________.
4. A ________ intersects two or more coplanar lines.

5. In the illustration, ∠4 and ∠6 are ________ interior angles.

6. In the illustration, ∠2 and ∠6 are ____________ angles.

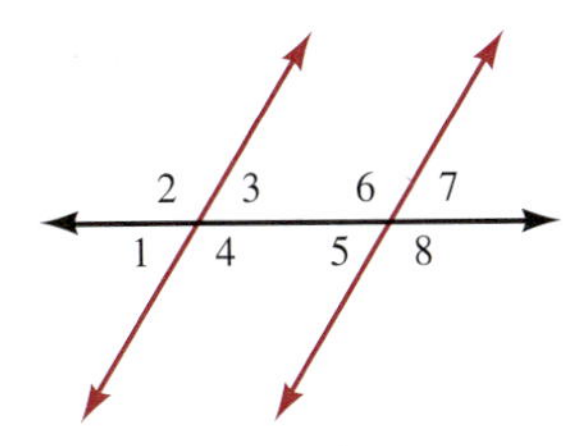

CONCEPTS

7. Which pairs of angles shown in the illustration above are alternate interior angles?

8. Which pairs of angles shown in the illustration above are corresponding angles?

9. Which angles shown in the illustration above are interior angles?

10. In the illustration below $l_1 \parallel l_2$. What can you conclude about l_1 and l_3?

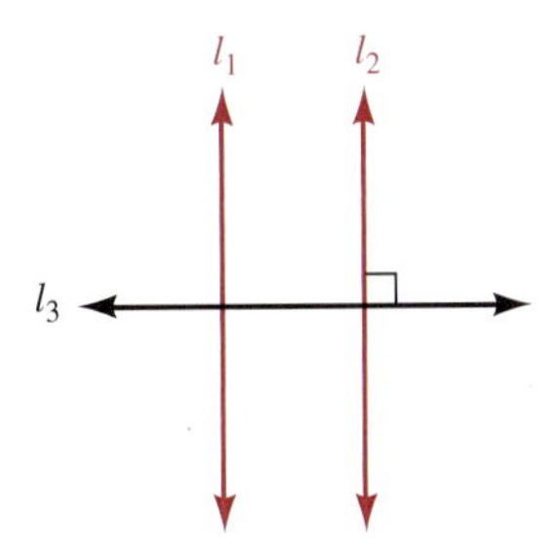

11. In the illustration, $l_1 \parallel l_2$ and $l_2 \parallel l_3$. What can you conclude about l_1 and l_3?

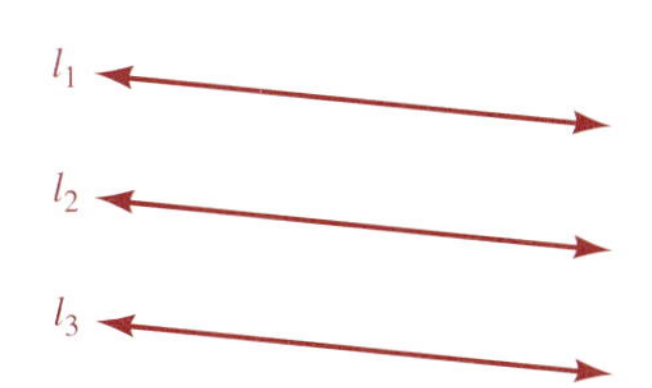

12. In the illustration, $\overline{AB} \parallel \overline{DE}$. What pairs of angles are congruent?

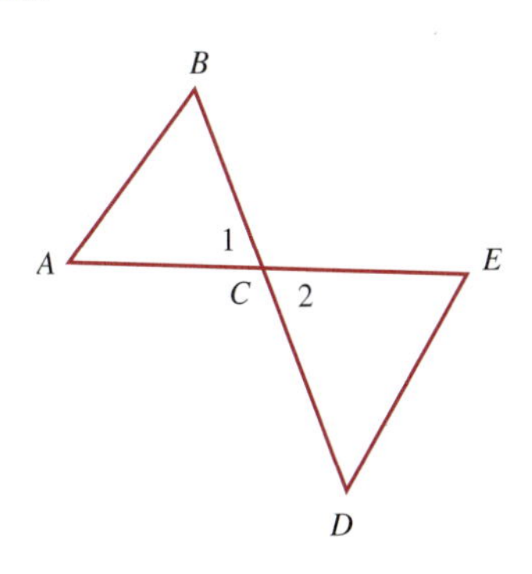

NOTATION *Fill in the blanks.*

13. The symbol ⏋ indicates ____________.

14. The symbol ∥ is read as "____________."

15. The symbol ⊥ is read as "________________."

16. The symbol l_1 is read as "__________."

PRACTICE

17. In the illustration, $l_1 \parallel l_2$ and $m(\angle 4) = 130°$. Find the measures of the other angles.

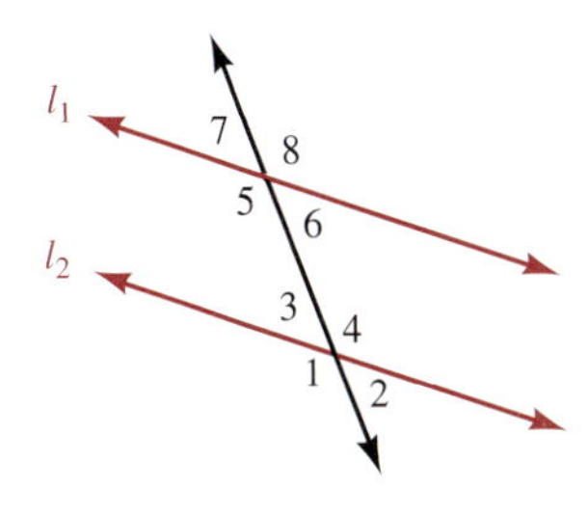

18. In the illustration, $l_1 \parallel l_2$ and $m(\angle 2) = 40°$. Find the measures of the other angles.

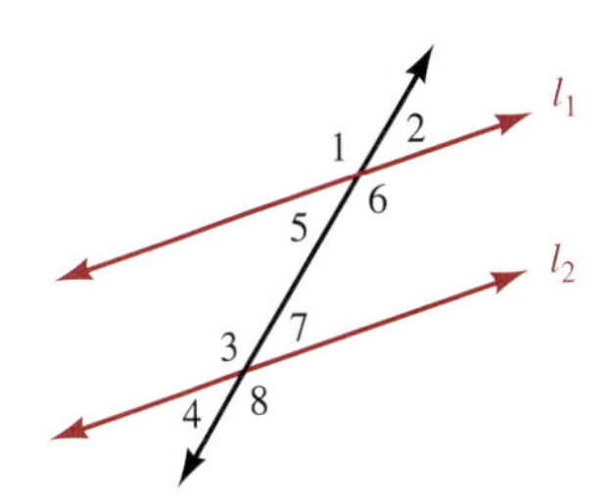

19. In the illustration, $l_1 \parallel \overrightarrow{AB}$. Find the measure of each angle.

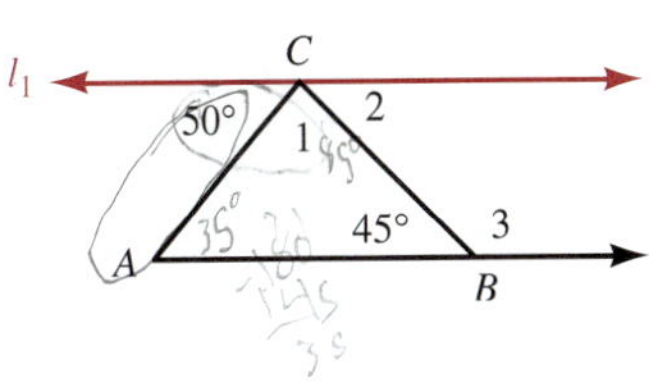

20. In the illustration, $\overline{AB} \parallel \overline{DE}$. Find $m(\angle B)$, $m(\angle E)$, and $m(\angle 1)$.

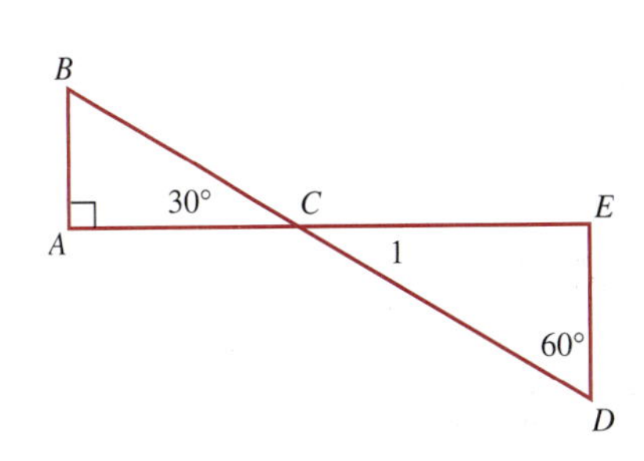

In Problems 21–24, $l_1 \parallel l_2$. Find x.

21.

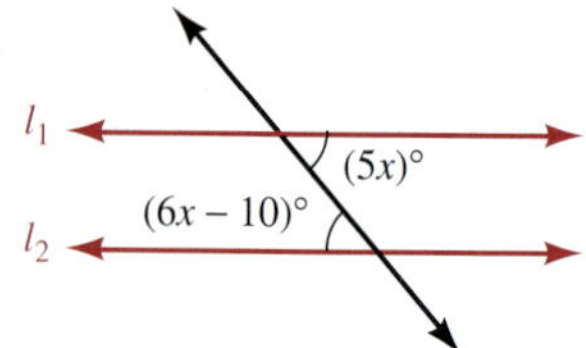

22.

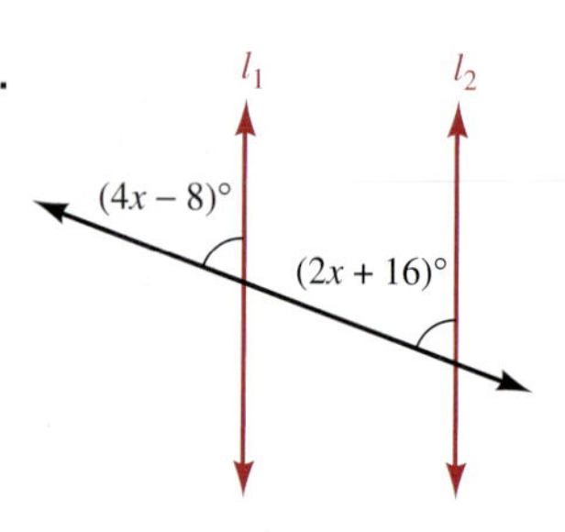

23.

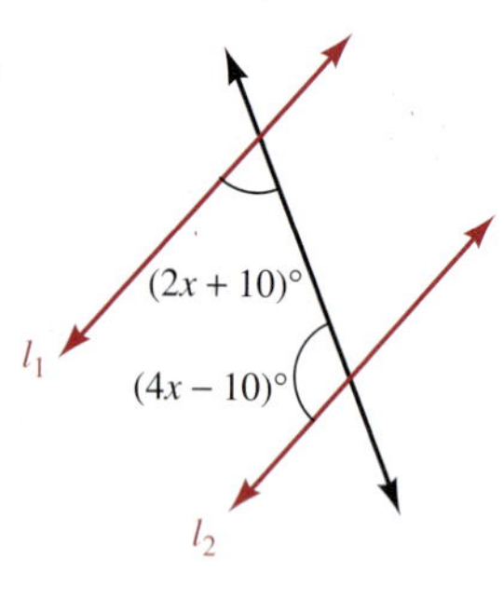

24.

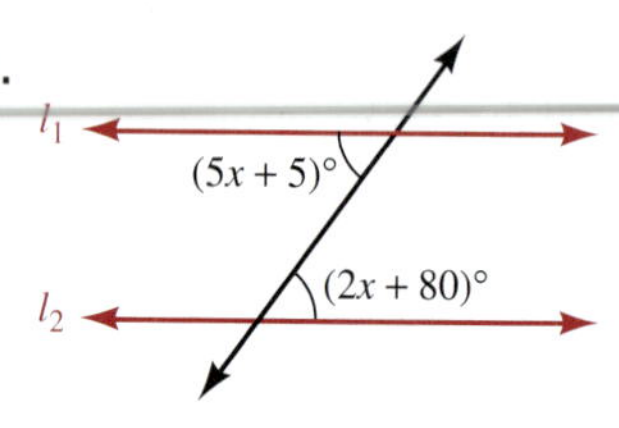

In Problems 25–28, find x.

25. $l_1 \parallel \overrightarrow{AC}$

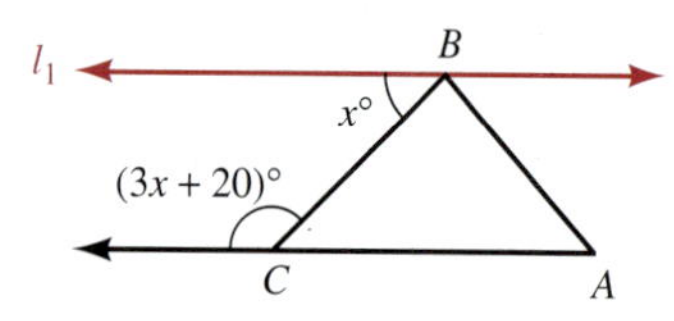

26. $\overline{AB} \parallel \overline{DE}$

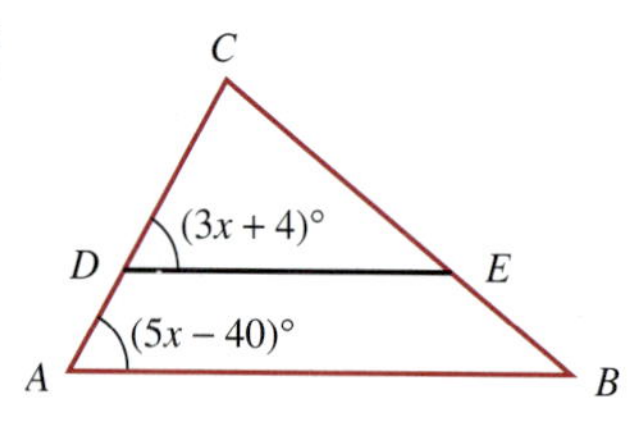

27. $\overline{AB} \parallel \overline{DE}$

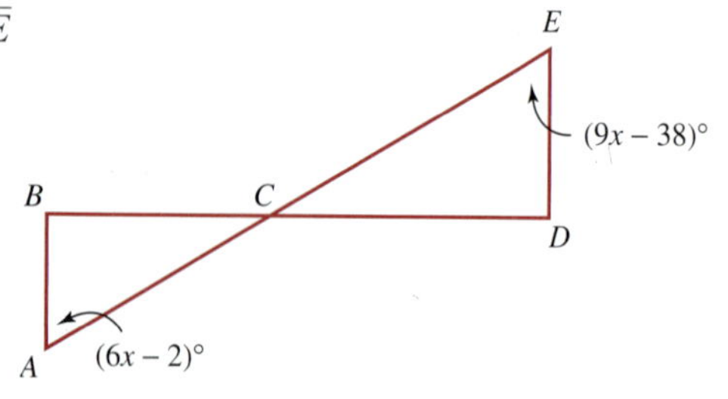

28. $\overline{AC} \parallel \overline{BD}$

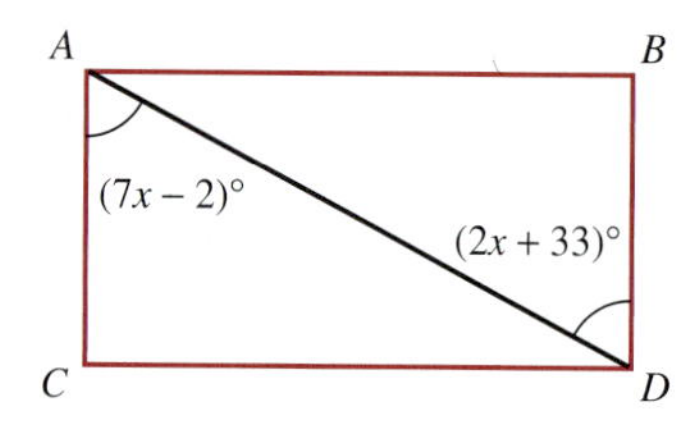

APPLICATIONS

29. PYRAMIDS The Egyptians used a device called a **plummet** to determine whether stones were properly leveled. A plummet, shown below, is made up of an A-frame and a plumb bob suspended from the peak of the frame. How could a builder use a plummet to tell that the stone on the left is not level and that the stones on the right are level?

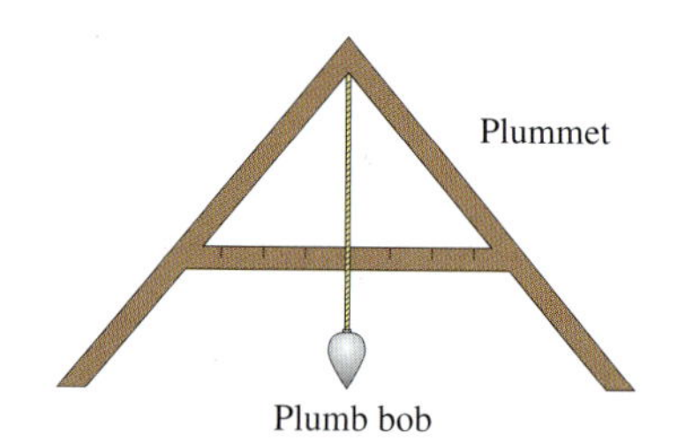

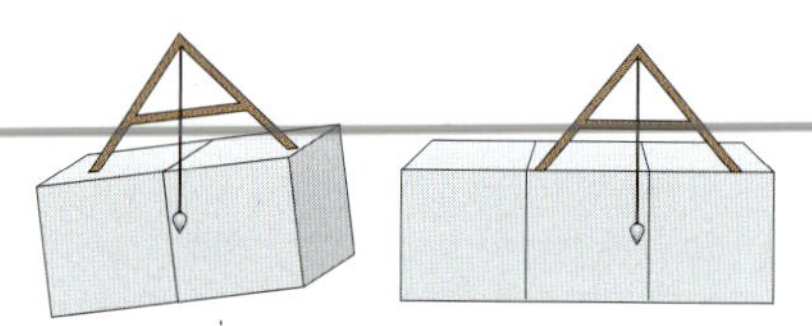

30. DIAGRAMMING SENTENCES English instructors have their students diagram sentences to help teach proper sentence structure. Below is a diagram of the sentence *The cave was rather dark and damp.* Point out pairs of parallel and perpendicular lines used in the diagram.

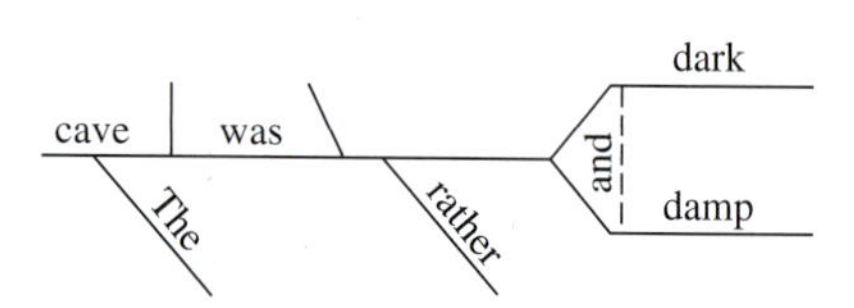

31. LOGOS Point out any perpendicular lines that can be found on the BMW company logo shown.

32. PAINTING SIGNS For many sign painters, the most difficult letter to paint is a capital E, because of all of the right angles involved. How many right angles are there?

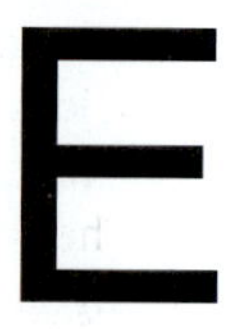

33. WALLPAPER Explain why the concepts of perpendicular and parallel are both important when hanging wallpaper.

34. TOOLS What geometric concepts are seen in the design of the rake?

WRITING

35. PARKING DESIGNS Using terms from this chapter, write a paragraph describing the parking layout shown.

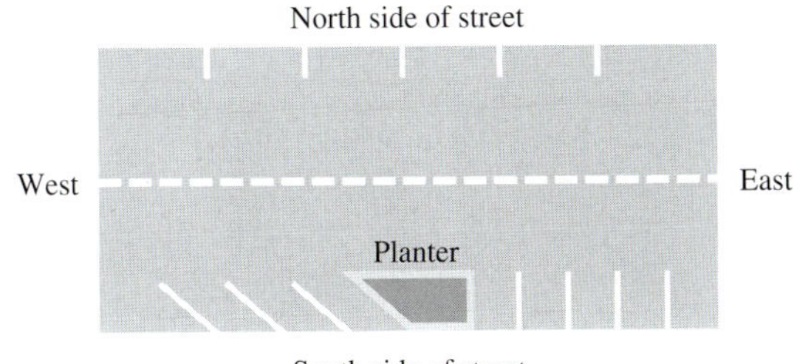

36. In your own words, explain what is meant by each of the following sentences.

a. The hikers were told that the path *parallels* the river.

b. John's quick rise to fame and fortune *paralleled* that of his older brother.

c. The judge stated that the case that was before her court was without *parallel.*

37. Why do you think that $\angle 4$ and $\angle 6$ shown in the illustration for Problem 7 are called alternate interior angles?

38. Why do you think that $\angle 4$ and $\angle 8$ shown in the illustration for Problem 8 are called corresponding angles?

39. Are pairs of alternate interior angles always congruent? Explain.

40. Are pairs of interior angles always supplementary? Explain.

REVIEW

41. Find 60% of 120.

42. 80% of what number is 400?

43. What percent of 500 is 225?

44. Simplify: $3.45 + 7.37 \cdot 2.98$.

45. Is every whole number an integer?

46. Multiply: $2\frac{1}{5} \cdot 4\frac{3}{7}$.

47. Express the phrase as a ratio in lowest terms: 4 ounces to 12 ounces.

48. Convert 5,400 milligrams to kilograms.

9.3 Polygons

- Polygons • Triangles • Properties of isosceles triangles
- The sum of the measures of the angles of a triangle • Quadrilaterals
- Properties of rectangles • The sum of the measures of the angles of a polygon

In this section, we will discuss figures called *polygons.* We see these shapes every day. For example, the walls in most buildings are rectangular. We also see rectangular shapes in doors, windows, and sheets of paper.

The gable ends of many houses are triangular, as are the sides of the Great Pyramid in Egypt. Triangular shapes are especially important because triangles are rigid and contribute strength and stability to walls and towers.

The designs used in tile or linoleum floors often use the shapes of a pentagon or a hexagon. Stop signs are in the shape of an octagon.

Polygons

Polygon

A **polygon** is a closed geometric figure with at least three line segments for its sides.

The figures in Figure 9-28 are **polygons.** They are classified according to the number of sides they have. The points where the sides intersect are called **vertices.** If a polygon has sides that are all the same length and angles that have the same measure, we call it a **regular polygon.**

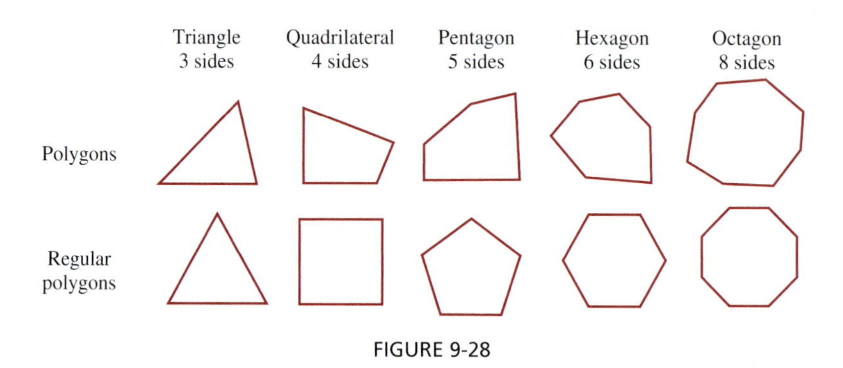

FIGURE 9-28

Self Check 1

Give the number of vertices of

a. a quadrilateral

b. a pentagon

Answers a. 4 **b.** 5

EXAMPLE 1 Give the number of vertices of **a.** a triangle and **b.** a hexagon.

Solution

a. From Figure 9-28, we see that a triangle has three angles and therefore three vertices.

b. From Figure 9-28, we see that a hexagon has six angles and therefore six vertices.

From the results of Example 1, we see that the number of vertices of a polygon is equal to the number of its sides.

Triangles

A **triangle** is a polygon with three sides. Figure 9-29 illustrates some common triangles. The slashes on the sides of a triangle indicate which sides are of equal length.

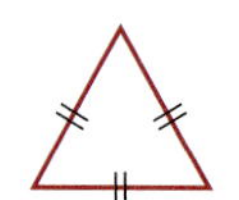

Equilateral triangle
(all sides equal length)

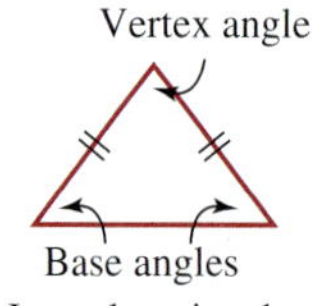

Isosceles triangle
(at least two sides of equal length)

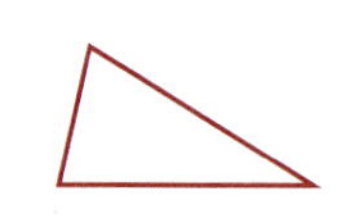

Scalene triangle
(no sides equal length)

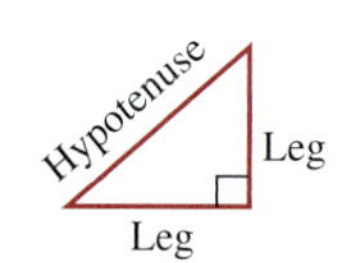

Right triangle
(has a right angle)

FIGURE 9-29

! COMMENT Since equilateral triangles have at least two sides of equal length, they are also isosceles. However, isosceles triangles are not necessarily equilateral.

Since every angle of an equilateral triangle has the same measure, an equilateral triangle is also **equiangular.**

In an isosceles triangle, the angles opposite the sides of equal length are called **base angles,** the sides of equal length form the **vertex angle,** and the third side is called the **base.**

The longest side of a right triangle is called the **hypotenuse,** and the other two sides are called **legs.** The hypotenuse of a right triangle is always opposite the 90° angle.

Properties of isosceles triangles

1. Base angles of an isosceles triangle are congruent.
2. If two angles in a triangle are congruent, the sides opposite the angles have the same length, and the triangle is isosceles.

EXAMPLE 2 Is the triangle in Figure 9-30 an isosceles triangle?

Solution $\angle A$ and $\angle B$ are angles of the triangle. Since $m(\angle A) = m(\angle B)$, we know that $m(\overline{AC}) = m(\overline{BC})$ and that $\triangle ABC$ (read as "triangle ABC") is isosceles.

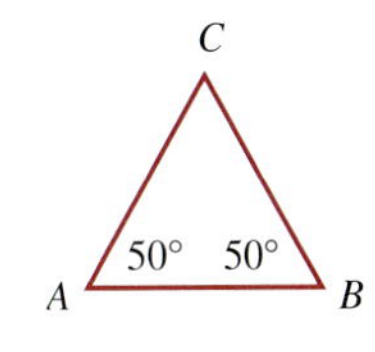

FIGURE 9-30

Self Check 2
In the figure below, $l_1 \parallel \overline{AB}$. Is the triangle an isosceles triangle?

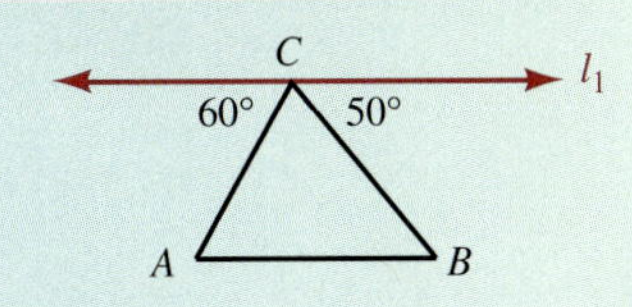

Answer no

The sum of the measures of the angles of a triangle

If you draw several triangles and carefully measure each angle with a protractor, you will find that the sum of the angle measures in each triangle is 180°.

> **Angles of a triangle**
> The sum of the angle measures of any triangle is 180°.

EXAMPLE 3 See Figure 9-31. Find x.

Solution Since the sum of the angle measures of any triangle is 180°, we have

$$x + 40 + 90 = 180$$

$$x + 130 = 180 \quad \text{Perform the addition: } 40 + 90 = 130.$$

$$x = 50 \quad \text{To undo the addition of 130, subtract 130 from both sides.}$$

FIGURE 9-31

Thus, x is 50.

Self Check 3
In the figure below, find y.

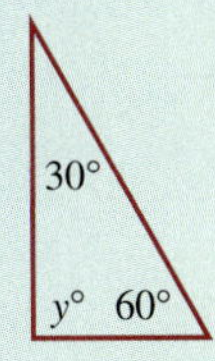

Answer 90

EXAMPLE 4 See Figure 9-32. If one base angle of an isosceles triangle measures 70°, how large is the vertex angle?

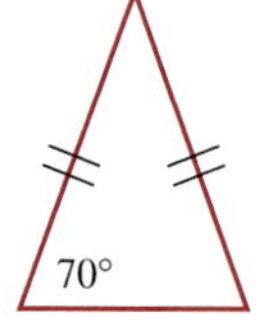

FIGURE 9-32

Solution Since one of the base angles measures 70°, so does the other. If we let x represent the measure of the vertex angle, we have

$x + 70 + 70 = 180$ The sum of the measures of the angles of a triangle is 180°.

$x + 140 = 180$ Perform the addition: $70 + 70 = 140$.

$x = 40$ To undo the addition of 140, subtract 140 from both sides.

The vertex angle measures 40°.

Quadrilaterals

A **quadrilateral** is a polygon with four sides. Some common quadrilaterals are shown in Figure 9-33.

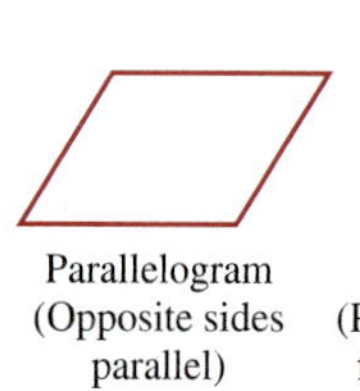
Parallelogram (Opposite sides parallel)

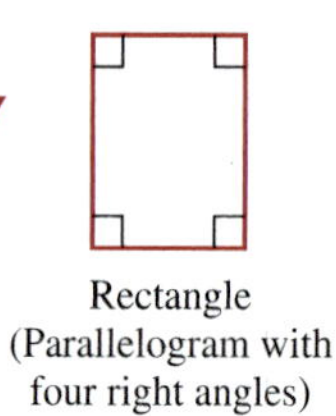
Rectangle (Parallelogram with four right angles)

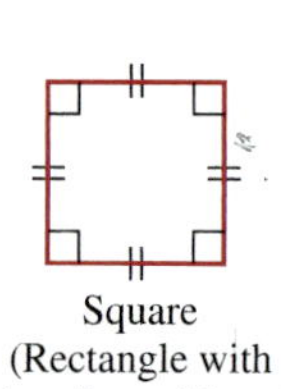
Square (Rectangle with sides of equal length)

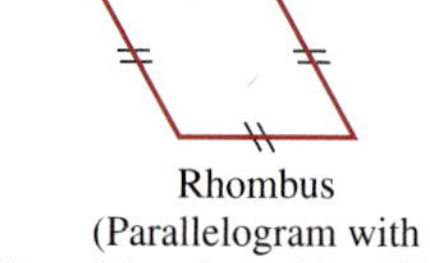
Rhombus (Parallelogram with sides of equal length)

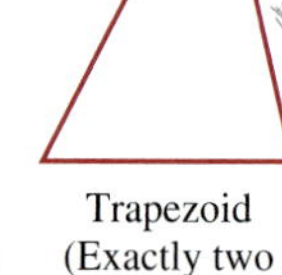
Trapezoid (Exactly two sides parallel)

FIGURE 9-33

Properties of rectangles

1. All angles of a rectangle are right angles.
2. Opposite sides of a rectangle are parallel.
3. Opposite sides of a rectangle are of equal length.
4. The diagonals of a rectangle are of equal length.
5. If the diagonals of a parallelogram are of equal length, the parallelogram is a rectangle.

EXAMPLE 5 **Squaring a foundation.** A carpenter intends to build a shed with an 8-by-12-foot base. How can he make sure that the rectangular foundation is "square"?

Solution See Figure 9-34. The carpenter can use a tape measure to find the lengths of diagonals AC and BD. If these diagonals are of equal length, the figure will be a rectangle and have four right angles. Then the foundation will be square.

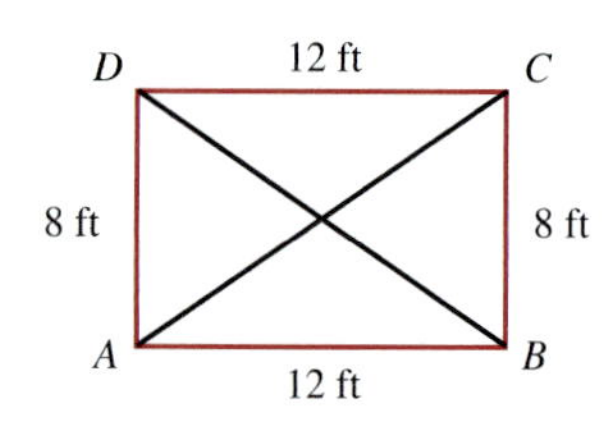

FIGURE 9-34

EXAMPLE 6 In rectangle $ABCD$ (Figure 9-35), the length of $\overline{AC}$ is 20 centimeters. Find each measure: **a.** $m(\overline{BD})$, **b.** $m(\angle 1)$, and **c.** $m(\angle 2)$.

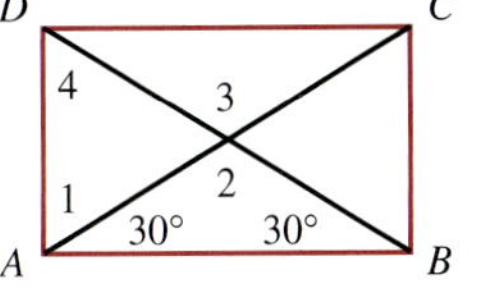

FIGURE 9-35

Solution

a. Since the diagonals of a rectangle are of equal length, $m(\overline{BD})$ is also 20 centimeters.

b. We let $m(\angle 1) = x$. Then, since the angles of a rectangle are right angles, we have

$$x + 30 = 90$$

$$x = 60 \quad \text{To undo the addition of 30, subtract 30 from both sides.}$$

Thus, $m(\angle 1) = 60°$.

c. We let $m(\angle 2) = y$. Then, since the sum of the angle measures of a triangle is 180°, we have

$$30 + 30 + y = 180$$

$$60 + y = 180 \quad \text{Simplify: } 30 + 30 = 60.$$

$$y = 120 \quad \text{To undo the addition of 60, subtract 60 from both sides.}$$

Thus, $m(\angle 2) = 120°$.

Self Check 6

In the rectangle $ABCD$, the length of $\overline{DC}$ is 16 centimeters. Find each measure:

a. $m(\overline{AB})$

b. $m(\angle 3)$

c. $m(\angle 4)$

Answers **a.** 16 cm **b.** 120° **c.** 60°

The parallel sides of a trapezoid are called **bases,** the nonparallel sides are called **legs,** and the angles on either side of a base are called **base angles.** If the nonparallel sides are the same length, the trapezoid is an **isosceles trapezoid.** In an isosceles trapezoid, the base angles are congruent.

EXAMPLE 7 **Cross section of a drainage ditch.** A cross section of a drainage ditch (Figure 9-36) is an isosceles trapezoid with $\overline{AB} \parallel \overline{CD}$. Find x and y.

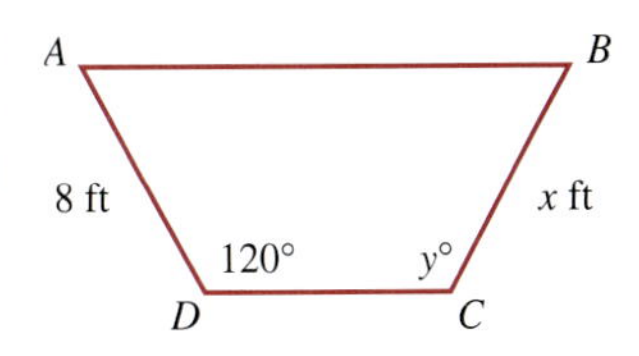

FIGURE 9-36

Solution Since the figure is an isosceles trapezoid, its nonparallel sides have the same length. So $m(\overline{AD})$ and $m(\overline{BC})$ are equal, and $x = 8$.

Since the base angles of an isosceles trapezoid are congruent, $m(\angle D) = m(\angle C)$. Thus, y is 120.

The sum of the measures of the angles of a polygon

We have seen that the sum of the angle measures of any triangle is 180°. Since a polygon with n sides can be divided into $n - 2$ triangles, the sum of the angle measures of the polygon, in degrees, is $(n - 2)180$.

Angles of a polygon

The sum S, in degrees, of the measures of the angles of a polygon with n sides is given by the formula

$$S = (n - 2)180$$

Self Check 8

Find the sum of the angle measures of a quadrilateral.

Answer 360°

EXAMPLE 8 Find the sum of the angle measures of a pentagon.

Solution Since a pentagon has 5 sides, we substitute 5 for n in the formula and simplify.

$$S = (n - 2)180$$
$$= (5 - 2)180 \quad \text{Substitute 5 for } n.$$
$$= (3)180 \quad \text{Perform the subtraction within the parentheses.}$$
$$= 540$$

The sum of the angles of a pentagon is 540°.

Self Check 9

The sum of the measures of the angles of a polygon is 720°. Find the number of sides the polygon has.

Answer 6

EXAMPLE 9 The sum of the measures of the angles of a polygon is 1,080°. Find the number of sides the polygon has.

Solution To find the number of sides the polygon has, we substitute 1,080 for S in the formula and then solve for n.

$$S = (n - 2)180$$
$$1{,}080 = (n - 2)180 \quad \text{Substitute 1,080 for } S.$$
$$1{,}080 = 180n - 360 \quad \text{Distribute the multiplication by 180.}$$
$$1{,}080 + 360 = 180n - 360 + 360 \quad \text{To undo the subtraction of 360, add 360 to both sides.}$$
$$1{,}440 = 180n \quad \text{Simplify.}$$
$$\frac{1{,}440}{180} = \frac{180n}{180} \quad \text{To undo the multiplication of 180, divide both sides by 180.}$$
$$8 = n \quad \text{Perform the division: } \frac{1{,}440}{180} = 8.$$

The polygon has 8 sides. It is an octagon.

Section 9.3 STUDY SET

VOCABULARY *Fill in the blanks.*

1. A ________ polygon has sides that are all the same length and angles that all have the same measure.
2. A polygon with four sides is called a ____________. A ________ is a polygon with three sides.
3. A ________ is a polygon with six sides.
4. A polygon with five sides is called a ________.
5. An eight-sided polygon is an ________.
6. The points where the sides of a polygon intersect are called ________.
7. A triangle with three sides of equal length is called an ________ triangle.
8. An ________ triangle has two sides of equal length.
9. The longest side of a right triangle is the ________.
10. The ______ angles of an isosceles triangle have the same measure.
11. A ____________ with a right angle is a rectangle.
12. A rectangle with all sides of equal length is a ______.
13. A ________ is a parallelogram with four sides of equal length.
14. A ________ has two sides that are parallel and two sides that are not parallel.

15. The legs of an ________ trapezoid have the same length.

16. The ________ of a polygon is the distance around it.

CONCEPTS *Give the number of sides each polygon has and classify it as a triangle, quadrilateral, pentagon, hexagon, or octagon. Then give the number of vertices it has.*

17. 

18. 

19. 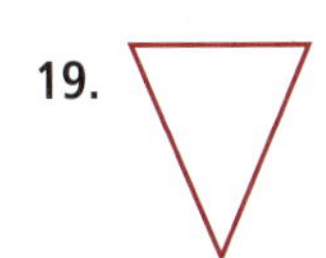

20. 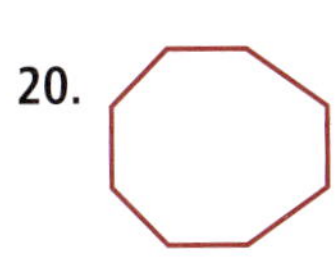

21. 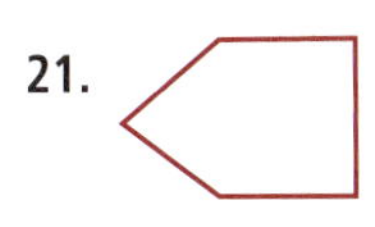

22. 

23. 

24. 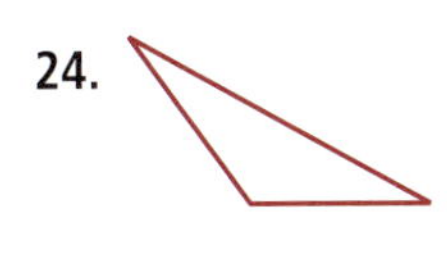

Classify each triangle as an equilateral triangle, an isosceles triangle, a scalene triangle, or a right triangle.

25. 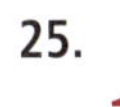 

26. 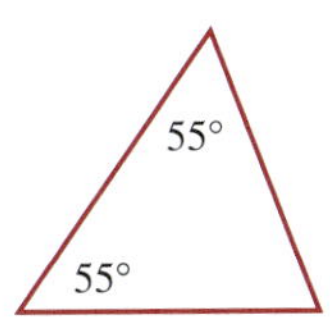

27. 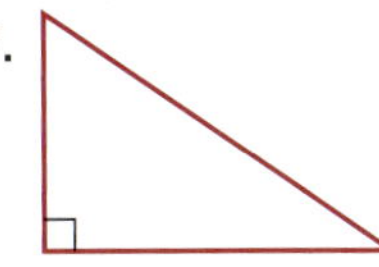

28. 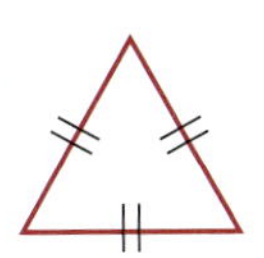

29. 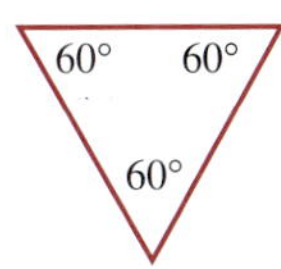

30. 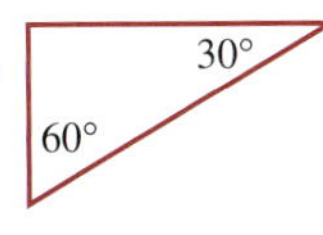

31. 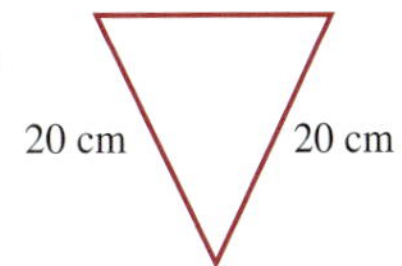

32. 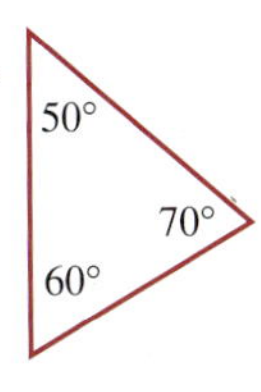

Classify each quadrilateral as a rectangle, a square, a rhombus, or a trapezoid. More than one name can be used for some figures.

33. 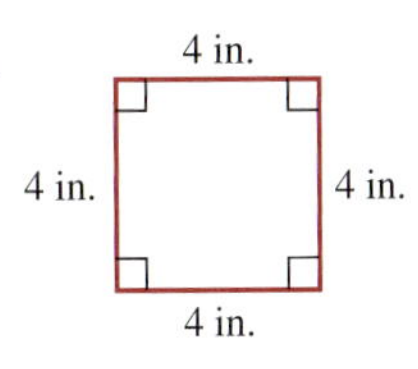

34. 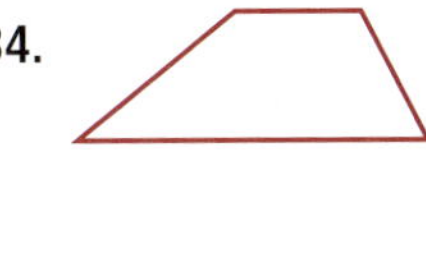

35. 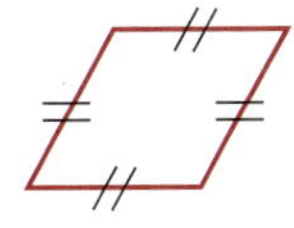

36. 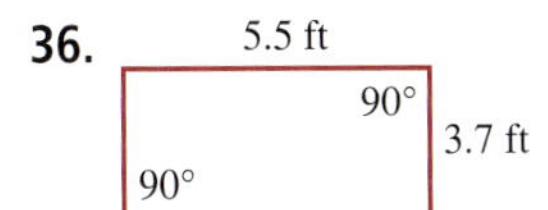

37. 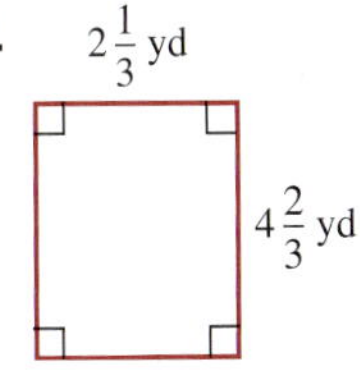

38. 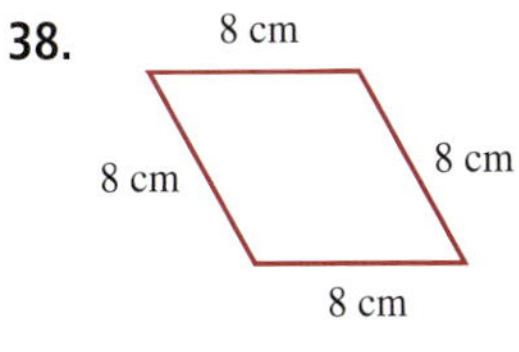

39.

40. 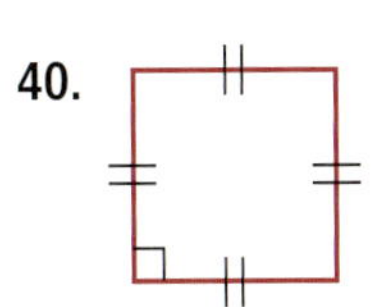

NOTATION *Fill in the blanks.*

41. The symbol $\triangle$ means ________.

42. The symbol $m(\angle 1)$ means the ________ of angle 1.

PRACTICE *The measures of two angles of $\triangle ABC$ shown in the following illustration are given. Find the measure of the third angle.*

43. $m(\angle A) = 30°$ and $m(\angle B) = 60°$
$m(\angle C) =$ _____

44. $m(\angle A) = 45°$ and $m(\angle C) = 105°$
$m(\angle B) =$ _____

45. $m(\angle B) = 100°$ and $m(\angle A) = 35°$
$m(\angle C) =$ _____

46. $m(\angle B) = 33°$ and $m(\angle C) = 77°$
$m(\angle A) =$ _____

47. $m(\angle A) = 25.5°$ and $m(\angle B) = 63.8°$
$m(\angle C) =$ ______

48. $m(\angle B) = 67.25°$ and $m(\angle C) = 72.5°$
$m(\angle A) =$ _______

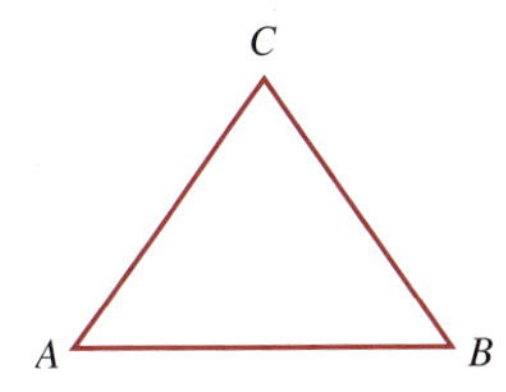

Refer to rectangle ABCD, shown below.

49. m(∠1) = _____

50. m(∠3) = _____

51. m(∠2) = _____

52. If $m(\overline{AC})$ is 8 cm, then $m(\overline{BD})$ = _____.

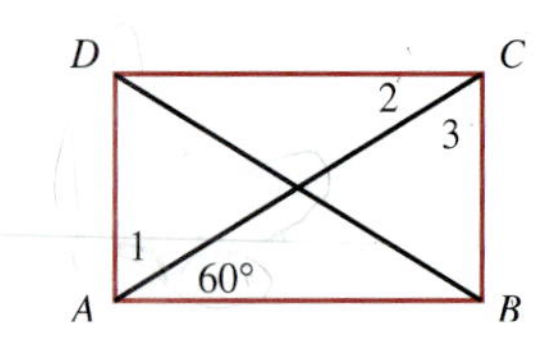

Find the sum of the angle measures of each polygon.

53. A hexagon

54. An octagon

55. A decagon (10 sides)

56. A dodecagon (12 sides)

Find the number of sides a polygon has if the sum of its angle measures is the given number.

57. 900°

58. 1,260°

59. 2,160°

60. 3,600°

APPLICATIONS

61. Give three uses of triangles in everyday life.

62. Give three uses of rectangles in everyday life.

63. Give three uses of squares in everyday life.

64. Give a use of a trapezoid in everyday life.

65. POLYGONS IN NATURE As we see in Illustration (a), a starfish has the approximate shape of a pentagon. What approximate polygon shape do you see in each of the other objects in the illustration?

b. Lemon **c.** Chili pepper **d.** Apple

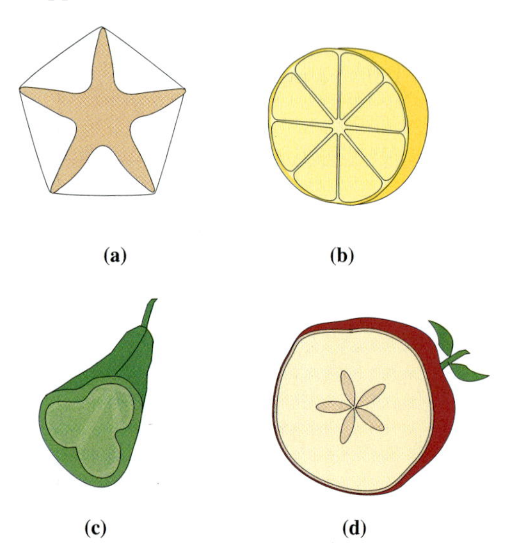

66. FLOWCHARTS A flowchart shows a sequence of steps to be performed by a computer to solve a given problem. When designing a flowchart, the programmer uses a set of standardized symbols to represent various operations to be performed by the computer. Locate a rectangle, a rhombus, and a parallelogram in the flow chart shown.

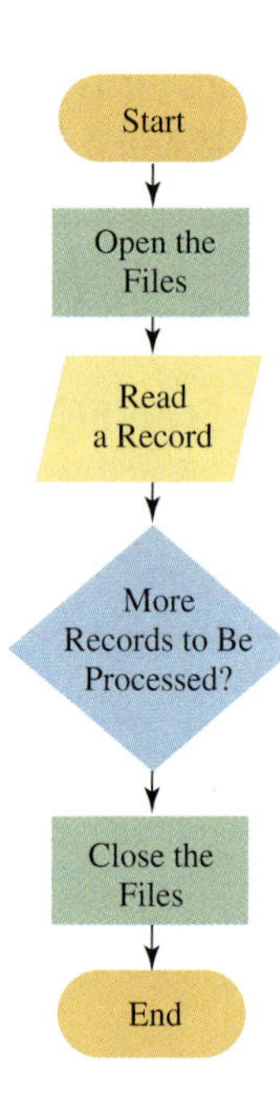

67. CHEMISTRY Polygons are used to represent the chemical structure of compounds graphically. In the illustration, what types of polygons are used to represent methylprednisolone, the active ingredient in an antiinflammatory medication?

Methylprednisolone

68. PODIUMS What polygon describes the shape of the upper portion of the podium?

69. EASELS Show how two of the legs of the easel form the equal sides of an isosceles triangle.

70. AUTOMOBILE JACKS Refer to the illustration. Show that no matter how high the jack is raised, it always forms two isosceles triangles.

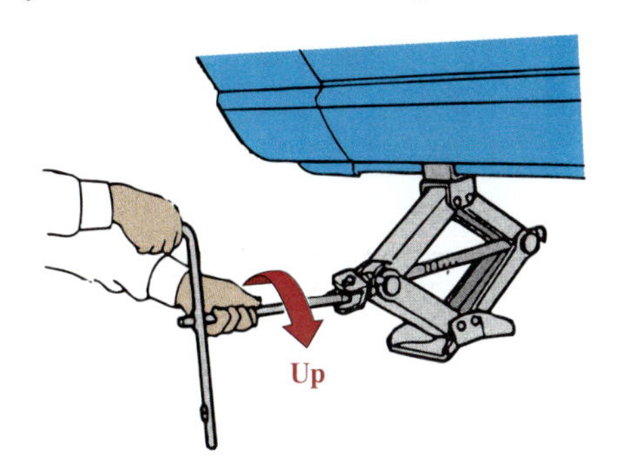

WRITING

71. Explain why a square is a rectangle.
72. Explain why a trapezoid is not a parallelogram.

REVIEW

73. Find 20% of 110.
74. Find 15% of 50.
75. What percent of 200 is 80?
76. 20% of what number is 500?
77. Simplify: $0.85 \div 2(0.25)$.
78. FIRST AID When checking an accident victim's pulse, a paramedic counted 13 beats during a 15-second span. How many beats would be expected in 60 seconds?

9.4 Properties of Triangles

- Congruent triangles • Similar triangles • The Pythagorean theorem

Proportions and triangles are often used to measure distances indirectly. For example, by using a proportion, Eratosthenes (275–195 B.C.) was able to estimate the circumference of the Earth with remarkable accuracy. On a sunny day, we can use properties of similar triangles to calculate the height of a tree while staying safely on the ground. By using a theorem proved by the Greek mathematician Pythagoras (about 500 B.C.), we can calculate the length of the third side of a right triangle whenever we know the lengths of two sides.

Congruent triangles

Triangles that have the same area and the same shape are called **congruent triangles.** In Figure 9-37 on the next page, triangles ABC and DEF are congruent.

$\triangle ABC \cong \triangle DEF$ Read as "Triangle ABC is congruent to triangle DEF."

Corresponding angles and corresponding sides of congruent triangles are called **corresponding parts.** The notation $\triangle ABC \cong \triangle DEF$ shows which vertices are corresponding parts.

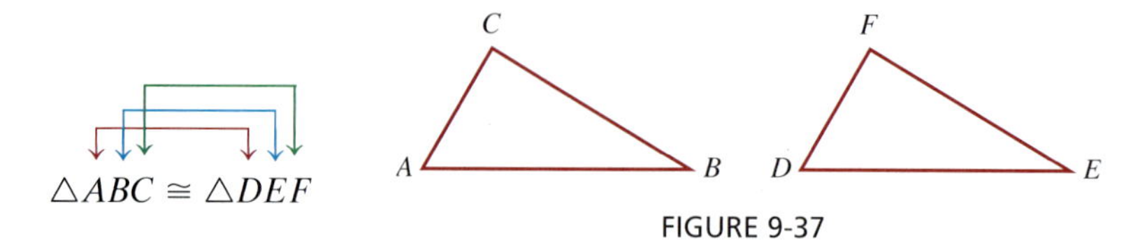

FIGURE 9-37

Corresponding parts of congruent triangles always have the same measure. In the congruent triangles shown in Figure 9-37,

$$m(\angle A) = m(\angle D), \quad m(\angle B) = m(\angle E), \quad m(\angle C) = m(\angle F),$$
$$m(\overline{BC}) = m(\overline{EF}), \quad m(\overline{AC}) = m(\overline{DF}), \quad m(\overline{AB}) = m(\overline{DE})$$

EXAMPLE 1 Name the corresponding parts of the congruent triangles in Figure 9-38.

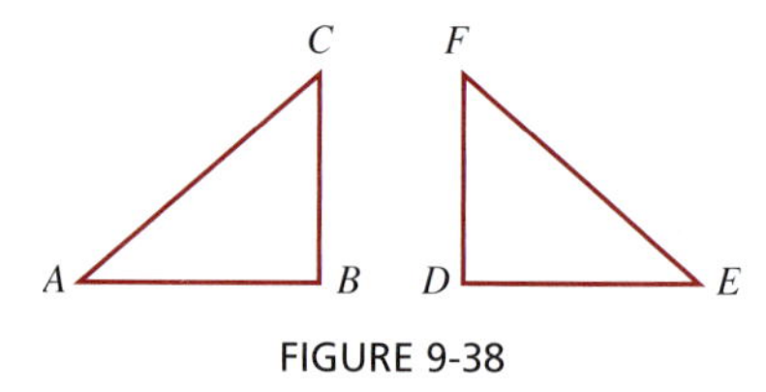

FIGURE 9-38

Solution The corresponding angles are

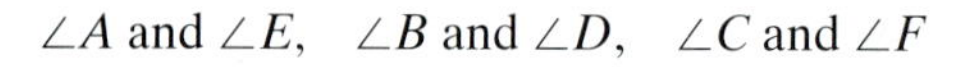

$\angle A$ and $\angle E$, $\angle B$ and $\angle D$, $\angle C$ and $\angle F$

Since corresponding sides are always opposite corresponding angles, the corresponding sides are

$\overline{BC}$ and $\overline{DF}$, $\overline{AC}$ and $\overline{EF}$, $\overline{AB}$ and $\overline{ED}$

We will discuss three ways of showing that two triangles are congruent.

SSS property

If three sides of one triangle are congruent to three sides of a second triangle, the triangles are congruent.

The triangles in Figure 9-39 are congruent because of the SSS property.

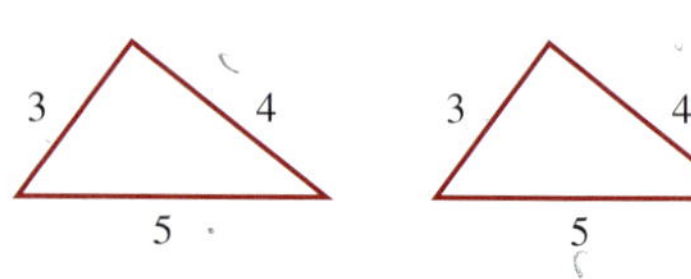

FIGURE 9-39

SAS property

If two sides and the angle between them in one triangle are congruent, respectively, to two sides and the angle between them in a second triangle, the triangles are congruent.

The triangles in Figure 9-40 are congruent because of the SAS property.

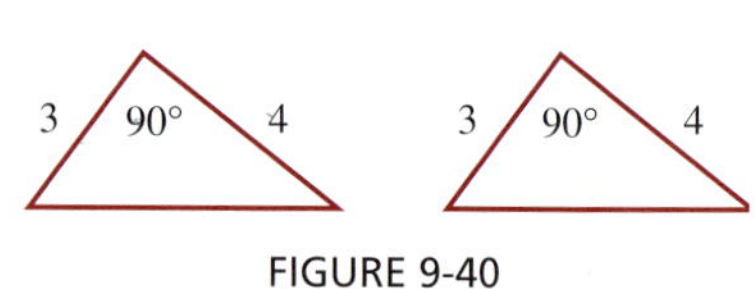

FIGURE 9-40

ASA property

If two angles and the side between them in one triangle are congruent, respectively, to two angles and the side between them in a second triangle, the triangles are congruent.

The triangles in Figure 9-41 are congruent because of the ASA property.

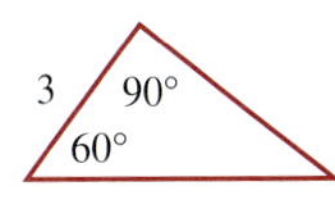

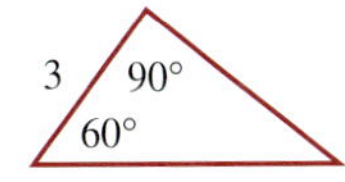

FIGURE 9-41

COMMENT There is no SSA property. To illustrate this, consider the triangles in Figure 9-42. Two sides and an angle of $\triangle ABC$ are congruent to two sides and an angle of $\triangle DEF$. But the congruent angle is *not* between the congruent sides.

We refer to this situation as SSA. Obviously, the triangles are not congruent, because they have different areas.

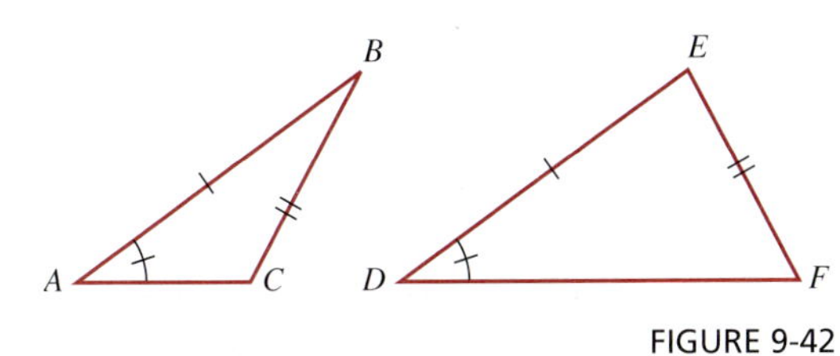

The slash marks indicate congruent parts. That is, the sides with one slash are the same length, the sides with two slashes are the same length, and the angles with one slash have the same measure.

FIGURE 9-42

EXAMPLE 2 Explain why the triangles in Figure 9-43 are congruent.

Solution Since vertical angles are congruent,

$$m(\angle 1) = m(\angle 2)$$

From the figure, we see that

$$m(\overline{AC}) = m(\overline{EC}) \quad \text{and} \quad m(\overline{BC}) = m(\overline{DC})$$

Since two sides and the angle between them in one triangle are congruent, respectively, to two sides and the angle between them in a second triangle, $\triangle ABC \cong \triangle EDC$ by the SAS property.

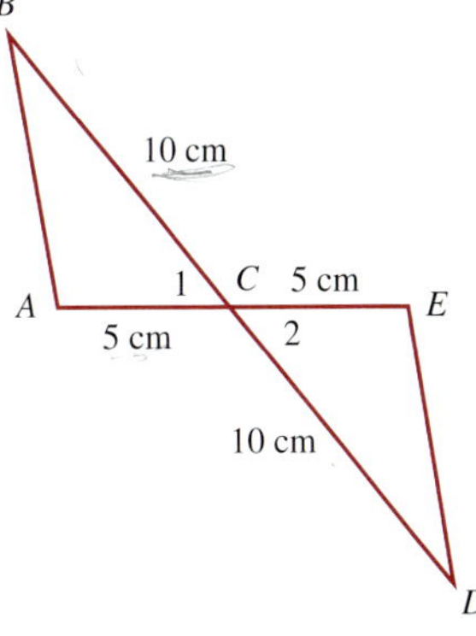

FIGURE 9-43

Similar triangles

If two angles of one triangle are congruent to two angles of a second triangle, the triangles will have the same shape. Triangles with the same shape are called **similar triangles.** In Figure 9-44, $\triangle ABC \sim \triangle DEF$ (read the symbol $\sim$ as "is similar to").

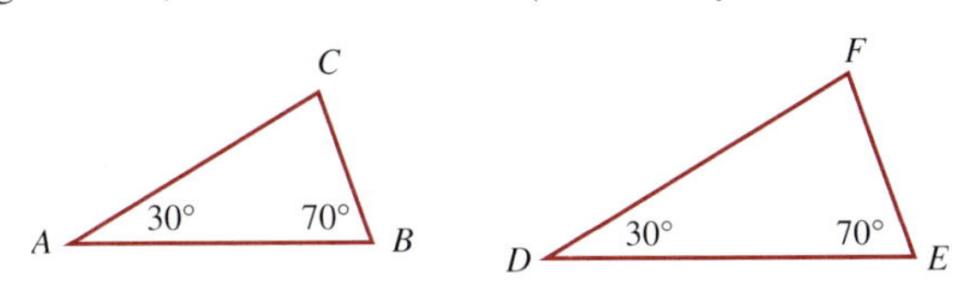

FIGURE 9-44

COMMENT Note that congruent triangles are always similar, but similar triangles are not always congruent.

Property of similar triangles

If two triangles are similar, all pairs of corresponding sides are in proportion.

In the similar triangles shown in Figure 9-44 on the previous page, the following proportions are true.

$$\frac{\overline{AB}}{\overline{DE}} = \frac{\overline{BC}}{\overline{EF}}, \quad \frac{\overline{BC}}{\overline{EF}} = \frac{\overline{CA}}{\overline{FD}}, \quad \text{and} \quad \frac{\overline{CA}}{\overline{FD}} = \frac{\overline{AB}}{\overline{DE}}$$

EXAMPLE 3 Tree height. A tree casts a shadow 18 feet long at the same time a woman 5 feet tall casts a shadow that is 1.5 feet long. (See Figure 9-45.) Find the height of the tree.

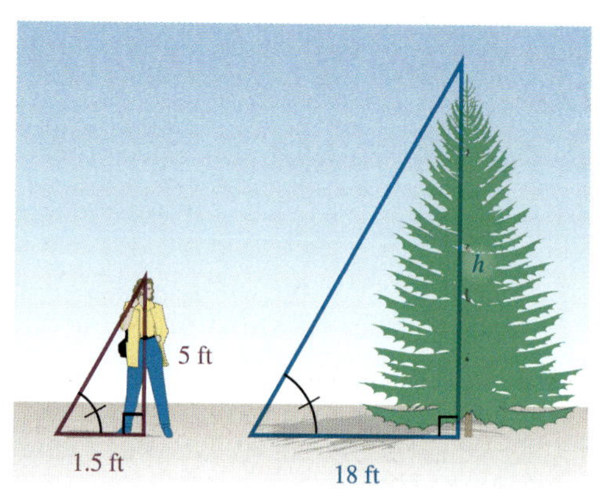

FIGURE 9-45

Solution The figure shows the triangles determined by the tree and its shadow and the woman and her shadow. Since the triangles have the same shape, they are similar, and the lengths of their corresponding sides are in proportion. If we let h represent the height of the tree, we can find h by solving the following proportion.

$\frac{h}{5} = \frac{18}{1.5}$ $\frac{\text{Height of the tree}}{\text{Height of the woman}} = \frac{\text{shadow of the tree}}{\text{shadow of the woman}}$

$1.5h = 5(18)$ In the proportion, the product of the extremes is equal to the product of the means.

$1.5h = 90$ Perform the multiplication: $5(18) = 90$.

$h = 60$ To undo the multiplication by 1.5, divide both sides by 1.5.

The tree is 60 feet tall.

The Pythagorean theorem

In the movie *The Wizard of Oz,* the scarecrow was in search of a brain. To prove that he had found one, he tried to recite the Pythagorean theorem. In words, the Pythagorean theorem can be stated as:

In a right triangle, the square of the hypotenuse is equal to the sum of squares of the other two sides.

Pythagorean theorem

If the length of the hypotenuse of a right triangle is c and the lengths of its legs are a and b, then

$$a^2 + b^2 = c^2$$

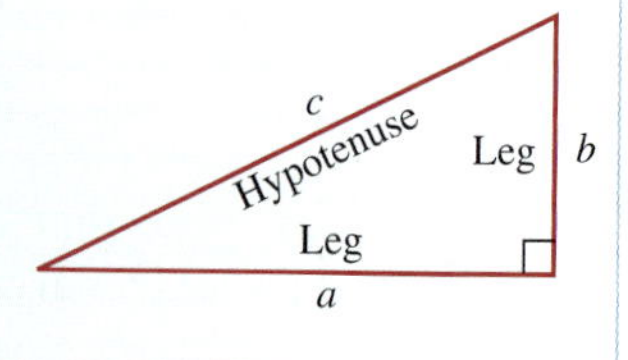

EXAMPLE 4 High-ropes adventure courses.

A builder of a high-ropes adventure course wants to secure the pole shown in Figure 9-46 by attaching a cable from the anchor stake 8 feet from its base to a point 6 feet up the pole. How long should the cable be?

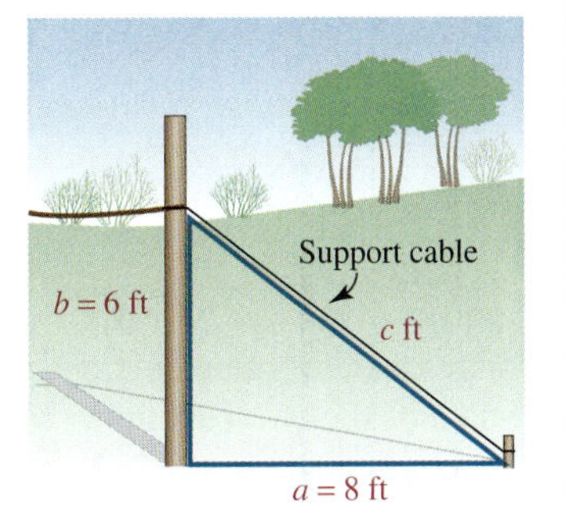

FIGURE 9-46

Solution The support cable, the pole, and the ground form a right triangle. If we let c represent the length of the cable (the hypotenuse), then we can use the Pythagorean theorem with $a = 8$ and $b = 6$ to find c.

$c^2 = a^2 + b^2$ This is the Pythagorean theorem.

$c^2 = 8^2 + 6^2$ Substitute 8 for a and 6 for b.

$c^2 = 64 + 36$ Evaluate the exponential expressions.

$c^2 = 100$ Simplify the right-hand side.

To find c, we must find a number that, when squared, is 100. There are two such numbers, one positive and one negative; they are the square roots of 100. Since c represents the length of a support cable, c cannot be negative. For this reason, we need only find the positive square root of 100 to get c.

$c^2 = 100$ This is the equation to solve.

$c^2 = \sqrt{100}$ The symbol $\sqrt{}$ is used to indicate the positive square root of a number.

$c = 10$ $\sqrt{100} = 10$, because $10^2 = 100$.

The support cable should be 10 feet long.

Self Check 4

A 26-foot ladder rests against the side of a building. If the base of the ladder is 10 feet from the wall, how far up the side of the building will the ladder reach?

Answer 24 ft

Finding the width of a television screen

CALCULATOR SNAPSHOT

The size of a television screen is the diagonal measure of its rectangular screen. (See Figure 9-47.) To find the width of a 27-inch screen that is 17 inches high, we use the Pythagorean theorem with $c = 27$ and $b = 17$.

$$c^2 = a^2 + b^2$$
$$27^2 = a^2 + 17^2$$
$$27^2 - 17^2 = a^2$$

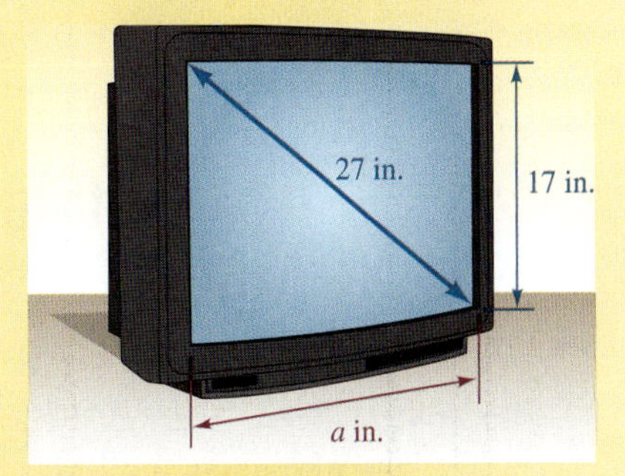

FIGURE 9-47

The variable a represents the width of a television screen, so it must be positive. To find a, we find the positive square root of the result when 17^2 is subtracted from 27^2. Using a radical symbol to indicate this, we have

$$\sqrt{27^2 - 17^2} = a$$

We can evaluate the expression on the left-hand side by entering these numbers and pressing these keys.

[( ] 27 [x^2] [−] 17 [x^2] [ )] [$\sqrt{}$] 20.97617696

To the nearest inch, the width of the television screen is 21 inches.

It is also true that

If the square of one side of a triangle is equal to the sum of the squares of the other two sides, the triangle is a right triangle.

Self Check 5
Is a triangle with sides of 9, 40, and 41 meters a right triangle?

Answer yes

EXAMPLE 5 Is a triangle with sides of 5, 12, and 13 meters a right triangle?

Solution We can use the Pythagorean theorem to answer this question. Since the longest side of the triangle is 13 meters, we must substitute 13 for c. It doesn't matter which of the two remaining side lengths we substitute for a and which we substitute for b.

$c^2 = a^2 + b^2$ This is the Pythagorean theorem.

$13^2 \stackrel{?}{=} 5^2 + 12^2$ Substitute 13 for c, 5 for a, and 12 for b.

$169 \stackrel{?}{=} 25 + 144$ Evaluate the exponential expressions.

$169 = 169$ Simplify the right-hand side.

Since the square of the longest side is equal to the sum of the squares of the other two sides, the triangle is a right triangle.

Self Check 6
Is a triangle with sides of 4, 5, and 6 inches a right triangle?

Answer no

EXAMPLE 6 Is a triangle with sides of 2, 2, and 3 feet a right triangle?

Solution We check to see whether the square of the longest side is equal to the sum of the squares of the other two sides.

$c^2 = a^2 + b^2$ This the Pythagorean theorem.

$3^2 \stackrel{?}{=} 2^2 + 2^2$ Substitute 3 for c, 2 for a, and 2 for b.

$9 \stackrel{?}{=} 4 + 4$ Evaluate the exponential expressions.

$9 \neq 8$ Simplify the right-hand side.

Since the square of the longest side is not equal to the sum of the squares of the other two sides, the triangle is not a right triangle.

THINK IT THROUGH Pythagorean Triples

CORBIS

"Fraternity and sorority membership nationwide is declining, down about 30% in the last decade." Chronicle of Higher Education, 2003

The first college social fraternity, Phi Beta Kappa, was founded in 1776 on the campus of The College of William and Mary. However, secret societies have existed since ancient times, and from these roots the essence of today's fraternities and sororities have their foundation. Pythagoras, the Greek mathematician of the 6th century BC, was the leader of a secret fraternity/sorority called the Pythagoreans. They were a community of men and women that studied mathematics, and in particular, the "magic 3-4-5 triangle." This right triangle is special because the sum of the squares of the lengths of its legs is equal to the square of the length of its hypotenuse: $3^2 + 4^2 = 5^2$ or $9 + 16 = 25$. Today, we call a set of three natural numbers a, b, and c that satisfy $a^2 + b^2 = c^2$ a Pythagorean triple. Show that each list of numbers is a Pythagorean triple.

1. 7, 24, 25 **2.** 8, 15, 17 **3.** 9, 40, 41

4. 11, 60, 61 **5.** 12, 35, 37 **6.** 88, 105, 137

Section 9.4 STUDY SET

VOCABULARY *Fill in the blanks.*

1. __________ triangles are the same size and the same shape.
2. All ____________ parts of congruent triangles have the same measure.
3. If two triangles are ________, they have the same shape.
4. The __________ is the longest side of a right triangle.

CONCEPTS *Determine whether each statement is true. If a statement is false, explain why.*

5. If three sides of one triangle are the same length as three sides of a second triangle, the triangles are congruent.
6. If two sides of one triangle are the same length as two sides of a second triangle, the triangles are congruent.
7. If two sides and an angle of one triangle are congruent, respectively, to two sides and an angle of a second triangle, the triangles are congruent.
8. If two angles and the side between them in one triangle are congruent, respectively, to two angles and the side between them in a second triangle, the triangles are congruent.
9. In a proportion, the product of the means is equal to the product of the extremes.
10. If two angles of one triangle are congruent to two angles of a second triangle, the triangles are similar.
11. Are the triangles shown below congruent?

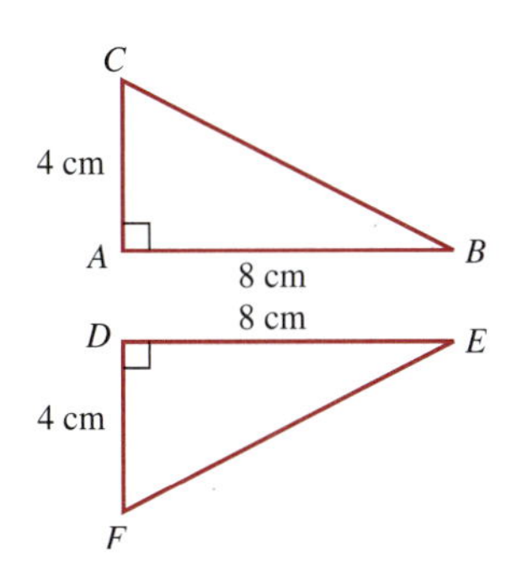

12. Are the triangles shown below congruent?

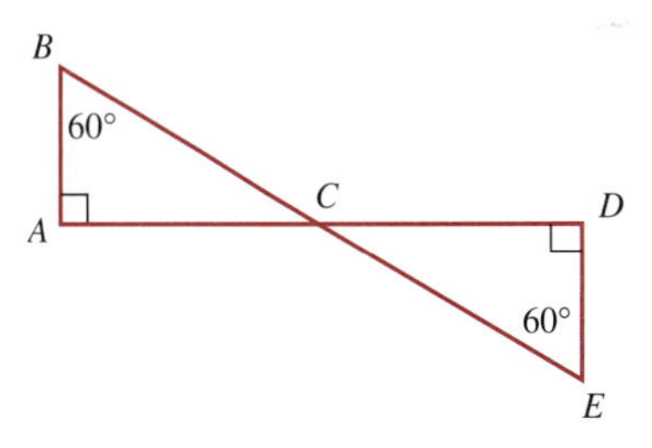

13. Are the triangles shown below similar?

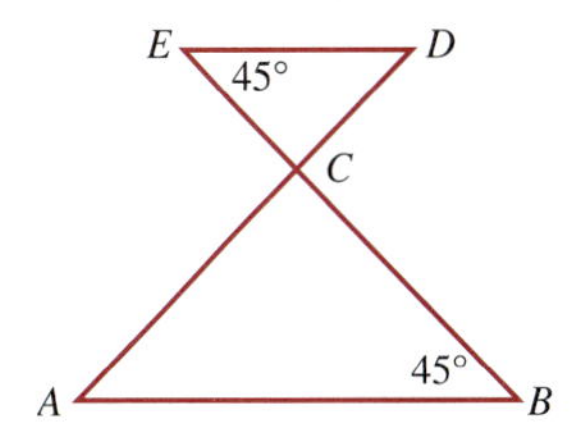

14. Are the triangles shown below similar?

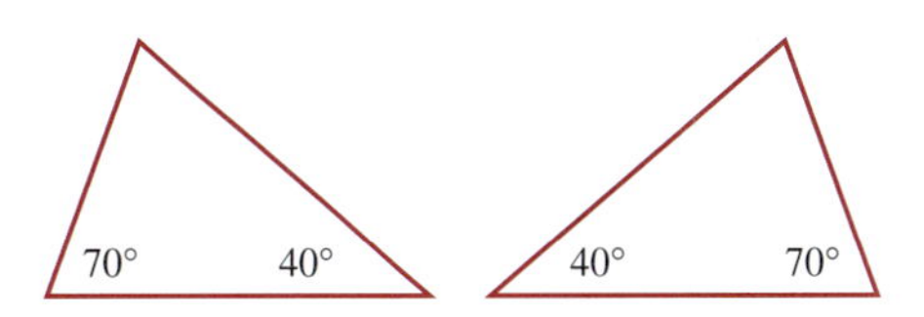

15. The Pythagorean theorem states that for a right triangle, $c^2 = a^2 + b^2$. What do the variables a, b, and c represent?
16. A triangle has sides of length 3, 4, and 5 centimeters. Substitute the lengths into $c^2 = a^2 + b^2$ and show that a true statement results. From the result, what can we conclude about the triangle?
17. Suppose that c represents the length of the hypotenuse of a right triangle and $c^2 = 25$. Fill in the blanks. To find c, we must find a number that, when squared, is ▢. Since c represents a positive number, we need only find the positive ________ ________ of 25 to get c.

$c^2 = 25$

$▢ = \sqrt{25}$ A radical symbol indicates the positive square root.

$c = ▢$

18. Solve: $\frac{h}{2.6} = \frac{27}{13}$.

NOTATION *Fill in the blanks.*

19. The symbol $\cong$ is read as "______________."

20. The symbol $\sim$ is read as "____________."

PRACTICE *Name the corresponding parts of the congruent triangles.*

21. Refer to the illustration.

$\overline{AC}$ corresponds to ____.

$\overline{DE}$ corresponds to ____.

$\overline{BC}$ corresponds to ____.

$\angle A$ corresponds to ____.

$\angle E$ corresponds to ____.

$\angle F$ corresponds to ____.

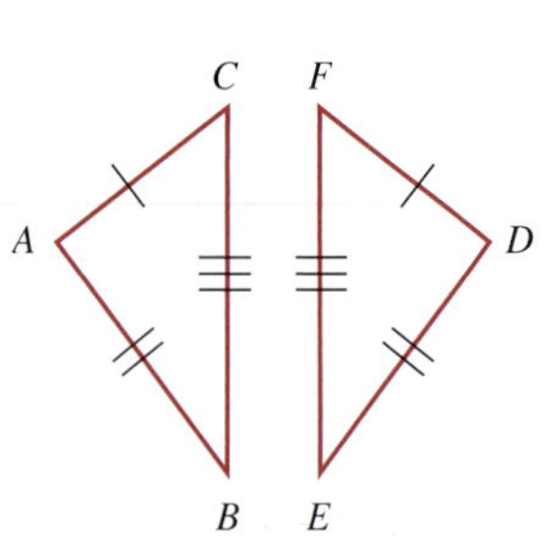

22. Refer to the illustration.

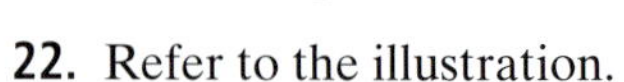

$\overline{AB}$ corresponds to ____.

$\overline{EC}$ corresponds to ____.

$\overline{AC}$ corresponds to ____.

$\angle D$ corresponds to ____.

$\angle B$ corresponds to ____.

$\angle 1$ corresponds to ____.

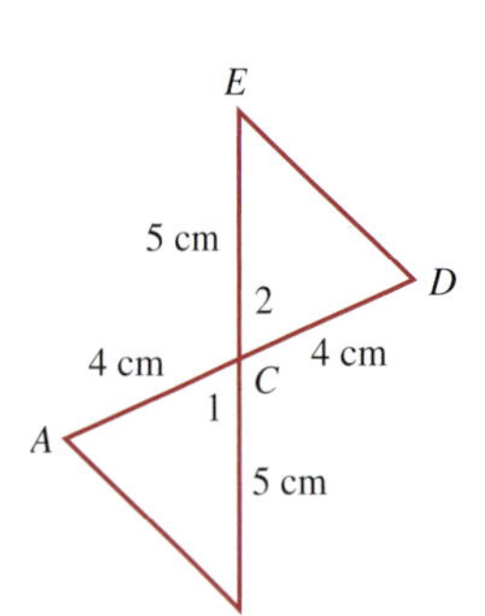

Determine whether each pair of triangles is congruent. If they are, explain why.

23.

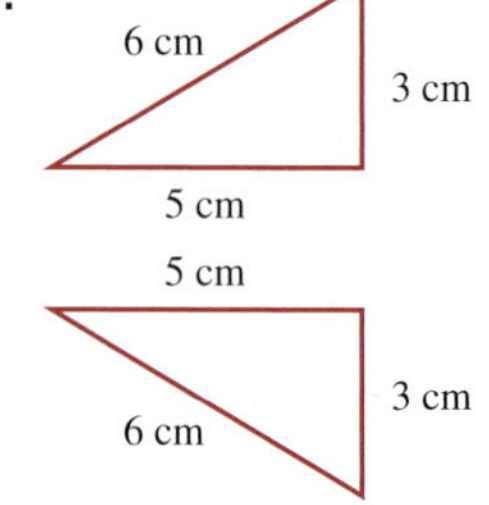

24.

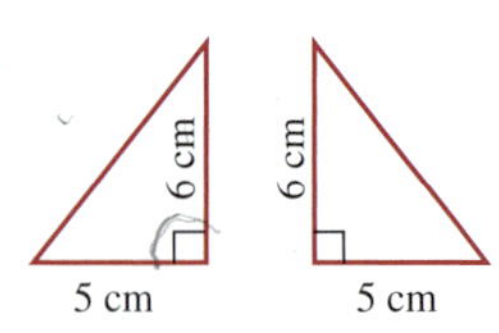

25.

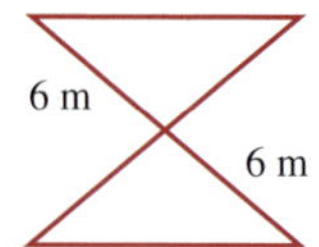

26.

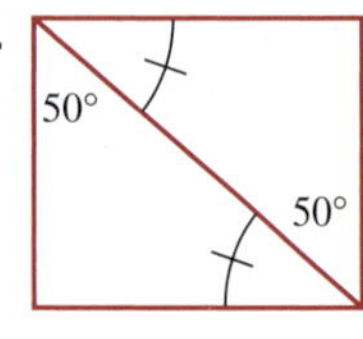

27.

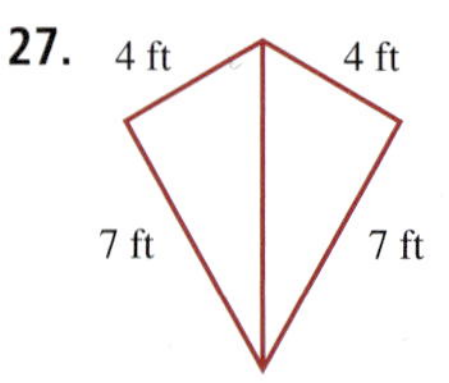

28.

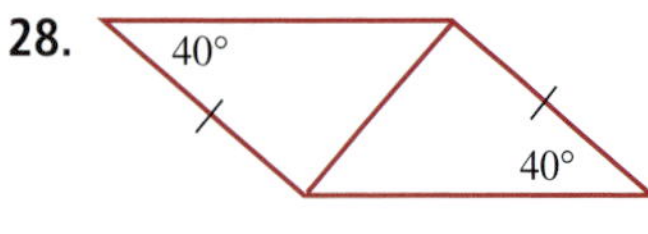

29.

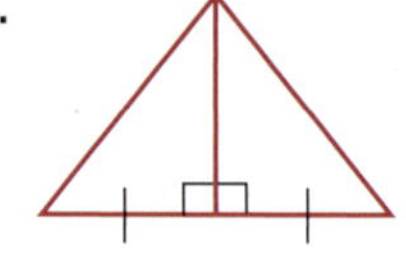

30.

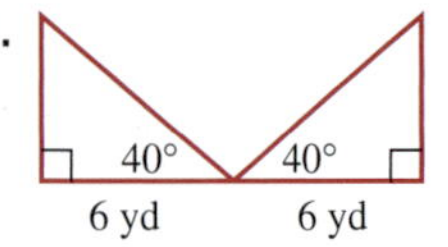

Find x.

31.

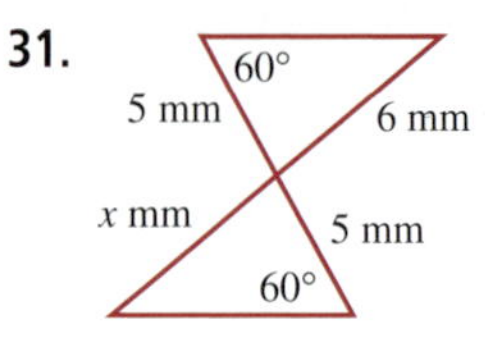

32.

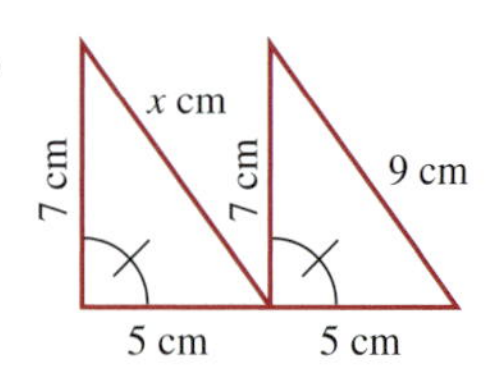

33.

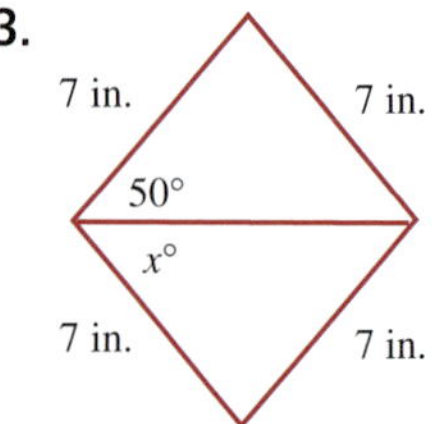

34.

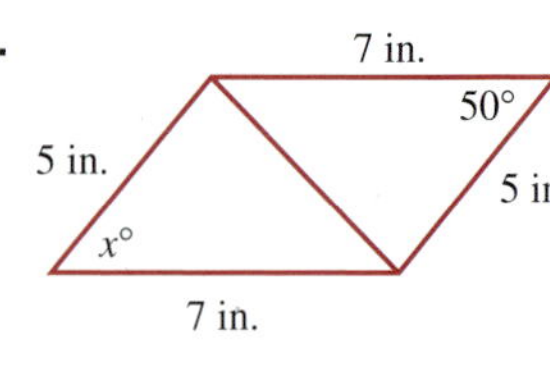

Determine whether the triangles are similar.

35.

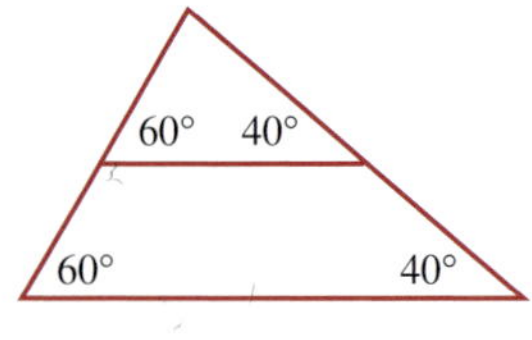

36.

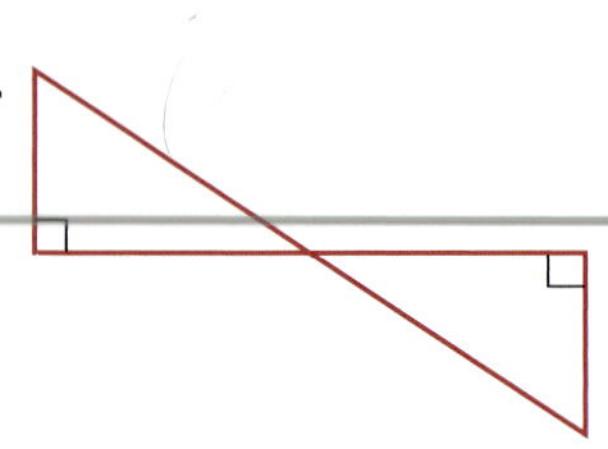

Refer to the illustration below and find the length of the unknown side.

37. $a = 3$ and $b = 4$. Find c.

38. $a = 12$ and $b = 5$. Find c.

39. $a = 15$ and $c = 17$. Find b.

40. $b = 45$ and $c = 53$. Find a.

41. $a = 5$ and $c = 9$. Find b.

42. $a = 1$ and $b = 7$. Find c.

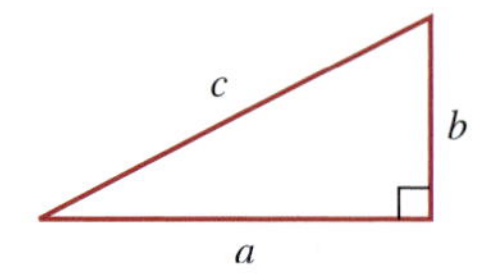

The lengths of the three sides of a triangle are given. Determine whether the triangle is a right triangle.

43. 8, 15, 17

44. 6, 8, 10

45. 7, 24, 26

46. 9, 39, 40

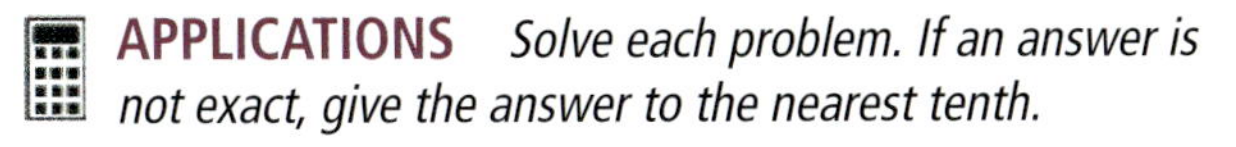

APPLICATIONS *Solve each problem. If an answer is not exact, give the answer to the nearest tenth.*

47. HEIGHT OF A TREE The tree below casts a shadow 24 feet long when a man 6 feet tall casts a shadow 4 feet long. Find the height of the tree.

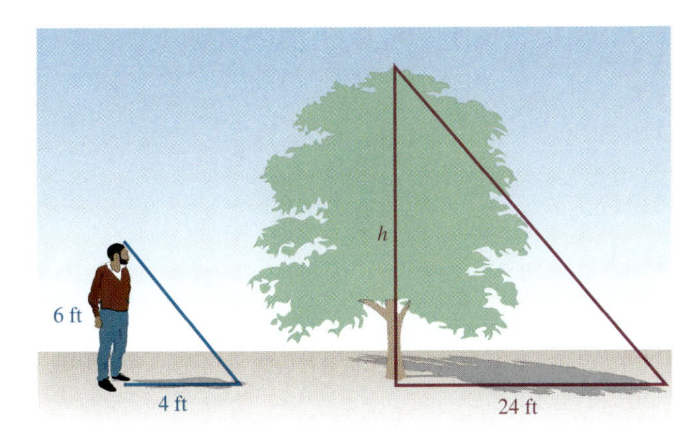

48. HEIGHT OF A BUILDING A man places a mirror on the ground and sees the reflection of the top of a building, as shown. Find the height of the building.

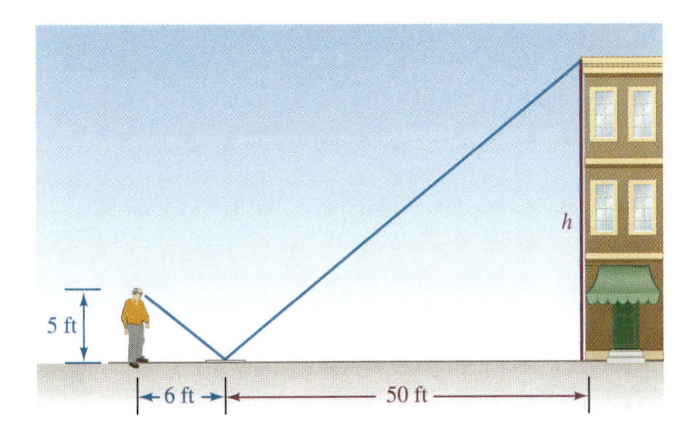

49. WIDTH OF A RIVER Use the dimensions in the illustration to find w, the width of the river.

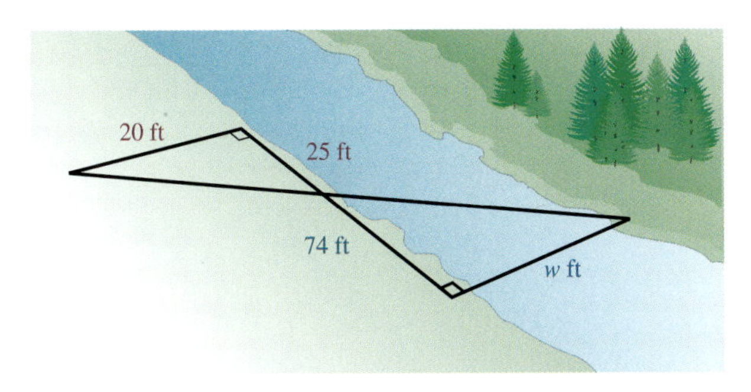

50. FLIGHT PATHS The airplane in the illustration in the next column ascends 200 feet as it flies a horizontal distance of 1,000 feet. How much altitude is gained as it flies a horizontal distance of 1 mile? (*Hint:* 1 mile = 5,280 feet.)

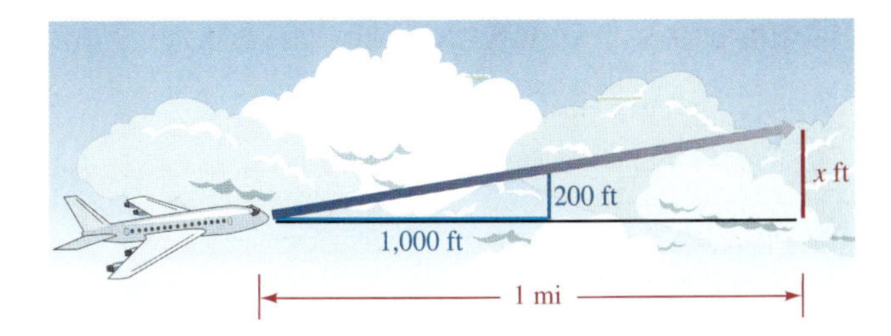

51. FLIGHT PATHS An airplane descends 1,200 feet as it flies a horizontal distance of 1.5 miles. How much altitude is lost as it flies a horizontal distance of 5 miles?

52. GEOMETRY If segment DE in the illustration is parallel to segment AB, $\triangle ABC$ will be similar to $\triangle DEC$. Find x.

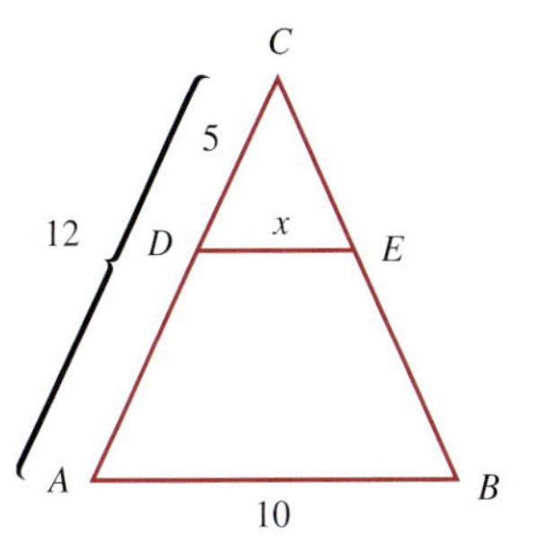

53. ADJUSTING LADDERS A 20-foot ladder reaches a window 16 feet above the ground. How far from the wall is the base of the ladder?

54. GUY WIRES A 30-foot tower is to be fastened by three guy wires attached to the top of the tower and to the ground at positions 20 feet from its base. How much wire is needed?

55. PICTURE FRAMES After gluing and nailing two pieces of picture frame molding together, a frame maker checks her work by making a diagonal measurement. If the sides of the frame form a right angle, what measurement should the frame maker read on the yardstick?

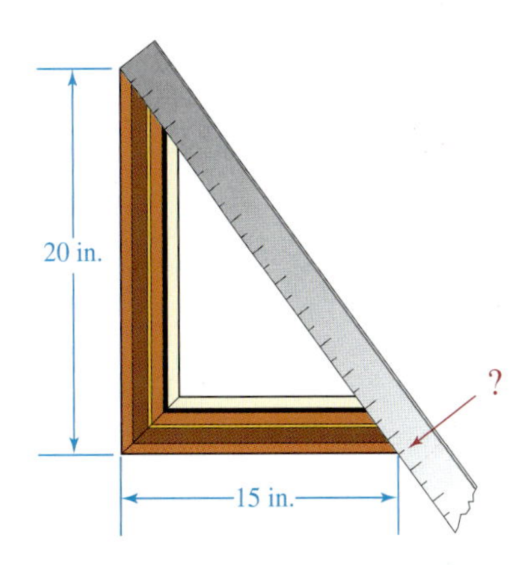

56. CARPENTRY The gable end of the roof shown below is divided in half by a vertical brace, 8 feet in height. Find the length of the roof line.

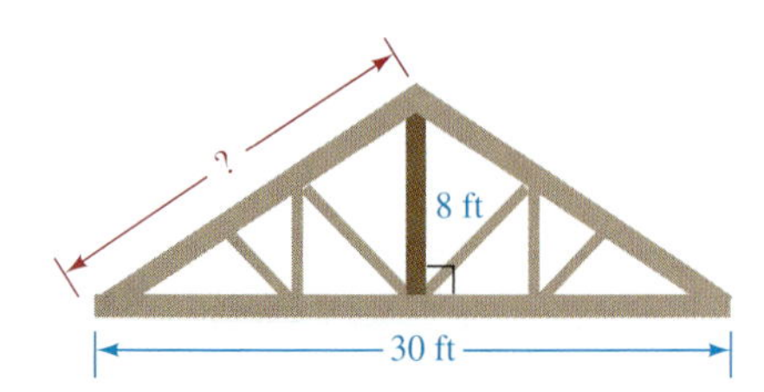

57. BASEBALL A baseball diamond is a square with each side 90 feet long. How far is it from home plate to second base?

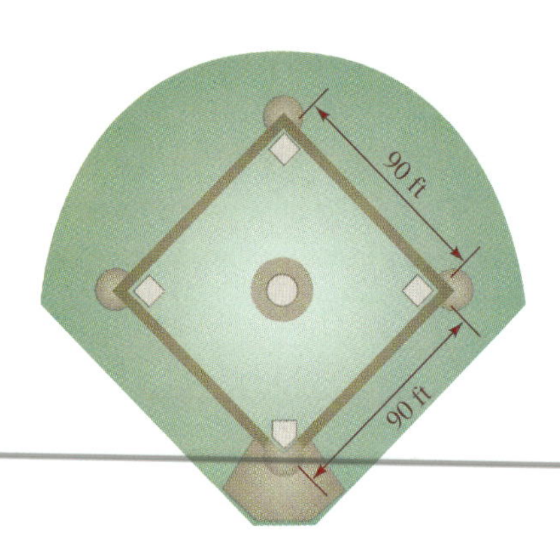

58. TELEVISION Find the size of the television screen shown.

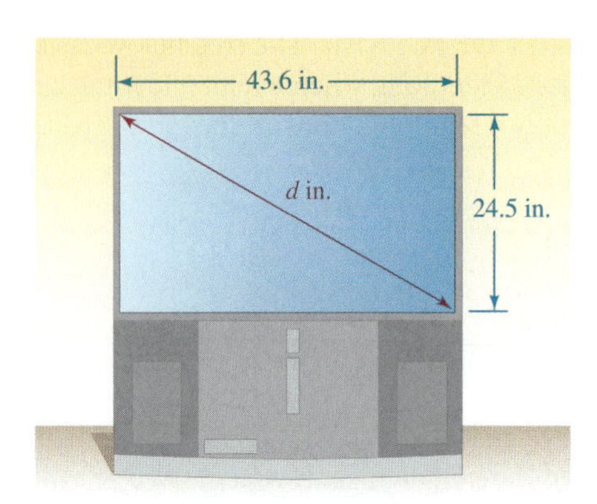

WRITING

59. Explain the Pythagorean theorem.

60. Explain the procedure used to solve the equation $c^2 = 64$. (Assume that c is positive.)

REVIEW *Estimate the answer to each problem.*

61. $\dfrac{0.95 \cdot 3.89}{2.997}$

62. 21% of 42

63. 32% of 60

64. $\dfrac{4.966 + 5.001}{2.994}$

65. 49.5% of 18.1

66. 98.7% of 0.03

9.5 Perimeters and Areas of Polygons

- Perimeters of polygons
- Perimeters of figures that are combinations of polygons
- Areas of polygons
- Areas of figures that are combinations of polygons

In this section, we will discuss how to find perimeters and areas of polygons. Finding perimeters is important when estimating the cost of fencing or estimating the cost of woodwork in a house. Finding areas is important when calculating the cost of carpeting, the cost of painting a house, or the cost of fertilizing a yard.

Perimeters of polygons

Recall that the **perimeter** of a polygon is the distance around it. Since a square has four sides of equal length s, its perimeter P is $s + s + s + s$, or $4s$.

Perimeter of a square

If a square has a side of length s, its perimeter P is given by the formula

$$P = 4s$$

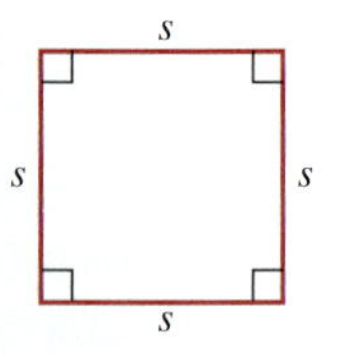

EXAMPLE 1 Find the perimeter of a square whose sides are 7.5 meters long.

Solution Since the perimeter of a square is given by the formula $P = 4s$, we substitute 7.5 for s and simplify.

$$P = 4s$$
$$= 4(7.5)$$
$$= 30$$

The perimeter is 30 meters.

Self Check 1
Find the perimeter of a square whose sides are 23.75 centimeters long.

Answer 95 cm

Since a rectangle has two lengths l and two widths w, its perimeter P is $l + l + w + w$, or $2l + 2w$.

Perimeter of a rectangle
If a rectangle has length l and width w, its perimeter P is given by the formula

$$P = 2l + 2w$$

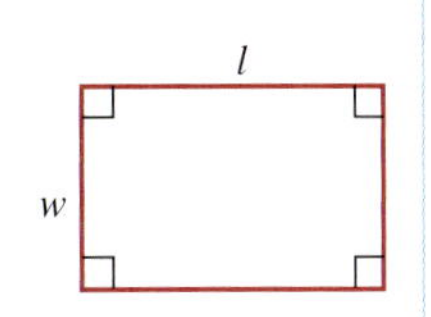

EXAMPLE 2 Find the perimeter of the rectangle in Figure 9-48.

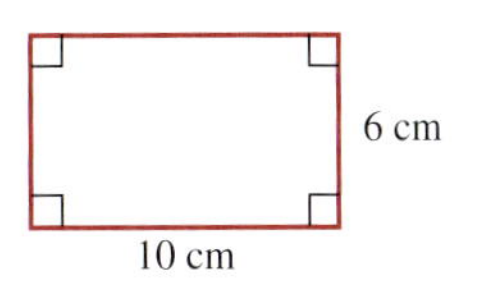

FIGURE 9-48

Solution Since the perimeter is given by the formula $P = 2l + 2w$, we substitute 10 for l and 6 for w and simplify.

$$P = 2l + 2w$$
$$= 2(10) + 2(6)$$
$$= 20 + 12$$
$$= 32$$

The perimeter is 32 centimeters.

Self Check 2
Find the perimeter of the isosceles trapezoid below.

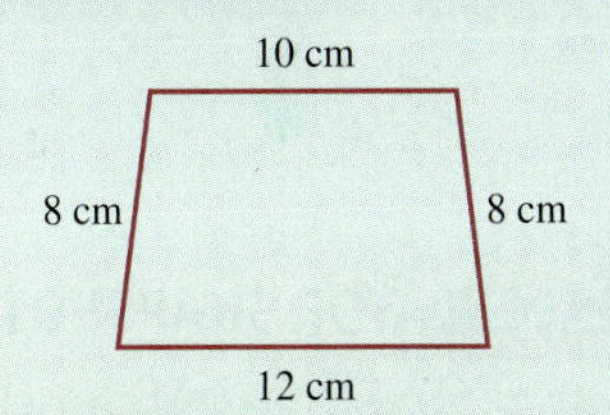

Answer 38 cm

EXAMPLE 3 Find the perimeter of the rectangle in Figure 9-49, in meters.

3 m

80 cm

FIGURE 9-49

Solution Since 1 meter = 100 centimeters, we can convert 80 centimeters to meters by multiplying 80 centimeters by the unit conversion factor $\frac{1 \text{ m}}{100 \text{ cm}}$.

$$80 \text{ cm} = 80 \text{ cm} \cdot \frac{1 \text{ m}}{100 \text{ cm}}$$ Multiply by 1: $\frac{1 \text{ m}}{100 \text{ cm}} = 1$.

$$= \frac{80}{100} \text{ m}$$ The units of centimeters divide out.

$$= 0.8 \text{ m}$$ Divide by 100 by moving the decimal point 2 places to the left.

Self Check 3
Find the perimeter of the triangle below, in inches.

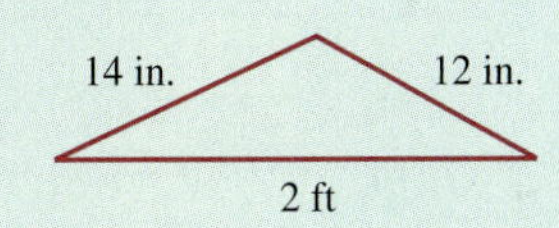

In practice, we don't find areas by counting squares in a figure. Instead, we use formulas for finding areas of geometric figures, as shown in Table 9-1.

Figure	Name	Formula for area
	Square	$A = s^2$, where s is the length of one side.
	Rectangle	$A = lw$, where l is the length and w is the width.
	Parallelogram	$A = bh$, where b is the length of the base and h is the height. (A height is always perpendicular to the base.)
	Triangle	$A = \frac{1}{2}bh$, where b is the length of the base and h is the height. The segment perpendicular to the base and representing the height is called an **altitude.**
	Trapezoid	$A = \frac{1}{2}h(b_1 + b_2)$, where h is the height of the trapezoid and b_1 and b_2 represent the lengths of the bases.

TABLE 9-1

EXAMPLE 5 Find the area of the square in Figure 9-54.

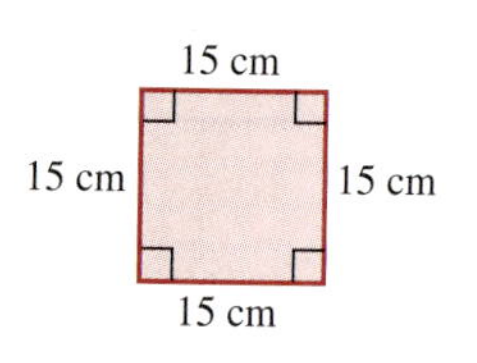

FIGURE 9-54

Solution We can see that the length of one side of the square is 15 centimeters. We can find its area by using the formula $A = s^2$ and substituting 15 for s.

$$A = s^2$$
$$= (\mathbf{15})^2 \quad \text{Substitute 15 for } s.$$
$$= 225 \quad \text{Evaluate the exponential expression: } 15 \cdot 15 = 225.$$

The area of the square is 225 cm^2.

Self Check 5

Find the area of the square shown below.

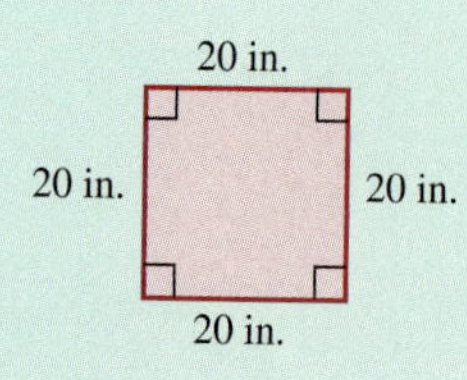

Answer 400 in.^{2}

EXAMPLE 6 Find the number of square feet in 1 square yard. (See Figure 9-55.)

Solution Since 3 feet = 1 yard, each side of 1 square yard is 3 feet long.

$$\begin{aligned} 1\text{ yd}^2 &= (\mathbf{1\ yd})^2 \\ &= (\mathbf{3\ ft})^2 && \text{Substitute 3 feet for 1 yard.} \\ &= 9\text{ ft}^2 && (3\text{ ft})^2 = (3\text{ ft})(3\text{ ft}) = 9\text{ ft}^2. \end{aligned}$$

There are 9 square feet in 1 square yard.

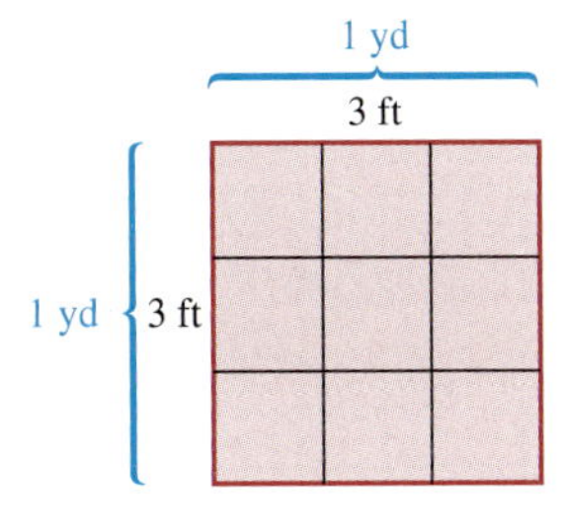

FIGURE 9-55

Self Check 6
Find the number of square centimeters in 1 square meter.

Answer 10,000 cm^2

EXAMPLE 7 **Field hockey.** Field hockey is a team sport in which players use sticks to try to hit a ball into their opponents' goal. Find the area of the rectangular field shown in Figure 9-56. Give the answer in square feet.

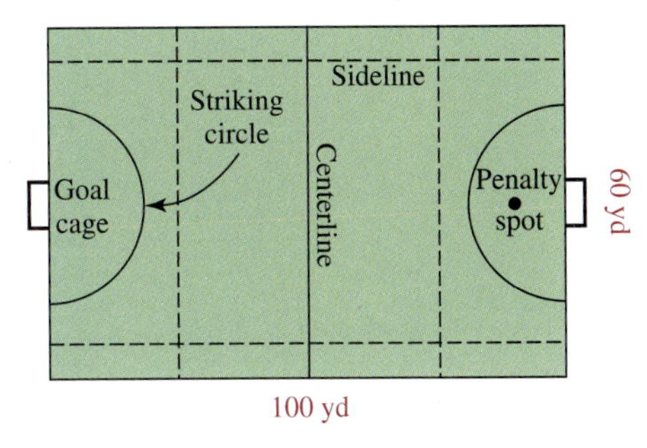

FIGURE 9-56

Solution To find the area in square yards, we substitute 100 for l and 60 for w in the formula for the area of a rectangle, and simplify.

$$\begin{aligned} A &= lw \\ &= 100(60) \\ &= 6{,}000 \end{aligned}$$

The area is 6,000 square yards. Since there are 9 square feet per square yard, we can convert this number to square feet by multiplying 6,000 square yards by $\frac{9\text{ ft}^2}{1\text{ yd}^2}$.

$$\begin{aligned} 6{,}000\text{ yd}^2 &= 6{,}000\text{ yd}^2 \cdot \frac{9\text{ ft}^2}{1\text{ yd}^2} && \text{Multiply by the unit conversion factor: } \frac{9\text{ ft}^2}{1\text{ yd}^2}. \\ &= 6{,}000 \cdot 9\text{ ft}^2 && \text{The units of square yards divide out.} \\ &= 54{,}000\text{ ft}^2 && \text{Multiply: } 6{,}000 \cdot 9 = 54{,}000. \end{aligned}$$

The area of the field is 54,000 ft^2.

Self Check 7
Find the area in square inches of a rectangle with dimensions of 6 inches by 2 feet.

Answer 144 in.^2

EXAMPLE 8 Find the area of the parallelogram in Figure 9-57.

Solution The length of the base of the parallelogram is

$$5\text{ feet} + 25\text{ feet} = 30\text{ feet}$$

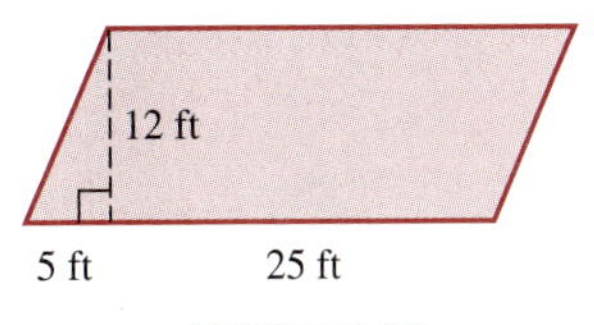

FIGURE 9-57

Self Check 8
Find the area of the parallelogram below.

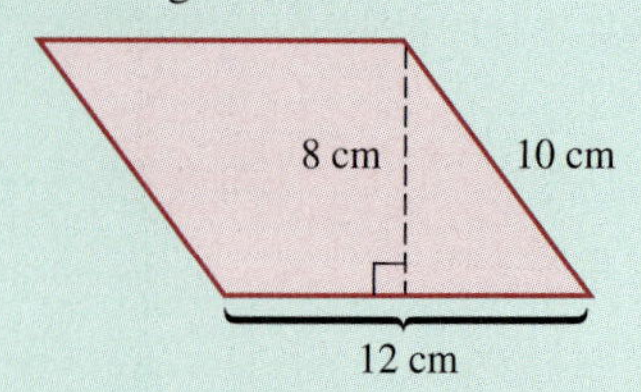

Answer 96 cm^2

The height is 12 feet. To find the area, we substitute 30 for b and 12 for h in the formula for the area of a parallelogram and simplify.

$$A = bh$$
$$= 30(12)$$
$$= 360$$

The area of the parallelogram is 360 ft^2.

Self Check 9
Find the area of the triangle below.

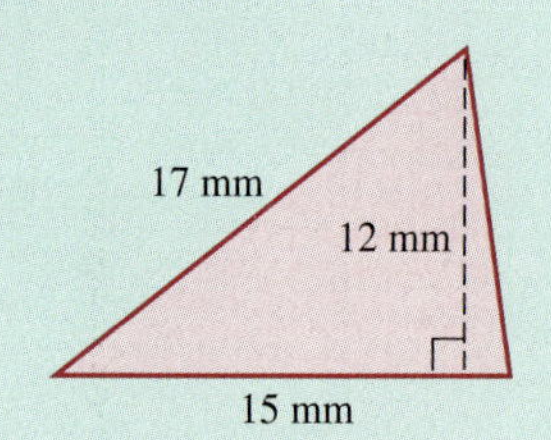

Answer 90 mm^2

EXAMPLE 9 Find the area of the triangle in Figure 9-58.

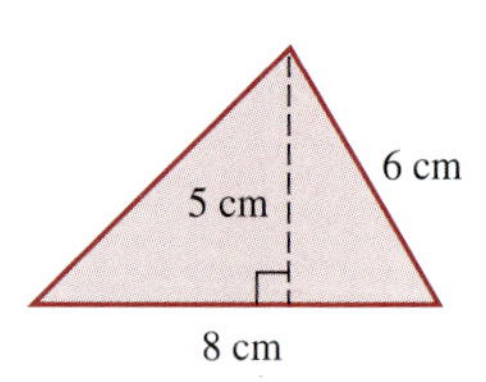

FIGURE 9-58

Solution We substitute 8 for b and 5 for h in the formula for the area of a triangle, and simplify. (The side having length 6 cm is additional information that is not used to find the area.)

$$A = \frac{1}{2}bh$$
$$= \frac{1}{2}(8)(5) \quad \text{The length of the base is 8 cm. The height is 5 cm.}$$
$$= 4(5) \quad \text{Perform the multiplication: } \tfrac{1}{2}(8) = 4.$$
$$= 20$$

The area of the triangle is 20 cm^2.

EXAMPLE 10 Find the area of the triangle in Figure 9-59.

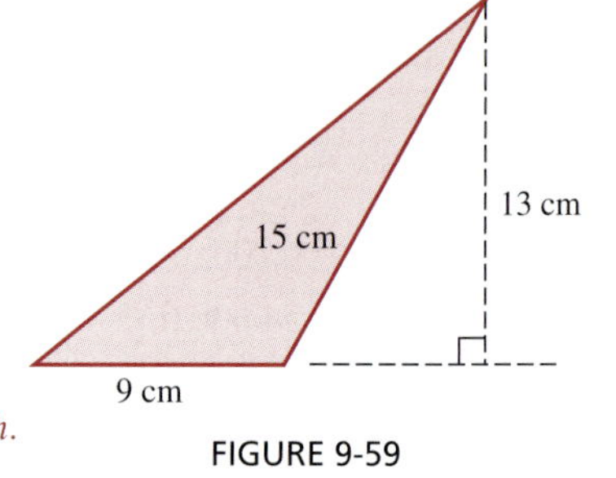

FIGURE 9-59

Solution In this case, the altitude falls outside the triangle.

$$A = \frac{1}{2}bh$$
$$= \frac{1}{2}(9)(13) \quad \text{Substitute 9 for } b \text{ and 13 for } h.$$
$$= \frac{1}{2}\left(\frac{9}{1}\right)\left(\frac{13}{1}\right) \quad \text{Write 9 as } \frac{9}{1} \text{ and 13 as } \frac{13}{1}.$$
$$= \frac{117}{2} \quad \text{Multiply the fractions.}$$
$$= 58.5 \quad \text{Perform the division.}$$

The area of the triangle is 58.5 cm^2.

EXAMPLE 11 Find the area of the trapezoid in Figure 9-60.

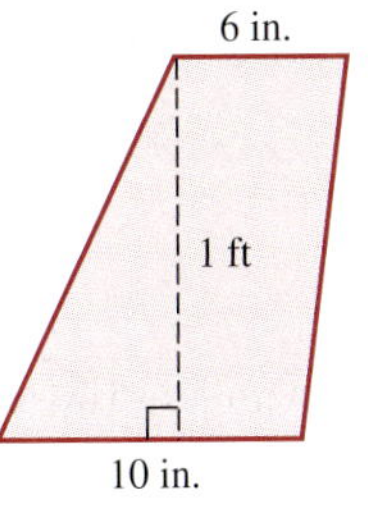

FIGURE 9-60

Solution In this example, $b_1 = 10$ and $b_2 = 6$. It is incorrect to say that $h = 1$, because the height of 1 foot must be expressed as 12 inches to be consistent with the units of the bases. Thus, we substitute 10 for b_1, 6 for b_2, and 12 for h in the formula for finding the area of a trapezoid and simplify.

$$A = \frac{1}{2}h(b_1 + b_2)$$

$$= \frac{1}{2}(12)(10 + 6)$$ The length of the lower base is 10 in. The length of the upper base is 6 in. The height is 12 in.

$$= \frac{1}{2}(12)(16)$$ Perform the addition within the parentheses.

$$= 6(16)$$ Perform the multiplication: $\frac{1}{2}(12) = 6$.

$$= 96$$

The area of the trapezoid is 96 in.^{2}

Self Check 11
Find the area of the trapezoid below.

Answer 54 m^2

Areas of figures that are combinations of polygons

EXAMPLE 12 **Carpeting a room.** A living room/dining room area has the floor plan shown in Figure 9-61. If carpet costs $29 per square yard, including pad and installation, how much will it cost to carpet the room? (Assume no waste.)

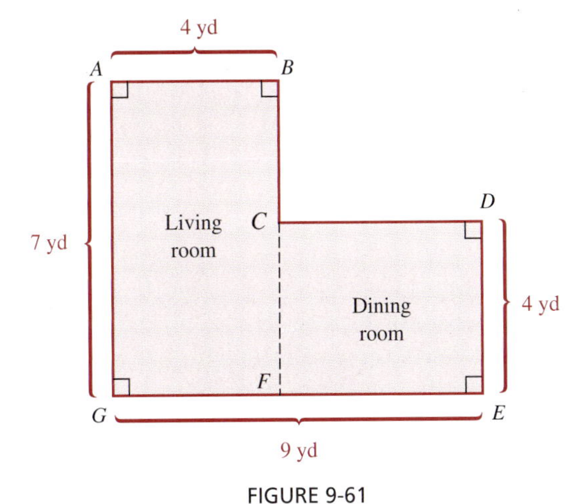

FIGURE 9-61

Solution First we must find the total area of the living room and the dining room.

$$A_{\text{total}} = A_{\text{living room}} + A_{\text{dining room}}$$

Since $\overline{CF}$ divides the space into two rectangles, the areas of the living room and the dining room are found by multiplying their respective lengths and widths.

$$\text{Area of living room} = lw$$
$$= 7(4)$$
$$= 28$$

The area of the living room is 28 yd^2.

To find the area of the dining room, we find its length by subtracting 4 yards from 9 yards to obtain 5 yards. We note that its width is 4 yards.

$$\begin{aligned}\text{Area of dining room} &= \boldsymbol{lw}\\ &= \mathbf{5}(\mathbf{4})\\ &= 20\end{aligned}$$

The area of the dining room is 20 yd^2.

The total area to be carpeted is the sum of these two areas.

$$\begin{aligned}A_{\text{total}} &= \boldsymbol{A}_{\text{living room}} + \boldsymbol{A}_{\text{dining room}}\\ &= \mathbf{28\ yd^2} + \mathbf{20\ yd^2}\\ &= 48\ \text{yd}^2\end{aligned}$$

At \$29 per square yard, the cost to carpet the room will be 48 · \$29, or \$1,392.

EXAMPLE 13 Area of one side of a tent.

Find the area of one side of the tent in Figure 9-62.

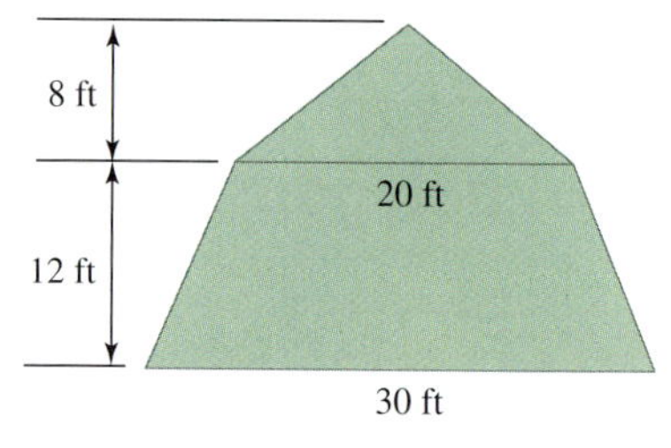

FIGURE 9-62

Solution Each side is a combination of a trapezoid and a triangle. Since the bases of each trapezoid are 30 feet and 20 feet and the height is 12 feet, we substitute 30 for b_1, 20 for b_2, and 12 for h into the formula for the area of a trapezoid.

$$\begin{aligned}A_{\text{trap.}} &= \frac{1}{2}\boldsymbol{h}(\boldsymbol{b_1} + \boldsymbol{b_2})\\ &= \frac{1}{2}(\mathbf{12})(\mathbf{30} + \mathbf{20})\\ &= 6(50)\\ &= 300\end{aligned}$$

The area of the trapezoid is 300 ft^2.

Since the triangle has a base of 20 feet and a height of 8 feet, we substitute 20 for b and 8 for h in the formula for the area of a triangle.

$$\begin{aligned}A_{\text{triangle}} &= \frac{1}{2}\boldsymbol{bh}\\ &= \frac{1}{2}(\mathbf{20})(\mathbf{8})\\ &= 80\end{aligned}$$

The area of the triangle is 80 ft^2.

The total area of one side of the tent is

$$\begin{aligned}A_{\text{total}} &= \boldsymbol{A}_{\text{trap.}} + \boldsymbol{A}_{\text{triangle}}\\ &= \mathbf{300\ ft^2} + \mathbf{80\ ft^2}\\ &= 380\ \text{ft}^2\end{aligned}$$

The total area is 380 ft^2.

Dorm Rooms

THINK IT THROUGH

"The United States has more than 4,000 colleges and universities, with 2 million students living in college dorms." Washingtonpost.com, 2004

The average dormitory room in a residence hall has about 180 square feet of floor space. The rooms are usually furnished with the following items having the given dimensions:

- 2 extra-long twin beds (39 in. W × 80 in. L × 24 in. H)
- 2 dressers (18 in. W × 36 in. L × 48 in. H)
- 2 bookcases (12 in. W × 24 in. L × 40 in. H)
- 2 desks (24 in. W × 48 in. L × 28 in. H)

How many square feet of floor space are left?

Section 9.5 STUDY SET

VOCABULARY *Fill in the blanks.*

1. The distance around a polygon is called the ________.
2. The perimeter of a polygon is measured in ________ units.
3. The measure of the surface enclosed by a polygon is called its ________.
4. If each side of a square measures 1 foot, the area enclosed by the square is 1 ________ foot.
5. The area of a polygon is measured in ________ units.
6. The segment that represents the height of a triangle is called an ________.

CONCEPTS *Sketch and label each of the figures described.*

7. Two different rectangles, each having a perimeter of 40 in.
8. Two different rectangles, each having an area of 40 in.^{2}
9. A square with an area of 25 m^2
10. A square with a perimeter of 20 m
11. A parallelogram with an area of 15 yd^2
12. A triangle with an area of 20 ft^2
13. A figure consisting of a combination of two rectangles whose total area is 80 ft^2
14. A figure consisting of a combination of a rectangle and a square whose total area is 164 ft^2

NOTATION *Fill in the blanks.*

15. The formula for the perimeter of a square is $P = \square$.
16. The formula for the perimeter of a rectangle is $P = \square$.
17. The symbol 1 in.^{2} means one ________ ________.
18. One square meter is expressed as 1 m$^{\square}$.
19. The formula for the area of a square is $A = \square$.
20. The formula for the area of a rectangle is $A = \square$.
21. The formula $A = \frac{1}{2}bh$ gives the area of a ________.
22. The formula $A = \frac{1}{2}h(b_1 + b_2)$ gives the area of a ________.

PRACTICE *Find the perimeter of each figure.*

23.

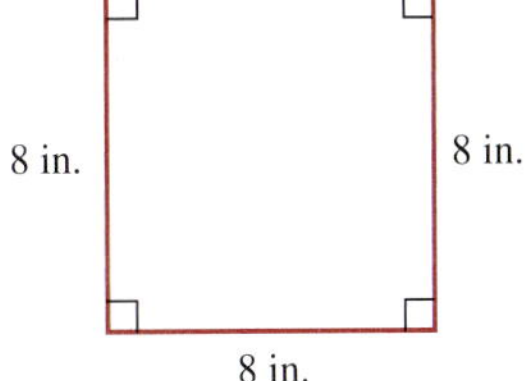

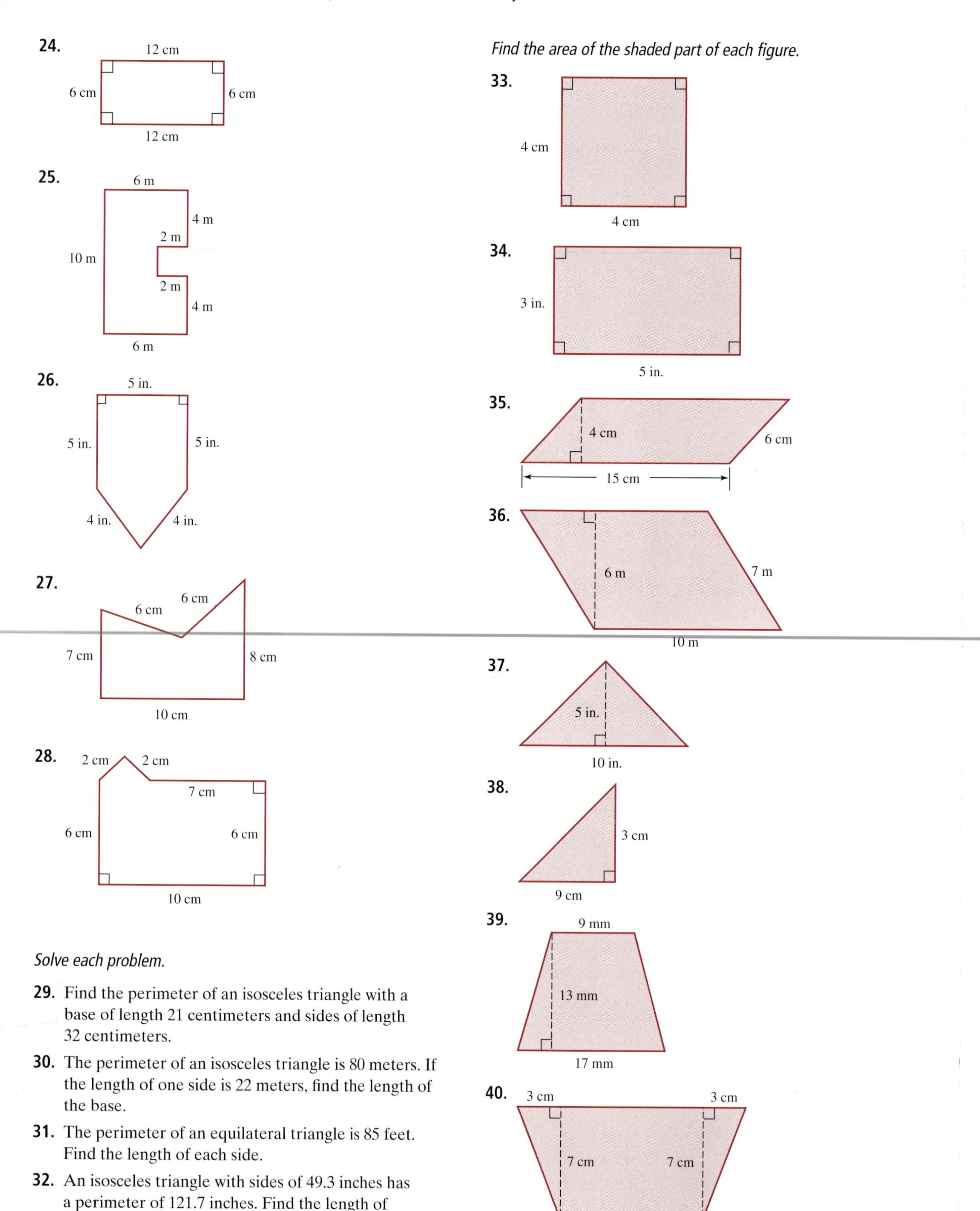

Find the area of the shaded part of each figure.

Solve each problem.

29. Find the perimeter of an isosceles triangle with a base of length 21 centimeters and sides of length 32 centimeters.

30. The perimeter of an isosceles triangle is 80 meters. If the length of one side is 22 meters, find the length of the base.

31. The perimeter of an equilateral triangle is 85 feet. Find the length of each side.

32. An isosceles triangle with sides of 49.3 inches has a perimeter of 121.7 inches. Find the length of the base.

41.

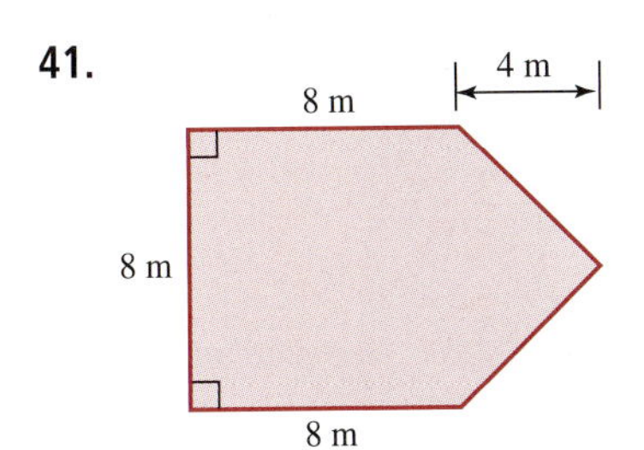

42.

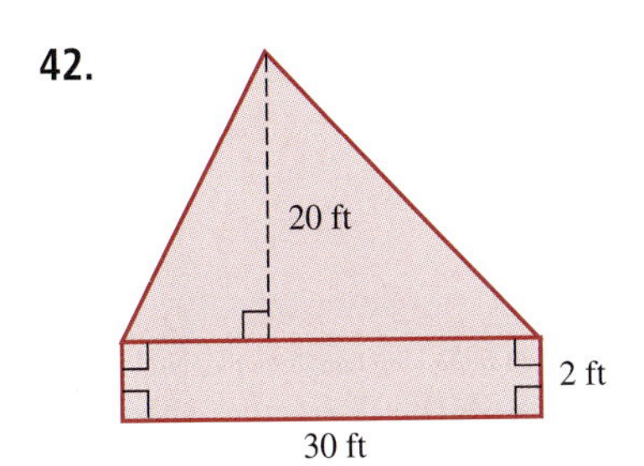

43.

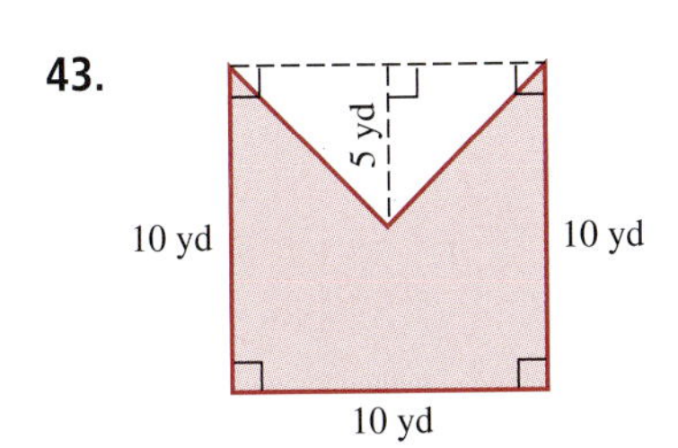

44.

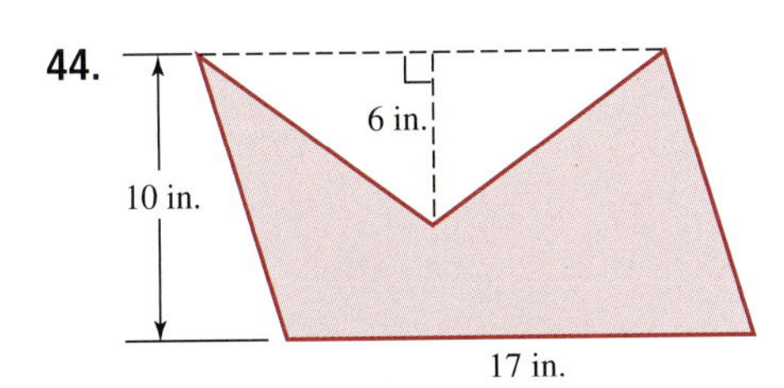

45.

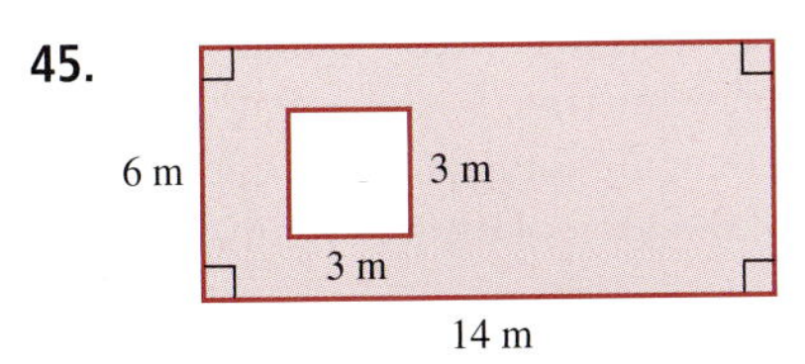

46.

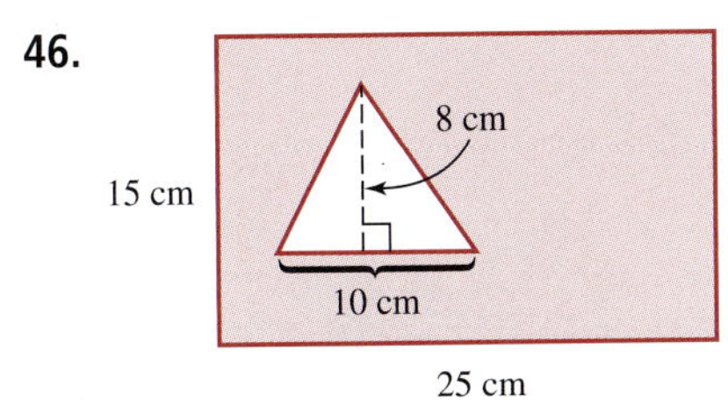

47. How many square inches are in 1 square foot?

48. How many square inches are in 1 square yard?

APPLICATIONS

49. FENCING A YARD A man wants to enclose a rectangular yard with fencing that costs $12.50 a foot, including installation. Find the cost of enclosing the yard if its dimensions are 110 ft by 85 ft.

50. FRAMING A PICTURE Find the cost of framing a rectangular picture with dimensions of 24 inches by 30 inches if framing material costs $8.46 per foot, including matting.

51. PLANTING A SCREEN A woman wants to plant a pine-tree screen around three sides of her backyard, shown below. If she plants the trees 3 feet apart, how many trees will she need?

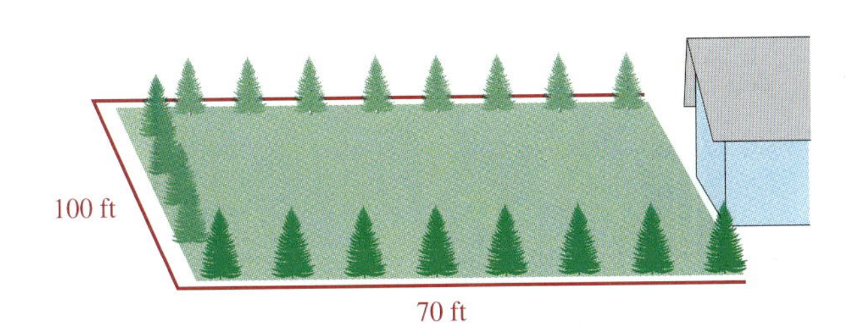

52. PLANTING MARIGOLDS A gardener wants to plant a border of marigolds around the garden shown, to keep out rabbits. How many plants will she need if she allows 6 inches between plants?

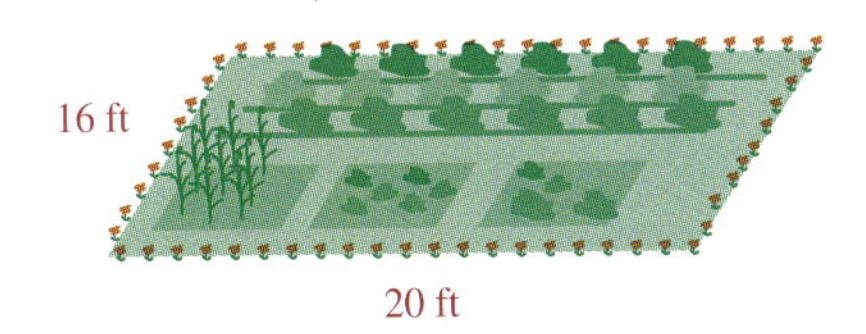

53. BUYING A FLOOR Which is more expensive: a ceramic-tile floor costing $3.75 per square foot or linoleum costing $34.95 per square yard?

54. BUYING A FLOOR Which is cheaper: a hardwood floor costing $5.95 per square foot or a carpeted floor costing $37.50 per square yard?

55. CARPETING A ROOM A rectangular room is 24 feet long and 15 feet wide. At $30 per square yard, how much will it cost to carpet the room? (Assume no waste.)

56. CARPETING A ROOM A rectangular living room measures 30 by 18 feet. At $32 per square yard, how much will it cost to carpet the room? (Assume no waste.)

57. TILING A FLOOR A rectangular basement room measures 14 by 20 feet. Vinyl floor tiles that are 1 ft^2 cost $1.29 each. How much will the tile cost to cover the floor? (Disregard any waste.)

58. PAINTING A BARN The north wall of a barn is a rectangle 23 feet high and 72 feet long. There are five windows in the wall, each 4 by 6 feet. If a gallon of paint will cover 300 ft^2, how many gallons of paint must the painter buy to paint the wall?

59. MAKING A SAIL If nylon is $12 per square yard, how much would the fabric cost to make a triangular sail with a base of 12 feet and a height of 24 feet?

60. PAINTING A GABLE The gable end of a warehouse is an isosceles triangle with a height of 4 yards and a base of 23 yards. It will require one coat of primer and one coat of finish to paint the triangle. Primer costs $17 per gallon, and the finish paint costs $23 per gallon. If one gallon covers 300 square feet, how much will it cost to paint the gable, excluding labor?

61. GEOGRAPHY Use the dimensions of the trapezoid that is superimposed over the state of Nevada to estimate the area of the "Silver State."

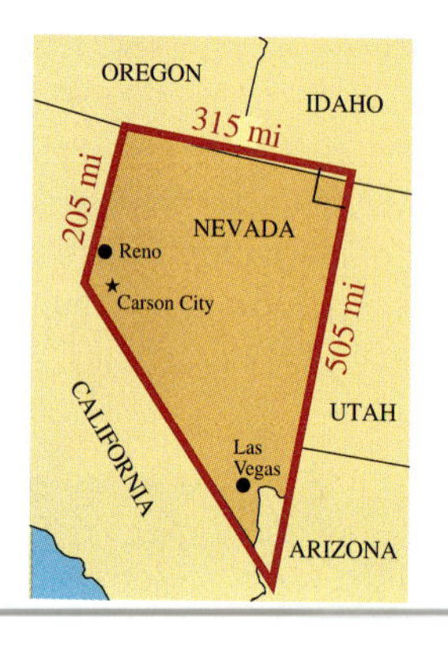

62. COVERING A SWIMMING POOL A swimming pool has the following shape. How many square meters of plastic sheeting will be needed to cover the pool? How much will the sheeting cost if it is $2.95 per square meter? (Assume no waste.)

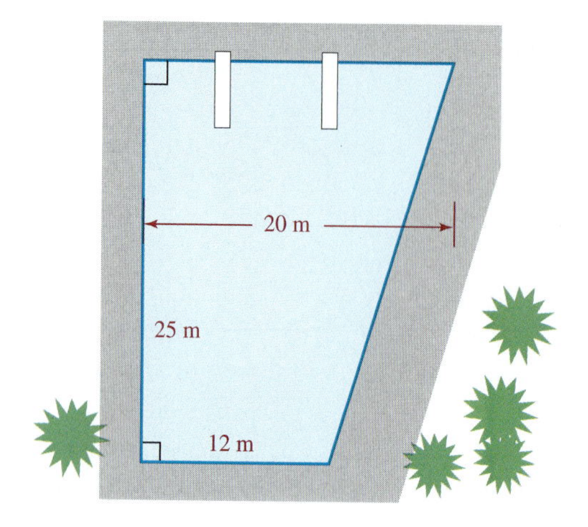

63. CARPENTRY How many sheets of 4-foot-by-8-foot sheetrock are needed to drywall the inside walls on the first floor of the barn shown in the next column? (Assume that the carpenters will cover each wall entirely and then cut out areas for the doors and windows.)

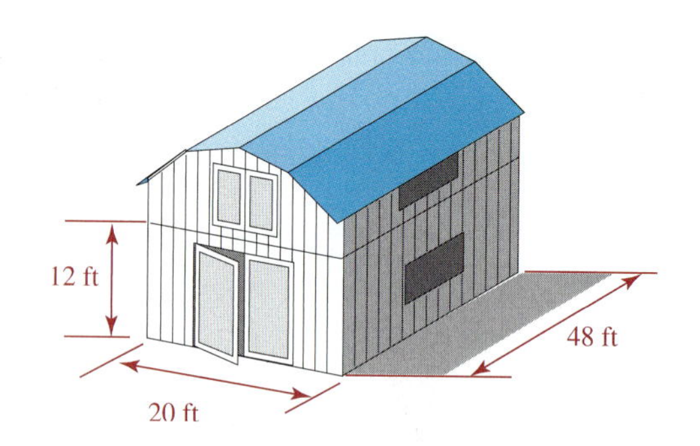

64. CARPENTRY If it costs $90 per square foot to build a one-story home in northern Wisconsin, estimate the cost of building the house with the floor plan shown.

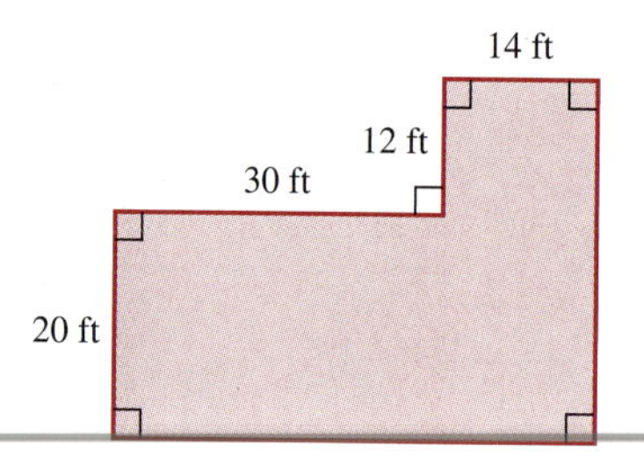

65. DRIVING SAFETY The illustration shows the areas on a highway that a truck driver cannot see in the truck's rear-view mirrors. Use the scale to determine the approximate dimensions of each blind spot. Then estimate the area of each of them.

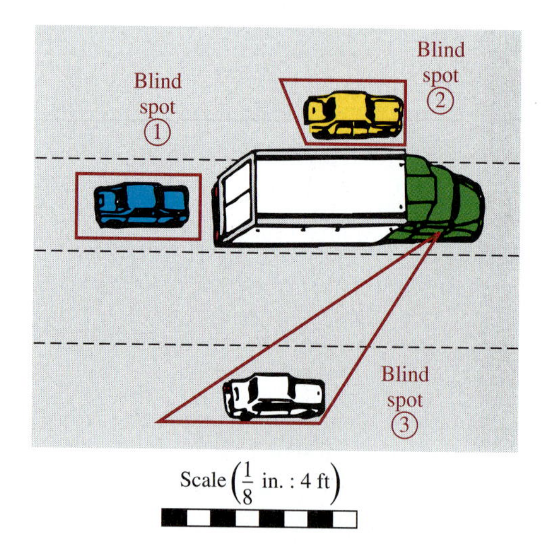

66. ESTIMATING AREA Estimate the area of the sole plate of the iron by thinking of it as a combination of a trapezoid and a triangle.

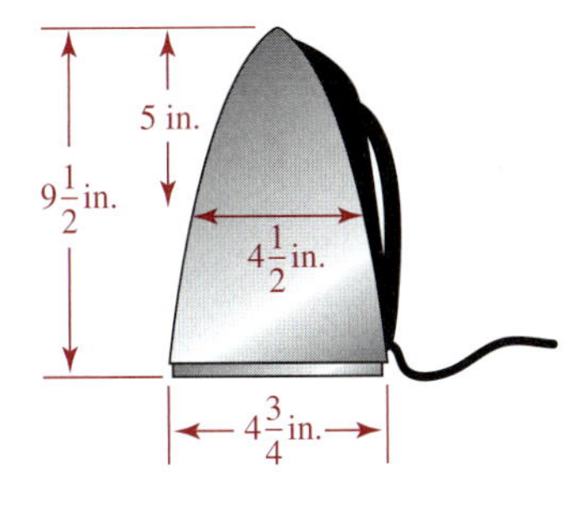

WRITING

67. Explain the difference between perimeter and area.

68. Why is area measured in square units?

REVIEW

Perform the calculations. Write all improper fractions as mixed numbers.

69. $\frac{3}{4} + \frac{2}{3}$

70. $\frac{7}{8} - \frac{2}{3}$

71. $3\frac{3}{4} + 2\frac{1}{3}$

72. $7\frac{5}{8} - 2\frac{5}{6}$

73. $7\frac{1}{2} \div 5\frac{2}{5}$

74. $5\frac{3}{4} \cdot 2\frac{5}{6}$

9.6 Circles

- Circles
- Circumference of a circle
- Area of a circle

In this section, we will discuss circles, one of the most useful geometric figures. In fact, the discovery of fire and the circular wheel were two of the most important events in the history of the human race.

Circles

Circle

A **circle** is the set of all points in a plane that lie a fixed distance from a point called its **center.**

A segment drawn from the center of a circle to a point on the circle is called a **radius.** (The plural of *radius* is *radii.*) From the definition, it follows that all radii of the same circle are the same length.

A **chord** of a circle is a line segment connecting two points on the circle. A **diameter** is a chord that passes through the center of the circle. Since a diameter D of a circle is twice as long as a radius r, we have

$$D = 2r$$

Each of the previous definitions is illustrated in Figure 9-63, in which O is the center of the circle.

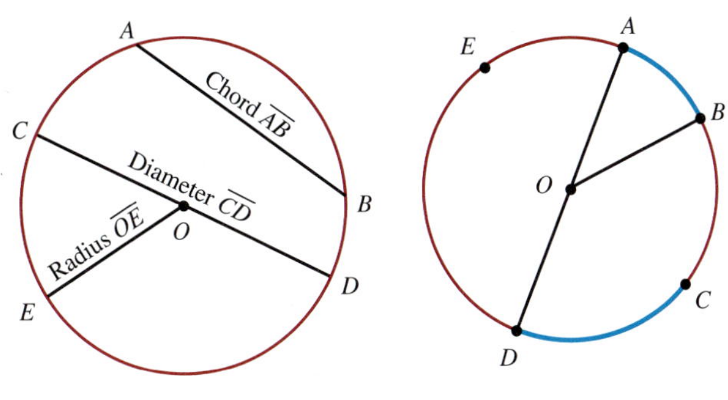

FIGURE 9-63

FIGURE 9-64

Any part of a circle is called an **arc.** In Figure 9-64 in the previous page, the part of the circle from point A to point B is $\overset{\frown}{AB}$, read as "arc AB." $\overset{\frown}{CD}$ is the part of the circle from point C to point D. An arc that is half of a circle is a **semicircle.**

Semicircle

A **semicircle** is an arc of a circle whose endpoints are the endpoints of a diameter.

If point O is the center of the circle in Figure 9-64, $\overline{AD}$ is a diameter and $\overset{\frown}{AED}$ is a semicircle. The middle letter E is used to distinguish semicircle $\overset{\frown}{AED}$ from semicircle $\overset{\frown}{ABCD}$.

An arc that is shorter than a semicircle is a **minor arc.** An arc that is longer than a semicircle is a **major arc.** In Figure 9-64,

$\overset{\frown}{AB}$ is a minor arc and $\overset{\frown}{ABCDE}$ is a major arc

Circumference of a circle

Since early history, mathematicians have known that the ratio of the distance around a circle (the circumference) divided by the length of its diameter is approximately 3. First Kings, Chapter 7, of the Bible describes a round bronze tank that was 15 feet from brim to brim and 45 feet in circumference, and $\frac{45}{15} = 3$. Today, we have a better value for this ratio, known as π (pi). If C is the circumference of a circle and D is the length of its diameter, then

$$\pi = \frac{C}{D}, \quad \text{where } \pi = 3.141592653589\ldots$$

$\frac{22}{7}$ and 3.14 are often used as estimates of π.

If we multiply both sides of $\pi = \frac{C}{D}$ by D, we have the following formula.

Circumference of a circle

The circumference of a circle is given by the formula

$C = \pi D$ where C is the circumference and D is the length of the diameter.

Since a diameter of a circle is twice as long as a radius r, we can substitute $2r$ for D in the formula $C = \pi D$ to obtain another formula for the circumference C.

$$C = 2\pi r$$

Self Check 1

To the nearest tenth, find the circumference of a circle that has a radius of 12 meters.

Answer 75.4 m

EXAMPLE 1 Find the circumference of a circle that has a diameter of 10 centimeters. (See Figure 9-65.)

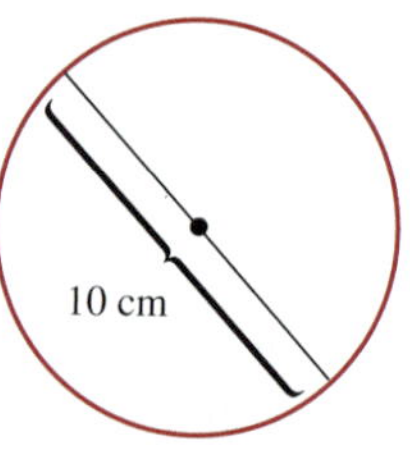

FIGURE 9-65

Solution We substitute 10 for D in the formula for the circumference of a circle.

$$\begin{aligned} C &= \pi D \\ &= \pi(10) \\ &\approx 3.14(10) \quad \text{Replace } \pi \text{ with an approximation: } \pi \approx 3.14. \\ &\approx 31.4 \end{aligned}$$

The circumference is approximately 31.4 centimeters.

Calculating revolutions of a tire

CALCULATOR SNAPSHOT

When the $\boxed{\pi}$ key on a scientific calculator is pressed (on some models, the $\boxed{\text{2nd}}$ key must be pressed first), an approximation of π is displayed. To illustrate how to use this key, consider the following problem. How many times does a 15-inch tire revolve when a car makes a 25-mile trip?

We first find the circumference of the tire.

$$C = \pi \mathbf{D}$$

$= \pi(\mathbf{15})$ Substitute 15 for D, the diameter of the tire.

$= 15\pi$ Normally, we rewrite a product such as $\pi(15)$ so that π is the second factor.

The circumference of the tire is 15π inches.

We then change 25 miles to inches using two unit conversion factors.

$$\frac{25}{1} \text{ miles} \cdot \frac{5{,}280 \text{ feet}}{1 \text{ mile}} \cdot \frac{12 \text{ inches}}{1 \text{ foot}} = 25(5{,}280)(12) \text{ in.}$$

The total distance of the trip is 25(5,280)(12) inches.

Finally, we divide the total distance of the trip by the circumference of the tire to get

$$\text{The number of revolutions of the tire} = \frac{25(5{,}280)(12)}{15\pi}$$

To approximate the value of $\frac{25(5{,}280)(12)}{15\pi}$ using a scientific calculator, we enter

$\boxed{(}$ 25 $\boxed{\times}$ 5280 $\boxed{\times}$ 12 $\boxed{)}$ $\boxed{\div}$ $\boxed{(}$ 15 $\boxed{\times}$ $\boxed{\pi}$ $\boxed{)}$ $\boxed{=}$

33613.52398

The tire makes about 33,614 revolutions.

EXAMPLE 2 **Architecture.** A Norman window is constructed by adding a semicircular window to the top of a rectangular window. Find the perimeter of the Norman window shown in Figure 9-66.

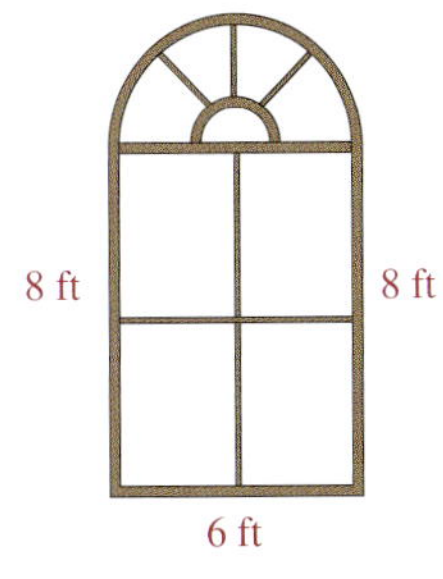

FIGURE 9-66

Solution The window is a combination of a rectangle and a semicircle. The perimeter of the rectangular part is

$P_{\text{rectangular part}} = 8 + 6 + 8 = 22$ Add only 3 sides.

The perimeter of the semicircle is one-half of the circumference of a circle that has a 6-foot diameter.

$$P_{\text{semicircle}} = \frac{1}{2}\pi \mathbf{D}$$

$= \frac{1}{2}\pi(\mathbf{6})$ Substitute 6 for D.

≈ 9.424777961 Use a calculator.

The total perimeter is the sum of the two parts.

$$P_{\text{total}} \approx 22 + 9.424777961$$
$$\approx 31.424777961$$

To the nearest hundredth, the perimeter of the window is 31.42 feet.

Area of a circle

If we divide the circle shown in Figure 9-67(a) into an even number of pie-shaped pieces and then rearrange them as shown in Figure 9-67(b), we have a figure that looks like a parallelogram. The figure has a base that is one-half the circumference of the circle, and its height is about the same length as a radius of the circle.

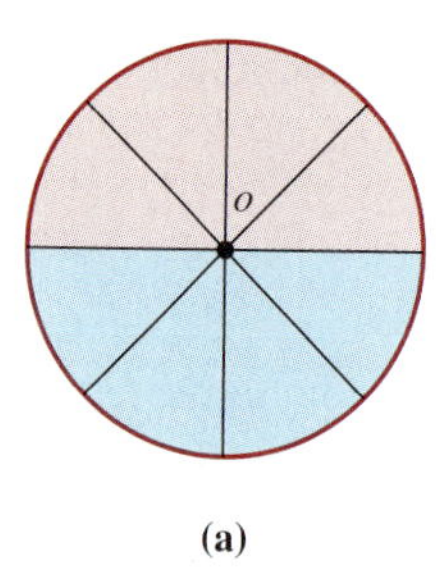

(a)

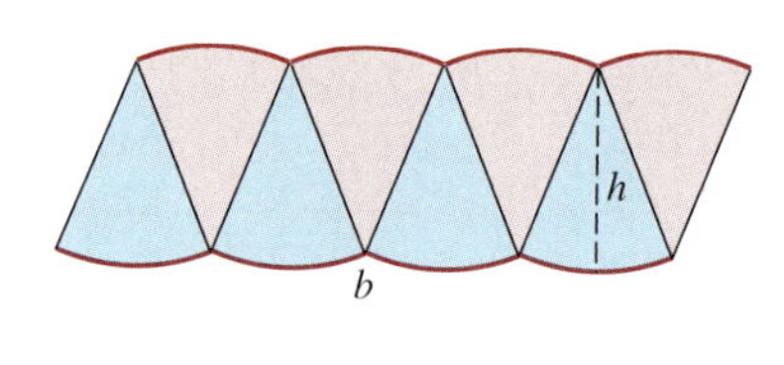

(b)

FIGURE 9-67

If we divide the circle into more and more pie-shaped pieces, the figure will look more and more like a parallelogram, and we can find its area by using the formula for the area of a parallelogram.

$A = bh$

$= \frac{1}{2}Cr$ Substitute $\frac{1}{2}$ of the circumference for b, and r for the height.

$= \frac{1}{2}(2\pi r)r$ Make a substitution: $C = 2\pi r$.

$= \pi r^2$ Simplify: $\frac{1}{2} \cdot 2 = 1$ and $r \cdot r = r^2$.

Area of a circle

The **area of a circle** with radius r is given by the formula

$$A = \pi r^2$$

Self Check 3

To the nearest tenth, find the area of a circle with a diameter of 12 feet.

Answer 113.1 ft^2

EXAMPLE 3 To the nearest tenth, find the area of the circle in Figure 9-68.

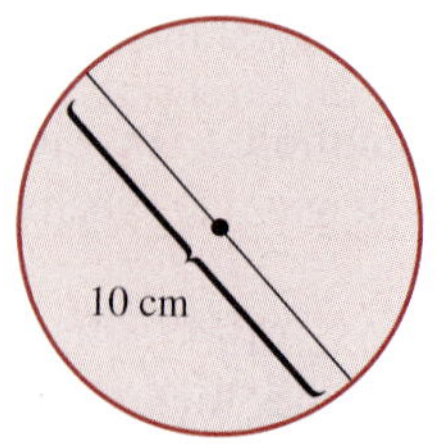

FIGURE 9-68

Solution Since the length of the diameter is 10 centimeters and the length of a diameter is twice the length of a radius, the length of the radius is 5 centimeters. To find the area of the circle, we substitute 5 for r in the formula for the area of a circle.

$A = \pi r^2$

$= \pi(5)^2$

$= 25\pi$

≈ 78.53981634 Use a calculator.

To the nearest tenth, the area is 78.5 cm^2.

Helicopter pads

CALCULATOR SNAPSHOT

Orange paint is available in gallon containers at \$19 each, and each gallon will cover 375 ft^2. To calculate how much the paint will cost to cover a circular helicopter pad 60 feet in diameter, we first calculate the area of the helicopter pad.

$$
\begin{aligned}
A &= \pi r^2 \\
&= \pi(30)^2 \quad \text{Substitute one-half of 60 for } r. \\
&= 30^2\pi
\end{aligned}
$$

The area of the pad is $30^2\pi$ ft^2. Since each gallon of paint will cover 375 ft^2, we can find the number of gallons of paint needed by dividing $30^2\pi$ by 375.

$$\text{Number of gallons needed} = \frac{30^2\pi}{375}$$

To approximate the value of $\frac{30^2\pi}{375}$ using a scientific calculator, we enter these numbers and press these keys.

30 [x^2] [×] [π] [=] [÷] 375 [=]   7.539822369

The result is approximately 7.54. Because paint comes in only full gallons, the painter will need to purchase 8 gallons. The cost of the paint will be 8(\$19), or \$152.

EXAMPLE 4 Find the shaded area in Figure 9-69.

Solution The figure is a combination of a triangle and two semicircles. By the Pythagorean theorem, the hypotenuse h of the right triangle is

$$h = \sqrt{6^2 + 8^2} = \sqrt{36 + 64} = \sqrt{100} = 10$$

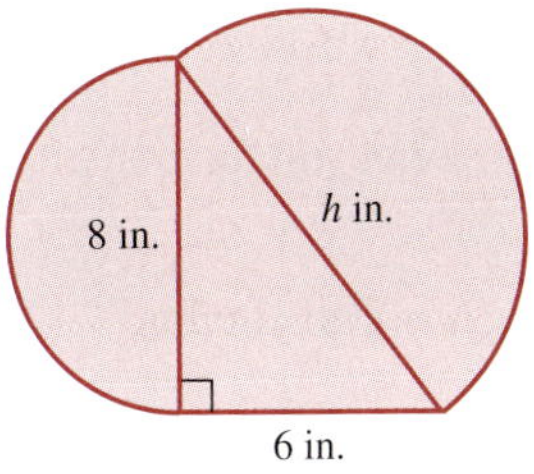

The area of the triangle is

$$A_{\text{right triangle}} = \frac{1}{2}bh = \frac{1}{2}(6)(8) = \frac{1}{2}(48) = 24$$

The area enclosed by the smaller semicircle is

$$A_{\text{smaller semicircle}} = \frac{1}{2}\pi r^2 = \frac{1}{2}\pi(4)^2 = \frac{1}{2}\pi(16) = 8\pi$$

The area enclosed by the larger semicircle is

$$A_{\text{larger semicircle}} = \frac{1}{2}\pi r^2 = \frac{1}{2}\pi(5)^2 = \frac{1}{2}\pi(25) = 12.5\pi$$

The total area is

$$A_{\text{total}} = 24 + 8\pi + 12.5\pi \approx 88.4026494 \quad \text{Use a calculator.}$$

To the nearest hundredth, the area is 88.40 $in.^2$.

8 in. | h in. | 6 in.

FIGURE 9-69

Section 9.6 STUDY SET

VOCABULARY *Fill in the blanks.*

1. A segment drawn from the center of a circle to a point on the circle is called a _______.
2. A segment joining two points on a circle is called a _______.
3. A _______ is a chord that passes through the center of a circle.
4. An arc that is one-half of a complete circle is a _______.
5. An arc that is shorter than a semicircle is called a _______ arc.

6. An arc that is longer than a semicircle is called a ______ arc.
7. The distance around a circle is called its ______.
8. The surface enclosed by a circle is called its ______.

CONCEPTS *Refer to the illustration.*

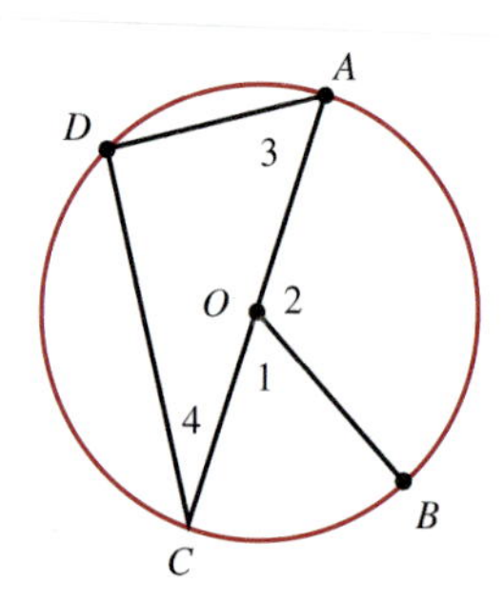

9. Name each radius.
10. Name a diameter.
11. Name each chord.
12. Name each minor arc.
13. Name each semicircle.
14. Name each major arc.
15. If you know the radius of a circle, how can you find its diameter?
16. If you know the diameter of a circle, how can you find its radius?
17. Suppose the two legs of the compass shown are adjusted so that the distance between the pointed ends is 1 inch. Then a circle is drawn.
    a. Find the radius of the circle.
    b. Find the diameter of the circle.
    c. Find the circumference of the circle.
    d. Find the area of the circle.

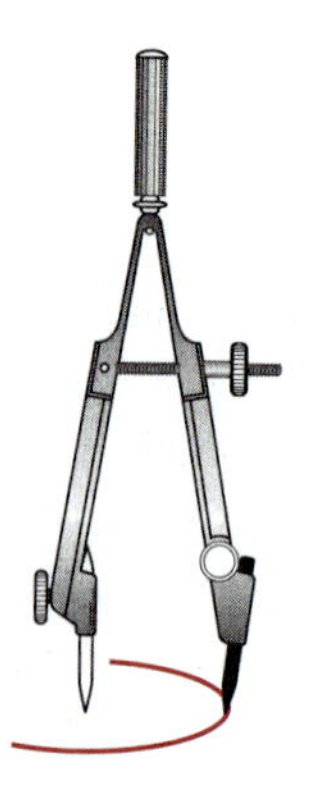

18. Suppose we find the distance around a can and the distance across the can using a measuring tape, as shown. Then we make a comparison, in the form of a ratio.

$$\frac{\text{The distance around the can}}{\text{The distance across the top of the can}}$$

After we do the indicated division, the result will be close to what number?

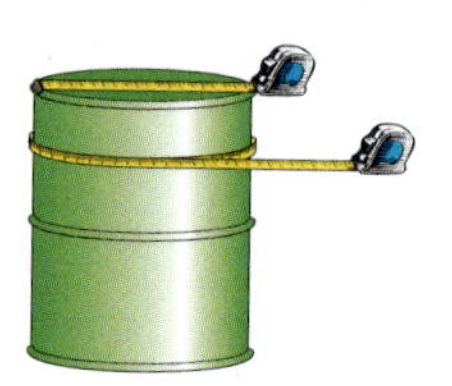

19. When evaluating $\pi(6)^2$, what operation should be performed first?
20. Round $\pi = 3.141592653589\ldots$ to the nearest hundredth.

NOTATION *Fill in the blanks.*

21. The symbol $\widehat{AB}$ is read as ______ ______.
22. To the nearest hundredth, the value of π is ______.
23. The formula for the circumference of a circle is $C =$ ______ or $C = 2\pi$ ______.
24. The formula $A = \pi r^2$ gives the area of a ______.
25. If C is the circumference of a circle and D is its diameter, then $\frac{C}{D} =$ ______.
26. If D is the diameter of a circle and r is its radius, then $D =$ ______ r.
27. Write $\pi(8)$ in a better form.
28. What does $2\pi r$ mean?

PRACTICE *Solve each problem. Answers may vary slightly depending on which approximation of π is used.*

29. To the nearest hundredth, find the circumference of a circle that has a diameter of 12 inches.
30. To the nearest hundredth, find the circumference of a circle that has a radius of 20 feet.
31. Find the diameter of a circle that has a circumference of 36π meters.
32. Find the radius of a circle that has a circumference of 50π meters.

Find the perimeter of each figure to the nearest hundredth.

33.

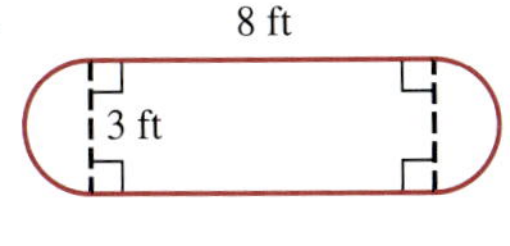

34.

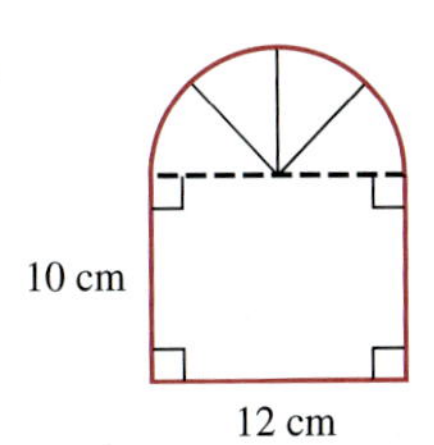

35.

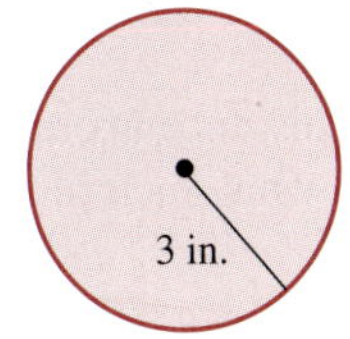

36.

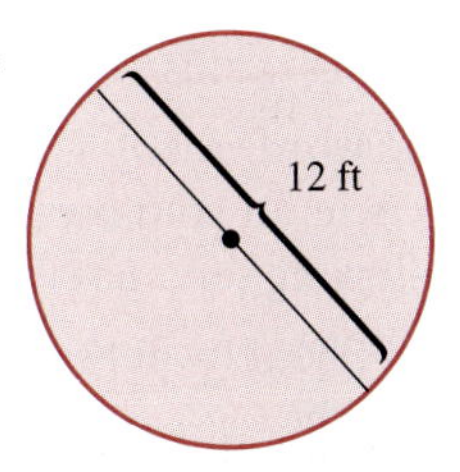

Find the area of each circle to the nearest tenth.

37.

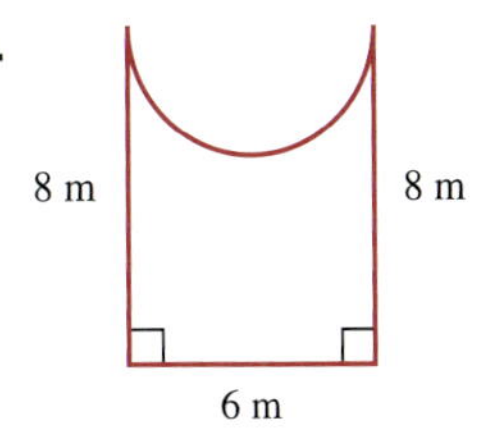

38.

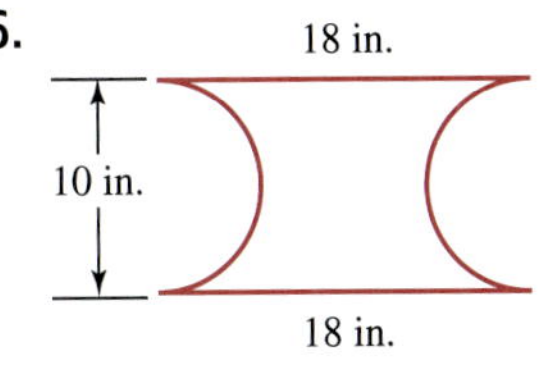

Find the total area of each figure to the nearest tenth.

39.

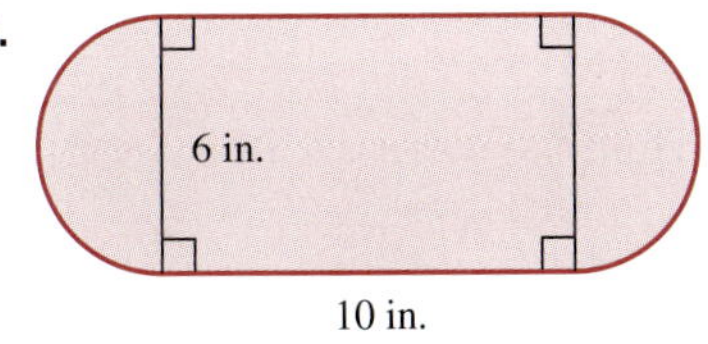

40.

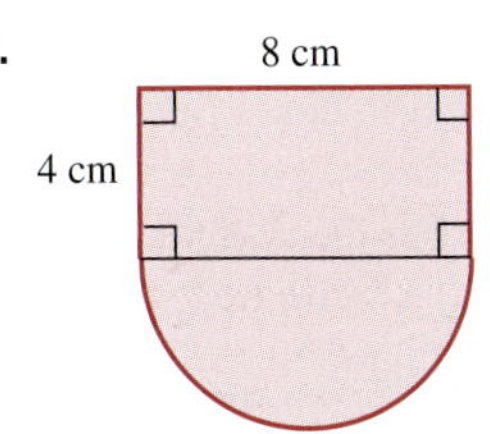

41.

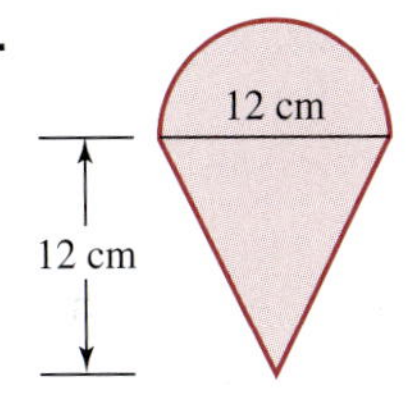

42.

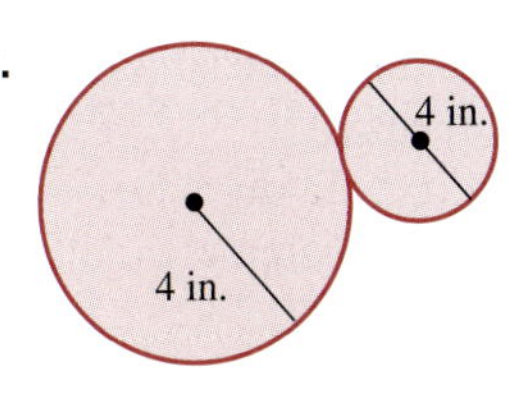

Find the area of each shaded region to the nearest tenth.

43.

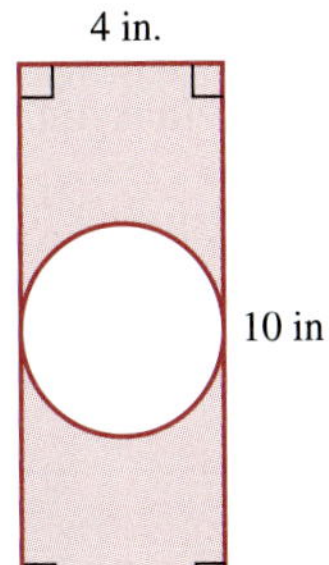

44.

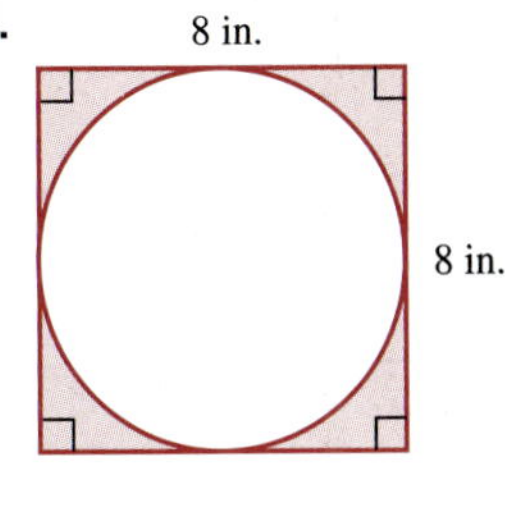

45.

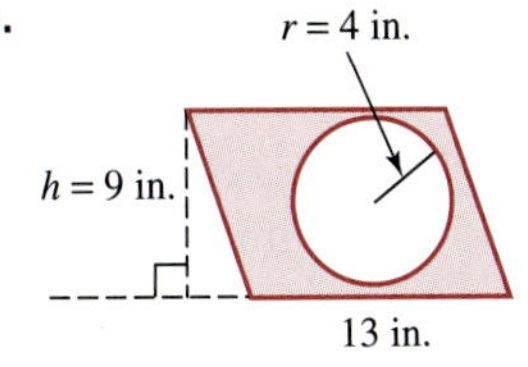

46.

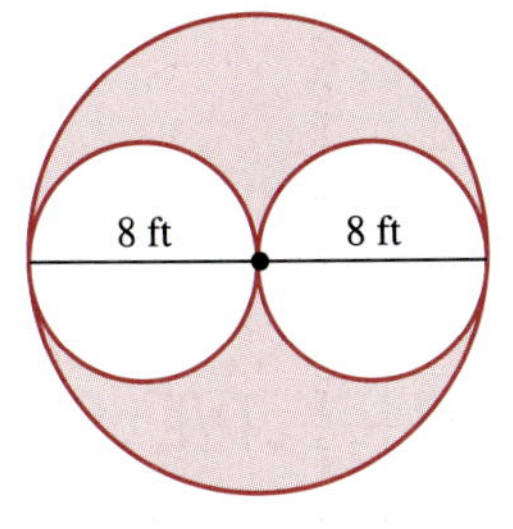

APPLICATIONS *Give each answer to the nearest hundredth. Answers may vary slightly depending on which approximation of π is used.*

47. AREA OF A LAKE Round Lake has a circular shoreline that is 2 miles in diameter. Find the area of the lake.

48. HELICOPTER How far does a point on the tip of a rotor blade travel when it makes one complete revolution?

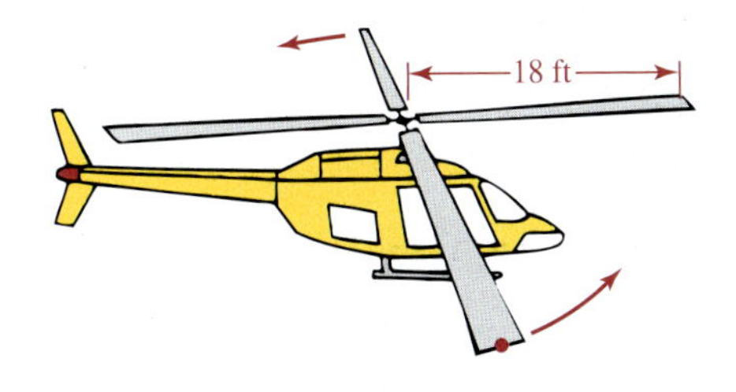

49. GIANT SEQUOIAS The largest sequoia tree is the General Sherman Tree in Sequoia National Park in California. In fact, it is considered to be the largest living thing in the world. According to the *Guinness Book of World Records,* it has a circumference of 102.6 feet, measured $4\frac{1}{2}$ feet above the ground. What is the diameter of the tree at that height?

50. TRAMPOLINES The distance from the center of the following trampoline to the edge of its steel frame is 7 feet. The protective padding covering the springs is 15 inches wide. Find the area of the circular jumping surface of the trampoline, in square feet.

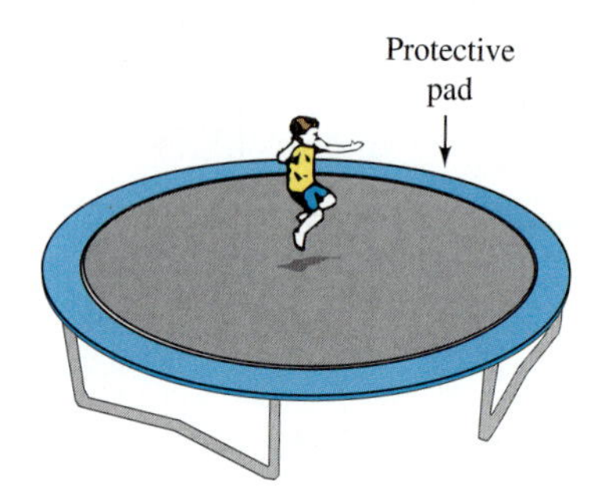

51. JOGGING Joan wants to jog 10 miles on a circular track $\frac{1}{4}$ mile in diameter. How many times must she circle the track?

52. FIXING THE ROTUNDA The rotunda at a state capitol is a circular area 100 feet in diameter. The legislature wishes to appropriate money to have the floor of the rotunda tiled. The lowest bid is \$83 per square yard, including installation. How much must the legislature spend?

53. BANDING THE EARTH A steel band is drawn tightly about the Earth's equator. The band is then loosened by increasing its length by 10 feet, and the resulting slack is distributed evenly along the band's entire length. How far above the Earth's surface is the band? (*Hint:* You don't need to know the Earth's circumference.)

54. CONCENTRIC CIRCLES Two circles are called **concentric circles** if they have the same center. Find the area of the band between two concentric circles if their diameters are 10 centimeters and 6 centimeters.

55. ARCHERY Find the areas of the entire target below and the bull's eye. What percent of the area of the target is the bull's eye?

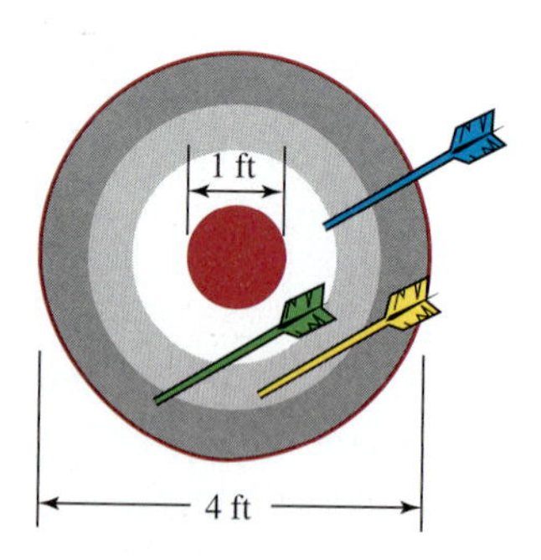

56. LANDSCAPE DESIGN How much of the following lawn does not get watered by the sprinklers at the center of each circle?

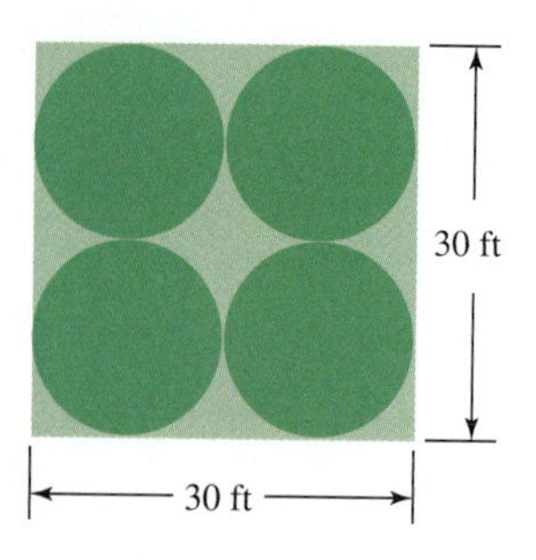

WRITING

57. Explain what is meant by the circumference of a circle.

58. Explain what is meant by the area of a circle.

59. Explain the meaning of π.

60. Distinguish between a major arc and a minor arc.

61. Explain what it means for a car to have a small turning radius.

62. The word *circumference* means the distance around a circle. In your own words, explain what each of the following sentences means.

a. A boat owner's dream was to *circumnavigate* the globe.

b. The teenager's parents felt that he was always trying to *circumvent* the rules.

c. The class was shown a picture of a circle *circumscribed* about an equilateral triangle.

REVIEW

63. Change $\frac{9}{10}$ to a percent.

64. Change $\frac{7}{8}$ to a percent.

65. UNIT COSTS A 24-ounce package of green beans sells for \$1.29. Give the unit cost in cents per ounce.

66. MILEAGE One car went 1,235 miles on 51.3 gallons of gasoline, and another went 1,456 on 55.78 gallons. Which car got the better gas mileage?

67. How many sides does a pentagon have?

68. What is the sum of the measures of the angles of a triangle?

9.7 Surface Area and Volume

- Volumes of solids • Surface areas of rectangular solids
- Volumes and surface areas of spheres • Volumes of cylinders
- Volumes of cones • Volumes of pyramids

In this section, we will discuss a measure of capacity called **volume.** Volumes are measured in cubic units, such as cubic inches, cubic yards, or cubic centimeters. For example,

- We buy gravel or topsoil by the cubic yard.
- We measure the capacity of a refrigerator in cubic feet.
- We often measure amounts of medicine in cubic centimeters.

We will also discuss surface area. The ability to compute surface area is necessary to solve problems such as calculating the amount of material necessary to make a cardboard box or a plastic beach ball.

Volumes of solids

A **rectangular solid** and a **cube** are two common geometric solids. (See Figure 9-70.)

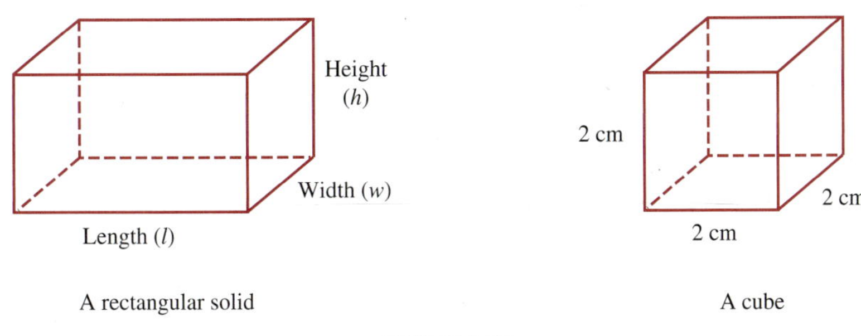

FIGURE 9-70

The **volume** of a rectangular solid is a measure of the space it encloses. Two common units of volume are cubic inches (in.^{3}) and cubic centimeters (cm^3). (See Figure 9-71.)

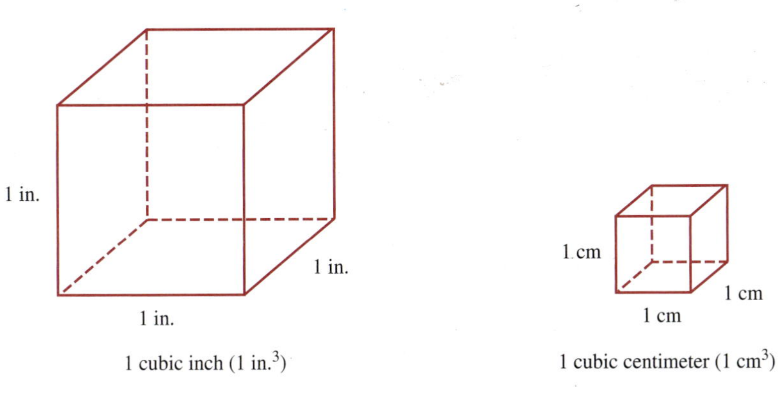

FIGURE 9-71

If we divide the rectangular solid shown in Figure 9-72 on the next page into cubes, each cube represents a volume of 1 cm^3. Because there are 2 levels with 12 cubes on each level, the volume of the rectangular solid is 24 cm^3.

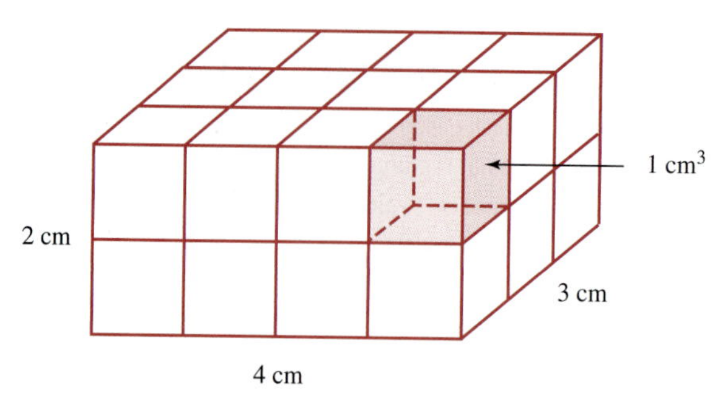

FIGURE 9-72

In practice, we don't find volumes by counting cubes. Instead, we use the formulas shown in Table 9-2.

Figure	Name	Volume	Figure	Name	Volume
s, s, s	Cube	$V = s^3$	h, r	Cylinder	$V = \pi r^2 h$ or $V = Bh$*
h, l, w	Rectangular solid	$V = lwh$	h, r	Cone	$V = \frac{1}{3}\pi r^2 h$ or $V = \frac{1}{3}Bh$*
h	Prism	$V = Bh$*	h	Pyramid	$V = \frac{1}{3}Bh$*
r	Sphere	$V = \frac{4}{3}\pi r^3$			

*B represents the area of the base that is shaded in the figure.

TABLE 9-2

! COMMENT The height of a geometric solid is always measured along a line perpendicular to its base. In each of the solids in Figure 9-73 on the next page, h is the height.

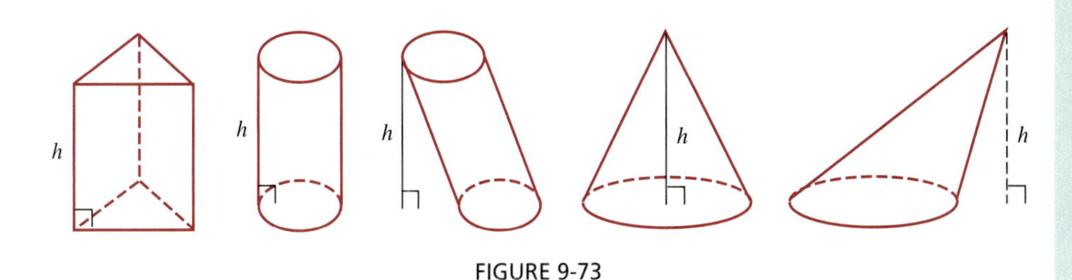

FIGURE 9-73

EXAMPLE 1 How many cubic inches are in 1 cubic foot? (See Figure 9-74.)

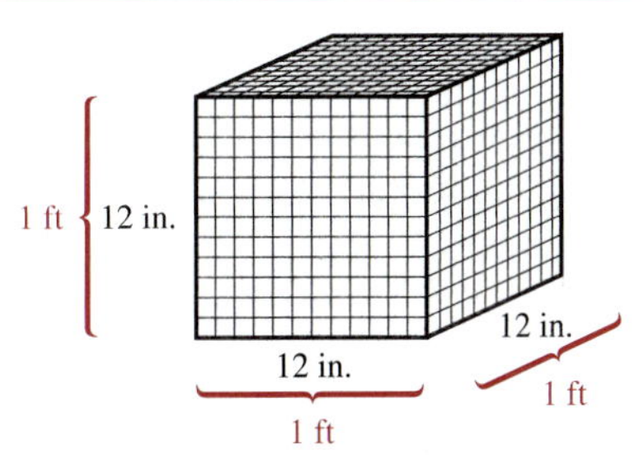

FIGURE 9-74

Solution Since a cubic foot is a cube with each side measuring 1 foot, each side also measures 12 inches. Thus, the volume in cubic inches is

$$V = s^3 \quad \text{This is the forumula for the volume of a cube.}$$
$$= (12)^3 \quad \text{Substitute 12 for } s.$$
$$= 1{,}728$$

There are 1,728 cubic inches in 1 cubic foot.

Self Check 1
How many cubic centimeters are in 1 cubic meter?

Answer $1{,}000{,}000 \text{ cm}^3$

EXAMPLE 2 An oil storage tank is in the form of a rectangular solid with dimensions of 17 by 10 by 8 feet. (See Figure 9-75.) Find its volume.

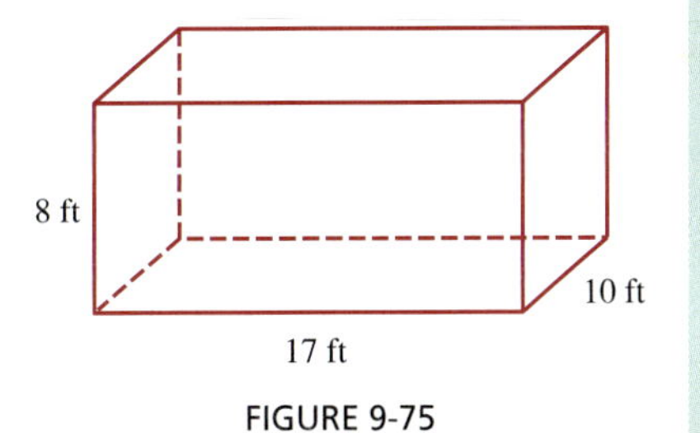

FIGURE 9-75

Solution To find the volume, we substitute 17 for l, 10 for w, and 8 for h in the formula $V = lwh$ and simplify.

$$V = lwh$$
$$= 17(10)(8)$$
$$= 1{,}360$$

The volume is $1{,}360 \text{ ft}^3$.

Self Check 2
Find the volume of a rectangular solid with dimensions of 8 by 12 by 20 meters.

Answer $1{,}920 \text{ m}^3$

EXAMPLE 3 Find the volume of the triangular prism in Figure 9-76.

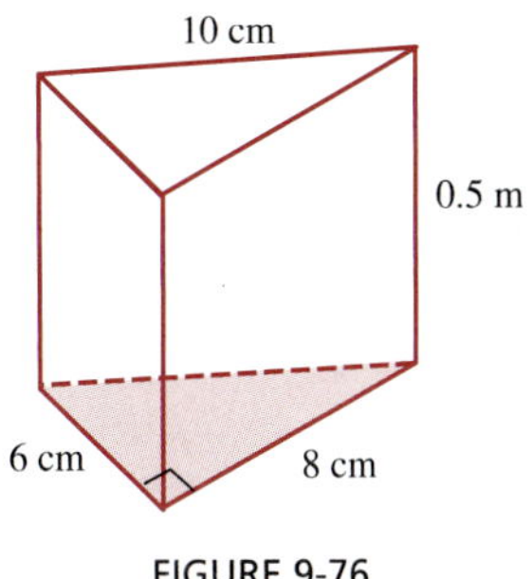

FIGURE 9-76

Solution The volume of the prism is the area of its base multiplied by its height. Since there are 100 centimeters in 1 meter, the height in centimeters is

$$0.5 \text{ m} = 0.5\,(1 \text{ m})$$
$$= 0.5\,(100 \text{ cm}) \quad \text{Substitute 100 centimeters for 1 meter.}$$
$$= 50 \text{ cm}$$

Self Check 3
Find the volume of the triangular prism below.

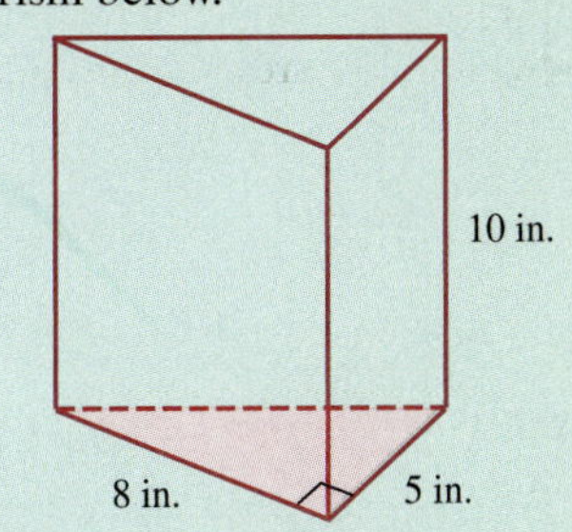

The area of the triangular base is $\frac{1}{2}(6)(8) = 24$ square centimeters. The height of the prism is 50 centimeters. Substituting into the formula for the volume of a prism, we have

$$V = Bh$$
$$= 24(50)$$
$$= 1{,}200$$

Answer 200 in.^3

The volume of the prism is $1{,}200 \text{ cm}^3$.

Surface areas of rectangular solids

The **surface area** of a rectangular solid is the sum of the areas of its six faces. Figure 9-77 shows how we can unfold the faces of a cardboard box to derive a formula for its surface area (SA).

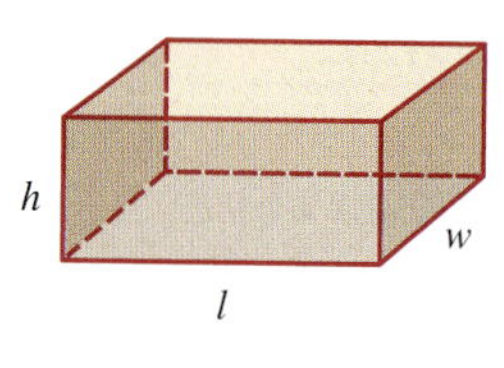

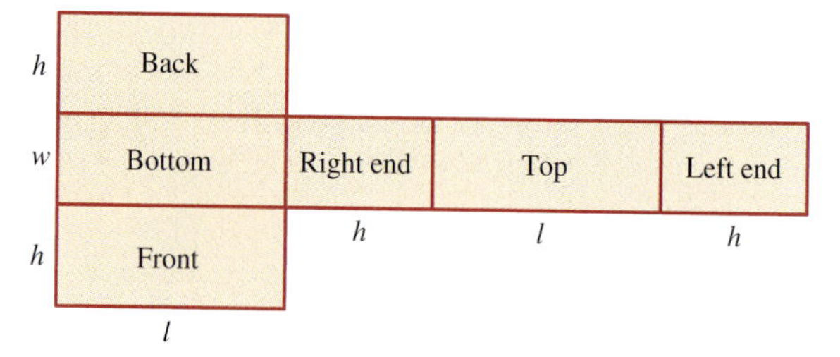

FIGURE 9-77

$$SA = A_{\text{bottom}} + A_{\text{back}} + A_{\text{front}} + A_{\text{right end}} + A_{\text{top}} + A_{\text{left end}}$$
$$= lw + lh + lh + hw + lw + hw$$
$$= 2lw + 2lh + 2hw \quad \text{Combine like terms.}$$

Surface area of a rectangular solid

The surface area of a rectangular solid is given by the formula

$$SA = 2lw + 2lh + 2hw$$

where l is the length, w is the width, and h is the height.

Self Check 4
Find the surface area of a rectangular solid with dimensions of 8 by 12 by 20 meters.

EXAMPLE 4 An oil storage tank is in the form of a rectangular solid with dimensions of 17 by 10 by 8 feet. (See Figure 9-78.) Find the surface area of the tank.

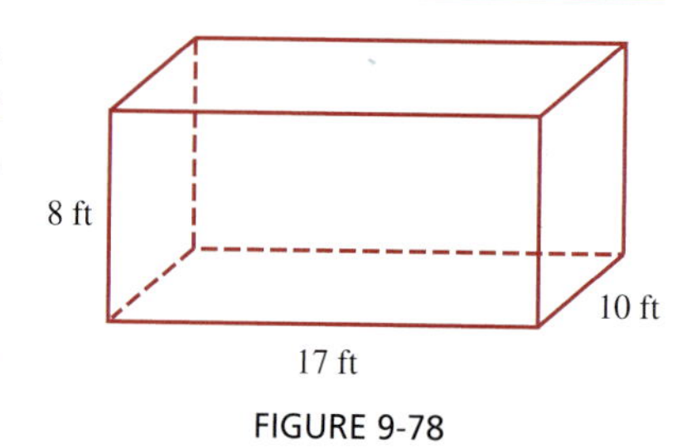

FIGURE 9-78

Solution To find the surface area, we substitute 17 for l, 10 for w, and 8 for h in the formula for surface area and simplify.

$$SA = 2lw + 2lh + 2hw$$
$$= 2(17)(10) + 2(17)(8) + 2(8)(10)$$
$$= 340 + 272 + 160$$
$$= 772$$

Answer 992 m^2

The surface area is 772 ft^2.

Volumes and surface areas of spheres

A **sphere** is a hollow, round ball. (See Figure 9-79.) The points on a sphere all lie at a fixed distance r from a point called its *center.* A segment drawn from the center of a sphere to a point on the sphere is called a *radius.*

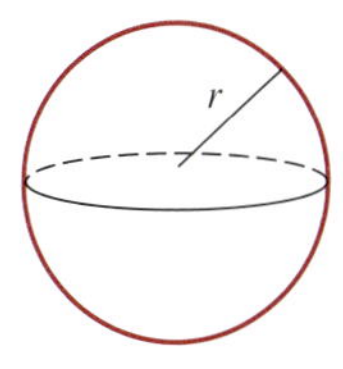

FIGURE 9-79

Filling a water tank — CALCULATOR SNAPSHOT

See Figure 9-80. To calculate how many cubic feet of water are needed to fill a spherical water tank with a radius of 15 feet, we substitute 15 for r in the formula for the volume of a sphere.

$$V = \frac{4}{3}\pi \mathbf{r}^3$$
$$= \frac{4}{3}\pi(\mathbf{15})^3$$

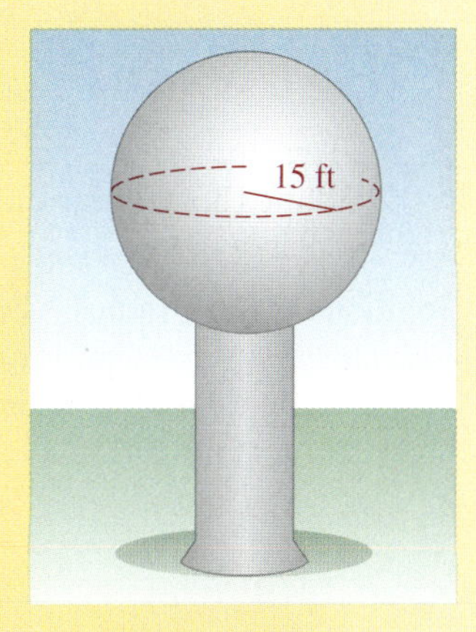

FIGURE 9-80

To approximate the value of $\frac{4}{3}\pi(15)^3$ using a scientific calculator, we enter

15 [y^x] 3 [=] [×] 4 [÷] 3 [=] [×] [π] [=]   14137.16694

To the nearest tenth, 14,137.2 ft^3 of water will be needed to fill the tank.

There is a formula to find the surface area of a sphere.

Surface area of a sphere

The surface area of a sphere with radius r is given by the formula

$$SA = 4\pi r^2$$

EXAMPLE 5 **Manufacturing beach balls.** A beach ball is to have a diameter of 16 inches. (See Figure 9-81.) How many square inches of material will be needed to make the ball? (Disregard any waste.)

Solution Since a radius r of the ball is one-half the diameter, $r = 8$ inches. We can now substitute 8 for r in the formula for the surface area of a sphere.

$$SA = 4\pi \mathbf{r}^2$$
$$= 4\pi(\mathbf{8})^2$$
$$= 4\pi(64)$$
$$= 256\pi \qquad \text{Simplify: } 4 \cdot 64 = 256.$$
$$\approx 804.2477193 \qquad \text{Use a calculator.}$$

A little more than 804 in.^2 of material is needed to make the ball.

FIGURE 9-81

Volumes of cylinders

A **cylinder** is a hollow figure like a piece of pipe. (See Figure 9-82.)

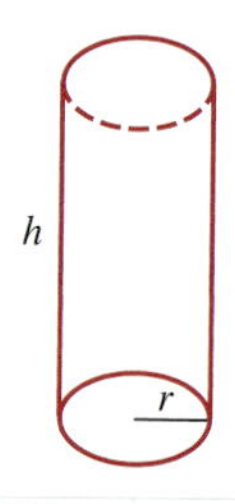

FIGURE 9-82

EXAMPLE 6 Find the volume of the cylinder in Figure 9-83.

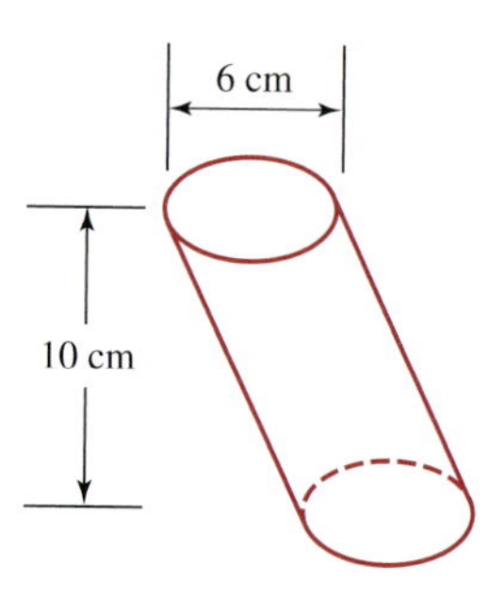

FIGURE 9-83

Solution Since a radius is one-half of the diameter of the circular base, $r = 3$ cm. From the figure, we see that the height of the cylinder is 10 cm. So we can substitute 3 for r and 10 for h in the formula for the volume of a cylinder.

$$V = \pi r^2 h$$
$$= \pi(3)^2(10)$$
$$= 90\pi \quad \text{Simplify: } (3)^2(10) = 9(10) = 90.$$
$$\approx 282.7433388 \quad \text{Use a calculator.}$$

To the nearest hundredth, the volume of the cylinder is 282.74 cm^3.

CALCULATOR SNAPSHOT

Volume of a silo

A silo is a structure used for storing grain. The silo in Figure 9-84 is a cylinder 50 feet tall topped with a **hemisphere** (a half-sphere). To find the volume of the silo, we add the volume of the cylinder to the volume of the dome.

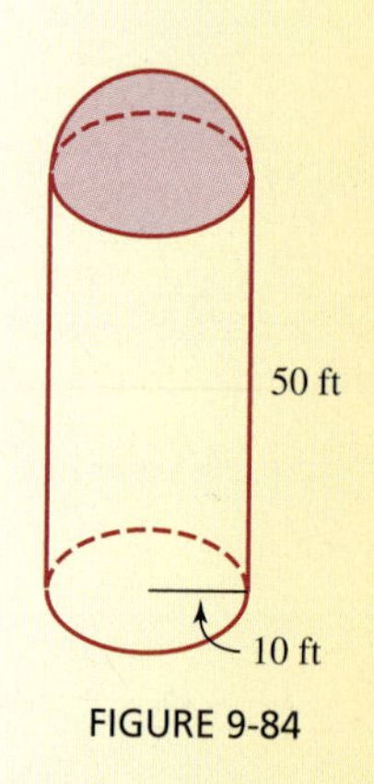

FIGURE 9-84

$$\text{Volume}_{\text{cylinder}} + \text{volume}_{\text{dome}} = (\text{area}_{\text{cylinder's base}})(\text{height}_{\text{cylinder}}) + \frac{1}{2}(\text{volume}_{\text{sphere}})$$
$$= \pi r^2 h + \frac{1}{2}\left(\frac{4}{3}\pi r^3\right)$$
$$= \pi r^2 h + \frac{2\pi r^3}{3} \qquad \frac{1}{2}\left(\frac{4}{3}\pi r^3\right) = \frac{1}{2}\cdot\frac{4}{3}\pi r^3 = \frac{4}{6}\pi r^3 = \frac{2\pi r^3}{3}$$
$$= \pi(10)^2(50) + \frac{2\pi(10)^3}{3} \qquad \text{Substitute 10 for } r \text{ and 50 for } h.$$

To approximate $\pi(10)^2(50) + \frac{2\pi(10)^3}{3}$ using a scientific calculator, we enter these numbers and press these keys.

[π] [×] 10 [x^2] [×] 50 [=] [+] [(] 2 [×] [π] [×] 10 [y^x] 3 [÷] 3 [)] [=] 17802.35837

The volume of the silo is approximately 17,802 ft^3.

EXAMPLE 7 Machining a block of metal. See Figure 9-85. Find the volume that is left when the hole is drilled through the metal block.

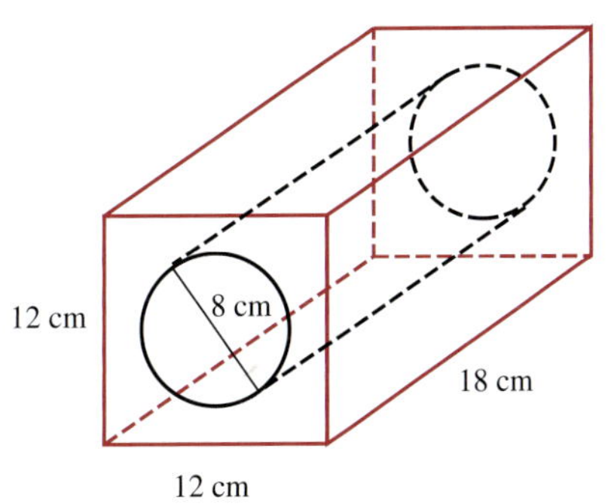

FIGURE 9-85

Solution We must find the volume of the rectangular solid and then subtract the volume of the cylinder. We will think of the rectangular solid and the cylinder as lying on their sides. Thus, the height is 18 cm when we find each volume.

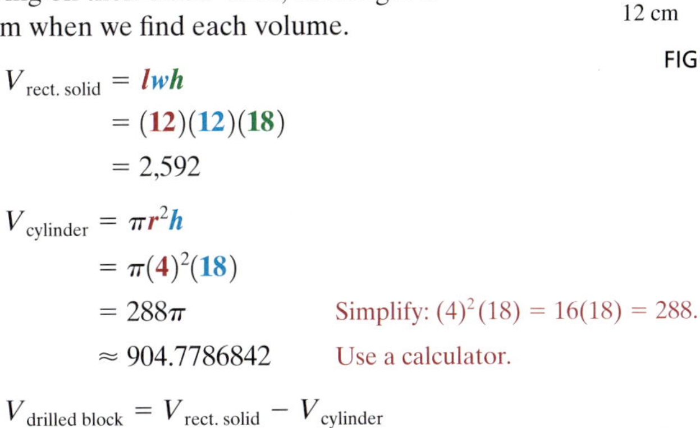

$$V_{\text{rect. solid}} = lwh$$
$$= (12)(12)(18)$$
$$= 2{,}592$$

$$V_{\text{cylinder}} = \pi r^2 h$$
$$= \pi(4)^2(18)$$
$$= 288\pi \qquad \text{Simplify: } (4)^2(18) = 16(18) = 288.$$
$$\approx 904.7786842 \qquad \text{Use a calculator.}$$

$$V_{\text{drilled block}} = V_{\text{rect. solid}} - V_{\text{cylinder}}$$
$$\approx 2{,}592 - 904.7786842$$
$$\approx 1{,}687.221316 \qquad \text{Use a calculator.}$$

To the nearest hundredth, the volume is 1,687.22 cm^3.

Volumes of cones

Two **cones** are shown in Figure 9-86. Each cone has a height h and a radius r, which is the radius of the circular base.

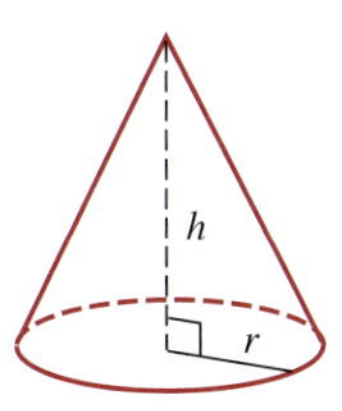

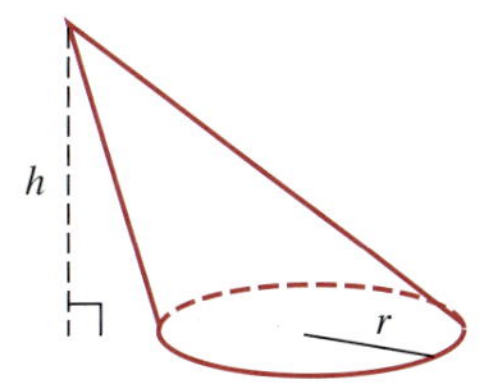

FIGURE 9-86

EXAMPLE 8 To the nearest tenth, find the volume of the cone in Figure 9-87.

Solution Since the radius is one-half of the diameter, $r = 4$ cm. We then substitute 4 for r and 6 for h in the formula for the volume of a cone.

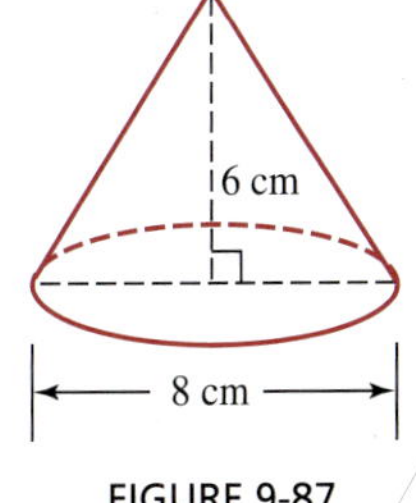

FIGURE 9-87

$$V = \frac{1}{3}\pi r^2 h$$
$$= \frac{1}{3}\pi(4)^2(6)$$
$$= \frac{1}{3}\pi(96) \qquad \text{Simplify: } (4)^2(6) = 16(6) = 96.$$
$$= 32\pi \qquad \text{Multiply: } \frac{1}{3}(96) = 32.$$
$$\approx 100.5309649$$

To the nearest tenth, the volume is 100.5 cubic centimeters.

Volumes of pyramids

Two **pyramids** with a height h are shown in Figure 9-88.

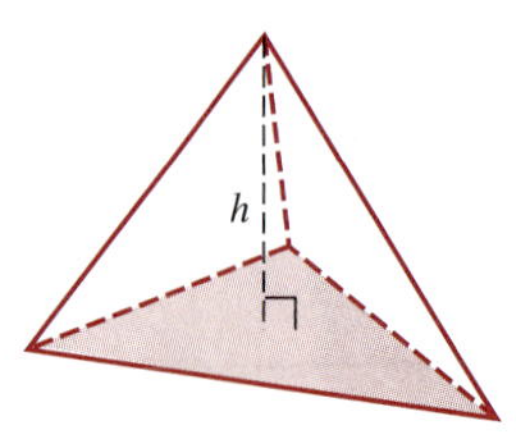

The base is a triangle.

(a)

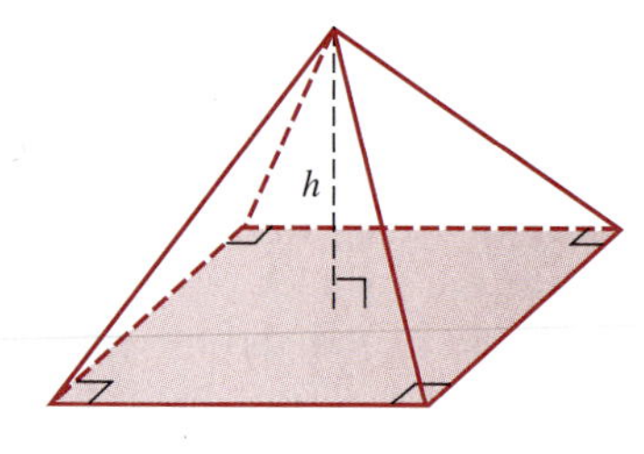

The base is a square.

(b)

FIGURE 9-88

Self Check 9
Find the volume of the pyramid shown below.

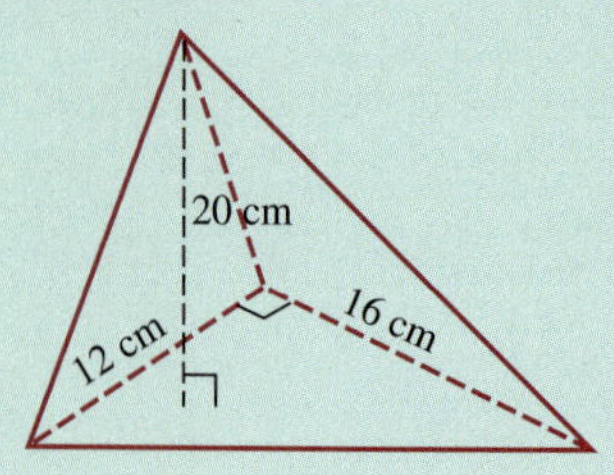

Answer 640 cm^3

EXAMPLE 9 Find the volume of a pyramid that has a square base with each side 6 meters long and a height of 9 meters.

Solution Since the base is a square with each side 6 meters long, the area of the base is 6^2 m^2, or 36 m^2. We can then substitute 36 for the area of the base and 9 for the height in the formula for the volume of a pyramid.

$$V = \frac{1}{3}Bh$$

$$= \frac{1}{3}(36)(9)$$

$$= 12(9) \quad \text{Multiply: } \frac{1}{3}(36) = 12.$$

$$= 108$$

The volume of the pyramid is 108 m^3.

Section 9.7 STUDY SET

VOCABULARY *Fill in the blanks.*

1. The space contained within a geometric solid is called its ________.
2. A ____________ solid is like a hollow shoe box.
3. A ______ is a rectangular solid with all sides of equal length.
4. The volume of a cube with each side 1 inch long is 1 ______ inch.
5. The ________ area of a rectangular solid is the sum of the areas of its faces.
6. The point that is equidistant from every point on a sphere is its ________.
7. A ________ is a hollow figure like a drinking straw.
8. A ___________ is one-half of a sphere.
9. A ______ looks like a witch's pointed hat.
10. A figure that has a polygon for its base and that rises to a point is called a ________.

CONCEPTS *Write the formula used for finding the volume of each solid.*

11. A rectangular solid
12. A prism
13. A sphere
14. A cylinder
15. A cone
16. A pyramid
17. Write the formula for finding the surface area of a rectangular solid.
18. Write the formula for finding the surface area of a sphere.

19. How many cubic feet are in 1 cubic yard?

20. How many cubic inches are in 1 cubic yard?

21. How many cubic decimeters are in 1 cubic meter?

22. How many cubic millimeters are in 1 cubic centimeter?

Which geometric concept (perimeter, circumference, area, volume, or surface area) should be applied to find each of the following?

23. **a.** The size of a room to be air conditioned

b. The amount of land in a national park

c. The amount of space in a refrigerator freezer

d. The amount of cardboard in a shoe box

e. The distance around a checkerboard

f. The amount of material used to make a basketball

24. **a.** The amount of cloth in a car cover

b. The size of a trunk of a car

c. The amount of paper used for a postage stamp

d. The amount of storage in a cedar chest

e. The amount of beach available for sunbathing

f. The distance the tip of a propeller travels

25. In the following illustration, the unit of measurement of length that was used to draw the figure was the inch.

a. What is the volume of the figure?

b. What is the area of the front of the figure?

c. What is the area of the base of the figure?

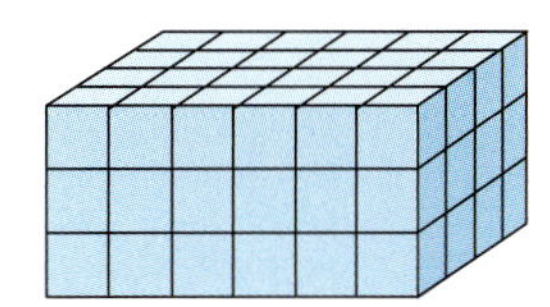

26. The cardboard box shown is a cube. Suppose the six faces were unfolded to lie flat on a table. Draw a picture of what this would look like.

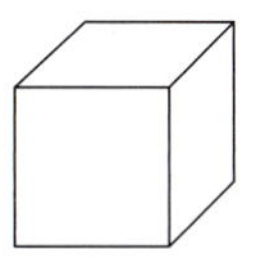

NOTATION *Fill in the blanks.*

27. The notation 1 in.^3 is read as ___ _____ _____.

28. One cubic centimeter is represented as 1 $\text{cm}^{\square}$.

PRACTICE *Find the volume of each solid. If an answer is not exact, round to the nearest hundredth. (Answers may vary slightly, depending on the approximation of π.)*

29. A rectangular solid with dimensions of 3 by 4 by 5 centimeters

30. A rectangular solid with dimensions of 5 by 8 by 10 meters

31. A prism whose base is a right triangle with legs 3 and 4 meters long and whose height is 8 meters

32. A prism whose base is a right triangle with legs 5 and 12 feet long and whose height is 10 feet

33. A sphere with a radius of 9 inches

34. A sphere with a diameter of 10 feet

35. A cylinder with a height of 12 meters and a circular base with a radius of 6 meters

36. A cylinder with a height of 4 meters and a circular base with a diameter of 18 meters

37. A cone with a height of 12 centimeters and a circular base with a diameter of 10 centimeters

38. A cone with a height of 3 inches and a circular base with a radius of 4 inches

39. A pyramid with a square base 10 meters on each side and a height of 12 meters

40. A pyramid with a square base 6 inches on each side and a height of 4 inches

Find the surface area of each solid. If an answer is not exact, round to the nearest hundredth.

41. A rectangular solid with dimensions of 3 by 4 by 5 centimeters

42. A cube with a side 5 centimeters long

43. A sphere with a radius of 10 inches

44. A sphere with a diameter of 12 meters

Find the volume of each figure. If an answer is not exact, round to the nearest hundredth. (Answers may vary slightly, depending on which approximation of π is used.)

45.

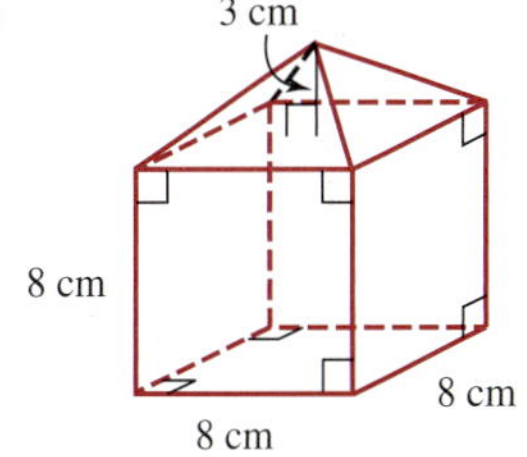

46.

47.

48.

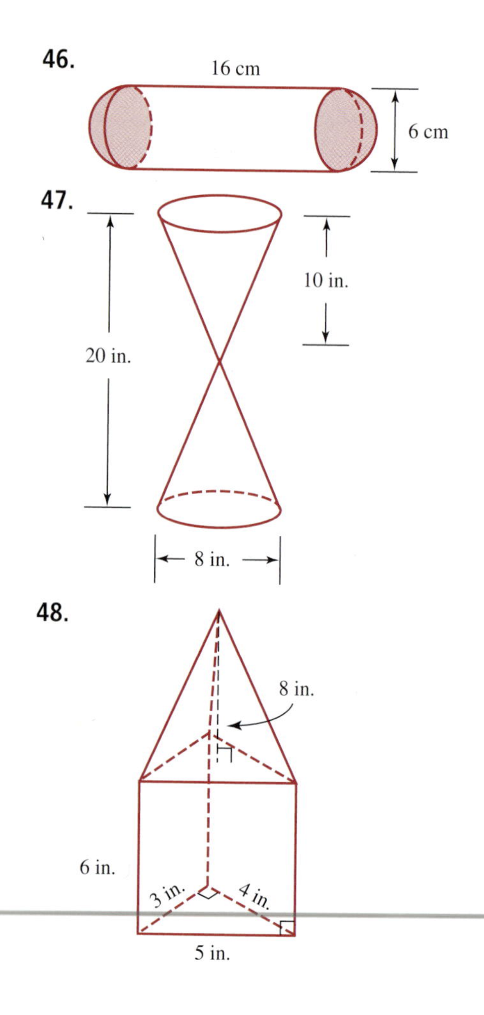

APPLICATIONS *Solve each problem. If an answer is not exact, round to the nearest hundredth.*

49. SUGAR CUBES A sugar cube is $\frac{1}{2}$ inch on each edge. How much volume does it occupy?

50. CLASSROOMS A classroom is 40 feet long, 30 feet wide, and 9 feet high. Find the number of cubic feet of air in the room.

51. WATER HEATERS Complete the ad for the high-efficiency water heater shown.

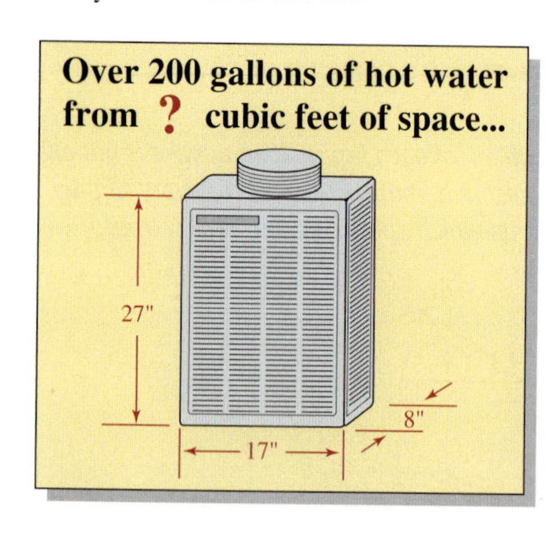

52. REFRIGERATOR CAPACITIES The largest refrigerator advertised in a J. C. Penney catalog has a capacity of 25.2 cubic feet. How many cubic inches is this?

53. OIL TANKS A cylindrical oil tank has a diameter of 6 feet and a length of 7 feet. Find the volume of the tank.

54. DESSERTS A restaurant serves pudding in a conical dish that has a diameter of 3 inches. If the dish is 4 inches deep, how many cubic inches of pudding are in each dish?

55. HOT-AIR BALLOONS The lifting power of a spherical balloon depends on its volume. How many cubic feet of gas will a balloon hold if it is 40 feet in diameter?

56. CEREAL BOXES A box of cereal measures 3 by 8 by 10 inches. The manufacturer plans to market a smaller box that measures $2\frac{1}{2}$ by 7 by 8 inches. By how much will the volume be reduced?

57. ENGINES The *compression ratio* of an engine is the volume in one cylinder with the piston at bottom-dead-center (B.D.C.), divided by the volume with the piston at top-dead-center (T.D.C.). From the data below, what is the compression ratio of the engine? Use a colon to express your answer.

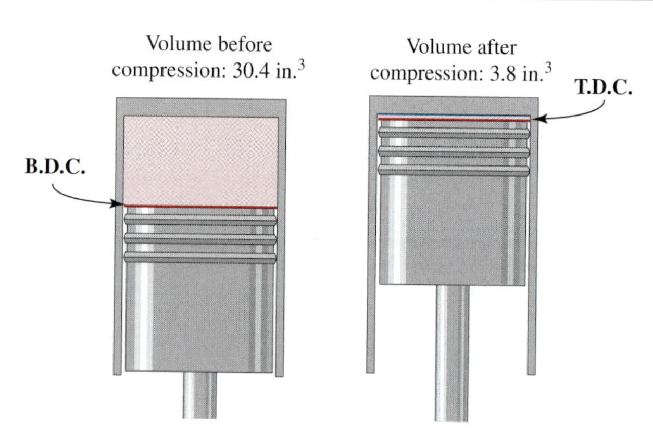

58. LINT REMOVER The following illustration shows a handy gadget; it uses a cylinder of sheets of sticky paper that can be rolled over clothing and furniture to pick up lint and pet hair. After the paper is full, that sheet is peeled away to expose another sheet of sticky paper. Find the area of the first sheet by using the formula $LSA = 2\pi rh$, where LSA represents the lateral surface area of the cylinder.

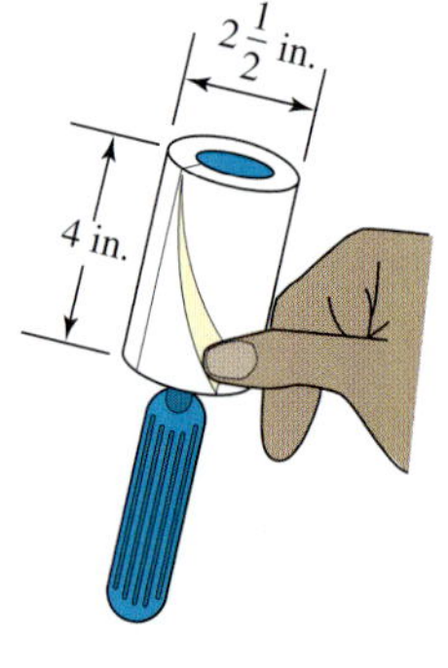

WRITING

59. What is meant by the *volume* of a cube?

60. What is meant by the *surface area* of a cube?

61. Are the units used to measure area different from the units used to measure volume? Explain.

62. The dimensions (length, width, and height) of one rectangular solid are entirely different numbers than the dimensions of another rectangular solid. Would it be possible for the rectangular solids to have the same volume? Explain.

REVIEW

63. Evaluate: $-5(5 - 2)^2 + 3$.

64. BUYING PENCILS Carlos bought 6 pencils at \$0.60 each and a notebook for \$1.25. He gave the clerk a \$5 bill. How much change did he receive?

65. Solve: $\frac{x + 7}{-4} = \frac{1}{4}$.

66. 38 is what percent of 40?

67. Express the phrase "3 inches to 15 inches" as a ratio in lowest terms.

68. Convert 40 ounces to pounds.

69. Convert 2.4 meters to millimeters.

70. State the Pythagorean theorem.

KEY CONCEPT

Formulas

A **formula** is a mathematical expression that is used to express a relationship between quantities. We have studied formulas used in mathematics, business, geometry, and science.

Write a formula describing the mathematical relationship between the given quantities.

1. Distance traveled (d), rate traveled (r), time traveling at that rate (t)
2. Sale price (s), original price (p), discount (d)
3. Perimeter of a rectangle (P), length of the rectangle (l), width of the rectangle (w)
4. Amount of interest earned (I), principal (P), interest rate (r), time the money is invested (t)

Use a formula to solve each problem.

5. Find the area (A) of the triangular lot shown.

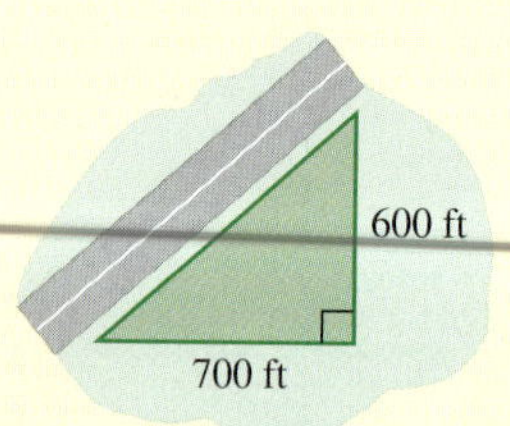

6. Find the volume (V) of the ice chest shown.

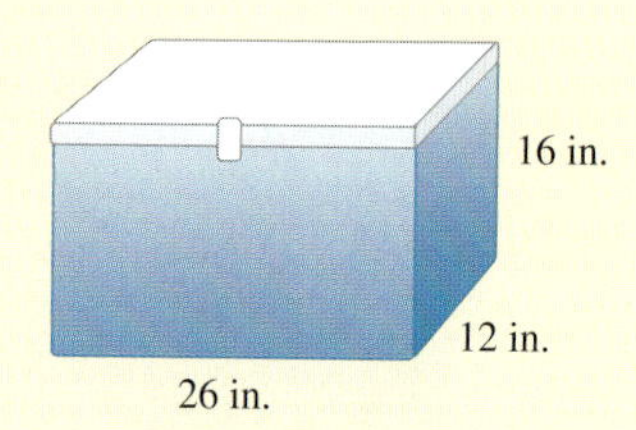

7. Find the retail price (p) of a cookware set that costs the store owner \$45.50 and is marked up \$35.
8. Find the profit (p) made by a school T-shirt sale if revenue was \$14,500 and costs were \$10,200.
9. Find the distance (d) that a rock falls in 3 seconds after being dropped from the edge of a cliff.
10. Find the temperature in degrees Celsius (C) if the temperature in degrees Fahrenheit is 59.

Sometimes we use the same formula to answer several related questions. The results can be displayed in a table.

11. Find the interest earned by each account.

Type of account	Principal	Annual rate earned	Time invested	Interest earned
Savings	\$5,000	5%	3 yr	
Passbook	\$2,250	2%	1 yr	
Trust found	\$10,000	6.25%	10 yr	

12. Complete the table.

Type of coin	Number	Value (¢)	Total value (¢)
Penny	15		
Nickel	n		
Dime	d		
Quarter	q		

ACCENT ON TEAMWORK

SECTION 9.1

WRITING DIGITS In the illustration, the digit 1 is drawn using one angle, and the digit 2 is drawn using two angles. Draw the digit 3 using three angles, the digit 4 using four angles, and so on for all of the digits up to and including 9.

SECTION 9.2

CONSTRUCTIONS

Step 1: See Illustration (a). Using a straightedge, draw $\overline{AB}$. Then place the sharp point of a compass at A and draw an arc.

Step 2: With the same compass setting, place the sharp point at B. As shown in Illustration (b), draw another arc that intersects the arc from step 1 at two points. Label these points C and D.

Step 3: Using a straightedge, draw a line through points C and D. Label the point where line CD intersects $\overline{AB}$ at point E. Does $m(\overline{AE}) = m(\overline{EB})$?

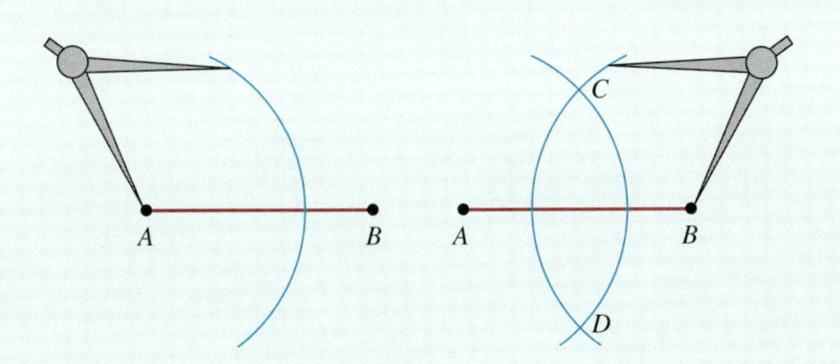

SECTION 9.3

TANGRAMS A tangram is a puzzle in which geometric shapes are arranged to form other shapes. Cut out the pieces in the illustration. Assemble them so that they form a square. There should be no gaps, overlaps, or holes.

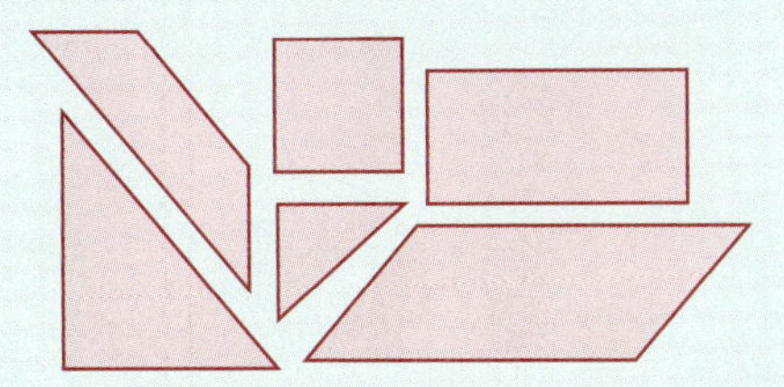

SECTION 9.4

CONGRUENT TRIANGLES Draw a triangle on a piece of paper. Then measure the lengths of its sides (with a ruler) and the angle measures (with a protractor). Choose a combination of any three measurements and tell them to your partner. Are the given facts sufficient for your partner to construct a triangle congruent to yours?

SECTION 9.5

AREAS Find the area of the shaded figure on the square grid shown below.

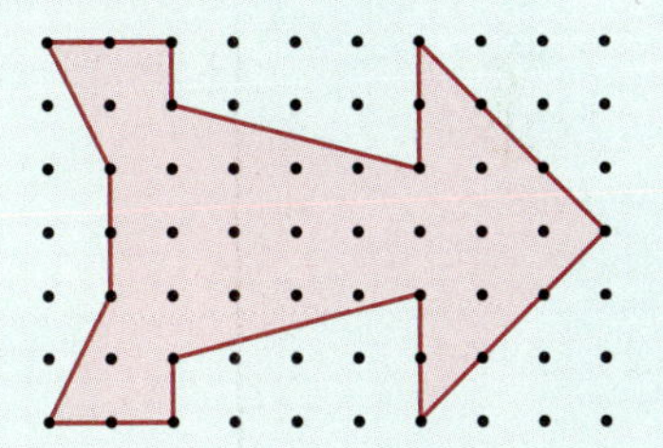

SECTION 9.6

PI Carefully measure the circumference and the diameter of different-size circles. Record the measurements in a table like the one below. Then use a calculator to find $\frac{C}{D}$. The result should be a number close to π.

Object	Circumference	Diameter	$\frac{C}{D}$
Jar	$3\frac{1}{2}$ in.	$1\frac{1}{8}$ in.	3.11

SECTION 9.7

PYRAMIDS Cut out, fold, and glue together the pattern shown below. Estimate the volume and surface area of the pyramid.

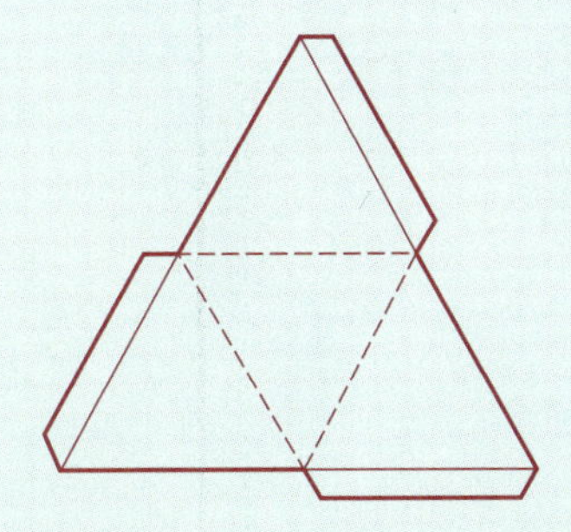

CHAPTER REVIEW

SECTION 9.1 Some Basic Definitions

CONCEPTS

In geometry, we study *points, lines,* and *planes.*

A *line segment* is a part of a line with two endpoints. A *ray* is a part of a line with one endpoint.

An *angle* is a figure formed by two rays with a common endpoint. The common endpoint is called the *vertex* of the angle.

A *protractor* is used to find the measure of an angle.

An *acute angle* is greater than 0° but less than 90°. A *right angle* measures 90°. An *obtuse angle* is greater than 90° but less than 180°. A *straight angle* measures 180°.

Two angles that have the same vertex and are side-by-side are called *adjacent angles.*

REVIEW EXERCISES

1. Identify a point, a line, and a plane on the table top.

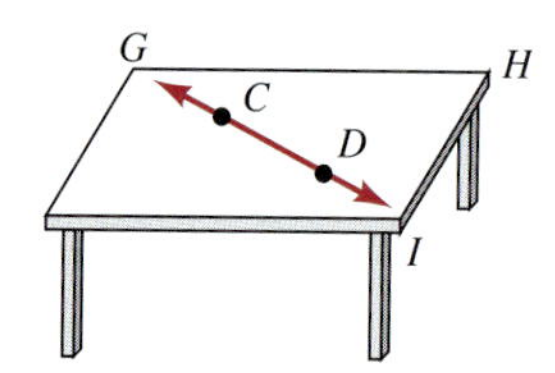

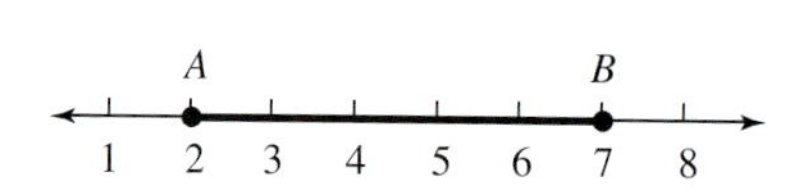

2. In the illustration shown, find $m(\overline{AB})$.
3. In the illustration below, give four ways to name the angle.

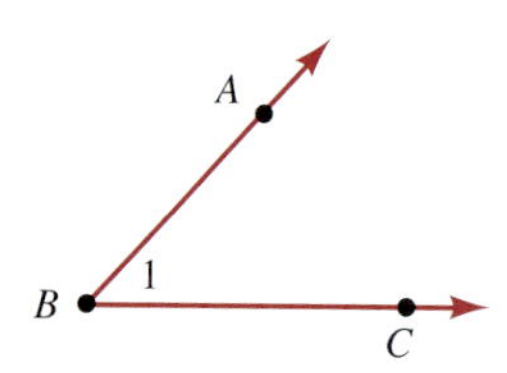

4. In the illustration above, use a protractor to find the measure of the angle.
5. In the illustration below, identify each acute angle, right angle, obtuse angle, and straight angle.

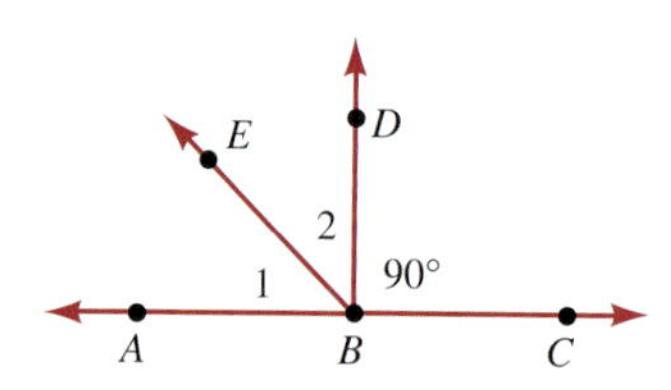

Use the measures to identify each angle as an acute angle, a right angle, an obtuse angle, or a straight angle.

6. $m(\angle A) = 150°$
7. $m(\angle B) = 90°$
8. $m(\angle C) = 180°$
9. $m(\angle D) = 25°$
10. The two angles shown below are adjacent angles. Find x.
11. Line AB is shown. Find y.

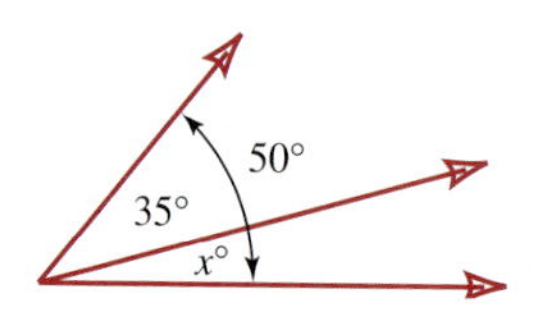

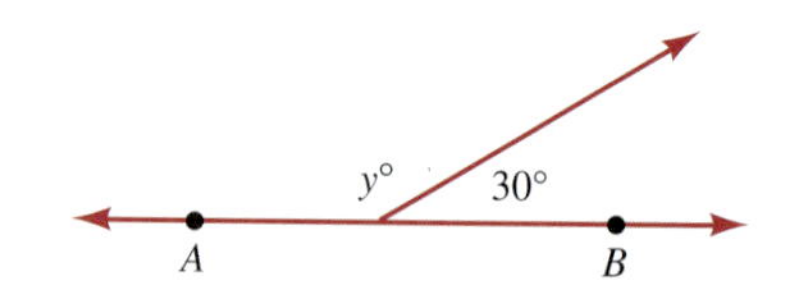

When two lines intersect, pairs of nonadjacent angles are called *vertical angles.*

Vertical angles have the same measure.

12. In the illustration, find **a.** m(∠1) and **b.** m(∠2).

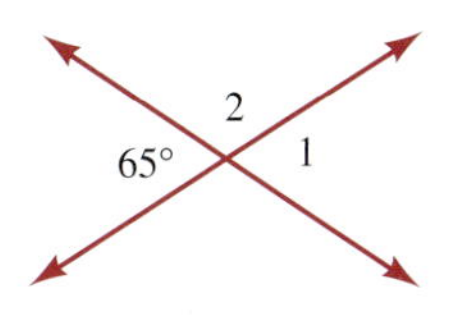

If the sum of two angles is 90°, the angles are *complementary.* If the sum of two angles is 180°, the angles are *supplementary.*

13. Find the complement of an angle that measures 50°.

14. Find the supplement of an angle that measures 140°.

15. Are angles measuring 30°, 60°, and 90° supplementary?

SECTION 9.2 *Parallel and Perpendicular Lines*

Parallel lines do not intersect. *Perpendicular* lines intersect and make right angles.

A line that intersects two or more *coplanar* lines is called a *transversal.*

When a transversal intersects two coplanar lines, *alternate interior angles* and *corresponding* angles are formed.

16. Which part of the illustration represents parallel lines?

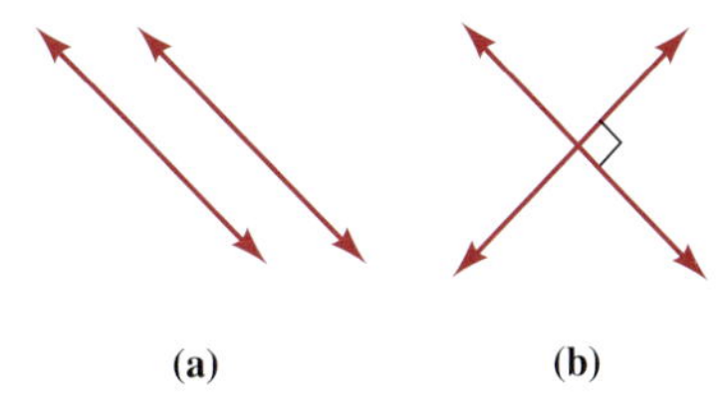

17. Identify all pairs of alternate interior angles shown in the illustration.

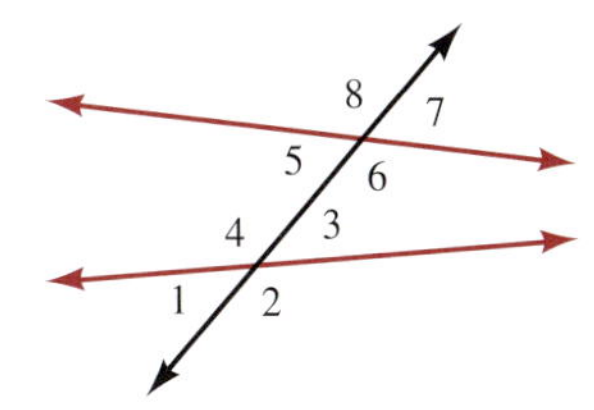

18. Identify all pairs of corresponding angles shown in the illustration above.

19. Identify all pairs of vertical angles shown in the illustration above.

If two parallel lines are cut by a transversal:

1. Alternate interior angles are congruent (have equal measures).
2. Corresponding angles are congruent.
3. Interior angles on the same side of the transversal are supplementary.

20. In the illustration below, $l_1 \parallel l_2$. Find the measure of each angle.

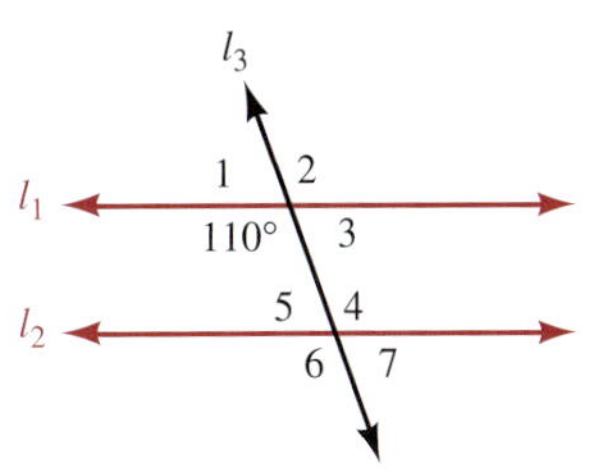

21. In the illustration, $\overline{DC} \parallel \overline{AB}$. Find the measure of each angle.

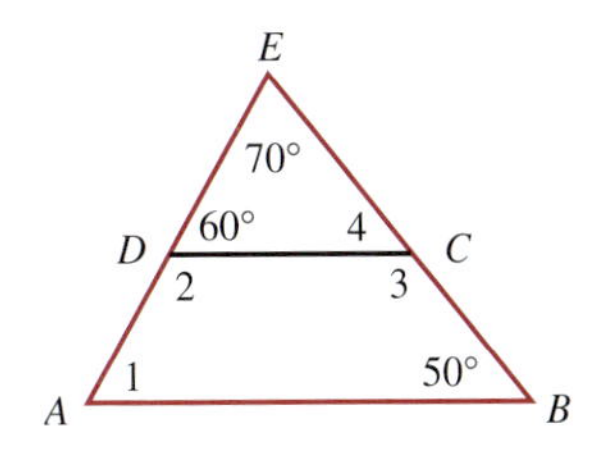

22. In the illustration, $l_1 \parallel l_2$. Find x.

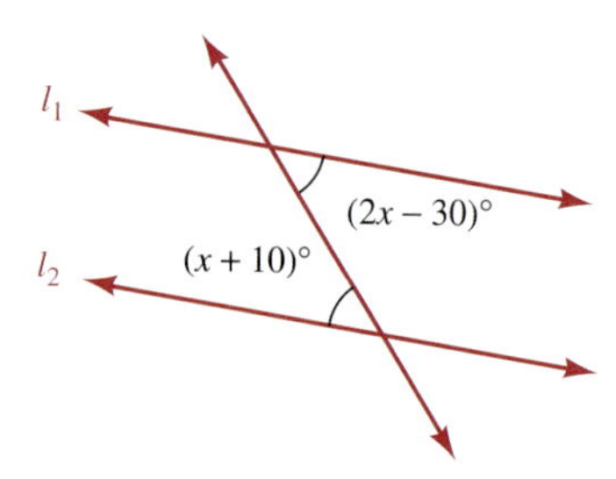

23. In the illustration, $l_1 \parallel l_2$. Find x.

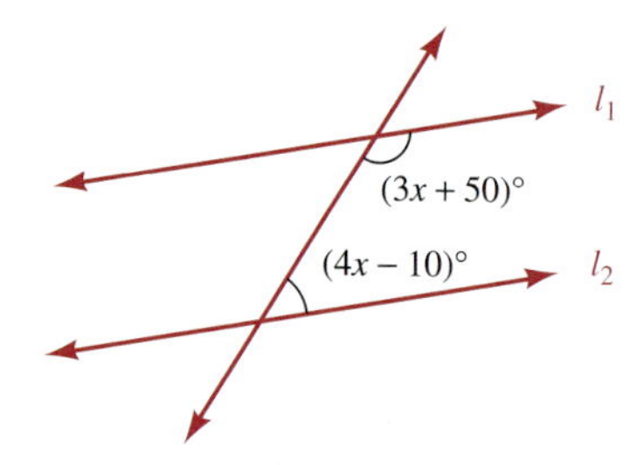

SECTION 9.3 Polygons

A *polygon* is a closed geometric figure. The points at which the sides intersect are called *vertices.*

A *regular polygon* has sides that are all the same length and angles that are all the same measure.

Polygons are classified as follows:

Number of sides	Name
3	Triangle
4	Quadrilateral
5	Pentagon
6	Hexagon
8	Octagon

An *equilateral triangle* has three sides of equal length.

An *isosceles triangle* has at least two sides of equal length.

A *scalene triangle* has no sides of equal length.

A *right triangle* has one right angle.

Identify each polygon as a triangle, a quadrilateral, a pentagon, a hexagon, or an octagon.

24.

25.

26.

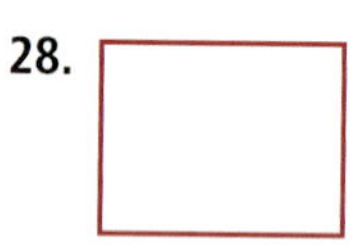

27.

28.

Give the number of vertices of each polygon.

29. Triangle

30. Quadrilateral

31. Octagon

32. Hexagon

Classify each of the triangles as an equilateral triangle, an isosceles triangle, a scalene triangle, or a right triangle.

33.

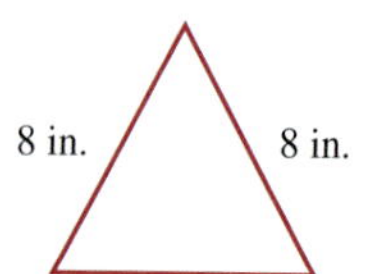

34.

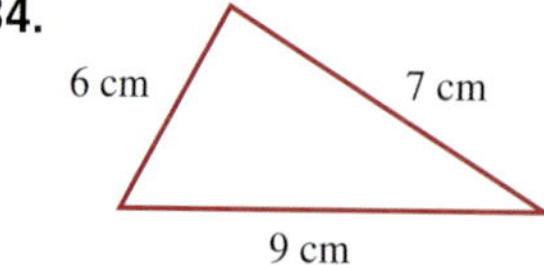

In an isosceles triangle, the angles opposite the sides of equal length are called *base angles.* The third angle is called the *vertex angle.* The third side is called the *base.*

Properties of isosceles triangles:

1. The base angles are congruent.
2. If two angles in a triangle are congruent, the sides opposite the angles are congruent, and the triangle is isosceles.

The sum of the measures of the angles of any triangle is 180°.

Quadrilaterals are classified as follows:

Property	Name
Opposite sides parallel	Parallelo-gram
Parallelogram with four right angles	Rectangle
Rectangle with all sides equal	Square
Parallelogram with sides of equal length	Rhombus
Exactly two sides parallel	Trapezoid

35.

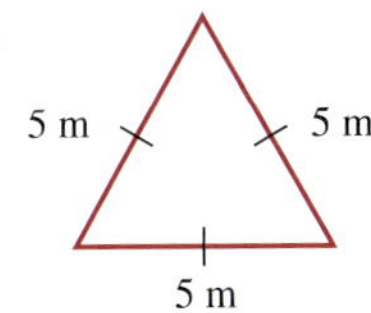

36.

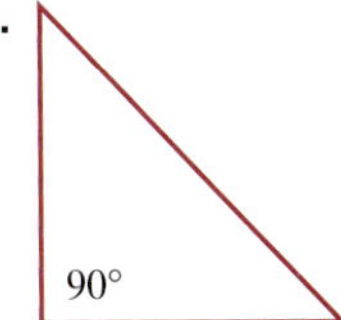

Determine whether each triangle is isosceles.

37.

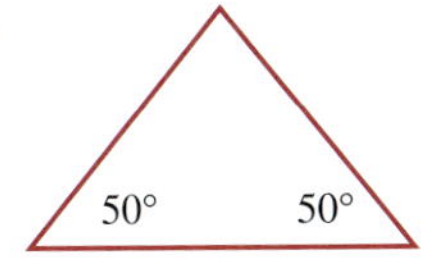

38.

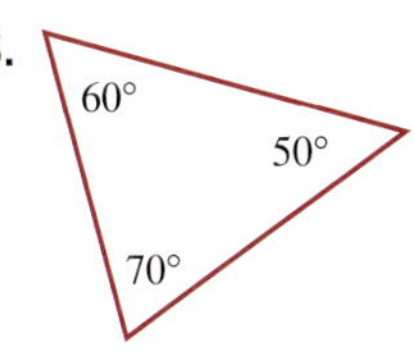

In each triangle, find x.

39.

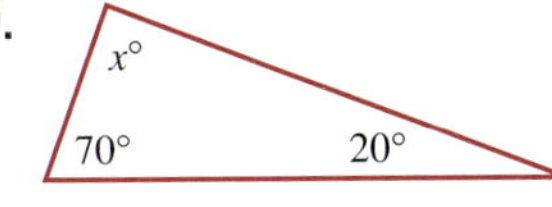

40.

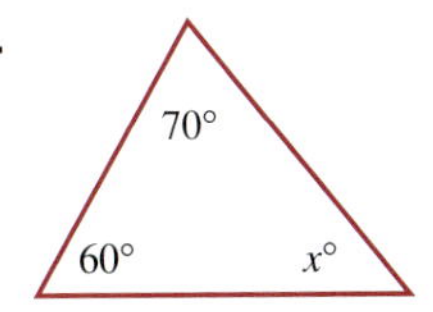

41. If one base angle of an isosceles triangle measures 65°, how large is the vertex angle?

42. If one base angle of an isosceles triangle measures 60°, what can you conclude about the triangle?

Classify each quadrilateral as a parallelogram, a rectangle, a square, a rhombus, or a trapezoid.

43.

44.

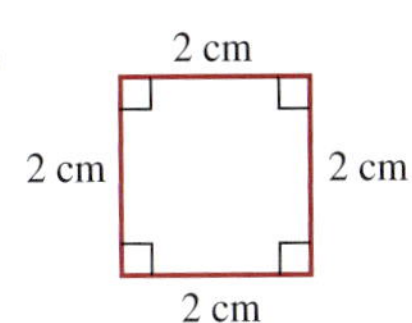

45.

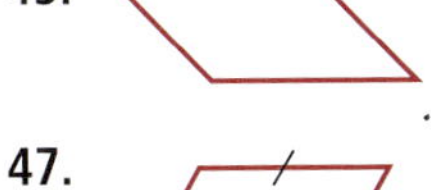

46.

47.

48.

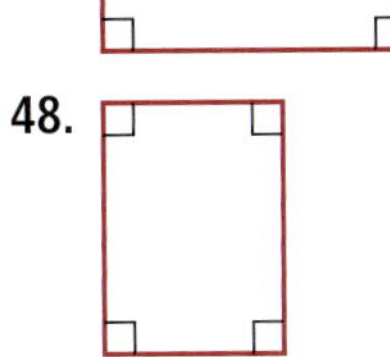

Properties of rectangles:

1. All angles are right angles.
2. Opposite sides are parallel.
3. Opposite sides are of equal length.
4. Diagonals are of equal length.
5. If the diagonals of a parallelogram are of equal length, the parallelogram is a rectangle.

In the illustration below, the length of diagonal $\overline{AC}$ of rectangle ABCD is 15 centimeters. Find each measure.

49. $m(\overline{BD})$ **50.** $m(\angle 1)$ **51.** $m(\angle 2)$

In the illustration below, ABCD is a rectangle. Classify each statement as true or false.

52. $m(\overline{AB}) = m(\overline{DC})$

53. $m(\overline{AD}) = m(\overline{DC})$

54. Triangle ABE is isosceles.

55. $m(\overline{AC}) = m(\overline{BD})$

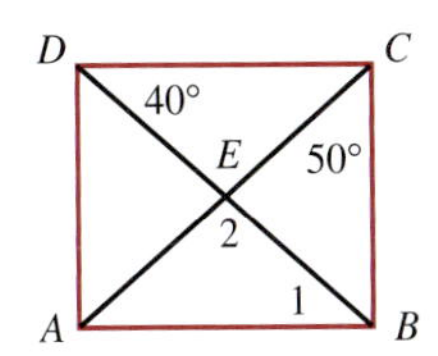

The parallel sides of a trapezoid are called *bases*. The nonparallel sides are called *legs*. If the legs of a trapezoid are of equal length, it is *isosceles*. In an isosceles trapezoid, the angles opposite the sides of equal length are *base angles*, and they are congruent.

In the illustration, ABCD is an isosceles trapezoid. Find each measure.

56. $m(\angle B)$

57. $m(\angle C)$

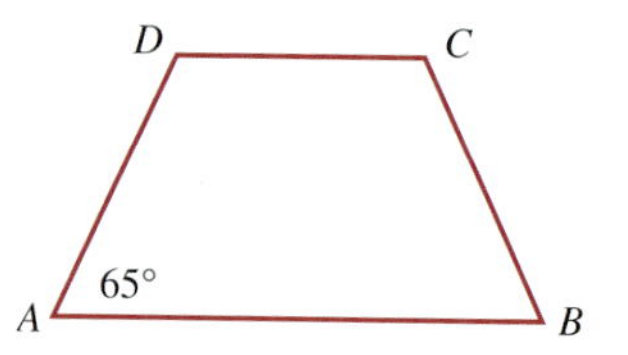

The sum of the measures of the angles of a polygon (in degrees) is given by the formula

$$S = (n - 2)180$$

Find the sum of the angle measures of each polygon.

58. Quadrilateral

59. Hexagon

SECTION 9.4 Properties of Triangles

If two triangles have the same size and the same shape, they are *congruent triangles*.

Corresponding parts of congruent triangles have the same measure.

60. See the illustration. Complete the list of corresponding parts.

a. $\angle A$ corresponds to ______

b. $\angle B$ corresponds to ______

c. $\angle C$ corresponds to ______

d. $\overline{AC}$ corresponds to ______

e. $\overline{AB}$ corresponds to ______

f. $\overline{BC}$ corresponds to ______

C F

A B E D

Three ways to show that two triangles are congruent are

1. The SSS property
2. The SAS property
3. The ASA property

Determine whether the triangles in each pair are congruent. If they are, tell why.

61.

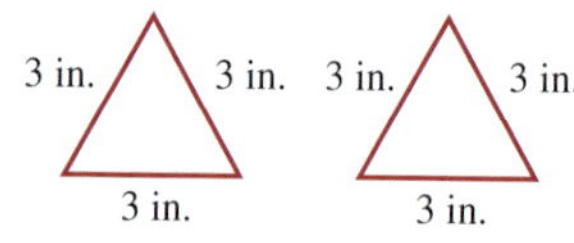

62.

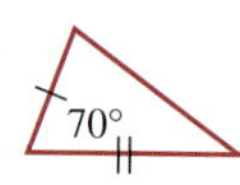

70°

63.

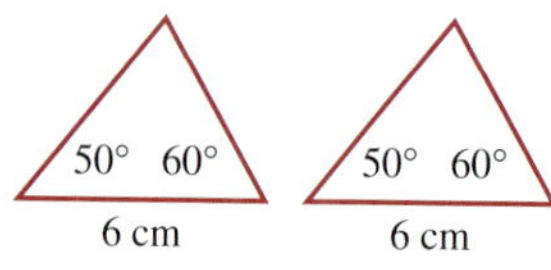

64.

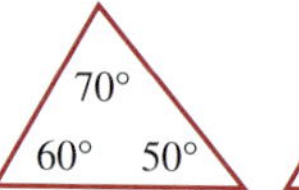

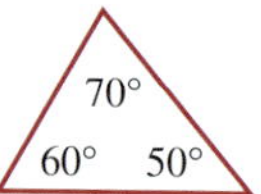

If two triangles have the same shape, they are said to be *similar*. If two angles of one triangle have the same measure as two angles of a second triangle, the triangles are similar.

Determine whether the triangles in each pair are similar.

65.

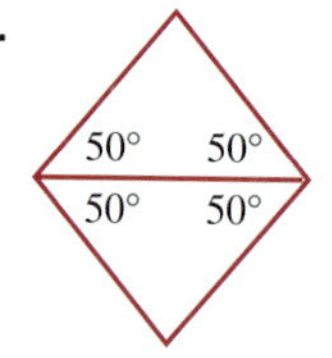

66.

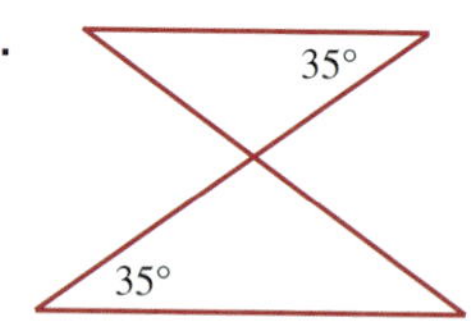

67. If a tree casts a 7-foot shadow at the same time a man 6 feet tall casts a 2-foot shadow, how tall is the tree?

The Pythagorean theorem: If the length of the *hypotenuse* of a right triangle is c, and the lengths of its legs are a and b, then

$$a^2 + b^2 = c^2$$

Refer to the illustration below and find the length of the unknown side.

68. If $a = 5$ and $b = 12$, find c.

69. If $a = 8$ and $c = 17$, find b.

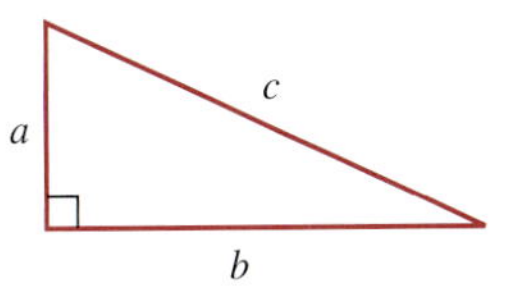

70. To the nearest tenth, find the height of the television screen shown.

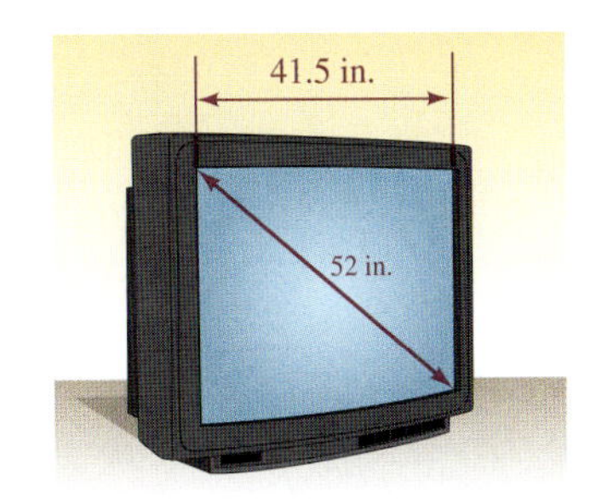

SECTION 9.5 Perimeters and Areas of Polygons

The *perimeter* of a polygon is the distance around it.

71. Find the perimeter of a square with sides 18 inches long.

72. Find the perimeter of a rectangle that is 3 meters long and 1.5 meters wide.

Find the perimeter of each polygon.

73.

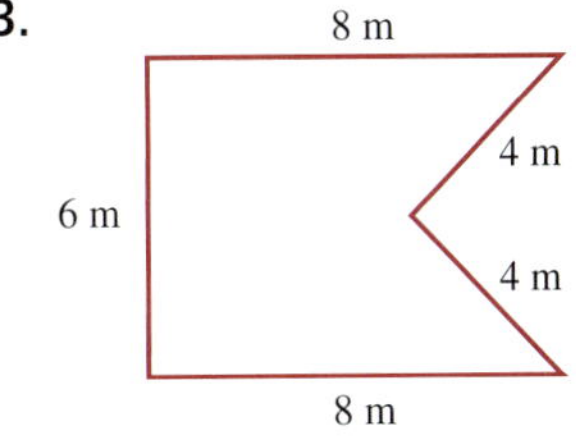

74.

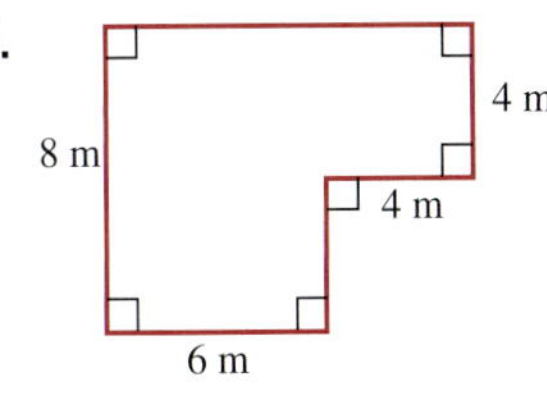

The *area* of a polygon is the measure of the surface it encloses.

Formulas for area:

Figure	Area
Square	s^2
Rectangle	lw
Parallelogram	bh
Triangle	$\frac{1}{2}bh$
Trapezoid	$\frac{1}{2}h(b_1 + b_2)$

Find the area of each polygon.

75.

3.1 cm

3.1 cm 3.1 cm

3.1 cm

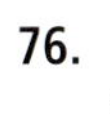

76.

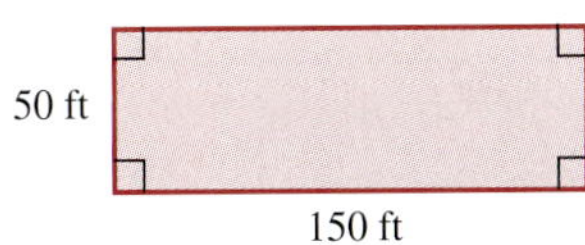

77.

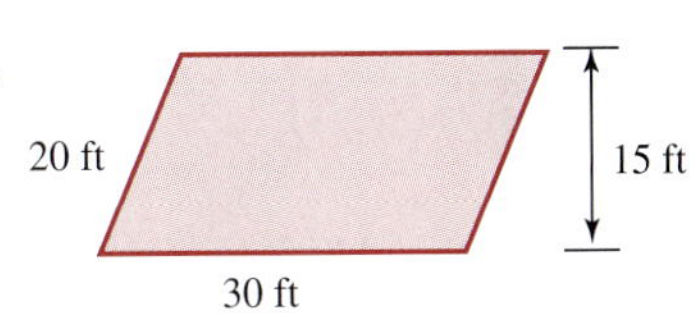

78.

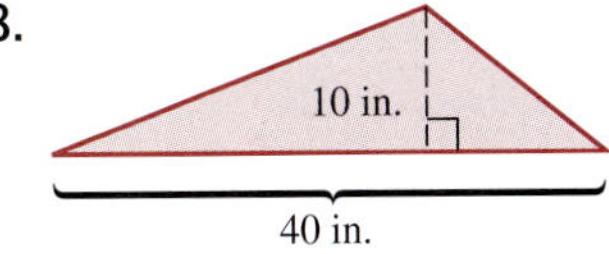

79.

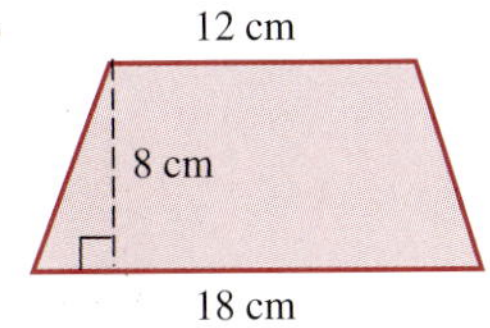

80.

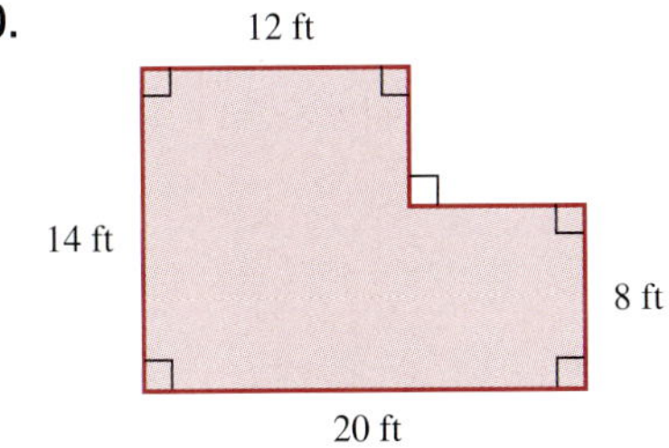

81.

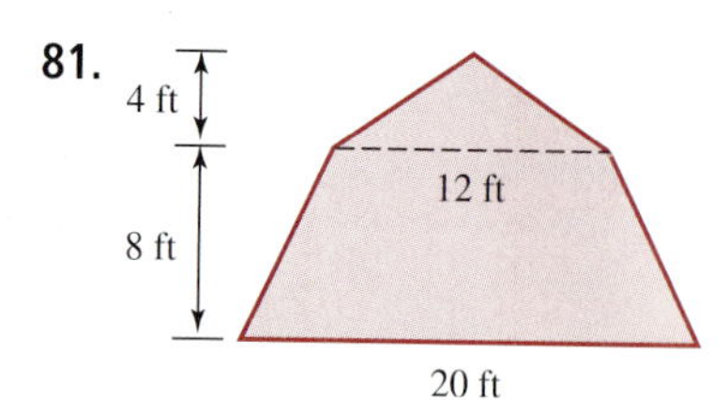

82.

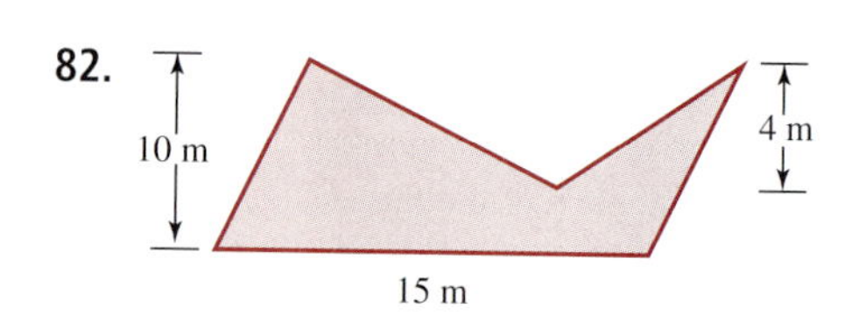

83. How many square feet are in 1 square yard?

84. How many square inches are in 1 square foot?

SECTION 9.6 Circles

A *circle* is the set of all points in a plane that lie a fixed distance from a point called its *center.* The fixed distance is the circle's *radius.*

A chord of a circle is a line segment connecting two points on the circle.

A diameter is a chord that passes through the circle's center.

The *circumference* (perimeter) of a circle is given by the formulas

$C = \pi D$ or $C = 2\pi r$

$\pi = 3.14159\ldots$

The *area* of a circle is given by the formula

$A = \pi r^2$

Refer to the illustration.

85. Name each chord.

86. Name each diameter.

87. Name each radius.

88. Name the center.

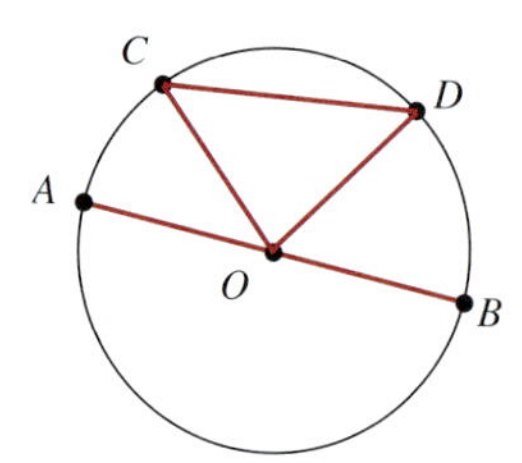

Find each answer to the nearest tenth.

89. Find the circumference of a circle with a diameter of 21 centimeters.

90. Find the perimeter of the figure shown.

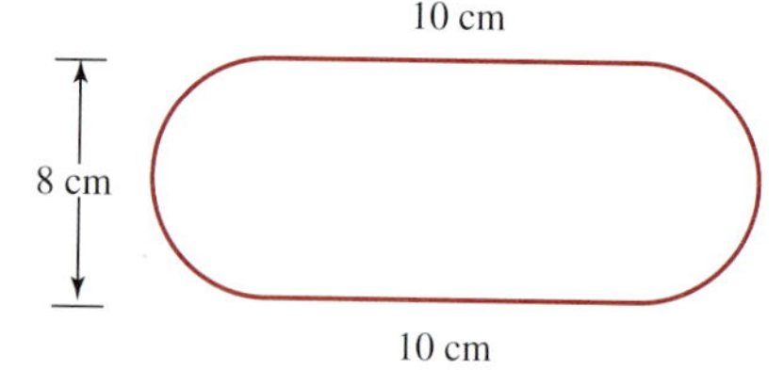

91. Find the area of a circle with a diameter of 18 inches.

92. Find the area of the figure shown for Problem 90.

SECTION 9.7 Surface Area and Volume

The *volume* of a solid is a measure of the space it occupies.

Find the volume of each solid to the nearest unit.

93.

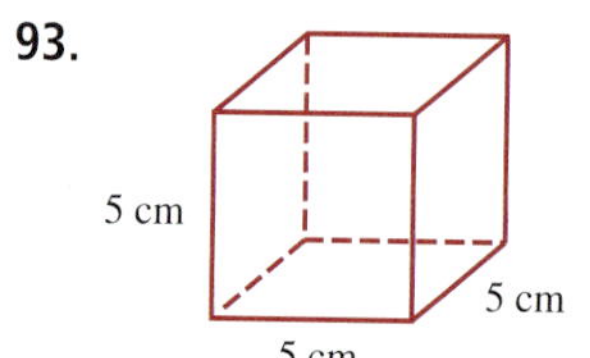

94.

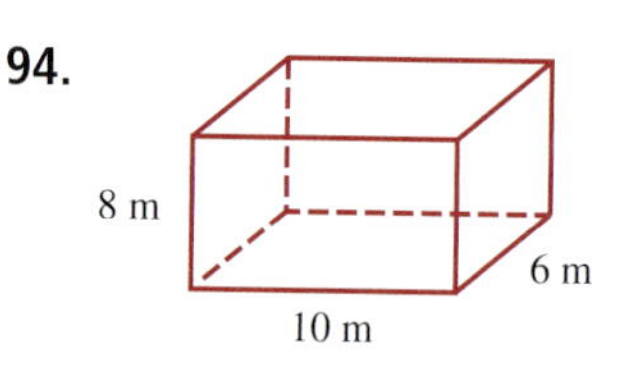

Formulas for volume:

Figure	Volume
Cube	s^3
Rectangular solid	lwh
Prism	Bh*
Sphere	$\frac{4}{3}\pi r^3$
Cylinder	$\pi r^2 h$
Cone	$\frac{1}{3}\pi r^2 h$
Pyramid	$\frac{1}{3}Bh$*

*B represents the area of the base.

The *surface area* of a rectangular solid is the sum of the areas of its six faces.

The surface area of a sphere is given by the formula

$$SA = 4\pi r^2$$

95.

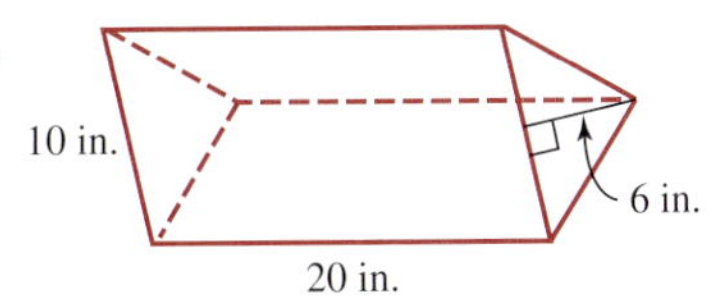

96.

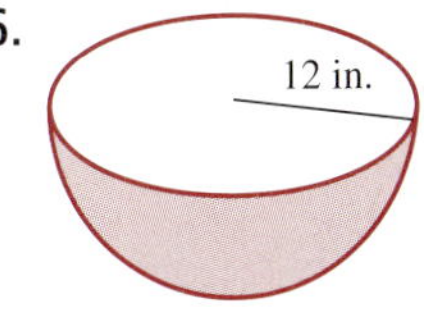

97.

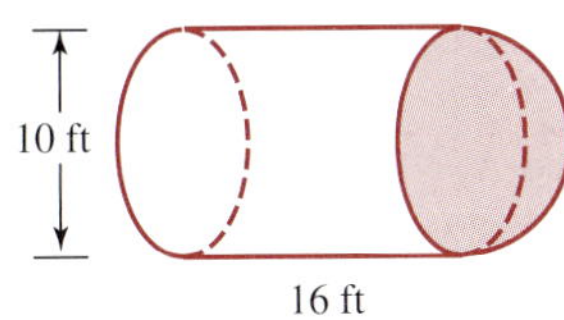

98.

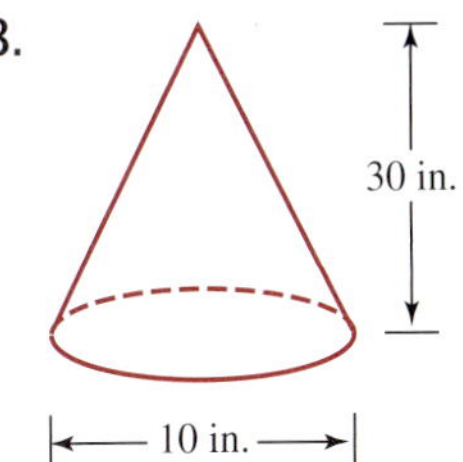

99.

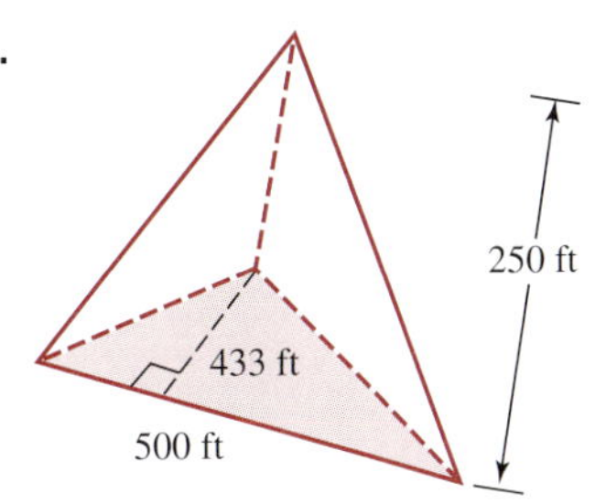

100.

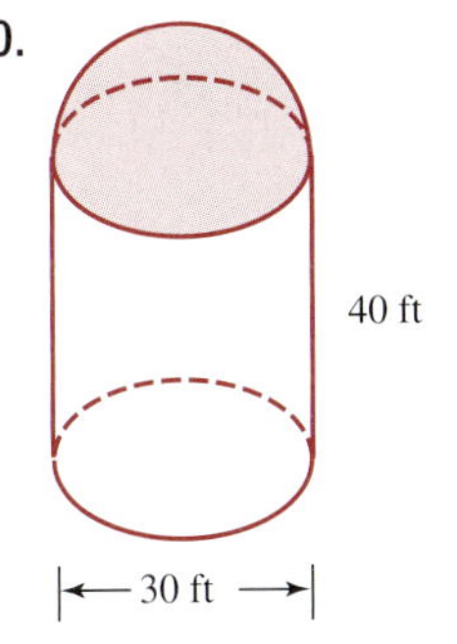

101. How many cubic inches are in 1 cubic foot?

102. How many cubic feet are in 2 cubic yards?

To the nearest tenth, find the surface area of each solid.

103.

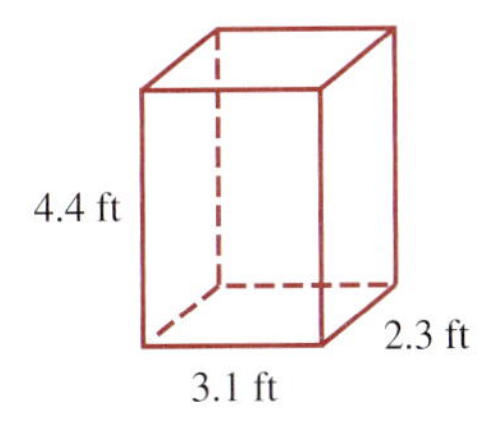

104.

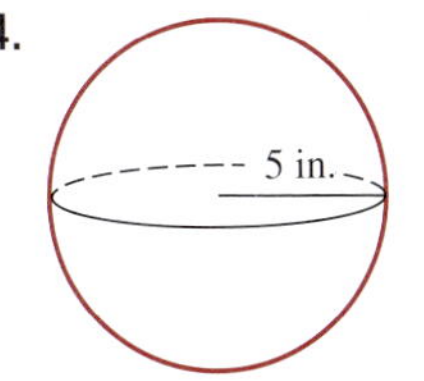

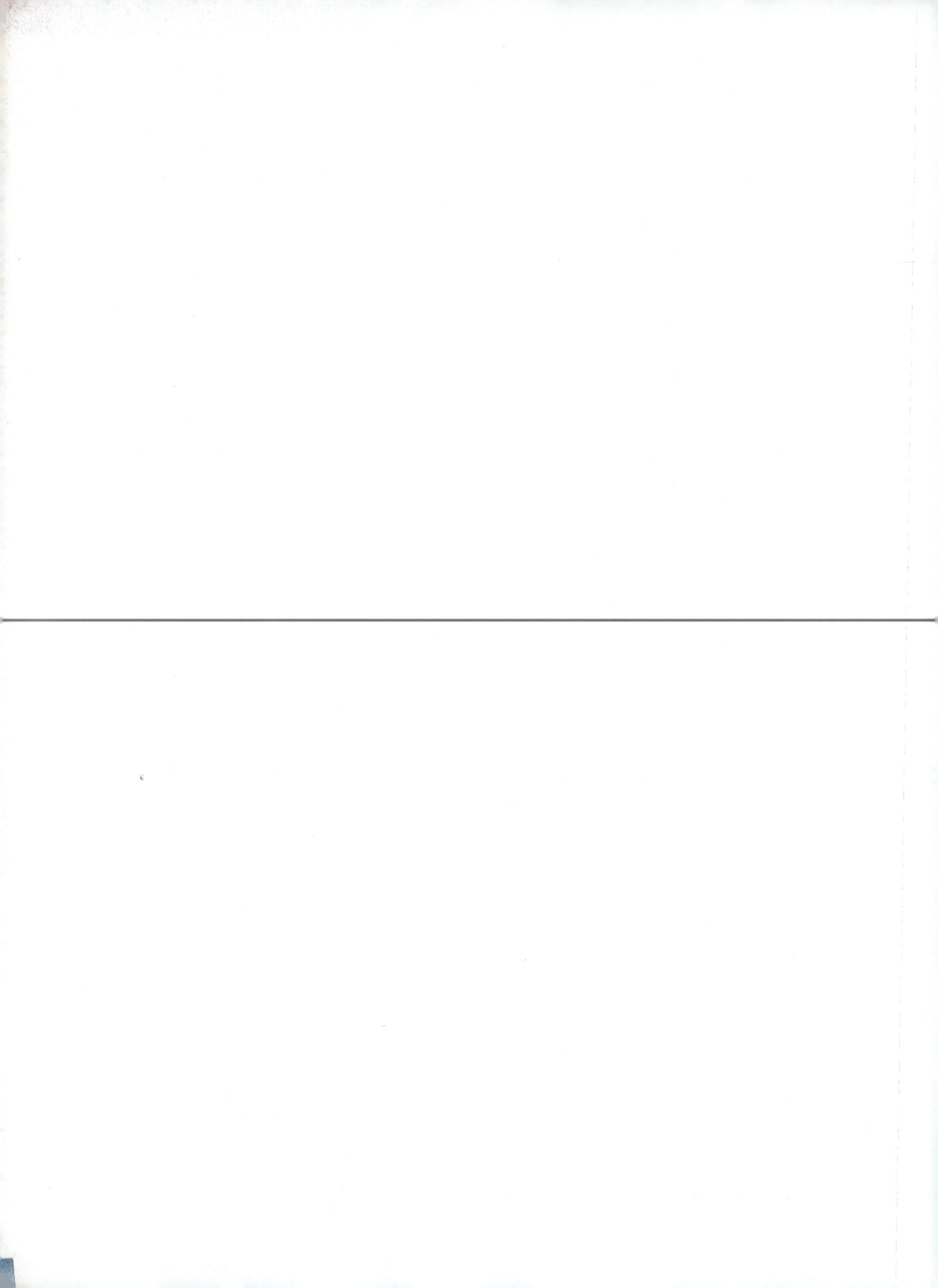

CHAPTER 9 TEST

1. Find $m(\overline{AB})$.

2. Which point is the vertex of $\angle ABC$?

Determine whether each statement is true or false.

3. An angle of 47° is an acute angle.

4. An angle of 90° is a straight angle.

5. An angle of 180° is a right angle.

6. An angle of 132° is an obtuse angle.

7. Find x.

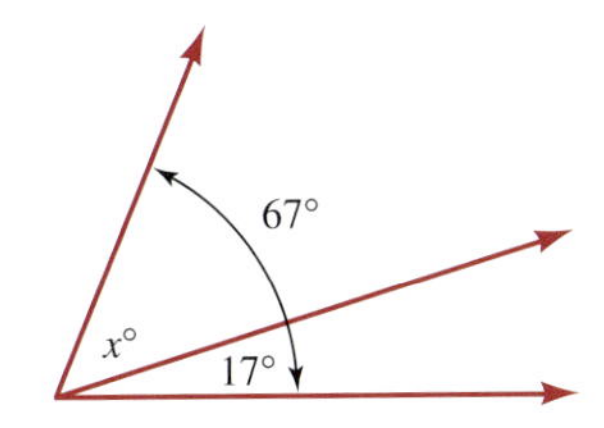

8. Find y.

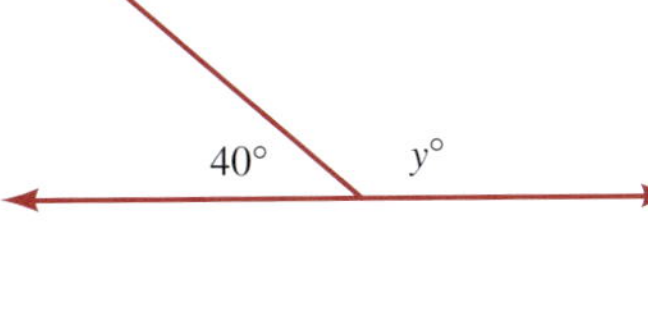

9. Find y.

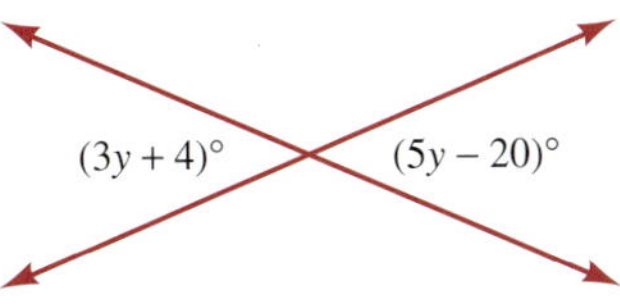

10. CALLIGRAPHY The illustration shows how the tip of the pen should be held at a 45° angle to the horizontal. What is x?

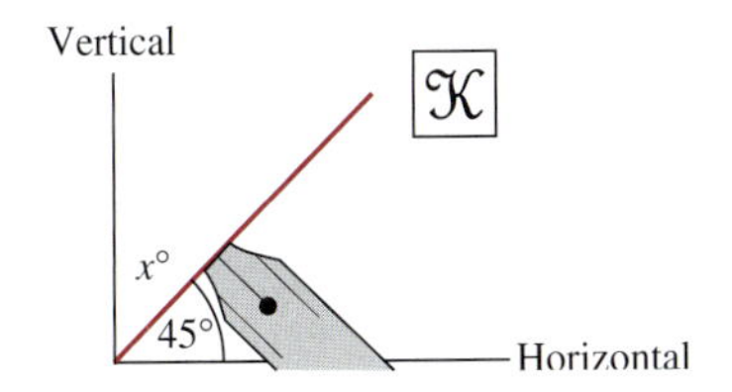

11. Find the complement of an angle measuring 67°.

12. Find the supplement of an angle measuring 117°.

Refer to the illustration below, in which $l_1 \parallel l_2$.

13. $m(\angle 1) =$ ____.

14. $m(\angle 2) =$ ____.

15. $m(\angle 3) =$ ____.

16. Find x.

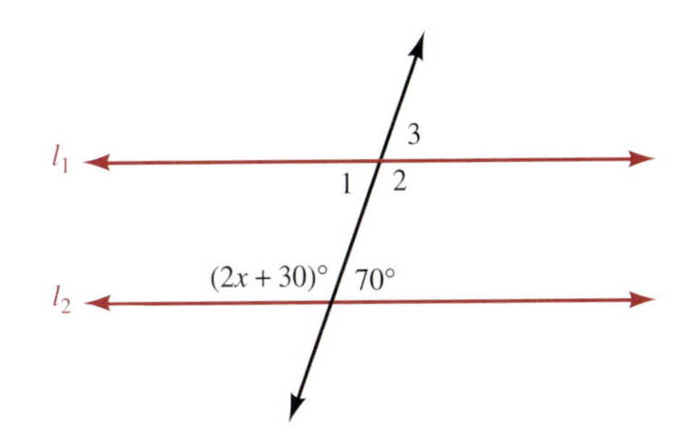

17. Complete the table.

Polygon	Number of sides
Triangle	
Quadrilateral	
Hexagon	
Pentagon	
Octagon	

18. Complete the table about triangles.

Property	Kind of triangle
All sides of equal length	
No sides of equal length	
Two sides of equal length	

Refer to the illustration.

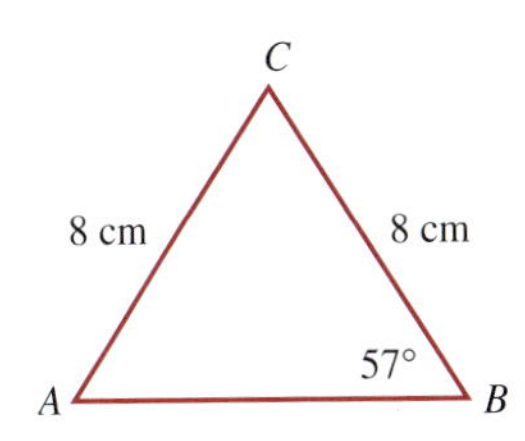

19. Find $m(\angle A)$.

20. Find $m(\angle C)$.

21. If the measures of two angles in a triangle are 65° and 85°, find the measure of the third angle.

22. Find the sum of the measures of the angles in a decagon (a ten-sided polygon).

23. In the illustration, $ABCD$ is a rectangle. Name three pairs of segments with equal lengths.

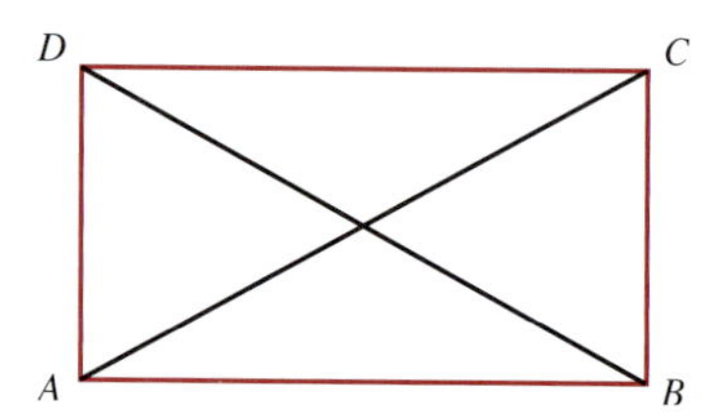

24. In the illustration, $ABCD$ is an isosceles trapezoid. Find x.

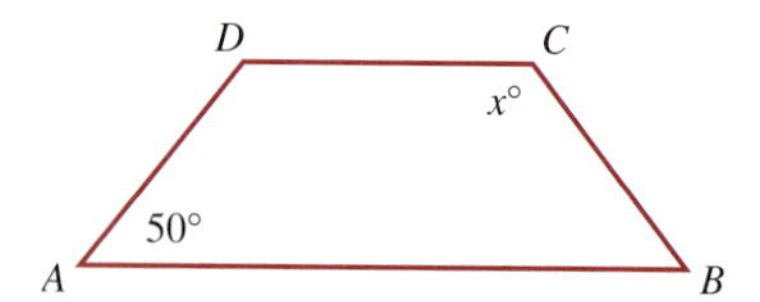

Refer to the illustration, in which $\triangle ABC \cong \triangle DEF$.

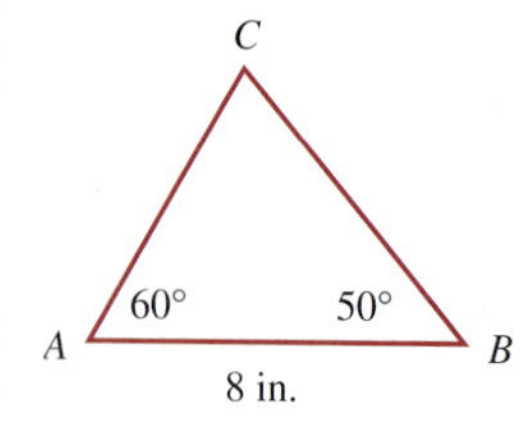

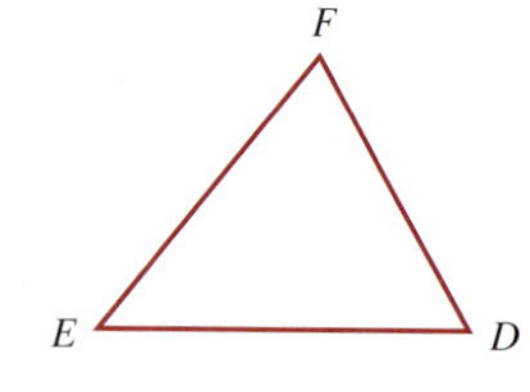

25. Find $m(\overline{DE})$.

26. Find $m(\angle E)$.

Refer to the illustration below, in which $m(\angle A) = m(\angle D)$ and $m(\angle C) = m(\angle F)$

27. Find x.

28. Find y.

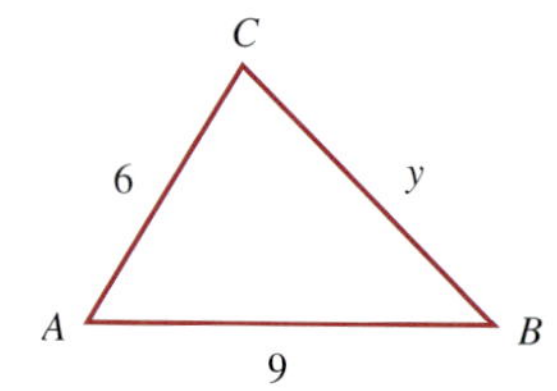

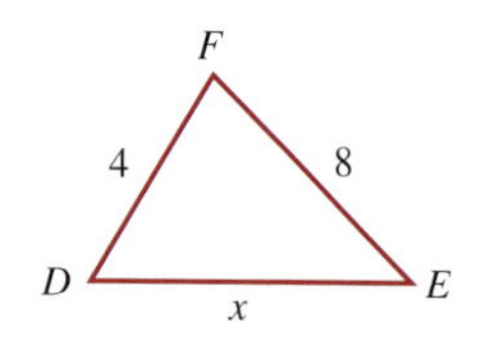

Give each answer to the nearest tenth.

29. A baseball diamond is a square with each side 90 feet long. What is the straight-line distance from third base to first base?

30. Find the area of a triangle with a base 44.5 centimeters long and a height of 17.6 centimeters.

31. Find the area of a trapezoid with a height of 6 feet and bases that are 12.2 feet and 15.7 feet long.

32. THE OLYMPICS A steel rod is to be bent to form the interlocking rings of the Olympic Games symbol. How many feet of steel rod will be needed to make the symbol if the diameter of each ring is to be 6 feet?

33. Find the area of a circle with a diameter that is 6 feet long.

34. Find the volume of a rectangular solid with dimensions 4.3 by 5.7 by 6.5 meters.

35. Find the volume of a sphere that is 8 meters in diameter.

36. Find the volume of a 10-foot-tall pyramid that has a rectangular base 5 feet long and 4 feet wide.

37. Give a real-life example in which the concept of perimeter is used. Do the same for area and for volume. Be sure to discuss the type of units used in each case.

38. Draw a cube. Explain how to find its surface area.

CHAPTERS 1–9 CUMULATIVE REVIEW EXERCISES

1. AMUSEMENT PARKS Use the data in the table to construct a bar graph on the illustration.

Fatal accidents on amusement park rides

Year	'93	'94	'95	'96	'97	'98	'99	'00	'01	'02
Number	4	2	4	3	4	7	6	1	3	2

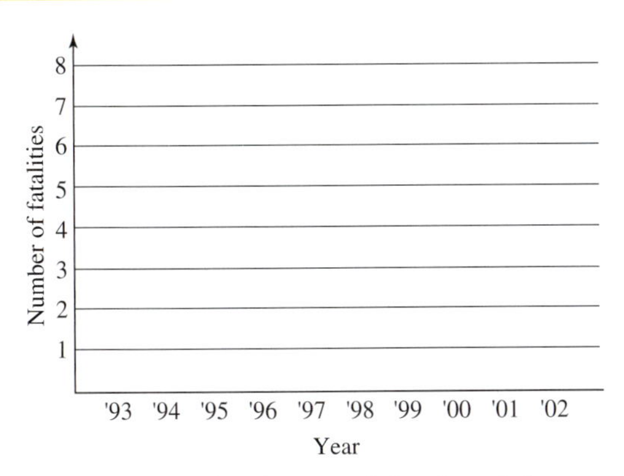

Source: U.S. Product Consumer Safety Commission

2. USED CARS The following ad appeared in *The Car Trader.* (O.B.O. means "or best offer.")

> 1969 Ford Mustang. New tires
> Must sell!!!! $10,500 O.B.O.

If offers of $8,750, $8,875, $8,900, $8,850, $8,800, $7,995, $8,995, and $8,925 were received, what was the selling price of the car?

3. Subtract: 35,021 − 23,999.
4. Divide: 1,353 ÷ 41.
5. Round 2,109,567 to the nearest thousand.
6. Prime factor 220.
7. Find all the factors of 24.
8. List the set of integers.
9. Evaluate: $-10(-2) - 2^3 + 1$.
10. Evaluate: $5 - 3[4^2 - (1 + 5 \cdot 2)]$.
11. Evaluate: $|-6 - (-3)|$.
12. Evaluate the expression $\frac{2x + 3y}{z - y}$ for $x = 2$, $y = -3$, and $z = -4$.
13. What is the difference between an equation and an expression?
14. Simplify: $4x - 2(3x - 4) - 5(2x)$.

Solve each equation and check each result.

15. $3(p + 15) + 4(11 - p) = 0$
16. $5t - 7 = 7t + 13$
17. $-x + 2 = 13$
18. $4 + \frac{x}{5} - 6 = -1$
19. SNAILS According to the *Guinness Book of World Records,* in the 1995 World Snail Racing Championships, a snail covered a 13-inch course in 2 minutes. What was the snail's rate in inches per minute?
20. SHOPPING What is the value of x coupons, each of which gives the shopper 50¢ off?
21. Translate into mathematical symbols: 5 less than a number.
22. LUMBER To find the number of board feet (b.f.) in a piece of lumber, use the formula

$$\text{b.f.} = \frac{\text{thickness (in.)} \cdot \text{width (in.)} \cdot \text{length (ft)}}{12}$$

Find the number of board feet in the piece of lumber shown. (*Hint:* The symbol ″ stands for inches and the symbol ′ stands for feet.)

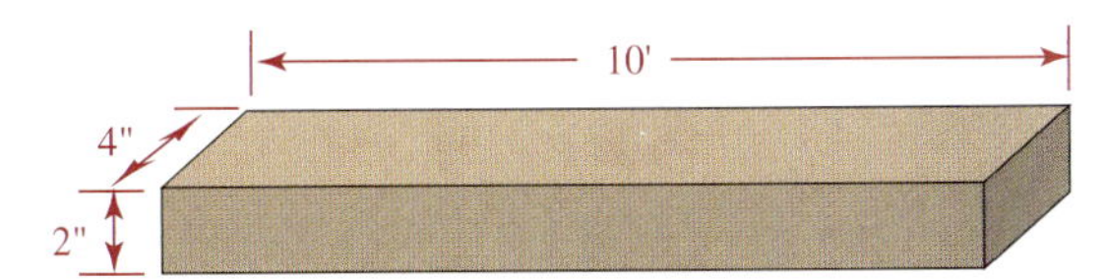

23. Simplify: $\frac{35a^2}{28a}$.
24. Add: $45\frac{2}{3} + 96\frac{4}{5}$.
25. Subtract: $\frac{x}{4} - \frac{3}{5}$.
26. BAKING A 5-pound bag of all-purpose flour contains $17\frac{1}{2}$ cups. A baker uses $3\frac{3}{4}$ cups. How much flour is left?
27. Multiply: $-\frac{6}{25}\left(2\frac{7}{24}\right)$.
28. Divide: $\frac{15}{8q^4} \div \frac{45}{8q^3}$.

29. PET MEDICINES A pet owner was told to use an eye dropper to administer medication to his sick kitten. The cup shown contains 8 doses of the medication. Determine the size of a single dose.

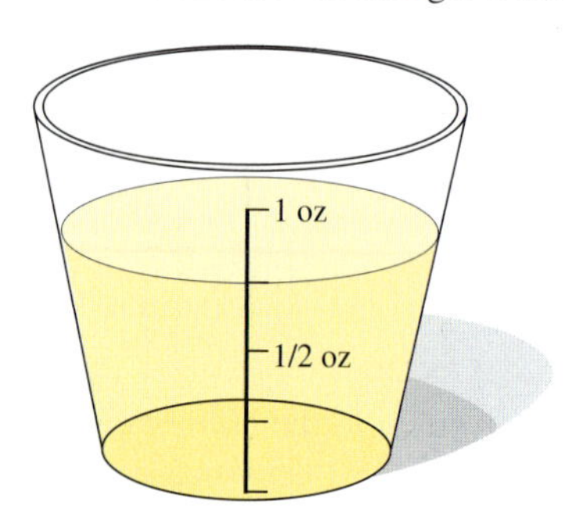

30. Solve: $\frac{x}{2} - \frac{1}{9} = \frac{1}{3}$.

31. Solve: $\frac{2}{3}q - 1 = -6$.

32. Evaluate: $\frac{3}{4} + \left(-\frac{1}{3}\right)^2\left(\frac{5}{4}\right)$.

33. Simplify: $\frac{7 - \frac{2}{3}}{4\frac{5}{6}}$.

34. GRAVITY Objects on the moon weigh only one-sixth as much as on Earth. If a rock weighs 3 ounces on the moon, how much does it weigh on Earth?

35. GLOBAL WARMING The graph below shows the annual mean global temperature change, as measured by satellites orbiting the Earth.

a. When was the greatest rise in temperature recorded? What was it?

b. When was the greatest decline in temperature recorded? What was it?

Change In Mean Global Surface Temperature

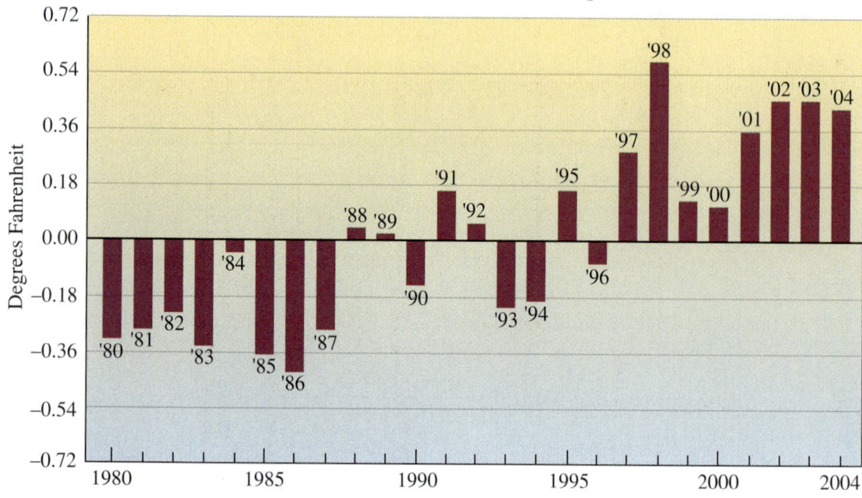

Source: National Oceanic and Atmospheric Administration

36. Graph: $\left\{-4\frac{5}{8}, \sqrt{17}, 2.89, \frac{2}{3}, -0.1, -\sqrt{9}, \frac{3}{2}\right\}$.

−5 −4 −3 −2 −1 0 1 2 3 4 5

37. Round the number pi to the nearest ten thousandth: $\pi = 3.141592654. . . .$

38. Place the proper symbol (> or <) in the blank: 154.34 ___ 154.33999.

39. Add: $3.4 + 106.78 + 35 + 0.008$.

40. Multiply: $-5.5(-3.1)$.

41. Multiply: $(89.9708)(1,000)$.

42. Divide: $\frac{0.0742}{1.4}$.

43. Evaluate: $-8.8 + (-7.3 - 9.5)$.

44. Evaluate: $\frac{7}{8}(9.7 + 15.8)$.

45. Change $\frac{2}{15}$ to a decimal.

46. Evaluate $\frac{(-1.3)^2 + 6.7}{-0.9}$ and round to the nearest hundredth.

47. DECORATIONS A mother has budgeted \$20 for decorations for her daughter's birthday party. She decides to buy a tank of helium for \$15.15 and some balloons. If the balloons sell for 5 cents apiece, how many balloons can she buy?

48. Solve: $1.7y + 1.24 = -1.4y - 0.62$.

49. INTERNET An Internet map Web site listed the directions for a trip as shown below. What is the total distance traveled in miles?

WEBQUEST

	DIRECTIONS	DISTANCE
START	1: Start out going Southeast on ANACAPA ST.	0.26 miles
	2: Turn LEFT onto E HALEY ST.	0.19 miles
	3: Turn RIGHT onto GARDEN ST.	0.15 miles
101	4: Merge onto US-101 N.	230.11 miles
EXIT	5: Take the exit toward MONTEREY PENINSULA.	0.21 miles
	6: Turn RIGHT onto S SANBORN RD.	0.81 miles
	7: Turn SLIGHT RIGHT onto E BLANCO RD.	1.59 miles
	8: Turn LEFT onto S MAIN ST/CA-68/SALINAS MONTEREY HWY. Continue to follow CA-68 W.	15.46 miles
	9: Turn SLIGHT LEFT to take the CA-1 S/CA-68	0.16 miles
END		
	Total Estimated Time: 4 hours, 14 minutes	Total Distance: ??? miles

50. Fill in the blank: $\sqrt{64} = 8$, because ____ $= 64$.

51. Evaluate: $2\sqrt{121} - 3\sqrt{64}$.

52. Evaluate: $\sqrt{\frac{49}{81}}$.

53. TABLE TENNIS The weights (in ounces) of 8 ping-pong balls that are to be used in a tournament are listed below.

0.85	0.87	0.88	0.88
0.85	0.86	0.84	0.85

Find the mean, median, mode, and range of the weights.

54. Graph each point: $(-4, -3)$, $(1.5, 1.5)$, $(-\frac{7}{2}, 0)$, $(0, 3.5)$.

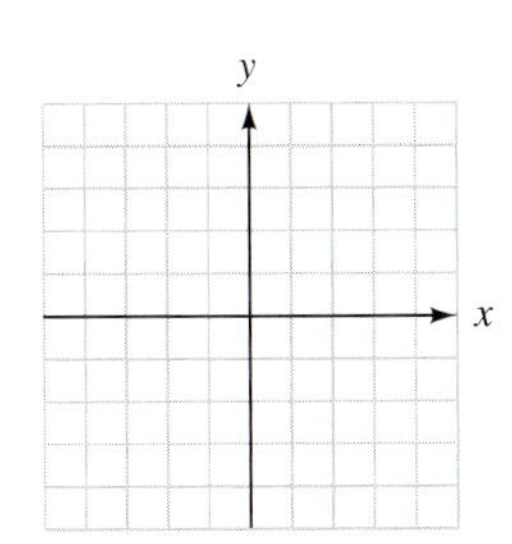

55. What are the coordinates of the origin?

56. Is $(-2, 1)$ a solution of $3x - y = -8$?

Graph each equation.

57. $y = -2x$

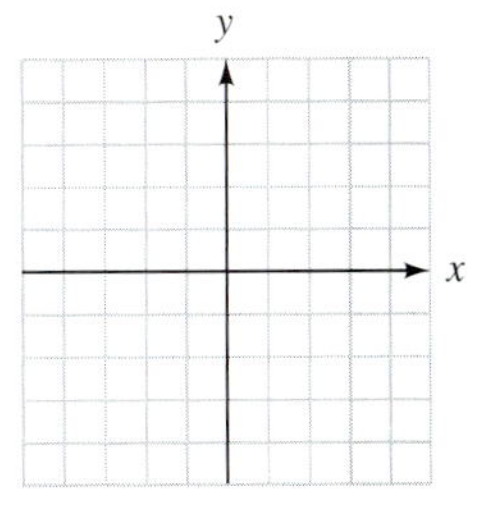

58. $2x - 3y = 14$

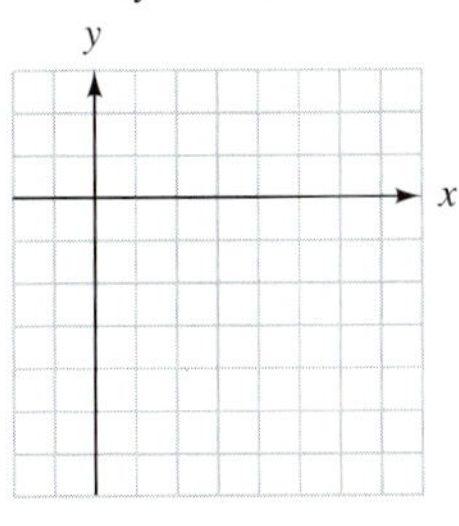

59. $x = 4$

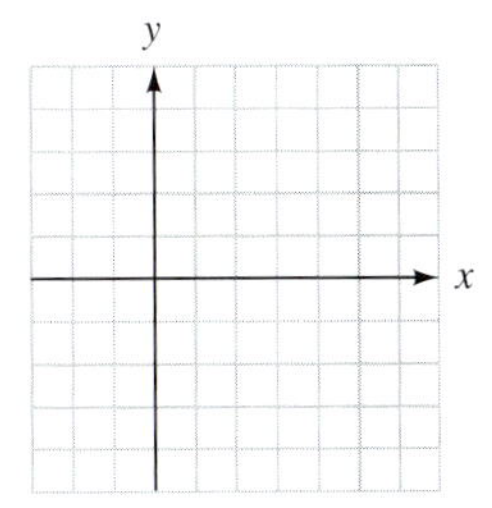

60. $y = x^2$

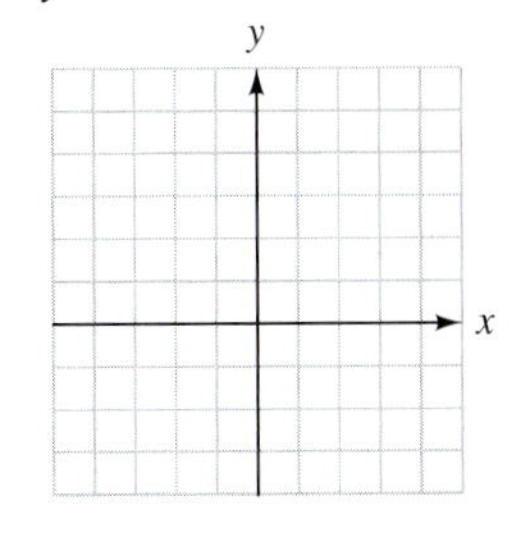

61. Evaluate $-3x^2 - 2x$ when $x = -2$.

62. a. Consider $(-3)^2$. What is the base and what is the exponent? Evaluate the expression.

b. Consider -3^2. What is the base and what is the exponent? Evaluate the expression.

Simplify each expression.

63. $s^4 \cdot s^5$

64. $(a^5)^7$

65. $-3h^9(-5h)$

66. $(2b^3c^6)^3$

67. $(y^5)^2(y^4)^3$

68. $x^m \cdot x^n$

69. Classify $3x^2 - 7x + 1$ as a monomial, a binomial, or a trinomial. Then give its degree.

Perform the operations.

70. $(5x^2 - 2x + 4) - (3x^2 - 5)$

71. $-3p(2p^2 + 3p - 4)$

72. $(3x + 5)(2x - 1)$

73. $(2y - 7)^2$

74. What percent of the figure in the illustration is shaded? What percent is not shaded?

75. Find 15% of 450.

76. 24.6 is 20.5% of what number?

77. Complete the table.

Percent	Decimal	Fraction
57%		
	0.001	
		$\frac{1}{3}$

78. STUDENT GOVERNMENTS In an election for Student Body President, 560 votes were cast. Stan Cisneros received 308 votes, and Amy Huang-Sims received 252 votes. Use a circle graph to show the percent of the vote received by each candidate.

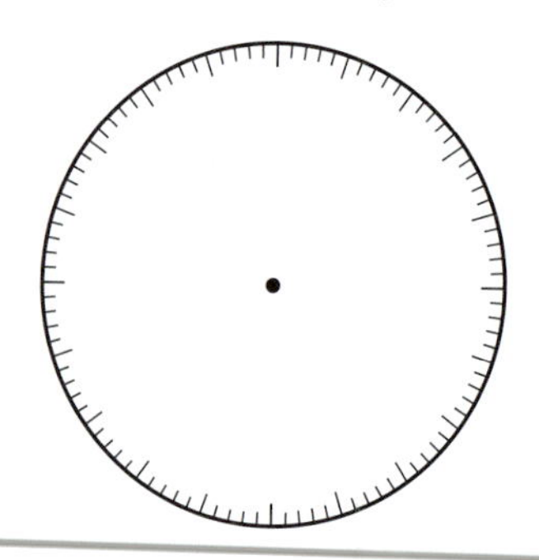

79. SHOPPING Find the regular price of the calculator shown below.

80. SALES TAXES If the sales tax rate is $6\frac{1}{4}\%$, how much sales tax will be added to the price of a new car selling for $18,550?

81. COLLECTIBLES A German Hummel porcelain figurine, which was originally purchased for $125, was sold by a collector ten years later for $750. What was the percent increase in the value of the figurine?

82. PAYING OFF A LOAN To pay for tuition, a college student borrows $1,500 for two months. If the annual interest rate is 9%, how much will the student have to repay when the loan comes due?

83. SAVING FOR RETIREMENT When he got married, a man invested $5,000 in an account that guaranteed to pay 8% interest, compounded monthly, for 50 years. At the end of 50 years, how much will his account be worth?

Write each phrase as a ratio.

84. 3 centimeters to 7 centimeters

85. 13 weeks to 1 year

86. COMPARISON SHOPPING A dry-erase whiteboard with an area of 400 in.^{2} sells for $24. A larger board, with an area of 600 in.^{2}, sells for $42. Which board is the better buy?

87. Solve the proportion: $\frac{5-x}{14} = \frac{13}{28}$.

88. INSURANCE CLAIMS In one year, an auto insurance company had 3 complaints per 1,000 policies. If a total of 375 complaints were filed that year, how many policies did the company have?

89. SCALE DRAWINGS Suppose the house plan shown below is drawn on a grid of $\frac{1}{4}$-inch squares. How long is the house?

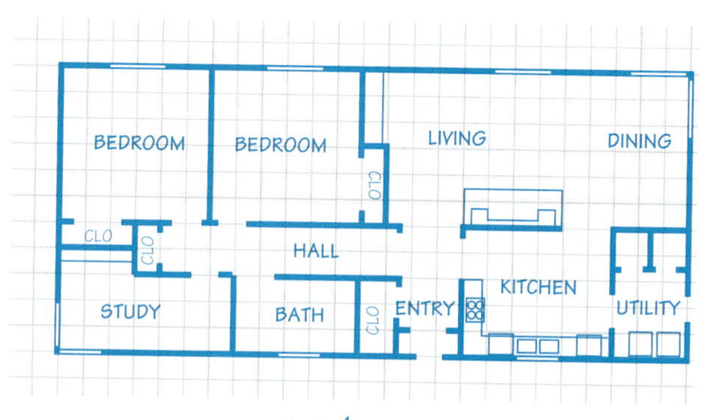

Scale $\frac{1}{4}$ in. : 3 ft

Make each conversion.

90. 168 inches = ▢ feet

91. 15 yards = ▢ inches

92. 212 ounces = ▢ pounds

93. 30 gallons = ▢ quarts

94. 25 cups = ▢ fluid ounces

95. 738 minutes = ☐ hours

96. 654 milligrams = ☐ centigrams

97. 500 milliliters = ☐ liter

98. 5,890 decimeters = ☐ dekameters

99. 75° C = ☐ F

100. THE AMAZON The Amazon River enters the Atlantic Ocean through a broad estuary, roughly estimated at 240,000 m in width. Convert the width to kilometers.

101. TENNIS A tennis ball weighs between 57 and 58 g. Express this range in centigrams.

102. OCEAN LINER When it was making the transatlantic cruises from England to America, the Queen Mary got 13 feet to the gallon.

a. How many meters a gallon is this?

b. The fuel capacity of the ship was 3,000,000 gallons. How many liters is this?

103. COOKING What is the weight of a 10-pound ham in kilograms?

104. How many degrees are in a right angle?

105. How many degrees are in an acute angle?

106. Find the supplement of an angle of 105°.

107. Find the complement of an angle of 75°.

Refer to the illustration, in which $l_1 \parallel l_2$. Find the measure of each angle.

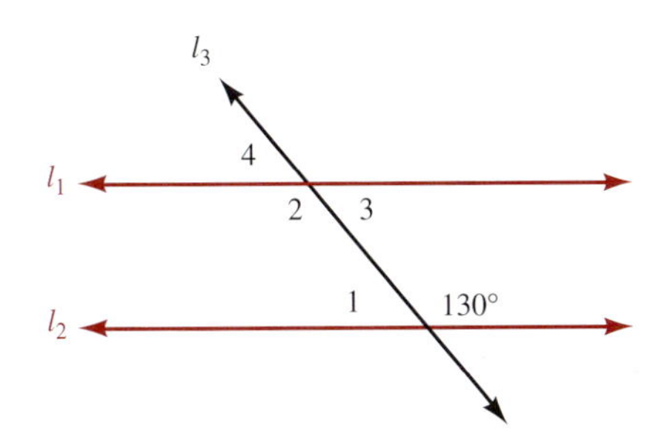

108. m(∠1) **109.** m(∠2)

110. m(∠3) **111.** m(∠4)

Refer to the illustration, in which $\overrightarrow{AB} \parallel \overline{DE}$ and $m(\overline{AC}) = m(\overline{BC})$. Find the measure of each angle.

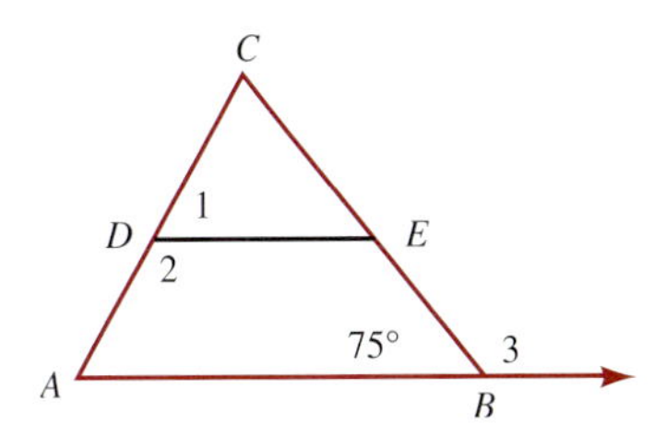

112. m(∠1) **113.** m(∠C)

114. m(∠2) **115.** m(∠3)

116. JAVELIN THROW In the illustration, determine x and y.

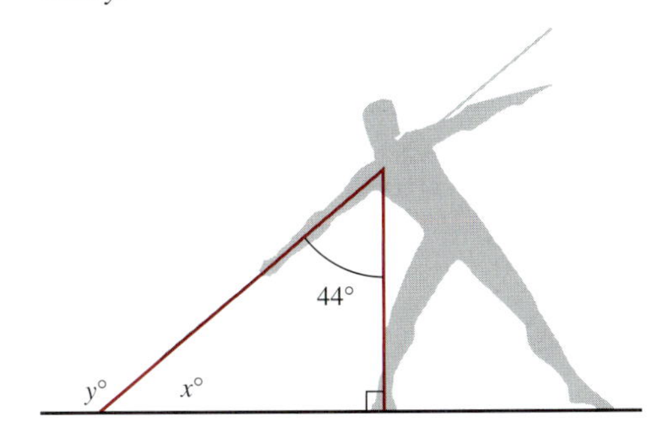

117. Find the sum of the angles of a pentagon.

118. If two sides of a right triangle measure 5 meters and 12 meters, how long is the hypotenuse?

If an answer is not exact, round to the nearest hundredth.

119. Find the perimeter and area of a rectangle with dimensions of 9 meters by 12 meters.

120. Find the area of a triangle with a base that is 14 feet long and an altitude of 18 feet.

121. Find the area of a trapezoid that has bases that are 12 inches and 14 inches long and a height of 7 inches.

122. Find the circumference and area of a circle with a diameter of 14 centimeters.

123. Find the area of the shaded region below, which is created using two semicircles.

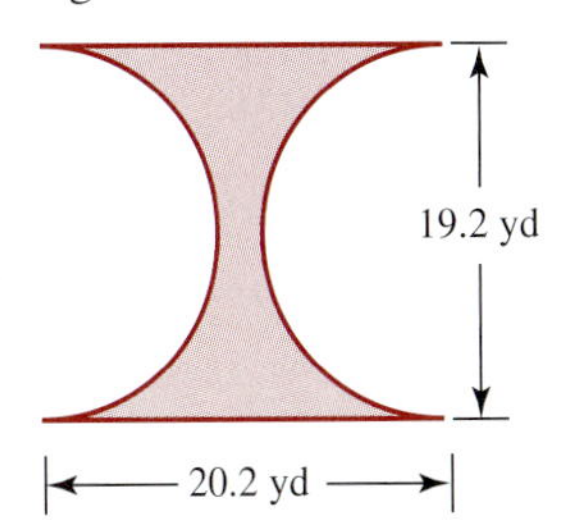

124. Find the volume of a rectangular solid with dimensions of 5 meters by 6 meters by 7 meters.

125. Find the volume of a sphere with a diameter of 10 inches.

126. Find the volume of a cone that has a circular base 8 meters in diameter and a height of 9 meters.

127. Find the volume of a cylindrical pipe that is 20 feet long and 6 inches in diameter.

128. Find the surface area of a block of ice that is in the shape of a rectangular solid with dimensions 15 in. $\times$ 24 in. $\times$ 18 in.

APPENDIX I

Inductive and Deductive Reasoning

- Inductive reasoning
- Deductive reasoning

To reason means to think logically. The objective of this appendix is to develop your problem-solving ability by improving your reasoning skills. We will introduce two fundamental types of reasoning that can be applied in a wide variety of settings. They are known as *inductive reasoning* and *deductive reasoning.*

Inductive reasoning

In a laboratory, scientists conduct experiments and observe outcomes. After several repetitions with similar outcomes, the scientist will generalize the results into a statement that appears to be true.

- If I heat water to 212° F, it will boil.
- If I drop a weight, it will fall.
- If I combine an acid with a base, a chemical reaction occurs.

When we draw general conclusions from specific observations, we are using **inductive reasoning.** The next examples show how inductive reasoning can be used in mathematical thinking. Given a list of numbers or symbols, called a *sequence,* we can often find a missing term of the sequence by looking for patterns and applying inductive reasoning.

EXAMPLE 1 **An increasing pattern.** Find the next number in the sequence 5, 8, 11, 14, ...

Solution The terms of the sequence are increasing. To discover the pattern, we find the *difference* between each pair of successive terms.

$8 - 5 = 3$ Subtract the first term from the second term.

$11 - 8 = 3$ Subtract the second term from the third term.

$14 - 11 = 3$ Subtract the third term from the fourth term.

The difference between each pair of numbers is 3. This means that each successive number is 3 greater than the previous one. Thus, the next number in the sequence is $14 + 3$, or 17.

Self Check 1
Find the next number in the sequence $-3, -1, 1, 3, \ldots$.

Answer 5

Self Check 2

Find the next number in the sequence $-0.1, -0.3, -0.5, -0.7 \ldots$

Answer -0.9

EXAMPLE 2 A decreasing pattern. Find the next number in the sequence $-2, -4, -6, -8, \ldots$

Solution The terms of the sequence are decreasing. Since each successive term is 2 less than the previous one, the next number in the pattern is $-8 - 2$, or -10.

Self Check 3

Find the next entry in the sequence Z, A, Y, B, X, C, ...

Answer W

EXAMPLE 3 An alternating pattern. Find the next letter in the sequence A, D, B, E, C, F, D, ...

Solution The letter A is the first letter of the alphabet, D is the fourth letter, B is the second letter, and so on. We can create the following letter–number correspondence.

$A \to 1$
$D \to 4$ (Add 3.)
$B \to 2$ (Subtract 2.)
$E \to 5$ (Add 3.)
$C \to 3$ (Subtract 2.)
$F \to 6$ (Add 3.)
$D \to 4$ (Subtract 2.)

The numbers in the sequence 1, 4, 2, 5, 3, 6, 4, . . . alternate in size. They change from smaller to larger, to smaller, to larger, and so on.

We see that 3 is added to the first number to get the second number. Then 2 is subtracted from the second number to get the third number. To get successive terms in the sequence, we alternately add 3 to one number and then subtract 2 from that result to get the next number.

If we apply this pattern, the next number in the numerical sequence is $4 + 3$, or 7. The next letter in the original sequence is G, because it is the seventh letter of the alphabet.

Self Check 4

Find the next geometric shape in the sequence below.

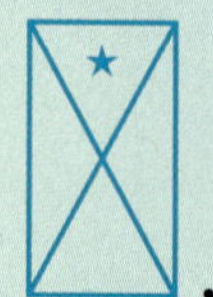

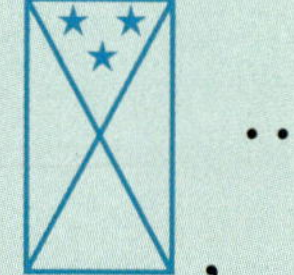
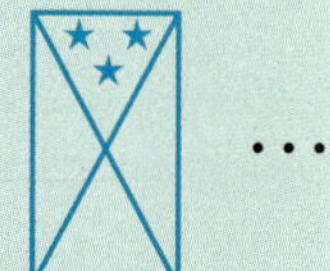

Answer

EXAMPLE 4 Two patterns. Find the next geometric shape in the sequence below.

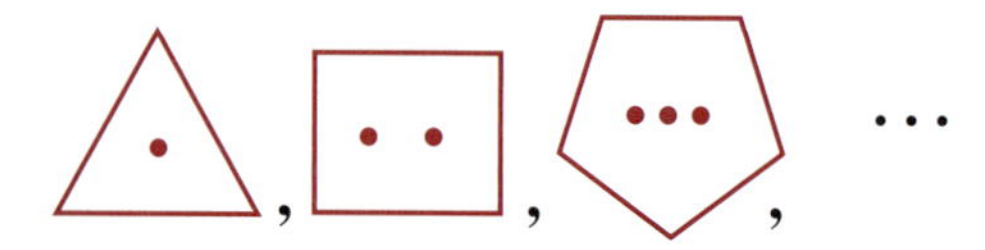

Solution This sequence has two patterns occurring at the same time. The first figure has three sides and one dot, the second figure has four sides and two dots, and the third figure has five sides and three dots. Thus, we would expect the next figure to have six sides and four dots, as shown in Figure A-1.

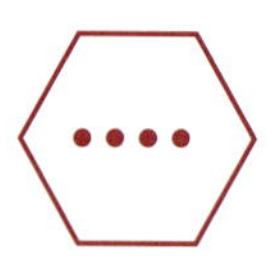

FIGURE A-1

EXAMPLE 5 **A circular pattern.** Find the next geometric shape in the sequence below.

Solution From figure to figure, we see that each dot moves from one point of the star to the next, in a counterclockwise direction. This is a circular pattern. The next shape in the sequence will be the one shown in Figure A-2.

FIGURE A-2

Self Check 5
Find the next geometric shape in the sequence below.

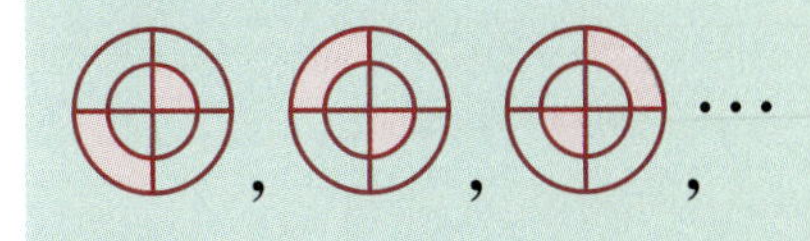

Answer

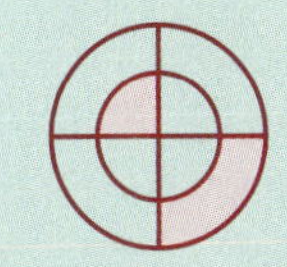

Deductive reasoning

As opposed to inductive reasoning, **deductive reasoning** moves from the general case to the specific. For example, if we know that the sum of the angles in any triangle is 180°, we know that the sum of the angles of $\triangle ABC$ is 180°. Whenever we apply a general principle to a particular instance, we are using deductive reasoning.

A deductive reasoning system is built on four elements.

1. **Undefined terms:** terms that we accept without giving them formal meaning
2. **Defined terms:** terms that we define in a formal way
3. **Axioms** or **postulates:** statements that we accept without proof
4. **Theorems:** statements that we can prove with formal reasoning

Many problems can be solved by deductive reasoning. For example, suppose that we plan to enroll in an early-morning algebra class and we know that Professors Perry, Miller, and Tveten are scheduled to teach algebra next semester. After some investigating, we find out that Professor Perry teaches only in the afternoon and Professor Tveten teaches only in the evenings. Without knowing anything about Professor Miller, we can conclude that he will be our teacher, since he is the only remaining possibility.

The following examples show how to use deductive reasoning to solve problems.

EXAMPLE 6 **Scheduling classes.** Four professors are scheduled to teach mathematics next semester, with the following course preferences.

1. Professors A and B don't want to teach calculus.
2. Professor C wants to teach statistics.
3. Professor B wants to teach algebra.

Who will teach trigonometry?

Solution The following chart shows each course, with each possible instructor.

Calculus	Algebra	Statistics	Trigonometry
A	A	A	A
B	B	B	B
C	C	C	C
D	D	D	D

Since Professors A and B don't want to teach calculus, we can cross them off the calculus list. Since Professor C wants to teach statistics, we can cross her off every other list. This leaves Professor D as the only person to teach calculus, so we can cross her off every other list. Since Professor B wants to teach algebra, we can cross him off every other list. Thus, the only remaining person left to teach trigonometry is Professor A.

Calculus	Algebra	Statistics	Trigonometry
~~A~~	A	A	A
~~B~~	B	~~B~~	~~B~~
~~C~~	~~C~~	C	~~C~~
D	~~D~~	~~D~~	~~D~~

Self Check 7

Of the 50 cars on a used-car lot, 9 are red, 31 are foreign models, and 6 are red foreign models. If a customer wants to buy an American model that is not red, how many cars does she have to choose from?

EXAMPLE 7 State flags.

The graph in Figure A-3 gives the number of state flags that feature an eagle, a star, or both. How many state flags have neither an eagle nor a star?

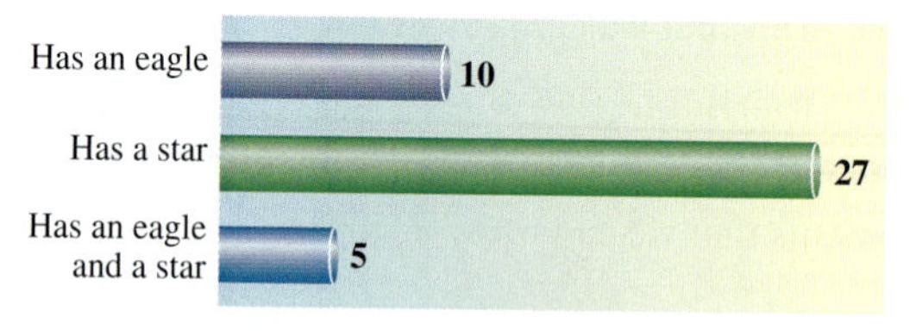

FIGURE A-3

Solution In Figure A-4(a), the intersection (overlap) of the circles is a way to show that there are 5 state flags that have both an eagle and a star. If an eagle appears on a total of 10 flags, then the left circle must contain 5 more flags outside of the intersection. See Figure A-4(b). If a total of 27 flags have a star, the right circle must contain 22 more flags outside the intersection.

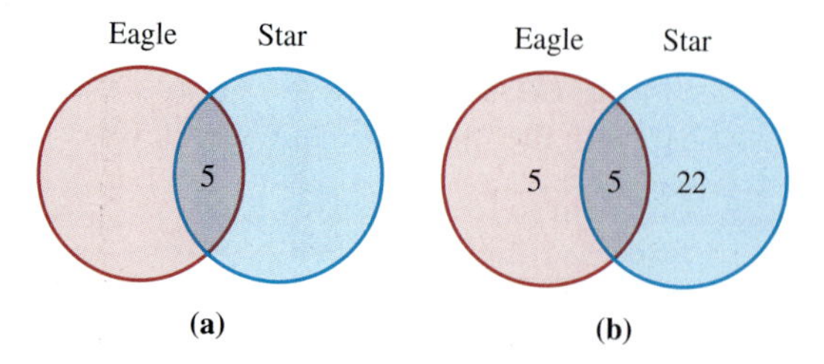

FIGURE A-4

From Figure A-4, we see that 5 + 5 + 22, or 32 flags have an eagle, a star, or both. To find how many flags have neither an eagle nor a star, we subtract this total from the number of state flags, which is 50.

$50 - 32 = 18$

There are 18 state flags that have neither an eagle nor a star.

Answer 16

Appendix I STUDY SET

VOCABULARY *Fill in the blanks.*

1. __________ reasoning draws general conclusions from specific observations.
2. __________ reasoning moves from the general case to the specific.

CONCEPTS *Determine whether the pattern shown is increasing, decreasing, alternating, or circular.*

3. 2, 3, 4, 2, 3, 4, 2, 3, 4, . . .
4. 8, 5, 2, −1, . . .
5. −2, −4, 2, 0, 6, . . .
6. 0.1, 0.5, 0.9, 1.3, . . .
7. a, c, b, d, c, e, . . .
8. , , , , . . .

9. ROOM SCHEDULING From the chart, determine what time(s) on a Wednesday morning a practice room in a music building is available. The symbol X indicates that the room has already been reserved.

	M	T	W	Th	F
9 A.M.	X		X	X	
10 A.M.	X	X			X
11 A.M.			X		X

10. QUESTIONNAIRES A group of college students were asked if they were taking a mathematics course and if they were taking an English course. The results are displayed.
    a. How many students were taking a mathematics course and an English course?
    b. How many students were taking an English course but not a mathematics course?
    c. How many students were taking a mathematics course?

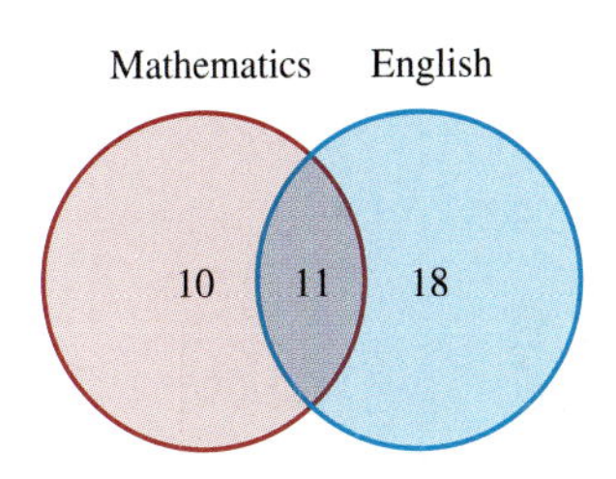

PRACTICE *Find the number that comes next in each sequence.*

11. 1, 5, 9, 13, . . .
12. 15, 12, 9, 6, . . .
13. −3, −5, −8, −12, . . .
14. 5, 9, 14, 20, . . .
15. −7, 9, −6, 8, −5, 7, −4, . . .
16. 2, 5, 3, 6, 4, 7, 5, . . .
17. 9, 5, 7, 3, 5, 1, . . .
18. 1.3, 1.6, 1.4, 1.7, 1.5, 1.8, . . .
19. −2, −3, −5, −6, −8, −9, . . .
20. 8, 11, 9, 12, 10, 13, . . .
21. 6, 8, 9, 7, 9, 10, 8, 10, 11, . . .
22. 10, 8, 7, 11, 9, 8, 12, 10, 9, . . .

Find the figure that comes next in each sequence.

23. 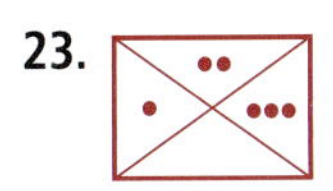, , 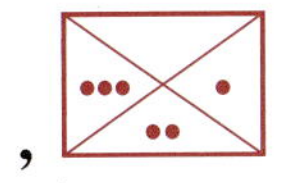, , . . .

24. , 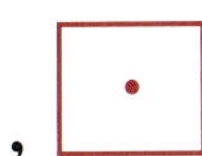, , 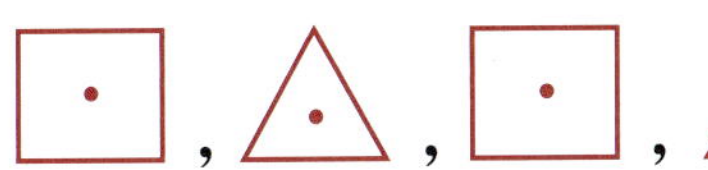, , . . .

Find the missing figure in each sequence.

25. 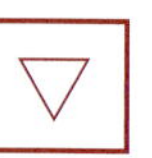, 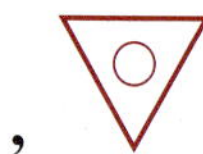, , ?, 

26. , 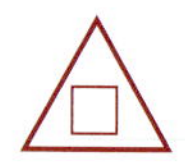, ?, , 

Find the next letter or letters in the sequence.

27. A, c, E, g, . . .

28. R, SS, TTT, . . .

29. d, h, g, k, j, n, . . .

30. B, N, C, N, D, . . .

What conclusion(s) can be drawn from each set of information?

31. Four people named John, Luis, Maria, and Paula have occupations as teacher, butcher, baker, and candlestick maker.

1. John and Paula are married.
2. The teacher plans to marry the baker in December.
3. Luis is the baker.

Who is the teacher?

32. In a zoo, a zebra, a tiger, a lion, and a monkey are to be placed in four cages numbered from 1 to 4, from left to right. The following decisions have been made.

1. The lion and the tiger should not be side by side.
2. The monkey should be in one of the end cages.
3. The tiger is to be in cage 4.

In which cage is the zebra?

33. A Ford, a Buick, a Dodge, and a Mercedes are parked side by side.

1. The Ford is between the Mercedes and the Dodge.
2. The Mercedes is not next to the Buick.
3. The Buick is parked on the left end.

Which car is parked on the right end?

34. Four divers at the Olympics finished first, second, third, and fourth.

1. Diver A beat diver B.
2. Diver C placed between divers B and D.
3. Diver B beat diver D.

In which order did they finish?

35. A green, a blue, a red, and a yellow flag are hanging on a flagpole.

1. The blue flag is between the green and yellow flags.
2. The red flag is next to the yellow flag.
3. The green flag is above the red flag.

What is the order of the flags from top to bottom?

36. Andres, Barry, and Carl each have two occupations: bootlegger, musician, painter, chauffeur, barber, and gardener. From the following facts, find the occupations of each man.

1. The painter bought a quart of spirits from the bootlegger.
2. The chauffeur offended the musician by laughing at his mustache.
3. The chauffeur dated the painter's sister.
4. Both the musician and the gardener used to go hunting with Andres.
5. Carl beat both Barry and the painter at monopoly.
6. Barry owes the gardener $100.

APPLICATIONS

37. JURY DUTY The results of a jury service questionnaire are shown. Determine how many of the 20,000 respondents have served on neither a criminal court nor a civil court jury.

Jury Service Questionnaire

997	Served on a criminal court jury
103	Served on a civil court jury
35	Served on both

38. POLLS The following Internet poll shows that 124 people voted for the first choice, 27 people voted for the second choice, and 19 people voted for both the first and the second choice. How many people clicked the third choice, "Neither"?

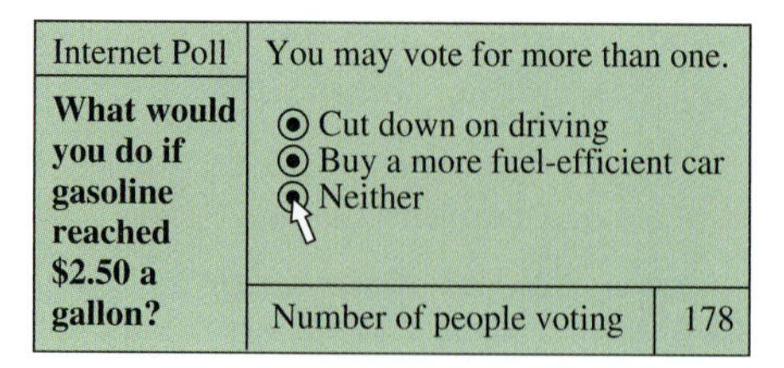

39. THE SOLAR SYSTEM The following graph shows some important characteristics of the 9 planets in our solar system. How many planets neither are rocky nor have moons?

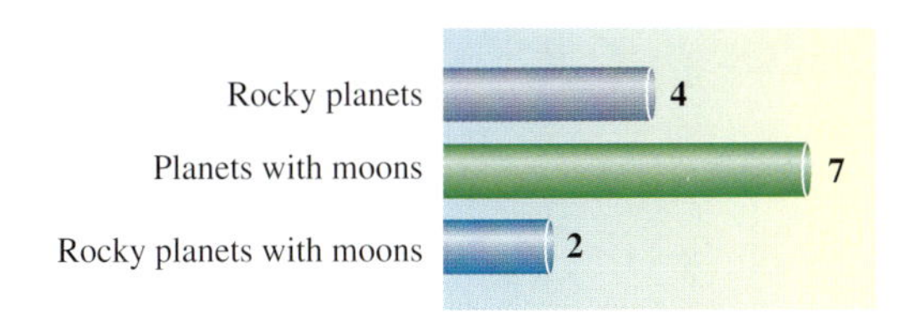

40. Write a problem in such a way that the following diagram can be used to solve it.

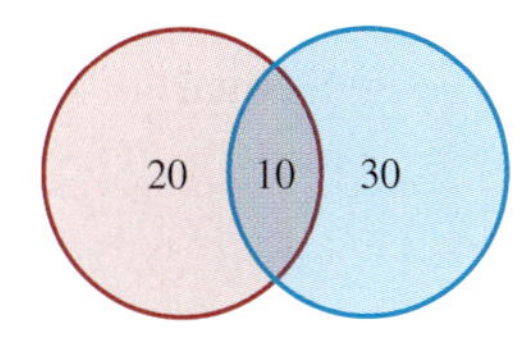

WRITING

41. Describe deductive reasoning.

42. Describe a real-life situation in which you might use deductive reasoning.

43. Describe inductive reasoning.

44. Describe a real-life situation in which you might use inductive reasoning.

APPENDIX REVIEW

APPENDIX I Inductive and Deductive Reasoning

CONCEPTS

Inductive reasoning draws general conclusions from specific observations.

Deductive reasoning draws specific conclusions by applying general rules.

REVIEW EXERCISES

1. Find the next number in the sequence 12, 8, 11, 7, 10, ...
2. Find the next number in the sequence 5, 9, 17, 33, ...
3. Find the missing geometric figure in the sequence.

4. Find the missing geometric figure in the sequence.

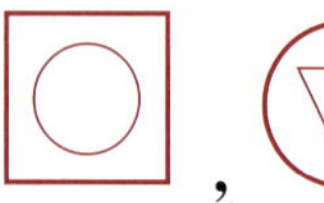

5. Find the next letter in the sequence.

   c, b, a, f, e, d, i, . . .

6. Four animals — a cow, a horse, a pig, and a sheep — are kept in a barn, each in a separate stall.
   1. The cow is in the first stall.
   2. The pig is between the horse and the sheep.
   3. The sheep cannot be next to the cow.

   What animal is in the last stall?

7. Jim, Sandra, Juan, and Mary all teach at the same college. One teaches mathematics, one teaches English, one teaches history, and one teaches music.
   1. Jim and Sandra eat lunch with the math teacher.
   2. Juan and Sandra carpool with the English teacher.
   3. Jim is married to the math teacher.
   4. Mary works in the same building as the history teacher.

   Who is the math teacher?

8. BUYING A CAR A new-car dealership has 103 cars for sale. A computer printout describing the cars in stock states the following.
   - 19 cars have a manual transmission (stick shift).
   - 53 cars have a CD player.
   - 12 cars have a manual transmission and a CD player.

   A customer wants to buy a car that has an automatic transmission. She says that she does not want a CD player in the car. How many cars does she have to look at?

APPENDIX II

Roots and Powers

n	n^2	$\sqrt{n}$	n^3	$\sqrt[3]{n}$	n	n^2	$\sqrt{n}$	n^3	$\sqrt[3]{n}$
1	1	1.000	1	1.000	51	2,601	7.141	132,651	3.708
2	4	1.414	8	1.260	52	2,704	7.211	140,608	3.733
3	9	1.732	27	1.442	53	2,809	7.280	148,877	3.756
4	16	2.000	64	1.587	54	2,916	7.348	157,464	3.780
5	25	2.236	125	1.710	55	3,025	7.416	166,375	3.803
6	36	2.449	216	1.817	56	3,136	7.483	175,616	3.826
7	49	2.646	343	1.913	57	3,249	7.550	185,193	3.849
8	64	2.828	512	2.000	58	3,364	7.616	195,112	3.871
9	81	3.000	729	2.080	59	3,481	7.681	205,379	3.893
10	100	3.162	1,000	2.154	60	3,600	7.746	216,000	3.915
11	121	3.317	1,331	2.224	61	3,721	7.810	226,981	3.936
12	144	3.464	1,728	2.289	62	3,844	7.874	238,328	3.958
13	169	3.606	2,197	2.351	63	3,969	7.937	250,047	3.979
14	196	3.742	2,744	2.410	64	4,096	8.000	262,144	4.000
15	225	3.873	3,375	2.466	65	4,225	8.062	274,625	4.021
16	256	4.000	4,096	2.520	66	4,356	8.124	287,496	4.041
17	289	4.123	4,913	2.571	67	4,489	8.185	300,763	4.062
18	324	4.243	5,832	2.621	68	4,624	8.246	314,432	4.082
19	361	4.359	6,859	2.668	69	4,761	8.307	328,509	4.102
20	400	4.472	8,000	2.714	70	4,900	8.367	343,000	4.121
21	441	4.583	9,261	2.759	71	5,041	8.426	357,911	4.141
22	484	4.690	10,648	2.802	72	5,184	8.485	373,248	4.160
23	529	4.796	12,167	2.844	73	5,329	8.544	389,017	4.179
24	576	4.899	13,824	2.884	74	5,476	8.602	405,224	4.198
25	625	5.000	15,625	2.924	75	5,625	8.660	421,875	4.217
26	676	5.099	17,576	2.962	76	5,776	8.718	438,976	4.236
27	729	5.196	19,683	3.000	77	5,929	8.775	456,533	4.254
28	784	5.292	21,952	3.037	78	6,084	8.832	474,552	4.273
29	841	5.385	24,389	3.072	79	6,241	8.888	493,039	4.291
30	900	5.477	27,000	3.107	80	6,400	8.944	512,000	4.309
31	961	5.568	29,791	3.141	81	6,561	9.000	531,441	4.327
32	1,024	5.657	32,768	3.175	82	6,724	9.055	551,368	4.344
33	1,089	5.745	35,937	3.208	83	6,889	9.110	571,787	4.362
34	1,156	5.831	39,304	3.240	84	7,056	9.165	592,704	4.380
35	1,225	5.916	42,875	3.271	85	7,225	9.220	614,125	4.397
36	1,296	6.000	46,656	3.302	86	7,396	9.274	636,056	4.414
37	1,369	6.083	50,653	3.332	87	7,569	9.327	658,503	4.431
38	1,444	6.164	54,872	3.362	88	7,744	9.381	681,472	4.448
39	1,521	6.245	59,319	3.391	89	7,921	9.434	704,969	4.465
40	1,600	6.325	64,000	3.420	90	8,100	9.487	729,000	4.481
41	1,681	6.403	68,921	3.448	91	8,281	9.539	753,571	4.498
42	1,764	6.481	74,088	3.476	92	8,464	9.592	778,688	4.514
43	1,849	6.557	79,507	3.503	93	8,649	9.644	804,357	4.531
44	1,936	6.633	85,184	3.530	94	8,836	9.695	830,584	4.547
45	2,025	6.708	91,125	3.557	95	9,025	9.747	857,375	4.563
46	2,116	6.782	97,336	3.583	96	9,216	9.798	884,736	4.579
47	2,209	6.856	103,823	3.609	97	9,409	9.849	912,673	4.595
48	2,304	6.928	110,592	3.634	98	9,604	9.899	941,192	4.610
49	2,401	7.000	117,649	3.659	99	9,801	9.950	970,299	4.626
50	2,500	7.071	125,000	3.684	100	10,000	10.000	1,000,000	4.642

APPENDIX III

Answers to Selected Exercises

Chapter 1 Check Your Knowledge (page 2)

1. natural, whole **2.** factors, product, quotient **3.** commutative, associative **4.** prime **5.** 7 thousands + 3 hundreds + 4 tens + 3 ones **6.** 27,500

7.

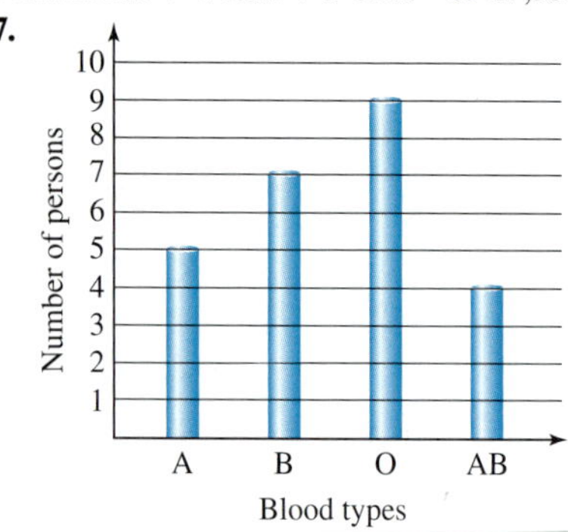

8.

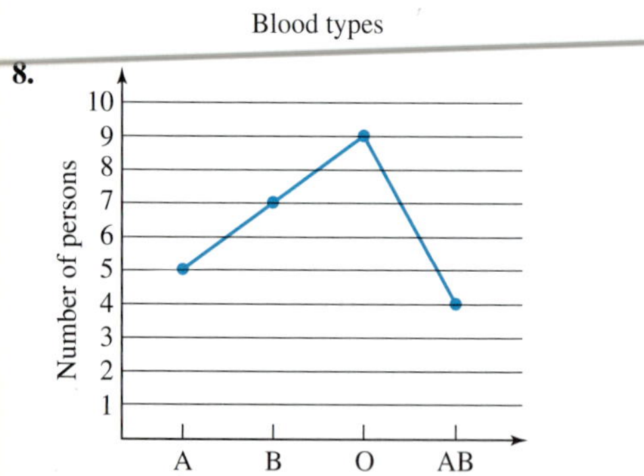

9. > **10.** 5,121 **11.** 58 **12.** 69° F **13.** 24,624 **14.** 57 R 34 **15.** 64 ft, 247 ft^2 **16.** $2 \cdot 5^2 \cdot 19$ **17.** 5 **18.** 103 **19.** 3 **20.** 19 **21.** 3 **22.** 2 **23.** 16 **24.** 48 **25.** 63 **26.** 89

Think It Through (page 9)

1. c **2.** b **3.** e **4.** d **5.** a

Study Set Section 1.1 (page 10)

1. set **3.** expanded **5.** number **7.** 3 **9.** 6 **11.** whole numbers

13

15.

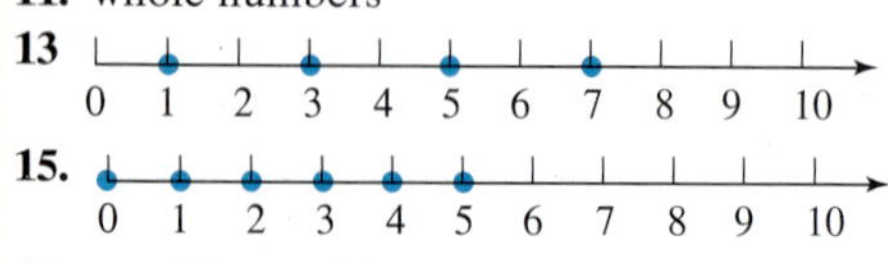

17. > **19.** > **21.** < **23.** > **25.** braces

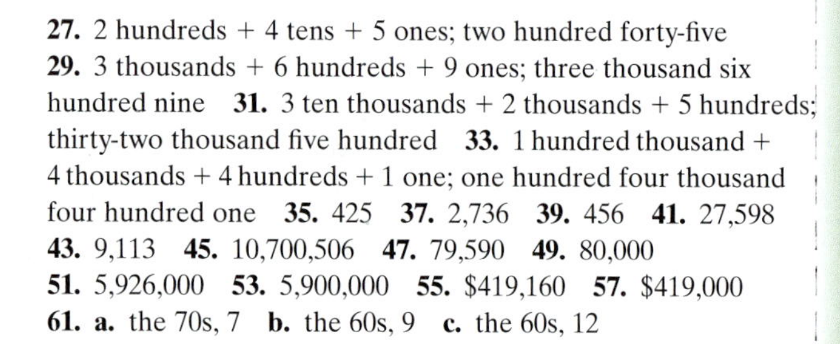

27. 2 hundreds + 4 tens + 5 ones; two hundred forty-five **29.** 3 thousands + 6 hundreds + 9 ones; three thousand six hundred nine **31.** 3 ten thousands + 2 thousands + 5 hundreds; thirty-two thousand five hundred **33.** 1 hundred thousand + 4 thousands + 4 hundreds + 1 one; one hundred four thousand four hundred one **35.** 425 **37.** 2,736 **39.** 456 **41.** 27,598 **43.** 9,113 **45.** 10,700,506 **47.** 79,590 **49.** 80,000 **51.** 5,926,000 **53.** 5,900,000 **55.** \$419,160 **57.** \$419,000 **61. a.** the 70s, 7 **b.** the 60s, 9 **c.** the 60s, 12

63.

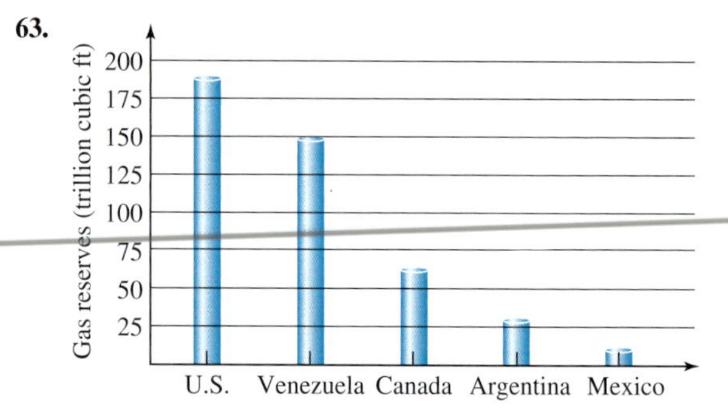

65.

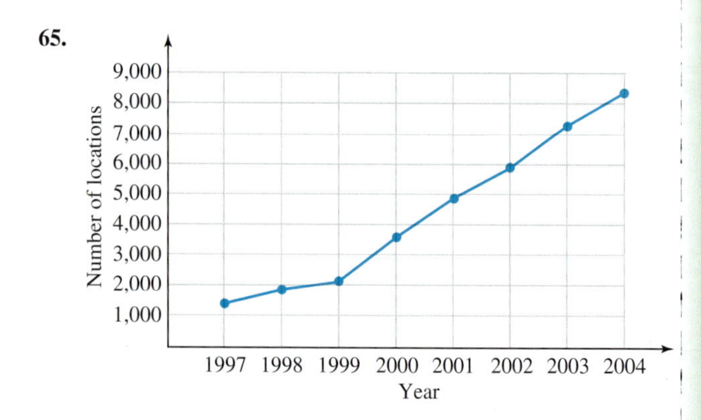

67.
a.

No. 201 — March 9, 20 05

Payable to Davis Chevrolet — \$ 15,601.00

Fifteen thousand six hundred one and 00/100 DOLLARS

45-365-02 — Don Smith

b.

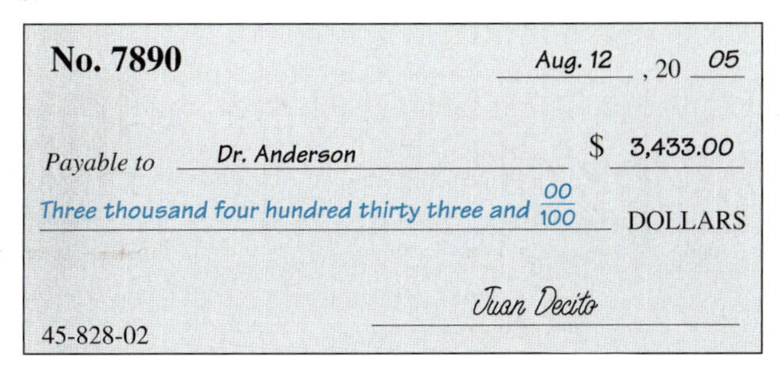

69. 1,865,593; 482,880; 1,503; 269; 43,449 **71. a.** 299,800,000 m/s **b.** 300,000,000 m/s

Study Set Section 1.2 (page 21)

1. sum, addends **3.** sum **5.** difference, subtrahend, minuend **7.** associative **9.** commutative property of addition **11.** associative property of addition **13. a.** $x + y = y + x$ **b.** $(x + y) + z = x + (y + z)$ **15.** 0 **17.** $4 + 3 = 7$ **19.** parentheses **21.** 47, 52 **23.** 38 **25.** 461 **27.** 111 **29.** 150 **31.** 363 **33.** 979 **35.** 1,985 **37.** 10,000 **39.** 15,907 **41.** 1,861 **43.** 5,312 **45.** 88 ft **47.** 68 in. **49.** 3 **51.** 25 **53.** 103 **55.** 65 **57.** 141 **59.** 0 **61.** 24 **63.** 118 **65.** 958 **67.** 1,689 **69.** 10,457 **71.** 303 **73.** 40 **75.** 110 **77.** \$18 **79.** 33 points **81.** \$213 **83.** 10,057 mi **85. a.** \$147,145 **b.** \$161,725 **87.** 91 ft **89.** \$6,233,000,000 **91.** 196 in. **95.** 3 thousands + 1 hundred + 2 tens + 5 ones **97.** 6,354,780 **99.** 6,350,000

Study Set Section 1.3 (page 34)

1. multiplication **3.** commutative, associative **5.** divisor, quotient **7.** $4 \cdot 8$ **9.** Multiply its length by its width. **11. a.** 25 **b.** 62 **c.** 0 **d.** 0 **13.** $5 \cdot 12 = 60$ **15. a.** $\times, \cdot, (\,)$ **b.** $\overline{)\ }$, $\div$, — **17.** square feet **19.** 84 **21.** 324 **23.** 180 **25.** 105 **27.** 7,623 **29.** 1,060 **31.** 2,576 **33.** 20,079 **35.** 2,919,952 **37.** 1,182,116 **39.** 84 in.^{2} **41.** 144 in.^{2} **43.** 8 **45.** 3 **47.** 12 **49.** 13 **51.** 73 **53.** 41 **55.** 205 **57.** 210 **59.** 8 R 25 **61.** 20 R 3 **63.** 30 R 13 **65.** 31 R 28 **67.** \$132 **69.** 406 mi **71.** 125,800 **73.** 312 **75.** yes **77.** 72 **79.** 4 **81.** 5 mi **83.** 440 ft **85.** \$41 **87.** 9 girls, 24 teams **89.** the square room; the square room **91.** 388 ft^2 **97.** 8 **99.** 872

Study Set Estimation (page 38)

1. no **3.** no **5.** no **7.** approx. 8,900 mi **9.** approx. 30 bags **11.** 1,800,000,000

Study Set Section 1.4 (page 44)

1. factors **3.** factor **5.** composite **7.** prime **9.** base, exponent **11.** $1 \cdot 27$ or $3 \cdot 9$ **13. a.** 44 **b.** 100 **15. a.** 1 and 11 **b.** 1 and 23 **c.** 1 and 37 **d.** They are prime numbers. **17.** yes **19.** 90 **21.** 605 **23.** no **25.** 2 and 5 **27.** 2 **29.** $3 \cdot 5 \cdot 2 \cdot 5$; $5 \cdot 3 \cdot 5 \cdot 2$; they are the same **31.** 13, 8, 7 **33.** 2 **35.** $7 \cdot 7 \cdot 7$ **37.** $3 \cdot 3 \cdot 3 \cdot 3 \cdot 3$ **39.** $5 \cdot 5 \cdot 11$ **41.** 10 **43.** 2^5 **45.** 5^4 **47.** $4^2(5^2)$ **49.** 1, 2, 5, 10 **51.** 1, 2, 4, 5, 8, 10, 20, 40 **53.** 1, 2, 3, 6, 9, 18 **55.** 1, 2, 4, 11, 22, 44 **57.** 1, 7, 11, 77 **59.** 1, 2, 4, 5, 10, 20, 25, 50, 100 **61.** $3 \cdot 13$ **63.** $3^2 \cdot 11$ **65.** $2 \cdot 3^4$ **67.** $2^2 \cdot 5 \cdot 11$ **69.** 2^6 **71.** $3 \cdot 7^2$ **73.** 81 **75.** 32 **77.** 144 **79.** 4,096 **81.** 72 **83.** 3,456 **85.** 12,812,904 **87.** 1,162,213 **89.** 1, 2, 4, 7, 14, 28; $1 + 2 + 4 + 7 + 14 = 28$ **91.** 2^2 square units; 3^2 square units; 4^2 square units **97.** 231,000 **99.** 0 **101.** $A = lw$

Think It Through (page 52)

10, 5, 12, 3, 4

Study Set Section 1.5 (page 52)

1. parentheses, brackets **3.** evaluate **5.** 3; square, multiply, subtract **7.** multiply, subtract **9.** $2 \cdot 3^2 = 2 \cdot 9$; $(2 \cdot 3)^2 = 6^2$ **11.** 4, 20 **13.** 36, 30 **15.** 27 **17.** 2 **19.** 15 **21.** 25 **23.** 5 **25.** 25 **27.** 18 **29.** 813 **31.** 5,239 **33.** 16 **35.** 5 **37.** 49 **39.** 24 **41.** 13 **43.** 10 **45.** 198 **47.** 18 **49.** 216 **51.** 17 **53.** 191 **55.** 3 **57.** 29 **59.** 14 **61.** 64 **63.** 192 **65.** 74 **67.** 137 **69.** 3 **71.** 21 **73.** 11 **75.** 1 **77.** 10,496 **79.** 2,845 **81.** $2(6) + 4(2) + 2(1)$; \$22 **83.** $24 + 6(5) + 10(10) + 12(20) + 2(50) + 100$; \$594 **85.** brick: $3(3) + 1 + 1 + 3 + 3(5)$; 29; aphid: $3[1 + 2(3) + 4 + 1 + 2]$; 42 **87.** 79° **89.** 5 **91.** 298 **97.** 7,300 **99.** 9,591

Study Set Section 1.6 (page 61)

1. equal, = **3.** check **5.** equivalent **7.** y, c **9.** addition of 6; subtract 6 from both sides **11.** 8, 8, 16, $\stackrel{?}{=}$, 24, 16 **13.** yes **15.** no **17.** yes **19.** yes **21.** yes **23.** yes **25.** no **27.** no **29.** no **31.** yes **33.** 10 **35.** 7 **37.** 3 **39.** 4 **41.** 13 **43.** 75 **45.** 740 **47.** 339 **49.** 3 **51.** 5 **53.** 9 **55.** 10 **57.** 1 **59.** 56 **61.** 84 **63.** 105 **65.** 4 **67.** 12 **69.** 8 **71.** 47 **75.** 94,683,948 **77.** 62 **79.** \$218,500 **81.** \$180 million **83.** 25 units **85.** \$190 **93.** 325,780 **95.** 90 **97.** 3

Study Set Section 1.7 (page 69)

1. division **3.** x **5.** y, z **7.** It is being multiplied by 4. Divide by 4. **9. a.** Subtract 5 from both sides. **b.** Add 5 to both sides. **c.** Divide both sides by 5. **d.** Multiply both sides by 5. **11.** 3, 3, 4, 12, 4 **13.** 1 **15.** 96 **17.** 3 **19.** 6 **21.** 1 **23.** 2 **25.** 14 **27.** 42 **29.** 75 **31.** 39 **33.** 50 **35.** 49 **37.** 10 **39.** 3 **41.** 2 **43.** 1 **45.** 40 **47.** 1,200 **51.** 390 words per minute **53.** 14 **55.** 96 **57.** 32 calls **59.** 55 lb **65.** 48 cm **67.** $2^3 \cdot 3 \cdot 5$ **69.** 72 **71.** 26 mpg

Key Concept (page 73)

1. Let x = the monthly cost to lease the van. **3.** Let x = the width of the field. **5.** Let x = the distance traveled by the motorist. **7.** $a + b = b + a$ **9.** $\frac{b}{1} = b$ **11.** $n - 1 < n$ **13.** $(r + s) + t = r + (s + t)$

Chapter Review (page 75)

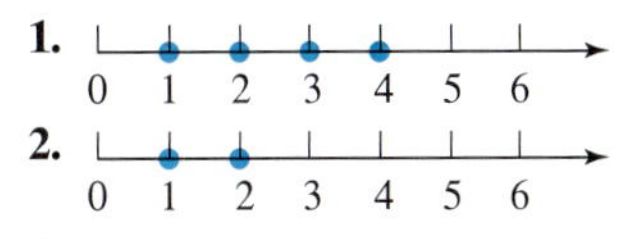

3.

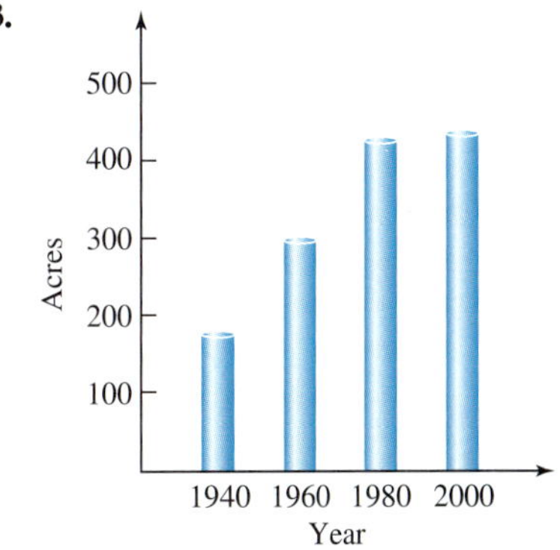

4.

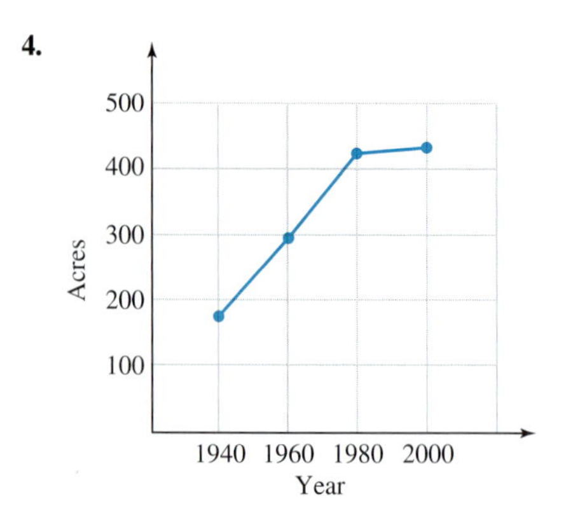

5. 6 **6.** 7 **7.** 5 hundred thousands + 7 ten thousands + 3 hundreds + 2 ones **8.** 3 ten millions + 7 millions + 3 hundred thousands + 9 thousands + 5 tens + 4 ones **9.** 3,207 **10.** 23,253,412 **11.** 16,000,000,000 **12.** > **13.** < **14.** 2,507,300 **15.** 2,510,000 **16.** 2,507,350 **17.** 2,500,000 **18.** 78 **19.** 137 **20.** 55 **21.** 149 **22.** 777 **23.** 2,332 **24.** 518 **25.** 6,000 **26.** 1,010 **27.** 24,986 **28.** commutative property of addition **29.** associative property of addition **30.** 96 in. **31.** 13 **32.** 4 **33.** 11 **34.** 54 **35.** 74 **36.** 2,075 **37.** $4 + 2 = 6$ **38.** $5 - 2 = 3$ **39.** \$45 **40.** \$785 **41.** \$23,541 **42.** 56 **43.** 56 **44.** 0 **45.** 7 **46.** 560 **47.** 210 **48.** 3,297 **49.** 178,704 **50.** 31,684 **51.** 455,544 **52.** associative property of multiplication **53.** commutative property of multiplication **54.** \$342 **55.** 108 ft, 288 ft^2 **56.** 720 **57.** 2 **58.** 15 **59.** undefined **60.** 0 **61.** 21 **62.** 37 **63.** 19 R 6 **64.** 23 R 27 **65.** 16, 25 **66.** 28 **67.** 1, 2, 3, 6, 9, 18 **68.** 1, 5, 25 **69.** prime **70.** composite **71.** neither **72.** neither **73.** composite **74.** prime **75.** odd **76.** even **77.** even **78.** odd **79.** $2 \cdot 3 \cdot 7$ **80.** $3 \cdot 5^3$ **81.** 6^4 **82.** $5^3 \cdot 13^2$ **83.** 125 **84.** 121 **85.** 200 **86.** 2,700 **87.** 49 **88.** 32 **89.** 75 **90.** 36 **91.** 38 **92.** 24 **93.** 8 **94.** 24 **95.** 53 **96.** 3 **97.** 19 **98.** 7 **99.** $3(6) + 2(5) = 28$ **100.** 201 **101.** no **102.** yes **103.** y **104.** t **105.** 9 **106.** 31 **107.** 340 **108.** 133 **109.** 9 **110.** 14 **111.** 120 **112.** 5 **113.** 7 **114.** 985 **115.** \$97,250 **116.** 185 **117.** 4 **118.** 3 **119.** 21 **120.** 14 **121.** 21 **122.** 36 **123.** 315 **124.** 425 **125.** 1 **126.** 144 **127.** 24 in. **128.** \$128

Chapter 1 Test (page 81)

1.

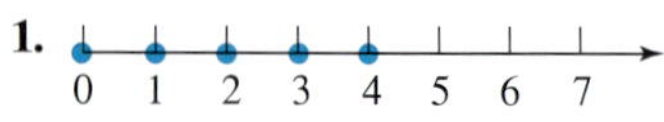

2. 5 thousands + 2 hundreds + 6 tens + 6 ones **3.** 7,507 **4.** 35,000,000

5.

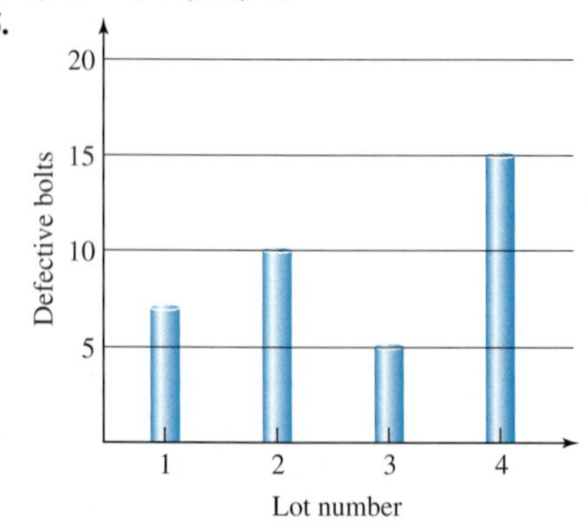

6.

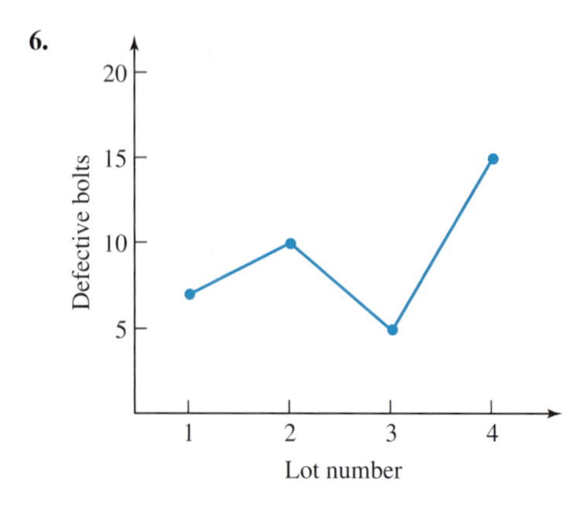

7. > **8.** < **9.** 1,491 **10.** 248 **11.** 58,105 **12.** 942 **13.** \$76 **14.** 1, 2, 4, 5, 10, 20 **15.** 424 **16.** 26,791 **17.** 72 **18.** 114 R 57 **19. a.** 0 **b.** 0 **20. a.** associative property of multiplication **b.** commutative property of addition **21.** 360 ft, 7,875 ft^2 **22.** 47 **23.** 3,456 **24.** $2^2 \cdot 3^2 \cdot 7$ **25.** 29 **26.** 44 **27.** 26 **28.** 39 **29.** yes **30.** 99 **31.** 30 **32.** 11 **33.** 81 **34.** 3,100 **35.** 194 yr **36.** To solve an equation means to find all the values of the variable that, when substituted into the equation, make a true statement.

Chapter 2 Check Your Knowledge (page 84)

1. absolute **2.** identity **3.** opposites (or negatives) **4.** unlike **5.** > **6.** 1 **7. a.** -14 **b.** 0 **c.** -8 **d.** 2 **8. a.** -6 **b.** 6 **c.** -12 **9. a.** -9 **b.** 100 **c.** -24 **10.** $3(-6) = -18$ **11. a.** -4 **b.** 30 **c.** undefined **12. a.** 7 **b.** -7 **c.** 12 **13. a.** -9 **b.** 9 **c.** 1 **14. a.** 16 **b.** -13 **c.** 29 **d.** 1 **15.** \$20 **16.** 15 **17.** $-\$230$

Think It Through (page 89)

\$84,621, \$1,073, \$3,325

Study Set Section 2.1 (page 92)

1. line **3.** graph **5.** inequality **7.** absolute value **9.** integers **11. a.** < **b.** > **c.** <, > **13.** yes **15.** $15 - 8$ **17.** $15 > 12$ **19. a.** -225 **b.** -10 **c.** -3 **d.** $-12{,}000$ **e.** -2 **21.** It is negative. **23.** -4 **25.** -8 and 2 **27.** -7 **29.** $6 - 4, -6, -(-6)$ (answers may vary) **31. a.** $-(-8)$ **b.** $|-8|$ **c.** $8 - 8$ **d.** $-|-8|$ **33.** 9 **35.** 8 **37.** 14 **39.** -20 **41.** -6 **43.** 203 **45.** 0 **47.** 11 **49.** 4 **51.** 12

53. −5 −4 −3 −2 −1 0 1 2 3 4 5

55. −5 −4 −3 −2 −1 0 1 2 3 4 5

57. < **59.** < **61.** > **63.** > **65.** ≥ **67.** ≥ **69.** ≤ **71.** ≤ **73.** 2, 3, 2, 0, -3, -7 **75.** peaks: 2, 4, 0; valleys: -3, -5, -2

77. a. −1 (1 below par) **b.** −3 (3 below par) **c.** Most of the scores are below par. **79. a.** −10° to −20° **b.** 40° **c.** 10° **81. a.** 200 yr **b.** A.D. **c.** B.C. **d.** the birth of Christ

83.

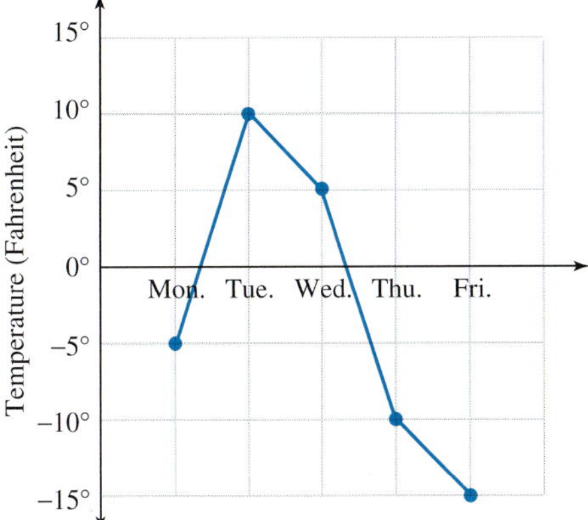

91. 23,500 **93.** 761 **95.** associative property of multiplication

Think It Through (page 101)

decrease expenses, increase income, decrease expenses, increase income, increase income, increase income, decrease expenses, decrease expenses, increase income, decrease expenses

Study Set Section 2.2 (page 103)

1. identity **3.** 3 **5.** −2 **7. a.** yes **b.** yes **9. a.** 7 **b.** 10 **11.** subtract, larger **13.** −18, −19 **15.** 5, 2 **17.** −5 should be within parentheses: −6 + (−5) **19.** 11 **21.** 23 **23.** 0 **25.** −99 **27.** −9 **29.** −10 **31.** 1 **33.** −7 **35.** −20 **37.** 15 **39.** 8 **41.** 2 **43.** −10 **45.** 9 **47.** 8 **49.** −21 **51.** 3 **53.** −10 **55.** −4 **57.** 7 **59.** −21 **61.** −7 **63.** 9 **65.** 0 **67.** 0 **69.** 5 **71.** 0 **73.** −3 **75.** −10 **77.** −1 **79.** −17 **81.** −8,346 **83.** −1,032 **85.** 3G, −3G **87.** no; $70 shortfall each month **89.** 5; 4% risk **91.** −1, 0 **93.** 7 ft over flood stage **95.** about −$2,500 **101.** 15 ft^2 **103.** 375 mi **105.** 5^3

Study Set Section 2.3 (page 111)

1. difference **3.** adding, opposite **5.** 6 **7.** $(-b)$ **9.** brackets **11.** −8 − (−4) **13.** 7 **15.** no; 8 − 3 = 5, 3 − 8 = −5 **17.** −3, 2, 0 **19.** −2, −10, 6, −4 **21.** 9 **23.** −13 **25.** −10 **27.** −1 **29.** 0 **31.** 8 **33.** 5 **35.** −4 **37.** −4 **39.** −20 **41.** 0 **43.** 0 **45.** −15 **47.** −9 **49.** 3 **51.** 9 **53.** −2 **55.** −10 **57.** −14 **59.** 3 **61.** −8 **63.** −18 **65.** −6 **67.** 10 **69.** −4 **71.** −2,447 **73.** 20,503 **75.** −1,676 **77.** −120 ft **79.** 16 points **81.** −8 **83.** 1,066 ft **85.** −4 yd

87. a.

Bottom: −12; Water level: 0; Platform: 25

b. 37 ft

89. No; he will be $244 overdrawn (−244). **95.** 5,990 **97.** 1, 2, 4, 5, 10, 20 **99.** 143 **101.** 3

Study Set Section 2.4 (page 119)

1. factors, product **3.** 3, exponent **5.** unlike **7.** commutative **9.** −9, the opposite of that number **11.** pos · pos, pos · neg, neg · pos, neg · neg **13. a.** negative **b.** positive **15. a.** 3 **b.** 12 **c.** 5 **d.** 9 **e.** 10 **f.** 25 **17. a.** 2, 4; 4, 16; 6, 64 **b.** even **19.** 6, −24 **21.** −5 should be in parentheses: −6(−5) **23.** 54 **25.** −15 **27.** −36 **29.** 56 **31.** −20 **33.** −120 **35.** 0 **37.** 6 **39.** 7 **41.** −23 **43.** −48 **45.** 40 **47.** −30 **49.** −60 **51.** −1 **53.** −18 **55.** 0 **57.** 0 **59.** 60 **61.** 16 **63.** −125 **65.** −8 **67.** 81 **69.** −1 **71.** 1 **73.** 49, −49 **75.** −144, 144 **77.** −59,812 **79.** 43,046,721 **81.** −25,728 **83.** 390,625 **85. a.** plan #1: −30 lb, plan #2: −28 lb **b.** plan #1; the workout time is double that of plan #2 **87. a.** high 2, low −3 **b.** high 4, low −6 **89.** −20° **91.** −20 ft **93.** −$24,330 **99.** 45 **101.** 2,100 **103.** is less than

Study Set Section 2.5 (page 125)

1. quotient, divisor **3.** absolute value **5.** positive **7.** 5(−5) = −25 **9.** 0(?) = −6 **11.** $\frac{-20}{5} = -4$ **13. a.** always true **b.** sometimes true **c.** always true **15.** −7 **17.** 2 **19.** 5 **21.** 3 **23.** −20 **25.** −2 **27.** 0 **29.** undefined **31.** −5 **33.** 1 **35.** −1 **37.** 10 **39.** −4 **41.** −3 **43.** 5 **45.** −4 **47.** −5 **49.** −4 **51.** −542 **53.** −16 **55.** −4° **57.** −1,000 ft **59.** −6 (6 games behind) **61.** −$15 **63.** −$1,740 **69.** 104 **71.** 2 · 3 · 5 · 7 **73.** yes **75.** 142

Study Set Section 2.6 (page 131)

1. order **3.** grouping **5.** 3; power, multiplication, subtraction **7.** multiplication; subtraction **9.** The base of the first exponential expression is 3; the base of the second is −3. **11.** 4, 20, −20, −28 **13.** 9, −36, −42 **15.** −7 **17.** 1 **19.** −21 **21.** −14 **23.** −7 **25.** −5 **27.** 12 **29.** −14 **31.** 30 **33.** 2 **35.** 15 **37.** −42 **39.** −5 **41.** −3 **43.** 4 **45.** 0 **47.** −14 **49.** 19 **51.** 4 **53.** −3 **55.** 25 **57.** −48 **59.** 44 **61.** 91 **63.** 3 **65.** −5 **67.** 17 **69.** 11 **71.** 8 **73.** 112 **75.** −1,707 **77.** −15 **79.** −5 **81.** −35 **83.** −200 **85.** −320 **87.** −9,000 **89.** −1,200 **91.** 19 **93.** 11 yd **95.** about a 60-cent gain **101.** 5,000 **103.** Add the lengths of all its sides. **105.** no

Study Set Section 2.7 (page 139)

1. solve **3.** x **5. a.** 3 **b.** (−3) **7. a.** multiplication by −2 **b.** addition of −6 **c.** mult. by −4, subtraction of 8 **d.** mult. by −5, addition of −6 **9.** simplify **11.** opposite **13. a.** subtraction of 3 **b.** addition of −6 **15.** −13, 7 **17.** 1, 1, −12, −4, −4, 3 **19.** $-10 \cdot x$ **21.** yes **23.** no **25.** −18 **27.** −14 **29.** 5 **31.** −1 **33.** −9 **35.** −14 **37.** −2 **39.** 0 **41.** −8 **43.** 5 **45.** 2 **47.** −1 **49.** 6 **51.** 0 **53.** 6 **55.** −52 **57.** −7 **59.** −4 **61.** −5 **63.** −2 **65.** 10 **67.** −6 **69.** −3 **71.** 3 **73.** −6 **75.** 54 **77.** 30 **79.** −14 **81.** −3 **83.** −2 **85.** −8 **87.** 15 **89.** −75, how many feet the cage was raised, the number of feet the cage was raised, addition, x, −75, −120, 120. The shark cage was raised 45 feet. 45 **91.** 34 **93.** −$435 **95.** 29 points **97.** $5 **99.** 18 ft **101.** zone −8 **105.** 5 · 5 · 5 · 5 · 5 · 5 **107.** 12 **109.** $\frac{16}{8}$

Key Concept (page 143)

1. −5 **3.** −30 **5.** +10 or 10 **7.** −205

9.

−4 −3 −2 −1 0 1 2 3 4

Negatives Positives

11. $x < y$ **13.** Like signs: Add their absolute values and attach their common sign to the sum. Unlike signs: Subtract their absolute values, the smaller from the larger, and attach the sign of the number with the larger absolute value to that result.

15. Like signs: The quotient is positive. Unlike signs: The quotient is negative.

Chapter Review (page 145)

1. **2.**

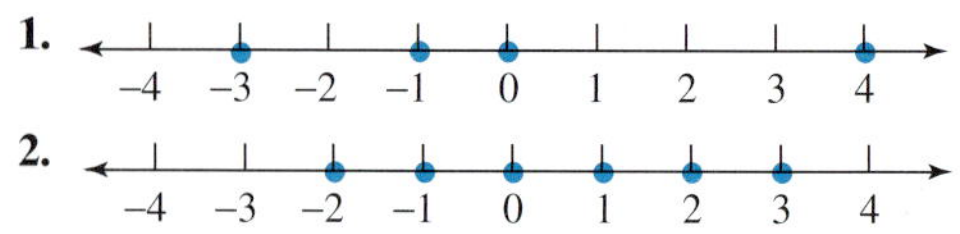

3. < **4.** < **5.** ≥ **6.** ≤ **7.** −33 ft **8.** −$1,200 **9.** −10 sec **10.** 4 **11.** 0 **12.** 43 **13.** −12 **14.** negative **15.** the opposite **16.** negative **17.** minus **18.** 12 **19.** −8 **20.** 8 **21.** 0 **22.** 2

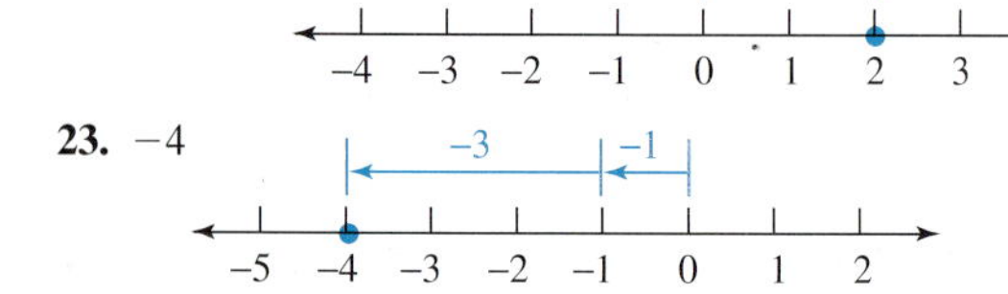

23. −4

−3 −1
−5 −4 −3 −2 −1 0 1 2

24. −10 **25.** −83 **26.** −8 **27.** −1 **28.** 112 **29.** −11 **30.** −3 **31.** −2 **32.** −4 **33.** −20 **34.** 0 **35.** 0 **36.** 11 **37.** −4 **38.** 65 ft **39.** −3 **40.** −21 **41.** 4 **42.** −112 **43.** −6 **44.** 6 **45.** −37 **46.** 30 **47.** adding, opposite **48.** −4 **49.** 15 **50.** 6 **51.** −8 **52.** −77 **53.** −1 **54.** −225 ft **55.** Alaska: 180°; Virginia: 140° **56.** −45 **57.** 18 **58.** −14 **59.** 376 **60.** −100 **61.** 1 **62.** −25 **63.** −150 **64.** −36 **65.** −36 **66.** 0 **67.** 1 **68.** −3, −6, −9 **69.** 25 **70.** −32 **71.** 64 **72.** −64 **73.** negative **74.** first expression: base of 2; second: base of −2; −4, 4 **75.** 5, −3, −15 **76.** −2 **77.** −5 **78.** −8 **79.** 101 **80.** 0 **81.** undefined **82.** 1 **83.** 10 **84.** −2 min **85.** −22 **86.** 4 **87.** −43 **88.** 8 **89.** 41 **90.** 0 **91.** −13 **92.** 32 **93.** 12 **94.** −16 **95.** −4 **96.** 1 **97.** −1 **98.** −4 **99.** −70 **100.** 20 **101.** −7,000 **102.** 1,100 **103.** yes **104.** no **105.** −10 **106.** 12 **107.** −8 **108.** 4 **109.** 15 **110.** −4 **111.** 3 **112.** −2 **113.** −12 **114.** 0 **115.** −$132 **116.** −46° **117.** 121 **118.** $8,200

Chapter 2 Test (page 151)

1. a. > **b.** < **c.** < **2.** {. . . , −3, −2, −1, 0, 1, 2, 3, . . .} **3.** Monroe **4.** −5

−2 −3
−6 −5 −4 −3 −2 −1 0 1 2 3 4 5 6

5. a. −34 **b.** −34 **c.** −8 **6. a.** −13 **b.** −1 **c.** −15 **d.** −150 **7. a.** −70 **b.** −48 **c.** 16 **d.** 0 **8.** (−4)(5) = −20 **9. a.** −8 **b.** undefined **c.** −5 **d.** 0 **10.** $3 million **11.** 154 ft **12. a.** 6 **b.** 7 **c.** −6 **d.** 132 **13. a.** 16 **b.** −16 **c.** −1 **14.** −27 **15.** 1 **16.** −34 **17.** 42 **18.** 4 **19.** −72° **20.** left, −7 **21.** −15 **22.** no **23.** −15 **24.** 16 **25.** −40 **26.** 0 **27.** −5 **28.** 2 **29.** −18 **30.** −$244 **31.** 18 **32.** −4 + (−4) + (−4) + (−4) + (−4) = −20 **33.** The absolute value of a number is the distance from the number to 0 on the number line. Distance is either positive or 0, but never negative. **34.** It is true because 12 = 12.

Cumulative Review Exercises (page 153)

1. 1, 2, 5, 9 **2.** 0, 1, 2, 5, 9 **3.** −2, −1 **4.** −2, −1, 0, 1, 2, 5, 9 **5.** 6 **6.** 3 **7.** 7,326,500 **8.** 7,330,000 **9.** CRF Cable **10.**

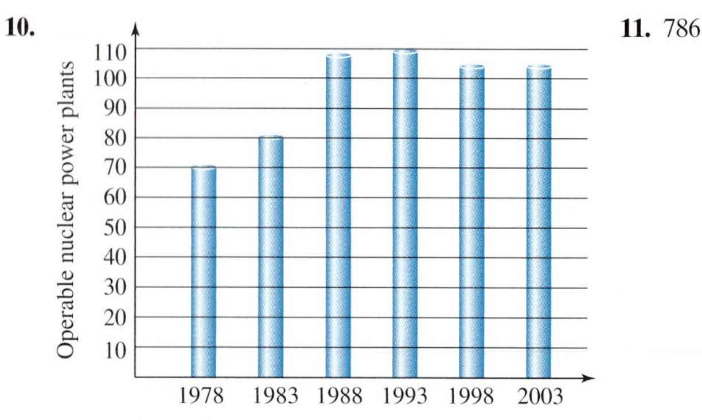

Source: *The World Almanac, 2005*

11. 786 **12.** 3,806 **13.** 4,684 **14.** 13,136 **15.** 104 ft, 595 ft^2 **16.** 65 **17.** 11,745 **18.** 13 **19.** 307,329 **20.** 467 **21.** 1,728 **22.** 1, 2, 3, 6, 9, 18 **23.** prime, odd **24.** composite, even **25.** even **26.** odd **27.** $2^3 \cdot 3^2 \cdot 7$ **28.** 11^4 **29.** 175 **30.** 38 **31.** 50 **32.** 2 **33.** no **34.** yes **35.** 13 **36.** 53 **37.** 27 **38.** 24 **39.**

−3 −2 −1 0 1 2 3

40.

−4 −3 −2 −1 0 1 2

41. true **42.** 9, −9 **43.** −5 **44.** −14 **45.** −8 **46.** −231 **47.** 24 **48.** −1,715 **49.** 2 **50.** −50 **51.** 26 **52.** −16 **53.** −3 **54.** 4 **55.** 3 **56.** −18 **57.** $126,037 **58.** −279° F

Chapter 3 Check Your Knowledge (page 156)

1. expression **2.** substitute **3.** simplify, solve **4.** like **5.** distributive **6. a.** $x + 7$ **b.** $y - 3$ **c.** $\frac{3}{4}z$ **7.** $\$(500 - x)$ **8. a.** 25*q*¢ or $0.25*q* **b.** $5*f* **9. a.** 25 **b.** −6 **10.** 36 mi **11. a.** $6x + 21$ **b.** $-3x + 7$ **c.** $-6x + 9$ **12. a.** factor **b.** term **13. a.** $3x$ **b.** $-x$ **14. a.** $15x$ **b.** $-5x$ **c.** $-5x + 1$ **15.** 6 **16.** 1 **17.** 10 **18.** 2 **19.** 6 **20.** 13, 4

Study Set Section 3.1 (page 162)

1. expression **3.** variable **5.** $10 + 3x$, $\frac{10 - x}{3}$ (answers may vary) **7.** Mr. Lamb; 15 mi **9.** $2p$, $3p$ **11.** 500, $500 + x$, $500 - x$ **13.** $\frac{h}{4}$ **15.** $450 - x$ **17.** $8x$ **19.** $\frac{10}{g}$ **21.** $x - 9$ **23.** $\frac{2}{3}p$ **25.** $6 + r$ **27.** $d - 15$ **29.** $1 - s$ **31.** $2p$ **33.** $s + 14$ **35.** $\frac{35}{b}$ **37.** $x - 2$ **39.** *c* increased by 7 **41.** 7 less than *c* **43. a.** $60m$ **b.** $3{,}600h$ **45. a.** $\frac{s}{12}$ **b.** $\frac{s}{52}$ **47. a.** $12f$ in. **b.** $\frac{f}{3}$ yd **49.** $j - 5$ **51.** $6s$ **53.** $\frac{p}{15}$ days **55.** $t + 2$ **57.** w = width, $w + 6$ = length **59.** g = gal drained out; $6 - g$ = gal remaining **61.** $3x + 5$ **63.** $10a + 12$ **65.** x = votes received by Nixon; $x + 118{,}550$ = votes received by Kennedy **67.** c = number of copies of *I Want to Hold Your Hand* sold; $c - 2{,}000{,}000$ = number of copies of *Hey Jude* sold **73.** −10 **75.** −4 **77.** {. . . , −3, −2, −1, 0, 1, 2, 3, . . .} **79.** 2

Think It Through (page 173)

4, 6, 6, 9, 8, 12, 10, 15

Study Set Section 3.2 (page 173)

1. formula **3.** substitute **5.** 2 − 8 + 10; it looks like subtraction **7. a.** x = length part 1; $x - 40$ = length part 2; $x + 16$ = length part 3 (answers may vary) **b.** 20 in. and 76 in. **9. a.** 22, 27, $p + 2$ **b.** $T = p + 2$ **11.** 48, $3t$, $3x$ **13. a.** health club instructor **b.** mechanic **c.** paleontologist **d.** realtor **e.** doctor **f.** economist **15. a.** $d = rt$ **b.** $C = \frac{5}{9}(F - 32)$ **c.** $d = 16t^2$ **17.** 17 **19.** 4 **21.** 40 **23.** −6 **25.** −6 **27.** 23 **29.** −8 **31.** 100 **33.** −28 **35.** 3 **37.** 44 **39.** −21 **41.** −3 **43.** −7 **45.** −18 **47.** 25 **49.** 21 **51.** −5 **53.** −29 **55.** −45 **57.** 21 **59.** 70 cents **61.** $8,200 **63.** $23 **65.** 300 mi **67.** −10°C **69.** 239 **71.** 64 ft **73.** 5,213, 5,079, 4,814; 2,053, 2,051, 1,921; 3,160, 3,028, 2,893 **75.** 30°, 15°, −5° **77.** 16, 16 ft; 64, 48 ft; 144, 80 ft; 256, 112 ft **79.** 4 **87.** 17, 37, 41 **89.** 7 **91.** division by 3 **93.** 3

Study Set Section 3.3 (page 182)

1. distributive, removed **3.** equivalent **5.** $x(y + z) = xy + xz$ **7.** $(w + 7)5$ **9.** 5, 6, 6, 2, 3 **11.** $-y - 9$ **13.** −5 **15.** −9, −9, $-45y$ **17. a.** x **b.** $x + 5$ **c.** $5x - 10y - 15$ **d.** $5x$ **19.** $12x$ **21.** $-30y$ **23.** $100t$ **25.** $12s$ **27.** $14c$ **29.** $-40h$ **31.** $-42xy$ **33.** $16rs$ **35.** $30xy$ **37.** $-30br$ **39.** $80c$ **41.** $-8e$ **43.** $4x + 4$ **45.** $16 - 4x$ **47.** $-6e - 6$ **49.** $-16q + 48$ **51.** $12 + 20s$ **53.** $42 + 24d$ **55.** $-25r + 30$ **57.** $-24 - 18d$ **59.** $9x - 21y + 6$ **61.** $9z + 9x + 15y$ **63.** $-x - 3$ **65.** $-4t - 5$ **67.** $3w + 4$ **69.** $-5x + 4y - 1$ **71.** $2(4x + 5)$ **73.** $(-4 - 3x)5$ **75.** $-3(4y - 2)$ **77.** $3(4 - 7t - 5s)$ **83.** 5 **85.** multiplication, division, subtraction, addition **87.** > **89.** carpeting, painting

Study Set Section 3.4 (page 189)

1. term **3.** combined **5.** distributive **7.** sum **9. a.** term **b.** factor **c.** factor **d.** factor **11. a.** 11 **b.** 8 **c.** −4 **d.** 1 **e.** −1 **f.** 102 **13.** 6, m; −75, t; 1, w; 4, bh **15.** It helps identify the like terms. **17.** $(2d + 25)$ mi **19.** To add the like terms, add 9 and 5 and keep the variable. **21.** 7 **23.** 2 **25.** 2, $5x$ **27.** $3x^2$, $5x$, 4 **29.** 5, $5t$, $-8t$, 4 **31.** 2 **33.** 5 **35.** $15t$ **37.** $4s$ **39.** x **41.** $4d$ **43.** $-4e$ **45.** cannot be simplified **47.** $-6z$ **49.** $-7x$ **51.** 0 **53.** 0 **55.** $4x$ **57.** cannot be simplified **59.** $-2y$ **61.** $3a$ **63.** $11t + 12$ **65.** $2w - 5$ **67.** $-7r + 11R$ **69.** $-50d$ **71.** $8x - 4y - 9$ **73.** $9x + 34$ **75.** $-22s + 23$ **77.** $19e - 21$ **79.** $2t + 8$ **81.** $3x + 8$ **83.** $10y - 32$ **85.** $288 **87.** 36 ft, 48 ft, 60 ft, 72 ft, 84 ft **93.** 2 **95.** $2^2 \cdot 5^2$ **97.** absolute value

Study Set Section 3.5 (page 196)

1. solve **3.** distributive **5.** combine **7.** When we substitute −5 for x, the result is a false statement: $-10 = -9$. **9.** $5k$ **11. a.** $4x$ **b.** $2x$ **13. a.** $2t - 8$ **b.** −4 **c.** −16 **15.** $2x$, 2, 2 **17.** 9, 45, 45, 45, $5x$, 5, 5 **19.** yes **21.** no **23.** 6 **25.** 3 **27.** −30 **29.** −28 **31.** 42 **33.** 37 **35.** 306 **37.** 735 **39.** 2 **41.** −14 **43.** −8 **45.** 5 **47.** −12 **49.** 4 **51.** 8 **53.** 10 **55.** 6 **57.** 0 **59.** −4 **61.** −10 **63.** 0 **65.** 1 **67.** 2 **69.** −11 **71.** −4 **73.** −7 **75.** 7 **77.** 3 **79.** 26 **81.** −3 **87.** −16 **89.** −3 **91.** 5 **93.** positive

Study Set Section 3.6 (page 203)

1. addition **3.** $5x$ **5.** $g - 100$ **7.** $3m$ **9.** $2w$ **11.** 30, 24, $5x$, $4(9 - x)$ **13. a.** 9 **b.** $9 - d$ **17.** 17 mo **19.** 11 yr **21.** $975 **23.** 61 **25.** 4 **27.** 400 gal **29.** 21 mi **31.** 10 ft **33.** 6 min **35.** 6 pairs of dress shoes, 3 pairs of athletic shoes **37.** 14 hr, 6 hr **43.** associative property of addition **45.** −100 **47.** addition **49.** $2^3 \cdot 5^2$

Key Concept (page 207)

1. parentheses, innermost, outermost **3.** multiplications, divisions **5.** −15 **7.** 0 **9.** 206 **11.** −15 **13.** 92 ft **15.** $3x$ **17.** no **19.** Undo the subtraction first. Then undo the multiplication.

Chapter Review (page 209)

1. Brandon is closer by 250 mi. **2.** $h + 7$ **3.** $n - 5$ **4.** $7x$ **5.** $\frac{6}{p}$ **6.** $s + (-15)$ **7.** $2l$ **8.** $D - 100$ **9.** $r + 2$ **10.** $\frac{45}{x}$ **11.** $\frac{c}{6}$ **12.** $(1{,}000 - x)$ dollars **13.** $2x$ = hours husband drove **14.** length = $w + 3$ **15.** $12x$ **16.** $\frac{d}{7}$ **17.** h = height of wall, length of upper base = $h - 5$, length of lower base = $2h - 3$ **18.** upper base = 5 ft, lower base = 17 ft **19.** 12 **20.** −8 **21.** 100 **22.** −4 **23.** 130, 114, $6x$, $55t$ **24.** $278 **25.** $15,230 **26.** 2002 **27.** 2004 **28.** They decreased. **29.** The pool is 2°C warmer. **30.** 144 ft **31.** 24 yr **32.** $-10x$ **33.** $42xy$ **34.** $60de$ **35.** $32s$ **36.** $2e$ **37.** $49xy$ **38.** $84k$ **39.** $100t$ **40.** $4y + 20$ **41.** $-30t - 45$ **42.** $-21 - 21x$ **43.** $-12e + 24x + 3$ **44.** $-6t + 4$ **45.** $-5 - x$ **46.** $-6t + 3s - 1$ **47.** $5a + 3$ **48.** $-4x$, 8 **49.** $-3y$, 1 **50.** factor **51.** term **52.** factor **53.** term **54.** yes **55.** no **56.** yes **57.** no **58.** $7x$ **59.** $-9t$ **60.** $-3z$ **61.** $5x$ **62.** $-12y$ **63.** $w - 5$ **64.** $-46d + 2a$ **65.** $10y + 15h - 1$ **66.** $13y + 48$ **67.** $-5t + 22$ **68.** $3x + 12$ **69.** $-50f + 84$ **70.** $-14m$ **71.** $14m$ **72.** 194 ft **73.** yes **74.** −18 **75.** 8 **76.** −3 **77.** 15 **78.** −3 **79.** 2 **80.** −3 **81.** −9 **82.** 10, 60; 25, 175; 1, x; 5, $5n$ **83.** $56 - c$ **84.** 6 hr **85.** 13 mi **86.** 2,200 **87.** 15 $3 drinks, 35 $4 drinks

Chapter 3 Test (page 215)

1. a. $r - 2$ **b.** $3xy$ **c.** $x + 100$ **2.** $51{,}000 - e$ = yearly earnings of husband **3. a.** −3 **b.** 26 **c.** 1 **4.** 165 mi **5.** $37,000 **6.** It would be 56 ft short of hitting the ground. **7.** 1 **8.** 250 ft **9.** 15°C **10. a.** $25x + 5$ **b.** $-42 + 6x$ **c.** $-6y - 4$ **d.** $6a + 9b - 21$ **11. a.** factor **b.** term **12. a.** $-28y + 10$ **b.** $-3t$ **13. a.** $8x^2$, $-4x$, −6 **b.** 8 **14. a.** $11x$ **b.** $24ce$ **c.** $5x$ **d.** $30y$ **e.** $-7x$ **f.** $9y$ **15.** $-16y - 3$ **16.** no **17.** −9 **18.** −3 **19.** 4 **20.** −10 **21.** −10 **22. a.** $10k$¢ **b.** $20(p + 2)$ dollars **23.** 3 hr **24.** 8 hr **25.** terms with exactly the same variables and exponents **26.** t = length trout, $t + 10$ = length salmon; s = length salmon, $s - 10$ = length trout **27.** No; we simplify expressions, and we solve equations. **28.** $2(x + 3) = 2x + 6$ (answers may vary)

Cumulative Review Exercises (page 217)

1. 358,600,000 gal **2.** 50,000 **3.** 54,604 **4.** 4,209 **5.** 23,115 **6.** 87 **7.** $683 + 459 = 1{,}142$ **8.** 2011 **9.** $4 \cdot 5 = 5 + 5 + 5 + 5 = 20$ **10.** 10,912 in.^{2} **11. a.** 1, 2, 3, 6, 9, 18 **b.** $3^2 \cdot 2$ **12.** 2, 3, 5, 7, 11, 13, 17, 19, 23, 29 **13.** It has factors other than 1 and itself. For example, $27 = 3 \cdot 9$. **14.** 22 **15.** 500 **16.** 6

17. −4 −3 −2 −1 0 1 2 3 4

18. 5 **19.** false

23. 1 **24.** true **25.** false **26.** $\frac{2}{9}$ **27.** $-\frac{8}{21}s$ **28.** $\frac{d}{21}$ **29.** $-\frac{5m}{9n}$ **30.** $\frac{9}{10}$ **31.** $-\frac{125}{8} = -15\frac{5}{8}$ **32.** $\frac{x^2}{9}$ **33.** $-\frac{8c^3}{125}$ **34.** 30 lb **35.** 60 in.^{2} **36.** 8 **37.** $-\frac{12}{11}$ **28.** $\frac{1}{x}$ **39.** $\frac{c}{ab}$ **40.** $\frac{25}{66}$ **41.** $-\frac{7}{2}$ **42.** $\frac{3}{32}$ **43.** $\frac{5}{2}$ **44.** $\frac{t}{2}$ **45.** $\frac{8}{5}$ **46.** $\frac{a^2}{b^2}$ **47.** $-\frac{6}{x}$ **48.** 12 **49.** $\frac{5}{7}$ **50.** $-\frac{6}{5}$ **51.** $\frac{2}{x}$ **52.** $\frac{7+t}{8}$ **53.** The denominators are not the same. **54.** 90 **55.** $\frac{5}{6}$ **56.** $\frac{1}{40}$ **57.** $-\frac{29}{24}$ **58.** $\frac{20}{7}$ **59.** $\frac{2x-15}{50}$ **60.** $\frac{y+21}{3y}$ **61.** $-\frac{23}{6}$ **62.** $\frac{47}{60}$ **63.** $\frac{7}{32}$ in. **64.** the second hour **65.** $2\frac{1}{6}$ **66.** $\frac{13}{6}$ **67.** $3\frac{1}{5}$ **68.** $-3\frac{11}{12}$ **69.** 1 **70.** $2\frac{1}{3}$ **71.** $\frac{75}{8}$ **72.** $-\frac{11}{5}$ **73.** $\frac{201}{2}$ **74.** $\frac{199}{100}$

75. Number line from -5 to 5 with points graphed at $-2\frac{2}{3}$, $\frac{8}{9}$, $\frac{59}{24}$

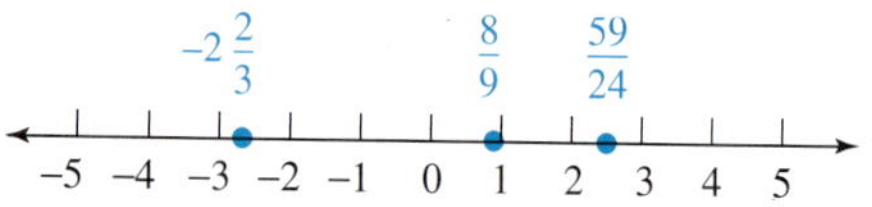

76. $-\frac{3}{10}$ **77.** $\frac{21}{22}$ **78.** 40 **79.** $-2\frac{1}{2}$ **80.** $48\frac{1}{8}$ in. **81.** $3\frac{23}{40}$ **82.** $6\frac{1}{6}$ **83.** $1\frac{1}{12}$ **84.** $1\frac{5}{16}$ **85.** $39\frac{11}{12}$ gal **86.** $182\frac{5}{18}$ **87.** $113\frac{3}{20}$ **88.** $31\frac{11}{24}$ **89.** $316\frac{3}{4}$ **90.** $20\frac{1}{2}$ **91.** $34\frac{3}{8}$ **92.** $\frac{8}{9}$ **93.** $\frac{19}{72}$ **94.** $-\frac{12}{17}$ **95.** $-\frac{2}{5}$ **96.** $\frac{97}{64}$ **97.** $-\frac{63}{32}$ **98.** 22 **99.** $\frac{14}{33}$ **100.** 24 **101.** 28 **102.** $-\frac{1}{3}$ **103.** $\frac{11}{2}$ **104.** $\frac{48}{5}$ **105.** $-\frac{18}{5}$ **106.** 16 **107.** $\frac{26}{3}$ **108.** 330

Chapter 4 Test (page 303)

1. a. $\frac{4}{5}$ **b.** $\frac{1}{5}$ **2. a.** $\frac{3}{4}$ **b.** $\frac{2n}{5}$ **3.** $-\frac{3}{20x}$ **4.** 40 **5.** $\frac{12}{a}$ **6.** $\frac{5x-24}{30}$ **7.** $\frac{21a}{24a}$ **8.** Number line from -2 to 3 with points graphed at $-1\frac{1}{7}$, $\frac{7}{6}$, $2\frac{4}{5}$

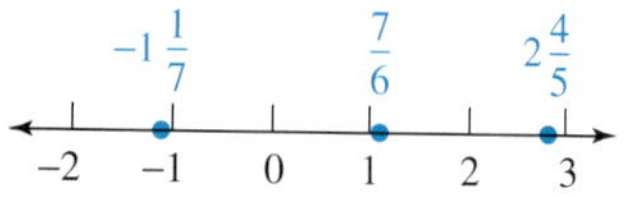

9. \$$1\frac{1}{2}$ million **10.** $\frac{2}{3}$ **11.** $261\frac{11}{36}$ **12.** $37\frac{5}{12}$ **13. a.** 0 lb **b.** $2\frac{3}{4}$ in. **c.** $3\frac{3}{4}$ in. **14.** $\frac{11}{7}$ **15.** $11\frac{1}{4}$ in. **16.** perimeter: $53\frac{1}{3}$ in.; area: $106\frac{2}{3}$ in.^{2} **17.** 60 **18.** 12 **19.** $\frac{13}{24}$ **20.** $-\frac{20}{21}$ **21.** $-\frac{5}{3}$ **22.** yes **23.** 42 **24.** $-\frac{36}{5}$ **25.** $\frac{1}{6}$ **26.** 8 **27.** 108, 36 **28.** numerator, fraction bar, denominator; equal parts of a whole, or a division **29.** When we multiply a number, such as $\frac{3}{4}$, and its reciprocal, $\frac{4}{3}$, the result is 1. **30. a.** simplifying a fraction: dividing the numerator and denominator of a fraction by the same number **b.** equivalent fractions: $\frac{1}{2} = \frac{2}{4}$ **c.** building a fraction: multiplying the numerator and denominator of a fraction by the same number

Cumulative Review Exercises (page 305)

1. 5,434,700 **2.** 5,430,000 **3.** 11,555, 10:30 A.M. **4.** hundred billions **5.** 8,136 **6.** 3,519 **7.** 299,320 **8.** 991 **9.** 450 ft **10.** 11,250 ft^2 **11.** $2^2 \cdot 3 \cdot 7$ **12.** $2 \cdot 3^2 \cdot 5^2$ **13.** $2^3 \cdot 3^2 \cdot 5$ **14.** $2^4 \cdot 3^2 \cdot 5^2$ **15.** 16 **16.** -35 **17.** 2 **18.** 2 **19.** $x + 15$ **20.** $x - 8$ **21.** $4x$ **22.** $\frac{x}{10}$ **23.** 7 **24.** 14 **25.** -52 **26.** -2 **27.** $-15x$ **28.** $28x^2$ **29.** $-6x + 8$ **30.** $-15x + 10y - 20$ **31.** $5x$ **32.** $7a^2$ **33.** $-x - y$ **34.** $-4x + 8$ **35.** -5 **36.** -5 **37.** -16 **38.** -5 **39.** 4 **40.** 4 **41.** 21 **42.** 21 ft by 84 ft **43.** $\frac{3}{4}$ **44.** $\frac{5x^3}{2y}$ **45.** $-\frac{4}{5}$ **46.** $\frac{1}{2p}$ **47.** $1\frac{5}{12}$ **48.** $\frac{20-3m}{5m}$ **49.** $\frac{23}{6}$ **50.** $-\frac{53}{8}$ **51.** $9\frac{11}{12}$ **52.** $5\frac{11}{15}$ **53.** $\frac{11}{16}$ in. **54.** 30 sec, 60 sec **55.** $\frac{2}{7}$ **56.** $-1\frac{9}{29}$ **57.** $-\frac{17}{15}$ **58.** 4 **59.** -15 **60.** $\frac{8}{3}$ **61.** An expression is a combination of numbers and/or variables with operation symbols. An equation contains an = symbol. **62.** a letter that is used to stand for a number

Chapter 5 Check Your Knowledge (page 308)

1. sum **2.** root **3.** $\frac{21}{250}$ **4.** \$85.80 **5. a.** 354.2782 **b.** 20,004.78 **6. a.** 7.875 ft^2 **b.** 11.5 ft **7.** 3.1 **8. a.** 0.15 **b.** 0.625 **c.** $0.\overline{1}$ **9.** 1.6

10. Number line from -2 to 2 with points graphed at -0.375, 0.25

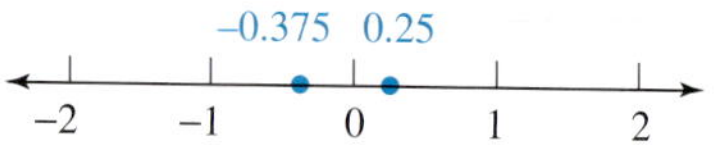

11. $\frac{5}{12}$ **12. a.** 6.9 **b.** 7 **c.** 9 **d.** 7

13. Number line from -2 to 2 with points graphed at -1.73, 1.41

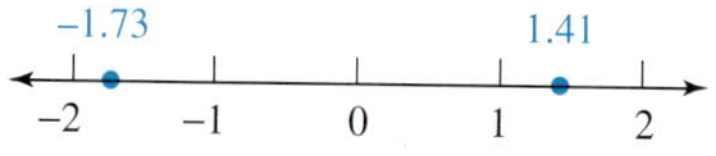

14. a. 19 **b.** $\frac{19}{20}$ **c.** -0.3 **15. a.** > **b.** < **c.** > **16.** 52.6 mph **17.** -9.43 **18.** -9 **19.** 0.8 **20.** 625

Study Set Section 5.1 (page 315)

1.

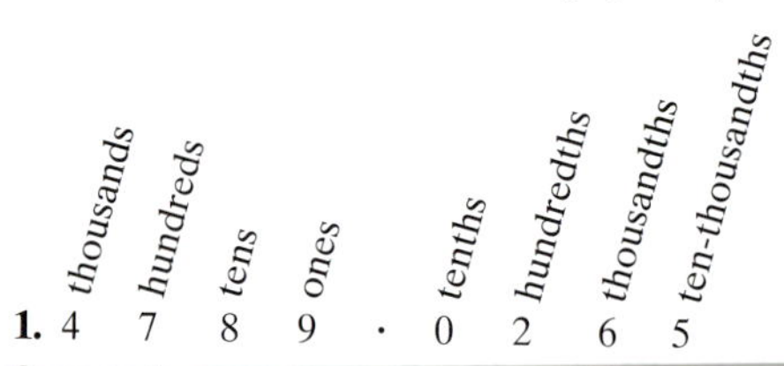

3. rounding **5. a.** thirty-two and four hundred fifteen thousandths **b.** 32 **c.** $\frac{415}{1{,}000}$ **d.** $30 + 2 + \frac{4}{10} + \frac{1}{100} + \frac{5}{1{,}000}$

7. Number line from -5 to 5 with points graphed at $-3\frac{1}{100}$, -0.7, $\frac{7}{10}$, 3.01

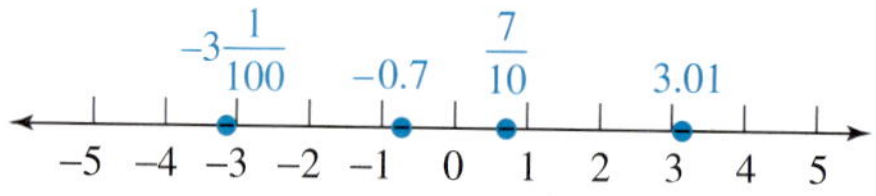

9. a. true **b.** false **c.** true **d.** true **11.** $\frac{47}{100}$, 0.47 **13.** Line segment with a mark labeled 0.3 **15.** 9,816.0245 **17.** fifty and one tenth; $50\frac{1}{10}$ **19.** negative one hundred thirty-seven ten-thousandths; $-\frac{137}{10{,}000}$ **21.** three hundred four and three ten-thousandths; $304\frac{3}{10{,}000}$ **23.** negative seventy-two and four hundred ninety-three thousandths; $-72\frac{493}{1{,}000}$ **25.** -0.39 **27.** 6.187 **29.** 506.1 **31.** 2.7 **33.** -0.14 **35.** 33.00 **37.** 3.142 **39.** 1.414 **41.** 39 **43.** 2,988 **45. a.** \$3,090 **b.** \$3,090.30 **47.** < **49.** > **51.** 132.64, 132.6401, 132.6499 **53.** \$1,025.78

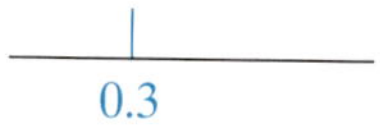

55.

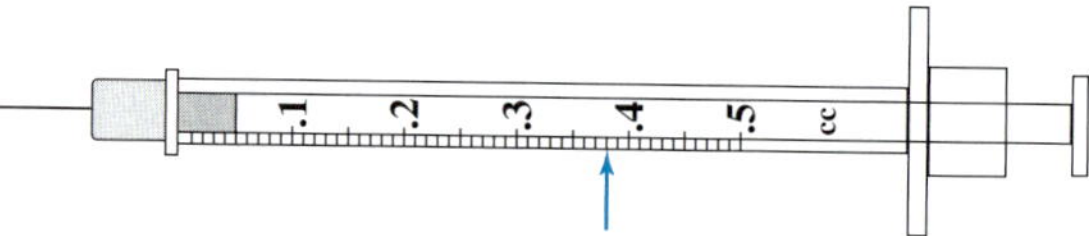

57. a. 0.30 **b.** 1,609.34 **c.** 453.59 **d.** 3.79 **59.** sand, silt, granule, clay **61.** Texas City, Houston, Westport, Galveston, White Plains, Crestline **63.** gold: Patterson; silver: Khorkina; bronze: Zhang **65. a.** Q1, 2004; \$0.25 **b.** Q2, 2002; $-$\$0.25 **73.** $164\frac{11}{20}$ **75.** $8R - 18$ **77.** 72 in.^{2} **79.** -1

Study Set Section 5.2 (page 322)

1. sum **3.** opposite **5.** point **7.** 39.9 **9.** 54.72 **11.** 15.9 **13.** 0.23064 **15.** 288.46 **17.** 58.04 **19.** 9.53 **21.** 70.29

23. 4.977 **25.** 0.19 **27.** −10.9 **29.** 38.29 **31.** −14.3 **33.** −0.0355 **35.** −16.6 **37.** 47.91 **39.** 2.598 **41.** 11.01 **43.** 4.1 **45.** 35.85 **47.** −57.47 **49.** 6.2 **51.** 15.2 **53.** 8.03 **55. a.** 53.044 sec **b.** 102.38 **57.** 103.4 in. **59.** 1.8, Texas **61.** 1.74, 2.32, 4.06, 2.90, 0, 2.90 **63.** 43.03 sec **65.** \$765.69, \$740.69 **67. a.** \$101.94 **b.** \$55.80 **69.** 8,156.9343 **71.** 1,932.645 **73.** 2,529.0582 **79.** $110\frac{23}{40}$ **80.** $-\frac{5}{6}$

Study Set Section 5.3 (page 331)

1. factors, product **3.** whole, sum **5. a.** $\frac{21}{1,1000}$ **b.** $\frac{21}{1,000}$ = 0.021. They are the same. **7.** 2.3 **9.** 0.08 **11.** −0.15 **13.** 0.98 **15.** 0.072 **17.** 12.32 **19.** −0.0049 **21.** −0.084 **23.** −8.6265 **25.** 9.6 **27.** −56.7 **29.** 12.24 **31.** −18.183 **33.** 0.024 **35.** −16.5 **37.** 42 **39.** 6,716.4 **41.** −0.56 **43.** 8,050 **45.** 980 **47.** −200 **49.** 0.01, 0.04, 0.09, 0.16, 0.25, 0.36, 0.49, 0.64, 0.81 **51.** 1.44 **53.** 1.69 **55.** −17.48 **57.** 14.24 **59.** 0.84 **61.** −3.872 **63.** 18.72 **65.** 86.49 **67.** 18.94 **69.** 7.6 **71.** 36.3 **73. a.** \$12.50, \$12,500, \$15.75, \$1,575 **b.** \$14,075 **75.** 0.75 in. **77.** 136.4 lb **79.** \$95.20, \$123.75, \$3.85x **81.** 160.6 m **83.** 0.000000136 in., 0.0000000136 in., 0.00000004 in. **85.** 15.29694 **87.** 631.2722 **89.** \$102.65 **95.** −12 **97.** the absolute value of negative three **99.** −1

Think It Through (page 339)

2.86

Study Set Section 5.4 (page 341)

1. dividend, divisor, quotient **3.** mean, median **5.** range **7.** whole, right, above **9.** true **11.** 10 **13.** Use multiplication to see whether 2.13 · 0.9 = 1.917. **15.** yes **17.** 3.3, 3.6, 2.3, 2.2 **19.** moving the decimal points in the divisor and dividend 2 places to the right **21.** 4.5 **23.** −9.75 **25.** 6.2 **27.** 32.1 **29.** 2.46 **31.** −7.86 **33.** 2.66 **35.** 7.17 **37.** 130 **39.** 1,050 **41.** 0.6 **43.** 0.6 **45.** 5.3 **47.** −2.4 **49.** 13.60 **51.** 0.79 **53.** 0.07895 **55.** −0.00064 **57.** 0.0348 **59.** 4.504 **61.** −0.96 **63.** 1,027.19 **65.** 9.1 **67.** 304.07 **69.** 280 **71.** 11 hr later: 6 P.M. **73.** 567 **75.** 1998: \$13.00; 2003: \$15.35 **77.** 0.37 mi **79.** 22.525 oz, 25 oz, 17.3 oz **81.** \$4.15, \$4.19, \$4.29 **83.** 7.24 **85.** −3.96 **91.** $\frac{7}{6}$ **93.** {. . . , −3, −2, −1, 0, 1, 2, 3, . . .} **95.** 12 **97.** 6

Study Set Estimation (page 345)

1. approx. \$240 **3.** approx. 2 cubic feet less **5.** approx. 30 **7.** approx. \$330 **9.** approx. \$520 **11.** not reasonable **13.** reasonable **15.** reasonable **17.** not reasonable

Study Set Section 5.5 (page 351)

1. repeating **3.** decimal **5. a.** 7 ÷ 8 **b.** numerator **7.** smaller

9.

$-3.8\overline{3}$ -0.75 $0.\overline{6}$ $1\frac{3}{4}$

−5 −4 −3 −2 −1 0 1 2 3 4 5

11. a. false **b.** true **c.** true **d.** false **13. a.** no **b.** It is a repeating decimal. **15.** 0.5 **17.** −0.625 **19.** 0.5625 **21.** −0.53125 **23.** 0.55 **25.** 0.775 **27.** −0.015 **29.** 0.002 **31.** $0.\overline{6}$ **33.** $0.\overline{45}$ **35.** $-0.58\overline{3}$ **37.** $0.0\overline{3}$ **39.** 0.23 **41.** 0.38 **43.** 0.152 **45.** 0.370 **47.** 1.33 **49.** −3.09 **51.** 3.75 **53.** −8.67 **55.** 12.6875 **57.** 203.73 **59.** < **61.** < **63.** $\frac{37}{90}$ **65.** $\frac{19}{60}$ **67.** $\frac{3}{22}$ **69.** $-\frac{1}{90}$ **71.** 1 **73.** 0.57 **75.** 5.27 **77.** 0.24 **79.** −2.55 **81.** 0.068 **83.** 7.11 **85.** −1.7 **87.** 4.25 **89.** 113.04 **91.** $0.\overline{2277}$ **93.** 37.2 **95.** 0.0625, 0.375, 0.5625, 0.9375 **97.** $\frac{3}{40}$ in. **99.** 23.4 sec, 23.8 sec, 24.2 sec, 32.6 sec **101.** 93.6 in.^{2} **107.** −1 **109.** $-T - 8t$ **111.** $6x^2$

Study Set Section 5.6 (page 359)

1. solve **3.** coefficient **5.** 2.1(1.7) − 6.3 = −2.73 **7.** $7.8x + 9.1 + 12.4$ **9. a.** \$0.25 **b.** \$0.01 **c.** \$2.50 **d.** \$0.99 **11.** the distributive property **13.** 2.3, 2.3, 0.6s, 0.6, 0.6 **15.** $10.1x$ **17.** $0.02h$ **19.** $-3.7r$ **21.** $-8.78x + 12.3$ **23.** $-0.5x + 3.9$ **25.** $-0.01x + 5$ **27.** 1.7 **29.** 7.11 **31.** −11.5 **33.** −0.1 **35.** −4.36 **37.** 1.3 **39.** −8.16 **41.** 22.44 **43.** −21.18 **45.** 0.4 **47.** −2.2 **49.** −2 **51.** 31 **53.** 1 **55.** 0.3 **57.** 0.6 **59.** −1.1 **61.** 2 **63.** 1.3 **65.** 5 **67.** 15, 30, 60, signatures, the number of signatures she needs to collect, \$0.30, 0.30x, 0.30x, 0.30x, 60, 0.30x. She needs to collect 150 signatures to make \$60. 150, 150, 45 **69.** \$8.6 million **71.** 3.27 **73.** 10.7 **75.** 12.4 mpg **77.** 200 **81.** $\frac{1}{12}$ **83.** $\frac{7}{8}$ **85.** −6

Study Set Section 5.7 (page 365)

1. root **3.** radical, positive **5.** radicand **7.** 25, 25 **9.** 7^2 **11.** $\frac{3}{4}$ **13.** $\sqrt{6}, \sqrt{11}, \sqrt{23}, \sqrt{27}$ **15. a.** 1 **b.** 0 **17. a.** 2.4 **b.** 5.76 **c.** 0.24

19.

$-\sqrt{5}$ $\sqrt{9}$

−5 −4 −3 −2 −1 0 1 2 3 4 5

21. a. 4, 5 **b.** 9, 10 **23.** −7, 8 **25.** 4 **27.** −11 **29.** −0.7 **31.** 0.5 **33.** 0.3 **35.** $-\frac{1}{9}$ **37.** $-\frac{4}{3}$ **39.** $\frac{2}{5}$ **41.** 31 **43.** −20 **45.** $-\frac{7}{20}$ **47.** −70 **49.** 2.56 **51.** −3.6 **53.** 1, 1.414, 1.732, 2, 2.236, 2.449, 2.646, 2.828, 3, 3.162 **55.** 37 **57.** 61 **59.** 3.87 **61.** 8.12 **63.** 4.904 **65.** −3.332 **67.** 4,899 **69.** −0.0333 **71. a.** 5 ft **b.** 10 ft **73.** 127.3 ft **75.** 42-inch **83.** subtraction and multiplication **85.** 16 **87.** {0, 1, 2, 3, 4, 5, 6, . . .} **89.** 30

The Real Numbers (page 368)

1.

-4 $-2\frac{1}{2}$ -0.1 $\frac{99}{100}$ $1.\overline{3}$ $\frac{13}{4}$ $\sqrt{17}$

−5 −4 −3 −2 −1 0 1 2 3 4 5

3. {1, 2, 3, 4, 5, . . .} **5.** {. . . , −3, −2, −1, 0, 1, 2, 3, . . .} **7.** nonterminating, nonrepeating decimals; a number that can't be written as a fraction **9.** false **11.** false **13.** true **15.** false **17.** true **19.** There is no real number that, when squared, yields a negative number.

Key Concept (page 369)

1. a. $5x - 8$ **b.** 3 **3. a.** $1.2y - 4.8$ **b.** −1.5

Chapter Review (page 371)

1. 0.67, $\frac{67}{100}$ **2.**

0.8

3. $10 + 6 + \frac{4}{10} + \frac{5}{100} + \frac{2}{1{,}000} + \frac{3}{10{,}000}$ **4.** two and three tenths, $2\frac{3}{10}$ **5.** negative fifteen and fifty-nine hundredths, $-15\frac{59}{100}$ **6.** six hundred one ten-thousandths, $\frac{601}{10{,}000}$ **7.** one one hundred thousandth, $\frac{1}{100{,}000}$

8. −2.7 −0.8 1.55
−5 −4 −3 −2 −1 0 1 2 3 4 5

9. Washington, Diaz, Chou, Singh, Gerbac **10.** true **11.** $<$ **12.** $>$ **13.** $=$ **14.** $<$ **15.** 4.58 **16.** 3,706.090 **17.** -0.1 **18.** 88.1 **19.** 66.7 **20.** 45.188 **21.** 15.17 **22.** 27.71 **23.** -7.7 **24.** 3.1 **25.** -4.8 **26.** -29.09 **27.** -25.6 **28.** 4.939 **29.** \$48.21 **30.** 8.15 in. **31.** -0.24 **32.** 2.07 **33.** -17.05 **34.** 197.945 **35.** 0.00006 **36.** 4.2 **37.** 90,145.2 **38.** 2,897 **39.** 0.04 **40.** 0.0225 **41.** 10.89 **42.** 0.001 **43.** -10.61 **44.** 25.82 **45.** 92.38 **46.** 68.62 in.^{2} **47.** 0.07 in. **48.** 1.25 **49.** -10.45 **50.** 1.29 **51.** 4.103 **52.** -2.9 **53.** 0.053 **54.** 63 **55.** 0.81 **56.** 12.9 **57.** -667.3 **58.** 20.22 **59.** \$8.34 **60.** 0.8976 **61.** -0.00112 **62.** 13.95 **63.** 14 **64.** 9.5 **65.** 7.3 microns, 7.2 microns, 6.9 microns, 1.1 microns **66.** \$1.45 billion **67.** 0.875 **68.** -0.4 **69.** -0.5625 **70.** 0.06 **71.** $0.\overline{54}$ **72.** $-0.\overline{6}$ **73.** 0.58 **74.** 1.03 **75.** $>$ **76.** $>$

77. −0.9 1.125 2.75
−5 −4 −3 −2 −1 0 1 2 3 4 5
$-0.\overline{3}$

78. $\frac{11}{15}$ **79.** -6.24

80. 93 **81.** 39.564 **82.** 33.49 **83.** 34.88 in.^{2} **84.** -18.41 **85.** 4.77 **86.** -5.34 **87.** 17 **88.** yes **89.** $2.8a - 12.4$ **90.** $3t - 1.4$ **91.** -0.6 **92.** 12 **93.** 9 **94. a.** radical **b.** 8^2, $(-8)^2$ **95.** 7 **96.** -4 **97.** 10 **98.** 0.3 **99.** $\frac{8}{5}$ **100.** 0.9 **101.** $-\frac{1}{6}$ **102.** 0 **103.** 9 and 10 **104.** It differs by 0.11.

105. $-\sqrt{2}$ $\sqrt{0}$ $\sqrt{3}$
−5 −4 −3 −2 −1 0 1 2 3 4 5

106. -30 **107.** 2.5 **108.** -27 **109.** 1.5 **110.** 4.36

Chapter 5 Test (page 377)

1. $\frac{79}{100}$, 0.79 **2.** Selway, Monroe, Paston, Covington, Cadia **3.** $\frac{271}{1{,}000}$ **4.** 33.050 **5.** \$208.75 **6. a.** 0.567909 **b.** 0.458 **7.** 1.02 in. **8.** 10.75 **9.** 6.121 **10.** 0.1024 **11.** 14.07 **12.** 125 mi^2 **13.** 0.004 in. **14.** 3.588 **15. a.** 0.34 **b.** $0.41\overline{6}$ **16.** -2.29 **17.** $1.\overline{18}$ **18.** 3.6, 3.6, 3.1 **19.** Half the families had more debt and half had less debt.

20. −0.8 0.375
−1 0 1

21. $\frac{41}{30}$ **22.** $4.96s + 2.3$

23. $5.2x - 18.7$ **24.** -7 **25.** 6.008 **26.** -0.425 **27.** 0.42 g **28.** 80 **29.** 12^2

30.

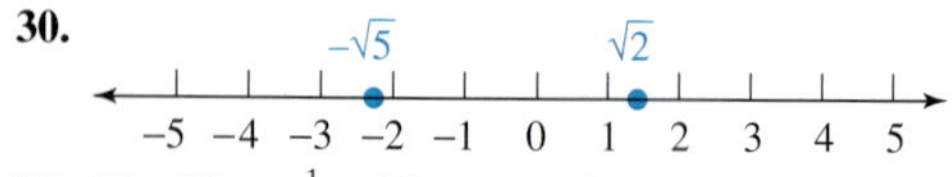

31. 11 **32.** $-\frac{1}{30}$ **33. a.** $>$ **b.** $>$ **c.** $>$ **d.** $<$ **34. a.** -0.2 **b.** 1.3

Cumulative Review Exercises (page 379)

1. \$788,000 **2.** $(x + y) + z = x + (y + z)$ **3.** 27 **4.** 1,000 **5.** $11 \cdot 5 \cdot 2^2$ **6.** 1, 2, 4, 5, 10, 20 **7.** $\{0, 1, 2, 3, 4, 5, \ldots\}$ **8.** -13 **9.** adding **10.** 8, -3, 36, -6, 6 **11.** $-15 = -5 \cdot 3$ **12.** -1 **13.** 9 **14.** 30 **15.** 35 **16.** 102 **17.** $3x$ ft **18.** $-\$1{,}100$ **19. a.** $(k + 1)$ in. **b.** $(m - 1)$ in. **20.** $2(4x + 5)$ **21.** 3 **22.** $-5w$ **23.** 5 **24.** $\frac{6}{13}$ **25.** equivalent fractions **26.** $\frac{5x}{7}$ **27.** $\frac{21}{128}$ **28.** $\frac{3y^2}{16}$ **29.** $\frac{28 + 2m}{7m}$ **30.** $19\frac{1}{8}$ **31.** $26\frac{7}{24}$ **32.** $-\frac{1}{3}$ **33.** -45 **34.** 8 **35.** 157.5 in.^{2}

36.

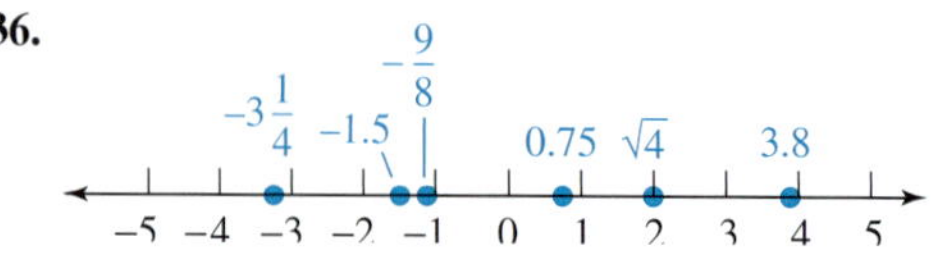

37. 0.001 in. **38.** $<$ **39.** -8.136 **40.** 5.6 **41.** 5,601.2 **42.** 0.0000897 **43.** 47.95 **44.** 33.6 hr **45.** 232.8°C **46.** 3.02, 3.005, 2.75 **47.** $0.14\overline{6}$ **48.** -9 **49.** 80 **50.** -6

Chapter 6 Check Your Knowledge (page 382)

1. quadrants **2.** origin **3.** line **4.** solution **5.** yes **6.** no

7.

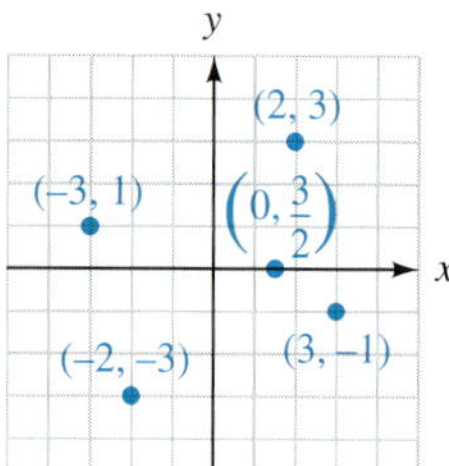

8. $A(3, 2)$; $B(2, -3)$; $C(-2, 0)$; $D(-1, 1)$; $E(0, -4)$

9. a. $-3, \frac{3}{2}, 1$ **b.** $(\frac{3}{2}, 0)$ **c.** $(0, -3)$

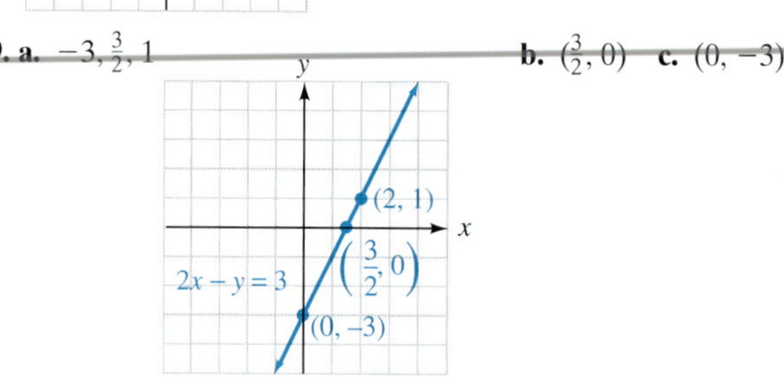

10. **11.**

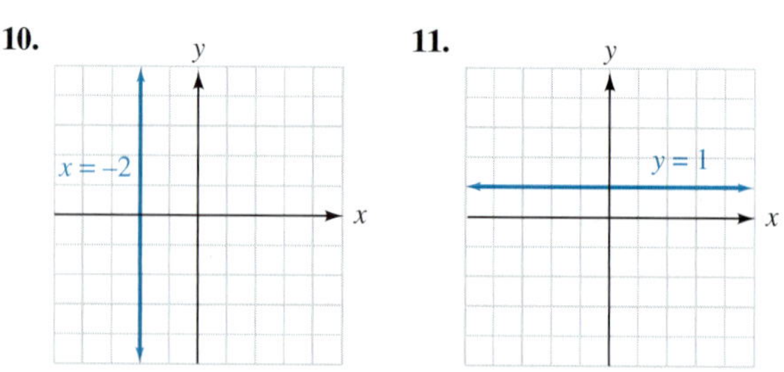

12. a. x^8 **b.** $6x^6$ **c.** $6x^3y^5$ **d.** x^6 **e.** $9x^2y^6$ **f.** x^{18} **13. a.** binomial **b.** 2nd degree **14.** -36

15. a.

y
x
$y = x^2 + 1$

16. $x^3 - x^2 - 2x + 8$ **17.** $-y^2 - 6y + 10$ **18.** $12x^2y^3z^2$ **19.** $2x^3 + 6x^2 - 10x$ **20.** $4y^2 - 49$ **21.** $8x^2 - 2x - 21$ **22.** $2x^3 - 5x^2 + 5x - 3$

Think It Through (page 389)

71,307,629

Study Set Section 6.1 (page 390)

1. ordered **3.** solution **5.** Cartesian **7.** origin

9.

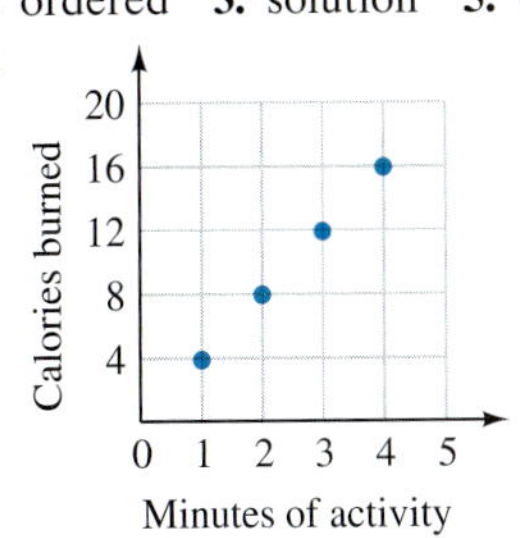

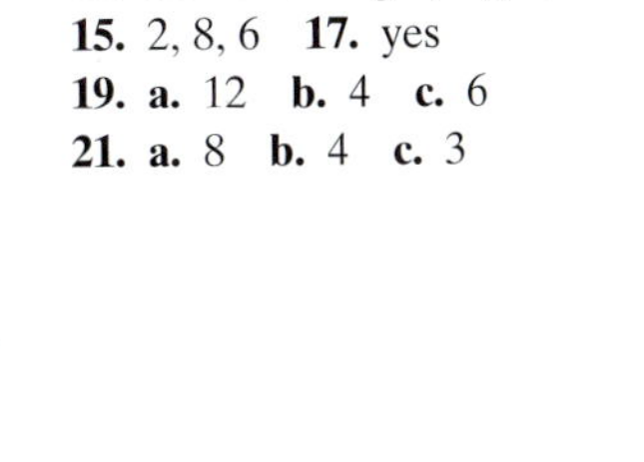

11. no **13.** origin, right, down **15.** 2, 8, 6 **17.** yes **19. a.** 12 **b.** 4 **c.** 6 **21. a.** 8 **b.** 4 **c.** 3

23.

x	y	(x, y)
0	−5	(0, −5)
4	0	(4, 0)
8	5	(8, 5)

25. (graph: points (−2, 4), (1, 3), (−3, −2), (3, −2))

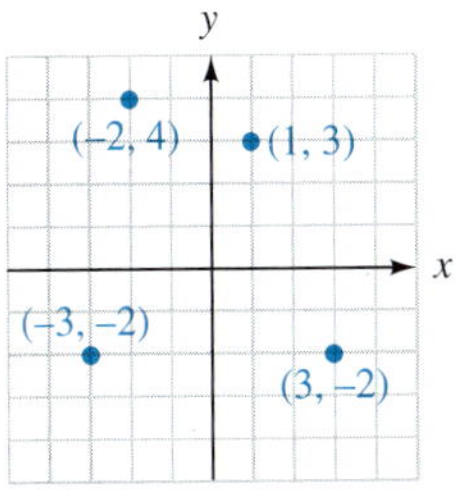

27.

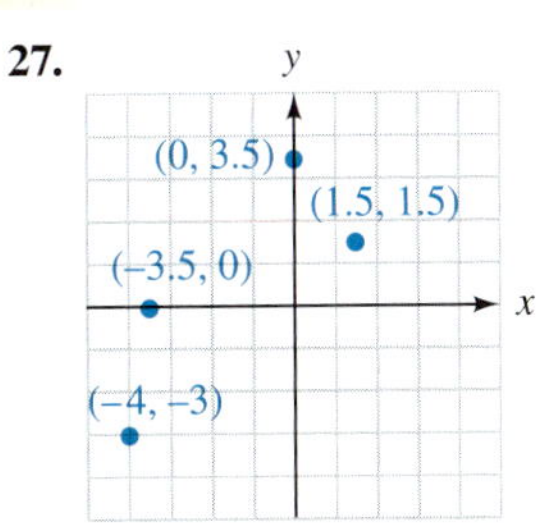

29. $A(2, 4)$, $B(-3, 3)$, $C(-2, -3)$, $D(4, -3)$ **31.** $A(-3, -4)$, $B(2.5, 3.5)$, $C(-2.5, 0)$, $D(2.5, 0)$ **33.** Rockford (5, B), Mount Carroll (1, C), Harvard (7, A), intersection (5, E) **35. a.** (2, −1) **b.** no **c.** yes **37.** New Delhi, Kampala, Coats Land, Reykjavik, Buenos Aires, Havana

39.

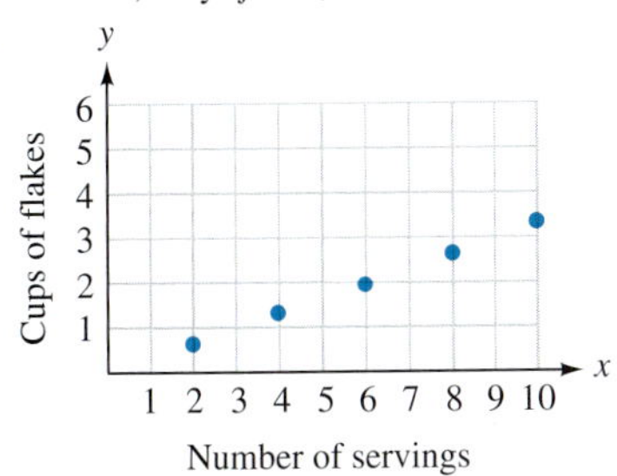

45. −16 **47.** 7 **49.** 21 **51.** −3 **53.** 65,536

Study Set Section 6.2 (page 400)

1. line **3.** x-intercept **5.** independent **7.** horizontal **9. a.** (0, 1) **b.** (−2, 0) **c.** yes **11.** Arrowheads were not drawn on both ends of the line. **13.** (−2, −2), (−1, −1), (0, 0), (1, 1), (2, 2), (3, 3) (answers may vary) **15.** 3, 6, 2 **17.** 0, (5, 0); −4, (−5, −4); 2, (10, 2) **19.** 3, (3, 3); −11, (−4, −11); 9, (6, 9) **21.** y-intercept is (0, 5); x-intercept is (5, 0) **23.** y-intercept is (0, 4); x-intercept is (5, 0)

25. 5, 5, 3

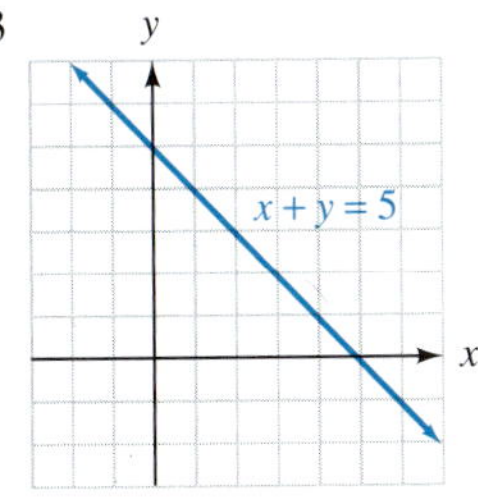

27. 4, 5, $\frac{16}{5}$

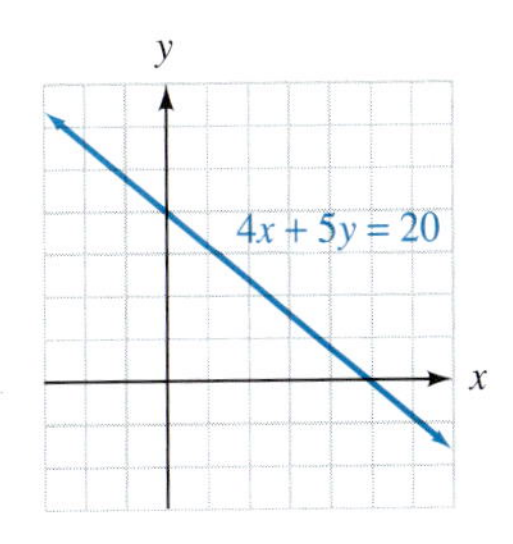

29. 2, −4, 4

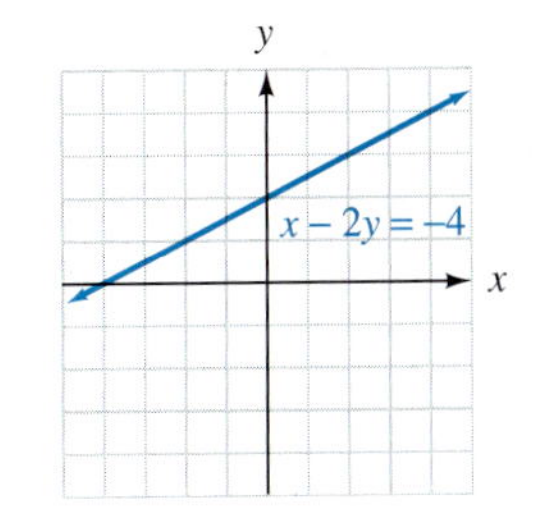

31.

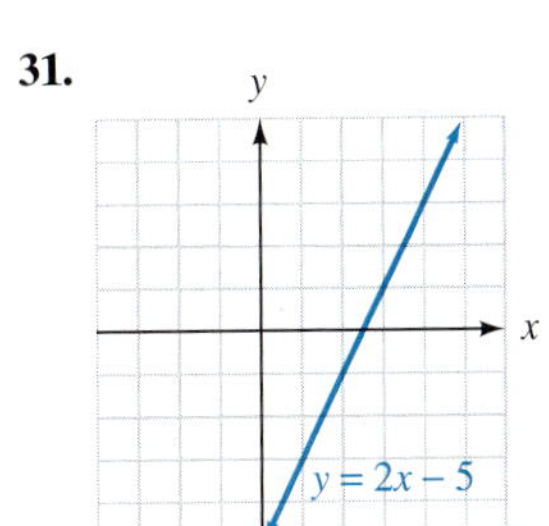

33.

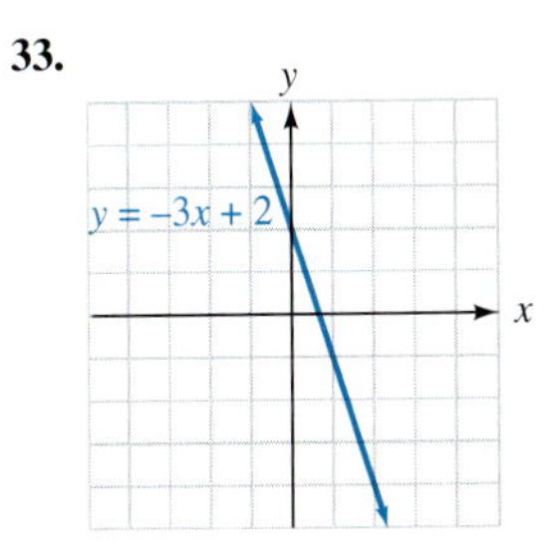

35.

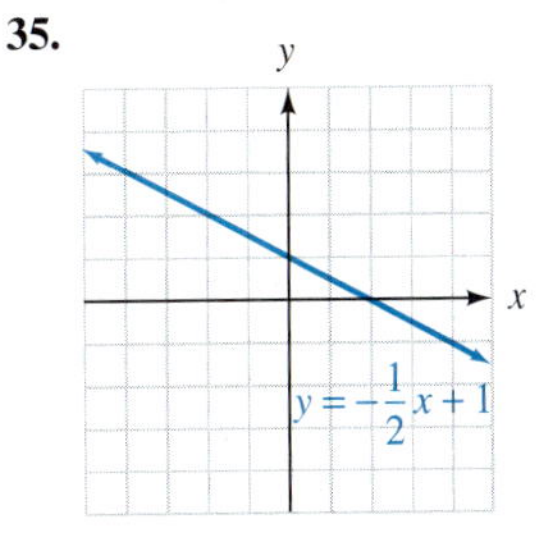

37.

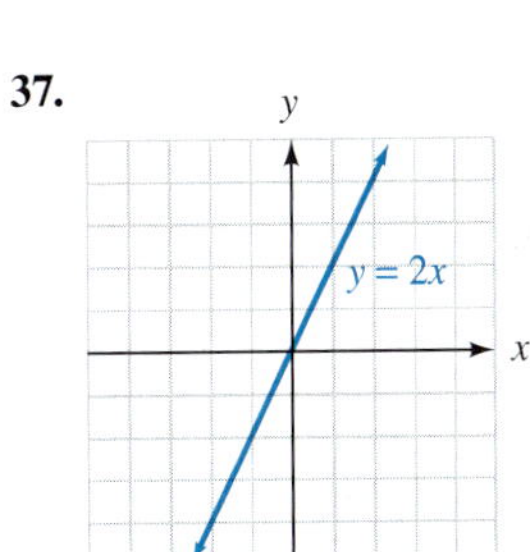

39.

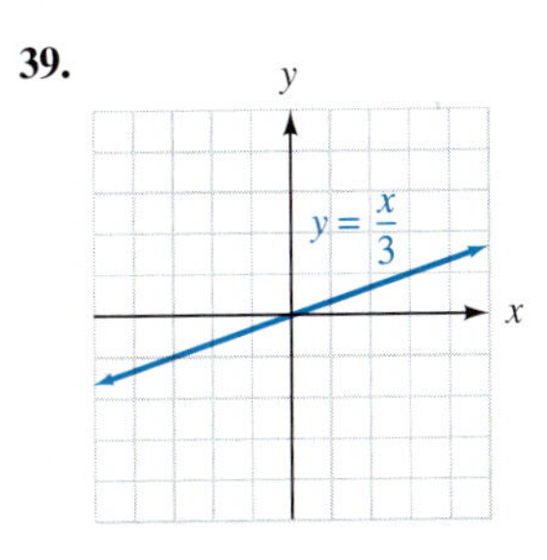

41.

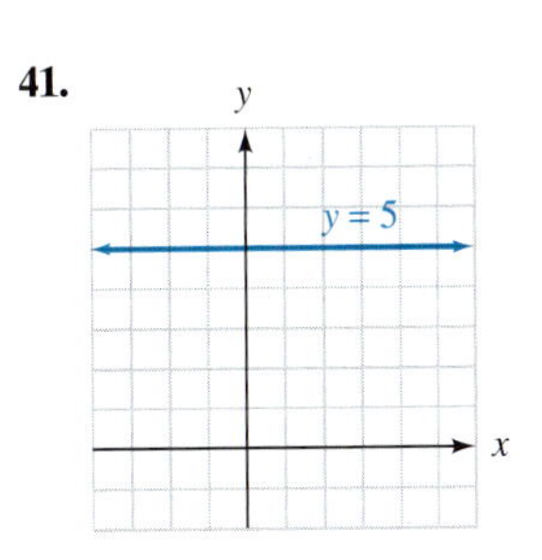

43.

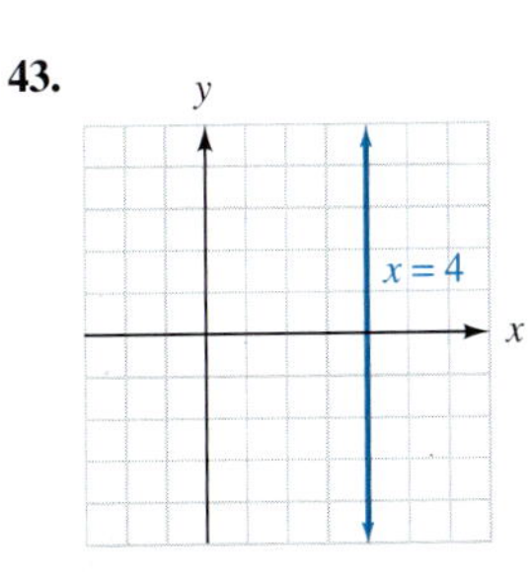

45.

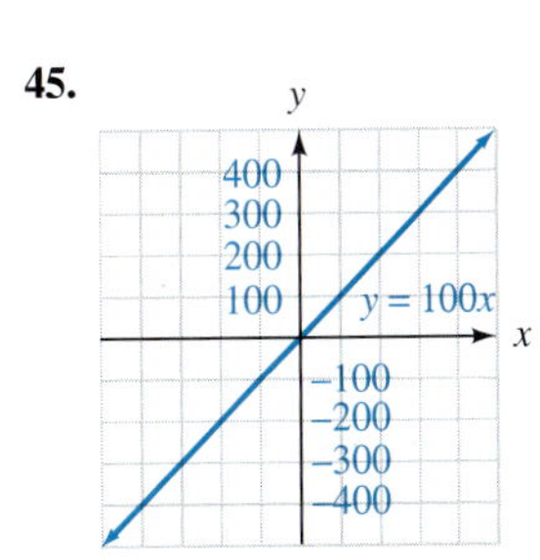

47.

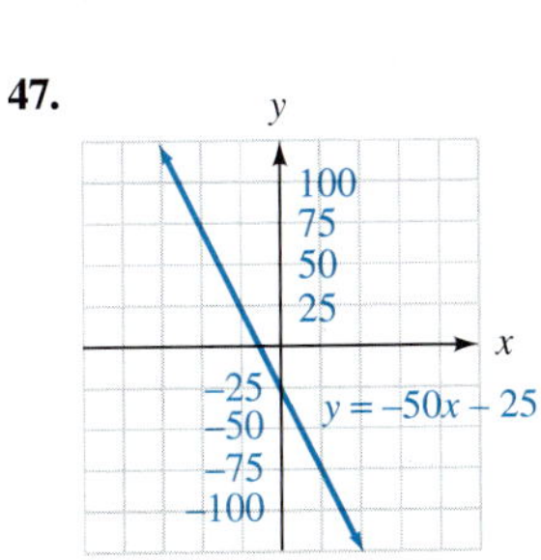

49. \$22.50

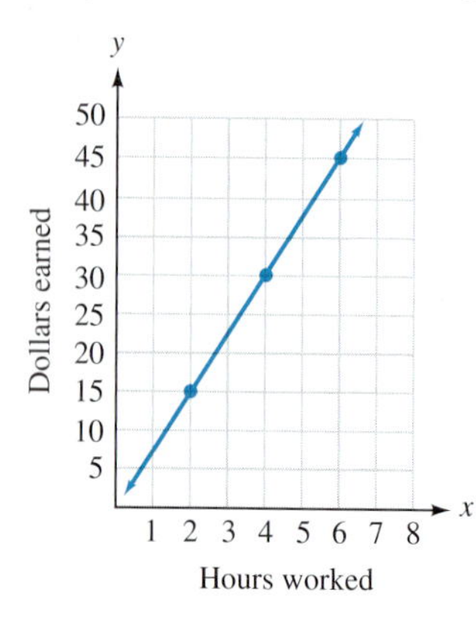

51. 2, 4, 6, 8, 10

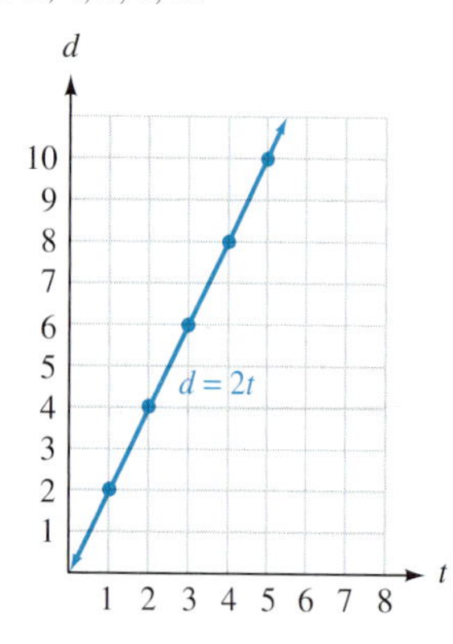

53. yes (2, 2) **61.** $2^2 \cdot 3^2 \cdot 5$ **63.** -5 **65.** $\frac{1}{4}$ sec

Think It Through (page 407)

$10^4 = 10{,}000$

Study Set Section 6.3 (page 409)

1. base, exponent **3.** like **5.** product **7. a.** x^7 **b.** x^2y^3 **c.** $3^4a^2b^3$ **9.** $x^2 \cdot x^6 = x^8$ (answers may vary) **11.** $(c^5)^2 = c^{10}$ (answers may vary) **13. a.** x^{m+n} **b.** x^{mn} **c.** a^nx^n **15. a.** 2 **b.** -10 **c.** x **17. a.** x^2; $2x$ **b.** x^3; $x + x^2$ **c.** x^4; $2x^2$ **19. a.** $4x^2$; $5x$ **b.** $12x^2$; $7x$ **c.** $12x^3$; $4x^2 + 3x$ **21.** 27 **23.** 5, 7 **25.** $x^4, x^3, 4, 3$ **27.** x^5 **29.** x^{10} **31.** f^{13} **33.** n^{32} **35.** l^{10} **37.** x^{11} **39.** 2^{12} **41.** 5^8 **43.** $8x^3$ **45.** $5t^{10}$ **47.** $-24x^5$ **49.** $-x^4$ **51.** $36y^8$ **53.** $-40t^{10}$ **55.** x^3y^3 **57.** b^8c^8 **59.** x^5y^2 **61.** a^4b^4 **63.** x^5y^7 **65.** $18x^3y^4$ **67.** $16x^4y^2$ **69.** $-24f^6t^4$ **71.** a^4b^3 **73.** $12x^4y^3$ **75.** x^8 **77.** m^{500} **79.** $8a^3$ **81.** x^4y^4 **83.** $27s^6$ **85.** $4s^4t^6$ **87.** x^{14} **89.** c^{30} **91.** $36a^{14}$ **93.** $216a^{15}$ **95.** x^{60} **97.** $32b^{25}$ **103.** $\frac{3}{4}$ **105.** 5 **107.** 7 **109.** 12

Study Set Section 6.4 (page 415)

1. monomial **3.** binomial **5.** binomial **7.** monomial **9.** monomial **11.** trinomial **13.** 3 **15.** 2 **17.** 1 **19.** 7 **21.** 2, 2, 4, 4, 16 **23.** 13 **25.** 6 **27.** 31 **29.** 4 **31.** 1

33.

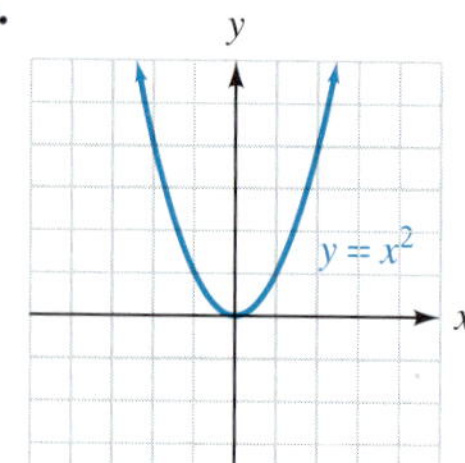

35.

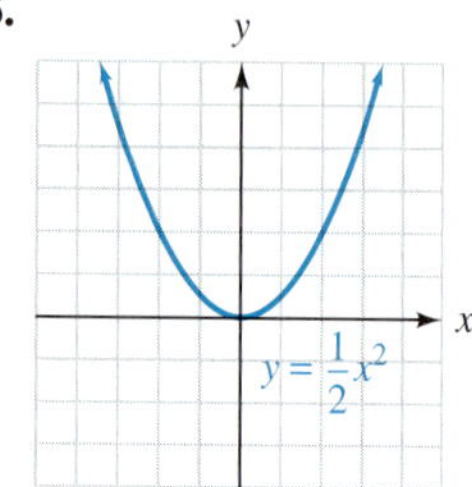

37.

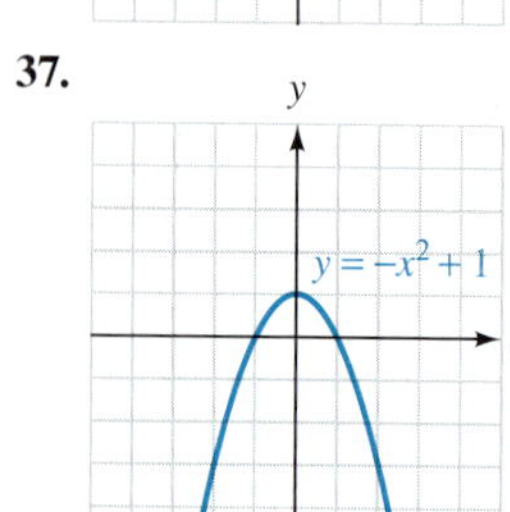

39.

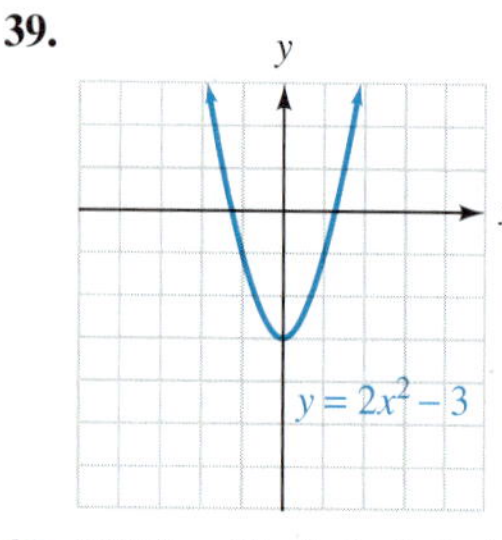

41. 0 ft **43.** 64 ft **45.** 63 ft **47.** 198 ft **49.** 0, 1, 4, 1, 4 **55.** 2 **57.** $\frac{13}{7}$ **59.** $\frac{3}{2}$ **61.** 16 **63.** 6

Study Set Section 6.5 (page 421)

1. like **3.** coefficients, variables **5.** yes, $7y$ **7.** no **9.** yes, $13x^3$ **11.** yes, $15x^2$ **13.** $2x^2, 7x, 5x^2$ **15.** $9y$ **17.** $-12t^2$ **19.** $14s^2$ **21.** $7x + 4$ **23.** $7x^2 - 7$ **25.** $12x^3 - 14.9x$ **27.** $8x^2 + 2x - 21$ **29.** $8y^2 + 4y - 2$ **31.** $6x^2 + x - 5$ **33.** $2n^2 + 5$ **35.** $5x^2 + x + 11$ **37.** $-7x^2 - 5x - 1$ **39.** $2x^2 + x + 12.9$ **41.** $16u^3$ **43.** $7x^5$ **45.** $1.6a + 8$ **47.** $-19x^2 - 5$ **49.** $7x^2 - 2x - 5$ **51.** $1.7y^2 + 3.1y - 9$ **53.** $7b + 4$ **55.** $p^2 - 2p$ **57.** $5x^2 + 6x - 8$ **59.** $-12x^2 - 13x + 36$ **61.** $-x^3 + x + 14$ **63.** \$92,000 **65.** \$112,800 **67.** \$211,000 **69.** $y = -800x + 8{,}500$ **71.** $y = -1{,}900x + 18{,}700$ **77.** 0.8 oz **79.** 54 ft

Study Set Section 6.6 (page 426)

1. monomial **3.** 3 **5.** numerical, factors **7.** term, like **9.** $2x, 5$ **11.** $12x^5$ **13.** $-6b^3$ **15.** $-6x^5$ **17.** $-\frac{1}{2}y^7$ **19.** $3x + 12$ **21.** $-4t - 28$ **23.** $3x^2 - 6x$ **25.** $-6x^4 + 2x^3$ **27.** $6x^3 + 8x^2 - 14x$ **29.** $-2p^3 + 3p^2 - 2p$ **31.** $3q^4 - 6q^3 + 21q^2$ **33.** $a^2 + 9a + 20$ **35.** $3x^2 + 10x - 8$ **37.** $6a^2 + 2a - 20$ **39.** $4x^2 + 12x + 9$ **41.** $4x^2 - 12x + 9$ **43.** $25t^2 + 10t + 1$ **45.** $6x^3 - x^2 + 1$ **47.** $x^3 - 1$ **49.** $x^3 - x^2 - 5x + 2$ **51.** $4x^2 + 11x + 6$ **53.** $12x^2 + 14x - 10$ **55.** $x^3 + 1$ **57.** $(x^2 - 4)$ ft^2 **59.** $R = -\frac{x^2}{100} + 30x$ **65.** four and ninety-one thousandths **67.** 0.109375 **69.** 134.657 **71.** 10

Key Concept (page 428)

1–6.

y-axis, x-axis; Q I, Q II, Q III, Q IV; (0, 4), P, (−3, 2), (−4, 0), (4, 0), (−3, −2), (3, −2), (0, −4)

7.

x	y	(x, y)
0	−2	(0, −2)
4	0	(4, 0)
2	−1	(2, −1)

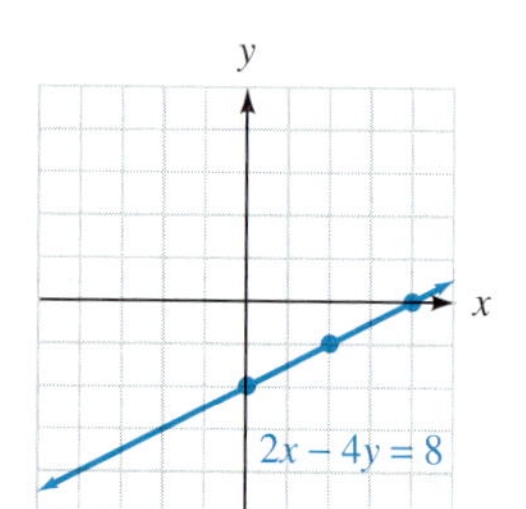

9.

x	y	(x, y)
−2	5	(−2, 5)
−1	2	(−1, 2)
0	1	(0, 1)
1	2	(1, 2)
2	5	(2, 5)

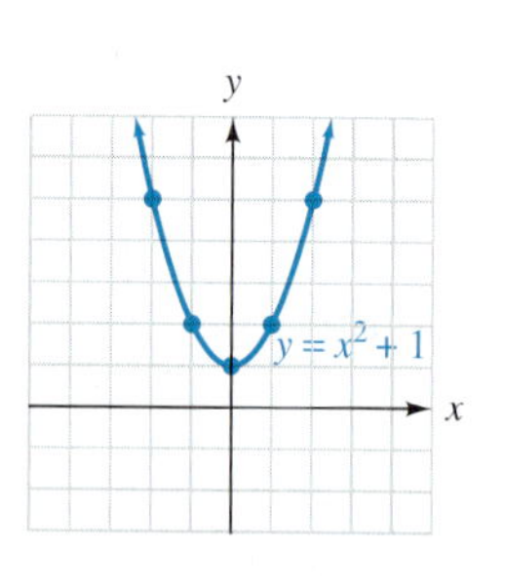

Chapter Review (page 430)

1. yes **2.** no **3.** $-3, -6$ **4.**

x	y	(x, y)
1	-5	$(1, -5)$
3	-11	$(3, -11)$
-2	4	$(-2, 4)$

5.

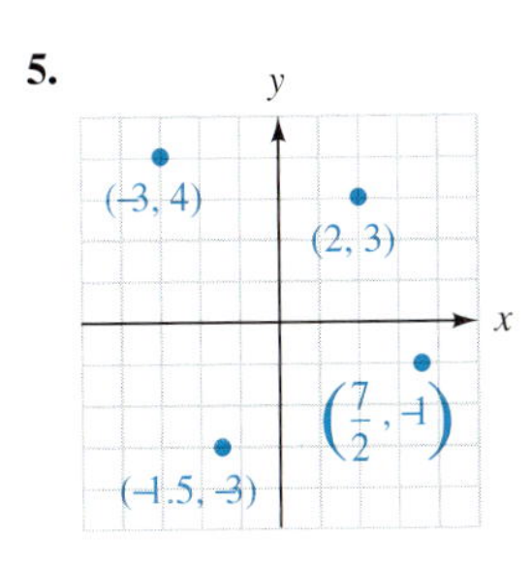

6. $A(4, 3)$, $B(-3, 3)$, $C(-4, 0)$, $D(-1.5, -3.5)$, $E(2.5, -1.5)$
7. III

9. $-5, \frac{5}{3}$

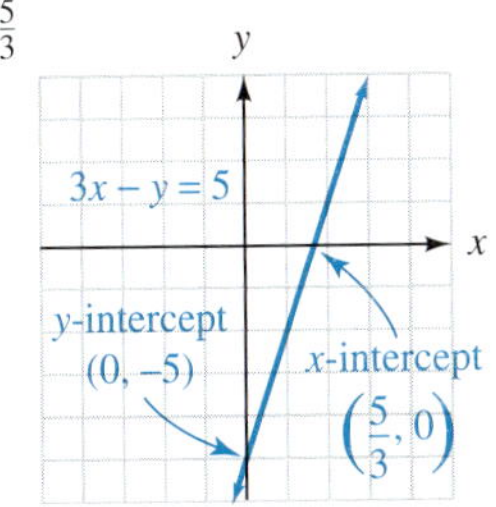

10.

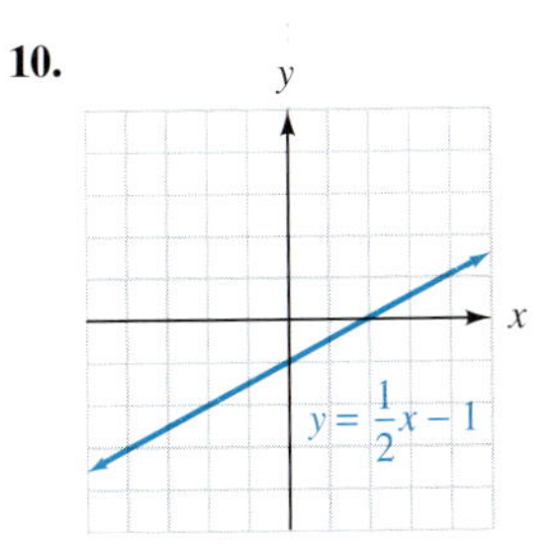

11.

y
x
y = 2

12.

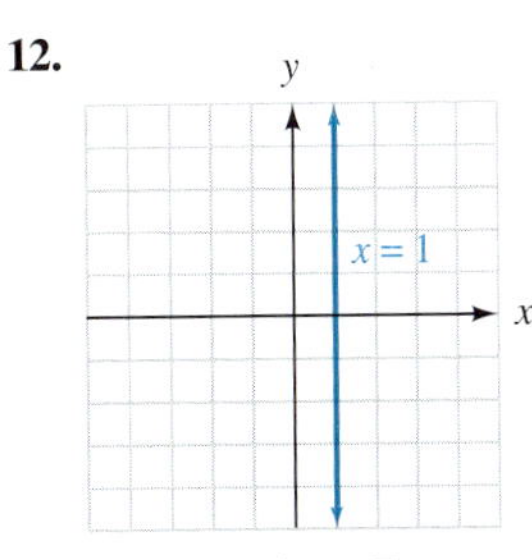

13. The line is made up of infinitely many points. Every point on the line is a solution of the equation, and every solution of the equation is on the line. **14.** $4h \cdot 4h \cdot 4h$ **15.** $5^2d^3m^4$ **16.** h^{10} **17.** t^8 **18.** w^7 **19.** 4^{12} **20.** $8b^7$ **21.** $-24x^4$ **22.** $24f^7$ **23.** $-a^2b^2$ **24.** x^2y^6 **25.** m^2n^2 **26.** $27m^3z^7$ **27.** $-20c^3d^6$ **28.** v^{12} **29.** $27y^3$ **30.** $25t^8$ **31.** $8a^{12}b^{15}$ **32.** c^{26} **33.** $108s^{12}$ **34.** c^{14} **35.** $8x^9$ **36.** trinomial **37.** monomial **38.** binomial **39.** 3 **40.** 4 **41.** 5 **42.** 7 **43.** 13

44.

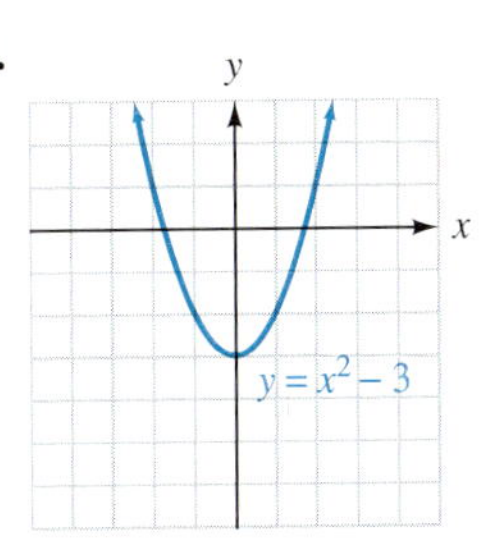

45.

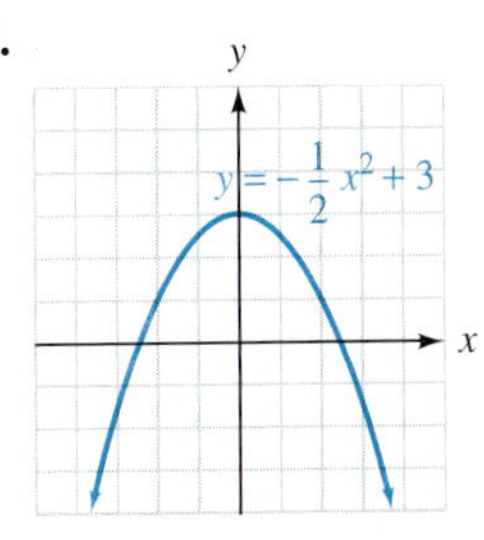

46. $5x^3$ **47.** $\frac{13}{2}p^2$ **48.** $9x + 4$ **49.** $2x^2 - 2x + 3$ **50.** $8x + 3$ **51.** $-2x^2 + x + 2$ **52.** $7p^3$ **53.** $-5y^2$ **54.** $1.1x - 8$ **55.** $z^2 - 4z + 6$ **56.** $2x - 7$ **57.** $8x^2 - 5x + 12$ **58.** $15x^5$ **59.** $-6z^4$ **60.** $6x^3 + 4x^2$ **61.** $-35t^5 + 30t^4 + 10t^3$ **62.** $6x^2 + x - 2$ **63.** $35t^2 - 2t - 24$ **64.** $15x^2 + 19x - 10$ **65.** $15x^2 - 5x - 10$ **66.** $6x^3 + x^2 + x + 2$ **67.** $6r^3 - 5r^2 - 12r + 9$ **68.** $15x^3 + 19x^2 - x + 15$ **69.** $15x^3 - 16x^2 - x + 2$

Chapter 6 Test (page 435)

1. yes **2.** no **3.** $-2, 4, -1$ **4.** $1, (0, 1); 2, (3, 2); 0, (-3, 0)$
5. $(30, 32), (30, 34), (31, 34), (38, 30)$
6.

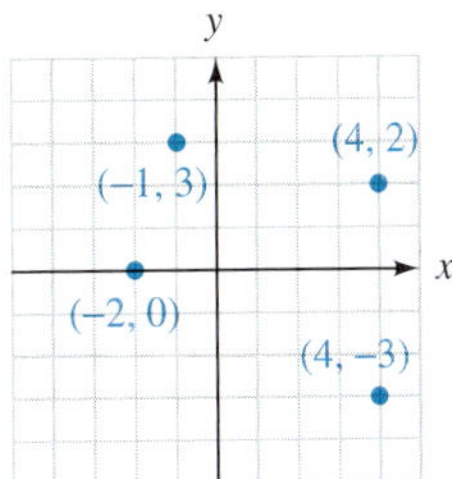

7. $A(0, 0)$, $B(2.5, 3.5)$, $C(-3, -2)$, and $D(0, -2)$
8.

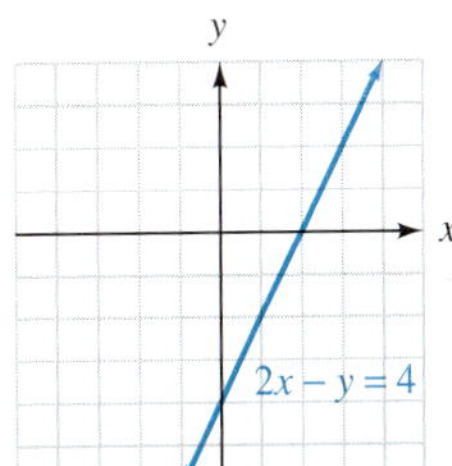

9. a. $(2, 0)$ **b.** $(0, -4)$

10.

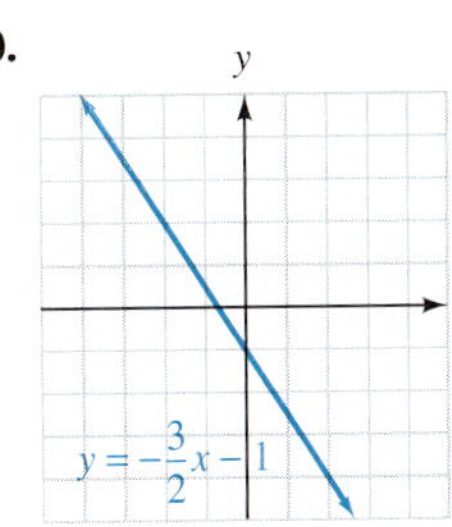

11.

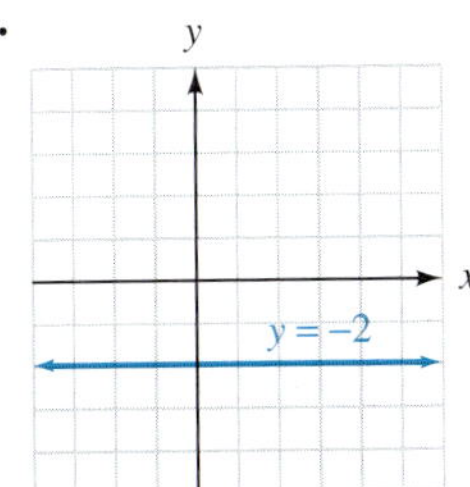

12.

y
x
x = 3

13. a. h^6 **b.** $-28x^5$ **c.** b^8 **d.** $24g^5k^{13}$ **14. a.** f^{15} **b.** $4a^4b^2$ **c.** x^{15} **d.** x^{15} **15.** binomial **16.** trinomial **17.** 6 **18.** 7 **19.** 25 **20.** 2

21.

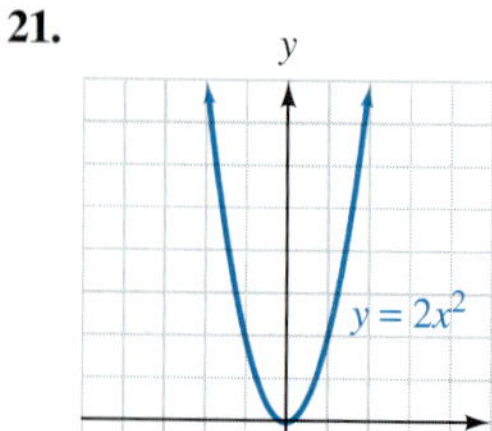

22.

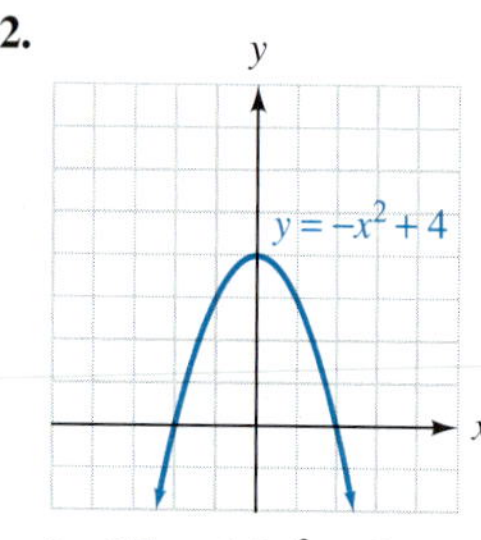

23. $5x^2 - 3x + 4$ **24.** $7x^2 + 2x - 2$ **25.** $-1.2p^2 + 3p$ **26.** $8d^2 - 9d + 12.5$ **27.** $-8x^5$ **28.** $3y^4 - 6y^3 + 9y^2$ **29.** $6x^2 - 7x - 20$ **30.** $2x^3 - 7x^2 + 14x - 12$ **31.** No; $(1, -2)$ lies in quadrant IV, and $(-2, 1)$ lies in quadrant II. **32.** There are infinitely many ordered pairs that satisfy this equation.

Cumulative Review Exercises (page 437)

1. 6,246,000 **2.** 6,000,000 **3.** 22 m **4.** 52 in. **5.** 180, 200, 250, 136 **6.** $\frac{80}{45}, \frac{120}{80}, \frac{140}{85}$ **7.** $2^3 \cdot 3 \cdot 5$ **8.** $3 \cdot 5^2 \cdot 7$ **9.** Visibility is 39 ft less. **10.** -10 **11.** -2 **12.** -9 **13.** $-6x$ **14.** $x + 10y$ **15.** 2 **16.** 4 **17.** -18 **18.** -2 **19.** $\frac{1}{b}$ **20.** $-\frac{17}{18}$ **21.** $20\frac{5}{18}$ **22.** $\frac{25 - mn}{5m}$ **23.** -32 **24.** 10 **25.** 34 **26.** To divide a number by a fraction, we multiply the number by the reciprocal of the fraction. **27.** 57.57 **28.** 351.053 **29.** 107.26 **30.** 2.1303 **31.** 4.67 **32.** $0.3\overline{5}$ **33.** 5.6 **34.** -4.8 **35.** 14.6 **36.** 120 **37.** 3.3 **38.** -73.5 **39.** 8.2, 8.0, 7.9, 1.4 **40.** 41,811.4 **41.** 800 **42.** 1,600 **43.** 11 **44.** $\frac{9}{2}$ **45.** 0.5 **46.** 29 **47.** yes

48.

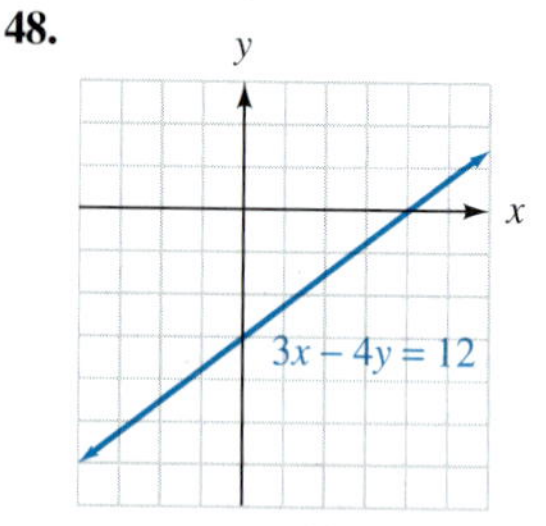

49. p^9 **50.** $-10q^8$ **51.** p^6q^6 **52.** $-27a^{15}$ **53.** $x^2 - 6x + 3$ **54.** $6x^2 + 7x - 3$

Chapter 7 Check Your Knowledge (page 441)

1. one hundred **2.** amount, percent, base **3.** discount **4.** principal **5.** compound, simple **6. a.** 0.75, 75% **b.** 0.625, 62.5% **c.** 1.45, 145% **7. a.** 35%, $\frac{7}{20}$ **b.** 398%, $\frac{199}{50}$ or $3\frac{49}{50}$ **c.** 10.5%, $\frac{21}{200}$ **8. a.** 0.25, $\frac{1}{4}$ **b.** 2 or 2.0, 2 or $\frac{2}{1}$ **c.** 0.005, $\frac{1}{200}$ **9.** 35% **10. a.** 37.5% or $37\frac{1}{2}$% **b.** $83\frac{1}{3}$% **11. a.** $66\frac{2}{3}$% **b.** 66.7% **12.** 325 **13.** 17% **14.** 52 **15.** 250% **16.** \$2.99, \$11.96 **17.** \$29.95, 17% **18.** \$29.94 **19.** \$46.00 **20.** \$1,676.47 **21.** 93% **22.** \$1,045.00 **23.** \$40.71 **24.** 13

Study Set Section 7.1 (page 448)

1. percent **3.** 100, simplify **5.** right **7. a.** 0.84, 84%, $\frac{21}{25}$ **b.** 16% **9.** $\frac{17}{100}$ **11.** $\frac{1}{20}$ **13.** $\frac{3}{5}$ **15.** $\frac{5}{4}$ **17.** $\frac{1}{150}$ **19.** $\frac{21}{400}$ **21.** $\frac{3}{500}$ **23.** $\frac{19}{1,000}$ **25.** 0.19 **27.** 0.06 **29.** 0.408 **31.** 2.5 **33.** 0.0079 **35.** 0.0025 **37.** 93% **39.** 61.2% **41.** 3.14% **43.** 843% **45.** 5,000% **47.** 910% **49.** 17% **51.** 16% **53.** 40% **55.** 105% **57.** 62.5% **59.** 18.75% **61.** $66\frac{2}{3}$% **63.** $8\frac{1}{3}$ **65.** 11.11% **67.** 55.56% **69. a.** 48, 59, 49, 49, 57 **b.** 8% **71. a.** $\frac{9}{22}$ **b.** 41% **73. a.** $\frac{5}{29}$ **b.** 17% **c.** 24% **75.** 5 ft **77.** 0.9944 **79.** as a decimal; 89.6% **81.** torso: 27.5%

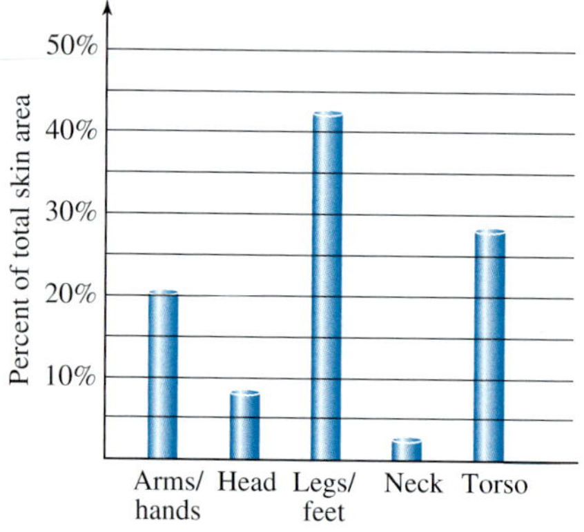

83. 92% **85.** 0.27% **93.** 9 **95.** 7, 11, 3 **97.** $x^2 + 3x + 2$

Think It Through (page 459)

36% are enrolled in college full time; 69% of full-time students read 5 or more assigned textbooks, manuals, or books during the current school year; 38% occasionally

Study Set Section 7.2 (page 459)

1. graph **3.** $x = 0.10 \cdot 50$ **5.** $48 = x \cdot 47$ **7. a.** 0.12 **b.** 0.056 **c.** 1.25 **d.** 0.0025 **9.** more **11. a.** 25 **b.** 100% **c.** 87 **13.** 44% **15. a.** multiply **b.** equals **c.** x (as a variable) **17.** 90 **19.** 80% **21.** 65 **23.** 0.096 **25.** 0.00125% **27.** 44 **29.** 43.5 **31.** 107.1 **33.** 99 **35.** 60 **37.** 31.25%

39.

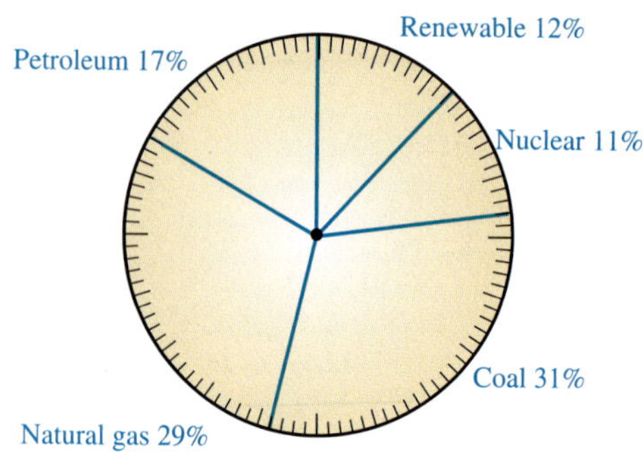

Source: Energy Information Administration

41. 120 **43.** 666 **45.** 38,000 = 38K **47.** 24 oz **49.** yes **51.** 30, 12 **53.** 2.7 in. **55.** 5% **57.** yes **63.** 18.17 **65.** 5.001 **67.** 34,546.4 **69.** -22.5

Think It Through (page 467)

48%, 59% 49%, 57%; greatest percent increase is medical assistant

Study Set Section 7.3 (page 469)

1. commission **3.** discount **5.** subtract, original **7.** \$42.75 **9.** 8% **11.** \$47.34, \$2.84, \$50.18 **13.** \$150 **15.** 8%, 1.2%, 1.45%, 6.2% **17.** 360 hr **19.** 96 calories

21. 1995–1996; 5% **23.** 10% **25.** 31% **27. a.** 25% **b.** 36% **29.** \$5,955 **31.** 1.5% **33.** \$12,000 **35.** \$39.95, 25% **37.** \$187.49 **39.** \$349.97, 13% **41.** \$3.60, 23%, \$11.88 **43.** \$76.50 **49.** −50 **51.** −24 **53.** $12d$ **55.** 13

57.

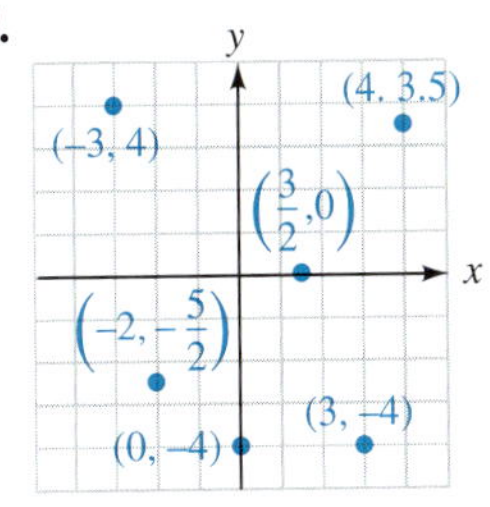

Study Set Estimation (page 474)

1. 164 **3.** \$60 **5.** \$54,000 **7.** 320 lb **9.** 130 **11.** 21 **13.** 18,000 **15.** 3,100

Study Set Section 7.4 (page 480)

1. principal **3.** interest **5.** simple **7. a.** 0.07 **b.** 0.098 **c.** 0.0625 **9.** \$1,800 **11. a.** compound interest **b.** \$1,000 **c.** 4 **d.** \$50 **e.** 1 year **13.** multiplication **15.** \$5,300 **17.** \$1,472 **19.** \$4,262.14 **21.** \$10,000, 0.0725, 2 yr, \$1,450 **23.** \$192, \$1,392, \$58 **25.** \$18.828 million **27.** \$755.83 **29.** \$1,271.22 **31.** \$570.65 **33.** \$30,915.66 **39.** $\frac{1}{2}$ **41.** no **43.** −3 **45.** 3 **47.** III

Key Concept (page 483)

1. $\frac{5}{12}$; simplifying a fraction **3.** $5x$; combining like terms **5.** x^5; when multiplying like bases, add the exponents **7.** $66\frac{2}{3}\%$; divide the numerator by the denominator, move the decimal point 2 places to the right, and insert a % symbol **9.** $2x^2 + 1$; combining like terms **11.** 1; a nonzero number divided by itself is 1 **13.** 30; the product of two negative numbers is positive **15.** x; 2 times x divided by 2 is x **17.** 17; perform multiplications before additions **19.** $2x + 10$; distributive property **21.** 1; the product of a number and its reciprocal is 1

Chapter Review (page 485)

1. 39%, 0.39, $\frac{39}{100}$ **2.** 111%, 1.11, $1\frac{11}{100}$ **3.** 61% **4.** $\frac{3}{20}$ **5.** $\frac{6}{5}$ **6.** $\frac{37}{400}$ **7.** $\frac{1}{1,000}$ **8.** 0.27 **9.** 0.08 **10.** 1.55 **11.** 0.018 **12.** 83% **13.** 62.5% **14.** 5.1% **15.** 600% **16.** 50% **17.** 80% **18.** 87.5% **19.** 6.25% **20.** $33\frac{1}{3}\%$ **21.** $83\frac{1}{3}\%$ **22.** 55.56% **23.** 266.67% **24.** 63% **25.** $0.1\% = \frac{1}{1,000}$ **26.** amount: 15, base: 45, percent: $33\frac{1}{3}\%$ **27.** $x = 32\% \cdot 96$ **28.** 200 **29.** 125 **30.** 1.75% **31.** 2,100 **32.** 121 **33.** 30 **34.** 14.4 gal nitro, 0.6 gal methane **35.** 68 **36.** 87% **37.** \$5.43

38.

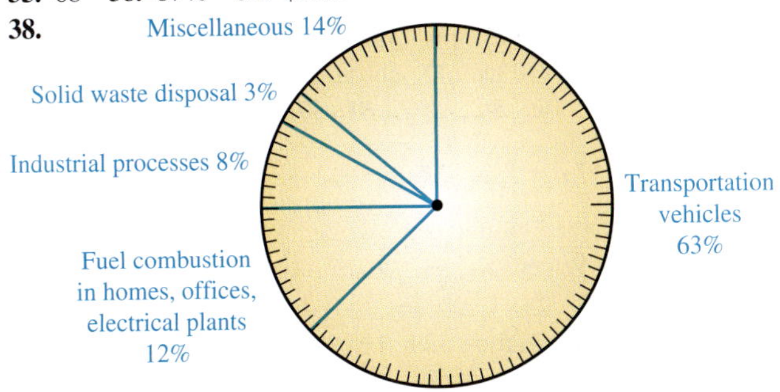

39. 139,531,200 mi² **40.** \$3.30, \$63.29 **41.** 4% **42.** \$40.20 **43.** original **44.** 25% **45.** 9.6% **46.** \$50, \$189.99, 26% **47.** \$6,000, 8%, 2 years, \$960 **48.** \$10,308.22 **49.** \$134.69 **50.** \$2,142.45 **51.** \$6,076.45 **52.** \$43,265.78

Chapter 7 Test (page 489)

1. 61%, $\frac{61}{100}$, 0.61 **2.** 199%, $\frac{199}{100}$, 1.99 **3. a.** 0.67 **b.** 0.123 **c.** 0.0975 **4. a.** 25% **b.** 62.5% **c.** 12% **5. a.** 19% **b.** 347% **c.** 0.5% **6. a.** $\frac{11}{20}$ **b.** $\frac{1}{10,000}$ **c.** $\frac{5}{4}$ **7.** 23.33% **8.** 60% **9.** $66\frac{2}{3}\%$ **10.** 25% **11. a.** 1.02 in. **b.** 32.98 in. **12.** 6.5% **13.** \$3.81 **14.** 93.9% **15.** 90 **16.** 21 **17.** 144 **18.** 27% **19.** \$35.92 **20.** \$41,440 **21.** \$11.95, \$3, 20% **22.** 22% **23.** \$150 **24.** \$5,079.60 **25.** The phrase "bringing crime down to 37%" is unclear. The question that arises is: 37% of what? **26.** Interest is money that is paid for the use of money.

Cumulative Review Exercises (page 491)

1.

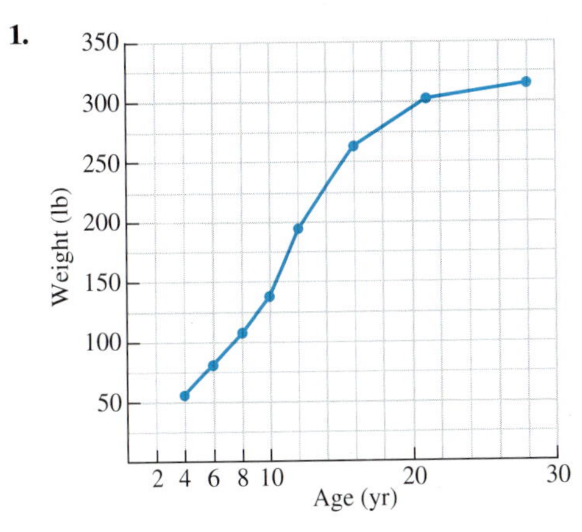

2. If a and b represent numbers, $ab = ba$. **3. a.** 1, 2, 4, 5, 8, 10, 20, 40 **b.** $5 \cdot 2^3$ **4.** \$2,106 **5.** 64 ft² **6.** −7 **7.** −4 **8.** 55 **9.** −2 **10.** $2t - 16$ **11.** 15°C **12.** 3 **13.** $\frac{4}{11}$ **14.** $\frac{2}{3}$ **15.** $-\frac{5}{21a}$ **16.** $\frac{2}{5}$ **17.** $\frac{28 + 2m}{7m}$ **18.** $20\frac{5}{18}$ **19.** −30 **20.** 8 **21.** 70.29 **22.** −8.6265 **23.** 752 **24.** 83.4 **25.** −2.33 **26.** 452.030 **27.** $0.7\overline{3}$ **28.** −11.1 **29.** −29 **30.** 3.5 hr **31.** $-2m^2 - 3m + 3$ **32.** $3x^2 + 10x - 8$ **33.** $4y^2 - 20y + 25$ **34.** y^7 **35.** h^{20} **36.** $8a^9b^{18}$ **37.** $-56g^9$

38.

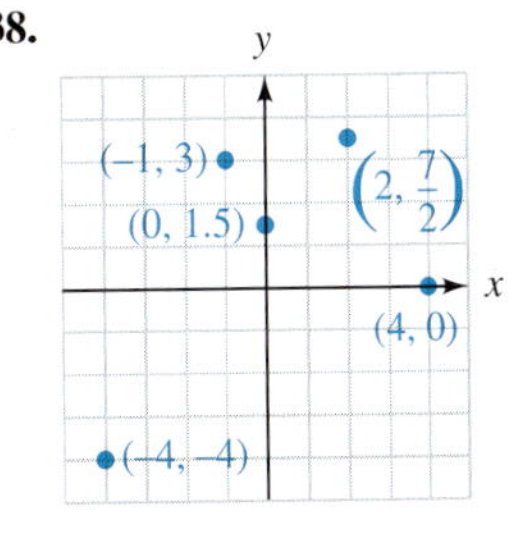

39.

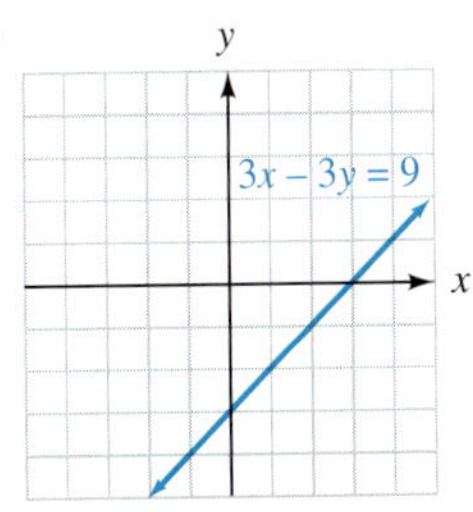

40.

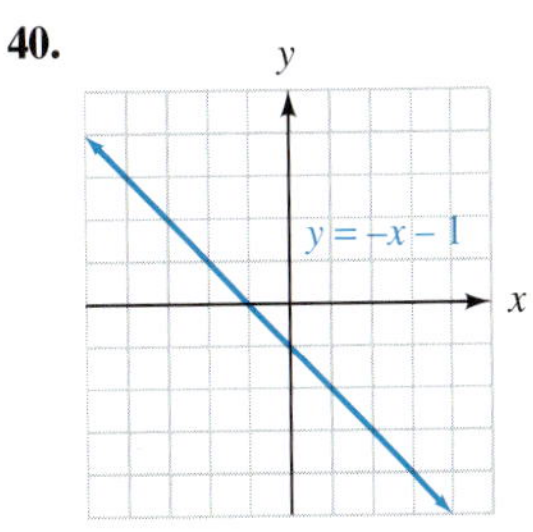

41. 29%, $\frac{29}{100}$; 0.473, $\frac{473}{1,000}$; 87.5%, 0.875 **42.** 125 **43.** \$12.00, \$87.18 **44.** 0.0018% **45.** \$1,450

Chapter 8 Check Your Knowledge (page 494)

1. ratio **2.** proportion **3.** length, metric **4.** Fahrenheit, Celsius **5.** $\frac{7}{5}$ **6.** $\frac{2}{9}$ **7.** 47 mpg **8.** the 12-ounce bag **9.** $\frac{4}{1}$ **10. a.** yes **b.** no **11.** no **12.** 68 **13.** −40 **14.** 25.74 **15.** −0.2 **16.** 500 cm **17.** $9\frac{2}{3}$ yd **18.** 26,400 ft **19.** 252 in. **20.** 109.4 yd **21.** 175.26 **22.** 2,500 **23.** $3\frac{1}{2}$ in. **24.** −40° C

Think It Through (page 498)

23 : 1; 24 : 1; 26 : 1; prealgebra has the lowest student-to-instructor ratio.

Study Set Section 8.1 (page 501)

1. ratio **3.** cost **5.** 3 **7.** 10 **9.** $\frac{11 \text{ minutes}}{60 \text{ minutes}} = \frac{11}{60}$ **11.** $\frac{13}{9}$, 13 to 9, 13 : 9 **13.** $\frac{5}{7}$ **15.** $\frac{1}{2}$ **17.** $\frac{2}{3}$ **19.** $\frac{5}{8}$ **21.** $\frac{2}{7}$ **23.** $\frac{1}{3}$ **25.** $\frac{1}{5}$ **27.** $\frac{3}{7}$ **29.** $\frac{3}{4}$ **31.** \$1,800 **33.** $\frac{1}{3}$ **35.** \$8,750 **37.** $\frac{1}{5}$ **39.** $\frac{32 \text{ ft}}{3 \text{ sec}}$ **41.** $\frac{21 \text{ made}}{25 \text{ attempts}}$ **43.** $\frac{375 \text{ students}}{2 \text{ yr}}$ **45.** $\frac{3 \text{ beats}}{2 \text{ measures}}$ **47.** 12 revolutions per min **49.** 1.5 errors per hr **51.** 7 presents per child **53.** 320 people per square mile **55.** \$0.07 per foot **57.** 1.2 cents per ounce **59.** \$68 per person **61.** \$0.8 billion per month **63.** $\frac{1}{1}$ **65.** $\frac{3}{2}$ **67.** $\frac{\frac{2}{3}}{3\frac{1}{2}}$ **69.** $\frac{12 \text{ hits}}{22 \text{ at-bats}} = \frac{6 \text{ hits}}{11 \text{ at-bats}}$ **71.** $\frac{5 \text{ compressions}}{2 \text{ breaths}}$ **73.** $\frac{329 \text{ complaints}}{100{,}000 \text{ passengers}}$ **75.** $\frac{1 \text{ faculty member}}{16 \text{ students}}$ **77.** \$1.89 per gal **79.** 7¢ per oz **81.** the 6-oz can **83.** the 50-tablet boxes **85.** the truck **87.** 440 gal per min **89.** 325 mi, 65 mph **91.** the second car **97.** 45.537 **99.** 192.7012 **101.** $1\frac{3}{4}$

Study Set Section 8.2 (page 511)

1. proportion **3.** cross **5.** *ad*, *bc* **7. a.** $\frac{5}{8} = \frac{15}{24}$ **b.** $\frac{3 \text{ teacher's aides}}{25 \text{ children}} = \frac{12 \text{ teacher's aides}}{100 \text{ children}}$ **9.** i, iv **11.** $18x$, 288, 18, 18 **13.** no **15.** yes **17.** no **19.** yes **21.** no **23.** yes **25.** 4 **27.** 6 **29.** 3 **31.** 36 **33.** 0 **35.** −17 **37.** $-\frac{3}{2}$ **39.** $\frac{83}{2}$ **41.** \$3,500 **43.** 5.625 **45.** \$218.75 **47.** \$11.76 **49.** the same **51.** 24 **53.** 975 **55.** $4\frac{2}{7}$, which is about $4\frac{1}{4}$ **57.** 19 sec **59.** 221 mi **61.** \$309 **63.** 10 ft **65.** 65.25 ft = 65 ft 3 in. **67.** 2.625 in. = $\frac{25}{8}$ in. **73.** 90% **75.** $\frac{1}{3}$ **77.** 2.6

Study Set Section 8.3 (page 521)

1. length **3.** 1 **5.** capacity **7.** 1 **9.** 5,280 **11.** 16 **13.** 8 **15.** 1 **17.** 24 **19.** $\frac{5}{8}$ in., $1\frac{3}{4}$ in., $2\frac{5}{16}$ in. **21. a.** $\frac{1 \text{ ton}}{2{,}000}$. **b.** $\frac{2 \text{ pt}}{1 \text{ qt}}$ **23. a.** iv **b.** i **c.** ii **d.** iii **25. a.** iii **b.** iv **c.** i **d.** ii **27.** 36, 36 **29.** 2, 4, 12 **31.** $2\frac{5}{8}$ in. **33.** $10\frac{3}{4}$ in. **35.** 48 in. **37.** 42 in. **39.** 2 ft **41.** 288 in. **43.** 2.5 yd **45.** $4\frac{2}{3}$ ft **47.** 15 ft **49.** $2\frac{1}{3}$ yd **51.** 3 mi **53.** 2,640 ft **55.** 5 lb **57.** 3.5 tons **59.** 24,800 lb **61.** 6 pt **63.** 2 gal **65.** 2 pt **67.** 4 hr **69.** 5 days **71.** 150 yd **73.** 2,880 in. **75.** 0.28 mi **77.** 61,600 yd **79.** 128 oz **81.** 4.95 tons **83.** 68 **85.** $71\frac{7}{8}$ gal = 71.875 gal **87.** 320 oz **89.** $6\frac{1}{8}$ days = 6.125 days **93.** 3,700 **95.** 3,673.26 **97.** 0.101 **99.** 0.1

Study Set Section 8.4 (page 532)

1. tens **3.** thousands **5.** hundredths **7.** metric **9.** 1 cm, 3 cm, 6 cm **11. a.** $\frac{1 \text{ km}}{1{,}000 \text{ m}}$ **b.** $\frac{100 \text{ cg}}{1 \text{ g}}$ **c.** $\frac{1{,}000 \text{ milliliters}}{1 \text{ liter}}$ **13 a.** iii **b.** i **c.** ii **15. a.** ii **b.** iii **c.** i **17.** 10 **19.** $\frac{1}{100}$ **21.** $\frac{1}{1{,}000}$ **23.** 1,000 **25.** 1,000 **27.** 1,000 **29.** $\frac{1}{100}$ **31.** 1 **33.** 1, 100 **35.** 1,000, 1, 1,000 **37.** 156 mm **39.** 28 cm **41.** 300 **43.** 570 **45.** 3.1 **47.** 7,680,000 **49.** 0.472 **51.** 4.532 **53.** 0.0325 **55.** 37.5 **57.** 125 **59.** 675,000 **61.** 6.383 **63.** 0.63 **65.** 69.5 **67.** 5.689 **69.** 5.762 **71.** 0.000645 **73.** 0.65823 **75.** 3,000 **77.** 2,000 **79.** 1,000,000 **81.** 0.5 **83.** 3,000 **85.** 5,000 **87.** 10 **89.** 0.5 km, 1 km, 1.5 km, 5 km, 10 km **91.** 12 cm, 8 cm **93.** 400,000 cg **95.** 40 dL **97.** 4 **99.** 3 g **105.** \$23.99 **107.** \$402 **109.** $11\frac{10}{21}$

Think It Through (page 540)

1. 216 mm × 279 mm **2.** 9 kilograms **3.** 22.5 milliliters

Study Set Section 8.5 (page 541)

1. Fahrenheit, Celsius **3. a.** meter **b.** meter **c.** inch **d.** mile **5. a.** liter **b.** liter **c.** gallon **7.** 0.3048, 1,371.6 **9.** 0.264 **11.** 91.4 **13.** 147.6 **15.** 39,370 **17.** 127 **19.** 1 **21.** 11,350 **23.** 17.5 **25.** 0.6 **27.** 0.1 **29.** 243.4 **31.** 710 **33.** 0.5 **35.** 10° **37.** 122° **39.** 14° **41.** −20.6° **43.** 5 mi **45.** 70 mph **47.** 1.9 km **49.** 1.9 cm **51.** 411 lb, 744 lb **53. a.** 226.8 g **b.** 0.24 L **55.** no **57.** the 3 quarts **59.** 62°C **61.** 28°C **63.** −5°C and 0°C **69.** $5y + 4$ **71.** $-7x$ **73.** x^3 **75.** $15b^2$

Key Concept (page 544)

1. teacher's aides needed to supervise 75 children; 2, 75, *x*; 2, *x*; 15, 75, 150, 15, 15 **3.** 10,800 ft

Chapter Review (page 546)

1. $\frac{1}{3}$ **2.** $\frac{1}{4}$ **3.** $\frac{3}{2}$ **4.** $\frac{2}{3}$ **5.** $\frac{37}{32}$ **6.** \$7.75 **7.** the 8-oz can **8.** 75 **9.** 15 **10.** no **11.** yes **12.** yes **13.** no **14.** 4.5 **15.** 16 **16.** 0 **17.** $-\frac{35}{9}$ **18.** 192.5 mi **19.** 300 **20.** 12 ft **21.** $1\frac{1}{2}$ in. **22.** $\frac{1 \text{ mi}}{5{,}280 \text{ ft}} = 1$, $\frac{5{,}280 \text{ ft}}{1 \text{ mi}} = 1$ **23.** 15 ft **24.** 216 in. **25.** 5.5 ft **26.** 306 in. **27.** 1.75 mi **28.** 1,760 yd **29.** 2 lb **30.** 275.2 oz **31.** 96,000 oz **32.** 2.25 tons **33.** 80 fl oz **34.** 0.5 gal **35.** 68 c **36.** 5.5 qt **37.** 40 pt **38.** 56 c **39.** 1,200 sec **40.** 15 min **41.** $8\frac{1}{3}$ days **42.** 360 min **43.** 108 hr **44.** 86,400 sec **45.** $484\frac{2}{3}$ yd **46.** 100 **47.** 4 cm **48.** $\frac{1 \text{ km}}{1{,}000 \text{ m}} = 1$, $\frac{1{,}000 \text{ m}}{1 \text{ km}} = 1$ **49.** 4.75 m **50.** 8,000 mm **51.** 0.03 km **52.** 2,000 dm **53.** 50 hm **54.** 25 hm **55.** 70 mg **56.** 8 g **57.** 5.425 kg **58.** 5,425,000 mg **59.** 7.5 g **60.** 0.05 kg **61.** 50 **62.** 1.5 L **63.** 3.25 kL **64.** 1,000 dL **65.** 40 cL **66.** 20 hL **67.** 400 mL **68.** 1,000 mL **69.** 164.04 ft **70.** the World Trade Center **71.** 3,106 km **72.** 198.12 cm **73.** 850.5 g **74.** 33 lb **75.** 11,000 g **76.** 910 kg **77.** about 2 lb **78.** LaCroix **79.** the 5-liter bottle **80.** 25°C **81.** 30°C

Chapter 8 Test (page 551)

1. $\frac{3}{4}$ **2.** $\frac{1}{6}$ **3.** the 2-pound can **4.** 22.5 kwh per day **5.** $\frac{1}{1}$, 1 : 1, 1 to 1 **6.** no **7.** yes **8.** yes **9.** 15 **10.** 63.24 **11.** 11 **12.** 3 **13.** \$3.43 **14.** $1\frac{2}{3}$ c **15.** 15 ft **16.** $8\frac{1}{3}$ yd **17.** 160 oz **18.** 3,200 lb **19.** 128 fl oz **20.** 115,200 min **21.** the one on the left **22.** the blue one **23.** the right side **24.** 0.5 km

25. 500 cm **26.** 0.08 kg **27.** 70,000 mL **28.** 7.5 g **29.** the 100-yd race **30.** Jim **31.** the 1-liter bottle **32.** 182°F **33.** A scale is a ratio (or rate) comparing the size of a drawing and the size of an actual object. For example, 1 inch to 6 feet (1 in.:6 ft). **34.** It is easier to convert from one unit to another in the metric system, because it is based on the number 10.

Cumulative Review Exercises (page 553)

1. 6 ten thousands + 4 thousands + 5 hundreds + 2 ones **2.** 20 R 3 **3.** 587, 278, −6,790 **4. a.** −8 **b.** undefined **c.** −8 **d.** 0 **e.** 8 **f.** 0 **5.** 15 shots **6.** −9, 9 **7.** −61 **8.** 1, 16 **9.** $60h$ **10.** $x - 7$ **11.** −2 **12.** $d = rt$ **13.** 79 **14.** $A = \frac{1}{2}bh$ **15.** $\frac{4}{5}$ **16.** $\frac{54t}{60t}$ **17.** $-\frac{1}{h}$ **18.** $\frac{1}{4}$ **19.** $\frac{3}{4}$ hp **20.** $\frac{4}{25}$ **21.** $\frac{10}{9}$ **22.** 55.1°F **23.** 304.07 **24.** $0.08\overline{3}$ **25.** 0.8 **26.** 23

27.

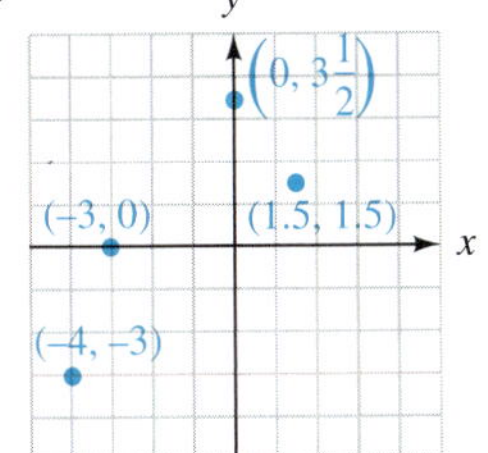

28. a.

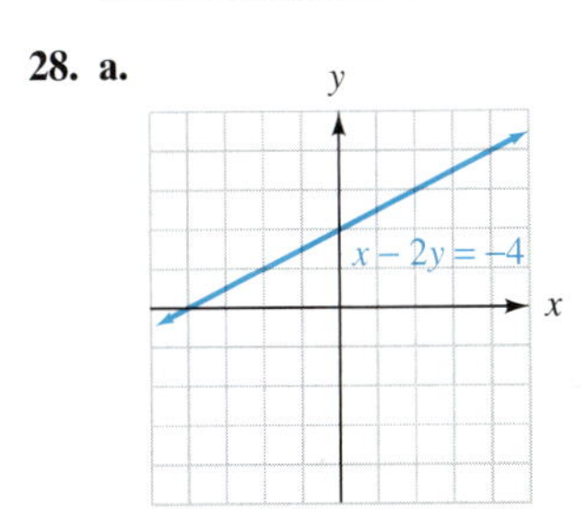

b.

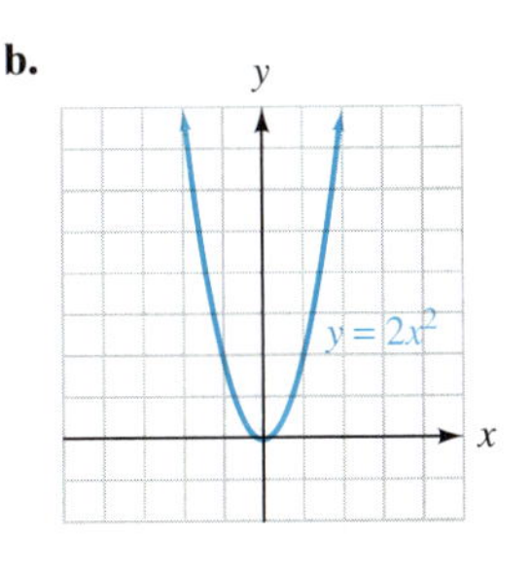

29. $2x^2 - 6x - 2$ **30.** $6x^2 - 11x - 10$ **31. a.** s^{13} **b.** s^{42} **c.** $27a^6b^{12}$ **d.** $-8w^8$ **32.** $66\frac{2}{3}\%$ **33.** $I = Prt$ **34.** 99%, $\frac{99}{100}$; 0.013, $\frac{13}{1,000}$; 31.25%, 0.3125 **35.** \$427.99; about 30% **36.** $\frac{1}{5}$ **37. a.** 960 hr **b.** 4,320 min **c.** 480 sec **38.** 2.5 lb **39.** 2,400 mm **40.** 0.32 kg **41. a.** 1 gal **b.** a meterstick **42.** the 45-kilogram bag

Chapter 9 Check Your Knowledge (page 556)

1. parallel, perpendicular **2.** quadrilateral, triangle **3.** right, hypotenuse **4.** congruent, similar **5.** perimeter, area **6.** radius, circumference, diameter **7.** volume **8. a.** III **b.** I **c.** IV **d.** II **9.** *B* **10.** 14 **11.** 36 **12.** 61° **13. a.** 50° **b.** 130° **c.** 130° **d.** 50° **e.** 75 **14.** 80° **15. a.** 13 ft **b.** 60 ft² **16.** 17.55 in.² **17.** 28π ft² ≈ 78.5 ft³ **18.** $10\frac{2}{3}\pi$ ft³ ≈ 33.5 ft³ **19.** 200 ft³ **20.** 160π ft³ ≈ 502.7 ft³

Study Set Section 9.1 (page 563)

1. segment **3.** midpoint **5.** protractor **7.** right **9.** 180° **11.** supplementary **13.** true **15.** false **17.** true **19.** true **21.** acute **23.** obtuse **25.** right **27.** straight **29.** true **31.** false **33.** yes **35.** yes **37.** no **39.** true **41.** true **43.** true **45.** true **47.** angle **49.** ray **51.** 3 **53.** 3 **55.** 1 **57.** *B* **59.** 40° **61.** 135° **63.** 10 **65.** 27.5 **67.** 30 **69.** 25 **71.** 60° **73.** 75° **75.** 130° **77.** 230° **79.** 100° **81.** 40°

83.

85. 65, 115 **87.** 30° **95.** 16 **97.** $\frac{7}{24}$

99.

101. 4

Study Set Section 9.2 (page 570)

1. coplanar **3.** perpendicular **5.** alternate **7.** ∠4 and ∠6, ∠3 and ∠5 **9.** ∠3, ∠4, ∠5, ∠6 **11.** They are parallel. **13. a.** a right angle **15.** is perpendicular to **17.** m(∠1) = 130°, m(∠2) = 50°, m(∠3) = 50°, m(∠5) = 130°, m(∠6) = 50°, m(∠7) = 50°, m(∠8) = 130° **19.** m(∠A) = 50°, m(∠1) = 85°, m(∠2) = 45°, m(∠3) = 135° **21.** 10 **23.** 30 **25.** 40 **27.** 12 **29.** If the stones are level, the plum bob string should pass through the midpoint of the crossbar of the A-frame. **41.** 72 **43.** 45% **45.** yes **47.** $\frac{1}{3}$

Study Set Section 9.3 (page 578)

1. regular **3.** hexagon **5.** octagon **7.** equilateral **9.** hypotenuse **11.** parallelogram **13.** rhombus **15.** isosceles **17.** 4, quadrilateral, 4 **19.** 3, triangle, 3 **21.** 5, pentagon, 5 **23.** 6, hexagon, 6 **25.** scalene triangle **27.** right triangle **29.** equilateral triangle **31.** isosceles triangle **33.** square, rhombus, rectangle **35.** rhombus **37.** rectangle **39.** trapezoid **41.** triangle **43.** 90° **45.** 45° **47.** 90.7° **49.** 30° **51.** 60° **53.** 720° **55.** 1,440° **57.** 7 sides **59.** 14 sides **65. b.** octagon **c.** triangle **d.** pentagon **67.** pentagon, hexagon **73.** 22 **75.** 40% **77.** 0.10625

Study Set Section 9.4 (page 587)

1. congruent **3.** similar **5.** true **7.** false **9.** true **11.** yes **13.** yes **15.** *a* and *b* represent the lengths of the legs; *c* represents the length of the hypotenuse. **17.** 25, square root, *c*, 5 **19.** is congruent to **21.** $\overline{DF}, \overline{AB}, \overline{EF}$, ∠D, ∠B, ∠C **23.** yes, SSS **25.** not necessarily **27.** yes, SSS **29.** yes, SAS **31.** 6 mm **33.** 50° **35.** yes **37.** 5 **39.** 8 **41.** $\sqrt{56}$ **43.** yes **45.** no **47.** 36 ft **49.** 59.2 ft **51.** 4,000 ft **53.** 12 ft **55.** 25 in. **57.** 127.3 ft **61.** $1\frac{1}{3}$ **63.** 20 **65.** 9

Think It Through (page 599)

$107\frac{2}{3}$ ft^2

Study Set Section 9.5 (page 599)

1. perimeter **3.** area **5.** square **7.** length 15 in. and width 5 in.; length 16 in. and width 4 in. (answers may vary) **9.** sides of length 5 m **11.** base 5 yd and height 3 yd (answers may vary) **13.** length 5 ft and width 4 ft; length 20 ft and width 3 ft (answers may vary) **15.** $4s$ **17.** square inch **19.** s^2 **21.** triangle **23.** 32 in. **25.** 36 m **27.** 37 cm **29.** 85 cm **31.** $28\frac{1}{3}$ ft **33.** 16 cm^2 **35.** 60 cm^2 **37.** 25 in.^{2} **39.** 169 mm^2 **41.** 80 m^2 **43.** 75 yd^2 **45.** 75 m^2 **47.** 144 **49.** \$4,875 **51.** 81 **53.** linoleum **55.** \$1,200 **57.** \$361.20 **59.** \$192 **61.** 111,825 mi^2 **63.** 51 **65.** spot 1: $l = 20$ ft, $w = 10$ ft, 200 ft^2; spot 2: $b_1 = 20$ ft, $b_2 = 16$ ft, $h = 10$ ft, 180 ft^2; spot 3: $b = 28$ ft, $h = 28$ ft, 392 ft^2 **69.** $1\frac{5}{12}$ **71.** $6\frac{1}{12}$ **73.** $1\frac{7}{18}$

Study Set Section 9.6 (page 607)

1. radius **3.** diameter **5.** minor **7.** circumference **9.** $\overline{OA}$, $\overline{OC}$, and $\overline{OB}$ **11.** $\overline{DA}$, $\overline{DC}$, and $\overline{AC}$ **13.** $\widehat{ABC}$ and $\widehat{ADC}$ **15.** Double the radius. **17. a.** 1 in. **b.** 2 in. **c.** 2π in. $\approx$ 6.28 in. **d.** π in.$^2 \approx$ 3.14 in.^{2} **19.** Square 6. **21.** arc AB **23.** πD, r **25.** π **27.** 8π **29.** 37.70 in.^{2} **31.** 36 m **33.** 25.42 ft **35.** 31.42 m **37.** 28.3 in.^{2} **39.** 88.3 in.^{2} **41.** 128.5 cm^2 **43.** 27.4 in.^{2} **45.** 66.7 in.^{2} **47.** 3.14 mi^2 **49.** 32.66 ft **51.** 12.73 times **53.** 1.59 ft **55.** 12.57 ft^2; 0.79 ft^2; 6.28% **63.** 90% **65.** 5.375¢ per oz **67.** five

Study Set Section 9.7 (page 618)

1. volume **3.** cube **5.** surface **7.** cylinder **9.** cone **11.** $V = lwh$ **13.** $V = \frac{4}{3}\pi r^3$ **15.** $V = \frac{1}{3}Bh$ or $V = \frac{1}{3}\pi r^2 h$ **17.** $SA = 2lw + 2lh + 2hw$ **19.** 27 ft^3 **21.** 1,000 dm^3 **23. a.** volume **b.** area **c.** volume **d.** surface area **e.** perimeter **f.** surface area **25. a.** 72 in.^{3} **b.** 18 in.^{2} **c.** 24 in.^{2} **27.** 1 cubic inch **29.** 60 cm^3 **31.** 48 m^3 **33.** 3,053.63 in.^{3} **35.** 1,357.17 m^3 **37.** 314.16 cm^3 **39.** 400 m^3 **41.** 94 cm^2 **43.** 1,256.64 in.^{2} **45.** 576 cm^3 **47.** 335.10 in.^{3} **49.** $\frac{1}{8}$ in.^{3} = 0.125 in.^{3} **51.** 2.125 **53.** 197.92 ft^3 **55.** 33,510.32 ft^3 **57.** 8 : 1 **63.** −42 **65.** −8 **67.** $\frac{1}{5}$ **69.** 2,400 mm

Key Concept (page 622)

1. $d = rt$ **3.** $P = 2l + 2w$ **5.** 210,000 ft^2 **7.** \$80.50 **9.** 144 ft **11.** \$750, \$45, \$6,250

Chapter Review (page 624)

1. points C and D, line CD, plane GHI **2.** 5 units **3.** $\angle ABC$, $\angle CBA$, $\angle B$, $\angle 1$ **4.** 48° **5.** $\angle 1$ and $\angle 2$ are acute, $\angle ABD$ and $\angle CBD$ are right angles, $\angle CBE$ is obtuse, and $\angle ABC$ is a straight angle. **6.** obtuse angle **7.** right angle **8.** straight angle **9.** acute angle **10.** 15 **11.** 150 **12.** $m(\angle 1) = 65°$, $m(\angle 2) = 11°$ **13.** 40° **14.** 40° **15.** no **16.** part a **17.** $\angle 4$ and $\angle 6$, $\angle 3$ and $\angle 5$ **18.** $\angle 1$ and $\angle 5$, $\angle 4$ and $\angle 8$, $\angle 2$ and $\angle 6$, $\angle 3$ and $\angle 7$ **19.** $\angle 1$ and $\angle 3$, $\angle 2$ and $\angle 4$, $\angle 5$ and $\angle 7$, $\angle 6$ and $\angle 8$ **20.** $m(\angle 1) = 70°$, $m(\angle 2) = 110°$, $m(\angle 3) = 70°$; $m(\angle 4) = 110°$, $m(\angle 5) = 70°$, $m(\angle 6) = 110°$, $m(\angle 7) = 70°$ **21.** $m(\angle 1) = 60°$, $m(\angle 2) = 120°$, $m(\angle 3) = 130°$, $m(\angle 4) = 50°$ **22.** 40 **23.** 20 **24.** octagon **25.** pentagon **26.** triangle **27.** hexagon **28.** quadrilateral **29.** 3 **30.** 4 **31.** 8 **32.** 6 **33.** isosceles **34.** scalene **35.** equilateral **36.** right triangle **37.** yes **38.** no **39.** 90 **40.** 50 **41.** 50° **42.** It is equilateral. **43.** trapezoid **44.** square **45.** parallelogram **46.** rectangle **47.** rhombus **48.** rectangle **49.** 15 cm **50.** 40° **51.** 100° **52.** true **53.** false **54.** true **55.** true **56.** 65° **57.** 115° **58.** 360° **59.** 720° **60.** $\angle D$, $\angle E$, $\angle F$, $\overline{DF}$, $\overline{DE}$, $\overline{EF}$ **61.** congruent, SSS **62.** congruent, SAS **63.** congruent, ASA **64.** not necessarily congruent **65.** yes **66.** yes **67.** 21 ft **68.** 13 **69.** 15 **70.** 31.3 in. **71.** 72 in. **72.** 9 m **73.** 30 m **74.** 36 m **75.** 9.61 cm^2 **76.** 7,500 ft^2 **77.** 450 ft^2 **78.** 200 in.^{2} **79.** 120 cm^2 **80.** 232 ft^2 **81.** 152 ft^2 **82.** 120 m^2 **83.** 9 ft^2 **84.** 144 in.^{2} **85.** $\overline{CD}$, $\overline{AB}$ **86.** $\overline{AB}$ **87.** $\overline{OA}$, $\overline{OC}$, $\overline{OD}$, $\overline{OB}$ **88.** O **89.** 66.0 cm **90.** 45.1 cm **91.** 254.5 in.^{2} **92.** 130.3 cm^2 **93.** 125 cm^3 **94.** 480 m^3 **95.** 600 in.^{3} **96.** 3,619 in.^{3} **97.** 1,518 ft^3 **98.** 785 in.^{3} **99.** 9,020,833 ft^3 **100.** 35,343 ft^3 **101.** 1,728 in.^{3} **102.** 54 ft^3 **103.** 61.8 ft^2 **104.** 314.2 in.^{2}

Chapter 9 Test (page 633)

1. 4 units **2.** B **3.** true **4.** false **5.** false **6.** true **7.** 50 **8.** 140 **9.** 12 **10.** 45 **11.** 23° **12.** 63° **13.** 70° **14.** 110° **15.** 70° **16.** 40 **17.** 3, 4, 6, 5, 8 **18.** equilateral triangle, scalene triangle, isosceles triangle **19.** 57° **20.** 66° **21.** 30° **22.** 1,440° **23.** $m(\overline{AB}) = m(\overline{DC})$, $m(\overline{AD}) = m(\overline{BC})$, and $m(\overline{AC}) = m(\overline{BD})$ **24.** 130° **25.** 8 in. **26.** 50° **27.** 6 **28.** 12 **29.** 127.3 ft **30.** 391.6 cm^2 **31.** 83.7 ft^2 **32.** 94.2 ft **33.** 28.3 ft^2 **34.** 159.3 m^3 **35.** 268.1 m^3 **36.** 66.7 ft^3 **38.** The surface area is 6 times the area of one face of the cube.

Cumulative Review Exercises (page 635)

1.

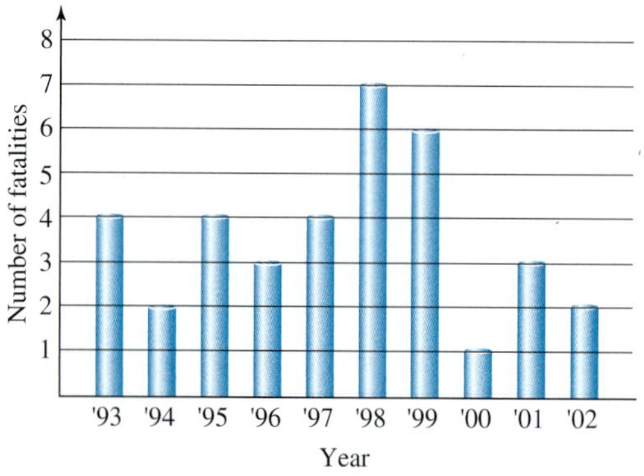

Source: U.S. Product Consumer Safety Commission

2. \$8,995 **3.** 11,022 **4.** 33 **5.** 2,110,000 **6.** $11 \cdot 5 \cdot 2^2$ **7.** 1, 2, 3, 4, 6, 8, 12, 24 **8.** $\{\ldots, -3, -2, -1, 0, 1, 2, 3, \ldots\}$ **9.** 13 **10.** −10 **11.** 3 **12.** 5 **13.** An equation contains an = sign; an expression does not. **14.** $-12x + 8$ **15.** 89 **16.** −10 **17.** −11 **18.** 5 **19.** 6.5 in./min **20.** $50x$¢ **21.** $x - 5$ **22.** $6\frac{2}{3}$ b.f. **23.** $\frac{5a}{4}$ **24.** $142\frac{7}{15}$ **25.** $\frac{5x - 12}{20}$ **26.** $13\frac{3}{4}$ cups **27.** $-\frac{11}{20}$ **28.** $\frac{1}{3q}$ **29.** $\frac{3}{32}$ fluid oz **30.** $\frac{8}{9}$ **31.** $-\frac{15}{2}$ **32.** $\frac{8}{9}$ **33.** $1\frac{9}{29}$ **34.** 18 oz **35. a.** 1998; about 0.6° F **b.** 1986; about −0.4° F

36.

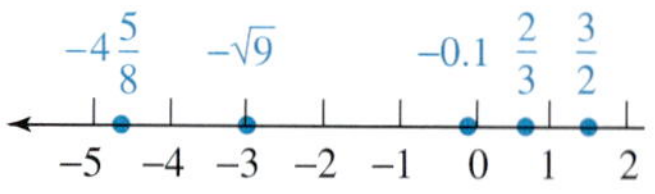

37. 3.1416 **38.** >

39. 145.188 **40.** 17.05 **41.** 89,970.8 **42.** 0.053 **43.** −25.6 **44.** 22.3125 **45.** $0.1\overline{3}$ **46.** −9.32 **47.** 97 **48.** −0.6 **49.** 248.94 **50.** 8^2 **51.** −2 **52.** $\frac{7}{9}$ **53.** 0.86 oz, 0.855 oz, 0.85 oz, 0.04 oz

54.

55. (0, 0) **56.** no

57.

58.

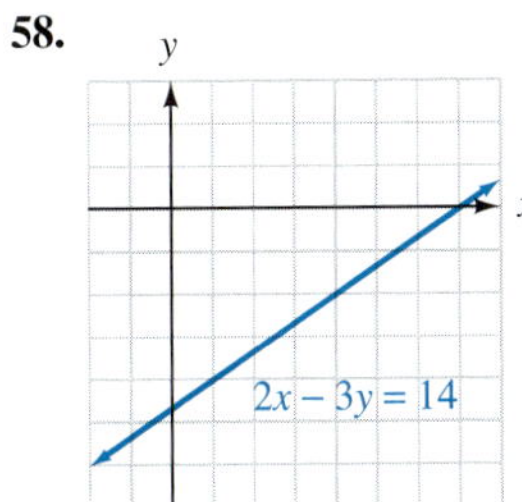

59.

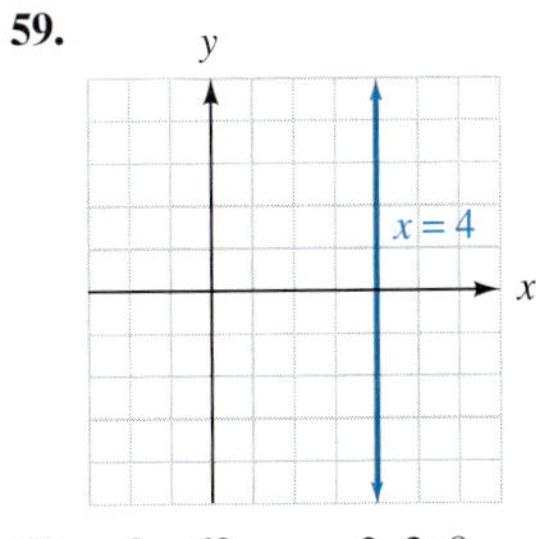

60.

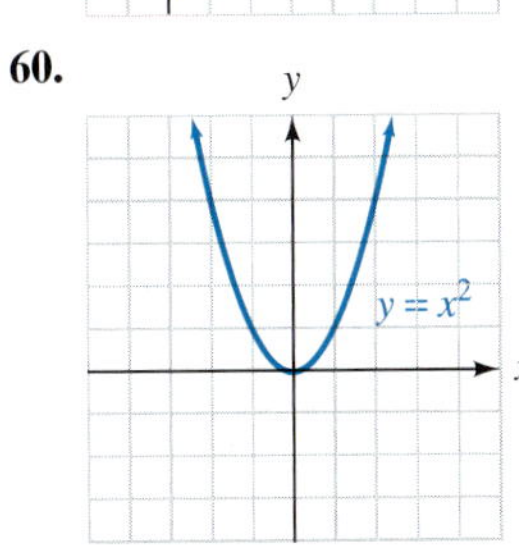

61. −8 **62. a.** −3, 2; 9 **b.** 3, 2; −9 **63.** s^9 **64.** a^{35} **65.** $15h^{10}$ **66.** $8b^9c^{18}$ **67.** y^{22} **68.** x^{m+n} **69.** trinomial; 2 **70.** $2x^2 - 2x + 9$ **71.** $-6p^3 - 9p^2 + 12p$ **72.** $6x^2 + 7x - 5$

73. $4y^2 - 28y + 49$ **74.** 93%, 7% **75.** 67.5 **76.** 120 **77.** 0.57, $\frac{57}{100}$; 0.1%, $\frac{1}{1,000}$; $33\frac{1}{3}\%$ $0.\overline{3}$

78.

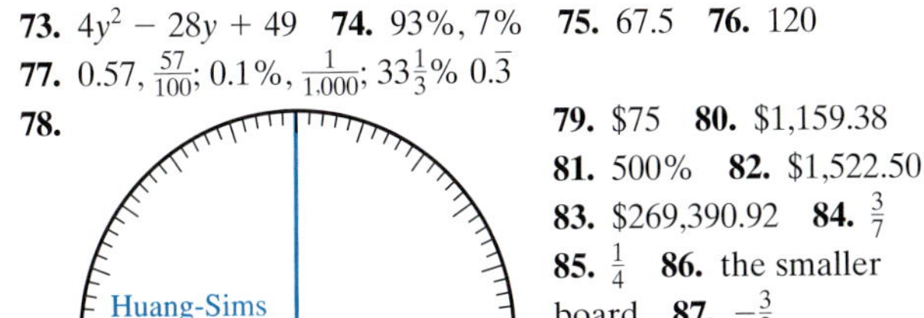

79. \$75 **80.** \$1,159.38 **81.** 500% **82.** \$1,522.50 **83.** \$269,390.92 **84.** $\frac{3}{7}$ **85.** $\frac{1}{4}$ **86.** the smaller board **87.** $-\frac{3}{2}$ **88.** 125,000 **89.** 75 ft **90.** 14 **91.** 540 **92.** 13.25 **93.** 120 **94.** 200 **95.** 12.3 **96.** 65.4 **97.** 0.5

98. 58.9 **99.** 167° **100.** 240 km **101.** between 5,700 and 5,800 cg **102. a.** about 4 m/gal **b.** 11,355,000 L **103.** about 4.5 kg **104.** 90 **105.** more than 0 but less than 90 **106.** 75° **107.** 15° **108.** 50° **109.** 130° **110.** 50° **111.** 50° **112.** 75° **113.** 30° **114.** 105° **115.** 105° **116.** 46, 134 **117.** 540° **118.** 13 m **119.** 42 m, 108 m² **120.** 126 ft² **121.** 91 in.² **122.** 43.98 cm, 153.94 cm² **123.** 98.31 yd² **124.** 210 m³ **125.** 523.60 in.³ **126.** 150.80 m³ **127.** 3.93 ft³ **128.** 2,124 in.²

Study Set Appendix I (page A-5)

1. inductive **3.** circular **5.** alternating **7.** alternating **9.** 10 A.M. **11.** 17 **13.** −17 **15.** 6 **17.** 3 **19.** −11 **21.** 9 **23.** **25.** **27.** I **29.** m **31.** Maria

33. the Mercedes **35.** green, blue, yellow, red **37.** 18,935 **39.** 0

Appendix Review (page A-8)

1. 6 **2.** 65 **3.** **4.**

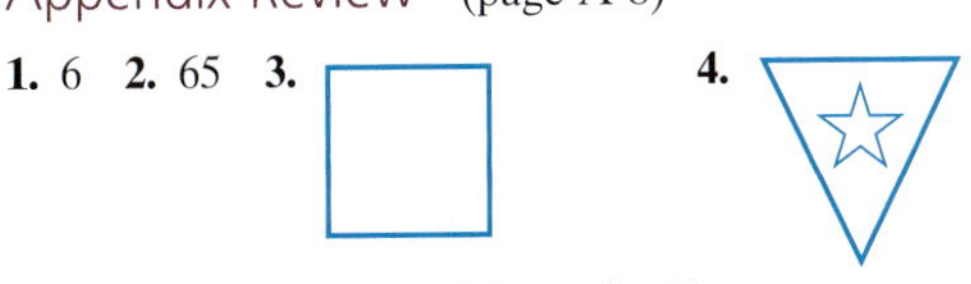

5. h **6.** the sheep **7.** Mary **8.** 43

INDEX